Mechanical Engineering Series

Frederick F. Ling
Series Editor

Mechanical Engineering Series

Frederick F. Ling
Series Editor

Giancarlo Genta

Dynamics of Rotating Systems

With 260 Figures

Springer

Giancarlo Genta
Politecnico di Torino
Torino, Italy

Series Editor
Frederick F. Ling
Ernest F. Gloyna Regents Chair in Engineering
Department of Mechanical Engineering
The University of Texas at Austin
Austin, TX 78712-1063, USA
and
Distinguished William Howard Hart
Professor Emeritus
Department of Mechanical Engineering,
Aeronautical Engineering and Mechanics
Rensselaer Polytechnic Institute
Troy, NY 12180-3590, USA

Library of Congress Cataloging-in-Publication Data
On file.

ISBN 0-387-20936-0 Printed on acid-free paper.

Printed in the United States of America. (EB)

9 8 7 6 5 4 3 2 1 SPIN 10949832

springeronline.com

To Franca and Alessandro

Series Preface

Mechanical engineering, an engineering discipline born of the needs of the industrial revolution, is once again asked to do its substantial share in the call for industrial renewal. The general call is urgent as we face profound issues of productivity and competitiveness that require engineering solutions, among others. The Mechanical Engineering Series is a series featuring graduate texts and research monographs intended to address the need for information in contemporary areas of mechanical engineering.

The series is conceived as a comprehensive one that covers a broad range of concentrations important to mechanical engineering graduate education and research. We are fortunate to have a distinguished roster of consulting editors, each an expert in one of the areas of concentration. The names of the consulting editors are listed on page ii of this volume. The areas of concentration are applied mathematics, biomechanics, computational mechanics, dynamic systems, and control, energetics, mechanics of materials, processing, thermal science, and tribology.

Preface

This book is the result of almost 30 years of work in the field of rotordynamics, which includes research, teaching, writing computer codes, and consulting. It is the outcome of an interdisciplinary research team that operated, and still operates, in the Mechanics Department and in the Interdepartmental Mechatronics Laboratory of Politecnico di Torino. The aim is mostly to write in a systematic way what has been the subject of a number of research papers, in such a way to give a consistent picture of the dynamic behavior of rotating machinery.

The author must then give credits to many colleagues and Ph.D. students who cooperated in various degree to this book: Much of the material they produced in their thesis work or in subsequent research found its way in these pages. An even greater number of students cooperated to this work in a more subtle way: with their thesis work and their questions, but mainly with their very presence that compels who tries to explain an involved subject to clarify his own ideas and to work out all details. To all of them goes the gratitude of the author.

A particular mention must be made to (in alphabetic order) Eugenio Brusa, Stefano Carabelli, and Andrea Tonoli, who not only worked during their doctoral thesis and subsequent research on these topics, but also who were very helpful in the actual writing of this book, reading and correcting the text, drawing figures, and doing much editorial work.

As usual, a deep gratitude goes to Franca for her support in the long work related with this book, and for the help in reading and correcting proofs.

As the title implies, this book is an attempt (only the reader can judge whether it is successful) to go beyond what is usually referred to as rotordynamics. The aim is that of dealing with the dynamic behavior of systems having in common the feature of rotating. This definition includes obviously those systems, like transmission shafts, turbine rotors, and gyroscopes, which are studied by rotordynamics, but also systems such as rotating blades (like in helicopter rotors) or flexible spinning spacecraft. Although rotordynamics usually deals only with the lateral behavior of rotors, some mention is made here also to torsional and axial vibration or to cases in which it is impossible to distinguish between them. However, the author imposed a limitation: No mention will be made of the dynamics of machines containing reciprocating parts, such as a crankshaft-connecting rod-piston mechanism. This arbitrary decision is based on the grounds that their vibration (mainly torsional vibration, but also axial and lateral vibration) is a very specialized topic, dealt with in many handbooks and textbooks and, above all, that to include it would have meant either to give a very insubstantial account or to double the size of the book.

Another area in which a decision about where to stop was needed is controlled rotors. A thorough study of the dynamics of many controlled rotors, like those running on active magnetic bearings or supplied with active dampers, would have implied a detailed study of their control systems (hardware and, in case of digital systems, software) sensors and actuators (with the critical issue of the power amplifiers). As is typical of mechatronic systems, only an integrated and interdisciplinary approach allows us to exploit the advantages of the potentialities modern technology has opened. As this would have lead too far from the main topics of this book, these areas will be touched only marginally.

The text is structured in two parts. The first one deals with what could be defined as *classic* or *basic rotordynamics*. The contents are basically well consolidated, although some incorrect statements can be found even in recent papers published on well-known journals. The basic assumptions are linearity, steady state operation, and at least some degree of axial symmetry.

The second part, containing topics that are usually considered as specialized aspects of rotordynamics, could be titled *advanced rotordynamics*. The mentioned assumptions are dropped, and more detailed models are built for rotors departing from the classic configurations studied in rotordynamics. The contents of this part are more research topics than consolidated applications.

The contents and the credits for the various chapters are the following:

- Chapter 1: Introduction. The basic concepts, graphical representation, and methods of rotordynamics are illustrated in a qualitative way. The expert reader, although familiar with these concepts, should not skip it altogether because the basic notation and the viewpoint that will be followed in the whole text are described.

Part 1: Basic topics

- Chapter 2: Jeffcott rotor. The so-called Jeffcott rotor is the simplest rotor model that can be conceived. Although unable to account for some typical phenomena linked with rotordynamics, like gyroscopic effect or centrifugal stiffening, it allows us to gain a good insight into the peculiarities of rotating systems. In particular, it is essential for understanding the role of damping in rotordynamics. The topics dealt with are as a whole standard, but the part on nonsynchronous damping, studied together with E. Brusa and published in [1], is less common.
- Chapter 3: Model with four degrees of freedom: Gyroscopic effect. A simple model in which a rigid body is substituted for the point mass of the Jeffcott rotor is then studied, to allow the study of gyroscopic effects. This model is representative for the behavior of any rigid rotor on compliant bearings and allows us to define a modal gyroscopic system, on which modal decomposition of rotors can be based under some assumptions.
- Chapter 4: Discrete multi-degrees-of-freedom rotors. The lateral behavior of a flexible rotor modeled as a discrete parameter beamlike (1-D approach) system is then studied. Older approaches, like the transfer matrices methods, are dealt with together with more modern ones, like the finite element method (FEM). Some work on reduction techniques by S. Carabelli and A. Tonoli [2] has been included.
- Chapter 5: Continuous systems: Transmission shafts. A short account on modeling simple rotors as continuous system is then included. This chapter can be considered more of academic rather than of practical relevance.
- Chapter 6: Anisotropy of rotors or supports. If either the rotor or the stator are not isotropic, it is still possible to obtain a closed-form solution for the linearized steady-state dynamics. Such systems are studied with particular reference to the backward whirling caused by unbalance in isotropic rotors on asymmetric supports and to the instability ranges of nonsymmetric rotors on isotropic supports.
- Chapter 7: Torsional and axial dynamics. The axial and torsional dynamics of rotors is briefly dealt with. Considering that the torsional and axial behavior is unaffected by the rotation of the system (at least if the basic assumptions of linearity and small displacements are made), just a brief account is reported.

- Chapter 8: Rotor-bearings interaction. The interaction between the behavior of the rotor and of the bearing is a complex subject, mainly because of the nonlinear behavior of the latter. The approach here followed is the classic one: The nonlinearity of the bearings is accounted for in computing their working conditions, and then the dynamic behavior is linearized assuming small displacements about the static equilibrium position (at speed). Rolling elements and lubricated and magnetic bearings are dealt with.

Part 2: Advanced topics

- Chapter 9: Anisotropy of rotors and supports. The assumption that either the stator or the rotor is isotropic is dropped. No closed-form solution is any more possible, although a truncated series solution can be attempted.
- Chapter 10: Nonlinear rotordynamics. Here another assumption, that of linearity, is dropped. The phenomena typical of nonlinear systems, like jumps and even chaotic behavior are discussed.
- Chapter 11: Nonstationary rotordynamics. The spin speed is no more assumed to be constant, or other parameters, like unbalance, are allowed to change. In particular, the acceleration of the rotor through a critical speed and the occurrence of a blade loss are dealt with in detail. The work performed with C. Delprete [3] has been thoroughly used.
- Chapter 12: Dynamic behavior of free rotors. Unconstrained rotating objects, like spinning celestial bodies or spacecraft, can be considered as rotors. The main aim of this section is to show that the assumption of constant angular momentum, typical of the dynamic study of free rotors, and that of constant angular velocity, typical of classic rotordynamics, coincide when the small displacement and rotations assumptions is made, so that the first can be approached with the methods of the latter. The chapter is based on the work performed with E. Brusa [4], [5].
- Chapter 13: Dynamics of rotating beams and blades. The effect of rotation, about an axis perpendicular to their longitudinal axis, on the dynamic behavior of beams and the blades-rotor interaction is studied using simple models. The well-known phenomena related to propeller and helicopter rotors' instability are dealt with, as well as other less-known phenomena regarding the effects of blade damping on the stability of a bladed rotor.
- Chapter 14: Dynamics of rotating discs and rings. Turbine and compressor discs are assumed, in classic rotordynamics, to behave as rigid

bodies. In this chapter, this assumption is dropped and the effects of the flexibility of the discs are dealt with using simple models, starting from that introduced about 80 years ago by Southwell [6].

- Chapter 15: Three-dimensional modeling of rotors. This chapter deals with numerical modeling, mostly based on the FEM, of complex rotors. The topics dealt with in Chapters 13 and 14 using simplified models are here treated with the aim of building more accurate models, yielding precise quantitative results. The work performed with A. Tonoli [7, 8] and the models developed by M. Silvagni in his Ph.D. thesis are included [9].

- Chapter 16: Dynamics of controlled rotors. Active vibration control is increasingly applied to rotors, either together with the use of active magnetic suspension or with techniques using active dampers or the control of more or less conventional bearings. As already stated, no attempt in modeling in detail the control, sensor or actuator dynamics is done, because it would lead too far from the central topics of this book. The work performed with S. Carabelli on sensor-actuator colocation [10] is reported.

- Appendix A: Vectors, matrices, and equations of motion. Some basic topics of system dynamics, particularly for the peculiar aspects linked with rotating systems, are summarized in this appendix, which owes much to the specific viewpoint of control theory for which the author is indebted to S. Carabelli. The results on circulatory and noncirculatory coupling published by Crandall [11] and relevant for rotordynamics are reported.

- Appendix B: An outline on rotor balancing. As many very good books have been written on rotor balancing, only a short account on the basic topics are dealt with.

- Appendix E: Bibliography. Some of the books specifically devoted to rotordynamics are listed in chronological order.

A CD-ROM comes with this book. It contains a simplified version of the DYNROT code and two short videos.

DYNROT Finite Element code was developed by the author starting in 1976. An initial version written in HPGL language for the early HP desktop computers together with G. Brussino, then student at the Politecnico di Torino, was followed by a version in BASIC written together with A. Gugliotta. In the subsequent years (almost 30), several versions have been developed, mostly using MatLab language with the help of countless researchers and students. The version reported here is DYNROT LIGHT, which contains only the basic elements and is based on the one-dimensional approach seen in Chapter 4 (while the full-blown DYNROT has capabilities also for studying bladed disks dynamics, with the possibility of time domain study of nonlinear and nonstationary rotors). A basic advantage of a MatLab code is its openness, so that the reader can understand how it works and can modify any points.

The two videos, *Gyroscopic Effect and Damping in Rotordynamics* and *Dynamic Behaviour of Rotors on Anisotropic Supports*, are based on experiments on simple demonstrators and were produced in an effort of showing students some physical evidence of what can be interpreted as mathematical divertissement more than the description of real-world machinery.

Giancarlo Genta

Torino, April 2004

Contents

Symbols

a	real part of a complex number, length, acceleration
b	imaginary part of a complex number, shaft bow, length
c	viscous damping coefficient, clearance
d	static offset
$\mathbf{f}(t)$	generalized forces vector
$\overline{\mathbf{f}}$	modal force vector
g	gravitational acceleration
h	thickness
i	imaginary unit ($i = \sqrt{-1}$), current
i_0	bias current
k	stiffness
l	length
m	mass
q_i	ith Lagrangian coordinate
$q(z)$	principal function
$\overline{q}$	complex conjugate of q
$\mathbf{q}(t)$	Vector of the generalized coordinates (real or complex)
r	complex coordinate ($r = x + iy$), radius
$\overline{\mathbf{r}}$	complex conjugate of $\mathbf{r}$
s	Laplace variable, complex frequency (as in $x = x_0 e^{st}$)
s'	complex frequency in the rotor fixed frame
$\mathbf{s}$	state vector (transfer matrices methods)
t	time, air gap
u_x, u_y, u_z,	displacements in xyz frame
uvw,	components of the displacement

xyz	reference frame
$\mathbf{z}$	state vector
A	area of the cross section
$\mathbf{A}$	dynamic matrix
B	magnetic field
$\mathbf{B}$	input gain matrix, matrix for discretization of stresses
$\mathbf{C}$	damping matrix, output gain matrix
$\overline{\mathbf{C}}$	modal damping matrix
E	Young's modulus
$\mathbf{E}$	stiffness matrix of the material
$\vec{F}$	force
$\mathcal{F}$	Rayleigh dissipation function
G	Shear modulus
$\mathbf{G}$	gyroscopic matrix
$\overline{\mathbf{G}}$	modal gyroscopic matrix
$\mathbf{H}$	circulatory matrix, angular momentum
I	area moment of inertia
$\mathbf{I}$	identity matrix
Im	imaginary part
$\mathbf{J}$	inertia tensor
J	moment of inertia
J_p	polar moment of inertia
J_t	transversal moment of inertia
K	stiffness, gain, constant
$\mathbf{K}$	stiffness matrix
$\overline{\mathbf{K}}$	modal stiffness matrix
$\mathcal{L}$	Lagrangian function
$\mathbf{M}$	mass matrix
$\overline{\mathbf{M}}$	modal mass matrix
$\vec{M}$	moment
$\mathbf{N}$	matrix of the shape functions
O	load factor (Ockvirk number)
Q	quality factor
Q_i	ith generalized force
Re	real part
R	radius
$\mathbf{R}$	rotation matrix
S	Sommerfeld number
$\mathcal{T}$	kinetic energy
$\mathbf{T}$	transfer matrix
$\mathbf{U}$	matrix of the right eigenvectors of the dynamic matrix
$\mathcal{U}$	potential energy
V	velocity, volume
α	phase angle, slenderness of a beam
β	phase angle, ratio J_t/ml^2, attitude angle, ratio ω/Ω
β_{ij}	generic element of the compliance matrix phase angle
γ	shear strain, ratio r_i/r_o

δ	ratio J_p/J_t , ratio r/l
δ'	ratio $(J_p - J_t)/m$
$\delta\mathcal{L}$	virtual work
δu	virtual displacement
ϵ	eccentricity
$\boldsymbol{\epsilon}$	strain vector
ζ	damping ratio, nondimensional coordinate ($\zeta = z/l$)
θ	rotation
κ	rotational stiffness, curvature
μ	viscosity, coefficent of the nonlinear restoring force
$\vec{\mu}$	magnetic dipole moment
ν	Poissons's ratio
$\xi\eta\zeta$	rotor-fixed reference frame
ρ	density, complex coordinate in the rotor fixed frame ($\rho = \xi + i\eta$)
ρ_p	radius of inertia $\rho_p = \sqrt{J_p/m}$
ρ_t	radius of inertia $\rho_t = \sqrt{J_t/m}$
σ	decay rate ($\sigma = \text{Re}(s)$), stress
$\boldsymbol{\sigma}$	stress vector
ϕ	complex coordinate ($\phi = \phi_y - i\phi_x$), phase
ϕ_x, ϕ_y, ϕ_z	rotation about x, y, z axes
χ	angular misalignment, shear factor, nondimensional coordinate $\chi = r/r_0$
ψ	angle
ω	frequency, whirl speed ($\omega = \text{Im}(s)$)
ω_n	natural frequency
$\mathbf{B}$	compliance matrix
Γ	rotational damping coefficient
$\boldsymbol{\Phi}$	matrix of the eigenvectors
Ω	rotational speed
Ω_{cr}	critical speed
	Subscripts
b	bearing
d	non synchronous rotating, deviatoric, damping
f	field
ff	fixed frame
i	inner
j	journal
m	mean
n	nonrotating, node
o	outer
p	pendulum
r	rotating, restoring force
rf	rotating frame
G	global
L	left
R	right

	Notation
λ	Scalar (real or complex)
$\overline{\lambda}$	complex conjugate of λ
$\mathbf{q}$	Vector (implicit notation: bold and lower case)
$\{q\}$	Vector (explicit notation: curl braces)
$\{q\}_n$	Vector of dimension n
$\mathbf{M}$	Matrix (implicit notation: bold and upper case)
$[M]$	Matrix (explicit notation: square braces)
$[M]_{n\times m}$	Matrix of dimension $n \times m$
$q(t)$, $\mathbf{q}(t)$	Time dependance
$\dot{q}(t)$, $\dot{\mathbf{q}}(t)$	Time derivative
$\ddot{q}(t)$, $\ddot{\mathbf{q}}(t)$	Time second order derivative
$(\lambda_i)_{i=1}^p$, $(\mathbf{q}_i)_{i=1}^p$, $(\mathbf{M}_i)_{i=1}^p$	Collection of p indexed elements
$(\{\mathbf{q}_i\}_n)_{i=1}^p$	Collection of p indexed vectors of dimension p

1

Introduction

Rotordynamics is that branch of systems dynamics dealing with mechanical devices in which at least one part, usually defined as *rotor*, rotates with significant angular momentum. Following the ISO definition, a rotor is a body suspended through a set of cylindrical hinges or bearings that allows it to rotate freely about an axis fixed in space. This definition is, however, too restrictive, because there are cases in which no material bearings constrain the position of the rotation axis in space: A spinning projectile or space vehicle thus can be also regarded as rotors; aside from the artificial rotors built by human technology, it is also possible to find natural rotors like spinning celestial bodies. The dynamic behavior of stars that spin at very high speeds, as neutron stars, display features that are typical of rotors.

Rotors provided with material bearings to constrain their spin axis in a more or less rigid way to a fixed position in space are usually referred to as *fixed rotors*, whereas those that are not constrained in any way are defined as *free rotors*. In the first case, the spin speed is usually considered as constant or at least as imposed by a driving device, whereas the speed of the latter is governed by the conservation of the angular momentum.

The parts of the machine that do not rotate are generally referred to with the general definition of *stator*.

Rotordynamic studies related to technological applications date back to the second half of the nineteenth century, when the increase of the rotational speed of many machine elements made it necessary to include rotation into the analysis of their dynamic behavior. However, the dynamics of rotating systems, as far as rigid rotors is concerned, was already well understood and the problem of the behavior of the spinning top had been successfully dealt with by several mathematicians and theoretical mechanicists.

The paper, *On the centrifugal force on rotating shafts*, published on *The Engineer* in 1869 by Rankine [12] is considered the first paper fully devoted to rotordynamics. It correctly states that a flexible rotating system has a speed, defined by the author as *critical speed*, at which very large vibration amplitudes are encountered. However, the author incorrectly predicts that stable running above the critical speed is impossible.

Early attempts to build turbines, mainly steam turbines, at the end of nineteenth century led to rotational speeds far higher than those common in other fields of mechanical engineering. At these speeds, some peculiar dynamic problems are usually encountered and must be dealt with to produce a successful design. De Laval had to solve the problem of correctly understanding the behavior of a rotor running at speeds in excess of the critical speed, i.e., in *supercritical* conditions, while designing his famous cream separator and then his steam turbine. From the beginning, the successful design of turbine engines depended on a thorough understanding of rotordynamics.

A theoretical explanation of supercritical running was supplied first by Föppl (1895)[13], Belluzzo (1905)[14], Stodola (1905)[15] (earlier works were published in German), and Jeffcott in his famous paper of 1919 [16]. Although the first turbine rotors were very simple and could be dealt with by using simple models, of the type now widely known as *Jeffcott rotor*, more complex machines required a more detailed modeling. Actually, although a simplified approach like the above-mentioned Jeffcott rotor can explain qualitatively many important features of real-life rotors, the most important being self-centering in supercritical conditions and the different roles of the damping of the rotor and of the nonrotating parts of the machine, it fails to explain other features, such as the dependence of the natural frequencies on the rotational speed. Above all, the simple Jeffcott rotor does not allow us to obtain a precise quantitative analysis of the dynamic behavior of complex systems, e.g., those encountered in gas- or steam-turbines, compressors, pumps, and many other types of machines.

To cope with the increasing complexity of rotating systems, graphical computation schemes were devised. They can be found in books and papers from the first quarter of the century, like the mentioned books by Belluzzo and Stodola, and were, without any doubt, the basic instrument for the dynamic analysis of turbine rotors. A large number of papers dealing with the flexural vibration of rotors and rotating shafts as well as with torsional vibration in reciprocating machinery appeared in the years after World War I. This trend continued for many years, fuelled by the increase of the size of power-generating machinery, of the specific power of engines of all types, and of the operational speed of machinery.

The availability of electromechanical calculators allowed us to develop tabular computational procedures, mainly based on the transfer matrices approach, which eventually substituted graphical computations. In particular, Holzer's method for the torsional vibrations of shafts and the Myklestadt-Prohl method for the computation of the critical speeds of turbine rotors were, and still are, widely used. These methods were immediately automatized when digital computers became available.

In the years after World War II, the fast diffusion of gas turbines and turbojet engines gave a new thrust to rotordynamic studies. Apart from papers published on journals and books specifically dedicated to this subject, rotordynamics was dealt with in a number of specialized conferences held every year. This lead to an explosion of the number of specialized papers and to the subdivision of the field in many very specialized branches with sometimes difficulties in recognizing clearly the unifying concepts laying under increasingly differentiating analytical procedures and jargon.

One example of these difficulties is shown by the peculiar approach developed for the study of the attitude dynamics of spinning spacecraft: Although the relevant phenomena are the same as those studied by mainstream rotordynamics, there is very little exchange of information between the specialists in the two fields and the different approaches make it difficult to recognize such similarities, particularly as far as the effects of damping are concerned.

The wide diffusion of the finite element method (FEM) deeply influenced also the field of rotordynamics. Strictly speaking, usual general purpose FEM codes cannot be used for rotordynamic analysis owing to the lack of consideration of gyroscopic effects. It is true that a gyroscopic matrix can be forced in the conventional formulation and that several manufacturers use commercial FEM codes to perform rotordynamic analysis, but the rotordynamic field is one of these applications in which purposely written, specialized FEM codes can give their best. Through FEM modeling, it is possible to study the dynamic behavior of machines containing high-speed rotors in greater detail and consequently to obtain quantitative predictions with an unprecedented degree of accuracy.

Correct quantitative prediction is particularly important as the trends of technology toward higher power density, lower weight, and faster machines tend to make worse all problems linked with the dynamic behavior of rotating machinery.

Higher speeds are often a goal in themselves, like in machine tools or other production machines in which spinning faster means directly increasing productivity. In applications related to power generation or utilization, a faster machine can develop or convert more power manipulating the same torque. As torque is usually the critical factor in dimensioning machine elements (shaft cross section, size of the conductors in electrical machinery, etc.), increasing the speed allows us to make power devices lighter. The use of materials able to withstand higher stresses allows us to reduce the mass and the size of machinery, but stronger materials (e.g., high strength steels or light alloys) are usually not stiffer and then these lighter machines are more compliant and more prone to vibrate. Another trend is toward higher operating temperature, both for increasing the thermodynamic efficiency and because higher power density means that the same amount of heat is generated in a smaller space, with less material, and hence lower thermal capacity. Higher temperatures lead to higher thermal stresses, lower damping, and often lower stiffness, which makes more severe strength and vibration problems.

Rotordynamic analysis is not restricted to the design stage: It can supply tools that are essential during testing and actual operation of machinery and is essential to gain a deep insight of the working conditions and to perform preventive maintenance.

The study of the *mechanical signature*, i.e., the vibration spectrum, of a rotating machine allows us to identify operating problems even before they become dangerous and to avoid the fact that the failure of a component precipitates failure or damages to other parts. Any deviation of the signature from its usual pattern provides a symptom, in many cases easily interpreted, of a problem that is developing and allows the required countermeasures in time to be taken.

Machines in general, but particularly those including fast rotating elements, are increasingly supplied with sensors to monitor the dynamic behavior, and they can even incorporate transducers, actuators, and control systems for controlling it in an active way. As it is commonly said, they are increasingly *intelligent* machines.

In spite of the progress that has been made in the recent years, or perhaps as a consequence of it, the rotordynamics field is still a field of very active research, both for the application of concepts that are well established from decades to new areas and new applications and for achieving a better understanding of the behavior of rotating systems. Nonlinear and nonstationary rotordynamics and the behavior of actively controlled rotating machines are only examples of areas in which much research work is still needed.

Every year many scientific conferences and seminars are devoted to rotordynamics, and the number of papers presented at them or published in scientific journals is in the thousands, and the state of the art is described in a large number of books.

1.1 Linear rotordynamics

The equations describing the motion of even a simple rigid body with mass m and principal moments of inertia J_ξ, J_η, and J_ζ referred to a reference frame $\xi\eta\zeta$ fixed to it in the three-dimensional space are actually complex, particularly when dealing with the rotational degrees of freedom, and they do not allow the direct use of any linear model. With reference to an inertial frame xyz and a rotating frame $\xi\eta\zeta$ fixed to the rigid body and coinciding with its principal axes of inertia, the six equations of motion under the action of the generic force $\vec{F}$ and moment $\vec{M}$ can be written in the form

$$\begin{cases} m\ddot{x} = F_x\,, \\ m\ddot{y} = F_y\,, \\ m\ddot{z} = F_z\,, \end{cases} \qquad \begin{cases} M_\xi = \dot{\Omega}_\xi J_\xi + \Omega_\eta \Omega_\zeta (J_\zeta - J_\eta)\,, \\ M_\eta = \dot{\Omega}_\eta J_\eta + \Omega_\xi \Omega_\zeta (J_\xi - J_\zeta)\,, \\ M_\zeta = \dot{\Omega}_\zeta J_\zeta + \Omega_\xi \Omega_\eta (J_\eta - J_\xi)\,. \end{cases} \tag{1.1}$$

The three equations for the rotational degrees of freedom, which are the well-known Euler equations, are clearly nonlinear in the angular velocity $\vec{\Omega}$.

However, a number of simplifications allow a linearized model to be obtained that retains the basic features of the dynamic behavior of rotating systems and allow us to describe it correctly, both in a qualitative and a quantitative manner.

The rotor has, in its undeformed configuration, a well-defined rotation axis, which coincides with one of the baricentrical principal axes of inertia. This condition is exactly verified only if the rotor is perfectly balanced and is only approximately true; however, in most cases the, *unbalance*, i.e., the deviation from this ideal condition, is small and it is possible to deal with it as a small perturbation. Also, all displacements and velocities, linear and angular, can be assumed to be small, with the exception of the rotation angle and angular velocity about the spin axis, which are not small but can be considered as imposed by the driving system.

1.1.1 Equation of motion

The two assumptions of small unbalance and small displacements allow the linearization of the equations of motion in a way that is consistent with what is usually done in the dynamics of structures (see, for example, [17, 18, 19]). However, even in the case of the discretised model[1] of a linear rotor that is axially symmetrical about its spin axis and rotates at a constant spin speed Ω, the linearized equation of motion (dynamic equilibrium equation) is of the following general form:

$$\mathbf{M}\ddot{\mathbf{q}}(\mathbf{t}) + (\mathbf{C}+\mathbf{G})\,\dot{\mathbf{q}}(t) + (\mathbf{K}+\mathbf{H})\,\mathbf{q}(t) = \mathbf{f}(t)\;, \tag{1.2}$$

---

[1] In the present text, time is always assumed to be continuous; hence, discretized is always to be intended with respect to space coordinates. Note that the expression "lumped parameters" is not a synonym of "discretized" because in the context of FEM, consistent parameter systems, which are discretized but not lumped parameters, are used.

Following a definition by Meirovitch, *lumped parameters depend on the spatial position only implicitly. On the other hand, distributed parameters depend explicitly on the spatial coordinates.* This definition, however, is more general and considers *consistent* FEM models as lumped parameters systems.

where $\mathbf{q}(t)$ is a vector containing the generalized coordinates, referred to an inertial frame, $\mathbf{M}$ is the symmetric mass matrix, $\mathbf{C}$ is the symmetric damping matrix, $\mathbf{G}$ is the skew-symmetric gyroscopic matrix, $\mathbf{K}$ is the symmetric stiffness matrix, $\mathbf{H}$ is the skew-symmetric circulatory matrix, and $\mathbf{f}(t)$ is a time-dependant vector in which all forcing functions are listed.

1.1.2 Rotating systems

When dealing with rotating systems, one of the forcing functions is usually that caused by the residual unbalance that, although small, cannot nevertheless be neglected. Unbalance forces are harmonic functions of time, with an amplitude proportional to Ω^2 and a frequency equal to Ω.

The gyroscopic matrix contains inertial, and hence conservative, terms that, in the case of rotordynamics, are strictly linked with the gyroscopic moments acting on the rotating parts of the machine. If the equation is written with reference to a noninertial frame, terms linked with Coriolis acceleration also are present in the gyroscopic matrix. The circulatory matrix contains nonconservative terms linked with the internal damping of rotating elements and, when using a linearized model for fluid bearings or seals, with the damping of the fluid film surrounding the rotor. It is well known that the presence of a circulatory matrix can cause instability, and rotors are no exception to this rule.

Equation (1.2) is that of a nonnatural, circulatory system and hence differs from the typical equations encountered in dynamics of structures, where all matrices are symmetric. It must be noted that in rotordynamics, the gyroscopic and circulatory matrices $\mathbf{G}$ and $\mathbf{H}$ are proportional to the spin speed Ω, and when Ω tends to zero, the skew-symmetric terms vanish and the equation reduces to that of a still structure. Also, the damping and stiffness matrices $\mathbf{C}$ and $\mathbf{K}$ may depend on the spin speed, often on its square Ω^2, and $\mathbf{H}$ can be a more complex function of Ω.

Some cases that, strictly speaking, could not be studied using the aforementioned assumptions can still be dealt with in the same way. Consider, for example, the rotor of an aircraft gas turbine during maneuvered flight. The direction of the axis of the rotor changes continuously in time, and no small-angle assumption can be considered for this motion. However, the motion of the rotor can be studied in a reference frame that is fixed to the aircraft, provided that the motion of the latter can be considered independent from the dynamic behavior of the first and the related inertia forces are added. This way of splitting the problem into its dynamic and quasistatic parts is possible if the characteristic times of the different phenomena under study are widely different. In the example given earlier, this is clearly the case if the frequencies that characterize the motion of the rotor of the turbine with respect to the aircraft are of several Hertz (periods of fractions of seconds), and the rotations of the airframe have characteristic times of the order of several seconds. On the contrary, the seismic actions on the rotor of a machine in a building may have frequencies of the same order of magnitude as those that characterize the rotor, and the problem may have to be studied without any uncoupling being possible.

Equation (1.2) has been obtained with the assumption of axial symmetry of the system about the spin axis. Actually, it still holds when the rotor is axially symmetrical, but it runs on a general stator, without any particular symmetry properties.

If, on the contrary, the rotor cannot be considered to be axially symmetrical, the study becomes very complicated, unless an axial symmetry assumption can be made on the nonrotating parts of the system. In the latter case, a rotating reference frame, i.e., one that rotates at the angular velocity of the rotor, can be used and an equation similar to Equation (1.2), although written with reference to a noninertial frame, is obtained.

If both stator and rotor are nonisotropic with respect to the rotation axis, the equation of motion that models its behavior has coefficients that are periodic in time, with a frequency equal to 2Ω. No closed-form solution of such equation is available, and even reaching an approximated solution is far more complicated than in the case in which either the rotor or the stator is axially symmetrical.

Most flexible rotors can be considered as beam-like structures. Under fairly wide assumptions, the lateral behavior of a beam can be considered as uncoupled from its axial and torsional behavior. The same uncoupling is usually assumed in rotordynamics, with the difference that no further uncoupling between bending in the principal planes is possible.

When the flexural behavior can be uncoupled from the axial and torsional ones, Equation (1.2) holds for the first one, and the torsional and axial equations of motion are usually those of a natural, noncirculatory system.

1.1.3 Complex coordinates

If both stator and rotor are isotropic with respect to the rotation axis, very simple models can be devised by introducing complex coordinates. Assume that the spin axis coincides with the z-axis of an inertial reference frame (Figure 1.1). The lateral displacement of any point of the rotor can be described in terms of the displacements in x- and y-directions or in terms of a displacement vector in the xy-plane. The latter can be expressed in the form of a complex number

$$r(t) = x(t) + iy(t) \ , \tag{1.3}$$

where i is the imaginary unit ($i = \sqrt{-1}$). The use of complex coordinates is somewhat equivalent to express the displacements in compact vector form

$$\mathbf{r}(t) = \left\{ \begin{array}{c} x(t) \\ y(t) \end{array} \right\} \ , \tag{1.4}$$

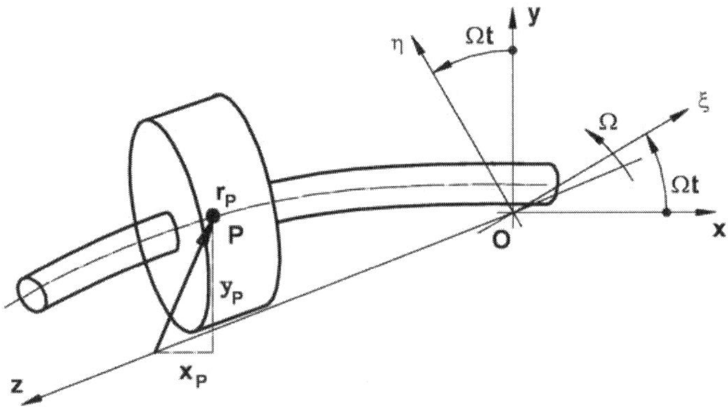

FIGURE 1.1. Inertial frame xyz and rotating frame $\xi\eta z$ for a rotor. Displacement vector of point P and its components.

but it allows a far more convenient analytical form to be used to obtain the solution of the equation of motion for rotors (see Section 2.2 for its application to the undamped Jeffcott rotor).

Remark 1.1 *In a similar way, complex coordinates for angular displacements can be defined (Chapter 3).*

In general, when using the complex-coordinates notation, the equation of motion of a multi-degrees-of-freedom axisymmetrical rotating system (Chapter 4) results in

$$\mathbf{M}'\ddot{\mathbf{q}}'(t)+(\mathbf{C}'+i\mathbf{G}')\,\dot{\mathbf{q}}'(t)+(\mathbf{K}'+i\mathbf{H}')\,\mathbf{q}'(t)=\mathbf{f}'(t)\ . \tag{1.5}$$

Substituting the expression of the complex coordinates $\mathbf{q}'(t)$ into Equation (1.5) and separating the real and imaginary parts

$$\mathbf{q}(t)=\left\{\begin{array}{c}\mathrm{Re}(\mathbf{q}'(t))\\ \mathrm{Im}(\mathbf{q}'(t))\end{array}\right\}, \tag{1.6}$$

the equation of motion (1.2) becomes

$$\begin{gathered}\left[\begin{array}{cc}\mathbf{M}' & \mathbf{0}\\ \mathbf{0} & \mathbf{M}'\end{array}\right]\ddot{\mathbf{q}}(t)+\left(\left[\begin{array}{cc}\mathbf{C}' & \mathbf{0}\\ \mathbf{0} & \mathbf{C}'\end{array}\right]+\left[\begin{array}{cc}\mathbf{0} & -\mathbf{G}'\\ \mathbf{G}' & \mathbf{0}\end{array}\right]\right)\dot{\mathbf{q}}(t)+\\ +\left(\left[\begin{array}{cc}\mathbf{K}' & \mathbf{0}\\ \mathbf{0} & \mathbf{K}'\end{array}\right]+\left[\begin{array}{cc}\mathbf{0} & -\mathbf{H}'\\ \mathbf{H}' & \mathbf{0}\end{array}\right]\right)=\mathbf{f}(t)\ .\end{gathered} \tag{1.7}$$

Remark 1.2 *The matrices that are symmetric when using the real coordinates approach are real in the equation written in terms of complex coordinates, whereas skew-symmetric matrices give way to symmetric, imaginary terms. When using the complex-coordinates approach, all relevant matrices, including the gyroscopic and circulatory ones, are then symmetric.*

Remark 1.3 *Although complex coordinates allow us to greatly simplify the equation of motion of axisymmetrical systems, they can be used as well in the study of general, nonisotropic, rotating systems.*

Remark 1.4 *When the spin speed reduces to zero, the gyroscopic and circulatory matrices vanish and the solution of the general equation of motion (1.7) is obtained as the sum of two equal and decoupled systems whose solution coincides with that of the system in complex coordinates of Equation (1.5).*

In the following chapters, the basics of rotordynamics are introduced by a two degrees-of-freedom point mass model, the so-called Jeffcott rotor (Chapter 2), a four degrees-of-freedom rigid body model (Chapter 3), and a multi-degrees-of-freedom flexural body model (Chapter 4) by taking full advantage of the complex coordinates notation.

1.1.4 Free vibration

The general solution of Equation (1.2) can be written by adding the complementary function (the general solution of the homogenous equation) to a particular integral of the complete equation. The former allows the free behavior of the system to be studied.

As usual, the solution for free vibration can be written in the form

$$\mathbf{q}(t) = \mathbf{q}_0 e^{st} , \tag{1.8}$$

where

$$s = \sigma + i\omega$$

is the *complex frequency*. The natural frequency of the free motion (*whirl frequency*) of the system is then the imaginary part ω of s, whereas its real part σ is the *decay rate* (i.e., the rate at which the amplitude decreases in time) changed in sign: A negative value of σ characterizes a motion that decays in time (stable motion), whereas a positive value characterizes an unstable motion, growing exponentially in time.

Spin speed dependence – Campbell diagram

As the spin speed can appear explicitly in the equation of motion, the natural frequencies of a machine containing a rotor can depend on the spin speed. When this occurs, the free behavior of the system is usually summarized by a plot of the natural frequencies $\omega_i = \mathrm{Im}(s_i)$ as functions of Ω. Because in many cases the frequencies of the exciting forces also depend on the speed, they can be reported on the same plot, obtaining

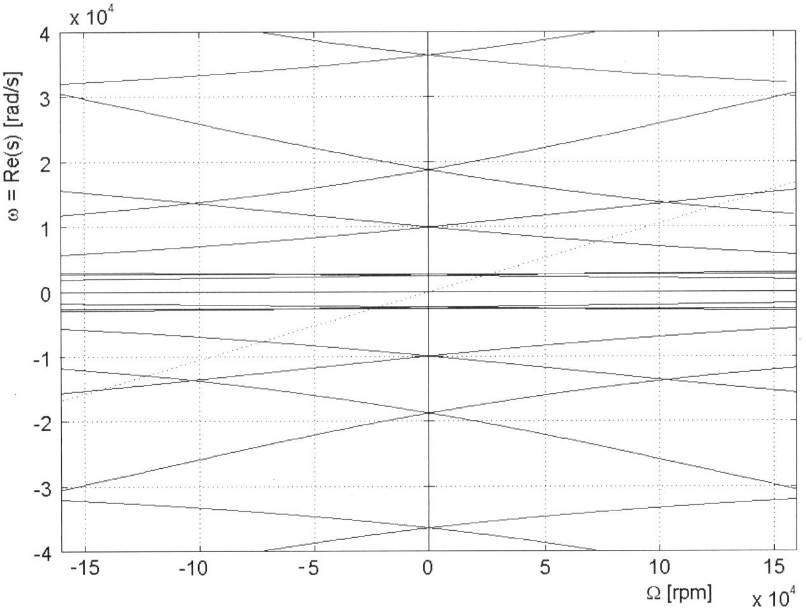

FIGURE 1.2. Four-quadrant Campbell diagram of a flexible rotor (the rotor of a small turbojet).

what is generally known as a *Campbell diagram*[2] (Figure 1.2). If damping is present, a second plot, in which the decay rates are reported as functions of the spin speed, can be plotted together with the Campbell diagram, as in Figure 1.4. Alternatively, the natural frequency can be plotted against the decay rate, obtaining what is generally referred to as a *roots locus* (Figure 1.3).

The Campbell diagram of Figure 1.2 has been represented over four quadrants. Clearly, it is symmetrical with respect to the ω axis, as the sign of the spin speed, i.e., the direction of the rotation of the machine, has no effect on the natural frequencies. It is also symmetrical with respect to the Ω axis: The solutions for s are either real or complex conjugate, and hence if ω is the imaginary part of an eigenvalue, $-\omega$ is the imaginary part of another eigenvalue.

Remark 1.5 *Both the Campbell diagram and the roots loci are symmetrical with respect to the abscissa axis.*

---

[2] A plot of the free vibration frequency as a function of the spin speed was extensively used in a paper by Campbell [20] describing the results of very extensive testing on rotating rubber membranes simulating turbine disks.

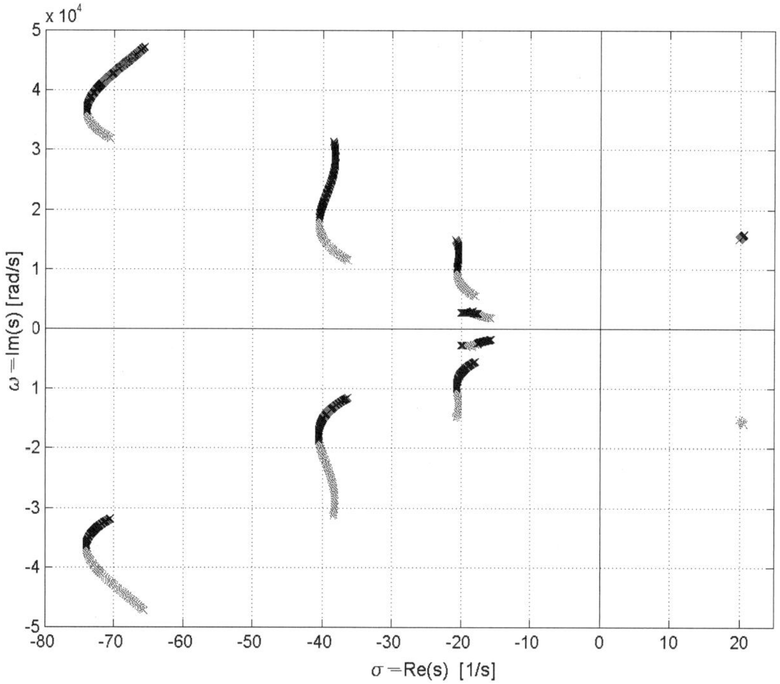

FIGURE 1.3. Roots locus of a flexible rotor depending on spin speed as the varying parameter.

Remark 1.6 *The intersections of the various branches of the Campbell diagram with the ω axis are the natural frequencies at standstill of the system. If the system is axially symmetrical (i.e., the characteristics in xz plane coincide with those in yz plane), the natural frequencies at standstill are pairs of coincident values. With increasing speed, the values are no more double: Two diverging branches start from each point on the ω axis of the Campbell diagram.*

Remark 1.7 *The Campbell diagram can be plotted only in the case of linear systems, because only in this case does the very concept of natural frequencies apply. However, in the case of nonlinear systems, the Campbell diagram of the linearized system may yield important information on the behavior of the system.*

As the Campbell diagram is symmetrical with respect to both the ω and the Ω axes, all of the information it conveys are contained in any one of its quadrants and it is customary to draw only one of them, usually the first quadrant. Such plot, together with the plot of the real part of the eigenvalues as functions of the speed, is reported in Figure 1.4.

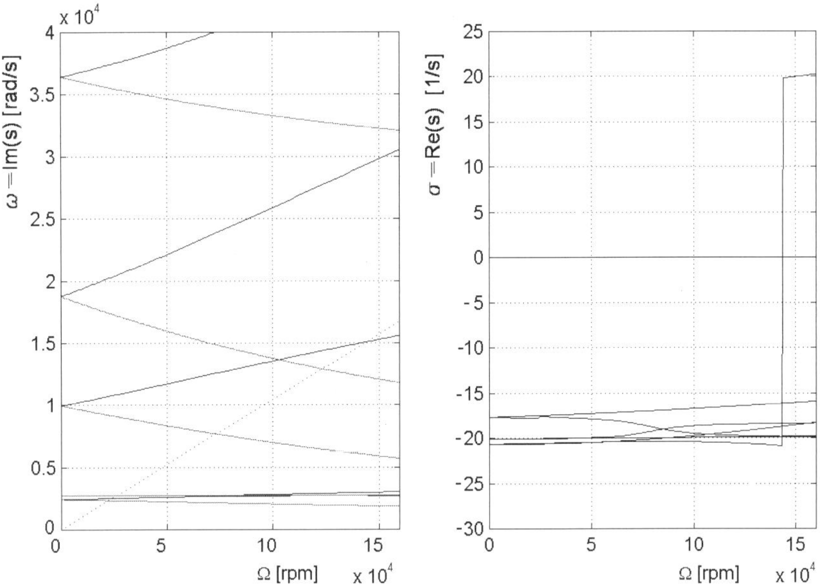

FIGURE 1.4. Standard Campbell diagram (only the first quadrant is represented) and decay rate plot of a flexible rotor.

Consider a pair of branches of the Campbell diagram of Figure 1.4 stemming from a single natural frequency at standstill (the rotor is axially symmetrical and the values on the ω-axis are double). As seen later in Chapters 2 and 3, they can be interpreted as the frequencies of two circular whirling motions, one (the darker in Figure 1.4) occurring in the same direction of the spin motion (*forward* or *direct whirling*) and one (the lighter) in the opposite direction (*backward* or *reverse whirling*).

It is then possible to plot the Campbell diagram in such a way as to differentiate between forward and backward whirling, using a positive value of ω, now interpreted as the speed of the free whirling motion, for the former and a negative value for the latter.

The Campbell diagram of Figure 1.4, plotted to include both a positive and negative value of the spin speed, takes now the aspect reported in Figure 1.5.

In the four quadrant representation of the Campbell diagram, the lines in the first one are related to forward whirling when spinning in forward (i.e., counterclockwise) direction, the second is for reverse whirling when spinning in clockwise direction, the third is for direct whirling in clockwise direction, and the fourth is for reverse whirling in counterclockwise direction. As already said, the direction of the spinning motion is immaterial and two quadrants are enough to convey all of the required information. The

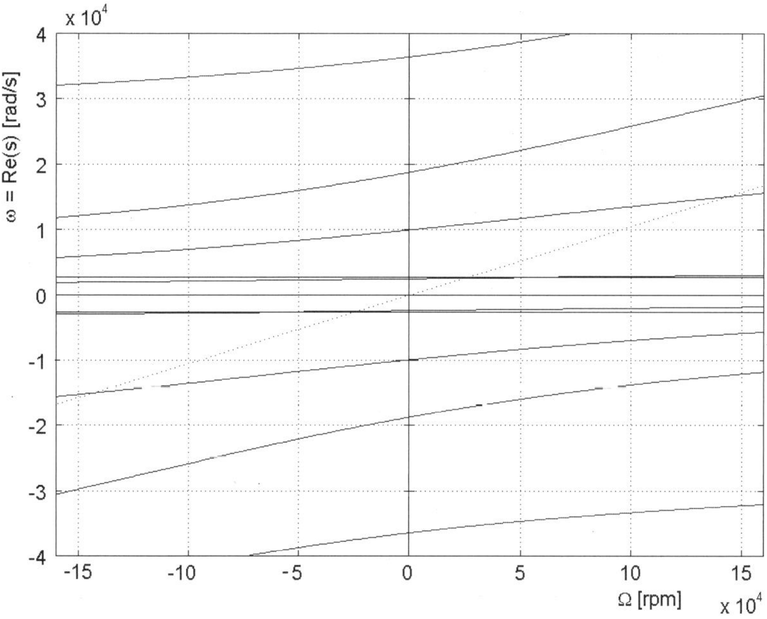

FIGURE 1.5. Campbell diagram of the same flexible rotor used for the previous plots, but represented using a positive vale of ω for forward whirling and a negative one for backward whirling. Both positive and negative values of Ω have been represented.

usual way is to represent the first and the fourth quadrants, but sometimes the first and second ones are plotted.

What has been said for the Campbell diagram applies also for the roots locus. Also here the branches for forward whirling can be represented with positive values of ω and those for backward whirling with negative values. A plot of the type shown in Figure 1.6 is thus obtained, which is not symmetrical with respect to the σ-axis.

The dependence of the dynamic response on the spin speed is also evident when the transfer functions of the rotating structure are used. Figures 1.7 and 1.8 show the transfer function from force to acceleration for the same rotor at standstill and at nominal rotating speed, respectively. Because of axial-symmetry, at standstill, the structure response is that of two equal and decoupled systems in the x- and y-directions: All of the eigenvalues of the overall system are complex conjugate and have double multiplicity. At the nominal spin speed, the gyroscopic effect couples the response in x- and y-directions: The eigenvalues are (have to be) complex conjugate but are no longer coincident.

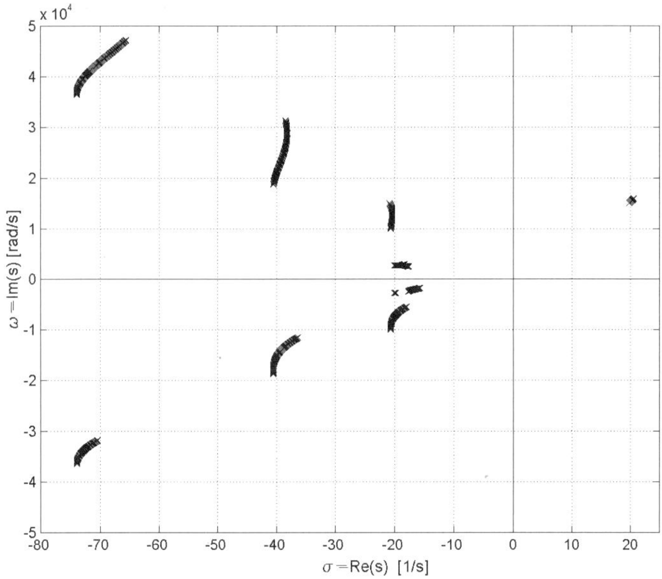

FIGURE 1.6. Roots locus in which forward and backward whirling are reported as positive and negative values of ω.

Each resonance peak in Figure 1.7 splits into two resonance peaks in Figure 1.8. This splitting of the resonances is typical of rotating systems that have a nonnegligible gyroscopic effect.

Critical speeds

Often rotors are subjected to forces that vary in time, and sometimes their time history is harmonic. This is the case, for example, of forces caused by the unbalance of the rotor, which can be described as a vector rotating with the same angular speed as the rotor and whose components in the fixed reference frame vary harmonically in time with circular frequency equal to the rotational speed Ω. In other cases the time history is less regular, but if it is periodical, it can always be represented as the sum of harmonic components.

In these cases, the frequency of the forcing function or of its harmonic components is often linked with the spin speed of the rotor and can be plotted on the Campbell diagram. In the case of the excitation caused by unbalance, for example, the forcing frequency can be represented on the $\Omega\omega$-plane of the Campbell diagram by the straight line $\omega = \Omega$, i.e., by the bisector of the first quadrant. In this case, the excitation is said to be *synchronous*. The relationship linking the frequency of the forcing function to the spin speed is often of simple proportionality and can be represented on the Campbell diagram by a straight line through the origin.

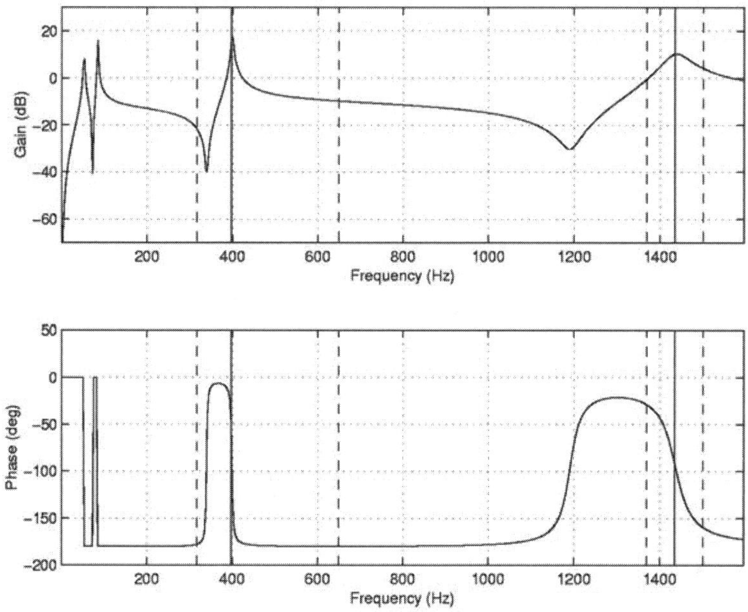

FIGURE 1.7. Frequency response of a rotor (on active magnetic bearings) at standstill. The transfer function between the acceleration at the location of a sensor and the force at the corresponding actuator is reported.

The spin speeds at which one of the forcing functions has a frequency coinciding with one of the natural frequencies of the system are usually referred to as *critical speeds* and can be identified on the Campbell diagram by the intersections of the curves related to the natural frequencies with those related to the forcing frequencies. A case in which the forcing frequencies are proportional to Ω is reported in Figure 1.9.

Not all of the intersections on the Campbell diagram are equally dangerous. If the frequency of a forcing function coincides with the natural frequency of a mode that is completely uncoupled from it (or, better, if the modal force corresponding to the forcing function and the resonant mode is vanishingly small), no resonance actually occurs. For example, if the frequency of the driving torque (i.e., of the torsional moment on the rotor) is coincident with a flexural natural frequency of the rotor and torsional and flexural behavior are completely uncoupled, no resonance takes place. In other cases, the resonance can be very weak and the damping of the system can be sufficient to avoid any measurable effect.

There are, however, cases in which a very strong resonance takes place and the rotor cannot operate at or near a critical speed without strong vibrations or even a catastrophic failure. In particular, the resonances caused by the coincidence of one of the flexural natural frequencies with the spin

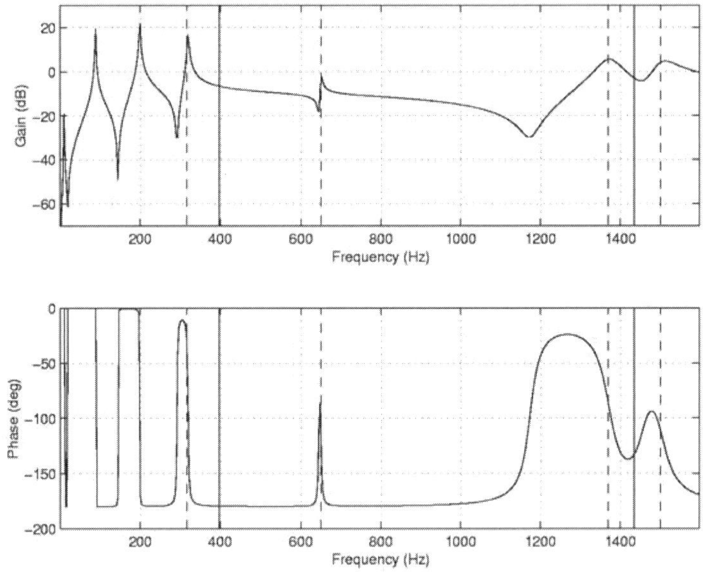

FIGURE 1.8. Frequency response of a rotor (on active magnetic bearings) at a speed of $\Omega = 33000$ rpm. Note the splitting of the resonance peaks caused by the gyroscopic effect.

speed are particularly dangerous; they can be detected on the Campbell diagram by the intersection of the curves related to the natural frequencies with the straight line $\omega = \Omega$. They are usually referred to as *flexural critical speeds*, without further indications, and other critical speeds related to bending behavior, which are usually less dangerous, are often said to be *secondary critical speeds*.

The response to a synchronous excitation, like unbalance, causes a synchronous whirling of the rotor; if the whole system is axi-symmetrical, a circular synchronous whirling occurs. The situation occurring in this condition is sketched in Figure 1.10: In (a), a whirling shaft is shown in its deflected configuration, and in (b), the situation occurring in xy plane is sketched. As the spin speed is equal to the whirl speed, the zone of the cross section of the shaft subjected to tensile stresses (shaded part close to point B) remains always under tensile loading, whereas that subjected to compression is always compressed. The situation is similar to that of the Moon, which always shows the same side to the Earth. In this condition, the material constituting the rotor is subjected to a constant stress state, which is good on one side as the rotor is not subjected to fatigue, but on the other side, the internal damping of the material cannot be effective in dissipating energy and hence in limiting the amplitude, particularly when the system is working at a speed close to a critical speed.

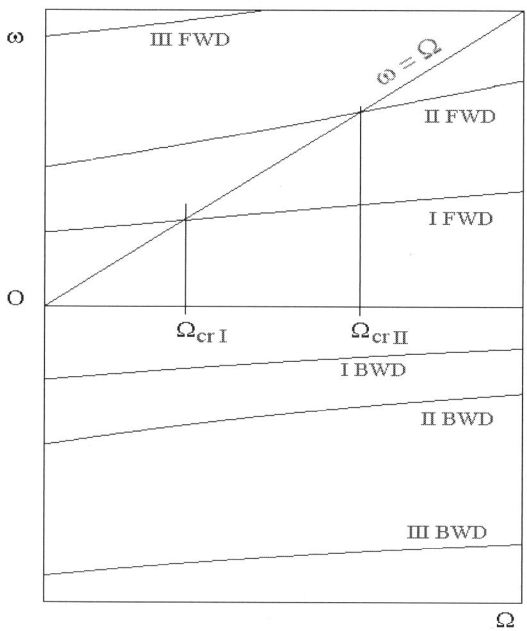

FIGURE 1.9. Intersections on the Campbell diagram to locate the critical speeds.

The flexural critical speeds can, in fact, be defined as the speeds at which the centrifugal forces due to the bending of the rotor are in indifferent equilibrium with the elastic restoring forces, and from this point of view, the situation is more similar to that characterizing elastic instability than that typical of vibratory phenomena. A rotor operating at a critical speed is then not subject to vibrations of any type but is a source of periodic excitation that can cause vibrations, often very strong, in the nonrotating parts of the machine.

In this condition, the amplitude of the vibration grows linearly in time and only the damping of the stator and the supports (because the damping of the rotor is completely ineffective in this case) and the unavoidable nonlinearities, which show up when the amplitude grows, can prevent the failure of the rotor. Actually, it is possible to design rotating machinery in such a way that operation at a critical speed is possible for a limited period of time, but it is, at any rate, necessary that the normal operating range is either below the first critical speed or between two critical speeds and that sustained operation in critical condition is avoided. An example of a machine operating above the first critical speed is that of the common domestic washing machine: In the transition between the washing and the spinning modes, the crossing of a critical speed is generally easily detected from strong vibrations, even without instruments.

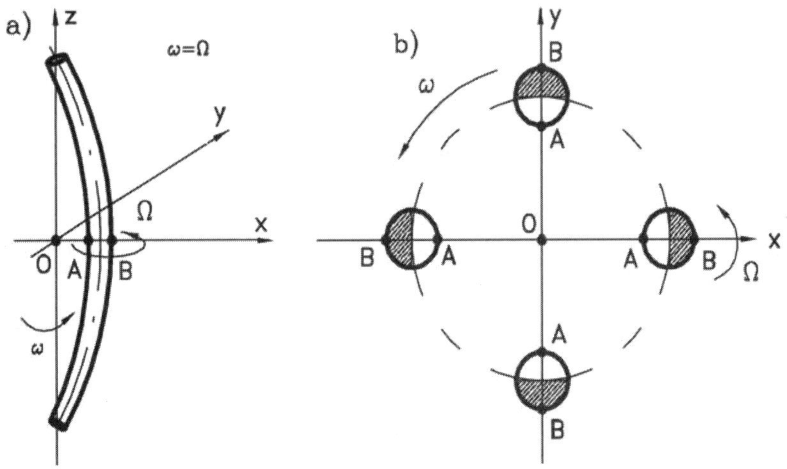

FIGURE 1.10. (a) Sketch of a shaft undergoing circular synchronous whirling, i.e., with whirl speed ω equal to the spin speed Ω; (b) situation on the xy-plane: The shaded area of the cross section is under tensile stress.

The speed range spanning from zero to the first critical speed is usually referred to as the *subcritical range*; above the first critical speed, the *supercritical range* starts. A growing number of machines work in the supercritical range, and then at least one of the critical speeds must be crossed during startup and shut-down procedures.

If the Campbell diagram related to flexural vibrations is made by straight lines parallel to the Ω-axis, i.e., if the natural frequencies are independent of the speed, the numerical values of the critical speeds coincide with those of the natural frequencies at standstill, as can be seen from Figure 1.11. Some confusion between the concepts of critical speed and the natural frequency that can still be found can probably be ascribed to this fact. Even if the numerical values are coincident, the two physical phenomena are different, particularly where the stressing of the rotor is concerned.

In addition to flexural critical speeds, torsional critical speeds can also be very dangerous, particularly in the case of reciprocating machinery. Many devices whose aim is to reduce the amplitude of the vibrations induced by critical speeds have been developed.

The very concept of critical speed has been defined with reference to a linear system, and it is impossible to define critical speeds in this sense in the case of nonlinear rotors. However, a more general definition of critical speed, as a spin speed in which strong vibrations are encountered, is often used. This definition, which also holds in the case of nonlinear rotors, has a certain degree of arbitrarity, because the amplitude of the vibration depends on the cause producing it. In the case of nonlinear rotors, the speed at which the maximum amplitude is reached, i.e., the critical speed following

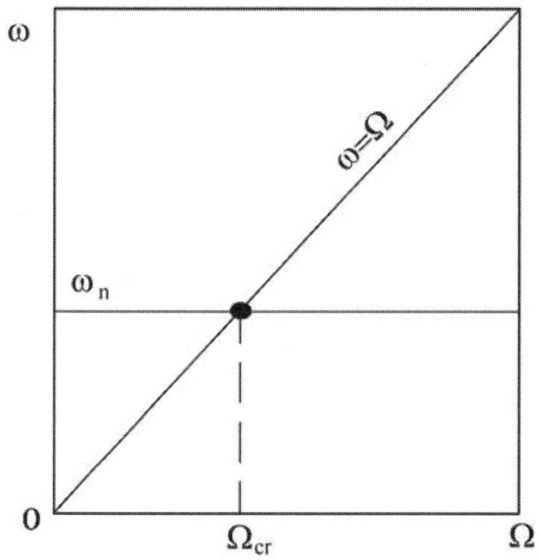

FIGURE 1.11. Critical speed of a rotor whose Campbell diagram is flat.

the last definition, also depends on the strength of the exciting causes (for example, the amount of unbalance in the case of flexural critical speeds). The critical speeds of linear systems are, on the contrary, characteristic of the system and are independent from the excitation.

Fields of instability

Rotors may develop an unstable behavior in well-defined velocity ranges. The velocities at which this unstable behavior occurs must not, however, be confused with the critical speeds of the rotor because the two phenomena are completely different. The term *unstable* can have several meanings, and different definitions of stability exist, one of the most common being that introduced by Liapunov.

An equilibrium configuration of a system is stable, following Liapunov criterion, if the system returns to a configuration close enough to the equilibrium one when displaced from it. Of course, this displacement must not be too large. If the system tends asymptotically in time to the equilibrium configuration, the latter is asymptotically stable.[3]

[3] The definition of stability introduced by Liapunov is based on a state space representation of the motion of the system. Consider a vector $\vec{X}(t)$ in the state space and use the symbol $|\vec{X}(t)|$ for its euclidean norm. Let $\vec{X}_0$ be an equilibrium position. Such a position is stable if, for any arbitrarily small positive quantity ϵ, a positive quantity δ exists such that the inequality

$$|\vec{X}(t) - \vec{X}_0| < \epsilon \qquad \text{for } 0 \leq t < \infty \,, \tag{1.9}$$

holds if $|\vec{X}(0) - \vec{X}_0| < \delta$.

It is possible to demonstrate that a linear system is asymptotically stable if all solutions for free motion defined by Equation (1.8) are such that all values of s have a negative real part; i.e., all $\sigma_i < 0$.

However, the theoretical definition of stability introduced by Liapunov may be difficult to apply in many engineering situations, and a technical definition of stability can be used: The behavior of a machine is considered *stable* when the amplitude of vibration in normal operation does not exceed a value considered acceptable. For rotating machinery, a definition of technical stability was stated by A. Muszynska [21] in the terms:

A rotating machine is stable if its rotor performs a pure rotational motion around an appropriate axis at a required rotational speed and this motion is not accompanied by other modes of vibrations of the rotor, its elements or other stationary parts of the machine, or, if such vibrations take place, their amplitudes do not exceed admitted, acceptable values. The stable rotating machine is immune to external perturbing forces, i.e., any random perturbation cannot drastically change its behavior. Such a perturbation causes only a transient decaying process leading to a previous regime of performance, or to a new one, which is included in the acceptable limits.

The amplitude of free vibration of damped linear systems decays exponentially in time, because of the energy dissipation caused by damping. In the case of rotors, however, there is a source of energy, the centrifugal field, that may in some cases cause an unbounded growth in time of the amplitude of free vibrations. The ranges of the spin speed in which this growth occurs, i.e., in which self-excited vibrations can develop, are usually called *instability fields* or *instability ranges*, and the speed at which the first of such field starts is the *threshold of instability*.

Instability ranges must not be confused with critical speeds: Critical speeds are a sort of resonance between a natural frequency and a forcing function acting on the rotor, and in instability ranges, true self-excited vibrations occur. They need the presence of some source of energy to sustain the vibration with increasing amplitude, and in this case, the energy can be supplied by the kinetic energy linked with rotation at the spin speed Ω. It is easy to verify that the kinetic energy stored in the rotor is greater by some orders of magnitude than the elastic potential energy the rotor can store without failure.

This means that any trajectory starting within a multidimensional sphere (if the state space has more than two dimensions, a circle in the case of a single-degree-of-freedom system, whose state space has two dimensions) of radius δ centered in the equilibrium point in the state space remains within a multidimensional sphere of radius ϵ for all values of time. If

$$\lim_{t \to \infty} |\vec{X}(t) - \vec{X}_0| = 0\,, \tag{1.10}$$

then the equilibrium position is asymptotically stable.

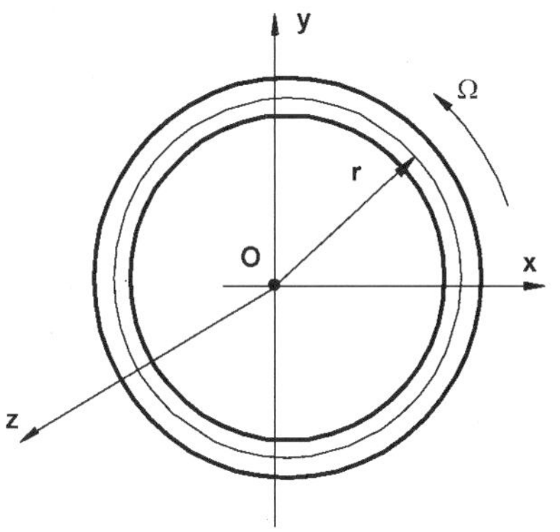

FIGURE 1.12. Thin ring of mean radius r rotating at the speed Ω.

Consider, for example, a thin ring with radius r and material density ρ rotating at speed Ω (Figure 1.12). As the hoop stress in the ring is simply

$$\sigma_h = \rho\Omega^2 r^2 ,$$

the relationship linking the kinetic energy with the stress is

$$\mathcal{T} = \frac{1}{2} m r^2 \Omega^2 = \frac{1}{2} \frac{m\sigma_h}{\rho} .$$

The maximum potential energy the ring can store in an axi-symmetrical tensile deformation is

$$\mathcal{U} = \frac{1}{2} V \frac{\sigma_U^2}{E} = \frac{1}{2} m \frac{\sigma_U^2}{\rho E} ,$$

where E, V, and σ_U are the Young's modulus, the volume of the material, and the ultimate strength, respectively. Note that the material has been assumed to be linear up to failure. The ratio between the kinetic energy and the potential energy corresponding to a deformation causing the failure of the ring (at failure $\sigma_h = \sigma_U$) is

$$\frac{\mathcal{T}}{\mathcal{U}} = \frac{E}{\sigma_U}.$$

By introducing the expression of the hoop stress into the last equation, it is easy to see that ratio E/σ_U is nothing other than the ratio v_s/v_U between the speed of sound in the material and the peripheral velocity v_U at failure of the ring, whose value is usually far greater than 10. The kinetic energy stored in the rotor is then at least one or two orders of magnitude greater than the energy needed to deform the rotor until failure occurs. Similar considerations would hold for other geometrical configurations or deformation patterns. It is then sufficient that a small portion of the kinetic energy of the rotor is transformed into deformation potential energy to cause failure.

Remark 1.8 *A structural system is inherently stable, as it can only dissipate energy, unless some mechanism that may supply energy to it is present. This is the case of the inverted pendulum (energy supplied by the gravitational field), aeroelastic vibrations (energy supplied by the aerodynamic field and ultimately by the kinetic energy of the aircraft), rotors (energy supplied by rotation), and active controlled systems (energy supplied by the controller). In all of those cases, instability may occur and the designer must study very carefully the conditions that assure a stable working of the system.*

It is easy, at least from a theoretical point of view, to predict the onset of unstable working conditions in a linear system by assuming that the free motion has a time history expressed by Equation (1.8) and by studying the sign of the decay rate, i.e., the sign of the real part of the complex frequency s.

At vanishingly small speeds, all values of $\sigma = \mathrm{Re}(s)$ are obviously negative (at least if the system does not contain active devices), because there is no external source of energy that can excite vibration. With increasing speed, the decay rate of some modes can decrease in absolute value showing a reduction of stability. If at a certain value of the spin speed Ω one of them vanishes and then becomes positive, that speed is the threshold of instability of the system.

A threshold of instability is clearly visible in the decay rate plot of Figure 1.4: One of the values of σ becomes positive at a speed slightly lower than 150,000 rpm.

Actually, it can be very difficult to estimate the decay rates with enough precision, because they are influenced by many factors that are very difficult to evaluate, one of them being damping. Often, it is only possible to perform a first-approximation theoretical or numerical study, whose results must be verified experimentally.

As a general rule, if the conditions for uncoupling of flexural, axial, and torsional behavior are met, only the first can give way to self-excited vibration. There are many mechanisms that may cause unstable conditions, including internal rotor damping caused by material damping and friction between the various components assembled by bolting, riveting, shrink fitting, and so on; rubbing between stator and rotor; fluid forces in journal bearings and seals; and energy dissipation caused by eddy currents. All of the mentioned mechanisms are potentially dangerous, but they do not necessarily always cause instability. The better known destabilizing effects are those caused by material damping in the rotor and lubricated journal bearings. The latter can cause the well-known oil whip phenomenon, consisting in very strong vibrations starting from a speed that is, in many cases, almost twice the first flexural critical speed.

To make it easier to distinguish between critical speeds and fields of instability, the following features may be listed:

Critical speeds

- They occur at well-defined values of the spin speed.
- The amplitude grows linearly in time if no damping is present. It can be maintained within reasonable limits, and as a consequence, a critical speed can be passed.
- The value of the speed is fixed, but that of the maximum amplitude depends on the amplitude of the perturbation causing it. In particular, the main flexural critical speeds do not depend on the amount of unbalance, but the amplitude increases with increasing unbalance.

Fields of instability

- Their span is usually large. Often, all speeds in excess of the threshold of instability give way to unstable behavior.
- The threshold of instability, if it exists, is usually located in the supercritical range.
- The amplitude grows exponentially in time. It grows in an uncontrollable way, and then working above the threshold of instability is impossible. When it falls within the working range, the system must be modified to raise it well above the maximum operating speed. Only possible nonlinearities of the system can prevent the amplitude from growing without limits, giving way to a limit cycle.

Some tables that can help in the identification of the causes of anomalies in the dynamic behavior of rotating machinery are here reported from Ehrich and Childs [22] (Tables 1.1-1.3).

1.1.5 Forced response

Response to unbalance

By introducing the forcing function caused by unbalance into the equation of motion (1.2), it is possible to obtain the response of the system in terms of the displacement vector $\mathbf{q}(t)$ and then to compute the stressing of the rotor and the vibration it induces on the nonrotating parts of the machine. If the system has axial symmetry, the response to unbalance is a circular whirling taking place at the spin speed of the machine. The rotor then does not vibrate, but simply spins in a deflected configuration; the material

	Forced or resonant vibrations	**Self-excited vibrations**
Relationship between frequency and speed	Frequency is equal to (i.e., synchronous with) the spin speed or a whole number or rational fraction of spin speed.	Frequency is nearly constant and essentially independent of spin speed or any external excitation or/and is at or near one of the shaft natural frequencies.
Relationship between amplitude and speed	Amplitude will peak in a narrow band of spin speed wherein the rotor's natural frequency is equal to the spin speed or to a whole number multiple or a rational fraction of the spin speed and is independent on external excitation.	Amplitude will suddenly increase at a threshold speed and continues at high or increasing levels as spin speed is increased.
Whirl direction	Almost always forward whirling.	Generally forward whirling, but backward whirling has been reported.
Rotor stressing	Static stressing in case of synchronous whirling.	Oscillatory stressing at frequency equal to $\omega - \Omega$.
Correcting actions	1. Introduce damping to limit peak amplitudes at critical speeds. 2. Tune the system's critical speeds to be outside the working range. 3. Eliminate all deviations from axial symmetry in the system as built or as induced during operation (e.g., balancing).	1. Increase damping to increase the threshold of instability above the operating speed range. 2. Raise the rotor natural frequencies as high as possible. 3. Identify and eliminate the instability mechanism.
Influence of damping	Addition of damping may reduce peak amplitude but does not affect the spin speed at which it occurs.	Addition of damping may raise the speed at which instability occurs but usually does not affect the amplitude after onset.
Influence of system geometry	Excitation level and hence amplitude are dependent on some lack of axial symmetry in the rotor mass distribution or geometry or external forces applied to the rotor. Amplitudes may be reduced by refining the system to make it more axisymmetric or balanced.	Amplitudes are independent of system axial symmetry. Given an infinitesimal deflection to an otherwise axisymmetric system, the amplitude will self-propagate for whipping speeds above the threshold of instability.

TABLE 1.1. Characterization of forced and self-excited rotor vibrations.

Mechanism	**Ratio** ω/Ω	**Direction**
Internal rotor damping	$0,2 < \omega/\Omega < 1$ ($\omega/\Omega = 0,5$)	Forward
Hydrodynamic bearings, labyrinth, or liquid seals	$\omega/\Omega < 0,5$ ($0,45 < \omega/\Omega < 0,48$)	Forward
Blade-tip clearance excitation	Dependent on fluid force levels	Forward
Centrifugal pump and compressor whirl	Dependent on fluid force levels	Forward
Propeller and turbomachinery whirl	Dependent on fluid force levels	Backward, if the vertex of the cone described in the whirl motion is after the rotor (referring to the direction of the fluid flow). Forward in the opposite case
Excitation due to fluid trapped in rotors	$0,5 < \omega/\Omega < 1,0$ ($0,7 < \omega/\Omega < 0,9$)	Forward

TABLE 1.2. Diagnostic table of self-excited vibrations of rotating machinery.

Mechanism	**Correcting action**
Internal rotor damping	Minimize number of separate elements in rotor; restrict span of rabbets and shrink-fitted parts; provide secure lock up on assembled elements.
Hydrodynamic bearings	Install tilting pad or rolling elements bearings.
Labyrinth seals	Add swirl brakes at seal inlets to reduce the inlet tangential velocity. Replace rotor-mounted labyrinth vanes with stator-mounted vanes. Replace labyrinth seals with honeycomb seals.
Liquid seals for pumps	Introduce swirl webs or brakes at seal inlet; roughen stator elements.
Blade-tip clearance	No ready measures that do not affect the unit operating efficiency.
Centrifugal pump and compressor whirl	Not well understood.
Propeller and turbomachinery whirl	Modify mode shapes to minimize angular motion of the plane of turbomachinery.
Excitation due to fluid trapped in rotors	Introduce drain holes to eliminate fluid accumulation or axial webs to inhibit rotation of fluid.

TABLE 1.3. Design correcting actions to reduce instability.

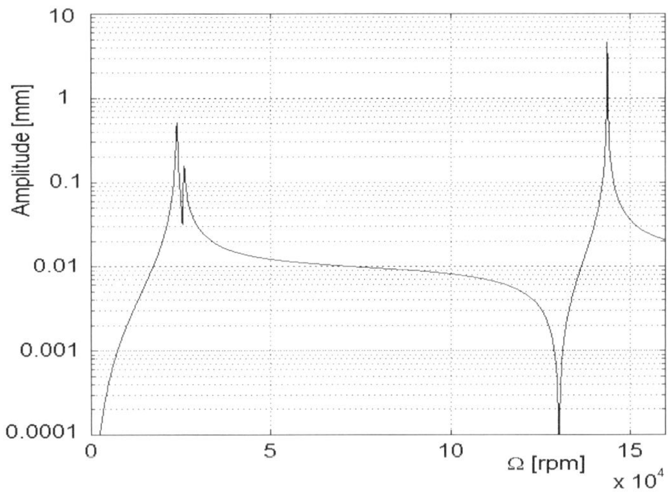

FIGURE 1.13. Amplitude of the unbalance response for a damped axi-symmetrical rotor. The plot refers to the same small turbojet engine used for Figure 1.2.

that constitutes the rotating parts of the machine is not subject to fatigue (at least, to the high number of stress cycles linked with high-frequency vibration) and its damping plays no role in its behavior. In the case of an undamped system (actually it is enough that the nonrotating damping vanishes) the amplitude goes to infinity in correspondence of the critical speed. Real-world machines, which are always damped, show a more or less pronounced vibration peak at the critical speed; to reduce its amplitude, the designer must increase nonrotating damping.

In the case of rotors running on nonisotropic bearings or stator, the whirling becomes elliptical, and in certain cases backward whirling may occur. If both stator and rotor are nonisotropic, the response becomes polyharmonic and the motion quite complex.

The unbalance response is usually reported in graphical form as a plot of the amplitude of the circular orbit (or, in case of elliptical whirling, of the semi-axes of the elliptical orbit) at a selected location as a function of the spin speed (Figure 1.13).

A more complete graphical representation of the unbalance response of a rotor, particularly useful in the case of the elliptical whirling of a rotor running on an anisotropic stator, is the plotting of the *orbital tube* (Figure 1.14). The orbits are plotted in a tridimensional graph, stacked along the spin-speed axis. The tube starts as a point at zero speed and then enlarges, taking an elliptical cross section. At very high speed, because of self-centering, it tends to a circular cylinder with radius equal to the eccentricity. The projection on the xy-plane (orbital view) directly gives the

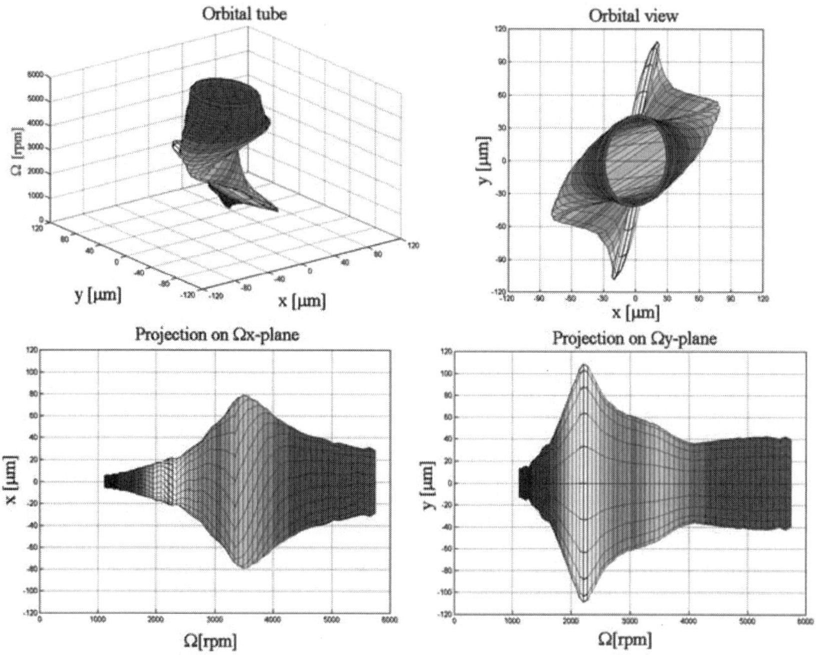

FIGURE 1.14. Orbital tube representation. (a) Three-dimensional view of the tube; (b) projection on the xy-plane; (c) and (d) projection on the Ωx- and Ωy-planes.

orbits at various speeds, superimposed on each other. The projections on the Ωx- and Ωy-planes give the peak-to-peak amplitude as a function of the spin speed.

Polyharmonic motion

Unbalance is not the only forcing function acting on a rotor, and seldom the measured response is harmonic. The synchronous component of the vibration of a rotating machine is always present and in most cases is the largest one, but it is usually accompanied by other components. In rotordynamic jargon, the synchronous component is usually referred to as the 1× component, whereas those with frequencies that are multiples of the spin frequency are called 2×, 3×, 4×, etc. components.

The experimental analysis of the vibrations caused by rotating machinery yields much useful information on the working conditions and allows the discovery of possible problems before consequences become too severe and, in some cases, even predicts their occurrence. The ultimate aim of the analysis is to diagnose the state of the machine to be able to perform preventive maintenance. To monitor the dynamic behavior of the machine, it is often

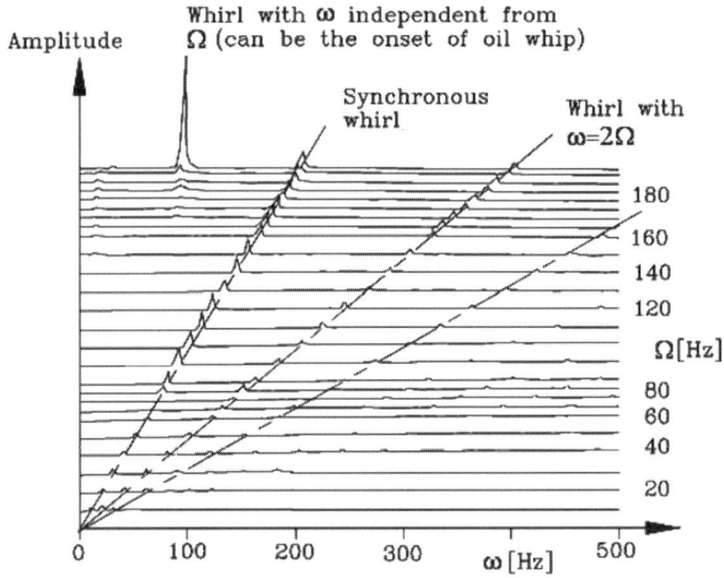

FIGURE 1.15. Cascade plot of a rotor.

sufficient to attach transducers that can measure the acceleration, velocity, or displacement in selected locations. In some machines, the transducers can be mounted permanently, and the signal they supply can be monitored continuously or at regular intervals or even only when working anomalies become apparent. Different kinds of transducers are available, but presently they are almost always connected to electronic data-acquisition systems that can perform various types of analysis and supply the relevant information in the form the user feels is more expedient. In particular, it is very useful to perform a harmonic analysis of the output of the transducers to obtain the acceleration or displacement spectrum. A very common way of representing these spectra is the so-called cascade plot: A tridimensional plot in which the spectra obtained at different spin speeds are reported in the planes with y constant in a tridimensional space xyz (Figure 1.15). Often, the frequency is reported in Hertz, and the spin speed is in revolutions per minute or revolutions per second.

Usually, the terms *cascade plot* and *waterfall plot* are used for two different types of diagrams. In the first, the spectra are plotted at different spin speeds; in the second, they are obtained at the same speed but at different times. The waterfall plot is not used to study the behavior of the machine in different working conditions but to follow the evolution in time of its dynamic behavior.

In the cascade plot, it is possible to identify, at each speed, the frequencies of the various motions of the rotor. If different transducers are placed in different radial planes, then by comparing the phases of the relevant signals, it is possible to assess the direction of the various whirl components. From the frequency and phase information, it is possible to study the causes that produce the vibration and to decide the proper correcting actions.

For example, the synchronous component is usually linked with unbalance and can be corrected by performing a more accurate balancing, whereas a component with frequency equal to twice the rotational speed is generally linked with rotor anisotropy and can be corrected by making it more symmetrical. As a general rule, balancing the rotor has very little effect on all of the nonsynchronous components.

Each machine produces a characteristic vibration spectrum, which is often referred to as the *mechanical signature* of the machine. Any alteration in time of the signature, as evidenced by a waterfall plot, is the symptom of an anomaly of the working conditions and must be considered very carefully. It can actually be linked with a problem that has occurred or is developing. Diagnosing problems before they actually occur, is important to reduce costs associated with maintenance and with the unavailability of the machine.

1.2 Nonlinear rotordynamics

Linearity is only an idealization of the actual behavior of any system, and real-world rotors always deviate, to a greater or lesser extent, from linearity. Classic rotordynamic analysis is typically a linearized analysis, and many of the concepts seen above, like the critical speeds or the instability threshold, are valid only within the limits imposed by linearity. Although rotors are often linear in their nominal conditions, at least if the displacements remain within the limits allowable in operation, other components like bearings, dampers, and seals can display a severe nonlinear behavior. The alternatives of considering these elements as rigid (e.g., in the case of bearings), to neglect them (e.g., seals) or to linearize their behavior, are very common, but in many cases the results obtained in this way may be an unsatisfactory approximation. Another possible cause of nonlinearity is the presence of a crack in some structural element, which causes a difference between local tensile and compressive characteristics. If nonlinearities are not neglected, many of the tools and concepts typical of rotordynamics, like the Campbell diagram and the critical speed analysis, lose their meaning and the attention must be concentrated on the unbalance response and on the spectral analysis of the vibration records. But this is only one aspect of the problem: When a system is highly nonlinear, new phenomena can develop, particularly those related with the presence of subharmonics and superharmonics and of chaotic vibrations. Chaotic motion was found in numerical simulations on simple mathematical rotor models, and results that indicate the possible presence of chaotic phenomena have been found in test results on turbojet engines.

If the system is isotropic with respect to the rotation axis, the unbalance response takes the form of circular whirling, even if this solution may be no more unique and other solutions may exist, each one with its own basin of attraction. As usual with nonlinear systems, it is possible that the response shows *jumps* from one equilibrium configuration to another one and that the configuration taken at any given speed during the spin-up of the system is different from that taken at the same speed when decelerating. If axial symmetry is not granted, as it may occur also in the case of isotropic systems operating in asymmetrical conditions, e.g., when a lateral load acts on the rotor, the whirling is no more circular and no closed-form solution can be found. The only way to study the unbalance response is by numerical integration in time of the nonlinear equations of motion.

At any rate, whether it is possible to reach a closed-form solution or not, it is no more possible to distinguish between a free and a forced behavior, and unbalance influences both the vibration frequency and the stability of the system.

A definition of critical speed as the speed at which the unbalance response of the rotor has a peak has been introduced to extend critical speed analysis to nonlinear rotors, but such a definition has the drawback that the critical speeds obtained in this way depend on the unbalance of the rotor. A similar consideration holds for the stability analysis: When the amplitude of the whirl motion grows, nonlinear phenomena become more and more important and a limit cycle may be reached. A rotor that is unstable in the small may reach stable working conditions because of nonlinearities; whether these conditions can be accepted or not depends on the resulting whirl amplitude and the design specifications of the specific machine.

1.3 Nonstationary rotordynamics

Rotordynamic analysis is usually performed at constant spin speed. A fast accelerating rotor may, however, behave in a way that is different from the behavior in steady-state running. In other cases, it may be the unbalance that changes quickly in time, as in the so-called blade-loss conditions of turbine rotors.

Even if the rotor can be assumed to behave in a linear way, in case of spin speed variations, not only the matrices appearing in Equation (1.2) become time-dependent, but they are also different because a number of terms linked with angular acceleration enter the equation of motion. In this condition, the solution must be obtained through numerical integration in time, and as a consequence, there is little advantage in linearizing the equation of motion.

Similarly, the blade-loss problem is studied through numerical integration in time, and nonlinearities add little to the complexity of the problem.

A further cause of variation of the angular velocity is present in free rotors: In this case, there is no device that keeps the spin speed constant and the amplitude of the motion may be larger. The motion of free rotors must then be studied under the condition that the angular momentum remains constant, i.e., that the spin speed and the whirl speed vary continuously in such a way that the total angular momentum is conserved.

1.4 Time domain versus frequency domain

Classic rotordynamics deals, as already stated, with linear (or at least linearized) systems, operating in steady-state condition. The analysis is based on a frequency domain approach and yields a comprehensive insight on the way any particular rotor behaves in the different conditions. The fact that the natural frequencies and the mode shapes depend on the speed makes things more complicated than in standard structural dynamics, but it does not change the basic approach. Frequency domain solutions allow us to perform parametric and optimization studies, in which a number of configurations are studied to refine the design or to assess its robustness against the variation of the values of some parameters or working conditions.

When the steady-state assumption is dropped, frequency domain solutions become impossible and the only way to solve the relevant equations is by numerical integration in time. The same occurs in the case of nonlinear rotors, even if in rotordynamics circular whirling is an exact solution for the nonlinear case (on the contrary, in structural dynamics, nonlinearities make it impossible for the system to vibrate with harmonic time history). However, even if circular whirling is possible, other solutions, which cannot be obtained through frequency domain analysis, may be present, and general solutions, describing the behavior of the system without the need to specify the particular working conditions, do not exist.

The time domain approach, based on the numerical integration in time of the differential equations of motion, although today very popular, has the notable drawback of allowing us to solve only particular cases, without yielding a complete insight on the relevant phenomena. Keeping this in mind, time domain modeling allows us to perform different types of transient analysis, such as blade loss or critical speed crossing, dropping the usual assumptions of linearity and axial symmetry of at least either the stator or the rotor.

The possible frequency and time domain approaches for the various types of models are summarized in Table 1.4.

An approach that allows us to perform time domain simulation is using standard multibody codes. In this case, the rotor is not dealt as such (i.e,.

Type of analysis	System	Frequency domain Eigenanalysis	Unbal. res.	Time domain
Linear, st. state	Isot. (R or S)(a)	YES	YES	Possible
	Nonisotropic	Approx.(b)	Approx.(b)	YES
Nonlin., st. state	Isotropic (all)	NO	YES(c)	YES
	Nonisot. (R or S)(a)	NO	NO	YES
Transient	Any	NO	NO	YES

TABLE 1.4. Frequency and time domain analysis for different types of rotors. (a) Either rotor or stator or both; (b) only a finite number of the infinite harmonics present in the response can be computed; (c) the closed-form solution is one of the possible solutions (perhaps not the stable one).

the usual assumptions that make rotordynamics a peculiar field of system dynamics are neglected) but as a general spinning body. This is, for example, the traditional way to analyze the dynamic behavior of spinning spacecraft. As the study is not limited to small displacements about an equilibrium position and the spin speed is not constrained to remain constant, free rotors can be studied in the correct condition of conservation of angular momentum. For fixed rotors, it is possible to model the drive system either as a device that imparts a constant (or controlled) velocity or a constant (or controlled) torque. This has a cost in terms of computational complexity, particularly if the system contains compliant bodies, and above all of the impossibility of gaining a deep general insight of the dynamic behavior of the system, comparable with that obtained once eigenvalues and eigenvectors are known.

Finally, when resorting to the time domain approach, there is no possibility of using the hysteretic damping model. Although for nonrotating damping it is possible to use the standard methods of structural dynamics for transforming hysteretic into viscous damping, in the case of rotating damping, things are much more complicated.

Part I

Basic topics

2
Jeffcott rotor

The simplest model that can be used to study the flexural behavior of rotors consists of a point mass attached to a massless shaft. As its dynamic behavior was deeply studied in a paper published by Jeffcott in 1919 [16], it is often referred to as *Jeffcott rotor*; however, this attribution is incorrect as August Föppl in 1895 published a paper [13] in which its behavior is correctly analyzed (he referred to it as *De Laval rotor*) and Stodola [15] and Belluzzo [14] described it in their books on turbomachinery of the first years of the twentieth century.

Although the Jeffcott rotor model is an oversimplification of real-world rotors, it retains some basic characteristics and allows us to gain a qualitative insight into important phenomena typical of rotordynamics, while being much simpler than more realistic models.

2.1 Undamped Jeffcott rotor

The three schemes sketched in Figure 2.1(a)-(c) yield the same results, as long as the system is

- Undamped, i.e., no damping effect is associated either to the springs or to the shaft,
- Axially symmetrical.

The overall stiffness k providing the restoring force can be considered the stiffness of the shaft, the supporting structure, or a combination of the two.

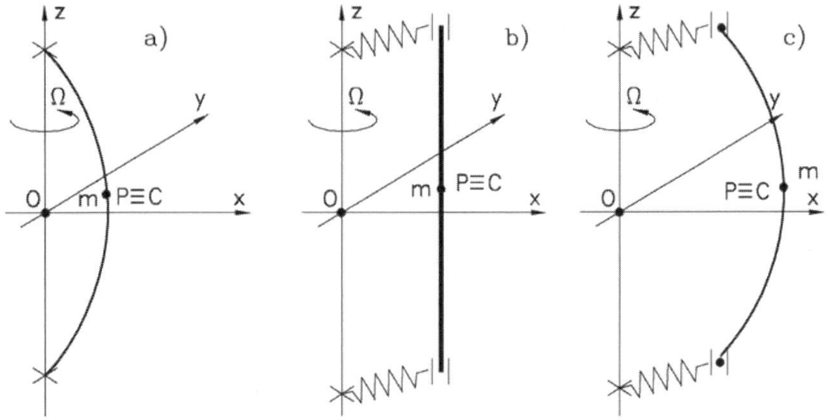

FIGURE 2.1. Perfectly balanced Jeffcott rotors. In (a), the rotor consists of a point mass on a flexible shaft running on stiff bearings. In (b), the shaft is stiff while the bearings are compliant. In (c), both the shaft and the bearings are considered as deformable bodies.

Point P, in which mass m is fixed, is always contained in the xy-plane. This statement is justified by the uncoupling between axial and radial motions and relies on the small displacement assumptions that are at the base of linear structural analysis. In the study of the flexural behavior, a model with only two degrees of freedom can then be used.

The mentioned schemes are, however, too much idealized: In practice, it never occurs that point P in which mass m is located exactly coincides with the elastic center C of the cross section of the shaft, i.e., with the point at which the elastic reaction of the shaft acts. However small may be the distance between points C and P, the presence of the eccentricity ϵ, as in Figure 2.2(a) and (b), causes a static unbalance $m\epsilon$ that can strongly affect the behavior of the system.

As in elementary rotordynamics, the spin speed Ω of the system is assumed to be constant; taking as initial time ($t = 0$) the instant in which vector $\overline{\text{P-C}}$ is parallel to the x-axis, the angle between $\overline{\text{P-C}}$ and the x-axis is $\theta = \Omega t$.

2.1.1 *Equations of motion*

Two choices for the generalized coordinates are possible: either coordinates x_C and y_C of point C, i.e., the geometric or elastic shaft center, or coordinates x_P and y_P of point P, i.e., the center of mass.

Using the first alternative, by far the most common one, the position and velocity of point P can be expressed as

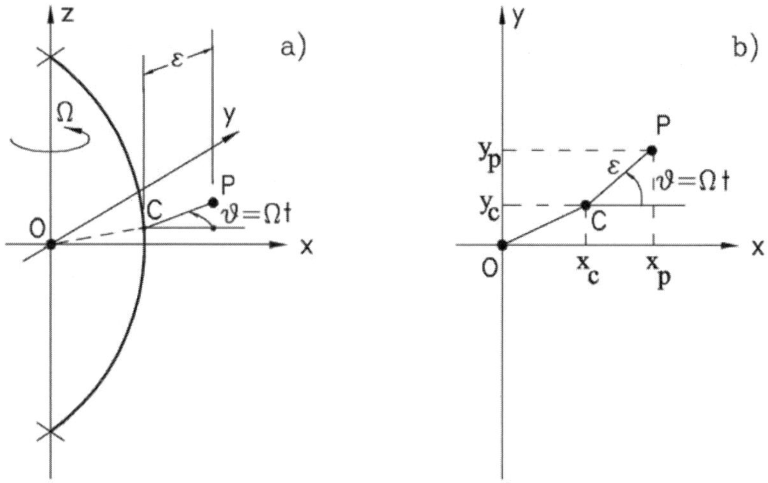

FIGURE 2.2. Unbalanced Jeffcott rotor, with unbalance $m\epsilon$. (a) Sketch of the system; (b) situation in the xy-plane.

$$\overline{\text{P-O}} = \mathbf{r}_\text{P}(t) = \left\{ \begin{array}{c} x_\text{P}(t) \\ y_\text{P}(t) \end{array} \right\} = \left\{ \begin{array}{c} x_\text{C}(t) + \epsilon \cos(\Omega t) \\ y_\text{C}(t) + \epsilon \sin(\Omega t) \end{array} \right\} , \tag{2.1}$$

$$\dot{\mathbf{r}}_\text{P}(t) = \left\{ \begin{array}{c} \dot{x}_\text{P}(t) \\ \dot{y}_\text{P}(t) \end{array} \right\} = \left\{ \begin{array}{c} \dot{x}_\text{C}(t) - \epsilon\Omega \sin(\Omega t) \\ \dot{y}_\text{C}(t) + \epsilon\Omega \cos(\Omega t) \end{array} \right\} . \tag{2.2}$$

The kinetic energy $\mathcal{T}$ and the potential energy $\mathcal{U}$ are, respectively,

$$\begin{gathered} \mathcal{T} = \tfrac{1}{2} m \left(\dot{x}_\text{P}^2 + \dot{y}_\text{P}^2 \right) = \\ = \tfrac{1}{2} m \left\{ \dot{x}_\text{C}^2 + \dot{y}_\text{C}^2 + \epsilon^2 \Omega^2 + 2\epsilon\Omega \left[-\dot{x}_\text{C} \sin(\Omega t) + \dot{y}_\text{C} \cos(\Omega t) \right] \right\} , \\ \mathcal{U} = \frac{1}{2} k \left(x_\text{C}^2 + y_\text{C}^2 \right) . \end{gathered} \tag{2.3}$$

A Lagrange equation can be written in the form

$$\frac{d}{dt} \left(\frac{\partial (\mathcal{T} - \mathcal{U})}{\partial \dot{q}_i} \right) - \frac{\partial (\mathcal{T} - \mathcal{U})}{\partial q_i} = Q_i , \tag{2.4}$$

where q_i are the Lagrangian coordinates, here x_C and y_C.

Assuming that an external force acts on point P in xy-plane (e.g., the weight of the rotor in case the axis of rotation is horizontal), forces Q_i can be easily obtained by assuming a virtual displacement of point C $[\delta x_\text{C}, \delta y_\text{C}]^T$. As the angular velocity is imposed by the driving system (i.e., angle $\theta = \Omega t$ does not depend on the generalized coordinates), the virtual displacement of point P is $[\delta x_\text{C}, \delta y_\text{C}]^T$ and the virtual work $\delta\mathcal{L}$ of a force F with components F_x and F_y acting on point P is

$$\delta\mathcal{L} = F_x \delta x_C + F_y \delta y_C \ . \tag{2.5}$$

The generalized forces can then be computed as $Q_i = \partial \delta\mathcal{L} / \partial \delta q_i$.

By performing the relevant derivatives and remembering that the spin speed Ω has been assumed to be constant, the following equations of motion are then obtained:

$$\begin{cases} m\ddot{x}_C(t) + kx_C(t) = m\epsilon\Omega^2 \cos(\Omega t) + F_x(t) \ , \\ m\ddot{y}_C(t) + ky_C(t) = m\epsilon\Omega^2 \sin(\Omega t) + F_y(t) \ , \end{cases} \tag{2.6}$$

where forces $F_x(t)$and $F_y(t)$ are considered as generic functions of time whereas unbalance forces are of the same amplitude but in time quadrature.

As usual, the general solution of Equation (2.6) can be obtained by adding the general solution of the homogeneous Equation (the complementary function)

$$\begin{cases} m\ddot{x}_C(t) + kx_C(t) = 0 \ , \\ m\ddot{y}_C(t) + ky_C(t) = 0 \ , \end{cases} \tag{2.7}$$

to the particular integral of the complete equation. Equations (2.7) yield the free motion of the perfectly balanced Jeffcott rotor, whereas Equations (2.6) yield the response to the static unbalance $m\epsilon$ and the response to a external force acting in the xy-plane. Note that, owing to linearity, it is possible to study separately the response to unbalance and that to a static force.

As already stated, it is possible to use the coordinates x_P and y_P of point P (center of gravity) as generalized coordinates. In this case, the position of point C can be expressed as

$$\overline{\text{C-O}} = \mathbf{r}_C(t) = \left\{ \begin{array}{c} x_C(t) \\ y_C(t) \end{array} \right\} = \left\{ \begin{array}{c} x_P(t) - \epsilon\cos(\Omega t) \\ y_P(t) - \epsilon\sin(\Omega t) \end{array} \right\} . \tag{2.8}$$

The kinetic and potential energies are

$$\mathcal{T} = \frac{1}{2} m \left(\dot{x}_P^2 + \dot{y}_P^2 \right) \ , \tag{2.9}$$

$$\mathcal{U} = \frac{1}{2} k \left(x_P^2 + y_P^2 + \epsilon^2 - 2\epsilon \left[x_P \cos(\Omega t) + y_P \sin(\Omega t) \right] \right) \ . \tag{2.10}$$

By introducing the Lagrangian function and the generalized forces F_x and F_y in the Lagrange equations, the following equations of motion of point P are then obtained:

$$\begin{cases} m\ddot{x}_P(t) + kx_P(t) = k\epsilon\cos(\Omega t) + F_x(t) \ , \\ m\ddot{y}_P(t) + ky_P(t) = k\epsilon\sin(\Omega t) + F_y(t) \ . \end{cases} \tag{2.11}$$

2.1.2 *Free whirling*

The equations of motion along each axis are coincident with the equation of the free motion of a system with a single degree of freedom. A purely mathematical approach to the solution of Equations (2.7) is to assume an exponential solution

$$\begin{cases} x_C(t) = x_{C_0}e^{st} \ , \\ y_C(t) = y_{C_0}e^{st} \ , \end{cases} \tag{2.12}$$

and solve for $s \in C$

$$\begin{cases} \left(ms^2 x_{C_0} + kx_{C_0}\right) e^{st} = 0 \ , \\ \left(ms^2 y_{C_0} + ky_{C_0}\right) e^{st} = 0 \ . \end{cases} \tag{2.13}$$

As $e^{st} \neq 0$ and a nontrivial solution is sought, i.e., $x_{C_0} \neq 0$ and $y_{C_0} \neq 0$, it implies that

$$\begin{cases} ms^2 + k = 0 \ , \\ ms^2 + k = 0 \ . \end{cases} \tag{2.14}$$

The absolute value of s that satisfies Equation (2.14) coincides with the natural frequency of the nonrotating system

$$\omega_n = \sqrt{k/m} \ ,$$

and the four solutions $\pm i\omega_n$ (actually two solutions, each one with multiplicity 2) are purely imaginary, owing to the conservative nature of the system.

Because of the symmetry of the system about x- and y-axes, the motion of point C is given by the combination of two harmonic motions taking place along axes x and y with the frequency ω_n coinciding with the natural frequency of the nonrotating shaft

$$\begin{cases} x_C(t) = X_1 e^{i\omega_n t} + X_2 e^{-i\omega_n t} \ , \\ y_C(t) = Y_1 e^{i\omega_n t} + Y_2 e^{-i\omega_n t} \ . \end{cases} \tag{2.15}$$

Constants X_1, X_2, Y_1, and Y_2 can be determined from the initial conditions on the positions $x_C(0)$ and $y_C(0)$, and on the velocity $\dot{x}_C(0)$, and $\dot{y}_C(0)$

$$\begin{aligned} x_C(0) &= X_1 + X_2 \ , & \dot{x}_C(0) &= i\,(X_1 - X_2)\,\omega_n \ , \\ y_C(0) &= Y_1 + Y_2 \ , & \dot{y}_C(0) &= i\,(Y_1 - Y_2)\,\omega_n \ . \end{aligned} \tag{2.16}$$

Substitution into Equations (2.15) gives the solution in terms of the initial conditions

$$\begin{cases} x_C(t) = x_C(0)\cos(\omega_n t) + \frac{1}{\omega_n}\dot{x}_C(0)\sin(\omega_n t) \ , \\ y_C(t) = y_C(0)\cos(\omega_n t) + \frac{1}{\omega_n}\dot{y}_C(0)\sin(\omega_n t) \ , \end{cases} \tag{2.17}$$

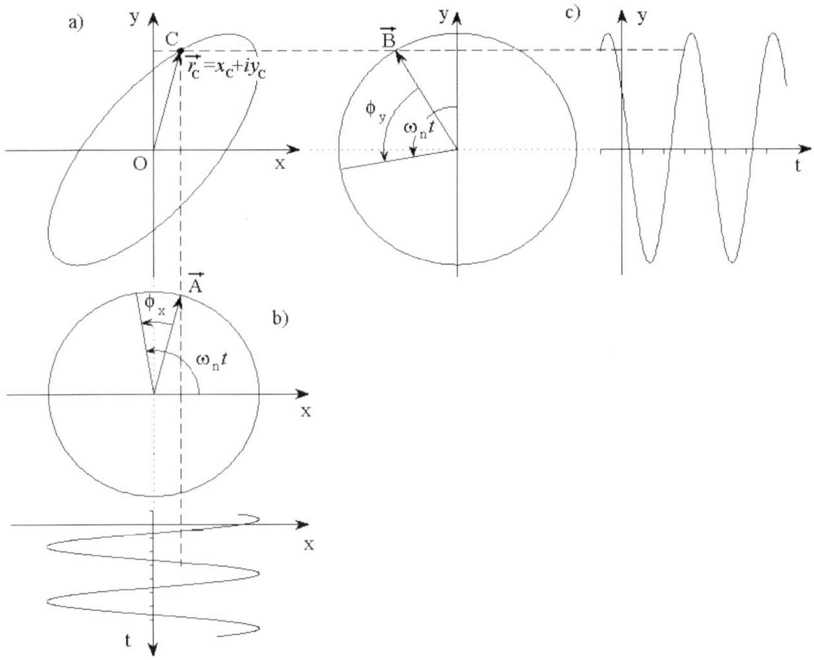

FIGURE 2.3. Response of the Jeffcott rotor to the free vibration. (a) Real (in x and y) and complex (in $r = x + iy$) representation of the orbit of point C; (b) and (c) representation of coordinates $x(t)$ and $y(t)$ of point C in the vector planes as projections on the x- and y-axes of rotating vectors $\vec{A}$ and $\vec{B}$.

which coincides with the responses of two decoupled harmonic oscillators. As usual, the solution can also be expressed in terms of amplitude and phase angles

$$\begin{cases} x_C(t) = X\cos(\omega_n t - \phi_x) \ , \\ y_C(t) = Y\cos(\omega_n t - \phi_y) \ , \end{cases} \tag{2.18}$$

where

$$\begin{array}{ll} x_C(0) = X\cos\phi_x \ , & \dot{x}_C(0) = \omega_n X \sin\phi_x \ , \\ y_C(0) = Y\cos\phi_y \ , & \dot{y}_C(0) = \omega_n Y \sin\phi_y \ . \end{array} \tag{2.19}$$

The equation of the motion as expressed by Equations (2.18) can be represented in a graphical form as shown in Figure 2.3. The trajectory of point C may be seen as that of vector $\mathbf{r}_C(t)$ whose coordinates $x_C(t)$ and $y_C(t)$ at time t are given by harmonic functions (2.18) of amplitudes X and Y, and phases ϕ_x and ϕ_y. They can also be thought of as the projection of rotating vectors $\vec{A}$ and $\vec{B}$ on the auxiliary axes x and y of Figure 2.3.

Remark 2.1 *It is worth noticing that although point* C *moves in the physical xy-plane, vectors $\vec{A}$ and $\vec{B}$ rotate in auxiliary vector planes where only one dimension has a physical meaning. This dimension is represented in Figure 2.3 by a continuous line, and the other dimension is auxiliary and is represented by a dotted line.*

Remark 2.2 *The trajectory of point* C *may be circular, elliptical, or rectilinear in any direction in the xy-plane depending on the initial conditions. Although in the vector planes [Figure 2.3(b) and (c)] the rotating vectors $\vec{A}$ and $\vec{B}$ rotate at constant angular velocity ω_n, in the physical xy-plane vector $\overline{O\text{-}C}$ rotates at nonconstant angular velocity but with a period equal to $2\pi/\omega_n$.*

Remark 2.3 *It can be appreciated that, if the initial conditions are such that the orbit is circular, the center line of the rotor rotates inflected but it does not vibrate in the plane containing the rotation axis. In this sense, the dynamic response is a whirling instead of a vibration. If the condition of synchronous whirling is added, no point of the rotor is subjected to vibration.*

Remark 2.4 *The natural frequency of the Jeffcott rotor is ω_n and is independent from the spin speed Ω. The flexural critical speed, defined as the speed at which the frequency of rotation is the same as the natural frequency, coincides with the natural frequency of the nonrotating system:*

$$\Omega_{cr} = \omega_n = \sqrt{\frac{k}{m}} \; . \tag{2.20}$$

Remark 2.5 *In the present case, the natural frequency of the rotor (i.e., the whirl speed) does not depend on the spin speed Ω: The Campbell diagram of a Jeffcott rotor is then made of horizontal straight lines. As already stated, the flexural critical speed, defined as the speed at which the natural frequency of the system is coincident with the frequency of rotation, coincides with the natural frequency of the nonrotating system, i.e., Equation (2.20).*

Matrix form and lambda matrix

The same result may be obtained recasting the system of Equations (2.7) in matrix form

$$\mathbf{r}_\mathrm{C}(t) = \left\{ \begin{array}{c} x_\mathrm{C}(t) \\ y_\mathrm{C}(t) \end{array} \right\} ,$$

$$\left[\begin{array}{cc} m & 0 \\ 0 & m \end{array} \right] \ddot{\mathbf{r}}_\mathrm{C}(t) + \left[\begin{array}{cc} k & 0 \\ 0 & k \end{array} \right] \mathbf{r}_\mathrm{C}(t) = 0 \; , \tag{2.21}$$

and using the general theory to solve the equations of motion [see appendix A, characteristic Equation (A.12)]

$$\det \begin{bmatrix} ms^2 + k & 0 \\ 0 & ms^2 + k \end{bmatrix} = 0 \ , \tag{2.22}$$

i.e.,

$$m^2 s^4 + 2kms^2 + k^2 = 0 \ . \tag{2.23}$$

A quadratic equation in s^2 is so obtained. Solving it in s^2, a double solution $s^2 = -k/m$ is found, accounting for the four solutions with double multiplicity that are also the eigenvalues of the associated eigenproblem. For what concerns the eigenvectors, the system is clearly decoupled and made of two single degree-of-freedom systems, and thus the eigenvectors may be chosen unitary and orthogonal

$$\mathbf{r}_1 = \begin{Bmatrix} 1 \\ 0 \end{Bmatrix} , \qquad \mathbf{r}_2 = \begin{Bmatrix} 0 \\ 1 \end{Bmatrix} , \tag{2.24}$$

and the resulting time solution

$$\mathbf{r}_{\mathrm{C}}(t) = \sum_{i=1}^{2} \mathbf{r}_i \left(C_i e^{s_i t} + C_i^* e^{s_i^* t} \right) \tag{2.25}$$

that may be expanded to Equations (2.15) and following.

2.1.3 Unbalance response

According to the equation of motion (2.6), when neglecting the external forces, the unbalance forces are represented as periodic forces in quadrature and with frequency equal to the rotational speed. The particular solution, i.e., the steady-state solution, of such a set of differential equations has the form

$$\begin{cases} x_{\mathrm{C}}(t) = x_{\mathrm{C}_0} \cos(\Omega t) \ , \\ y_{\mathrm{C}}(t) = y_{\mathrm{C}_0} \sin(\Omega t) = y_{\mathrm{C}_0} \cos(\Omega t + \pi/2) \ . \end{cases} \tag{2.26}$$

By introducing it into the equation of motion, the latter transforms into the algebraic equation

$$\begin{cases} \left(k - m\Omega^2\right) x_{\mathrm{C}_0} = m\epsilon\Omega^2 \ , \\ \left(k - m\Omega^2\right) y_{\mathrm{C}_0} = m\epsilon\Omega^2 \ , \end{cases} \tag{2.27}$$

which yields

$$x_{\mathrm{C}_0} = y_{\mathrm{C}_0} = \epsilon \frac{m\Omega^2}{k - m\Omega^2} = \epsilon \frac{\Omega^2}{\Omega_{cr}^2 - \Omega^2} = \epsilon \frac{\frac{\Omega^2}{\Omega_{cr}^2}}{1 - \frac{\Omega^2}{\Omega_{cr}^2}} \ , \tag{2.28}$$

where Ω_{cr} is the critical speed defined by Equation (2.20).

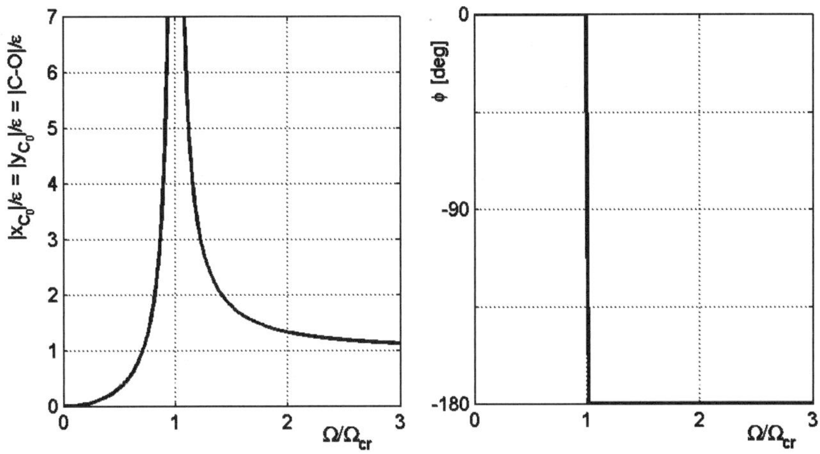

FIGURE 2.4. Amplitude and phase of the unbalance response of an undamped Jeffcott rotor.

The amplitudes of the two harmonic motions in xz- and yz-planes are equal and they are out of phase of 90°: The response is then a circular whirling occurring at speed Ω, i.e., a circular synchronous whirling. Vector $\overline{\text{C-O}}$ rotates with velocity Ω in the xy-plane remaining in line with vector $\overline{\text{P-C}}$. The value of velocity Ω that causes the denominator of the expressions for x_{C_0} and y_{C_0} in Equation (2.28) to vanish, i.e., which causes the amplitude to reach an infinite value, is coincident with the flexural critical speed of the rotor.

The amplitude of the circular orbit of point C caused by the presence of the unbalance $m\epsilon$, i.e., the *unbalance response*, coincides with either x_{C_0}or y_{C_0}. It can be expressed as a function of the speed in the nondimensional form derived from Equation (2.28)

$$\frac{|\overline{\text{C-O}}|}{\epsilon} = \frac{\sqrt{(x_{C_0}\cos\Omega t)^2 + (y_{C_0}\sin\Omega t)^2}}{\epsilon} = \tag{2.29}$$

$$= \frac{x_{C_0}}{\epsilon} = \frac{y_{C_0}}{\epsilon} = \frac{\Omega^2/\Omega_{cr}^2}{1-\Omega^2/\Omega_{cr}^2} \tag{2.30}$$

and plotted as in Figure 2.4.

In the subcritical range, i.e., $\Omega < \Omega_{cr}$, the whirling amplitude grows from zero tending to infinity at the critical speed Ω_{cr}. Both x_{C_0} and y_{C_0} remain positive, showing that the phase of the response with respect to the excitation is 0; i.e., points O, C and P are aligned in that order [Figure 2.5(a)].

In the supercritical range, i.e., $\Omega > \Omega_{cr}$, the values of x_{C_0} and y_{C_0} are negative, and their absolute values decrease monotonically with the speed. When the speed tends to infinity, the amplitudes tend to $-\epsilon$. The response is out of phase of 180° with respect to the excitation, and points O, C,

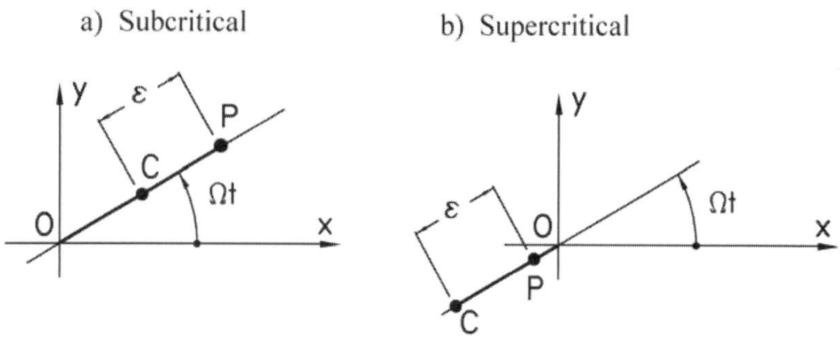

FIGURE 2.5. Unbalance response of an undamped Jeffcott rotor: situation in xy-plane for the cases of (a) subcritical running and (b) supercritical working of the undamped Jeffcott rotor.

and P are aligned in the order OPC [Figure 2.5(b)]. When the speed tends to infinity, point P tends to point O. This phenomenon is usually referred to as *self-centering* and indicates that the rotor tends to rotate about its center of mass instead of its geometrical center.

Remark 2.6 *The coincidence of the critical speed, computed from the unbalance response or from the free behavior of the system, with the natural frequency of the undamped nonrotating system, is a peculiar characteristic of the Jeffcott model or, more generally, of all those rotors in which the natural frequency does not depend on the speed, and must not be considered a general feature.*

Remark 2.7 *When the dynamic behavior is dominated by the stiffness, as it occurs in subcritical range, i.e.,* $\Omega < \sqrt{k/m}$*, rotation takes place about a point close to the geometrical center, whereas when the behavior is dominated by the inertia as in supercritical range, i.e.,* $\Omega > \sqrt{k/m}$*, the system rotates about a point close to the center of mass.*

2.1.4 Response to external forces in the frequency domain

The frequency domain approach is an alternative method of solving differential equations using the Laplace transform. It allows us to compute the system response to a generic but Laplace-transformable force function, i.e., those functions for which the following expression exists:

$$\hat{F}(s) = \int_0^{\infty} F(t)e^{st}dt = \mathcal{L}\left\{F(t)\right\}.$$

The differential time domain equation (2.6) with external force only

$$\begin{cases} m\ddot{x}_{\mathrm{C}}(t) + kx_{\mathrm{C}}(t) = F_x(t)\ , \\ m\ddot{y}_{\mathrm{C}}(t) + ky_{\mathrm{C}}(t) = F_y(t)\ , \end{cases} \tag{2.31}$$

is Laplace transformed as follows:

$$\begin{cases} (ms^2+k)\hat{x}_C(s) = \hat{F}_x(s) + x_{C_{t=0}}s + \dot{x}_{C_{t=0}} \ , \\ (ms^2+k)\hat{y}_C(s) = \hat{F}_y(s) + y_{C_{t=0}}s + \dot{y}_{C_{t=0}} \ , \end{cases} \tag{2.32}$$

where

$$\hat{x}_C(s) = \mathcal{L}\{x_C(t)\} = \int_0^\infty x_C(t)e^{st}dt \tag{2.33}$$

with $s \in C$. The system response in then obtained algebraically as

$$\begin{cases} \hat{x}_C(s) = \dfrac{1}{ms^2+k}\left[\hat{F}_x(s) + x_C(0)s + \dot{x}_C(0)\right] \ , \\ \hat{y}_C(s) = \dfrac{1}{ms^2+k}\left[\hat{F}_y(s) + y_C(0)s + \dot{y}_C(0)\right] \ , \end{cases} \tag{2.34}$$

where the first term on the right-hand side is the forced solution and the other two are the free response to nonzero initial conditions.

The so-called *transfer function* from input forces to output positions are defined as follows in the complex variables and for zero initial conditions:

$$\begin{cases} \dfrac{\hat{x}_C(s)}{\hat{F}_x(s)} = \dfrac{1}{ms^2+k} \ , \\ \dfrac{\hat{y}_C(s)}{\hat{F}_y(s)} = \dfrac{1}{ms^2+k} \ . \end{cases} \tag{2.35}$$

Unbalance force

The transfer function of a system is a complex function and may be represented in terms of its module and phase as a function of the complex variable s. In particular with $s = i\omega$, these diagrams are called Bode's diagrams, which give, frequency by frequency, the change of amplitude and phase of the output function, here a displacement, with respect to the input function, here a force.

Given the module of the transfer function

$$\left|\frac{\hat{x}_C(i\omega)}{\hat{F}_x(i\omega)}\right| = \left|\frac{1}{-m\omega^2+k}\right| \tag{2.36}$$

and the amplitude of the sinusoidal unbalance force $m\epsilon\Omega^2$ whose frequency is Ω, the amplitude of the sinusoidal displacement at the frequency of the spin speed $\omega = \Omega$ is obtained as

$$|\hat{x}_C(i\Omega)| = \left|\frac{m\epsilon\Omega^2}{-m\Omega^2+k}\right| \ , \tag{2.37}$$

which is the same result already obtained in equation (2.28).

Static force

The particular solution of the equation in which only a static (and constant in time) force F is applied is

$$\begin{cases} x_C(t) = x_0 \ , \\ y_C(t) = y_0 \ . \end{cases} \tag{2.38}$$

By introducing Equation (2.38) into the equation of motion (2.6), the following solution is obtained:

$$\begin{cases} x_0 = \dfrac{F_x}{k} \ , \\ y_0 = \dfrac{F_y}{k} \ . \end{cases} \tag{2.39}$$

Under the effect of a static force, the rotor takes the same static deformation that characterizes the nonrotating system. As will be seen later, this is because, so far, damping has not been included in the model.

2.2 Complex coordinates in rotordynamics

The simplicity of the Jeffcott rotor is used to introduce the complex coordinates' formalism for rotordynamics. As the center of the shaft, i.e., point C, moves in a plane, it is very expedient to express vector $\overline{\text{C-O}}$ by means of a complex number

$$r_C(t) = x_C(t) + iy_C(t), \tag{2.40}$$

usually referred to as a complex coordinate.[1] By multiplying the second of Equations (2.6) by the imaginary unit i and adding the two equations, a single complex equation is obtained

$$m\ddot{r}_C(t) + kr_C(t) = m\epsilon\Omega^2 e^{i\Omega t} + F_n(t) \ , \tag{2.41}$$

where the nonrotating force is $F_n(t) = F_x(t) + iF_y(t)$.

2.2.1 *Free whirling*

The solution of the homogeneous differential equation expressing the free motion of the system is

$$r_C(t) = r_{C_0} e^{st} \ , \tag{2.42}$$

[1] It is common to use the letter z instead of r for the complex coordinate. Here r is used to avoid confusion with the z-coordinate of the reference frame xyz.

where r_{C_0} is, generally speaking, a complex number $r_{C_0} = x_{C_0} + iy_{C_0}$. By introducing this solution into the homogeneous equation of motion, a homogeneous algebraic equation is obtained

$$\left(ms^2 + k\right) r_{C_0} = 0 \tag{2.43}$$

that has nontrivial solution $r_{C_0} \neq 0$ only if $s_{1,2} = \pm i\sqrt{k/m}$. The general solution s of the characteristic equation (2.43) can be split in its real and imaginary parts

$$s = \sigma + i\omega \ . \tag{2.44}$$

From the comparison with the values of $s_{1,2}$, it follows that

$$\sigma_{1,2} = 0 \ , \qquad \omega_{1,2} = \pm\sqrt{k/m} = \pm\omega_n \ . \tag{2.45}$$

The general solution of the homogenous equation of motion is then

$$r_C(t) = R_1 e^{i\omega_n t} + R_2 e^{-i\omega_n t} \tag{2.46}$$

$$= r_f(t) + r_b(t) \tag{2.47}$$

with R_1 and R_2 *non-necessarily conjugate* complex numbers (because the dynamic solution $r_C(t)$ is a complex valued function), and $r_f(t)$ and $r_b(t)$ rotating vectors in the physical plane.

Using real coordinates, the motion was split into the composition of two vibrations along the axes: Among the four solutions, two were related to the motion in x direction and two to the motion in y direction [see Figure 2.3 and mode shapes of Equation (2.24)]. In the present case, the solution is obtained as the sum of two vectors rotating in forward and in backward directions.

The motion expressed by Equation (2.46) is then the superimposition of a circular forward or direct whirl motion (i.e., occurring in the same direction as the spin speed that will always be considered positive) and a circular backward whirl motion. They both occur at an angular velocity, often called whirl speed, equal to the natural frequency ω_n of the nonrotating system.

The result of the superimposition of the two motions depends on the initial conditions, i.e., on the values of complex constants R_1 and R_2. If, for example, R_2 is equal to 0 a circular forward whirling occurs, and if the two constants are equal and real, a harmonic vibration along a direction at 45° with respect to the x-axis takes place.

Remark 2.8 *Vector* $\mathbf{r}_C(t)$ *of Equation (2.25) and complex number* $r_C(t)$ *are different mathematical representations of the same physical phenomenon. The result obtained obviously does not depend on which solution is used.*

When real coordinates are used, ω_n is the natural frequency of two harmonic motions in two planes [see Equation (2.18)] and their signs have no physical meaning. Vector $e^{i\omega_n t}$ can be used to express the motion in the form of Equations (2.15) and interpreted as a rotating vector in the auxiliary vector planes of Figure 2.3 whose real components only have a physical meaning.

When complex coordinates are used, ω_n is a true angular velocity, and the relevant vector rotates in the physical space. The deflected shape rotates about the undeformed configuration with angular velocity ω_n, although no rotation of a material object takes place with that speed. As a consequence, its sign states the direction of rotation of the deflected configuration.

2.2.2 Unbalance response

The particular integral of Equation (2.41) for the unbalance response is

$$r_C(t) = r_{C_0} e^{i\Omega t} .$$

By introducing it into Equation (2.41), the latter transforms into the algebraic equation

$$(-m\Omega^2 + k) r_{C_0} = m\epsilon\Omega^2 ,$$

which yields

$$r_{C_0} = \epsilon \frac{\Omega^2}{\Omega_{cr}^2 - \Omega^2} , \tag{2.48}$$

where Ω_{cr} is the critical speed defined by Equation (2.20). The value of the amplitude r_0 so obtained is real, and as a consequence, vector $(\overline{\text{C-O}})$, i.e., r_C, rotates with velocity Ω in the xy-plane remaining in line with vector $(\overline{\text{P-C}})$. The value of Ω that causes the denominator of the expression for r_0 in Equation (2.48) to vanish, i.e., the amplitude to reach an infinite value, is coincident with the flexural critical speed of the rotor. This result obviously coincides with that obtained earlier using real coordinates.

If instead of using Equation (2.6), which describes the motion of point C, Equation (2.11) describing the motion of point P was used, and introducing the complex coordinate

$$r_P = x_P + iy_P , \tag{2.49}$$

the relevant equation of motion would have been

$$m\ddot{r}_P + kr_P = k\epsilon e^{i\Omega t} + F_n . \tag{2.50}$$

The response to unbalance is in this case $r_P(t) = r_{P_0} e^{i\Omega t}$, where the amplitude r_{P_0} is

$$r_{P_0} = \epsilon \frac{k}{k - m\Omega^2} = \epsilon \frac{\Omega_{cr}^2}{\Omega_{cr}^2 - \Omega^2} . \tag{2.51}$$

Remembering that $r_{P_0} = r_0 + \epsilon$, this result coincides with that of Equation (2.48).

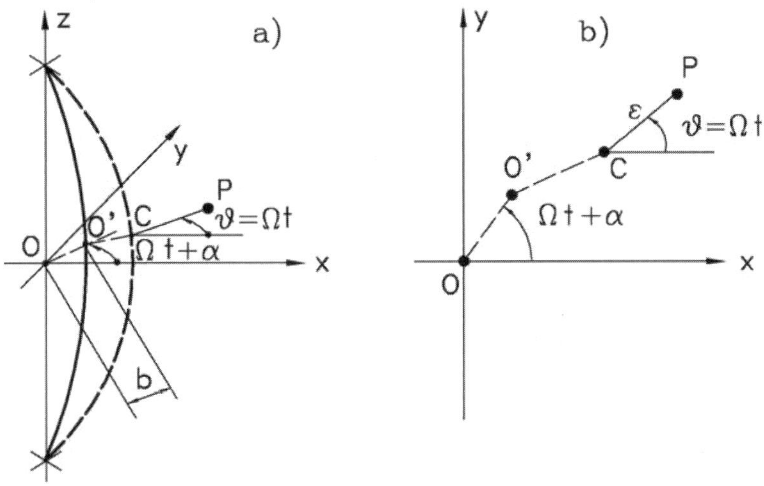

FIGURE 2.6. Jeffcott rotor with shaft bow. (a) Sketch. Full line: undeflected configuration; dashed line: deflected configuration. (b) Situation in xy-plane.

Self-centering becomes even more clear when using complex coordinates: When the speed tends to infinity, the amplitude of the orbit of point C tends to $-\epsilon$. The sign of the solution determines the equilibrium configurations as shown in Figure 2.5(a) and (b). When the solution is positive, in the subcritical field, points O, C, and P are aligned in the mentioned order and the center of mass of the rotor lies outside the deformed configuration of the shaft. In the supercritical field, however, point P lies between points C and O, and when the speed tends to infinity, the amplitude r_0 tends to $-\epsilon$, or point P tends to point O.

2.3 Jeffcott rotor with shaft bow

Eccentricity is not the only imperfection whose effects can be studied by means of the Jeffcott rotor model: The shaft of Figure 2.1(a) can be slightly bent in its undeflected configuration (Figure 2.6, full line). Note that the direction of the bow b is generally not coincident with the direction of the eccentricity ϵ: In the figure, it is assumed that the eccentricity has the direction of x-axis when $t = 0$, whereas the bow of the shaft has a direction that at the same time makes an angle α with the same axis.

As usual, as the system is linear, the effects of unbalance and shaft bow can be studied separately and then the results can be added to each other.

At time t, the position of point O' can be expressed as

$$(\overline{\text{O'} - \text{O}}) = \left\{ \begin{array}{c} x_{\text{O'}} \\ y_{\text{O'}} \end{array} \right\} = \left\{ \begin{array}{c} b\cos(\Omega t + \alpha) \\ b\sin(\Omega t + \alpha) \end{array} \right\} , \tag{2.52}$$

where b is the bow of the shaft. The elastic reaction of the shaft tends to bring back point C (deflected position of point O') toward O'. The potential energy is then

$$\mathcal{U} = \frac{1}{2}k(\overline{\text{C} - \text{O'}})^2 = \frac{1}{2}k\left\{[x_{\text{C}} - b\cos(\Omega t + \alpha)]^2 + [y_{\text{C}} - b\sin(\Omega t + \alpha)]^2\right\} , \tag{2.53}$$

i.e.,

$$\mathcal{U} = \frac{1}{2}k\left[x_{\text{C}}^2 + y_{\text{C}}^2 + b^2 - 2bx_{\text{C}}\cos(\Omega t + \alpha) - 2by_{\text{C}}\sin(\Omega t + \alpha)\right] . \tag{2.54}$$

By introducing Equation (2.54) instead of (2.3) into the Lagrange equations, the latter yields[2]

$$\left\{ \begin{array}{l} m\ddot{x}_{\text{C}} + k\left[x_{\text{C}} - b\cos(\Omega t + \alpha)\right] = m\epsilon\Omega^2\cos(\Omega t) + F_x , \\ m\ddot{y}_{\text{C}} + k\left[y_{\text{C}} - b\sin(\Omega t + \alpha)\right] = m\epsilon\Omega^2\sin(\Omega t) + F_y , \end{array} \right. \tag{2.55}$$

i.e.,

$$\left\{ \begin{array}{l} m\ddot{x}_{\text{C}} + kx_{\text{C}} = kb\cos(\Omega t + \alpha) + m\epsilon\Omega^2\cos(\Omega t) + F_x , \\ m\ddot{y}_{\text{C}} + ky_{\text{C}} = kb\sin(\Omega t + \alpha) + m\epsilon\Omega^2\sin(\Omega t) + F_y , \end{array} \right. \tag{2.56}$$

or, in complex coordinates,

$$m\ddot{r}_{\text{C}} + kr_{\text{C}} = kbe^{i\alpha}e^{i\Omega t} + m\epsilon\Omega^2 e^{i\Omega t} + F_n . \tag{2.57}$$

The particular solution of the equation containing only the term linked with shaft bow is

$$r_{\text{C}} = b\frac{\Omega_{cr}^2}{\Omega_{cr}^2 - \Omega^2}e^{i\alpha}e^{i\Omega t} . \tag{2.58}$$

i.e., is a synchronous circular whirling. The response to a shaft bow is reported in nondimensional form in Figure 2.7.

Apart from the phase α, this result is similar to the unbalance response, with the difference being that now the response is larger at lower speed (up to the critical speed) and smaller at higher speeds, tending to zero for $\Omega \to \infty$. Note that also here a sort of self-centering occurs at high speed, as the effect of shaft bow disappears. At low speed, the effect of shaft bow can be more important than that of unbalance, whereas the opposite occurs in the supercritical range. Recalling that the response here computed is because of shaft bow alone, when the speed tends to infinity and the response tends to zero, the rotor spins about point C, which coincides with point P and the rotor is self-centered.

---

[2] The term due to shaft bow is similar to the term due to the motion of the contraints in the equation of motion for vibrating systems, written with reference to an inertial frame.

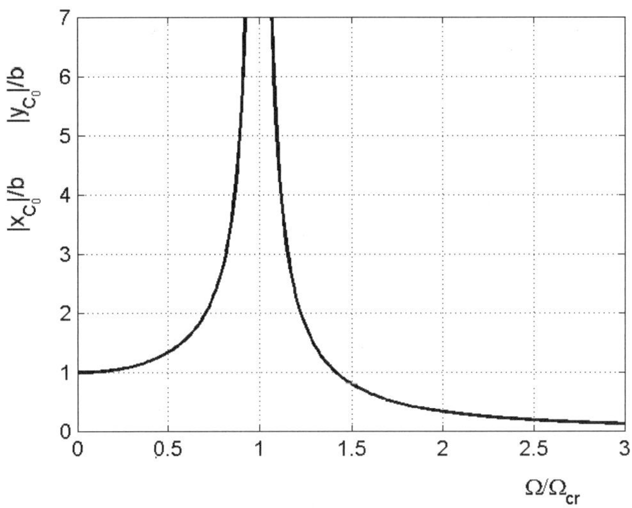

FIGURE 2.7. Nondimensional response of an undamped Jeffcott rotor to a shaft bow.

2.4 Jeffcott rotor with viscous damping

2.4.1 Equations of motion

When considering a damped rotor, it is very important to distinguish between the damping effects that can be associated with the stationary parts of the machine, usually referred to as *nonrotating damping*, and those directly associated with the rotor, or *rotating damping*. The former usually has a stabilizing effect that the designer can use to achieve the required stability in the whole working range of the machine. The latter, on the contrary, can reduce the amplitude of vibration in subcritical conditions but shows destabilizing effects in the supercritical range. Designers must then be very careful when studying machines operating in the supercritical range, taking into account that all mechanisms increasing energy dissipation within the rotor (such as material damping, friction in threaded, riveted, shrink-fitted connections in built-up rotors, splined shafts, intershaft dampers in multishaft machines, and so on) can cause severe instability problems.

The model of Figure 2.1 can be extended to the damped system by adding to the right-hand side of Equation (2.6) the generalized forces caused by damping. Because the generalized coordinates chosen are the displacements in the x- and y-directions of point C, such forces are the components in the directions of the x- and y-axes of damping forces applied to the point mass

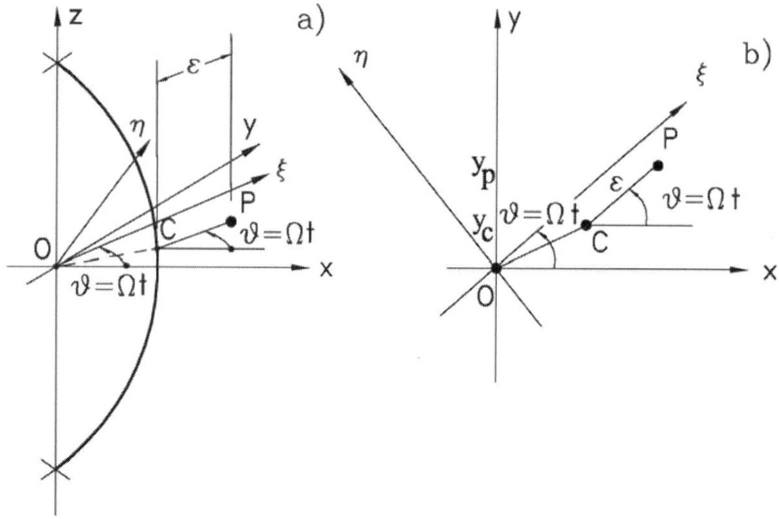

FIGURE 2.8. Definition of the rotating frame $O\xi\eta z$.

in point C. The force caused by viscous nonrotating damping is proportional to the speed of point C in a nonrotating reference frame

$$\vec{F}_n = \left\{ \begin{array}{c} F_{n_x} \\ F_{n_y} \end{array} \right\} = -c_n \left\{ \begin{array}{c} \dot{x}_C \\ \dot{y}_C \end{array} \right\} , \tag{2.59}$$

where c_n is the nonrotating damping coefficient.

The force caused by viscous rotating damping is proportional to the speed of point C as observed by a reference frame rotating with the same speed of the rotor. It follows that for the study of rotating damping, a rotating reference frame $O\xi\eta z$ must be introduced (Figure 2.8). The origin and z-axis of the rotating frame are the same as those of the fixed reference frame $Oxyz$ of Figure 2.1, but axes ξ and η rotate in the xy-plane at the same speed of the rotor. When the rotational speed is constant, the angle between the two reference frames is simply given by Ωt.

The force caused by rotating viscous damping can be expressed in the rotating reference frame $O\xi\eta z$ as proportional to the speed of point C, expressed in the rotating frame

$$\vec{F}_{r_{\xi\eta}} = \left\{ \begin{array}{c} F_{n_\xi} \\ F_{n_\eta} \end{array} \right\} = -c_r \left\{ \begin{array}{c} \dot{\xi}_C \\ \dot{\eta}_C \end{array} \right\} . \tag{2.60}$$

The coordinates ξ_C and η_C of point C in the rotating frame can be expressed as functions of the coordinates of the same point in the fixed frame as

$$\left\{ \begin{array}{c} \xi_C \\ \eta_C \end{array} \right\} = \mathbf{R} \left\{ \begin{array}{c} x_C \\ y_C \end{array} \right\} , \tag{2.61}$$

where the rotation matrix $\mathbf{R}$ is simply

$$\mathbf{R} = \left[\begin{array}{cc} \cos(\Omega t) & \sin(\Omega t) \\ -\sin(\Omega t) & \cos(\Omega t) \end{array} \right] . \tag{2.62}$$

By differentiating Equation (2.61), the velocity of point C can be expressed as

$$\left\{ \begin{array}{c} \dot{\xi}_C \\ \dot{\eta}_C \end{array} \right\} = \mathbf{R} \left\{ \begin{array}{c} \dot{x}_C \\ \dot{y}_C \end{array} \right\} + \dot{\mathbf{R}} \left\{ \begin{array}{c} x_C \\ y_C \end{array} \right\} , \tag{2.63}$$

where

$$\dot{\mathbf{R}} = \mathbf{\Omega} \left[\begin{array}{cc} -\sin(\Omega t) & \cos(\Omega t) \\ -\cos(\Omega t) & -\sin(\Omega t) \end{array} \right] . \tag{2.64}$$

From Equation (2.60), the force caused by rotating viscous damping can be expressed in the O$\xi\eta z$ frame as a function of the displacement and velocity of point C in the $Oxyz$ frame

$$\vec{F}_{r_{\xi\eta}} = \left\{ \begin{array}{c} F_{n_\xi} \\ F_{n_\eta} \end{array} \right\} = -c_r \left(\mathbf{R} \left\{ \begin{array}{c} \dot{x}_C \\ \dot{y}_C \end{array} \right\} + \dot{\mathbf{R}} \left\{ \begin{array}{c} x_C \\ y_C \end{array} \right\} \right) \tag{2.65}$$

The same force can be expressed in the fixed reference frame simply by multiplying Equation (2.65) by the rotation matrix $\mathbf{R}^{-1} = \mathbf{R}^T$

$$\vec{F}_{r_{xy}} = \mathbf{R}^T \vec{F}_{r_{\xi\eta}} = -c_r \left(\mathbf{R}^T\mathbf{R} \left\{ \begin{array}{c} \dot{x}_C \\ \dot{y}_C \end{array} \right\} + \mathbf{R}^T\dot{\mathbf{R}} \left\{ \begin{array}{c} x_C \\ y_C \end{array} \right\} \right) . \tag{2.66}$$

As usual with rotation matrices, $\mathbf{R}^T\mathbf{R} = \mathbf{I}$. With simple computations, it follows that

$$\mathbf{R}^T\dot{\mathbf{R}} -\Omega \left[\begin{array}{cc} 0 & 1 \\ -1 & 0 \end{array} \right] . \tag{2.67}$$

The force caused by rotating damping is then simply

$$\vec{F}_{r_{xy}} = -c_r \left\{ \begin{array}{c} \dot{x}_C \\ \dot{y}_C \end{array} \right\} - c_r\Omega \left[\begin{array}{cc} 0 & 1 \\ -1 & 0 \end{array} \right] \left\{ \begin{array}{c} x_C \\ y_C \end{array} \right\} . \tag{2.68}$$

Introducing the Expressions (2.59) and (2.68) for the forces caused by nonrotating and rotating damping on the right-hand side of the equation of motion (2.6) and writing the results in matrix form, it follows that

$$\begin{bmatrix} m & 0 \\ 0 & m \end{bmatrix}\begin{Bmatrix} \ddot{x}_C \\ \ddot{y}_C \end{Bmatrix}+\begin{bmatrix} c_n+c_r & 0 \\ 0 & c_n+c_r \end{bmatrix}\begin{Bmatrix} \dot{x}_C \\ \dot{y}_C \end{Bmatrix}+$$
$$+\left(\begin{bmatrix} k & 0 \\ 0 & k \end{bmatrix}+\Omega\begin{bmatrix} 0 & c_r \\ -c_r & 0 \end{bmatrix}\right)\begin{Bmatrix} x_C \\ y_C \end{Bmatrix}= \tag{2.69}$$
$$=\begin{Bmatrix} m\epsilon\Omega^2\cos(\Omega t) \\ m\epsilon\Omega^2\sin(\Omega t) \end{Bmatrix}+\begin{Bmatrix} F_x \\ F_y \end{Bmatrix}.$$

The presence of rotating damping gives thus way to skew-symmetric terms in the stiffness matrix, i.e., to a circulatory matrix. The circulatory matrix vanishes when Ω tends to zero, as it should be expected: When the spin speed vanishes, the rotor becomes a stationary system and its equation of motion becomes that of a natural system.

Instead of writing explicitly the forces due to nonrotating and rotating damping, the equation of motion (2.69) could be obtained by introducing the Rayleigh dissipation function into the Lagrange equation. Its expression is

$$\mathcal{F}=\frac{1}{2}c_n\left(\dot{x}_C^2+\dot{y}_C^2\right)+\frac{1}{2}c_r\left(\dot{\xi}_C^2+\dot{\eta}_C^2\right). \tag{2.70}$$

By introducing Equation (2.63) into Equation (2.70), it follows that

$$\mathcal{F}=\tfrac{1}{2}c_n\left(\dot{x}_C^2+\dot{y}_C^2\right)+$$
$$+\tfrac{1}{2}c_r\left(\begin{Bmatrix} \dot{x}_C \\ \dot{y}_C \end{Bmatrix}^T\mathbf{R}^T+\begin{Bmatrix} x_C \\ y_C \end{Bmatrix}^T\dot{\mathbf{R}}^T\right)\left(\mathbf{R}\begin{Bmatrix} \dot{x}_C \\ \dot{y}_C \end{Bmatrix}+\dot{\mathbf{R}}\begin{Bmatrix} x_C \\ y_C \end{Bmatrix}\right). \tag{2.71}$$

By performing the products and remembering that $\mathbf{R}^T\mathbf{R}=\mathbf{I}$ and $\dot{\mathbf{R}}^T\dot{\mathbf{R}}=\Omega^2\mathbf{I}$, it follows that

$$\mathcal{F}=\frac{1}{2}(c_n+c_r)\left(\dot{x}_C^2+\dot{y}_C^2\right)+\frac{1}{2}c_r\Omega^2\left(x_C^2+y_C^2\right)+c_r\Omega\begin{Bmatrix} \dot{x}_C \\ \dot{y}_C \end{Bmatrix}^T\mathbf{R}^T\dot{\mathbf{R}}\begin{Bmatrix} x_C \\ y_C \end{Bmatrix}, \tag{2.72}$$

i.e.,

$$\mathcal{F}=\frac{1}{2}(c_n+c_r)\left(\dot{x}_C^2+\dot{y}_C^2\right)+\frac{1}{2}c_r\Omega^2\left(x_C^2+y_C^2\right)+c_r\Omega\left(\dot{x}_Cy_C-\dot{y}_Cx_C\right). \tag{2.73}$$

By introducing Equation (2.73) into the Lagrange equation

$$\frac{d}{dt}\left(\frac{\partial(\mathcal{T}-\mathcal{U})}{\partial\dot{q}_i}\right)-\frac{\partial(\mathcal{T}-\mathcal{U})}{\partial q_i}+\frac{\partial\mathcal{F}}{\partial\dot{q}_i}=Q_i \tag{2.74}$$

the equation of motion (2.69) is readily obtained.

Also in this case it is possible to resort to complex coordinates: By multiplying the second Equation (2.69) by the imaginary unit i and adding it to the first one, the following equation is obtained

$$m\ddot{r}_C + (c_r + c_n)\dot{r}_C + (k - ic_r\Omega)r_C = m\epsilon\Omega^2 e^{i\Omega t} + F_n \ . \tag{2.75}$$

Equation (2.75) is formally identical to the equation of motion of a mass m suspended on a spring with complex stiffness $k - i\Omega c_r$ and a viscous damper with damping coefficient $c_n + c_r$ on which a force with harmonic time history with frequency Ω and amplitude $m\epsilon\Omega^2$ and a constant force are acting.

Remark 2.9 *What is skew-symmetric in real coordinates is imaginary when complex coordinates are used: This is a general rule.*

2.4.2 Some considerations on rotating damping

As it is clear from the equations seen in the previous section, nonrotating damping, i.e., the damping associated with the nonrotating parts of the machine, has the same role in rotordynamics as in the dynamics of natural, noncirculatory systems. This consideration holds in general.

Rotating damping, i.e., the damping associated with the rotor, performs two tasks: one is that of dissipating energy (in the equation of motion of linear rotors, this is seen by its presence in the damping matrix) and the other is that of transferring energy from the rotation of the system to its vibration (this can be seen from its presence into the circulatory matrix).

To better understand this feature, some examples can be mentioned. First consider a rotating dashpot in which a spherical pendulum oscillates [23] [Figure 2.9(a)]. The force the pendulum receives from the fluid is proportional to a damping coefficient c and to the velocity of the pendulum with respect to the fluid. Assume that at a certain instant the pendulum moves in radial direction [Figure 2.9(b)]. Vector $(\overline{\text{P-O}})$ can be expressed using the complex notation as $r = x + iy$, whereas the absolute velocity is $\dot{r} = \dot{x} + i\dot{y}$. If the fluid was not moving ($\Omega = 0$), the only force acting on the pendulum would be $-c\dot{r}$. If the dashpot rotates, the absolute velocity must be added to the velocity of the fluid with respect to the inertial frame. In point P, such velocity is equal, in modulus, to Ω multiplied by the modulus of r and has a direction that is perpendicular to the radius. By remembering that to rotate a vector by 90° in complex notation, it must be multiplied by i, the velocity vector that must be added is $i\Omega r$. As $\dot{r}$ is the absolute velocity of point P, the velocity of the same point relative to the fluid is

$$V_{\mathrm{P}_f} = \dot{r} - i\Omega r \ .$$

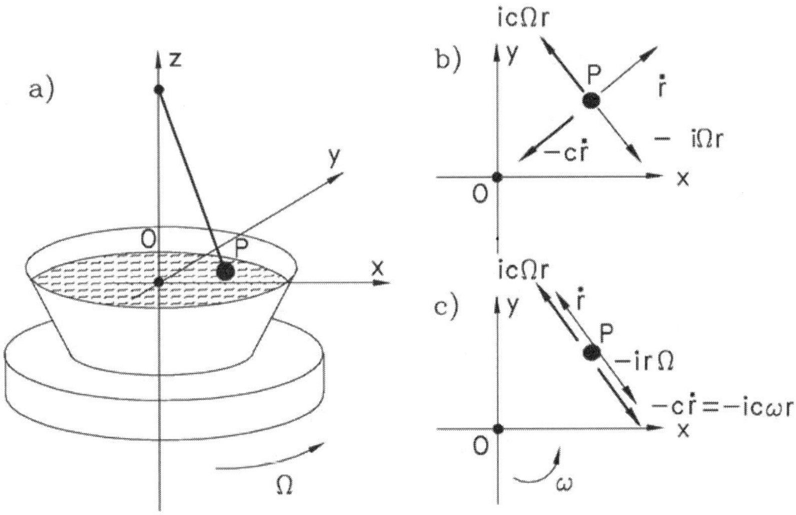

FIGURE 2.9. Example of rotating damping: a spherical pendulum in a rotating dashpot. (a) Sketch of the system; (b) situation in xy-plane when the pendulum moves radially toward the center; (c) situation in xy-plane when the pendulum moves on a circular orbit with whirl frequency ω (thin arrows: velocities, thick arrows: forces).

As the force caused by the fluid is assumed to be of the viscous type, the pendulum receives a force from the rotating fluid whose value is

$$F_{P_f} = -cV_{P_f} = -c(\dot{r} - i\Omega r) \; .$$

In the equation of motion a term made by an imaginary coefficient multiplying the displacement [$i\Omega r$ in Equation (2.75)] is thus present. As an imaginary term in complex coordinates is equivalent to a skew-symmetric term in real coordinates, the presence of a circulatory matrix is thus justified. The meaning of the imaginary unit in the formula is clear: The relevant component of the force is rotated by 90° with respect to the direction of the radius (the direction of vector r is radial).

Assume now that the pendulum performs a circular forward whirling motion with whirl speed ω [Figure 2.9(c)]. The absolute velocity $\dot{r}$ has now a direction perpendicular to the radius and its value is $|\dot{r}| = \omega|r|$. The velocity of the fluid with respect to the inertial frame is again perpendicular to the radius, and its value is Ωr. The total force received by the pendulum from the rotating fluid is thus $-icr(\omega - \Omega)$: If the spin speed is higher than the whirl speed, such force produces a counter clockwise moment that tends to sustain the whirl motion, transferring energy from rotation to whirling, a mechanism that can cause instability. On the contrary, if the spin speed is lower than the whirl speed, the moment caused by the damping force

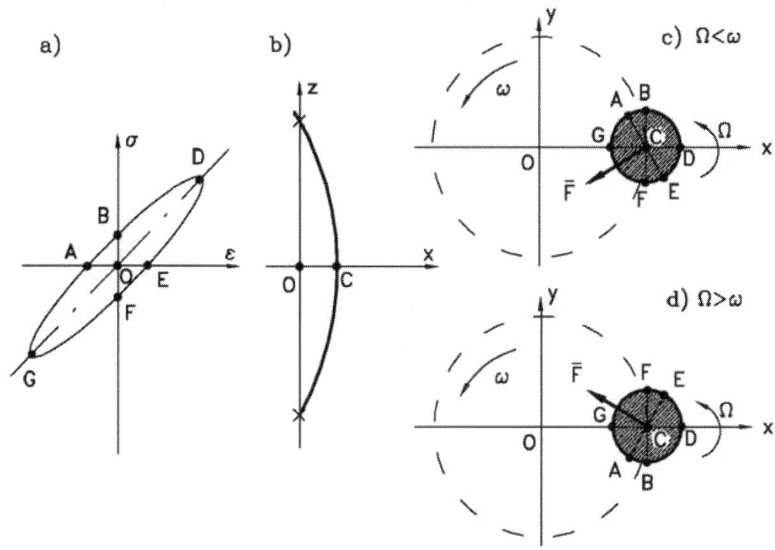

FIGURE 2.10. Shaft with hysteretic damping. (a) Hysteresis cycle of the material in $\sigma\epsilon$-plane; (b) sketch of the whirling system, seen when line $(\overline{\text{C-O}})$ is aligned with x-axis; (c) and (d) situation in xy-plane in circular whirling, respectively, with $\Omega < \omega$ (subcritical whirling) and $\Omega > \omega$ (supercritical whirling).

is in a clockwise direction and tends to oppose the whirl motion, i.e., to subtract energy from it.

A second example of rotating damping is that of the structural damping of a shaft. Internal damping can be modeled by modifying the stress-strain curve, which in a linear elastic material is a straight line, into an elliptical hysteresis cycle [Figure 2.10(a)]. During a stress cycle, the point on the $\sigma\epsilon$ plane moves on the ellipse in clockwise direction. The area under the line ABD, gone through during loading, is the energy supplied to the material, whereas the area under the line DE is the energy given back by the material during unloading. The area of the ellipse is then the energy dissipated by hysteresis. Actually, for structural materials, the ellipse is so narrow that it can hardly be distinguished from its larger axis OD.

Consider a compliant shaft running on stiff bearings and performing a generic circular whirling motion at the whirl speed ω [24]. Point C whirls on a circle centered on O [Figure 2.10(b)]. The situation in xy-plane is shown in Figure 2.10(c) and (d): The circle ABDEFG is the outline of the shaft, and the letters indicate the corresponding points on the hysteresis cycle of the material with reference to Figure 2.10(a).; the dashed circle is the orbit of C during circular whirling.

Due to the combination of the whirl and the spin motions, the material is subject to a strain cycle whose period, seen in the inertial frame, is that of a rotation of the deflected configuration about point O, i.e., corresponding to the whirl speed ω. As the shaft rotates with speed Ω, it goes through the hysteresis cycle at the frequency $\omega-\Omega$: If the whirling motion is synchronous ($\omega = \Omega$) the material doesn't go through the hysteresis cycle and rotating damping plays no role in the behavior of the system, as already stated.

Points B and F on the hysteresis cycle [Figure 2.10(a)] are those in which the strain vanishes. The corresponding points B and F on the cross section of the shaft in Figure 2.10(c) and (d) define the neutral line BF of strains. The fibers on the side of D with respect to BF are stretched and those on the other side are shortened. The points in which the stress vanishes are A and E; line AE is then the neutral line of stress, and the fibers on the side of D with respect to AE are under tensile stress whereas those on the other side are compressed. The elastic reaction of the shaft, that in absence of damping would be directed toward point O, is perpendicular to the neutral line of stresses and then has a small component perpendicular to line $(\overline{\text{C-O}})$.

If the rotation is slower than the whirl motion [$\Omega < \omega$, Figure 2.10(c)], the elastic reaction of the shaft has a moment about point O that opposes the whirl motion: Rotating damping tends to quench the whirl motion.

On the contrary, if the rotation is faster than the whirl motion [$\Omega > \omega$, Figure 2.10(d)], the elastic reaction of the shaft has a moment about O with the same direction of the whirl motion and tends to drive it, with the effect of transferring energy into the whirl motion and exciting it.

2.4.3 Free whirling

Equation (2.75) can be solved, as usual, by adding a particular integral to the complementary function. The solution of the homogeneous equation yielding the behavior of a perfectly balanced rotor is the usual one,

$$r_{\text{C}} = r_{\text{C}_0} e^{st} \ , \tag{2.76}$$

where both the amplitude r_{C_0} and the complex frequency $s = \sigma + i\omega$ are expressed by complex numbers. By introducing this solution into Equation (2.75), the following characteristic equation is obtained

$$ms^2 + (c_r + c_n)s + k - i\Omega c_r = 0 \, . \tag{2.77}$$

The roots of this quadratic equation with complex coefficients are

$$s = \sigma + i\omega = -\frac{c_r + c_n}{2m} \pm \sqrt{\frac{(c_r + c_n)^2 - 4m(k - i\Omega c_r)}{4m^2}} \, . \tag{2.78}$$

The real and imaginary parts of the complex frequency can be separated by resorting to the formula

$$\pm\sqrt{a+ib} = \pm\left[\sqrt{\frac{\sqrt{a^2+b^2}+a}{2}} + i\ \mathrm{sgn}(b)\sqrt{\frac{\sqrt{a^2+b^2}-a}{2}}\right] , \tag{2.79}$$

where sgn(b) indicates the sign of the imaginary part b. From Equation (2.78) it follows that the solutions are[3]

$$\begin{aligned} \sigma_{1,2} &= -\tfrac{c_r+c_n}{2m} \pm \tfrac{1}{\sqrt{2}}\sqrt{\sqrt{\Gamma^2+\left(\tfrac{\Omega c_r}{m}\right)^2}-\Gamma} , \\ \omega_{1,2} &= \pm\tfrac{\mathrm{sgn}(\Omega)}{\sqrt{2}}\sqrt{\sqrt{\Gamma^2+\left(\tfrac{\Omega c_r}{m}\right)^2}+\Gamma} , \end{aligned} \tag{2.80}$$

where subscripts $_{1,2}$ indicate the solutions with positive and negative signs, respectively, and

$$\Gamma = \frac{k}{m} - \frac{(c_r+c_n)^2}{4m^2} . \tag{2.81}$$

Two solutions of the characteristic Equation (2.77) can be found for each value of the spin speed Ω. They are not conjugate, because Equation (2.77) has complex coefficients.

The general solution of the homogeneous equation of motion is then

$$r = R_1 e^{(\sigma_1+i\omega_1)t} + R_1 e^{(\sigma_2+i\omega_2)t} . \tag{2.82}$$

The two complex constants R_1 and R_2 are to be determined from the initial conditions.

Because of the not null value of $\sigma_{1,2}$, the motion of point C has an amplitude varying exponentially with time. The resulting motion is then given by the superposition of two logarithmic spirals. The analysis of the stability and of the whirl direction of the two spiral motions can be made by taking the signs of the two solutions of Equation (2.80) into account.

If σ is negative, the amplitude decays in time and point C tends to point O. The rotor has a stable behavior because the whirl motion tends to reduce its amplitude. If σ is positive, the amplitude grows exponentially: The motion is unstable, as any small perturbation can trigger this self-excited whirling.

Solution $\sigma_1 + i\omega_1$

The sign of ω_1 in Equation (2.80) is always the same as the rotating speed Ω, and the corresponding logarithmic spiral is in forward direction. The

---

[3]In equation (2.79) the term in sgn(b) is included, to take into account the possibility that the spin speed is negative. In general, however, the spin speed is assumed to be positive (first and fourth quadrants of the Campbell diagram) and the term sgn(b) = sgn(Ω) is neglected.

sign of σ_1 can be positive or negative, leading to a self-excited or damped motion, respectively. With simple computations, the condition for stability can be shown to be

$$\Omega < \sqrt{\frac{k}{m}}\left(1 + \frac{c_n}{c_r}\right) . \tag{2.83}$$

The rotor is then stable in the subcritical range, i.e., when the speed is lower than the critical speed $\sqrt{k/m}$ of the undamped system. In the supercritical field, stability depends on the value of ratio c_n/c_r between the nonrotating and the rotating damping. If there is no nonrotating damping $(c_n = 0)$, the motion is unstable in the whole supercritical range. Increasing nonrotating damping, the threshold of instability becomes higher.

Solution $\sigma_2 + i\omega_2$

The sign of ω_2 in Equation (2.80) is opposite to the rotating speed Ω. As σ_2 is always negative, the corresponding spiral motion is in backward direction and it is stable. Usually this motion damps out quite fast, and this solution is of little practical interest.

Campbell diagram and root locus

Equation (2.80) can be rewritten in the following nondimensional form:

$$\begin{cases} \sigma^* = -\left(\zeta_r + \zeta_n\right) \pm \sqrt{\sqrt{\Gamma^{*2} + \Omega^{*2}\zeta_r^2} - \Gamma^*} \ , \\ \omega^* = \pm\mathrm{sgn}(\Omega^*)\sqrt{\sqrt{\Gamma^{*2} + \Omega^{*2}\zeta_r^2} + \Gamma^*} \ , \end{cases} \tag{2.84}$$

where

$$\begin{array}{ll} \omega^* = \omega/\sqrt{k/m} \ , & \sigma^* = \sigma/\sqrt{k/m} \ , \\ \Omega^* = \Omega/\sqrt{k/m} \ , & \Gamma^* = [1 - (\zeta_n + \zeta_r)^2]/2 \ , \\ \zeta_r = c_r/2\sqrt{km} \ , & \zeta_n = c_n/2\sqrt{km} \ . \end{array} \tag{2.85}$$

The nondimensional values of σ^* and ω^* are then functions of the spin speed Ω^* and of only two parameters ζ_n and ζ_r. A plot of σ^* and ω^* as functions of the nondimensional spin speed Ω^* is shown in Figure 2.11 (Campbell diagram). Four values of ζ_r have been considered, and rotating and nonrotating damping have been assumed to be equal $(\zeta_n = \zeta_r)$.

The whirl speed ω^* is little affected by the presence of damping: A value of $\zeta_n = 0.1$ is already high, and in most cases, the curve $\omega^*(\Omega^*)$ cannot be distinguished from that of the undamped case.

The roots locus is shown in Figure 2.12. It is the representation on the complex plane of the solutions $s_{1,2} = \sigma_{1,2} + i\omega_{1,2}$ of the characteristic equation (2.77) for a number of values of the whirl speed Ω. As solutions $s_{1,2}$ are not conjugate, the locus is not symmetrical with respect to the real axis. As already evidenced in the Campbell diagram, for high enough values

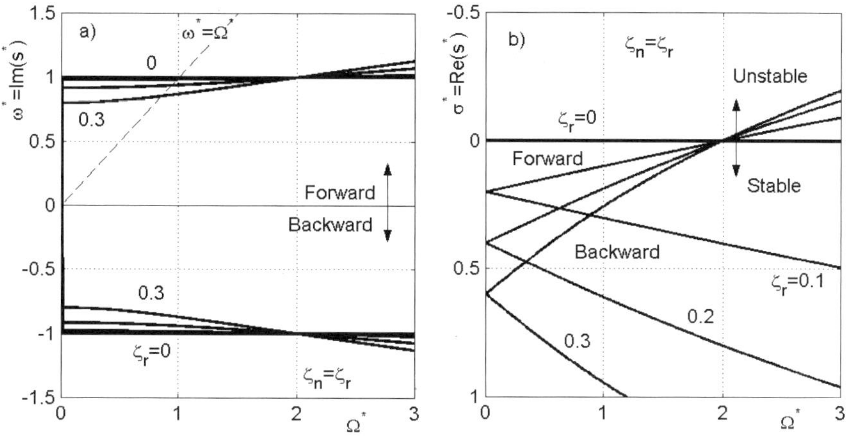

FIGURE 2.11. (a) Campbell diagram and (b) decay rate plot of a damped Jeffcott rotor with $\zeta_n = \zeta_r$, for three different values of ζ_n.

of the rotating speed Ω, the forward solution enters the first quadrant and the system becomes unstable. The locus representing the backward solution remains in the third quadrant for all values of the rotating speed: This solution does not give way to instability.

Free whirling in real coordinates

The same results could be obtained without resorting to the complex coordinates. By introducing the Solution (2.12) into the homogeneous equation associated with Equation (2.69), the following characteristic equation can be obtained:

$$\det \begin{bmatrix} s^2 m + s\,(c_n + c_r) + k & \Omega c_r \\ -\Omega c_r & s^2 m + s\,(c_n + c_r) + k \end{bmatrix} = 0 \,. \qquad (2.86)$$

It is a fourth-degree equation with real coefficients: It has four solutions that are either real or complex conjugate pairs. They are still expressed by Equation (2.80), but this time, all combinations of the double signs are possible. Clearly by resorting to real coordinates all information regarding the direction of whirl motion is lost and it is impossible to state from the study of the eigenvalues that backward whirling is always stable, whereas forward whirling can be unstable. It is still possible to assess the direction of each mode, but this information must be obtained from the study of the eigenvectors and eigenvalues.

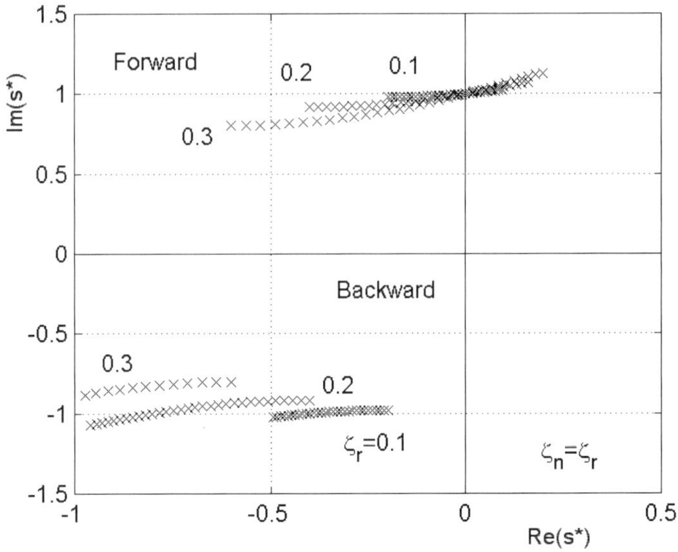

FIGURE 2.12. Roots locus of the same Jeffcott rotors as in Figure 2.11.

2.4.4 Unbalance response

If the rotor is not perfectly balanced, it is necessary to resort to the nonhomogeneous Equation (2.75). By resorting to complex coordinates, the particular integral, i.e., the response to static unbalance $m\epsilon$, is

$$r_C = r_{C_0} e^{i\Omega t} . \tag{2.87}$$

The amplitude of the unbalance response is then obtained by introducing the particular integral into the equation of motion (2.75), obtaining

$$r_{C_0}(-m\Omega^2 + i\Omega c_n + k) = m\epsilon\Omega^2 . \tag{2.88}$$

Rotating damping does not enter Equation (2.88): Unbalance produces a forward synchronous excitation, i.e., an excitation that rotates in the xy-plane at the same speed Ω as the rotor, and the latter rotates in the deflected configuration but is not subject to deformations that change in time.

The amplitude r_{C_0} is expressed by a complex number. By separating the real and the imaginary part, it follows that

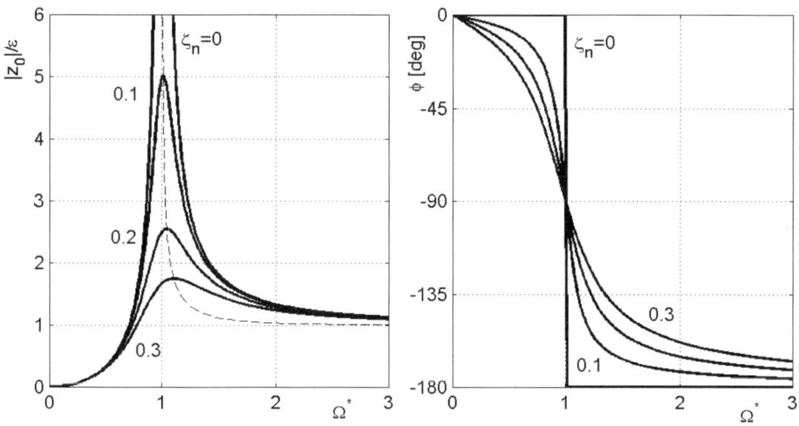

FIGURE 2.13. Unbalance response of a damped Jeffcott rotor: nondimensional amplitude and phase as functions of the nondimensional spin speed.

$$
\begin{aligned}
\mathrm{Re}(r_{C_0}) &= \frac{m\epsilon\Omega^2(k-m\Omega^2)}{(k-m\Omega^2)^2+\Omega^2c_n^2} = \epsilon\frac{{\Omega^*}^2(1-{\Omega^*}^2)}{(1-{\Omega^*}^2)^2+4\zeta_n^2{\Omega^*}^2},\\
\mathrm{Im}(r_{C_0}) &= -\frac{m\epsilon\Omega^3c_n}{(k-m\Omega^2)^2+\Omega^2c_n^2} = -\epsilon\frac{2{\Omega^*}^3\zeta_n}{(1-{\Omega^*}^2)^2+4\zeta_n^2{\Omega^*}^2},\\
|r_{C_0}| &= \epsilon\frac{{\Omega^*}^2}{\sqrt{(1-{\Omega^*}^2)^2+4\zeta_n^2{\Omega^*}^2}},\\
\Phi &= \arctan\left(\frac{-2\Omega^*\zeta_n}{1-{\Omega^*}^2}\right).
\end{aligned}
\tag{2.89}
$$

The amplitude and phase of r_{C_0} are plotted in nondimensional form as functions of the speed in Figure 2.13. The different curves have been obtained with different values of the nonrotating damping ζ_n. The equation yielding the unbalance response is identical to that yielding the amplitude of the response of a vibrating system with a single degree of freedom to a harmonic excitation whose amplitude is proportional to the square of the frequency. The damped resonance peak lies on the right of the undamped resonance and not on the left, as in the case of vibrating systems forced by an excitation whose amplitude is independent from the frequency.

The situation in the xy-plane is shown in Figure 2.14. As angle Φ is always negative, point C always lags the line forming an angle Ωt with the x-axis, i.e., lags ξ-axis. The delay is exactly 90° at the critical speed. The plot of the real and imaginary parts of the complex amplitude r_{C_0} yields directly the trajectory of point C in the $\xi\eta$-plane (full line in Figure 2.15).

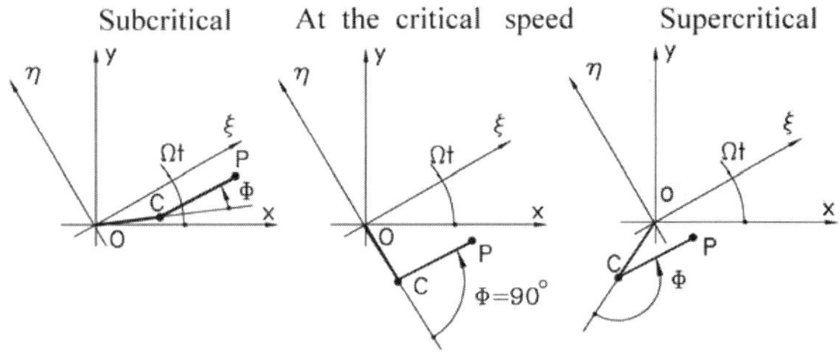

FIGURE 2.14. Configurations taken by the system when working in the subcritical and supercritical range and at the critical speed.

By adding vector ϵ to vector r_{C_0}, the trajectory of point P is obtained (dashed line in Figure 2.15).

The plot of the trajectory of C is very similar to a Nyquist diagram, with the important difference that the latter is plotted in the Argand plane, and the former gives the actual position of point C in the $\xi\eta$-plane in the physical space.

Remark 2.10 *The term* trajectory *has been used with an extended meaning: All of these figures were obtained neglecting the angular acceleration, and the points of the trajectories refer to steady-state conditions at different speeds and are not successive positions during an acceleration of the rotor.*

However, if the acceleration is very slow and dynamic effects linked to it can be neglected, the curves of Figure 2.15 can be assumed to be at least a good approximation for the actual trajectories in the $\xi\eta$-plane. From them, it is clear that self-centering is strictly linked with an increase of phase Φ from 0° to −180°, i.e., with a rotation of point C in the $\xi\eta$-plane. The typical situations in the subcritical regime, at the critical speed, and in the supercritical regime are shown in Figure 2.14.

If point C was constrained to remain on the ξ-axis, no self-centering would occur, as in the case of the system sketched in Figure 2.16, where mass m is allowed to move along a rotating guide and is constrained to the center of rotation O by a spring of stiffness k. The system is only apparently similar to the Jeffcott rotor of Figure 2.1: If the length of the guide was infinite, it would behave similarly to a Jeffcott rotor in the subcritical field, but in supercritical conditions, point P would not come back to the self-centered position and would remain at infinity (or, in real life, at the end of the guide).

As a final result, the motion of point C can be considered as the superimposition of a backward inward-spiral motion, which usually decays very quickly in time, a forward spiral motion, whose amplitude can be de-

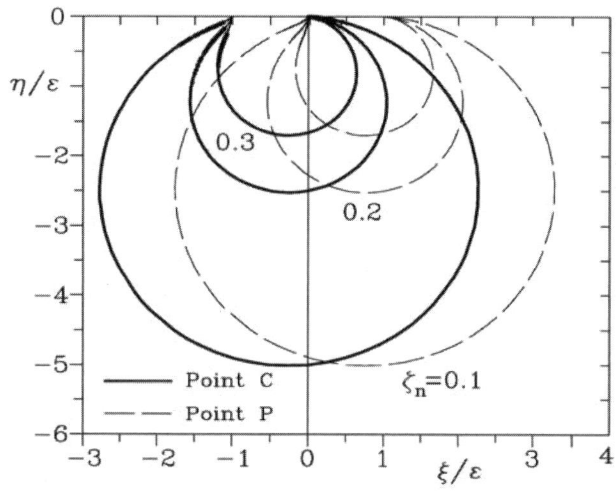

FIGURE 2.15. Nondimensional trajectories of points C and P in the rotating plane $O\xi\eta$.

creasing, constant, or increasing in time depending on the stability of the system, and a circular synchronous motion, with constant amplitude. As the frequency of the forward and backward free motions is the same, they can add to each other, giving way to elliptical whirling or straight vibration depending on the initial conditions; however, the decay rate of the forward motion is smaller than that of the backward component, and with time, the trajectory of the free whirling becomes more and more similar to that related to forward whirling alone. The amplitude of the circular forced whirling depends on the spin speed and the eccentricity ϵ.

Maximum amplitude of the response

If nonrotating damping is very low, the maximum amplitude of the unbalance response is obtained at a speed that is very close to the critical speed of the undamped system. With increasing damping, the speed of maximum amplitude increases: By differentiating the third Equation (2.89) with respect to Ω^* and equating the derivative to zero, it follows that

$$\Omega^*_{(z_{\max})} = \frac{1}{\sqrt{1-2\zeta_n^2}} . \tag{2.90}$$

The corresponding peak amplitude is

$$|r_{C_0}|_{\max} = \epsilon \frac{1}{2\zeta_n\sqrt{1-\zeta_n^2}} , \tag{2.91}$$

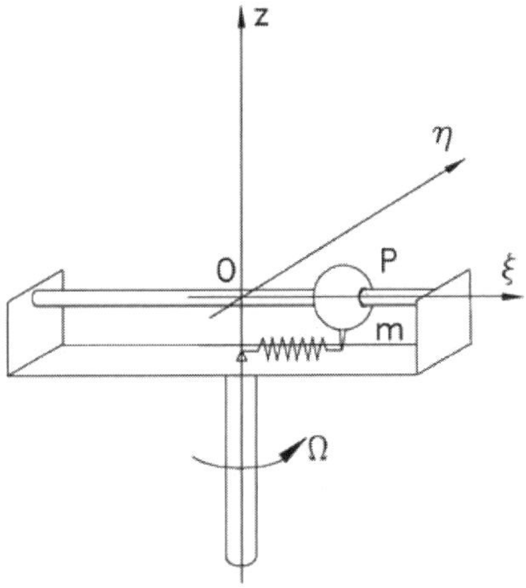

FIGURE 2.16. Rotating system in which point mass P is constrained to move along ξ-axis: It cannot undergo self-centering.

i.e., if the damping is low enough to neglect ζ_n^2 with respect to unity

$$|r_{C_0}|_{\max} \approx \epsilon \frac{1}{2\zeta_n} = \epsilon \frac{\sqrt{km}}{c_n} . \tag{2.92}$$

The ratio $|r_{C_0}|_{\max}/\epsilon \approx 1/2\zeta_n$ is usually referred to as the *quality factor*, and symbol Q is used for it. It gives immediately an idea of the magnitude of the response of the system while crossing a critical speed.

2.4.5 Response to a static force constant in time

The response to a static force F_n that is constant in time is a static displacement r_{C_n}, which can be computed from the equation

$$(k - i\Omega c_r)\, r_{C_n} = F_n , \tag{2.93}$$

i.e.,

$$r_{C_n} = \frac{F_n}{k - i\Omega c_r} = \frac{F_n}{k^2 + \Omega^2 c_r^2} (k + i\Omega c_r) . \tag{2.94}$$

The static deflection depends in this case on the spin speed, and the displacement does not occur in the same direction as the force: The presence of rotating damping causes a lateral deflection of the system, which grows with increasing speed.

2.4.6 Shaft bow

If a bow is present, Equation (2.75) becomes

$$m\ddot{r}_C + (c_r + c_n)\dot{r}_C + (k - ic_r\Omega)r_C = m\epsilon\Omega^2 e^{i\Omega t} + kbe^{i\alpha}e^{i\Omega t} + F_n . \tag{2.95}$$

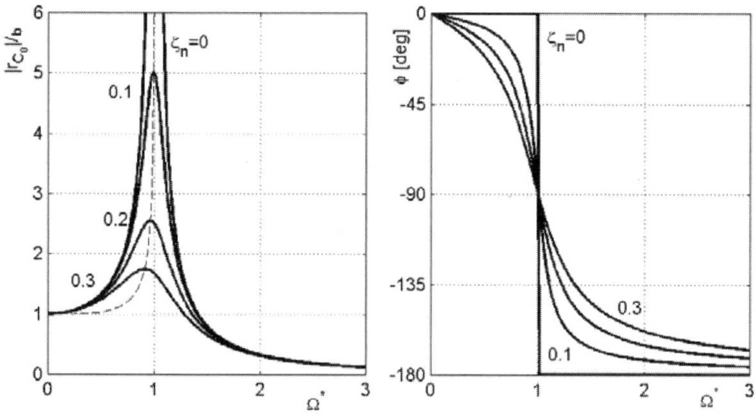

FIGURE 2.17. Response of a damped Jeffcott rotor to a shaft bow: nondimensional amplitude and phase as functions of the nondimensional spin speed.

The particular solution of the equation containing only the nonhomogeneous term for the shaft bow is of the type of the second Equation (2.87), i.e., $r = r_0 e^{i\Omega t}$, where

$$r_{C_0} = \frac{kbe^{i\alpha}}{(k - m\Omega^2) + i\Omega c_n} . \tag{2.96}$$

The response for some values of the nonrotating damping ratio ζ_n is reported in nondimensional form in Figure 2.17.

The response of the rotor is in general a combination of the response to bow and to unbalance, which add to each other with their own phases. At low speed, the first one dominates, whereas in the supercritical range, the second is more important. As an example, the response of a Jeffcott rotor with equal bow and eccentricity ($b = \epsilon$) and with a phasing of 90° is reported in Figure 2.18. Note that at very low speed, the phasing of the response is 90°, i.e., that of the shaft bow, whereas at very high speed, it tends to 180°, i.e., the system is self-centered.

The speed at which the response to a shaft bow is maximum is

$$\Omega^*_{(r_{\max})} = \sqrt{1 - 2\zeta_n^2} , \tag{2.97}$$

and the corresponding peak amplitude is

$$|r_{C_0}|_{\max} = b \frac{1}{2\zeta_n \sqrt{1 - \zeta_n^2}} . \tag{2.98}$$

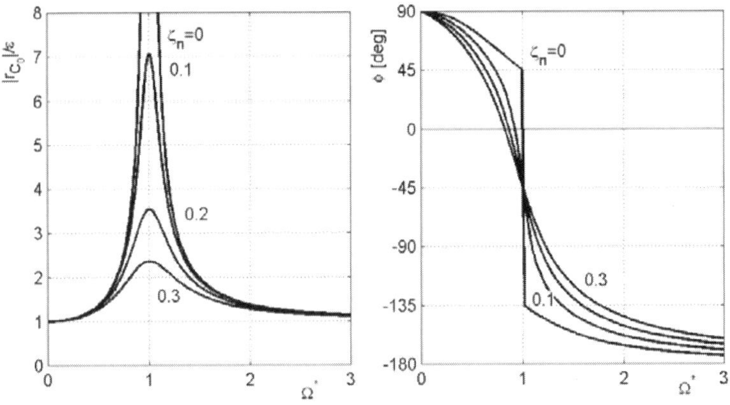

FIGURE 2.18. Response of a damped Jeffcott rotor to a combination of shaft bow and unbalance, with $b = \epsilon$ and $\alpha = 90°$. Nondimensional amplitude and phase as functions of the nondimensional spin speed.

2.4.7 *Frequency response*

The previous sections dealt with the response of a Jeffcott rotor to excitations caused by unavoidable imperfections of the shaft, such as unbalance or bow. This type of excitation is synchronous, i.e., is characterized by a frequency equal to the spin speed and can be modeled as a force that is nonrotating in the rotating reference frame $O\xi\eta\zeta$.

Rotors can be subjected to excitations whose frequency is unrelated to the rotational speed. Consider a perfectly balanced Jeffcott rotor excited by a force F_n whose time history is harmonic with an arbitrary frequency ω. The equation of motion in complex coordinates is still Equation (2.41) with vanishing unbalance and

$$F_n(t) = F_0 e^{i\omega t} \ . \tag{2.99}$$

Note that Equation (2.99) describes a force that rotates in xy-plane with angular velocity ω.

The particular integral of the equation of motion is

$$r_C(t) = r_{C_\omega} e^{i\omega t} \ . \tag{2.100}$$

The complex amplitude r_{C_ω} can be readily obtained by substituting the solution (2.100) into the equation of motion (2.41)

$$\left[-m\omega^2 + i\omega(c_n + c_r) + (k - i\Omega c_r)\right] r_{C_\omega} = F_0 \ . \tag{2.101}$$

The amplitude of the response can be expressed in terms of the frequency response $G(i\omega)$

$$r_{C_\omega}(i\omega) = G(i\omega) F_0 \ , \tag{2.102}$$

where

$$G(i\omega) = \frac{1}{k - m\omega^2 + i\left[\omega(c_n + c_r) - \Omega c_r\right]} \ . \tag{2.103}$$

Rotating damping causes the frequency response to depend explicitly on the spin speed Ω. Note that the response is that of a linear system with equivalent damping

$$c_{eq} = c_n + c_r \left(1 - \frac{\Omega}{\omega}\right) \ . \tag{2.104}$$

At standstill what matters is the sum of rotating and nonrotating damping and the response to a static force [Equation (2.94)] is obtained. With increasing speed, damping decreases, and when $\omega = \Omega$, i.e., when the system is excited at a frequency equal to the spin speed (synchronous whirling), the total damping reduces to nonrotating damping. With further increase of the speed, damping reduces even more and when

$$\frac{\Omega}{\omega} = 1 + \frac{c_n}{c_r} \ ,$$

the system behaves as if it was undamped. At higher speeds, it behaves as a system with negative damping[4] and its response is unstable in a field close to the natural frequency.

The response to a harmonic forcing function at standstill and for three values of the speed ($0.5\omega_n$, ω_n, and $1.8\omega_n$) of a Jeffcott rotor with damping ratios $\zeta_n = \zeta_r = 0.1$ is reported in Figure 2.19. Although the rotor is heavily damped, the response shows a resonance peak whose amplitude increases with the speed.

In the case of arbitrary excitation $F_n(t)$, the response of the system can be more easily obtained by resorting to a Laplace domain representation. The equation of motion (2.41) is transformed as

$$\left[ms^2 + (c_n + c_r)s + k - i\Omega c_r\right] r_C(s) = m\epsilon\Omega^2 \frac{1}{s - i\Omega} + F_n(s) \ , \tag{2.105}$$

where s is the complex variable of the Laplace transform. The system response can then be expressed by introducing the transfer function $G(s)$

$$r_C(s) = G(s) \left[\frac{m\epsilon\Omega^2}{s - i\Omega} + F_n(s)\right] \ , \tag{2.106}$$

[4] This statement must be interpreted in the sense that it has the same frequency response of a system with negative damping.

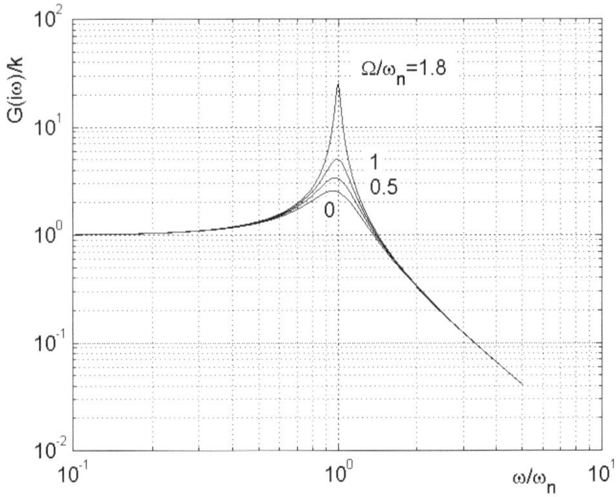

FIGURE 2.19. Nondimensional frequency response of a perfectly balanced Jeffcott rotor to a nonsynchronous harmonic excitation at four different values of the spin speed $\Omega = 0$, $\Omega = 0.5\omega_n$, $\Omega = \omega_n$, and $\Omega = 1.8\omega_n$. System with damping ratios $\zeta_n = \zeta_r = 0.1$.

where

$$G(s) = \frac{1}{ms^2 + (c_n + c_r)s + k - i\Omega c_r} . \tag{2.107}$$

The transfer function allows us to characterize completely the system by applying the standard methods of system theory. The response to unbalance and external forces can thus be studied, and the computation of the poles of the transfer function allows us to obtain the natural frequencies and to study the stability. The equation yielding the poles

$$ms^2 + (c_n + c_r)s + k - i\Omega c_r = 0 \tag{2.108}$$

obtained by equating to zero the denominator of the transfer function, coincides with Equation (2.77) yielding the natural frequencies of the system.

2.5 Jeffcott rotor with structural damping

2.5.1 *Equation of motion*

Rotating and nonrotating damping have been assumed to be of the viscous type. This can be a realistic model for nonrotating damping, particularly when dampers of the squeeze-film type are used, but for the rotating damping, a structural (or hysteretic) damping model is usually better suited.

It is well known that, in the case of a vibrating system with a single degree of freedom with mass m, stiffness k, and loss factor η, vibrating at the frequency ω, hysteretic damping is equivalent to a viscous damping with damping coefficient $c_{eq} = \eta k / |\omega|$ [25]. It is also well known that in most cases, i.e., when damping is small enough, the correct solution can be approximated by introducing an equivalent damping coefficient independent from the vibration frequency, $c_{eq} = \eta k / |\omega_n|$.

In the case of rotors, however, this approach can be used only for the nonrotating damping. For the rotor, the expression of the equivalent damping must be modified into $c_{eq} = \eta k / \omega_m$, where ω_m is the frequency at which the material goes through the hysteresis cycle. Although for nonrotating damping it coincides with $|\omega|$, modulus of the imaginary part of the complex whirl frequency $s = \sigma + i\omega$, it takes the value $|\omega - \Omega|$ for rotating damping.

If both rotating and nonrotating damping are of structural type, they can be accounted for in Equation (2.75) in terms of equivalent viscous damping

$$c_{n_{eq}} = \frac{\eta_n k_n}{|\omega|} , \qquad c_{r_{eq}} = \frac{\eta_r k_r}{|\omega - \Omega|} , \tag{2.109}$$

and the equation of the motion in terms of complex coordinates becomes

$$m\ddot{r}_C + \left(\frac{\eta_n k_n}{|\omega|} + \frac{\eta_r k_r}{|\omega - \Omega|} \right) \dot{r}_C + \left(k - i\Omega \frac{\eta_r k_r}{|\omega - \Omega|} \right) r_C = m\epsilon\Omega^2 e^{i\Omega t} + F_n , \tag{2.110}$$

where the unbalance forces and the nonrotating static forces F_n have been added similarly to Equation (2.75).

Remark 2.11 *Equation (2.110) must not be taken at face value, because it holds only when function $r_C(t)$ is harmonic [26]. This is a general problem with hysteretic damping and will not be discussed further here; however, the following equations written in the frequency domain are correct.*

2.5.2 Free whirling

The solution of the homogeneous dynamic Equation (2.110) can be found in the same form used for the case of the undamped rotor and of the rotor with viscous damping

$$r_C = r_{C_0} e^{st} . \tag{2.111}$$

By introducing the solution into the equation of the motion, the following characteristic equation is obtained:

$$m(\sigma + i\omega)^2 + \left(\frac{\eta_n k_n}{|\omega|} + \frac{\eta_r k_r}{|\omega - \Omega|} \right) (\sigma + i\omega) + k - i\Omega \frac{\eta_r k_r}{|\omega - \Omega|} = 0 . \tag{2.112}$$

By separating the real and the imaginary parts of Equation (2.112), it follows that

$$\begin{cases} m\left(\sigma^2-\omega^2\right)+\sigma\left(\frac{\eta_n k_n}{|\omega|}+\frac{\eta_r k_r}{|\omega-\Omega|}\right)+k=0\ , \\ 2m\sigma\omega+\eta_n k_n\frac{\omega}{|\omega|}+\eta_r k_r\frac{\omega-\Omega}{|\omega-\Omega|}=0\ . \end{cases} \tag{2.113}$$

The second Equation (2.113) can be used to compute the decay rate

$$\sigma=-\frac{1}{2m\omega}\left[\eta_n k_n \mathrm{sgn}(\omega)+\eta_r k_r \mathrm{sgn}(\omega-\Omega)\right]\ . \tag{2.114}$$

Three cases can thus be identified:

- Subcritical forward whirling $\omega-\Omega>0$ and $\omega>0$: The expression in braces is always positive, and the decay rate is negative. The system is *stable.*
- Supercritical forward whirling $\omega-\Omega<0$ and $\omega>0$: The sign of the expression in braces depends on whether $\eta_n k_n>\eta_r k_r$. The following two cases can be identified
  - $\eta_n k_n>\eta_r k_r$: The expression is positive, and the decay rate is negative. The system is *stable.*
  - $\eta_n k_n<\eta_r k_r$: The expression is negative, and the decay rate is positive. The system is *unstable.*
- Backward whirling $\omega-\Omega<0$ and $\omega<0$: The expression in braces is always negative, but the decay rate is negative. The system is *stable.*

If only rotating damping is acting ($\eta_n=0$), the system is stable in backward and forward subcritical whirling, whereas it is unstable in forward supercritical whirling; i.e., the threshold of instability coincides with the critical speed. Rotating hysteretic damping is stabilizing in subcritical conditions and destabilizing in supercritical condition, exactly like rotating viscous damping.

Remark 2.12 *This result has been well known for many years (see, for example, [27]); however, incorrect statements suggesting that rotating hysteretic damping is always destabilizing can be found in the literature (e.g., [28], and [29]). The misunderstanding comes also from a misquotation of the book by Dimentberg [27].*

By introducing the value of the decay rate so computed into the first Equation (2.113), an equation allowing us to compute the whirl speed ω can be written as

$$\omega^5-\Omega\omega^4-\frac{k}{m}\omega^3+\frac{k}{m}\Omega\omega^2+\frac{\left(k_r\eta_r\mp k_n\eta_n\right)^2}{4m^2}\omega+\frac{k_r^2\eta_r^2-k_n^2\eta_n^2}{4m^2}\Omega=0\ , \tag{2.115}$$

where the upper signs hold in case of backward or subcritical whirling and the lower ones in case of supercritical whirling.

To avoid solving Equation (2.115), a simplified equation is usually obtained by neglecting the real part σ of the complex frequency in the computation of the term proportional to $(\sigma + i\omega)$. Equation (2.112) becomes

$$m(\sigma + i\omega)^2 + k + i\left(\frac{\omega}{|\omega|}\eta_n k_n + \frac{\omega - \Omega}{|\omega - \Omega|}\eta_r k_r\right) = 0\,. \tag{2.116}$$

The term neglected operating in this way

$$\left(\frac{\eta_n k_n}{|\omega|} + \frac{\eta_r k_r}{|\omega - \Omega|}\right)\sigma \tag{2.117}$$

is quite small with respect to k, at least if the real part σ of the complex frequency is small and the whirl frequency ω is different enough from Ω. Note that at the threshold of instability ($\sigma = 0$), the imaginary part of the whirl frequency vanishes and then Equation (2.116) yields exact results.

Equation (2.116) can be then rewritten in terms of the sign of the whirl speed ω and of the speed difference $\omega - \Omega$

$$m(\sigma + i\omega)^2 + k + i\left[\mathrm{sgn}(\omega)\eta_n k_n + \mathrm{sgn}(\omega - \Omega)\eta_r k_r\right] = 0\,. \tag{2.118}$$

With reference to Figure 2.20, the $\Omega\omega$-plane has been split in four regions according to the values assumed by functions $\mathrm{sgn}(\omega)$ and $\mathrm{sgn}(\omega - \Omega)$. Different hatches have been used in the same figure to indicate the regions where forward, backward, subcritical, or supercritical motions occur.

Because of the presence of the sign functions, Equation (2.118) is nonlinear and the solution requires a minimum amount of iteration to be found. The complex frequency $\sigma + i\omega$ can be obtained from Equation (2.118) as the square root of a complex number

$$\sigma + i\omega = \pm\sqrt{-\frac{k}{m} - \frac{i}{m}\left[\mathrm{sgn}(\omega)\eta_n k_n + \mathrm{sgn}(\omega - \Omega)\eta_r k_r\right]}\,. \tag{2.119}$$

The computation of the explicit values of σ and ω can be done by means of Equation (2.79) introduced in the case of damping of viscous type. The real and imaginary parts a and b are in this case

$$a = -\frac{k}{m}; \qquad b = -\frac{1}{m}\left[\mathrm{sgn}(\omega)\eta_n k_n + \mathrm{sgn}(\omega - \Omega)\eta_r k_r\right]. \tag{2.120}$$

As shown in Figure 2.20, four different cases are possible for the sign functions. For each of the four cases, the general solution of Equation (2.79) gives way to two values of the complex frequency $\sigma + i\omega$. The pair of solutions found for each case must be checked against the assumptions for functions $\mathrm{sgn}(\omega)$ and $\mathrm{sgn}(\omega - \Omega)$ relative to that case to assess the only one that is consistent with the assumptions and therefore acceptable.

Each of the four possible cases are discussed in the following in order to analyze the stability and the whirl frequency and direction.

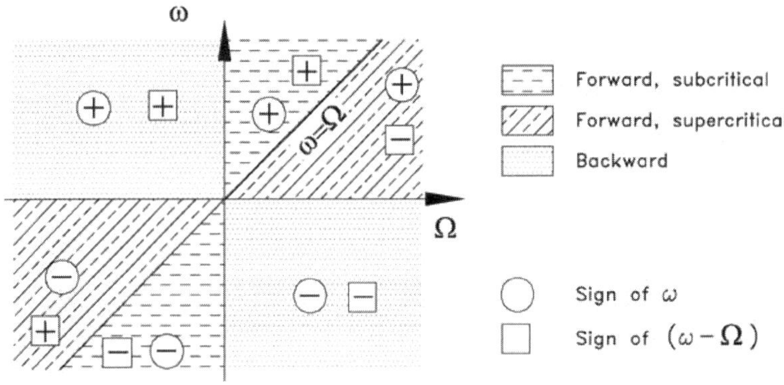

FIGURE 2.20. Signs of the terms in ω and $\omega - \Omega$ in the characteristic equation for systems with hysteretic damping.

Case 1. $\text{sgn}(\omega) = 1$; $\text{sgn}(\omega - \Omega) = 1$. *(first quadrant, subcritical; second quadrant).*

From Equation (2.79),

$$\left\{ \begin{array}{l} \sigma = \pm \dfrac{1}{\sqrt{2m}} \sqrt{\sqrt{k^2 + (k_n \eta_n + k_r \eta_r)^2} - k} \; , \\ \\ \omega = \mp \dfrac{1}{\sqrt{2m}} \sqrt{\sqrt{k^2 + (k_n \eta_n + k_r \eta_r)^2} + k} \; . \end{array} \right. \tag{2.121}$$

Among the two possible signs affecting the solution, only the lower signs are acceptable as they give way to positive whirl speed in accordance with the assumption $\text{sgn}(\omega) = 1$ at the base of this case. The solution of this case is then

$$\left\{ \begin{array}{l} \sigma = -\dfrac{1}{\sqrt{2m}} \sqrt{\sqrt{k^2 + (k_n \eta_n + k_r \eta_r)^2} - k} \\ \\ \omega = \dfrac{1}{\sqrt{2m}} \sqrt{\sqrt{k^2 + (k_n \eta_n + k_r \eta_r)^2} + k} \end{array} \right. \quad \Omega < \omega \; . \tag{2.122}$$

Taking the assumption that $\text{sgn}(\omega - \Omega) = 1$ into account, this solution is valid for $\Omega < \omega$, i.e., in subcritical conditions.

Case 2. $\mathrm{sgn}(\omega) = -1$; $\mathrm{sgn}(\omega - \Omega) = -1$ *(third quadrant, subcritical; fourth quadrant).*

A discussion similar to that performed in the Case 1 can be done in this case. The result is

$$\begin{cases} \sigma = -\dfrac{1}{\sqrt{2m}}\sqrt{\sqrt{k^2 + (k_n\eta_n + k_r\eta_r)^2} - k} \\ \omega = -\dfrac{1}{\sqrt{2m}}\sqrt{\sqrt{k^2 + (k_n\eta_n + k_r\eta_r)^2} + k} \end{cases} \quad \Omega > \omega \, . \tag{2.123}$$

As in Case 1, this solution is valid for subcritical conditions.

Case 3. $\mathrm{sgn}(\omega) = 1$; $\mathrm{sgn}(\omega - \Omega) = -1$ *(first quadrant, supercritical).*

The complex frequency is

$$\begin{cases} \sigma = -\mathrm{sgn}(k_n\eta_n - k_r\eta_r)\dfrac{1}{\sqrt{2m}}\sqrt{\sqrt{k^2 + (k_n\eta_n + k_r\eta_r)^2} - k} \\ \omega = \dfrac{1}{\sqrt{2m}}\sqrt{\sqrt{k^2 + (k_n\eta_n + k_r\eta_r)^2} + k} \end{cases} \quad \Omega > \omega > 0 \, . \tag{2.124}$$

Case 4. $\mathrm{sgn}(\omega) = -1$; $\mathrm{sgn}(\omega - \Omega) = 1$ *(third quadrant, supercritical).*

The complex frequency is

$$\begin{cases} \sigma = -\mathrm{sgn}(k_n\eta_n - k_r\eta_r)\dfrac{1}{\sqrt{2m}}\sqrt{\sqrt{k^2 + (k_n\eta_n + k_r\eta_r)^2} - k} \\ \omega = -\dfrac{1}{\sqrt{2m}}\sqrt{\sqrt{k^2 + (k_n\eta_n + k_r\eta_r)^2} + k} \end{cases} \quad \Omega < \omega < 0 \, . \tag{2.125}$$

Stability

Case 1 and Case 2 have a negative real part σ of the complex frequency, the corresponding motion is then stable. With reference to Figure 2.20, the two cases include the backward motion and the forward subcritical motion.

Case 3 and Case 4 have a real part σ whose sign is a function of the difference $k_n\eta_n - k_r\eta_r$. In the case

$$\eta_n k_n > \eta_r k_r \, , \tag{2.126}$$

the system is stable at any speed. If, on the contrary, it is not satisfied, the threshold of instability coincides with the critical speed, and no supercritical running is possible.

2.5.3 Mixed damping

The case in which there are different types of damping can be studied in the same way. If rotating damping is of the structural type and nonrotating damping is viscous, Equation (2.112) becomes

$$m\left(\sigma+i\omega\right)^2+\left(c_n+\frac{\eta_r k_r}{|\omega-\Omega|}\right)\left(\sigma+i\omega\right)+k-i\Omega\frac{\eta_r k_r}{|\omega-\Omega|}=0\,. \quad (2.127)$$

By neglecting the term

$$\sigma\frac{\eta_r k_r}{|\omega-\Omega|} \quad (2.128)$$

and introducing the complex frequency s, Equation (2.127) becomes

$$ms^2+k+c_n s+i\,\mathrm{sgn}(\omega-\Omega)\eta_r k_r=0\,. \quad (2.129)$$

By operating in the same way as seen for the case of systems with viscous damping, and taking into account the sign function, three cases can be identified:

- Case 1: forward subcritical whirling

$$s=-\frac{c_n}{2m}-\frac{1}{\sqrt{2}}\sqrt{\sqrt{\Gamma^2+\left(\frac{\eta_r k_r}{m}\right)^2}-\Gamma}+i\frac{1}{\sqrt{2}}\sqrt{\sqrt{\Gamma^2+\left(\frac{\eta_r k_r}{m}\right)^2}+\Gamma}\,, \quad (2.130)$$

where

$$\Gamma=\frac{k}{m}-\frac{{c_n}^2}{4m^2}\,. \quad (2.131)$$

- Case 2: forward supercritical whirling

$$s=-\frac{c_n}{2m}+\frac{1}{\sqrt{2}}\sqrt{\sqrt{\Gamma^2+\left(\frac{\eta_r k_r}{m}\right)^2}-\Gamma}+i\frac{1}{\sqrt{2}}\sqrt{\sqrt{\Gamma^2+\left(\frac{\eta_r k_r}{m}\right)^2}+\Gamma}\,. \quad (2.132)$$

- Case 3: backward whirling

$$s=-\frac{c_n}{2m}-\frac{1}{\sqrt{2}}\sqrt{\sqrt{\Gamma^2+\left(\frac{\eta_r k_r}{m}\right)^2}-\Gamma}-i\frac{1}{\sqrt{2}}\sqrt{\sqrt{\Gamma^2+\left(\frac{\eta_r k_r}{m}\right)^2}+\Gamma}\,. \quad (2.133)$$

Again, there is no stability problem in forward subcritical and backward whirling, whereas in the case of forward supercritical whirling, a minimum value of the nonrotating damping can be found

$$c_n > \eta_r k_r \sqrt{\frac{m}{k}} .$$

Also in this case, the system is either always stable or the threshold of instability coincides with the critical speed.

2.5.4 Unbalance response

In the case of the unbalance response, rotating damping has no influence on the behavior of the system. If nonrotating damping is of the structural type, the amplitude and phase of the unbalance response are

$$\begin{aligned} |r_{C_0}| &= \epsilon \frac{\Omega^{*^2}}{\sqrt{(1-\Omega^{*^2})^2+\eta_n^2}} , \\ \Phi &= \arctan\left(\frac{-\eta_n}{1-\Omega^{*2}}\right) . \end{aligned} \tag{2.134}$$

2.5.5 Dependence of the loss factor on frequency

Note that the assumption of constant loss factor is only a rough approximation. The very model of hysteretic damping loses its validity when the frequency at which the hysteresis cycle is gone through ($|\omega|$ for nonrotating damping and $|\omega - \Omega|$ for rotating damping) tends to zero.

If the dependence of the loss factor from the frequency is known, the free whirling frequencies can be computed iteratively. Because the whirl speed is little influenced by the value of the damping, the procedure can converge quickly. For the computation of the unbalance response, however, the dependence of damping from the frequency introduces no computational difficulty. Other effects, such as the dependence of damping from the value of the maximum stress, introduce nonlinearities into the model and make the solution very difficult.

2.6 Jeffcott rotor with nonsynchronous damping

In the sections above, damping was assumed to be either stationary or rotating at the same speed Ω of the rotor. There may be cases in which the energy dissipating element rotates at a speed different from both 0 and Ω; i.e., damping is rotating in a nonsynchronous way. It can be either rotating in the same direction of the rotor (corotating damping) or in the opposite direction (counter-rotating damping).

Consider a damped Jeffcott rotor running at constant speed and assume that a damper, with damping coefficient c_d, rotating at a speed $\Omega_d = \alpha\Omega$, is present together with the usual nonrotating and rotating damping c_n and c_r. A positive value of α characterizes corotating damping and a negative value characterizes counter-rotating damping. By resorting to complex coordinates, the equation of motion (2.75) can be written in the form

$$m\ddot{r}_C + c\dot{r}_C + (k - i\Omega c_r - i\Omega_d c_d)\, r_C = m\epsilon\Omega^2 e^{i\Omega t}, \tag{2.135}$$

where c is the total damping ($c = c_n + c_r + c_d$). The same equation can be written using real coordinates

$$\begin{bmatrix} m & 0 \\ 0 & m \end{bmatrix} \begin{Bmatrix} \ddot{x}_C \\ \ddot{y}_C \end{Bmatrix} + \begin{bmatrix} c & 0 \\ 0 & c \end{bmatrix} \begin{Bmatrix} \dot{x}_C \\ \dot{y}_C \end{Bmatrix} +$$
$$+ \begin{bmatrix} k & \Omega c_r + \Omega_d c_d \\ -\Omega c_r - \Omega_d c_d & k \end{bmatrix} \begin{Bmatrix} x_C \\ y_C \end{Bmatrix} = m\epsilon\Omega^2 \begin{Bmatrix} \cos(\Omega t) \\ \sin(\Omega t) \end{Bmatrix} . \tag{2.136}$$

The homogeneous equation associated with Equation (2.135) allows us to study the stability of the system. By assuming a solution of the type $r_C = r_{C_0} e^{st}$, the following characteristic equation is obtained:

$$ms^2 + cs + k - i\Omega c_r - i\Omega_d c_d = 0 . \tag{2.137}$$

Equation (2.135) can be written in nondimensional form using the same nondimensional quantities used for Equation (2.84) to which the obvious relationship

$$\Omega_d^* = \frac{\Omega_d}{\Omega_{cr}} = \frac{\Omega_d}{\sqrt{k/m}} \tag{2.138}$$

is added. The solutions of the characteristic equation can be shown to be

$$\omega^* = \pm\sqrt{\Gamma^* + \sqrt{\Gamma^{*2} + \Xi^{*2}}} , \tag{2.139}$$

$$\begin{cases} \sigma^* = -\zeta \pm \sqrt{-\Gamma^* + \sqrt{\Gamma^{*2} + \Xi^{*2}}} & \text{if} \quad \Xi \geq 0 , \\ \sigma^* = -\zeta \mp \sqrt{-\Gamma^* + \sqrt{\Gamma^{*2} + \Xi^{*2}}} & \text{if} \quad \Xi < 0 , \end{cases} \tag{2.140}$$

where

$$\Gamma^* = \frac{(1 - \zeta^2)}{2} , \quad \zeta = \frac{c}{2\sqrt{km}} , \text{ and } \Xi^* = \Omega^* \zeta_r + \Omega_d^* \zeta_d . \tag{2.141}$$

The upper and lower signs indicate forward and backward whirling, respectively.

From Equation (2.140), it follows immediately that when $\Xi^* \geq 0$, the backward mode is always stable. The condition for stability of the forward mode is

$$\zeta > \sqrt{-\Gamma^* + \sqrt{\Gamma^{*2} + \Xi^{*2}}}\,, \text{ i.e.,} \quad \Omega^* < \frac{\zeta}{\zeta_r} - \Omega_d^* \frac{\zeta_d}{\zeta_r}\,. \tag{2.142}$$

By introducing the ratio $\alpha = \Omega_d^*/\Omega^*$ between the speed of rotation of the damper c_d and the spin speed, the condition for stability becomes

$$\Omega^* < \frac{\zeta}{\zeta_r + \alpha\zeta_d}\,. \tag{2.143}$$

A particular case is that of counter-rotating damping with $\alpha = -1$. The nondimensional threshold of instability, which for synchronous damping ($\zeta_d = 0$) has a value ζ/ζ_r, is now equal to $\zeta/(\zeta_r - \zeta_d)$.

If $\Xi^* < 0$ the forward mode is always stable. The condition for stability of the backward mode is

$$\zeta > \sqrt{-\Gamma^* + \sqrt{\Gamma^{*2} + \Xi^{*2}}}\,, \text{ i.e.,} \quad \Omega^* < -\frac{\zeta}{\zeta_r} - \Omega_d^* \frac{\zeta_d}{\zeta_r} \tag{2.144}$$

or

$$\Omega^* < \frac{\zeta}{-\zeta_r - \alpha\zeta_d} \quad \text{or} \quad \Omega^* \frac{\zeta_r}{\zeta} < -\frac{1}{1 + \alpha\frac{\zeta_d}{\zeta_r}}\,. \tag{2.145}$$

A stability chart is shown in Figure 2.21. Forward modes are always stable for counter-rotating damping with $|\Omega_d^*| > \Omega^*\zeta_r/\zeta_d$, whereas for counter-rotating damping with a speed lower (in absolute value) than the mentioned value and for corotating damping, the condition for stability is expressed by Equation (2.142). Backward modes are always stable for corotating or for counter-rotating damping with $|\Omega_d^*| < \Omega^*\zeta_r/\zeta_d$, whereas for counter-rotating damping with a speed higher (in absolute value) than the mentioned value, the condition for stability is expressed by Equation (2.144).

It is possible to find a value of α that assures a stable working of the machine at any speed for each value of rotating damping. A fairly large zone about $\alpha\zeta_d/\zeta_r = 1$, where the value of the threshold of stability is very high, can be found. No threshold of instability exists if $\alpha\zeta_d/\zeta_r = 1$.

Note that nonsynchronous damping can make the system unstable even at standstill (i.e., for $\Omega = 0$), the condition for stability being

$$|\Omega_d^*| < \frac{\zeta_n + \zeta_r + \zeta_d}{\zeta_d}\,. \tag{2.146}$$

In the case of nonrotating systems, the sign of Ω_d has no meaning because the direction of the rotation of the damper is immaterial.

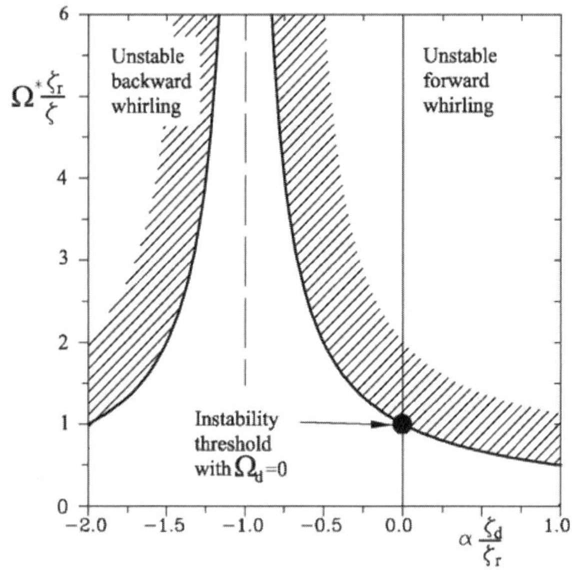

FIGURE 2.21. Stability chart for the forward and backward whirl modes. Nondimensional parameter $\Omega^*\zeta_r/\zeta$ as a function of $\alpha\zeta_d/\zeta_r$.

2.7 Effect of the compliance of the bearings

The solutions obtained in the previous sections were based on the assumption that both the rotating and nonrotating damping work for the whole displacement of point C. Actually this is not the case in most real-life applications, as the compliance of rotor systems can be usually ascribed partially to the rotor itself and partially to the stator, and the two forms of damping are usually acting separately: Nonrotating damping can be ascribed to the stator while rotating damping can be associated with the rotor (Figure 2.22)

Assuming that damping is of the viscous type and neglecting the mass of the bearings, the simplest model of the system can be written in the form

$$\begin{bmatrix} 0 & 0 \\ 0 & m \end{bmatrix}\begin{Bmatrix} \ddot{r}_B \\ \ddot{r}_C \end{Bmatrix}+\begin{bmatrix} c_n+c_r & -c_r \\ -c_r & c_r \end{bmatrix}\begin{Bmatrix} \dot{r}_B \\ \dot{r}_C \end{Bmatrix}+\left(\begin{bmatrix} k_n+k_r & -k_r \\ -k_r & k_r \end{bmatrix}+\right.$$

$$\left.-i\Omega\begin{bmatrix} c_r & -c_r \\ -c_r & c_r \end{bmatrix}\right)\begin{Bmatrix} r_B \\ r_C \end{Bmatrix}=\begin{Bmatrix} 0 \\ m\epsilon\Omega^2 \end{Bmatrix}e^{i\Omega t}\,, \tag{2.147}$$

where r_B and r_C are the complex displacements of the bearings (point B) and of point C, k_n and k_r and c_n and c_r are the stiffness and the damping of the stator and the rotor, respectively.

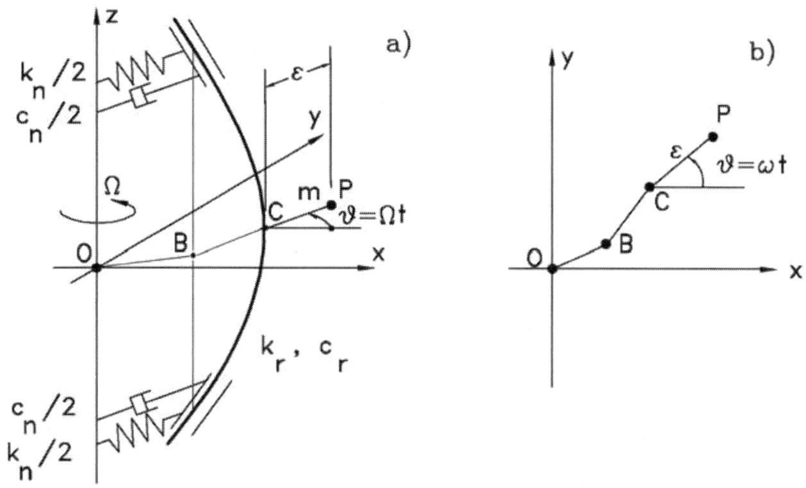

FIGURE 2.22. Compliant Jeffcott rotor on deformable bearings, as in Figure 2.1(c), but with damping added to both rotor and stator.

If the system is undamped, the first equation can be solved in r_B, yielding

$$r_B = r_C \frac{k_r}{k_n + k_r} . \tag{2.148}$$

By introducing Equation (2.148) into the second Equation (2.147) and neglecting damping, it follows that

$$m\ddot{r}_C + \frac{k_n k_r}{k_n + k_r} r_C = m\epsilon\Omega^2 e^{i\Omega t} . \tag{2.149}$$

The system behaves as a Jeffcott rotor with a stiffness equal to that of the two springs k_n and k_r in series. As already stated, the system of Figure 2.1(c) is equivalent to those of Figures 2.1(a) and (b). The critical speed is then

$$\Omega_{cr} = \sqrt{\frac{k_n k_r}{m (k_n + k_r)}} . \tag{2.150}$$

This is not the case for the damped system, which cannot be dealt with as a single degree-of-freedom system (two in real coordinates). As the mass of the bearings has been neglected, the system is a third-order one, in complex coordinates.

2.7.1 Unbalance response

The unbalance response can be obtained in the usual way, assuming that

$$\left\{ \begin{array}{c} r_B \\ r_C \end{array} \right\} = \left\{ \begin{array}{c} r_{B_0} \\ r_{C_0} \end{array} \right\} e^{i\Omega t}$$

and obtaining

$$\left(-\Omega^2 \begin{bmatrix} 0 & 0 \\ 0 & m \end{bmatrix} + i\Omega \begin{bmatrix} c_n & 0 \\ 0 & 0 \end{bmatrix} + \begin{bmatrix} k_n + k_r & -k_r \\ -k_r & k_r \end{bmatrix} \right) \left\{ \begin{array}{c} r_{B_0} \\ r_{C_0} \end{array} \right\} =$$

$$= m\epsilon\Omega^2 \left\{ \begin{array}{c} 0 \\ 1 \end{array} \right\} . \tag{2.151}$$

As expected, thc damping of the shaft does not enter the equation for synchronous whirling. It is easy to expect that the system behaves as an undamped system in both cases with $c_n = 0$ and $c_n \longrightarrow \infty$: In the former case because there is no damping, whereas in the latter because the bearings cannot move and the only compliant part of the system is the shaft, whose damping does not enter the equation for synchronous whirling. Moreover, the first system has a critical speed expressed by Equation (2.150), whereas the critical speed of the second is $\sqrt{k_r/m}$.

For all intermediate values of damping ($0 < c_n < \infty$), the amplitude will be limited, with a value of c_n yielding a minimum peak amplitude of the response of point C.

Solving the first Equation (2.151) in r_{B_0}

$$r_{B_0} = r_{C_0} \frac{k_r}{k_n + k_r + i\Omega c_n} , \tag{2.152}$$

and introducing it in the second one, it yields

$$r_{C_0} = m\epsilon\Omega^2 \frac{k_n + k_r + i\Omega c_n}{k_n k_r - m\Omega^2 (k_n + k_r) + i\Omega c_n (k_r - m\Omega^2)} . \tag{2.153}$$

By introducing the following nondimensional parameters:

$$\Omega^* = \frac{\Omega}{\Omega_1} \quad , \quad \alpha = \frac{k_n}{k_r} \quad , \quad \zeta = \frac{c_n}{2\sqrt{mk_n}} ,$$

where $\Omega_1 = \sqrt{k_n/m}$ is the critical speed of a Jeffcott rotor with a stiffness equal to that of the bearings alone, it follows that

$$\left\{ \begin{array}{l} \dfrac{r_{B_0}}{\epsilon} = \Omega^{*2} \dfrac{1}{1 - \Omega^{*2} (1 + \alpha) + 2i\Omega^* \zeta (1 - \alpha\Omega^{*2})} , \\ \dfrac{r_{C_0}}{\epsilon} = \Omega^{*2} \dfrac{1 + \alpha + 2i\Omega^* \zeta\alpha}{1 - \Omega^{*2} (1 + \alpha) + 2i\Omega^* \zeta (1 - \alpha\Omega^{*2})} . \end{array} \right. \tag{2.154}$$

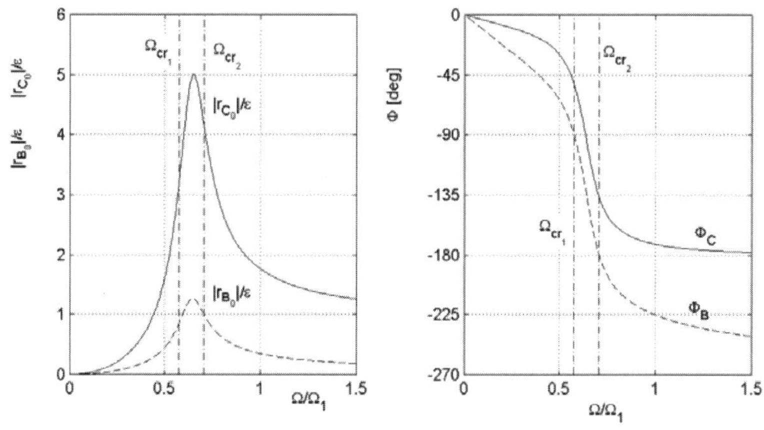

FIGURE 2.23. Jeffcott rotor on compliant bearings: nondimensional unbalance response for $\alpha = 2$ and $\zeta = 1$

The amplitudes and the phases of the response are then

$$\begin{cases} \left|\frac{r_{B_0}}{\epsilon}\right| = \Omega^{*2}\frac{1}{\sqrt{\left[1-\Omega^{*2}\left(1+\alpha\right)\right]^2 + 4\Omega^{*2}\zeta^2\left(1-\alpha\Omega^{*2}\right)^2}}\,, \\ \Phi_B = \arctan\left[\frac{-2\Omega^*\zeta\left(1-\alpha\Omega^{*2}\right)}{1-\Omega^{*2}\left(1+\alpha\right)}\right]\,, \\ \left|\frac{r_{C_0}}{\epsilon}\right| = \Omega^{*2}\frac{\sqrt{\left(1+\alpha\right)^2 + 4\Omega^{*2}\zeta^2\alpha^2}}{\sqrt{\left[1-\Omega^{*2}\left(1+\alpha\right)\right]^2 + 4\Omega^{*2}\zeta^2\left(1-\alpha\Omega^{*2}\right)^2}}\,, \\ \Phi_C = \arctan\left[\frac{-2\Omega^*\zeta}{\left(1+\alpha\right)\left[1-\Omega^{*2}\left(1+\alpha\right)\right] + 4\Omega^{*2}\zeta^2\alpha\left(1-\alpha\Omega^{*2}\right)}\right]\,. \end{cases} \tag{2.155}$$

The nondimensional response depends only on the nondimensional speed Ω^* and two nondimensional parameters, namely, α and ζ. As an example, the unbalance response for $\alpha = 2$ and $\zeta = 1$ is reported in Figure 2.23. Note that $\zeta = 1$ is a high value of the bearing damping, and in practice, it requires the presence of a purposely added damper.

By differentiating the third Equation (2.155) with respect to Ω^* and equating the derivative to zero, it is possible to compute the speed at which the peak of the response of point C is located. By performing the relevant derivatives, the following cubic equation in Ω^{*2} can be found:

$$\begin{aligned} &-32\zeta^4\alpha^3\Omega^{*6} + 16\zeta^2\alpha^2\left[2\zeta^2 - \left(1+\alpha\right)\right]\Omega^{*4} + \\ &+\left\{2\left(1+\alpha\right)^2\left[\zeta^2 - \left(1+\alpha\right)\right] + 12\zeta^2\alpha^2\right\}\Omega^{*2} + 2\left(1+\alpha\right)^2 = 0\,. \end{aligned} \tag{2.156}$$

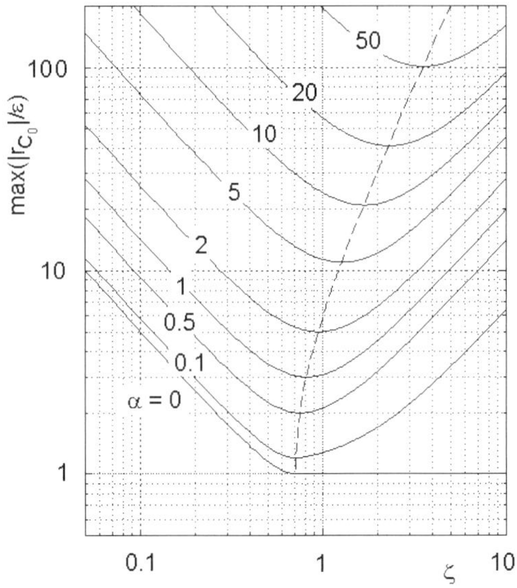

FIGURE 2.24. Jeffcott rotor on compliant bearings: peak amplitude of the orbit of point C as a function of the damping ratio of the bearings, for various values of the stiffness ratio α.

By solving Equation (2.156) for various values of α and ζ, it is possible to obtain the plot of Figure 2.24 in which the peak amplitude is reported as a function of the damping ratio of the bearings, for various values of the stiffness ratio α. Note that $\alpha = 0$ means that $k_r \longrightarrow \infty$, i.e., that the shaft is rigid while the bearings are compliant (system of Figure 2.1(a) with damping added).

Remark 2.13 *For each value of α, there is an optimum value of the damping, i.e., a value of the damping that minimizes the peak of the unbalance response. Such a value increases with increasing α: The more compliant is the shaft, the higher the damping of the bearings must be.*

2.7.2 Free whirling

The solution for free whirling is

$$\left\{ \begin{array}{c} r_B \\ r_C \end{array} \right\} = \left\{ \begin{array}{c} r_{B_0} \\ r_{C_0} \end{array} \right\} e^{st} .$$

Operating as usual, the algebraic equation for free whirling is thus

$$\left[\begin{array}{cc} k_n + k_r + s\,(c_n + c_r) - i\Omega c_r & -k_r - (s - i\Omega)\,c_r \\ -k_r - (s - i\Omega)\,c_r & -m\omega^2 + k_r + (s - i\Omega)\,c_r \end{array} \right] \left\{ \begin{array}{c} r_{B_0} \\ r_{C_0} \end{array} \right\} = \mathbf{0} . \tag{2.157}$$

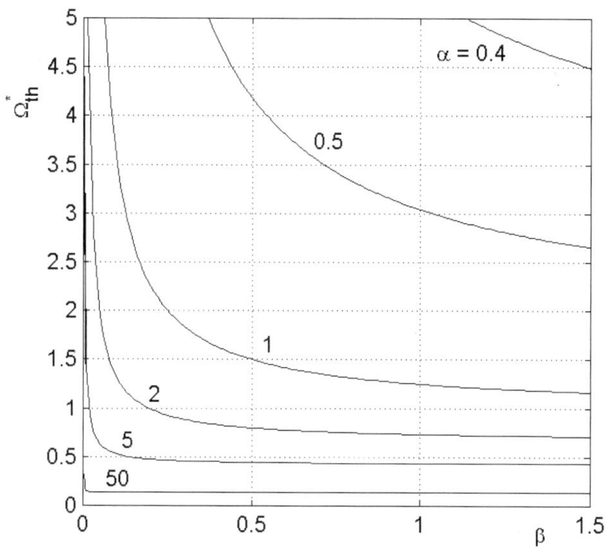

FIGURE 2.25. Jeffcott rotor on compliant bearings: nondimensional threshold of instability Ω^*_{th} as a function of ratio β for various values of the stiffness ratio α and for a fixed value of the damping ratio $\zeta = 1$.

The relevant eigenproblem can be written in the following nondimensional form:

$$\det \begin{bmatrix} 1+\alpha+2\alpha\zeta\left[s^*\left(1+\beta\right)-i\Omega^*\beta\right] & -1-2\alpha\zeta\beta\left(s^*-i\Omega^*\right) \\ -1-2\alpha\zeta\beta\left(s^*-i\Omega^*\right) & \alpha s^{*2}+1+2\alpha\zeta\beta\left(s^*-i\Omega^*\right) \end{bmatrix} = 0\ , \tag{2.158}$$

where $\beta = c_r/c_n$ is the ratio between the rotating and nonrotating damping and $s^* = s/\Omega_1$ is the nondimensional whirl frequency. The solution depends now on three nondimensional parameters, and it is impossible to plot a chart that summarizes the behavior of the system, like the one of Figure 2.24. As an example, the nondimensional threshold of instability Ω^*_{th} is plotted as a function of ratio β for various values of the stiffness ratio α and for a fixed value of the damping ratio $\zeta = 1$ in Figure 2.25.

Remark 2.14 *The threshold of instability decreases with increasing ratio α, i.e., with increasing compliance of the rotor.*

2.8 Rotating coordinates

In the previous sections, the equations of motion were written with reference to the inertial frame Oxy. There is, however, no difficulty in writing them with reference to the rotating frame $O\xi\eta$ defined in Figure 2.8. By solving Equation (2.61) in x_C and y_C and remembering that $\mathbf{R}^{-1} = \mathbf{R}^T$, it follows that

$$\left\{ \begin{array}{c} x_C \\ y_C \end{array} \right\} = \mathbf{R}^T \left\{ \begin{array}{c} \xi_C \\ \eta_C \end{array} \right\} , \tag{2.159}$$

where $\mathbf{R}^T$ is

$$\mathbf{R}^T = \left[\begin{array}{cc} \cos(\Omega t) & -\sin(\Omega t) \\ \sin(\Omega t) & \cos(\Omega t) \end{array} \right] . \tag{2.160}$$

By differentiating Equation (2.159), the velocity and the acceleration of point C can be expressed as

$$\left\{ \begin{array}{c} \dot{x}_C \\ \dot{y}_C \end{array} \right\} = \mathbf{R}^T \left\{ \begin{array}{c} \dot{\xi}_C \\ \dot{\eta}_C \end{array} \right\} + \dot{\mathbf{R}}^T \left\{ \begin{array}{c} \xi_C \\ \eta_C \end{array} \right\} , \tag{2.161}$$

$$\left\{ \begin{array}{c} \ddot{x}_C \\ \ddot{y}_C \end{array} \right\} = \mathbf{R}^T \left\{ \begin{array}{c} \ddot{\xi}_C \\ \ddot{\eta}_C \end{array} \right\} + 2\dot{\mathbf{R}}^T \left\{ \begin{array}{c} \dot{\xi}_C \\ \dot{\eta}_C \end{array} \right\} + \ddot{\mathbf{R}}^T \left\{ \begin{array}{c} \xi_C \\ \eta_C \end{array} \right\} , \tag{2.162}$$

where

$$\dot{\mathbf{R}}^T = \mathbf{\Omega} \left[\begin{array}{cc} -\sin(\Omega t) & -\cos(\Omega t) \\ \cos(\Omega t) & -\sin(\Omega t) \end{array} \right] , \tag{2.163}$$

$$\ddot{\mathbf{R}}^T = \mathbf{\Omega}^2 \left[\begin{array}{cc} -\cos(\Omega t) & \sin(\Omega t) \\ -\sin(\Omega t) & -\cos(\Omega t) \end{array} \right] = -\Omega^2 \mathbf{R}^T . \tag{2.164}$$

The force caused by a rotating viscous damping, with damping coefficient c_r, can be expressed in the $O\xi\eta z$-frame. By introducing Equations (2.159), (2.161), and (2.162) into the equation of motion (2.69), the latter yields

$$\left[\begin{array}{cc} m & 0 \\ 0 & m \end{array} \right] \mathbf{R}^T \left\{ \begin{array}{c} \ddot{\xi}_C \\ \ddot{\eta}_C \end{array} \right\} + \left(\left[\begin{array}{cc} c_n + c_r & 0 \\ 0 & c_n + c_r \end{array} \right] \mathbf{R}^T + 2 \left[\begin{array}{cc} m & 0 \\ 0 & m \end{array} \right] \dot{\mathbf{R}}^T \right) \times$$

$$\times \left\{ \begin{array}{c} \dot{\xi}_C \\ \dot{\eta}_C \end{array} \right\} + \left(\left[\begin{array}{cc} k & 0 \\ 0 & k \end{array} \right] \mathbf{R}^T + \Omega \left[\begin{array}{cc} 0 & c_r \\ -c_r & 0 \end{array} \right] \mathbf{R}^T + \left[\begin{array}{cc} c_n + c_r & 0 \\ 0 & c_n + c_r \end{array} \right] \times$$

$$\times \dot{\mathbf{R}}^T - \Omega^2 \left[\begin{array}{cc} m & 0 \\ 0 & m \end{array} \right] \mathbf{R}^T \right) \left\{ \begin{array}{c} \xi_C \\ \eta_C \end{array} \right\} = \left\{ \begin{array}{c} m\epsilon\Omega^2 \cos(\Omega t) \\ m\epsilon\Omega^2 \sin(\Omega t) \end{array} \right\} + \left\{ \begin{array}{c} F_x \\ F_y \end{array} \right\} . \tag{2.165}$$

By premultiplying Equation (2.165) by $\mathbf{R}$, it follows that

$$\left[\begin{array}{cc} m & 0 \\ 0 & m \end{array} \right] \left\{ \begin{array}{c} \ddot{\xi}_C \\ \ddot{\eta}_C \end{array} \right\} + \left(\left[\begin{array}{cc} c_n + c_r & 0 \\ 0 & c_n + c_r \end{array} \right] + 2\Omega \left[\begin{array}{cc} 0 & -m \\ m & 0 \end{array} \right] \right) \times$$

$$\times \left\{ \begin{array}{c} \dot{\xi}_C \\ \dot{\eta}_C \end{array} \right\} + \left(\left[\begin{array}{cc} k & 0 \\ 0 & k \end{array} \right] - \Omega^2 \left[\begin{array}{cc} m & 0 \\ 0 & m \end{array} \right] + \Omega \left[\begin{array}{cc} 0 & -c_n \\ c_n & 0 \end{array} \right] \right) \left\{ \begin{array}{c} \xi_C \\ \eta_C \end{array} \right\} =$$

$$= \left\{ \begin{array}{c} m\epsilon\Omega^2 \\ 0 \end{array} \right\} + \left\{ \begin{array}{c} F_x \cos(\Omega t) + F_y \sin(\Omega t) \\ F_y \cos(\Omega t) - F_x \sin(\Omega t) \end{array} \right\} . \tag{2.166}$$

Equation (2.166) contains two skew symmetric matrices, one linked with Coriolis acceleration (it is often referred to as gyroscopic matrix because it adds to the damping matrix, even if it has little to do with gyroscopic moments), and a circulatory matrix similar (but not equal) to the one present in the equation written in the inertial frame. Note that a matrix proportional to Ω^2 linked with the centrifugal acceleration acting on point C is also present. This terms is like a negative stiffness proportional to the square of the spin speed.

The same result could be obtained in a simpler way by defining a complex coordinate ρ in the rotating frame

$$\rho = \xi + i\eta = re^{-i\Omega t} . \tag{2.167}$$

The position, velocity, and acceleration of point C can then be expressed as functions of complex coordinate ρ

$$\begin{cases} r = \rho e^{i\Omega t} , \\ \dot{r} = (\dot{\rho} + i\Omega\rho)\, e^{i\Omega t} , \\ \ddot{r} = \left(\ddot{\rho} + 2i\Omega\dot{\rho} - \Omega^2\rho\right) e^{i\Omega t} . \end{cases} \tag{2.168}$$

By introducing Equation (2.168) into the equation of motion of the damped system (2.75) and rearranging the various terms, it follows that

$$m\ddot{\rho} + (c_n + c_r + 2im\Omega)\dot{\rho} + (k - m\Omega^2 + i\Omega c_n)\rho = m\epsilon\Omega^2 + F_n e^{-i\Omega t} . \tag{2.169}$$

Again the main differences between Equation (2.75) written in the inertial frame and Equation (2.169) are the presence in the latter of the terms $2im\Omega\dot{\rho}$ linked with Coriolis acceleration and $m\Omega^2\rho$ caused by the centrifugal acceleration. As already stated, the skew-symmetric terms present in the equation written in real coordinates correspond to imaginary terms in the complex-coordinates equations. Note also that the forces caused by unbalance are constant, in both direction and modulus, in the rotating frame, whereas a stationary force is seen as a force rotating backward with velocity Ω.

To show the differences and similarity between the fixed-frame and the rotating-frame approach, the relevant homogeneous equations for an undamped, perfectly balanced, Jeffcott rotor, written in complex coordinates, are reported in the following table:

Fixed (inertial) frame	**Rotating frame**

Equation of motion

$m\ddot{r} + kr = 0\,.$	$m\ddot{\rho} + 2m\Omega i\dot{\rho} + (k - m\Omega^2)\rho = 0\,.$

Solution

$r = r_0 e^{st}\,.$	$\rho = \rho_0 e^{s't}\,.$

Characteristic equation

$ms^2 + k = 0\,.$	$ms'^2 + 2im\Omega s' + k - m\Omega^2 = 0\,.$

Solution in the Laplace domain

$s = \pm i\sqrt{\frac{k}{m}}\,.$	$s' = i\left(\Omega \pm \sqrt{\frac{k}{m}}\right)\,.$

Campbell diagram

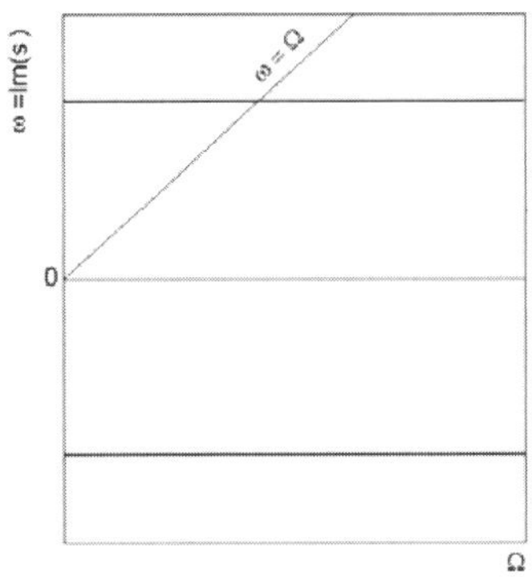

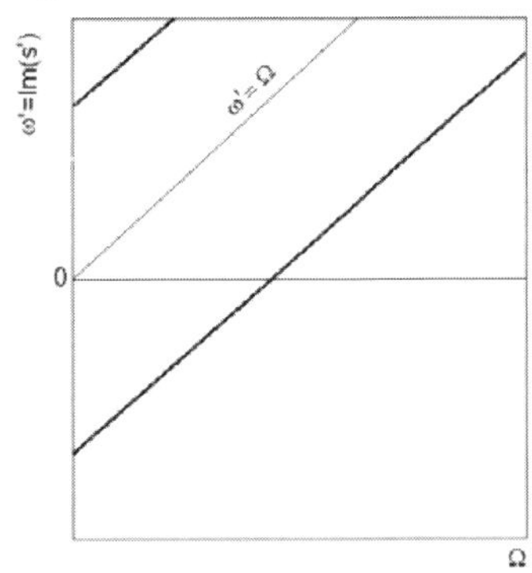

Conversion into inertial coordinates:

$$s = s' - i\Omega = \pm\sqrt{\frac{k}{m}}\,.$$

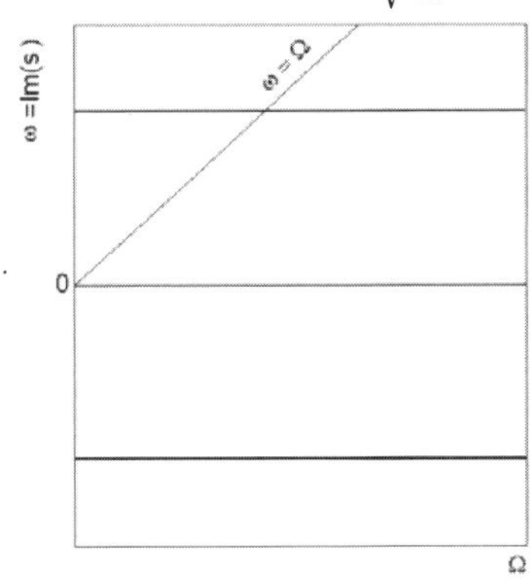

Remark 2.15 *In the Campbell diagram plotted with reference to the rotating frame $s'(\Omega)$, the critical speed is seen as an intersection with the Ω-axis. In the rotating frame, the critical speed is then a sort of resonance with a static force (unbalance force in the rotating frame are constant in amplitude and direction), and hence it is a condition similar to elastic instability.*

2.9 Stability in the supercritical field

The steady-state unbalance response of an undamped or damped Jeffcott rotor was computed in the preceding sections without reaching any conclusion about the stability of the equilibrium position. As the system is linear, the results about the stability of the free motion apply as well for the unbalance response: A simple way to state the stability of the equilibrium position is by observing that the complete solution of the equation of motion can be obtained by adding the solution for free whirling to the unbalance response. When the first one leads to a stable behavior, the overall behavior of the system is stable. The equilibrium position is then stable in the whole supercritical range up to the previously computed threshold of instability.

Another way to reach the same conclusion is by studying the motion in the small of point C in the vicinity of the above-mentioned equilibrium position that can be obtained in terms of rotating coordinates from Equation (2.169). This procedure is typical of the study of the behavior in the small of nonlinear systems, but it can be applied for linear ones as well. By equating to zero all derivatives of the generalized coordinates with respect to time and neglecting static forces, the equilibrium position ρ_0 is simply

$$\rho_0 = \frac{m\epsilon\Omega^2}{k - m\Omega^2 + i\Omega c_n} \,. \tag{2.170}$$

Note that Equation (2.170) coincides with that obtained from Equations (2.87) and (2.88). The motion in the small in the vicinity of point ρ_0 can be expressed as

$$\rho(t) = \rho_1(t) + \rho_0 \,. \tag{2.171}$$

By introducing Equation (2.171) into the equation of motion (2.169) and remembering the expression for the equilibrium position (2.170), the following equation for the motion about the equilibrium position is obtained:

$$m\ddot{\rho}_1 + (c_n + c_r + 2im\Omega)\dot{\rho}_1 + (k - m\Omega^2 + i\Omega c_n)\rho_1 = 0 \,. \tag{2.172}$$

The solution of Equation (2.172) is of the usual type $\rho_1 = \rho_{1_0} e^{s't}$, yielding a spiral motion about the equilibrium position whose amplitude can be either decreasing or increasing in time depending on the sign of the real part of the complex frequency s'. By introducing this solution into Equation (2.172), the following characteristic equation allowing the computation of the complex frequency is readily obtained:

$$ms'^{\,2} + [(c_n + c_r) + 2im\Omega]\, s' + k - m\Omega^2 + i\Omega c_n = 0 \,. \tag{2.173}$$

The solution of the characteristic equation is

$$s' = \sigma + i\omega' = -i\Omega - i\frac{c_r + c_n}{2m} \pm \sqrt{\frac{(c_r + c_n)^2 - 4m(k - i\Omega c_r)}{4m^2}}. \quad (2.174)$$

By comparing Equation (2.174) with Equation (2.78), it is immediately clear that the real parts of the two expressions are equal and the imaginary parts differ by a term equal to $-\Omega$. This result is expected because the former is expressed in a frame of reference that rotates at an angular velocity equal to Ω with respect to the one in which Equation (2.78) is expressed. Because the real part of the complex frequency is the same as that given by Equation (2.78), when Condition (2.83) for stability is satisfied, the motion about the equilibrium position is, in the rotating reference frame, a decaying spiral and the equilibrium position is stable.

2.10 Drag torque at constant speed

If the system has damping properties, energy is dissipated during whirling motion. If no driving torque is applied to the rotor, the spin speed decays in time: The assumption of constant spin speed postulates then that some energy is supplied by a driving system to compensate for the energy losses. The latter can be seen as a form of resistance to motion caused by rotordynamic effects, which must be added to the bearing drag and all other forms of drag (e.g., aerodynamic drag) that can contribute to the overall resistance to motion.

Consider a Jeffcott rotor supplied with both rotating and nonrotating damping, running at constant spin speed Ω. Assume also that, owing to unbalance, it performs a circular whirling motion with amplitude r_0 and frequency Ω, i.e., $r = r_{C_0}e^{i\Omega t}$, where $|r_{C_0}|$ is given by the third Equation (2.89). The force caused by nonrotating damping is thus equal to $-c_n\dot{r}_C$, and the power it dissipates is

$$P_d = c_n|\dot{r}_C|^2 = c_n\Omega^2 r_{C_0}^2. \quad (2.175)$$

Rotating damping does not have any effect, as the system is whirling in a synchronous way. The power supplied by the motor is simply $P_m = M_m\Omega$. By equating the power supplied by the motor to that dissipated by damping and remembering the expression of the amplitude of the whirling motion caused by unbalance [Equation (2.89)], it follows that

$$M_m = c_n\Omega r_{C_0}^2 = \frac{2k\epsilon^2{\Omega^*}^5\zeta_n}{(1 - {\Omega^*}^2)^2 + 4\zeta_n^2{\Omega^*}^2}. \quad (2.176)$$

Equation (2.176) yields then the torque needed to drive the rotor at constant speed against the drag provided by nonrotating damping caused

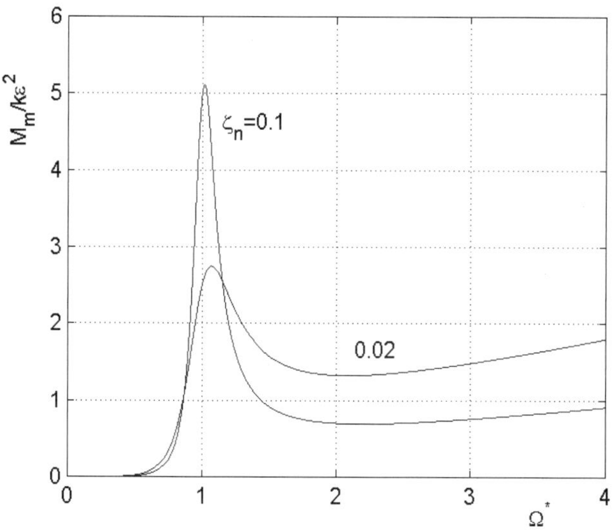

FIGURE 2.26. Nondimensional plot of the drag torque caused by nonrotating damping as a function of Ω^* for two different values of the damping ratio ζ_n: $\zeta_n = 0.1$ and $\zeta_n = 0.2$.

by the unbalance of the rotor. A nondimensional plot of this drag torque as a function of speed is shown in Figure 2.26 for two different values of the damping ratio ζ. Note that the drag depends on the square of the eccentricity and is strong only in case of much unbalanced rotors. In many cases, it is neglected, because it is considered as vanishingly small owing to the assumption of small unbalance that is underlying the linearization of the model. It has a peak at the critical speed, whose value is

$$M_{m_{max}} = \frac{k\epsilon^2}{2\zeta_n} . \tag{2.177}$$

If the rotor is slowly accelerated through the critical speed (no account has been taken for the effect of angular acceleration in the above formulas) and the driving torque is smaller than the value given by Equation (2.177), the rotor stalls, i.e., it fails to go through the critical speed. All of the power supplied by the driving system is dissipated by nonrotating damping, and acceleration is no longer possible. Obviously, the value so computed must be added to that required to overcome all other forms of drag, such as aerodynamic or bearing drag.

In the high supercritical range, the torque needed to operate at constant speed grows linearly with Ω. From Equation (2.176), it follows that

$$\lim_{\Omega\to\infty} M_m = 2k\epsilon^2\Omega^*\zeta_n . \tag{2.178}$$

3

Model with four degrees of freedom: Gyroscopic effect

In the Jeffcott model, the rotor was assumed to be a point mass, and consequently, no allowance was taken for its moments of inertia. This assumption, while greatly simplifying the analysis, precludes the possibility of studying phenomena that considerably influence the dynamic behavior of rotors, and in particular that cause the natural frequencies of bending modes to depend on the spin speed, i.e., cause the Campbell diagram to be different from a number of straight lines running in a horizontal direction. The simplest model to evaluate this effect is shown in Figure 3.1(a) or (b): either a rigid body attached to a compliant massless shaft or a rigid rotor spinning on compliant bearings. A more complex system is that of Figure 3.1(c) in which both the bearings and the rotor are compliant. As it was the case for the Jeffcott rotor, the three cases of Figure 3.1 coincide if damping is neglected.

The present model is not much different from the Jeffcott rotor of Figure 2.1, with the only difference that a rigid body with nonvanishing moments of inertia is located in point C instead of a point mass. One of the principal axes of inertia coincides, in the undeformed position, with z-axis and its ellipsoid of inertia has axial symmetry with respect to the same axis. The principal moments of inertia of the rigid body will be referred to as the *polar* moment of inertia J_p about the rotation axis and *transversal* moment of inertia J_t about any axis in the rotation plane.

If $J_p > J_t$, the body is usually referred to as a *disc*; the limiting case is that of an infinitely thin disc in which $J_p = 2J_t$. If $J_p < J_t$, the rotor is usually referred to as a *long rotor*. A particular case is when $J_p = J_t$; i.e., the inertia ellipsoid degenerates into a sphere; as it will be shown later, this configuration must be avoided in the case of fast spinning rotors.

Assume also that, owing to small errors, the position of point P, in which the center of gravity of the rigid body is located, does not coincide with that of point C, the center of the shaft.[1] The distance between the two points is the eccentricity ϵ. Moreover, the axis of symmetry of the rigid body does not coincide exactly with the rotation axis, the angle between them being a small angular error χ. These two errors, which are assumed to be small, cause a *static unbalance* and a *couple unbalance*, respectively.

Strictly speaking, the system has six degrees of freedom, and six generalized coordinates must be defined for the study of its dynamic behavior. The uncoupling among axial, flexural, and torsional behavior seen for the Jeffcott rotor will be shown to hold also in this case, and a model with four degrees of freedom is adequate for the study of the flexural behavior at constant speed, at least under wide simplifying assumptions.

3.1 Generalized coordinates and equations of motion

3.1.1 Kinematics

The generalized coordinates will be defined with reference to the frames shown in Figure 3.2(a).

- Frame OXYZ: Inertial frame, with origin in O, and Z-axis coinciding with the rotation axis of the rotor.

- Frame OΞHZ with origin in O, and Z-axis coinciding with that of the preceding frame: Axes Ξ, H rotate in the XY-plane with angular velocity Ω, in the case of constant speed operation. It will be referred to as a *rotating frame*.

- Frame C$X'Y'Z'$ with origin in C: Its axes remain parallel with those of frame OXYZ. $X'Y'$-plane remains parallel to XY-plane.

- Frame Cxyz with origin in C: Its z-axis coincides with the rotation axis of the rigid body in the deformed position, and x- and y-axes are defined by the following rotations:

[1] Point C is the geometric center of the cross section of the shaft where the rigid body is located. It is an elastic center in the sense that the elastic reaction of the shaft tends to put point C on the line connecting the centers of the bearings in the undeflected postion.

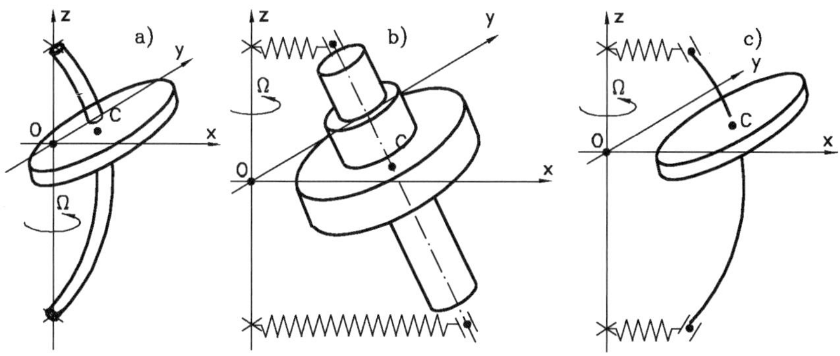

FIGURE 3.1. Rotor with four degrees of freedom. (a) Rigid body on a massless compliant shaft, (b) rigid rotor on compliant bearings, and (c) rotor in which both the bearings and the shaft are compliant. In (a) and (c), the disc is assumed to be a rigid body.

- - Rotate the axes of $CX'Y'Z'$ frame about the X'-axis of an angle $\phi_{X'}$ until the Y'-axis enters the rotation plane of the rigid body in its deformed configuration. Let the axes so obtained be the y- and z^*-axes. The rotation matrix allowing one to express the components of a vector in $CX'yz^*$-frame from those in $CX'Y'Z'$-frame (or in the inertial frame, because the directions of the axes coincide) is

$$\mathbf{R}_1 = \begin{bmatrix} 1 & 0 & 0 \\ 0 & \cos(\phi_{X'}) & \sin(\phi_{X'}) \\ 0 & -\sin(\phi_{X'}) & \cos(\phi_{X'}) \end{bmatrix} . \tag{3.1}$$

  - Rotate the frame obtained after the mentioned rotation, about the y-axis until X'-axis also enters the rotation plane of the rigid body in its deformed configuration. Let the axis so obtained be the x-axis and the rotation angle be ϕ_y. After the two mentioned rotations, z-axis coincides, apart from the angular error χ, with the symmetry axis of the rigid body in its deformed configuration. Frame $Cxyz$ is centered in the center of the shaft of the rigid body and follows it in its whirling motion. However, it does not rotate with the spin speed Ω. It will then be referred to as *whirling frame*. Let the matrix expressing this second rotation be

$$\mathbf{R}_2 = \begin{bmatrix} \cos(\phi_y) & 0 & -\sin(\phi_y) \\ 0 & 1 & 0 \\ \sin(\phi_y) & 0 & \cos(\phi_y) \end{bmatrix} . \tag{3.2}$$

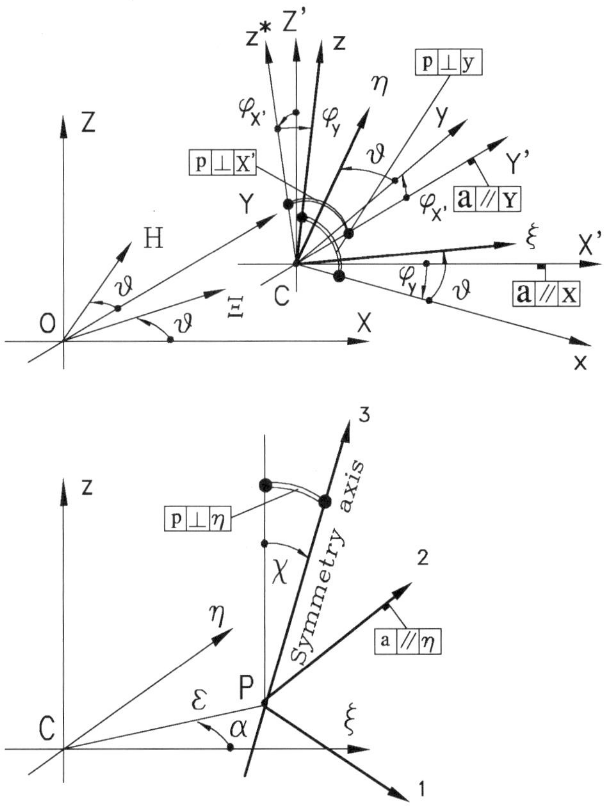

FIGURE 3.2. Reference frames used in the study of a rotor with four degrees of freedom. The symbol in the boxes indicate whether the axis (a) or plane (p, indicated with the double arc) is parallel or perpendicular to an axis.

- Frame $C\xi\eta z$ with origin in C: It is obtained from frame $Cxyz$ by rotating x- and y-axes in xy-plane of an angle equal to rotation angle θ of the rotor corresponding to the spin speed. If rotation occurs with constant spin speed Ω, angle θ is equal to Ωt. Frame $C\xi\eta z$ is actually fixed to the rigid body, although not being centered in its center of gravity owing to the eccentricity ϵ, and not being principal of inertia owing to the angular error χ. It will be referred to as *rotating and whirling frame* . The matrix allowing one to express a vector in the $C\xi\eta z$-frame from the components in the $Cxyz$-frame is

$$\mathbf{R}_3 = \begin{bmatrix} \cos(\theta) & \sin(\theta) & 0 \\ -\sin(\theta) & \cos(\theta) & 0 \\ 0 & 0 & 1 \end{bmatrix}, \tag{3.3}$$

from	to	translation	rotation
$OXYZ$	$CX'Y'Z'$	$\begin{bmatrix} X_C & Y_C & 0 \end{bmatrix}^T$	-
$CX'Y'Z'$	$CX'yz^*$	-	$\mathbf{R}_1$
$CX'yz^*$	$Cxyz$	-	$\mathbf{R}_2$
$Cxyz$	$C\xi\eta z$	-	$\mathbf{R}_3$
$C\xi\eta z$	P123	$\begin{bmatrix} \epsilon_\xi & \epsilon_\eta & 0 \end{bmatrix}^T$	$\mathbf{R}_4$

TABLE 3.1. Summary of the transformations between the various reference frames in terms of displacement vectors and rotation matrices.

where $\theta = \Omega t$ in constant-speed operation.

- Frame P123, principal axes of inertia of the rigid body: As already stated, the rotor is assumed to be slightly unbalanced. As the angular position of the rotor in the $C\xi\eta z$-frame is immaterial, the principal axis of inertia corresponding to the moment of inertia J_p will be assumed to lie in a plane parallel to ξz-plane. As the static unbalance cannot be assumed to lie in the same plane as the couple unbalance, the eccentricity cannot be assumed to lie along the ξ-axis, as was the case of the Jeffcott rotor. The conditions of unbalance are summarized in Figure 3.2(b), where the static unbalance is shown to lead the couple unbalance of a phase angle α. A fourth rotation matrix $\mathbf{R}_4$ allowing passage from the rotor system of reference $C\xi\eta z$ to the principal axes of the rigid body P123 is thus defined

$$\mathbf{R}_4 = \begin{bmatrix} \cos(\chi) & 0 & -\sin(\chi) \\ 0 & 1 & 0 \\ \sin(\chi) & 0 & \cos(\chi) \end{bmatrix} . \tag{3.4}$$

The transformations needed to pass from one reference frame to the others are reported in Table 3.1

Take the X-, Y- and Z-coordinates of point C and angles $\phi_{X'}$, ϕ_y, and θ as generalized coordinates of the rigid body. A small displacement assumption on coordinates X, Y, Z, $\phi_{X'}$, and ϕ_y will allow great simplification of the problem. Coordinate θ, on the contrary, cannot be considered small.

The velocity of the center of gravity (point P) and the angular velocity expressed in the principal system of inertia must be computed to compute the kinetic energy of the rigid body. The position of point P is easily obtained

$$\overline{(\mathrm{P}-\mathrm{O})} = \begin{Bmatrix} X \\ Y \\ Z \end{Bmatrix} + \mathbf{R}_1^T \mathbf{R}_2^T \mathbf{R}_3{}^T \begin{Bmatrix} \epsilon_\xi \\ \epsilon_\eta \\ 0 \end{Bmatrix} . \tag{3.5}$$

The small-displacement assumption allows the linearization of the trigonometric functions of angles $\phi_{X'}$ and ϕ_y and the neglect of some terms, which are of the same order of magnitude as those that are neglected when truncating the series for the trigonometric functions after the first term.[2] Equation (3.5) then reduces to

$$(\overline{\mathrm{P}-\mathrm{O}}) = \left\{ \begin{array}{c} X + \epsilon\cos(\theta + \alpha) \\ Y + \epsilon\sin(\theta + \alpha) \\ Z + \epsilon\left[\phi_{X'}\sin(\theta + \alpha) - \phi_y\cos(\theta + \alpha)\right] \end{array} \right\}, \tag{3.6}$$

where the unbalance has been expressed in terms of amplitude and phase (ϵ and α) instead of using the components $\epsilon_\xi = \epsilon\cos(\alpha)$ and $\epsilon_\eta = \epsilon\sin(\alpha)$.

3.1.2 Equations of motion in real coordinates

The equations of motion of the rotor are obtained following a Lagrangian approach from the expressions of the kinetic and potential energies. The generalized coordinates of the system are five: X, Y, Z, $\phi_{X'}$ and ϕ_y; θ is not considered as a Lagrangian coordinate because the angular velocity has been assumed to be constant. Only the lateral and axial behavior of the system is then considered, and the torsional dynamics is not studied here.

Kinetic energy

The kinetic energy of the system is easily computed as the sum of the translational kinetic energy of its center of mass plus the rotational kinetic energy

$$\mathcal{T} = \mathcal{T}_t + \mathcal{T}_t = \frac{1}{2} m V_{\mathrm{P}}^2 + \frac{1}{2} \Omega_{123}'^T \, \mathbf{J} \, \Omega_{123}' , \tag{3.7}$$

where V_{P} is the velocity of the center of mass, Ω'_{123} is the angular velocity vector expressed in the 123 reference frame and

$$\mathbf{J} = \begin{bmatrix} J_t & 0 & 0 \\ 0 & J_t & 0 \\ 0 & 0 & J_p \end{bmatrix} \tag{3.8}$$

is the inertia tensor of the rigid body expressed in the same reference frame.

---

[2] All terms of order higher than two in the expression of the kinetic energy give way to nonlinear terms in the equations of motion. By linearizing the expressions of the displacements and the velocities, the expression of the kinetic energy is then simplified in such a way to yield linearized equations of motion. The results so obtained were checked against those obtained by computing the kinetic energy using the complete expression of the displacements and then making the simplifications at the end of the computations. The two procedures yield, as expected, identical results.

The velocity of the center of gravity of the rotor is easily computed by performing the derivatives of vector $\overline{(\mathrm{P}-\mathrm{O})}$ with respect to time. Remembering that $\dot{\theta}=\Omega$ and angle α are constant, it follows that

$$V_{\mathrm{P}}=\left\{\begin{array}{c}\dot{X}-\epsilon\Omega\sin(\theta+\alpha)\\ \dot{Y}+\epsilon\Omega\cos(\theta+\alpha)\\ \dot{Z}+\epsilon\left[(\Omega\phi_{X'}-\dot{\phi}_y)\cos(\theta+\alpha)+(\Omega\phi_y+\dot{\phi}_{X'})\sin(\theta+\alpha)\right]\end{array}\right\}. \tag{3.9}$$

In the third line of Equation (3.9), there are two terms: The first is the velocity in the axial direction caused by the displacement of point C in the same direction, and the second is the velocity in the axial direction caused by the eccentricity and rotations of the cross section of the shaft. It is easy to verify that the last term causes a coupling between bending and axial behavior of the rotor; however, if the eccentricity is small, it is negligible when compared with the first one. In the following developments, all terms containing the product of the eccentricity or the angular error by a small quantity will be neglected and no axial-flexural coupling will be obtained. The translational kinetic energy is then

$$\mathcal{T}_t=\frac{1}{2}m\left\{\dot{X}^2+\dot{Y}^2+\dot{Z}^2+\epsilon^2\Omega^2+2\epsilon\Omega\left[-\dot{X}\sin(\Omega t+\alpha)+\dot{Y}\cos(\Omega t+\alpha)\right]\right\}. \tag{3.10}$$

The angular velocity Ω'_{123} can be computed as the sum of three angular velocity vectors acting in different directions: $\dot{\phi}_{X'}$ along the X'-axis, $\dot{\phi}_y$ along the y-axis, and $\dot{\theta}$ along the z-axis. Using the relevant rotation matrices, the components of the angular velocity along the principal axes of inertia of the rotor Ω'_{123} are

$$\Omega'_{123}=\mathbf{R}_4\left[\mathbf{R}_3\left(\mathbf{R}_2\left\{\begin{array}{c}\dot{\phi}_{X'}\\0\\0\end{array}\right\}_{X'Y'Z'}+\left\{\begin{array}{c}0\\ \dot{\phi}_y\\0\end{array}\right\}_{xyz'}\right)+\left\{\begin{array}{c}0\\0\\ \dot{\theta}\end{array}\right\}_{\xi\eta z}\right]. \tag{3.11}$$

By remembering the small-displacement assumptions, Equation (3.11) reduces to

$$\Omega'_{123}=\left\{\begin{array}{c}\dot{\phi}_{X'}\cos(\theta)+\dot{\phi}_y\sin(\theta)-\chi\Omega\\ -\dot{\phi}_{X'}\sin(\theta)+\dot{\phi}_y\cos(\theta)\\ \dot{\phi}_{X'}\left[\chi\cos(\theta)+\phi_y\right]+\dot{\phi}_y\chi\sin(\theta)+\Omega\end{array}\right\}. \tag{3.12}$$

As the components of Ω'_{123} are referred to the principal axes of inertia, neglecting the small terms, the rotational kinetic energy can be computed as

$$\mathcal{T}_r = \frac{1}{2}\left\{J_t(\dot{\phi}_{X'}^2 + \dot{\phi}_y^2 + \chi^2\Omega^2) + J_p(\Omega^2 + 2\Omega\dot{\phi}_{X'}\phi_y) + \right.$$
$$\left. +2\Omega\chi(J_p - J_t)\left[\dot{\phi}_{X'}\cos(\theta) + \dot{\phi}_y\sin(\theta)\right]\right\} . \tag{3.13}$$

Potential energy

The only forces and moments acting on the rigid body in P that will be considered are those caused by the elastic reaction of the shaft. Because the behavior of the shaft is assumed to be linear, they are linked to the generalized coordinates by the stiffness matrix of the shaft; moreover, the lateral behavior of the rotor is again uncoupled from the axial and torsional behavior and only the forces and displacements in xy-plane and the related rotations need to be considered. The situation in the xz-plane is similar to that in the yz-plane, but if the same elements of the stiffness matrix are used, owing to the axial symmetry of the shaft, the different sign convention in the two coordinate planes compels the use of opposite signs for the elements with subscripts 12 and 21 in the stiffness matrices related to the two planes:

$$\mathbf{K}_{xz} = \begin{bmatrix} K_{11} & K_{12} \\ K_{12} & K_{22} \end{bmatrix} , \qquad \mathbf{K}_{yz} = -\begin{bmatrix} K_{11} & -K_{12} \\ -K_{12} & K_{22} \end{bmatrix} . \tag{3.14}$$

The stiffness matrices $\mathbf{K}_{xz}$ and $\mathbf{K}_{yz}$ can be obtained in different ways, e.g., using the formulas reported on many stress analysis handbooks or through the finite element method (see Chapter 4). They can be also obtained by inverting the compliance matrix

$$\mathbf{K} = \mathbf{B}^{-1} = \begin{bmatrix} \beta_{11} & \beta_{12} \\ \beta_{12} & \beta_{22} \end{bmatrix}^{-1} , \tag{3.15}$$

where the various β_{ij} are the coefficients of influence of the system. In the present case, β_{11} is the displacement of point P caused by a unit force applied in the same point; $\beta_{12} = \beta_{21}$ is the rotation in the same point caused by a unit force or the displacement caused by a unit torque and β_{22} is the rotation caused by a unit torque. The computation of the compliance and the stiffness matrices will be shown in detail in the examples at the end of this chapter.

The potential energy is thus

$$\mathcal{U} = \frac{1}{2}\left\{\begin{array}{c} X \\ \phi_y \end{array}\right\}^T \mathbf{K}_{xz}\left\{\begin{array}{c} X \\ \phi_y \end{array}\right\} + \frac{1}{2}\left\{\begin{array}{c} Y \\ \phi_{X'} \end{array}\right\}^T \mathbf{K}_{yz}\left\{\begin{array}{c} Y \\ \phi_{X'} \end{array}\right\} . \tag{3.16}$$

By introducing the Lagrangian function $\mathcal{T} - \mathcal{U}$ into Lagrange equation (2.4) and performing the relevant derivatives, the equations of motion of the undamped rotor with four degrees of freedom are readily obtained

$$\begin{cases} m\ddot{X} + K_{11}X + K_{12}\phi_y = m\epsilon\Omega^2\cos(\Omega t + \alpha)\,, \\ m\ddot{Y} + K_{11}Y - K_{12}\phi_{X'} = m\epsilon\Omega^2\sin(\Omega t + \alpha)\,, \\ J_t\ddot{\phi}_{X'} + J_p\Omega\dot{\phi}_y - K_{12}Y + K_{22}\phi_{X'} = -\chi\Omega^2(J_t - J_p)\sin(\Omega t)\,, \\ J_t\ddot{\phi}_y - J_p\Omega\dot{\phi}_{X'} + K_{12}X + K_{22}\phi_y = \chi\Omega^2(J_t - J_p)\cos(\Omega t)\,. \end{cases} \tag{3.17}$$

Equations (11.28) can be written in the following matrix form:

$$\begin{bmatrix} m & 0 & 0 & 0 \\ 0 & J_t & 0 & 0 \\ 0 & 0 & m & 0 \\ 0 & 0 & 0 & J_t \end{bmatrix} \begin{Bmatrix} \ddot{X} \\ \ddot{\phi}_y \\ \ddot{Y} \\ \ddot{\phi}_{X'} \end{Bmatrix} + \Omega \begin{bmatrix} 0 & 0 & 0 & 0 \\ 0 & 0 & 0 & -J_p \\ 0 & 0 & 0 & 0 \\ 0 & J_p & 0 & 0 \end{bmatrix} \begin{Bmatrix} \dot{X} \\ \dot{\phi}_y \\ \dot{Y} \\ \dot{\phi}_{X'} \end{Bmatrix} +$$

$$+ \begin{bmatrix} K_{11} & K_{12} & 0 & 0 \\ K_{12} & K_{22} & 0 & 0 \\ 0 & 0 & K_{11} & -K_{12} \\ 0 & 0 & -K_{12} & K_{22} \end{bmatrix} \begin{Bmatrix} X \\ \phi_y \\ Y \\ \phi_{X'} \end{Bmatrix} = \Omega^2 \begin{Bmatrix} m\epsilon\cos(\Omega t + \alpha) \\ \chi(J_t - J_p)\cos(\Omega t) \\ m\epsilon\sin(\Omega t + \alpha) \\ -\chi(J_t - J_p)\sin(\Omega t) \end{Bmatrix}. \tag{3.18}$$

If $-\phi_{X'}$ is used instead of $\phi_{X'}$ as the generalized coordinate for rotations about the X-axis, the stiffness matrix assumes a more regular pattern with all terms positive and the skew-symmetric gyroscopic matrix, containing the polar moments of inertia, is replaced by its transpose.

3.1.3 Equations of motion in complex coordinates

Also, in the current case, it is possible to define a set of complex coordinates that allow the equations of motion to be written in a more compact form

$$\begin{cases} r = X + iY\,, \\ \phi = \phi_y - i\phi_{X'}\,. \end{cases} \tag{3.19}$$

Note that by defining the complex coordinate for rotation in this way, the real part of the coordinates refers to the behavior in xz-plane and the imaginary part refers to the behavior in yz-plane. Moreover, the (−) sign in the definition of ϕ takes care of the different conventions of sign in the two inflection planes. All alternative definitions of ϕ, e.g.,

$$\phi = \phi_{X'} + i\phi_y\,,$$

would lead to more complicated equations with a mixing between the features occurring in the two inflection planes: Almost all advantages of complex coordinates would be lost.

Multiplying the second Equation (11.28) by the imaginary unit i and adding it to the first one, and multiplying the third equation by $-i$ and adding it to the fourth one, reduces the equations of motion to

$$\begin{cases} m\ddot{r} + K_{11}r + K_{12}\phi = m\epsilon\Omega^2 e^{i(\Omega t+\alpha)} , \\ J_t\ddot{\phi} - i\Omega J_p\dot{\phi} + K_{12}r + K_{22}\phi = \chi\Omega^2(J_t - J_p)e^{i\Omega t} , \end{cases} \tag{3.20}$$

or, in the more compact form,

$$\mathbf{M}\ddot{\mathbf{q}} - i\Omega\mathbf{G}\dot{\mathbf{q}} + \mathbf{K}\mathbf{q} = \Omega^2\mathbf{f}e^{i\Omega t} . \tag{3.21}$$

The vector of complex coordinates, mass matrix, gyroscopic matrix, stiffness matrix, and vector of unbalances in Equations (3.21) are, respectively,

$$\mathbf{q} = \left\{ \begin{array}{c} r \\ \phi \end{array} \right\} , \qquad \mathbf{M} = \left[\begin{array}{cc} m & 0 \\ 0 & J_t \end{array} \right] , \qquad \mathbf{G} = \left[\begin{array}{cc} 0 & 0 \\ 0 & J_p \end{array} \right] ,$$

$$\mathbf{K} = \left[\begin{array}{cc} K_{11} & K_{12} \\ K_{12} & K_{22} \end{array} \right] , \qquad \mathbf{f} = \left\{ \begin{array}{c} m\epsilon e^{i\alpha} \\ \chi(J_t - J_p) \end{array} \right\} . \tag{3.22}$$

Note that all matrices are symmetric when using the complex coordinates notation, and that when using real coordinates, the gyroscopic matrix is skew symmetric.

The presence of damping can be accounted for in a very easy way if the viscous or hysteretic damping models can be accepted. As in the case of the Jeffcott model, it is important to distinguish between nonrotating and rotating damping and to introduce separately the two damping matrices. In the case of viscous damping, the equation of motion of the damped system is

$$\mathbf{M}\ddot{\mathbf{q}} + (\mathbf{C}_n + \mathbf{C}_r - i\Omega\mathbf{G})\dot{\mathbf{q}} + (\mathbf{K} - i\Omega\mathbf{C}_r)\mathbf{q} = \Omega^2\mathbf{f}e^{i\Omega t} , \tag{3.23}$$

where the structure of matrices $\mathbf{C}_n$ and $\mathbf{C}_r$ is similar to that of the stiffness matrix.

Again rotating damping causes the presence of a circulatory matrix, which in real coordinates is skew-symmetric and in complex coordinates is symmetric but imaginary.

3.1.4 Static and couple unbalance

The conditions of unbalance have been summarized in Figure 3.2(b), with the principal axis of inertia corresponding to the moment of inertia J_p of the rigid body lying in a plane parallel to ξz-plane and with the static unbalance leading the couple unbalance of a phase angle α. This is generally enough for the study of an isotropic rotor because the angular position of the rotor in the $C\xi\eta z$-frame is immaterial. However, in many cases, the position of ξ- and η-axes is fixed with respect to the rotor and does not depend on the couple unbalance, which in general can be assumed to lead the plane parallel to ξz-plane by an angle β.

The vector of the generalized forces caused by unbalance then becomes

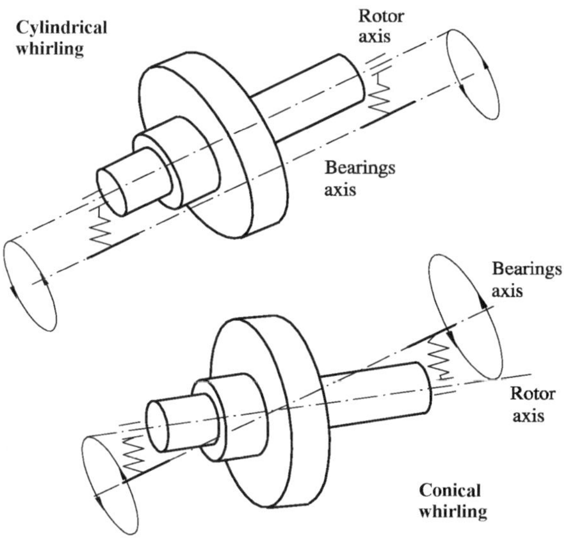

FIGURE 3.3. Cylindrical and conical whirling of a rigid rotor on compliant bearings (the label *bearings axis* refers to the undeformed position).

$$\mathbf{f} = \left\{ \begin{array}{c} m\epsilon e^{i\alpha} \\ \chi(J_t - J_p)e^{i\beta} \end{array} \right\}. \tag{3.24}$$

3.2 Uncoupled gyroscopic system

3.2.1 Complex coordinates

As can be seen from equation (3.20), the two equations uncouple if $K_{12} = 0$, i.e., if the translational degrees of freedom are elastically uncoupled from the rotational ones. Translational motion occurs with the axis of the rotor remaining parallel to itself and is referred to as *cylindrical whirling*; rotational motion occurs about the center of mass and is called *conical whirling* (Figure 3.3). This can occur, for instance, in the case of a rigid rotor on two equal bearings with the center of mass exactly at midspan, which is a situation not uncommon in practice. The equation of motion reduces to

$$\left\{ \begin{array}{l} m\ddot{r} + K_{11}r = m\epsilon\Omega^2 e^{i(\Omega t+\alpha)} , \\ J_t\ddot{\phi} - i\Omega J_p\dot{\phi} + K_{22}\phi = \chi\Omega^2(J_t - J_p)e^{i(\Omega t+\beta)} . \end{array} \right. \tag{3.25}$$

The first Equation (3.25) is identical to the equation of motion of a Jeffcott rotor and needs no further study. The second equation describes the behavior of the simplest gyroscopic system that can be conceived. The free whirling of the undamped system can be studied using the homogeneous equation associated with the second Equation (3.25); by introducing a solution of the type $\phi = \phi_0 e^{st}$, it yields the following algebraic linear equation:

$$\phi_0 \left(J_t s^2 + isJ_p\Omega + K_{22} \right) = 0 \,. \tag{3.26}$$

The characteristic equation yielding the whirling frequencies is

$$\left(J_t s^2 + isJ_p\Omega + K_{22} \right) = 0 \,. \tag{3.27}$$

whose solutions are

$$s = \frac{iJ_p\Omega \pm \sqrt{-J_p^2\Omega^2 - 4J_tK_{22}}}{2J_t} = i\frac{J_p\Omega \pm \sqrt{J_p^2\Omega^2 + 4J_tK_{22}}}{2J_t} \,. \tag{3.28}$$

By remembering that $s = \sigma + i\omega$, it follows that

$$\begin{cases} \omega = \dfrac{J_p\Omega \pm \sqrt{J_p^2\Omega^2 + 4J_tK_{22}}}{2J_t} \,, \\ \sigma = 0 \,, \end{cases} \tag{3.29}$$

or, in nondimensional form,

$$\begin{cases} \omega^* = \dfrac{\delta\Omega^*}{2} \pm \sqrt{\dfrac{\delta^2\Omega^{*2}}{4} + 1} \,, \\ \sigma^* = 0 \,, \end{cases} \tag{3.30}$$

where

$$\delta = \frac{J_p}{J_t}$$

can have values spanning from 0 to 2. The nondimensional whirl frequency $\omega^* = \omega/\omega_0$ and spin speed $\Omega^* = \Omega/\omega_0$ have been obtained using the natural frequency of the system at standstill $\omega_0 = \sqrt{K_{22}/J_t}$. Equation (3.30) has been plotted for different values of δ in Figure 3.4.

Remark 3.1 *As it was shown for the case of the Jeffcott rotor, the eigenvalues would have been 4 if real coordinates had been used. Owing to axial symmetry, they are two identical pairs, given by Equation (12.22).*

The main features of the Campbell diagram are the following:

- The curves are not horizontal straight lines, except if $\delta = 0$, i.e., if the gyroscopic effect vanishes.
- The frequency of backward whirling decreases in absolute value with increasing speed, and tends to 0 for $\Omega \to \infty$.

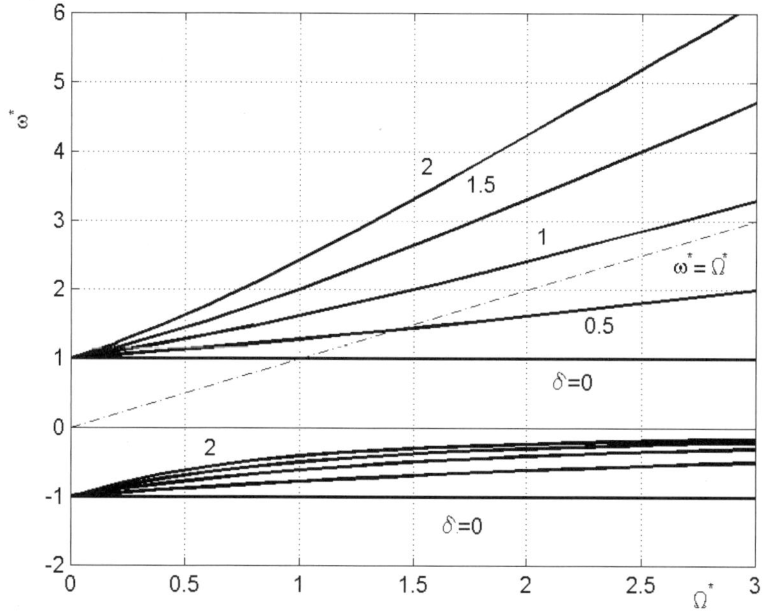

FIGURE 3.4. Nondimensional Campbell diagram for the rotational motions of a rotor with four degrees of freedom in which translational and rotational motions are uncoupled. Five values of $\delta = J_p/J_t$ have been considered: $\delta = 0$, 0.5, 1, 1.5, and 2.

- The frequency of forward whirling increases with increasing speed, and the relevant curve has an inclined asymptote with equation $\omega = \delta\Omega$.

- If $\delta > 1$, i.e., the rotor is disk-like, there is no intersection of the curve with the line $\omega = \Omega$, i.e., there is no critical speed.

- If $\delta < 1$, i.e., the rotor is a long rotor, the curve intersects the line $\omega = \Omega$ and there is a critical speed linked with conical motion. The value of this critical speed can be easily obtained by introducing $\omega = \Omega = \Omega_{cr}$ into Equation (12.21), obtaining

$$\Omega_{cr} = \sqrt{\frac{K_{22}}{J_t - J_p}} = \sqrt{\frac{K_{22}}{J_t}} \frac{1}{\sqrt{1-\delta}} \,. \tag{3.31}$$

- If $\delta = 1$, i.e., the rotor is a spherical rotor, the curve does not intersect the line $\omega = \Omega$, but tends asymptotically to it for $\Omega \to \infty$. There is no critical speed linked with conical motion, but the situation is in

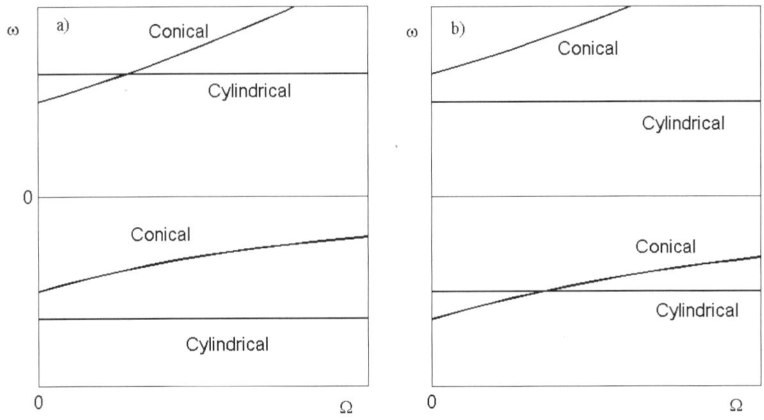

FIGURE 3.5. Campbell diagram for a rotor with complete uncoupling between conical and cylindrical whirling. Cases with the natural frequency at standstill of the cylindrical mode (a) higher and (b) lower than that of the conical mode.

a way similar to that of a field of instability. Although not giving way to any self-excited vibration, the response to unbalance grows steadily with increasing speed and no self-centered configuration can be achieved at high speed.

The complete Campbell diagram for a rotor with complete uncoupling between conical and cylindrical whirling can thus be obtained by superimposing the plot of Figure 3.4 with the Campbell diagram of the Jeffcott rotor, obtaining a plot of the type shown in Figure 3.5(a) or (b). The first plot refers to a case in which the natural frequency at standstill of the cylindrical mode is higher than that of the conical mode, i.e., $\sqrt{K_{11}/m} > \sqrt{K_{22}/J_t}$. In this case, the two branches of the plot intersect in the first quadrant. The second figure refers to a case in which the natural frequency at standstill of the cylindrical mode is lower than that of the conical mode, i.e., $\sqrt{K_{11}/m} < \sqrt{K_{22}/J_t}$, and the intersection occurs in the fourth quadrant.

3.2.2 Real coordinates

If real coordinates are used, the equations of motion are

$$\begin{cases} m\ddot{X} + K_{11}X = m\epsilon\Omega^2 \cos(\Omega t + \alpha)\,, \\ m\ddot{Y} + K_{11}Y = m\epsilon\Omega^2 \sin(\Omega t + \alpha)\,, \\ J_t\ddot{\phi}_{X'} + J_p\Omega\dot{\phi}_y + K_{22}\phi_{X'} = -\chi\Omega^2(J_t - J_p)\sin(\Omega t)\,, \\ J_t\ddot{\phi}_y - J_p\Omega\dot{\phi}_{X'} + K_{22}\phi_y = \chi\Omega^2(J_t - J_p)\cos(\Omega t)\,. \end{cases} \tag{3.32}$$

The first two are the same equations already seen for the Jeffcott rotor, and the other two can be written in matrix form as

$$\begin{bmatrix} J_t & 0 \\ 0 & J_t \end{bmatrix} \begin{Bmatrix} \ddot{\phi}_y \\ \ddot{\phi}_{X'} \end{Bmatrix} + \Omega \begin{bmatrix} 0 & -J_p \\ J_p & 0 \end{bmatrix} \begin{Bmatrix} \dot{\phi}_y \\ \dot{\phi}_{X'} \end{Bmatrix} +$$

$$+ \begin{bmatrix} K_{22} & 0 \\ 0 & K_{22} \end{bmatrix} \begin{Bmatrix} \phi_y \\ \phi_{X'} \end{Bmatrix} = \chi(J_t - J_p)\Omega^2 \begin{Bmatrix} \cos(\Omega t) \\ -\sin(\Omega t) \end{Bmatrix} . \quad (3.33)$$

By assuming a solution of the type $\phi_X = \phi_{X0}e^{st}$, $\phi_y = \phi_{y0}e^{st}$, the homogeneous equation yields the characteristic equation

$$\det \begin{bmatrix} s^2 J_t + K_{22} & -s\Omega J_p \\ s\Omega J_p & s^2 J_t + K_{22} \end{bmatrix} = 0 , \quad (3.34)$$

i.e.,

$$s^4 J_t^2 + s^2 \left(2J_t K_{22} + \Omega^2 J_p^2\right) + K_{22}^2 = 0 . \quad (3.35)$$

Its solution is

$$s^2 = -\frac{K_{22}}{J_t} - \frac{\Omega^2 \delta^2}{2} \pm \Omega\delta \sqrt{\frac{K_{22}}{J_t} + \frac{\Omega^2 \delta^2}{4}} , \quad (3.36)$$

or in terms of s,

$$s = \pm i \left(\frac{\Omega\delta}{2} \pm \sqrt{\frac{K_{22}}{J_t} + \frac{\Omega^2\delta^2}{4}} \right) , \quad (3.37)$$

which coincides with Equation (3.30). The only difference is that the solutions found through the real coordinates approach are twice in number and their sign loses any meaning.

3.3 Free whirling of the coupled, undamped system

If no uncoupling is possible, i.e., $K_{12} \neq 0$, the free whirling of the undamped system can be studied using the homogeneous equation associated with Equation (3.20). By introducing a solution of the type

$$\begin{Bmatrix} r \\ \phi \end{Bmatrix} = \begin{Bmatrix} r_0 \\ \phi_0 \end{Bmatrix} e^{st} ,$$

into the equation of motion, the following algebraic linear equations are readily obtained:

$$\begin{cases} r_0 \left(ms^2 + K_{11}\right) + \phi_0 K_{12} = 0 , \\ r_0 K_{12} + \phi_0 \left(J_t s^2 + isJ_p\Omega + K_{22}\right) = 0 . \end{cases} \quad (3.38)$$

The characteristic equation allowing computation of the whirl frequency is

$$\det\begin{bmatrix} ms^2 + K_{11} & K_{12} \\ K_{12} & J_t s^2 + isJ_p\Omega + K_{22} \end{bmatrix} = 0\,. \tag{3.39}$$

As the solutions of the eigenproblem (3.39) are imaginary, it is handy to use directly ω as an unknown, instead of $s = i\omega$. In this case, Equation (3.39) yields

$$\omega^4 - \Omega\omega^3 \frac{J_p}{J_t} - \omega^2 \left(\frac{K_{11}}{m} + \frac{K_{22}}{J_t}\right) + \Omega\omega \frac{K_{11} J_p}{m J_t} + \frac{K_{11}K_{22} - K_{12}^2}{m J_t} = 0\,. \tag{3.40}$$

Equation (3.40) has four real roots, two of which are positive. The Campbell diagram of the system is of the type shown in Figure 3.6: At each speed Ω, four whirl modes, occurring at different frequencies, are possible. Two of them occur in the forward direction and two in the backward direction. The plot refers to a rigid rotor on two equal compliant bearings: When the rotor is at midspan ($a = l/2$), $K_{12} = 0$ and conical and cylindrical modes are uncoupled: The plot is of the type of Figure 3.5(a) as the natural frequency at standstill of the cylindrical mode is higher than that of the conical mode. Otherwise the modes mix with each other and it is no more possible to speak of conical or cylindrical whirling. It is usually still possible to speak of modes that are mostly cylindrical and modes that are mostly conical; however, following a branch of the plot, the mode can transform from one type to the other (e.g., the mode related to the first forward mode is conical at low speed, to become more and more cylindrical with increasing speed).

Because all solutions of Equation (3.40) are real, the corresponding eigenvectors $\mathbf{q}_i$ are also real.

If the initial conditions are taken in such a way that r_0 and ϕ_0 are real, at time $t = 0$, both Y and $\phi_{X'}$ vanish: The z-axis is contained in a plane also containing the Z-axis and rotating about the latter with a constant angular velocity equal to the whirl speed ω. The axis of the rotor describes a cone whose axis is the rotor axis in its undeformed position. Because the deflected shape is contained in a plane, the motion could have been studied in such a plane from the beginning, using a model with only two degrees of freedom. This model, however, cannot be used to study the damped system, and a more general model was preferred from the beginning.

Actually, the free whirling of the system can be more complex because the motion is the combination of all four circular whirling motions that occur at different frequencies

$$\mathbf{q} = \sum_{k=1}^{4} \mathbf{q}_k e^{i\omega_k t}\,. \tag{3.41}$$

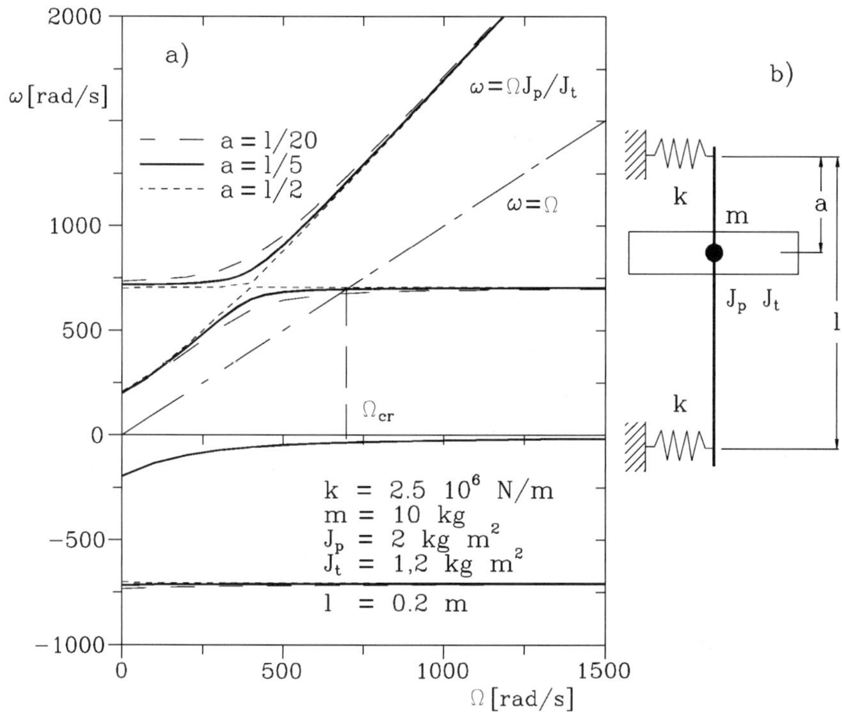

FIGURE 3.6. Campbell diagram of a system made by a rigid rotor on two identical elastic supports.

Remark 3.2 *The gyroscopic terms couple the behavior in the planes passing through the rotation axis and make it impossible for the system to perform elliptical or rectilinear motions.*

An equation of the type of Equation (3.41) can yield elliptical motions only in the case in which two of the eigenfrequencies ω_k have the same modulus and opposite sign, as was the case for the Jeffcott rotor and for the horizontal branches of a rotor with uncoupling between conical and cylindrical whirling. The introduction of gyroscopic moments causes forward whirl frequencies to be different from backward whirl frequencies and causes all whirl motions to be circular. Obviously the four circular modes can add to each other, yielding Lissajous curves.

To avoid solving Equation (3.40) in ω, the Campbell diagram can be obtained solving the same equation in Ω. The equation is linear in this unknown, and a closed-form solution can be easily found

$$\Omega = \frac{mJ_t\omega^4 - (J_tK_{11} + mK_{22})\omega^2 + K_{11}K_{22} - K_{12}^2}{\omega J_p(m\omega^2 - K_{11})} \; . \tag{3.42}$$

From Figure 3.6, it is clear that the Campbell diagram has three horizontal asymptotes: Two for backward motions and one for forward whirling. Another asymptote has the equation

$$\omega = \Omega\frac{J_p}{J_t} \; . \tag{3.43}$$

The intersections between the curve $\omega(\Omega)$ with the bisector of the first quadrant $\omega = \Omega$ yield the conditions for forward synchronous whirling, i.e., the critical speeds. By introducing condition $\Omega = \omega$ into Equation (3.40), the following quadratic equation in Ω^2 is obtained:

$$\Omega^4 m(J_p - J_t) - \Omega^2\left[(J_p - J_t)K_{11} - mK_{22}\right] - K_{11}K_{22} + K_{12}^2 = 0 \, . \tag{3.44}$$

By neglecting the negative solutions corresponding to the intersections of curve $\omega(\Omega)$ with the straight line $\omega = \Omega$ lying in the third quadrant of the $\Omega\omega$-plane, the following values of the critical speed are obtained:

$$\Omega_{cr} = \sqrt{\frac{K_{11}(J_p - J_t) - mK_{22} \pm \sqrt{\left[K_{11}(J_p - J_t) + mK_{22}\right]^2 - 4m(J_p - J_t)K_{12}^2}}{2m(J_p - J_t)}} \; . \tag{3.45}$$

Introducing ratio

$$\delta' = \frac{J_p - J_t}{m} \; ,$$

which has the dimensions of the square of a length, Equation (3.45) can be written in the more compact form

$$\Omega_{cr} = \frac{1}{\sqrt{2m}}\sqrt{K_{11} - \frac{K_{22}}{\delta'} \pm \sqrt{\left(K_{11} + \frac{K_{22}}{\delta'}\right)^2 - 4\frac{K_{12}^2}{\delta'}}} \; . \tag{3.46}$$

Note that

$$\delta' = \frac{J_p}{m}\left(\frac{\delta - 1}{\delta}\right) = \rho_p^2\left(\frac{\delta - 1}{\delta}\right) \; , \tag{3.47}$$

where ρ_p is the radius of inertia of the rigid body related to its polar moment of inertia ($\rho_p = \sqrt{J_p/m}$).

If $J_p < J_t$ ($\delta < 1$ or $\delta' < 0$), as happens in the case of long rotors, there are two real solutions and, as a consequence, two values of the critical speed.

If, on the contrary, $J_p > J_t$ or $\delta > 1$, as happens in the case of discs, one of the solutions is imaginary and only one critical speed exists.

The same behavior of the critical speeds as functions of the values of J_p and J_t was already clear from Figure 3.6 for the case with uncoupled translational and rotational motions ($a/l = 0.5$).

Example 3.1 *Cantilever rotor with four degrees of freedom (the Stodola-Green rotor).*

An interesting example is the so-called Stodola-Green rotor: a disc attached at the end of a prismatic cantilever beam. The simplest model is shown in Figure 3.7(a), where the moments of inertia of the disc are neglected and the Jeffcott rotor model is used. The model studied here [Figure 3.7(b)] on the contrary takes into account also the latter and is based on a model with four degrees of freedom.

To compute the stiffness matrix, the compliance of the system is evaluated first. As in the previous cases, a unit force and a unit bending moment are applied at the free end of the beam. Apply first a force F; the displacement x and the rotation ϕ_y of the same point can be computed by using the formulas available in various handbooks

$$x = \frac{Fl^3}{3EI}, \quad \phi_y = \frac{Fl^2}{2EI}, \tag{3.48}$$

where E and I are, respectively, the Young's modulus of the material and the area moment of inertia of the cross section of the shaft.

Then apply a unit bending moment M_y. The displacement x of the center of mass and the rotation ϕ_y are

$$x = \frac{M_y l^2}{2EI}, \quad \phi_y = \frac{M_y l}{EI}. \tag{3.49}$$

The compliance and stiffness matrices are then

$$\mathbf{B} = \frac{l}{6EI}\begin{bmatrix} 2l^2 & 3l \\ 3l & 6 \end{bmatrix}, \tag{3.50}$$

$$\mathbf{K} = \mathbf{B}^{-1} = \frac{EI}{l^3}\begin{bmatrix} 12 & -6l \\ -6l & 4l^2 \end{bmatrix}. \tag{3.51}$$

The natural frequency at standstill of the Jeffcott rotor model (i.e., of a system with negligible moments of inertia) is

$$\omega_0 = \sqrt{\frac{3EI}{ml^3}}. \tag{3.52}$$

If the moments of inertia are taken into account, the natural frequencies depend on the spin speed and can be computed using Equation (3.40), which can be written in nondimensional form by introducing parameters

$$\delta = \frac{J_p}{J_t}, \quad \beta = \frac{J_t}{ml^2} = \left(\frac{\rho_t}{l}\right)^2, \tag{3.53}$$

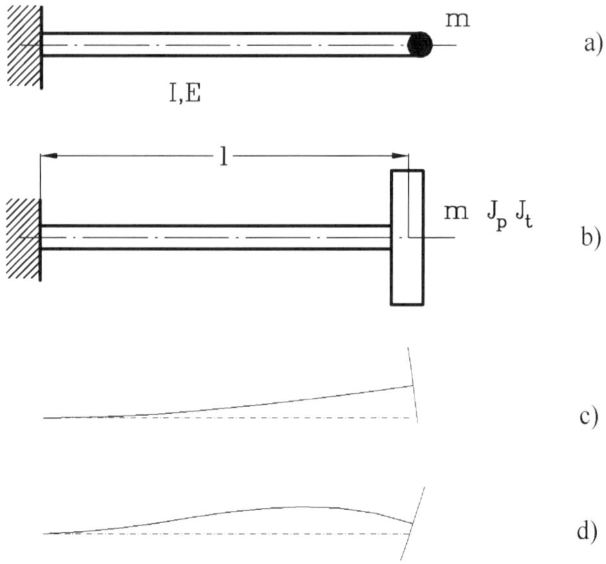

FIGURE 3.7. The so-called Stodola-Green rotor: a disc attached at the end of a prysmatic cantilever beam. (a) Model in which the moments of inertia of the disc are neglected (Jeffcott model); (b) model with four degrees of freedom; and (c) and (d) mode shapes at standstill of model (b).

where $\rho_t = \sqrt{J_t/m}$ *is the radius of inertia of the disc corresponding to the moment of inertia* J_t*, and the nondimensional whirl and spin speed*

$$\omega^* = \frac{\omega}{\omega_0} \quad , \qquad \Omega^* = \frac{\Omega}{\omega_0} \,. \tag{3.54}$$

Parameter β *is vanishing small for a point mass rotor [the Jeffcott rotor of Figure 3.7(a)], and the speeds are made nondimensional with the whirl speed of the same Jeffcott rotor.*

The free response of the system can thus be studied using the homogeneous equation

$$\left(-\omega^{*2}\begin{bmatrix} 1 & 0 \\ 0 & \beta \end{bmatrix} + \omega^*\Omega^*\begin{bmatrix} 0 & 0 \\ 0 & \delta\beta \end{bmatrix} + \begin{bmatrix} 4 & -2 \\ -2 & \frac{4}{3} \end{bmatrix}\right)\left\{ \begin{array}{c} r_0 \\ \phi_0 l \end{array} \right\} = \mathbf{0} \,. \tag{3.55}$$

The characteristic equation takes the form

$$\omega^{*4} - \delta\Omega^*\omega^{*3} - 4\left(1 + \frac{1}{3\beta}\right)\omega^{*2} + 4\delta\Omega^*\omega^* + \frac{4}{3\beta} = 0\,. \tag{3.56}$$

The nondimensional natural frequencies of the system depend then on only two parameters, namely, ratio β*, assessing the importance of the transversal moment of inertia with respect to the mass, and the product* $\delta\Omega^*$*, assessing the importance of the gyroscopic effect.*

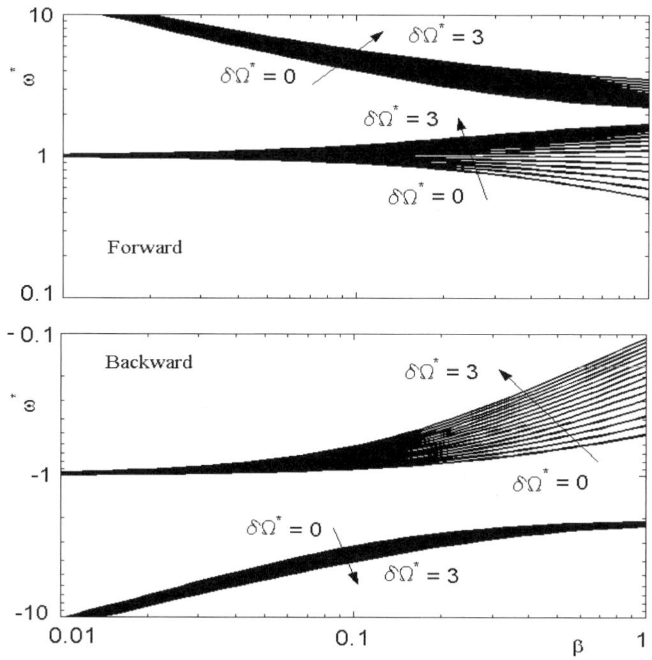

FIGURE 3.8. Nondimensional natural frequency of the Stodola-Green rotor as a function of ratio β. The various curves refer to values of the nondimensional speed $\delta\Omega^*$ spanning from 0 to 3, with increments of 0.2.

The dependence of the frequencies on such parameters is summarized in the plot of Figure 3.8. If $\beta \rightarrow 0$*, the behavior of the rotor tends to that of the Jeffcott model: The nondimensional frequency of the first mode tends to 1 and does not depend on the speed, whereas the frequency of the second mode tends to infinity (there is only one mode). With increasing* β*, the effect of the speed increases and the frequency of the second mode (both forward and backward) is increasingly lower. Note that* $\beta = 1$ *is a very high value and corresponds to a very large disc, with a radius of inertia equal to the length of the beam. Values lower than 0.1 are more realistic for actual systems.*

The nondimensional Campbell diagram is reported in Figure 3.9. Note that the overall pattern is different not only from that of Figure 3.5 but also from that of Figure 3.6. Here the coupling between the two modes is strong, and actually it is impossible to speak of conical and cylindrical whirling. The two curves for the two modes are far from each other, and they do not tend to cross, neither in the first nor in the fourth quadrant.

The mode shapes, computed for $\beta = 0.1$ *at varying speed, are reported in Figure 3.10. For vanishing small speed, they coincide with those sketched in Figure 3.7(c)*

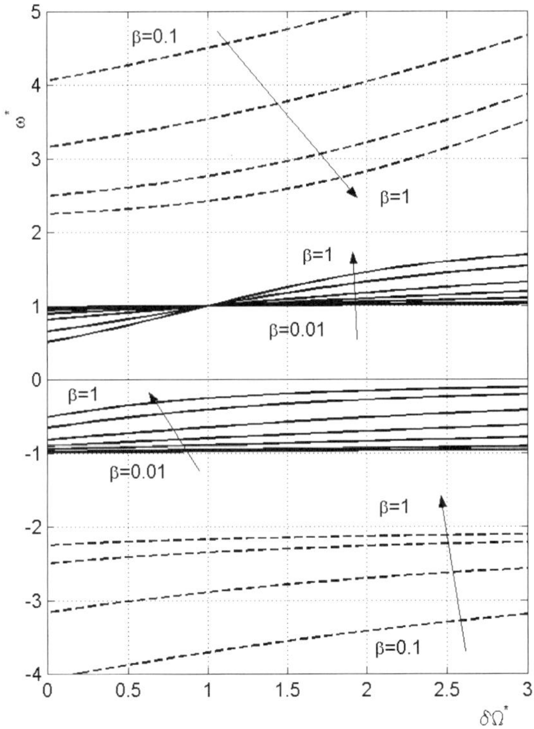

FIGURE 3.9. Nondimensional Campbell diagram of the Stodola-Green rotor. Note that the spin speed has been corrected multiplying it by ratio δ, to summarize the behavior of the system with only four sets of lines (for the four modes), with parameter β. The curves have been computed for $\beta = 0.01$, 0.02, 0.05, 0.1, 0.2, 0.5, and 1. Full lines: first mode, forward and backward; dashed lines: second mode, forward and backward.

and (d), and obviously they are identical from forward and backward motions. Note that at very high speed, the first forward mode and the second backward one tends to be cylindrical and the other two to be conical.

Example 3.2 *Compliant rotor on two bearings located at two opposite sides.*

Consider a rotor made by a rigid gyroscopic body attached to a flexible uniform shaft running on rigid bearings (Figure 3.11, inset sketch). The stiffness matrix can be computed using the finite element method, by modeling the shaft with two beam elements and then applying a static reduction technique (see Chapter 4). Alternatively, here the compliance matrix is obtained and then inverted.

Apply a unit force F in the center of mass of the rigid body. The displacement x of the same point and the rotation ϕ_y can be computed by using the formulas

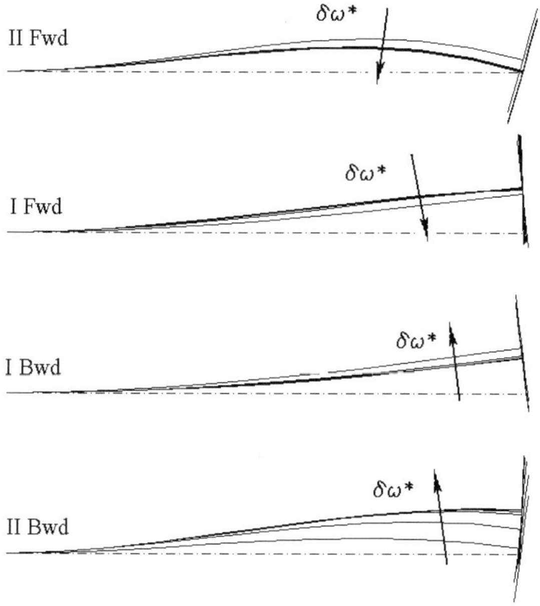

FIGURE 3.10. Mode shapes at varying speed (for $\delta\Omega^*$ spanning from 0 and 50) for a Stodola-Green rotor with $\beta = 0.05$.

available in various handbooks

$$\left\{\begin{array}{l} x = \dfrac{F}{3EIl}a^2(l-a)^2 , \\ \\ \phi_y = \dfrac{F}{3EIl}a(l-a)(l-2a) . \end{array}\right. \tag{3.57}$$

Then apply a unit bending moment M_y. The displacement x of the center of mass and the rotation ϕ_y are

$$\left\{\begin{array}{l} x = \dfrac{M_y}{3EIl}a(l-a)(l-2a) , \\ \\ \phi_y = \dfrac{M_y}{3EIl}\left(l^2 - 3al + 3a^2\right) . \end{array}\right. \tag{3.58}$$

The compliance matrix is then obtained by assembling in matrix form the displacements and rotations caused by a unit force F and a unit moment M_y

$$\mathbf{B} = \frac{1}{3EIl}\left[\begin{array}{cc} a^2(l-a)^2 & a(l-a)(l-2a) \\ a(l-a)(l-2a) & l^2 - 3al + 3a^2 \end{array}\right] . \tag{3.59}$$

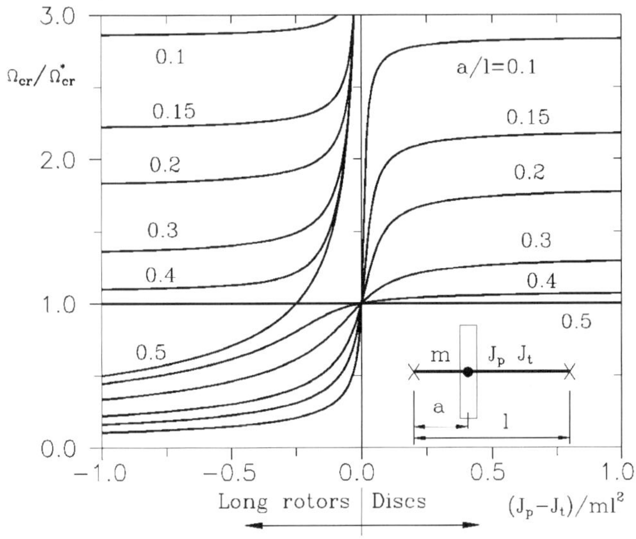

FIGURE 3.11. Influence of the gyroscopic moment on the critical speeds of a flexible rotor on two bearings located at opposite sides of the center of mass (inset the sketch of the system).

The stiffness matrix in xz plane is then

$$\mathbf{K} = \mathbf{B}^{-1} = \frac{3EIl}{a(l-a)} \begin{bmatrix} \frac{l^2 - 3al + 3a^2}{a^2(l-a)^2} & \frac{2a-l}{a(l-a)} \\ \frac{2a-l}{a(l-a)} & 1 \end{bmatrix} . \tag{3.60}$$

The Campbell diagram and the critical speeds can be computed as shown in the previous sections. The influence of the gyroscopic moment on the critical speeds can be shown by plotting the values of the critical speed or, better, of ratio $\Omega_{cr}/\Omega_{cr}^*$, where Ω_{cr}^* is the critical speed computed neglecting gyroscopic effects (i.e., using the Jeffcott rotor model), as functions of δ, δ' or, better, of nondimensional parameter $\delta'/l^2 = (J_p - J_t)/ml^2$. The graph, plotted for various values of a/l, is shown in Figure 3.11. If the rotor is at midspan, i.e., if $a/l = 0.5$, the translational and rotational motions are uncoupled. The gyroscopic moment has no effect on critical speeds in the case of the discs and causes a second critical speed to be present in the case of long rotors; one of the critical speeds can be computed using the Jeffcott rotor model. In all other cases, a disc-type rotor causes an increase in the critical speed, which is sometimes explained by saying that the gyroscopic moment causes a stiffening of the system, which is only a phenomenological explanation because no actual stiffening takes place. In the case of a long rotor, the critical speed decreases and a second critical speed, usually higher than the first, occurs. Note that the values of $|(J_p - J_t)|/ml^2$ are in practice small, particularly in the case of long rotors, and Figure 3.11 holds only for a narrow *zone about the ordinate axis. On the right of this narrow zone, the discs are too thin and they can no longer be considered rigid bodies, and on the left the rotor becomes too long and thin to be considered rigid again.*

Remark 3.3 *If the ellipsoid of inertia is close to be a sphere, i.e., J_p is close to J_t, the value of the critical speed is sensitive to the values of the parameters of the system. This situation must be avoided, particularly for rotors that spin at a speed far higher than the first critical speed.*

Example 3.3 *The considerations drawn from Figure 3.11, although obtained for a particular case, are, however, qualitatively applicable in general.*

3.4 Response to unbalance and shaft bow

In the previous sections, the shaft has been assumed to be perfectly straight in its undeflected configuration. This cannot be exactly the case in an actual situation, and as it has been seen for the Jeffcott rotor, a certain bow may be present. However, when dealing with shaft bow in the context of the present model, not only the bow b but also the angular misalignment χ_b caused by the predeformations of the shaft must be accounted for (Figure 3.12). The angular misalignment χ_b must not be confused with angle χ defined in Figure 3.2(b): The former defines the position toward which the tangent to deflected configuration of the shafts is recalled when no forces are applied to it, and the second defines the direction of the principal axis of the rotor at rest. The bow and the angular misalignment do not lie, in general, in the ξz-plane, but form angles α_b and β_b with it.

The elastic restoring forces act to bring point C toward O' (which rotates in xy-plane with velocity Ω) instead of O, and then the reaction forces do not depend on the displacements X and Y and on angles $\phi_{X'}$ and ϕ_y, but on $X - b\cos(\Omega t + \alpha_b)$, $Y - b\sin(\Omega t + \alpha_b)$, $\phi_{X'} - \chi_b\sin(\Omega t + \beta_b)$, and $\phi_y - \chi_b\cos(\Omega t + \beta_b)$.

By introducing these values of the generalized displacements into the Expression (3.16) of the potential energy of the system, and performing the relevant computations, the equation of motion of the damped system in complex coordinates becomes

$$\mathbf{M}\ddot{\mathbf{q}} + (\mathbf{C}_n + \mathbf{C}_r - i\Omega\mathbf{G})\dot{\mathbf{q}} + (\mathbf{K} - i\Omega\mathbf{C}_r)\mathbf{q} =$$

$$\Omega^2\left\{\begin{array}{c} m\epsilon e^{i\alpha} \\ \chi(J_t - J_p)e^{i\beta}\end{array}\right\}e^{i\Omega t} + \mathbf{K}\left\{\begin{array}{c} be^{i\alpha_b} \\ \chi_b e^{i\beta_b}\end{array}\right\}e^{i\Omega t}. \tag{3.61}$$

The particular integral of Equation (3.61) can be expressed in the form $r = r_0e^{i\Omega t}$, $\phi = \phi_0e^{i\Omega t}$, i.e.,

$$\mathbf{q} = \mathbf{q}_0e^{i\Omega t} = \left\{\begin{array}{c} r_0 \\ \phi_0\end{array}\right\}e^{i\Omega t}. \tag{3.62}$$

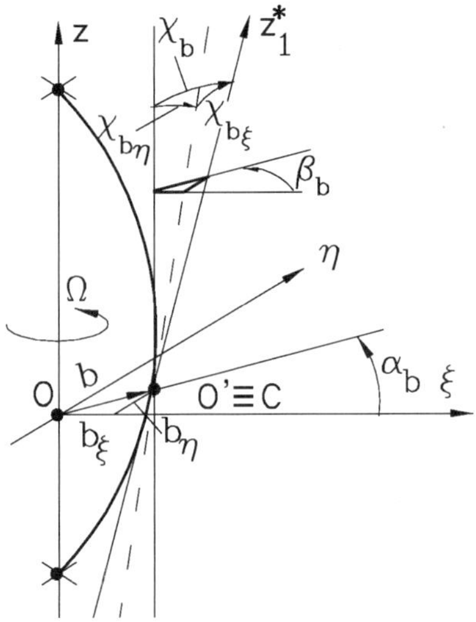

FIGURE 3.12. Undeflected configuration of a bowed shaft, in the rotating frame at standstill. Point O' is the undeflected position of the center of the shaft (as the system is shown in the undeflected configuration, point C coincides with point O').

By introducing the particular integral into the equation of motion, the following algebraic equation is obtained

$$\left[-\Omega^2(\mathbf{M}-\mathbf{G})+i\Omega\mathbf{C}_n+\mathbf{K}\right]\mathbf{q}_0=\Omega^2\left\{\begin{array}{c} m\epsilon e^{i\alpha} \\ \chi(J_t-J_p)e^{i\beta}\end{array}\right\}+\mathbf{K}\left\{\begin{array}{c} be^{i\alpha_b} \\ \chi_b e^{i\beta_b}\end{array}\right\}. \tag{3.63}$$

The amplitudes z_o and ϕ_o of the response are, in general, complex numbers, even in the case of undamped systems (i.e., even in the case the matrix of the coefficients is real) because static and couple unbalances and the bow do not lie in the same plane, i.e., the phase angles α, β, α_b, and β_b are, in general, not equal to zero. As it was the case for the Jeffcott rotor, the response to unbalance is similar to the response to an excitation proportional to the square of the frequency, and it starts from 0 at standstill. The response to shaft bow, on the contrary, is similar to the response to an excitation with constant amplitude, taking a nonzero value at standstill and decreasing to zero, after one or two resonance peaks (the number of peaks in the response depends on the number of critical speeds) at high speed.

To study separately the effects of static and couple unbalance, consider first an undamped system with eccentricity ϵ and vanishing phase α. The amplitude of the response is

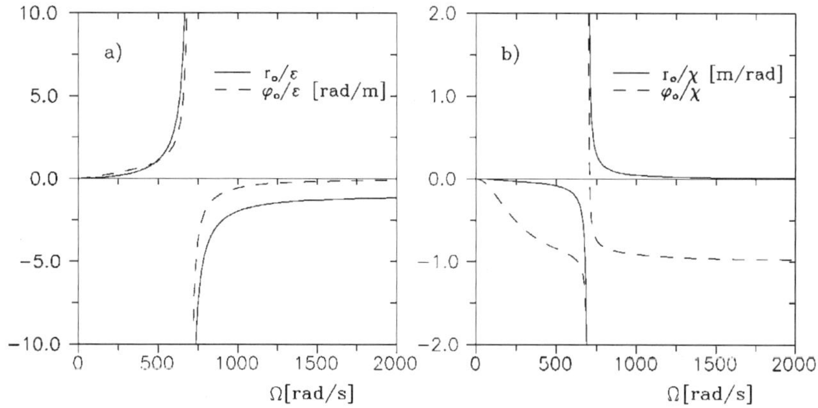

FIGURE 3.13. Amplitude of the unbalance response for the same rotor of Figure 3.6 with $l/a = 4$. (a) Response to static unbalance; (b) response to couple unbalance.

$$\begin{cases} z_0 = m\epsilon\Omega^2 \dfrac{(J_p - J_t)\Omega^2 + K_{22}}{\Delta}, \\ \phi_0 = -m\epsilon\Omega^2 \dfrac{K_{12}}{\Delta}, \end{cases} \tag{3.64}$$

where

$$\Delta = -m(J_p - J_t)\Omega^4 + [K_{11}(J_p - J_t) - mK_{22}]\,\Omega^2 + K_{11}K_{22} - K_{12}^2 .$$

In the case of couple unbalance with phase $\beta = 0$, the response is

$$\begin{cases} z_0 = \chi\Omega^2(J_p - J_t)\dfrac{K_{12}}{\Delta}, \\ \phi_0 = \chi\Omega^2(J_p - J_t)\dfrac{m\Omega^2 - K_{11}}{\Delta}. \end{cases} \tag{3.65}$$

Equations (3.64) and (3.65) are plotted in Figure 3.13 for the same rotor as in Figure 3.6 with $l/a = 4$. As is clear from the figure, in the current case, self-centering occurs at speeds in excess of the critical speed in the case of static unbalance, whereas a certain self-centering also occurs in the subcritical field in the case of couple unbalance.

This is because the system has only one critical speed, because $J_p > J_t$, and the natural frequency at standstill related to the rotational mode is lower than that linked with the translational mode.

3.5 Frequency response

The previous sections dealt with the response of a four degrees-of-freedom rotor to excitations caused by unavoidable imperfections of the shaft, such as unbalance or bow. As it was the case for the Jeffcott rotor, also in the present case, this type of excitation is synchronous, i.e., is characterized by a frequency equal to the spin speed and can be modeled as a force that is nonrotating in the rotating reference frame $O\xi\eta\zeta$.

Consider now a harmonic excitation whose frequency is unrelated to the rotational speed. Consider a perfectly balanced rotor with four degrees of freedom excited by a force F_n and a bending moment M_n whose time history is harmonic with an arbitrary frequency ω. The equation of motion in complex coordinates is still Equation (3.61) with vanishing unbalance and bow, but with forcing function

$$\mathbf{f}_n(t) = \left\{ \begin{array}{c} F_{n_0} \\ M_{n_0} \end{array} \right\} e^{i\omega t} . \tag{3.66}$$

Note that Equation (3.66) describes a force and a moment that rotate in xy-plane with angular velocity ω.

The particular integral of the equation of motion is

$$\mathbf{q}(t) = \mathbf{q}_0 e^{i\omega t} = \left\{ \begin{array}{c} r_0 \\ \phi_0 \end{array} \right\} e^{i\omega t} . \tag{3.67}$$

By substituting the Solution (3.67) into the equation of motion (3.61), an algebraic equation, which can be used to compute the amplitude of the response, is obtained

$$\left\{-\omega^2\mathbf{M}+\omega\left[\Omega\mathbf{G}+i\left(\mathbf{C}_n+\mathbf{C}_r\right)\right]+\mathbf{K}-i\Omega\mathbf{C}_r\right\}\mathbf{q}_0 = \left\{ \begin{array}{c} F_{n_0} \\ M_{n_0} \end{array} \right\} . \tag{3.68}$$

The amplitude of the response can be expressed in terms of the frequency response $\mathbf{G}(i\omega)$

$$\mathbf{q}_0(i\omega) = \mathbf{G}(i\omega)\mathbf{f}_0 , \tag{3.69}$$

where

$$\mathbf{G}(i\omega) = \left[\mathbf{K} - \omega^2\mathbf{M}+\omega\Omega\mathbf{G}+i\omega\left(\mathbf{C}_n+\mathbf{C}_r\right)-i\Omega\mathbf{C}_r\right]^{-1} . \tag{3.70}$$

The effects of rotating damping are the same already seen for the Jeffcott rotor. Also in the present case, an equivalent, speed-dependent damping can be defined as

$$\mathbf{C}_{eq} = \mathbf{C}_n + \mathbf{C}_r\left(1 - \frac{\Omega}{\omega}\right) . \tag{3.71}$$

With increasing speed, the system behaves as if the damping was decreased, and it is possible that, at high speed, instability occurs, owing to the destabilizing effect of rotating damping.

3.6 Unbalance response: modal computation

The frequency domain Equation (3.63) for synchronous whirling is formally identical to the equation of a vibrating system with mass matrix equal to $\mathbf{M} - \mathbf{G}$ excited by harmonic forces having frequency Ω [30]. If the system is undamped, it is possible to perform a modal analysis of a vibrating system having the same stiffness matrix $\mathbf{K}$ as the rotating system, and a mass matrix equal to $\mathbf{M} - \mathbf{G}$. The eigenvectors $\mathbf{q}_i$ of such a system then uncouple the modes, and it is possible to compute the unbalance response (also the response to a shaft bow, but in this section the latter will be neglected) by superimposing the modal responses to the various modal forces.

Remark 3.4 *Modal uncoupling is strictly linked to the m- and k-orthogonality of the eigenvectors and, hence, with the symmetry of the mass and stiffness matrix. As the gyroscopic matrix is symmetrical only when using the complex coordinates approach, the modes are uncoupled only when the equations are written in terms of complex coordinates.*

Remark 3.5 *As matrix* $\mathbf{M} - \mathbf{G}$ *may be negative defined, not all its eigenvalues in* Ω^2 *are necessarily positive. In particular, if* $J_p > J_p$, *one of the two eigenvalues is negative and only one real critical speed exists, as explained above.*

Let $\mathbf{\Phi}$ be the matrix of the eigenvectors (because the system has just two complex degrees of freedom, there are only two eigenvectors)

$$\mathbf{\Phi} = \begin{bmatrix} \mathbf{q}_1 & \mathbf{q}_2 \end{bmatrix} , \tag{3.72}$$

The modal transformation can be expressed as

$$\begin{gathered} \overline{\mathbf{M}} = \mathbf{\Phi}^T \left(\mathbf{M} - \mathbf{G}\right) \mathbf{\Phi} , \\ \overline{\mathbf{K}} = \mathbf{\Phi}^T \left(\mathbf{K}\right) \mathbf{\Phi} , \qquad \overline{\mathbf{f}} = \mathbf{\Phi}^T \mathbf{f} , \end{gathered} \tag{3.73}$$

where both $\overline{\mathbf{M}}$ and $\overline{\mathbf{K}}$ are diagonal matrices, but in the present case, not all elements of the former are positive. Vector $\overline{\mathbf{f}}$ contains the modal unbalances related to the various modes. By dividing the modal equations in the frequency domain by the modal masses, the corresponding values of the modal stiffness are the squares of the critical speeds and the two algebraic equations yielding the modal responses η_{0_i}, corresponding to Equations (3.64) and (3.65), are

$$\left(\Omega_{cr_i}^2 - \Omega^2\right) \eta_{0_i} = \Omega^2 \overline{f}_i \, . \tag{3.74}$$

The modal response

$$\eta_{0_i} = \overline{f}_i \frac{\Omega^2}{\left(\Omega_{cr_i}^2 - \Omega^2\right)} \tag{3.75}$$

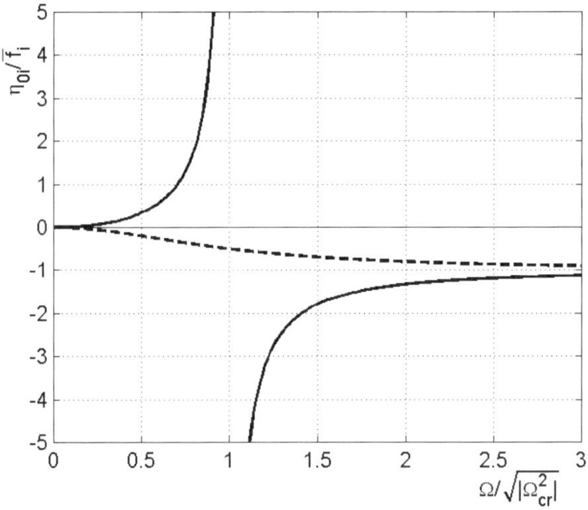

FIGURE 3.14. Modal response for the modes with positive (full line) and negative (dashed line) eigenvalue.

corresponds to the unbalance response of a Jeffcott rotor with $m\epsilon = \overline{f}_i$ and critical speed Ω_{cr_i}.

The modes characterized by a negative modal mass, or better, by a negative value of $\Omega^2_{cr_i}$, do not *resonate*, i.e., do not have a critical speed. However, mathematically the solution of the equation yielding the critical speed exists, but it is imaginary. The so-called imaginary critical speed

$$\Omega_{im_i} = \sqrt{\left|\Omega^2_{cr_i}\right|} \tag{3.76}$$

has a physical meaning. The nondimensional modal responses for a mode with positive eigenvalue (real critical speed) and negative eigenvalue (imaginary critical speed) are reported in Figure 3.14. In the former case, the plot coincides with the response of a Jeffcott rotor (Figure 2.4), whereas in the latter, there is no resonance, but the mode self-centers at high speed. At the imaginary critical speed Ω_{im}, the response is half of the self-centered response.

Here this property was shown for a system with two degrees of freedom, but it holds in general (see Chapter 4).

Remark 3.6 *The modes whose imaginary critical speed lies within the working range contribute to self-centering together with the modes having a real critical speed.*

Remark 3.7 *The response close to a critical speed is dominated by the resonating mode, and it can be obtained considering only one mode.*

The last properties supply the theoretical foundations on which the modal balancing procedure is based (see Appendix B).

The considerations seen above apply, strictly speaking, only in case of undamped systems. The presence of damping (nonrotating damping, because rotating damping has no effect on the unbalance response of isotropic systems) couples all equations of motion even if modal coordinates are used and makes modal analysis impossible. However, if the system is lightly damped, the modal approach still supplies a very good approximation of the response.

3.7 Modal uncoupling of gyroscopic systems

3.7.1 Configuration-space approach

The equation of motion of a gyroscopic (and possibly circulatory, if rotating damping is present) systems cannot be uncoupled in the usual way. It is still possible to use the eigenvectors of the undamped natural system (i.e., of a system obtained by neglecting the gyroscopic, damping and circulatory matrices) to perform the modal transformation, but the resulting modal equations are generally not uncoupled.

This procedure can be performed either by using real or complex coordinates: Here the latter approach will be followed to simplify the relevant computations. The starting equation for the modal transformation in complex coordinates is Equation (3.23). Let $\mathbf{\Phi}$ be the eigenvector matrix of the undamped natural system. In the present case (rotor with two complex degrees of freedom), the eigenvalues are

$$\omega_{n1} = \sqrt{\frac{K_{22} + \beta K_{11} + D}{2\beta m}}\,,\qquad \omega_{n2} = \sqrt{\frac{K_{22} + \beta K_{11} - D}{2\beta m}}\,, \tag{3.77}$$

where

$$\beta = J_t/m\ , \qquad D = \sqrt{(K_{22} - \beta K_{11})^2 + 4\beta K_{12}{}^2}\ .$$

Note that the two values of ω_n are the natural frequencies of the undamped system at standstill.

After normalizing them in such a way that the modal masses have unit values, the eigenvectors are

$$\mathbf{\Phi} = \begin{bmatrix} \dfrac{A}{\sqrt{m}\sqrt{A^2 + 4\beta K_{12}^2}} & \dfrac{B}{\sqrt{m}\sqrt{B^2 + 4\beta K_{12}^2}} \\ \dfrac{2K_{12}}{\sqrt{m}\sqrt{A^2 + 4\beta K_{12}^2}} & \dfrac{2K_{12}}{\sqrt{m}\sqrt{B^2 + 4\beta K_{12}^2}} \end{bmatrix}, \tag{3.78}$$

where

$$A = -K_{22} + \beta K_{11} + D \; ,$$
$$B = -K_{22} + \beta K_{11} - D \; .$$

By performing the usual modal transformation, it follows that

$$\overline{\mathbf{M}}\ddot{\boldsymbol{\eta}} + (\overline{\mathbf{C}}_n + \overline{\mathbf{C}}_r - i\Omega\overline{\mathbf{G}})\dot{\boldsymbol{\eta}} + (\overline{\mathbf{K}} - i\Omega\overline{\mathbf{C}}_r)\boldsymbol{\eta} = \Omega^2\overline{\mathbf{f}}e^{i\Omega t} \; , \tag{3.79}$$

where $\boldsymbol{\eta}$ are the modal complex coordinates related to the two modes, and

$$\overline{\mathbf{M}}=\begin{bmatrix} 1 & 0 \\ 0 & 1 \end{bmatrix} , \qquad \overline{\mathbf{K}}=\begin{bmatrix} \omega_{n1}^2 & 0 \\ 0 & \omega_{n2}^2 \end{bmatrix} , \tag{3.80}$$

are the modal mass matrix and the modal stiffness matrix, respectively. They are both diagonal.

The modal gyroscopic matrix

$$\overline{\mathbf{G}}=\frac{4K_{12}^2 Jp}{m}\begin{bmatrix} \frac{1}{A^2+4K_{12}^2\beta} & \frac{1}{\sqrt{(A^2+4K_{12}^2\beta)(B^2+4K_{12}^2\beta)}} \\ \frac{1}{\sqrt{(A^2+4K_{12}\beta)(B^2+4K_{12}^2)}} & \frac{1}{B^2+4K_{12}^2\beta} \end{bmatrix} \tag{3.81}$$

is not diagonal, because the eigenvectors are m-orthogonal and k-orthogonal but are not g-orthogonal.

Also the nonrotating and rotating modal damping matrices $\overline{\mathbf{C}}_n$ and $\overline{\mathbf{C}}_r$ are not diagonal, but their explicit expression is too complicated to be reported here. As the coupling effect of damping in gyroscopic system is the same as in natural damped systems, nothing will be added here.

The point for rotating system is the coupling caused by gyroscopic terms that can be avoided only if the gyroscopic matrix is a linear combination of the mass and stiffness matrices (case similar to proportional damping described by Rayleigh) or follow the same rule defined by Caughey for the damping matrices (generalized proportional damping, see [17]). However, these instances are even less likely for the gyroscopic matrix than for the damping matrix, so gyroscopic coupling is always present.

3.7.2 *State-space, complex-coordinates approach*

The equations of motion can be written with reference to the state space. Using again complex coordinates and defining the complex state vector as

$$\mathbf{z} = \begin{bmatrix} \dot{\mathbf{q}}^T & \mathbf{q}^T \end{bmatrix}^T \; , \tag{3.82}$$

the state space equation corresponding to the configuration space Equation (3.23) is

$$\dot{\mathbf{z}} = \begin{bmatrix} -\mathbf{M}^{-1}(\mathbf{C}_n + \mathbf{C}_r - i\Omega\mathbf{G}) & -\mathbf{M}^{-1}(\mathbf{K} - i\Omega\mathbf{C}_r) \\ \mathbf{I} & \mathbf{0} \end{bmatrix}\mathbf{z}+\Omega^2\begin{bmatrix} \mathbf{M}^{-1} \\ \mathbf{0} \end{bmatrix}\mathbf{f}e^{i\Omega t} \; . \tag{3.83}$$

Note that the dynamic matrix $\mathbf{A}$ depends on the speed and is complex, owing to the presence of gyroscopic and circulatory terms. Its eigenvalues yield the complex whirling frequencies of the system.

It is well known that if the matrix $\mathbf{U}$ of the complex right eigenvectors of the dynamic matrix $\mathbf{A}$ is obtained, it is possible to uncouple the equations of motion in the form

$$\dot{\overline{\mathbf{z}}} = \mathbf{U}^{-1}\mathbf{A}\mathbf{U}\overline{\mathbf{z}} + \mathbf{U}^{-1}\mathbf{B}\mathbf{u} \; , \tag{3.84}$$

where modal dynamic matrix $\overline{\mathbf{A}} = \mathbf{U}^{-1}\mathbf{A}\mathbf{U}$ is a complex diagonal matrix and $\overline{\mathbf{z}} = \mathbf{U}^{-1}\mathbf{z}$ are the modal states of the system. Equation (3.84) is a set of $2n$ uncoupled equations, if the number of complex degrees of freedom is n and hence in the present case is a set of four equations. Note that also in the case of a damped gyroscopic system, this uncoupling can be performed, but because the dynamic matrix of the system is a function of the spin speed Ω, it requires us to recalculate the modal dynamic matrix and the modal input gain matrix for each value of the speed. This implies the solution of an eigenproblem of order $2n$ for each value of the speed, which is however exactly what is needed for plotting the Campbell diagram.

In the case of an undamped gyroscopic system, the eigenvalues of the dynamic matrix $\mathbf{A}$ are imaginary and the eigenvectors have the first n elements that are imaginary and the other ones that are real.

Although Equation (3.84) is all what is needed for uncoupling the equations of motion (written with reference to the state space) of any damped gyroscopic system, a further elaboration aimed to obtain uncoupled equations in the configuration space, i.e., aimed to split the multi-degrees-of-freedom rotor into a number of uncoupled single-degree-of-freedom rotors, may be useful. If the eigenvalues are ordered in such a way that the forward-backward pairs are placed one after the other, the diagonal matrix $\overline{\mathbf{A}}$ had the form

$$\overline{\mathbf{A}} = \begin{bmatrix} s_{1+} & 0 & 0 & \dots & 0 \\ 0 & s_{1-} & 0 & \dots & 0 \\ 0 & 0 & s_{2+} & \dots & 0 \\ \dots & \dots & \dots & \dots & \dots \\ 0 & 0 & 0 & \dots & s_{n-} \end{bmatrix} . \tag{3.85}$$

The system with n complex degrees of freedom can be split into n gyroscopic systems with a single degree of freedom, whose equations of motion are

$$\ddot{\eta}_i + \left(\overline{C}_i - i\Omega\overline{G}_i\right)\dot{\eta}_i + \left(\overline{K}_i - i\Omega\overline{C}_{ri}\right)\eta_i = \Omega^2\overline{f}_{ri}e^{i\omega t} \; . \tag{3.86}$$

The modal parameters can be easily computed by solving the eigenproblem related to the homogeneous Equation (3.86) and equating the solution to the ith pair of eigenvalues (s_{i+} and s_{i-}, where the signs + and − identify the forward and backward modes) of the original system

$$\begin{bmatrix} 1 & \Omega\,\mathrm{Im}\,(s_{i-}) & \mathrm{Re}\,(s_{i-}) & 0 \\ 0 & -\Omega\,\mathrm{Re}\,(s_{i-}) & \mathrm{Im}\,(s_{i-}) & -\Omega \\ 1 & \Omega\,\mathrm{Im}\,(s_{i+}) & \mathrm{Re}\,(s_{i+}) & 0 \\ 0 & -\Omega\,\mathrm{Re}\,(s+) & \mathrm{Im}\,(s_{i+}) & -\Omega \end{bmatrix} \begin{Bmatrix} \overline{K}_i \\ \overline{G}_i \\ \overline{C}_i \\ \overline{C}_{ri} \end{Bmatrix} = \begin{Bmatrix} \mathrm{Im}\,(s_{i-})^2 - \mathrm{Re}\,(s_{i-})^2 \\ -2\,\mathrm{Im}\,(s_{i-})\,\mathrm{Re}\,(s_{i-}) \\ \mathrm{Im}\,(s_{i+})^2 - \mathrm{Re}\,(s_{i+})^2 \\ -2\,\mathrm{Im}\,(s_{i+})\,\mathrm{Re}\,(s_{i+}) \end{Bmatrix} \tag{3.87}$$

Equation (3.87) is easily solved in closed form, yielding

$$\begin{cases} \overline{K}_i = -\,\mathrm{Im}\,(s_{i-})\,\mathrm{Im}\,(s_{i+}) + \mathrm{Re}\,(s_{i-})\,\mathrm{Re}\,(s_{i+}) \ , \\ \overline{G}_i = \frac{1}{\Omega}\left[\mathrm{Im}\,(s_{i-}) + \mathrm{Im}\,(s_{i+})\right] \ , \\ \overline{C}_i = -\,\mathrm{Re}\,(s_{i-}) - \mathrm{Re}\,(s_{i+}) \ , \\ \overline{C}_{ri} = -\frac{1}{\Omega}\left[\mathrm{Re}\,(s_{i-})\,\mathrm{Im}\,(s_{i-}) + \mathrm{Re}\,(s_{i+})\,\mathrm{Im}\,(s_{i+})\right] \ . \end{cases} \tag{3.88}$$

Note that by definition $\mathrm{Im}\,(s_{i+})$ is positive and $\mathrm{Im}\,(s_{i-})$, is negative. In the case of the undamped system, the eigenvalues are imaginary and $\overline{K}_i = -\,\mathrm{Im}\,(s_{i-})\,\mathrm{Im}\,(s_{i+})$.

The modal parameters of the system can thus be used to write a state-space equation

$$\dot{\overline{\overline{\mathbf{z}}}} = \overline{\overline{\mathbf{A}}}\,\overline{\overline{\mathbf{z}}} + \overline{\overline{\mathbf{B}}}\mathbf{u} \ , \tag{3.89}$$

where the structure of the dynamic matrix is

$$\overline{\overline{\mathbf{A}}} = \begin{bmatrix} \overline{\overline{\mathbf{A}}}_1 & \mathbf{0} & \dots & \mathbf{0} \\ \mathbf{0} & \overline{\overline{\mathbf{A}}}_2 & \dots & \mathbf{0} \\ \dots & \dots & \dots & \dots \\ \mathbf{0} & \mathbf{0} & \dots & \overline{\overline{\mathbf{A}}}_n \end{bmatrix} , \tag{3.90}$$

with

$$\overline{\overline{\mathbf{A}}}_i = \begin{bmatrix} s_{i+} + s_{i-} & -s_{i+}s_{i-} \\ 1 & 0 \end{bmatrix} = \begin{bmatrix} -\left(\overline{C}_i - i\omega\overline{G}_i\right) & -\left(\overline{K}_i - i\omega\overline{C}_{ri}\right) \\ 1 & 0 \end{bmatrix} . \tag{3.91}$$

The transformation matrix to obtain the state variables $\overline{\overline{\mathbf{z}}}$ from $\overline{\mathbf{z}}$ is the matrix of the eigenvectors of matrix $\overline{\overline{\mathbf{A}}}$, or better, because there are n uncoupled systems, there are n eigenvector matrices of the various matrices $\overline{\overline{\mathbf{A}}}_i$. Let $\mathbf{U}_{1i}$ = eigenvectors $(\overline{\overline{\mathbf{A}}}_i)$, and it is easy to verify that

$$\mathbf{U}_{1i} = \begin{bmatrix} s_{i+} & s_{i-} \\ 1 & 1 \end{bmatrix} . \tag{3.92}$$

To transform the whole system in terms of the modal coordinates $\overline{\overline{\mathbf{z}}}$ defined by Equation (3.89), the augmented eigenvector matrix $\mathbf{U}_{1i}$ can be defined as

$$\mathbf{U}_1 = \begin{bmatrix} \mathbf{U}_{11} & \mathbf{0} & \dots & \mathbf{0} \\ \mathbf{0} & \mathbf{U}_{12} & \dots & \mathbf{0} \\ \dots & \dots & \dots & \dots \\ \mathbf{0} & \mathbf{0} & \dots & \mathbf{U}_{1n} \end{bmatrix} . \tag{3.93}$$

The transformation from the physical complex coordinates and those of the uncoupled modal system is then

$$\overline{\overline{\mathbf{z}}} = \mathbf{U}_1\overline{\mathbf{z}} = \mathbf{U}_1\mathbf{U}^{-1}\mathbf{z} \ . \tag{3.94}$$

The input gain matrix $\overline{\overline{\mathbf{B}}}$ is then

$$\overline{\overline{\mathbf{B}}} = \mathbf{U}_1\mathbf{U}^{-1}\mathbf{B} \ . \tag{3.95}$$

The modal force to be included into Equation (3.86) includes not only the inputs $\mathbf{u}$ but also their derivatives $\dot{\mathbf{u}}$. This is because all rows of matrix $\overline{\overline{\mathbf{B}}}$ can be nonzero. By differentiating the first row of the i-th state space modal equation with respect to time and substituting the expression of the second state space coordinate into the first equation, the configuration space equation of the ith modal system is

$$\ddot{\eta}_i + \left(\overline{C}_i - i\Omega\overline{G}_i\right)\dot{\eta}_i + \left(\overline{K}_i - i\Omega\overline{C}_{ri}\right)\eta_i = -\left(\overline{K}_i - i\Omega\overline{C}_{ri}\right)\overline{\overline{\mathbf{B}}}_{2i}\mathbf{u} + \overline{\overline{\mathbf{B}}}_{1i}\dot{\mathbf{u}} \ , \tag{3.96}$$

where $\overline{\overline{\mathbf{B}}}_{2i}$ and $\overline{\overline{\mathbf{B}}}_{1i}$ are the second and the first row of the input gain matrix $\overline{\overline{\mathbf{B}}}$ related to the ith mode.

Equation (3.96) can then be used to compute the modal forces caused by unbalance or the modal control forces caused by active devices.

3.7.3 State-space, real-coordinates approach

If no damping is present, a more traditional way, based on the real coordinates approach, to uncouple the equations of motion of a gyroscopic system is as follows. The modal equations are written in nonstandard form:

$$\begin{bmatrix} \mathbf{M}' & \mathbf{0} \\ \mathbf{0} & \mathbf{K}' \end{bmatrix} \dot{\mathbf{z}} + \begin{bmatrix} \Omega\mathbf{G}' & \mathbf{K}' \\ -\mathbf{K}' & \mathbf{0} \end{bmatrix} \mathbf{z} = \Omega^2 \begin{bmatrix} \mathbf{I} \\ \mathbf{0} \end{bmatrix} \mathbf{f}' e^{i\Omega t} \ , \tag{3.97}$$

where $\mathbf{M}'$, $\mathbf{K}'$, $\mathbf{G}'$, and $\mathbf{f}'$ are the mass, stiffness, and gyroscopic matrices and the unbalance force vector for the real coordinates approach (their expression can be found in Equation (3.18); their size is 4×4), and the state vector $\mathbf{z}$ contains the generalized velocities and the displacements in the direction of x- and y-axes.

Usually symbols $\mathbf{M}^*$ and $\mathbf{G}^*$ are used for the two matrices

$$\mathbf{M}^* = \begin{bmatrix} \mathbf{M}' & \mathbf{0} \\ \mathbf{0} & \mathbf{K}' \end{bmatrix} , \qquad \mathbf{G}^* = \begin{bmatrix} \Omega\mathbf{G}' & \mathbf{K}' \\ -\mathbf{K}' & \mathbf{0} \end{bmatrix} . \tag{3.98}$$

The first one is symmetrical, and the second one is skew-symmetrical (and a function of Ω). The eigenproblem

$$\det\left(s - \mathbf{M}^{*-1}\mathbf{G}^*\right) = 0 \tag{3.99}$$

yielding the whirl complex frequencies of the system has then a number of pure imaginary complex conjugate solutions (in the present case, there are four pairs, corresponding to the four whirling speeds of the system). Note that because real coordinates have been used, no information on the direction of whirling can be obtained from eigenvalues.

Also, the eigenvectors are conjugate. By introducing the whirling frequency $\omega = is$ and separating the real and the imaginary parts of the solutions, a new eigenproblem, this time with real symmetric, positive-defined matrices, can be obtained

$$\det\left(-\omega^2 \mathbf{M}^{*-1} + \mathbf{K}^*\right) = 0 \; , \tag{3.100}$$

where

$$\mathbf{K}^* = \mathbf{G}^{*T} \mathbf{M}^{*-1} \mathbf{G}^* \; . \tag{3.101}$$

Equation (3.100) can be interpreted as the eigenproblem related to a natural undamped system with mass matrix $\mathbf{M}^*$ and stiffness matrix $\mathbf{K}^*$. Such a system can be uncoupled by the matrix of the eigenvectors obtained from the Eigenproblem (3.100).

Remark 3.8 *This approach for uncoupling gyroscopic systems cannot be used for free rotors: If the stiffness matrix $\mathbf{K}'$ is singular, also matrix $\mathbf{M}^*$ is such and cannot be inverted. Equation (3.100) cannot then be used to uncouple the system.*

Remark 3.9 *The methods described in the present section allow modal uncoupling of the equations of motion of a gyroscopic system, but they have the drawback of needing a new modal analysis for each value of the spin speed Ω, which is considered. Apart from this, the first approach, being based on complex coordinates, can be used only for systems that possess axial symmetry, but works also in the case of damped (and circulatory) systems. The second one does not require any symmetry assumption, but it cannot be applied to damped systems and free rotors.*

Example 3.4 *Modal analysis of a cantilever rotor with four degrees of freedom.*

Consider the same Stodola-Green rotor studied in Example 3.4, but study its dynamic behavior through modal analysis, assuming a value of ratio $\beta = J_t/ml^2 = 0.8$.

The homogeneous equation of motion of the undamped system written with reference to the complex coordinates in the time domain is

$$m\begin{bmatrix} 1 & 0 \\ 0 & \beta \end{bmatrix}\begin{Bmatrix} \ddot{r} \\ \ddot{\phi}l \end{Bmatrix} + i\delta\Omega m\begin{bmatrix} 0 & 0 \\ 0 & \beta \end{bmatrix}\begin{Bmatrix} \dot{r} \\ \dot{\phi}l \end{Bmatrix} + k\begin{bmatrix} 4 & -2 \\ -2 & \frac{4}{3} \end{bmatrix}\begin{Bmatrix} r \\ \phi \end{Bmatrix} = \mathbf{0} \; , \tag{3.102}$$

where

$$k = \frac{3EI}{l^3} \; . \tag{3.103}$$

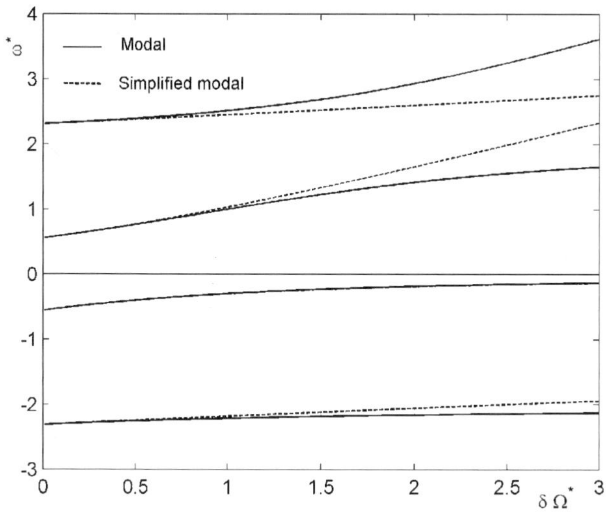

FIGURE 3.15. Nondimensional Campbell diagram of a Stodola-Green rotor with $\beta = 0.8$. The nondimensional quantities are $\omega^* = \omega/\sqrt{k/m}$; $\Omega^* = \Omega/\sqrt{k/m}$.

Standard modal analysis

The modal analysis of the natural system is easily performed. The natural frequencies and the eigenvectors are

$$\omega_{n1} = 0.5579\sqrt{\frac{k}{m}}, \qquad \omega_{n2} = 2.3142\sqrt{\frac{k}{m}}, \tag{3.104}$$

$$\mathbf{\Phi} = \frac{1}{\sqrt{m}}\begin{bmatrix} 0.5184 & 0.8552 \\ 0.9561 & -0.5796 \end{bmatrix}, \tag{3.105}$$

and coincide with the natural frequencies and the mode shapes of the system at standstill.

The modal mass matrix and the modal stiffness matrix are respectively

$$\overline{\mathbf{M}}=\begin{bmatrix} 1 & 0 \\ 0 & 1 \end{bmatrix}, \qquad \overline{\mathbf{K}}=\frac{k}{m}\begin{bmatrix} 0.3112 & 0 \\ 0 & 5.3555 \end{bmatrix}. \tag{3.106}$$

The modal gyroscopic matrix

$$\overline{\mathbf{G}}= \delta \begin{bmatrix} 0.7313 & -0.4433 \\ -0.4433 & 0.2687 \end{bmatrix} \tag{3.107}$$

is not diagonal, as expected.

The Campbell diagram is reported in Figure 3.15. The solutions obtained using a nonmodal procedure and a modal computation taking into account the coupling between the equations of motion caused by the gyroscopic matrix are completely superimposed (full lines).

The same computation was repeated by neglecting gyroscopic coupling (i.e., the out-of-diagonal terms of the modal gyroscopic matrix). The results are reported on the same figure, dashed lines. It is clear that the errors introduced by neglecting modal coupling are unacceptable except for very low speed.

Uncoupling in the space state, complex coordinates.

The dynamic matrix is

$$\mathbf{A} = \begin{bmatrix} i\Omega\delta \begin{bmatrix} 0 & 0 \\ 0 & 1 \end{bmatrix} & \frac{k}{m} \begin{bmatrix} 4 & -2 \\ -\frac{2}{\beta} & \frac{4}{3\beta} \end{bmatrix} \\ \begin{bmatrix} 1 & 0 \\ 0 & 1 \end{bmatrix} & \begin{bmatrix} 0 & 0 \\ 0 & 0 \end{bmatrix} \end{bmatrix} . \tag{3.108}$$

Note that the dynamic matrix $\mathbf{A}$ *depends on the speed and is complex. Its eigenvalues and eigenvectors must be computed for each value of the speed. For instance, if* $\delta\Omega = 2\sqrt{k/m}$, *the eigenvalues and the eigenvectors are*

$$s_1 = -2.1636i\sqrt{\frac{k}{m}} , \qquad s_2 = -0.1857i\sqrt{\frac{k}{m}} , \tag{3.109}$$

$$s_3 = 1.4131i\sqrt{\frac{k}{m}} , \qquad s_2 = 2.9362i\sqrt{\frac{k}{m}} ,$$

$$\mathbf{U} = \begin{bmatrix} 0.8593 & 0.0822 & 0.5767 & -0.3760 \\ -0.2926 & 0.1630 & 0.5777 & 0.8687 \\ 0.3972i & 0.4427i & -0.4081i & 0.1281i \\ -0.1352i & 0.8779i & -0.4088i & -0.2959i \end{bmatrix} . \tag{3.110}$$

The modal dynamic matrix $\overline{\mathbf{A}} = \mathbf{U}^{-1}\mathbf{A}\mathbf{U}$ *can thus be obtained; its elements coincide with the complex whirling frequencies. When* $\delta\Omega = 2\sqrt{k/m}$, *its value is*

$$\overline{\mathbf{A}} = i \begin{bmatrix} -2.1636 & 0 & 0 & 0 \\ 0 & -0.1857 & 0 & 0 \\ 0 & 0 & 1.4131 & 0 \\ 0 & 0 & 0 & 2.9362 \end{bmatrix} . \tag{3.111}$$

The equations of motion are thus uncoupled. The system can then be reduced to two uncoupled undamped single-degree-of-freedom (if complex coordinates are used) gyroscopic systems, whose modal mass, stiffness, and gyroscopic terms are functions of the spin speed. For instance, if $\delta\Omega = 2\sqrt{k/m}$, *they are*

$$\overline{\mathbf{M}}_1 = 1 , \qquad \overline{\mathbf{G}}_1 = 0.6137\,\delta , \qquad \overline{\mathbf{K}}_1 = 0.2624\frac{k}{m} ,$$

$$\overline{\mathbf{M}}_2 = 1 , \qquad \overline{\mathbf{G}}_2 = 0.3863\,\delta , \qquad \overline{\mathbf{K}}_2 = 6.3526\frac{k}{m} .$$

The modal stiffness and the modal gyroscopic term are plotted as functions of the nondimensional speed in Figure 3.16.

Uncoupling in the space-state, real coordinates

Matrices $\mathbf{M}^*$ *and* $\mathbf{G}^*$ *have eight rows and eight columns, because each one of them is made by four matrices referred to the real coordinates approach, whose size is 4×4. They are*

$$\mathbf{M}^* = \begin{bmatrix} \mathbf{M}' & \mathbf{0} \\ \mathbf{0} & \mathbf{K}' \end{bmatrix} , \qquad \mathbf{G}^* = \begin{bmatrix} \Omega\mathbf{G}' & \mathbf{K}' \\ -\mathbf{K}' & \mathbf{0} \end{bmatrix} , \tag{3.112}$$

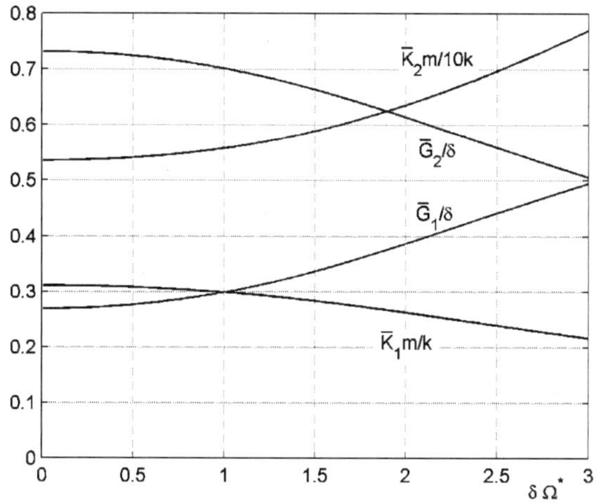

FIGURE 3.16. Modal stiffness and modal gyroscopic terms for the two modal systems equivalent to the Stodola-Green rotor as functions of the nondimensional speed.

where

$$\mathbf{M}' = m \begin{bmatrix} 1 & 0 & 0 & 0 \\ 0 & \beta & 0 & 0 \\ 0 & 0 & 1 & 0 \\ 0 & 0 & 0 & \beta \end{bmatrix}, \qquad \mathbf{K}' = k \begin{bmatrix} 4 & -2 & 0 & 0 \\ -2 & \frac{4}{3} & 0 & 0 \\ 0 & 0 & 4 & -2 \\ 0 & 0 & -2 & \frac{4}{3} \end{bmatrix}, \tag{3.113}$$

$$\mathbf{G}' = \delta m \beta \begin{bmatrix} 0 & 0 & 0 & 0 \\ 0 & 0 & 0 & -1 \\ 0 & 0 & 0 & 0 \\ 0 & 1 & 0 & 0 \end{bmatrix}. \tag{3.114}$$

Matrix

$$\mathbf{K}^* = \mathbf{G}^{*T}\mathbf{M}^{*-1}\mathbf{G}^* \tag{3.115}$$

is readily computed, and the natural system characterized by matrices $\mathbf{M}^*$ *and* $\mathbf{K}^*$ *can be decomposed modally in eight single-degrees-of-freedom systems. By normalizing the eigenveectors to obtain unit modal masses, the modal stiffness can be computed at the various speeds. If* $\delta\Omega = 2\sqrt{k/m}$*, for instance, the modal stiffness matrix is*

$$\overline{\mathbf{K}^*} = \frac{k}{m} diag \left[\begin{array}{cccccccc} 0.0345 & 1.997 & 4.681 & 8.621 & 0.0345 & 1.997 & 4.681 & 8.621 \end{array} \right].$$

As the system is axially symmetrical, each modal system has multiplicity 2. At standstill, forward and backward modes are equal and two modal masses with multiplicity 4 are found. The values of the modal stiffnesses are reported as functions of the speed in Figure 3.17.

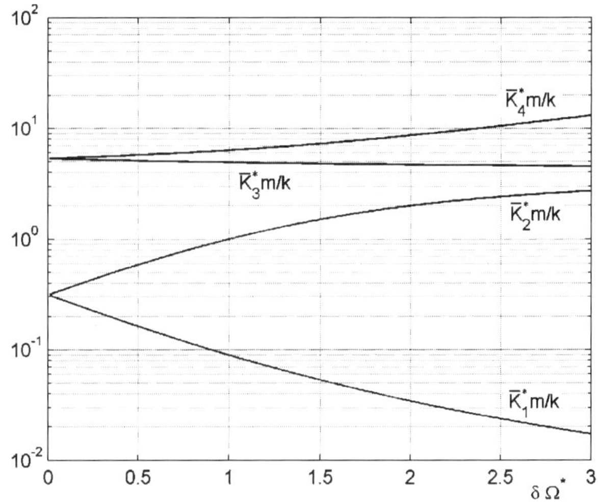

FIGURE 3.17. Modal stiffnesses of a Stodola-Green rotor as functions of the speed. Modal uncoupling referred to real coordinates. Each value has a multiplicity 2.

Example 3.5 *Rigid rotor on two bearings located at two opposite sides.*

Consider a rigid rotor running on two bearings, located at the two opposite sides of the center of mass (Figure 3.18).

The mass and the gyroscopic matrices are

$$\mathbf{M} = \begin{bmatrix} m & 0 \\ 0 & J_t \end{bmatrix}, \qquad \mathbf{G} = \begin{bmatrix} 0 & 0 \\ 0 & J_p \end{bmatrix}, \tag{3.116}$$

which hold also for the following three examples and will not be repeated.

To compute the compliance matrix, apply a unit force F in point G. The reaction forces F_1 and F_2 the bearings apply on the rotor in xz-plane are

$$\begin{cases} F_1 = -F\dfrac{b}{l}, \\ F_2 = -F\dfrac{a}{l}. \end{cases} \tag{3.117}$$

The corresponding displacements x_1 and x_2 are then

$$\begin{cases} x_1 = -\dfrac{F_1}{K_1} = \dfrac{F}{K_1}\dfrac{b}{l}, \\ x_2 = -\dfrac{F_2}{K_2} = \dfrac{F}{K_2}\dfrac{a}{l}. \end{cases} \tag{3.118}$$

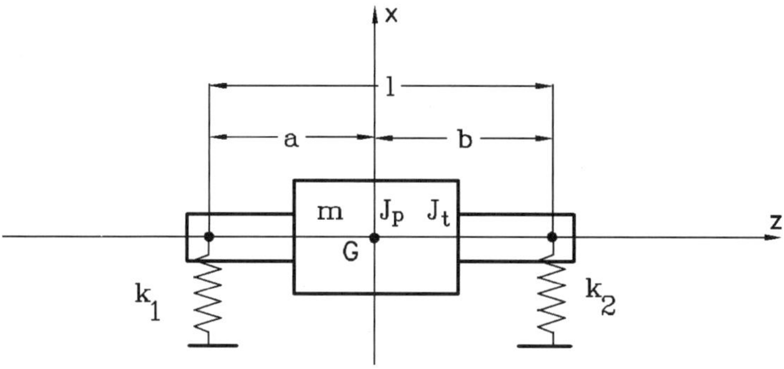

FIGURE 3.18. Sketch of a rigid rotor on compliant bearings. The center of mass is between the bearings.

The displacement x of the center of mass and the rotation ϕ_y are

$$\begin{cases} x = x_1 + \frac{a}{l}(x_2 - x_1) = \frac{F}{l^2}\left(\frac{a^2}{K_2} + \frac{b^2}{K_1}\right) = \frac{F}{K_1 K_2 l^2}\left(a^2 K_1 + b^2 K_2\right) , \\ \phi_y = \frac{x_2 - x_1}{l} = \frac{F}{l^2}\left(\frac{a}{K_2} - \frac{b}{K_1}\right) = \frac{F}{K_1 K_2 l^2}\left(a K_1 - b K_2\right) . \end{cases} \tag{3.119}$$

Then apply a unit bending moment M_y in point G. The reaction forces F_1 and F_2 the bearings apply on the rotor in xz-plane are

$$\begin{cases} F_1 = \frac{M_y}{l} , \\ F_2 = -\frac{M_y}{l} . \end{cases} \tag{3.120}$$

The corresponding displacements x_1 and x_2 are then

$$\begin{cases} x_1 = -\frac{F_1}{K_1} = -\frac{M_y}{l K_1} , \\ x_2 = -\frac{F_2}{K_2} = \frac{M_y}{l K_2} . \end{cases} \tag{3.121}$$

The displacement x of the center of mass and the rotation ϕ_y are then

$$\begin{cases} x = x_1 + \frac{a}{l}(x_2 - x_1) = \frac{M_y}{l^2}\left(\frac{a}{K_2} - \frac{b}{K_1}\right) = \frac{M_y}{K_1 K_2 l^2}\left(a K_1 - b K_2\right) , \\ \phi_y = \frac{x_2 - x_1}{l} = \frac{M_y}{l^2}\left(\frac{1}{K_2} + \frac{1}{K_1}\right) = \frac{M_y}{K_1 K_2 l^2}\left(K_1 + K_2\right) . \end{cases} \tag{3.122}$$

The compliance matrix is then obtained by assembling in matrix form the displacements and rotations caused by a unit force F and a unit moment M_y

$$\mathbf{B} = \frac{1}{K_1 K_2 l^2}\begin{bmatrix} a^2 K_1 + b^2 K_2 & a K_1 - b K_2 \\ a K_1 - b K_2 & K_1 + K_2 \end{bmatrix} . \tag{3.123}$$

The stiffness matrix in xz-plane is then

$$\mathbf{K} = \mathbf{B}^{-1} = \begin{bmatrix} K_1 + K_2 & -aK_1 + bK_2 \\ -aK_1 + bK_2 & a^2K_1 + b^2K_2 \end{bmatrix} . \tag{3.124}$$

Remark 3.10 *As expected, if $aK_1 = bK_2$, translational and rotational motions uncouple and the system behaves as a Jeffcott rotor plus a single-degree-of-freedom gyroscopic system.*

Another important particular case is that with two identical bearings, $K_1 = K_2 = K$. The stiffness matrix then simplifies as

$$\mathbf{K} = K \begin{bmatrix} 2 & -a+b \\ -a+b & a^2+b^2 \end{bmatrix} . \tag{3.125}$$

The Campbell diagram of Figure 3.6 and the unbalance response of Figure 3.13 refer to this case.

The equation for the undamped response to static unbalance can be written in the following nondimensional form:

$$\left(-\Omega^{*2}\begin{bmatrix} 1 & 0 \\ 0 & 1-\delta \end{bmatrix} + \begin{bmatrix} 1 & \frac{b-a}{2\rho_t} \\ \frac{b-a}{2\rho_t} & \frac{b^2+a^2}{2\rho_t^2} \end{bmatrix}\right)\left\{ \begin{array}{c} \frac{r_0}{\epsilon} \\ \phi_0\frac{\rho_t}{\epsilon} \end{array} \right\} = \Omega^{*2}\left\{ \begin{array}{c} 1 \\ 0 \end{array} \right\} , \tag{3.126}$$

where the nondimensional spin speed and the radius of inertia are

$$\Omega^* = \Omega/\sqrt{2K/m} , \qquad \rho_t = \sqrt{J_t/m} .$$

Unbalance response

The unbalance response for the system of Figure 3.6 with $l/a = 4$ but with $\rho_t/l = 0.5$ and four different values of δ is plotted in Figure 3.19. The first plot refers to the case with $\delta = 1.6$: The rotor is a disc, and only one critical speed is present. There is only one resonance peak. The last plot refers to the case with $\delta = 0.6$: a long rotor with two critical speeds and two resonance peaks. The other two plots are for almost identical rotors with δ very close to 1 ($\delta = 1.001$ and $\delta = 0.999$). Note that the two responses are different in the high supercritical regime. As expected, because δ is very close to 1, the response is very sensitive to the values of the parameters and a very large response at high speed can be expected.

Unbalance response; modal computation

Using the same data as above ($a/l = 0.25$, $\rho_t/l = 0.5$, and $\delta = 1.6$), the eigenvalues and the corresponding eigenvectors are

$$\Omega^{*2} = \left\{ \begin{array}{c} 0.8584 \\ -1.9417 \end{array} \right. , \quad \mathbf{q}_1 = \left\{ \begin{array}{c} 0.9621 \\ -0.2726 \end{array} \right\} , \quad \mathbf{q}_2 = \left\{ \begin{array}{c} -0.1676 \\ 0.9859 \end{array} \right\} . \tag{3.127}$$

There is only one critical speed: $\Omega^*_{cr} = 0.9265$; the imaginary critical speed is $\Omega^*_{im} = 1.3934$.

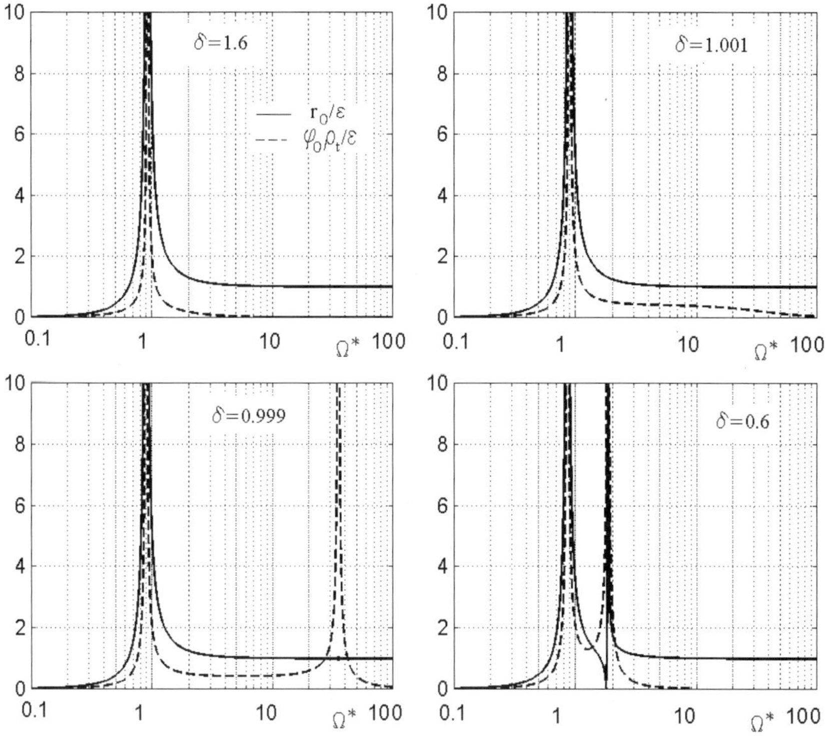

FIGURE 3.19. Undamped response to static unbalance in nondimensional form for the system of Figure 3.6 with $l/a = 4$, $\rho_t/l = 0.5$ and four different values of δ:1.6, 1.001, 0.999, and 0.6.

The eigenvectors can be normalized in such a way that the first modal mass is equal to 1 *and the second one to* −1. *As the matrix that is nonpositive defined is the mass matrix, it is better that the nonpositive modal matrix is* $\overline{\mathbf{M}}$*; however, a more standard normalization in which the modal mass matrix is an identity matrix could be used, but it would result in an imaginary eigenvector.*

The matrix of the normalized eigenvectors and the modal matrices are

$$\mathbf{\Phi} = \begin{bmatrix} 1.0250 & -0.2249 \\ -0.2904 & 1.3232 \end{bmatrix} , \tag{3.128}$$

$$\overline{\mathbf{M}} = \begin{bmatrix} 1 & 0 \\ 0 & -1 \end{bmatrix} , \qquad \overline{\mathbf{K}} = \begin{bmatrix} 0.8584 & 0 \\ 0 & 1.9417 \end{bmatrix} . \tag{3.129}$$

The unbalance force vector for a unit eccentricity and the corresponding modal unbalances are

$$\mathbf{f} = \begin{Bmatrix} 1 \\ 0 \end{Bmatrix} , \qquad \overline{\mathbf{f}} = \begin{Bmatrix} 1.0250 \\ -0.2249 \end{Bmatrix} . \tag{3.130}$$

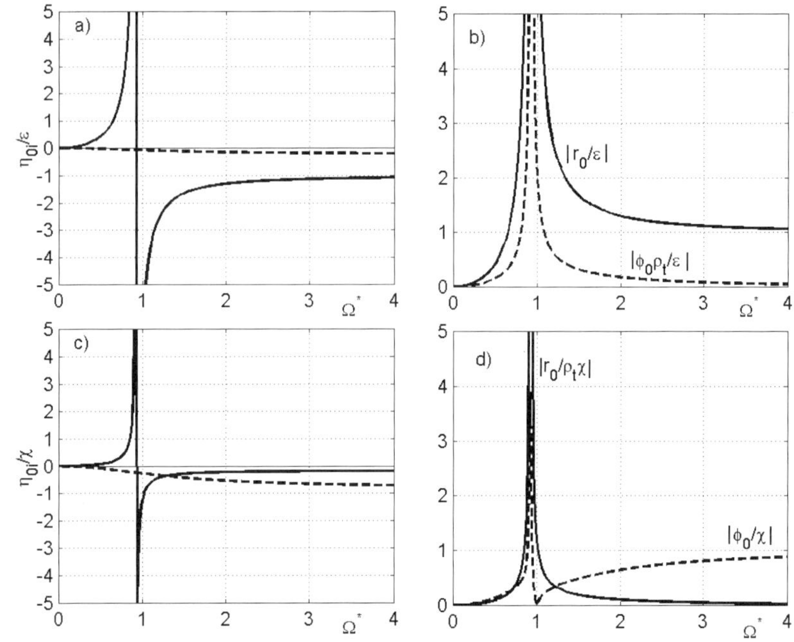

FIGURE 3.20. Nondimensional unbalance response of the system of Figure 3.19. Response to a static unbalance in modal form (a, full line first mode, dashed line second mode) and in physical coordinates (b). Response to a couple unbalance in modal form (c, full line first mode, dashed line second mode) and in physical coordinates (d).

The modal responses are plotted in Figure 3.20(a) and then recombined to yield the responses in terms of physical coordinates in Figure 3.20(b). Note that this result coincides with that reported in Figure 3.19(a), although here a linear scale has been used for the spin speed.

In case of a unit couple unbalance (or better, a unit value of the product $\chi\rho_t$*), the unbalance force vector and the corresponding modal unbalances are*

$$\mathbf{f} = \left\{ \begin{array}{c} 0 \\ 1-\delta \end{array} \right\} = \left\{ \begin{array}{c} 0 \\ -0.6 \end{array} \right\}, \qquad \overline{\mathbf{f}} = \left\{ \begin{array}{c} 0.1742 \\ -0.7939 \end{array} \right\}. \tag{3.131}$$

The responses are plotted in Figures 3.20(c) and (d).

Example 3.6 *Rigid overhung rotor on two bearings.*

A case similar to that of the previous example is shown in Figure 3.21, but now the two bearings are located on the same side of the center of mass.

Again the compliance matrix is obtained by applying a unit force F *and a unit bending moment* M_y *in point* G*. The reaction forces* F_1 *and* F_2 *the bearings apply*

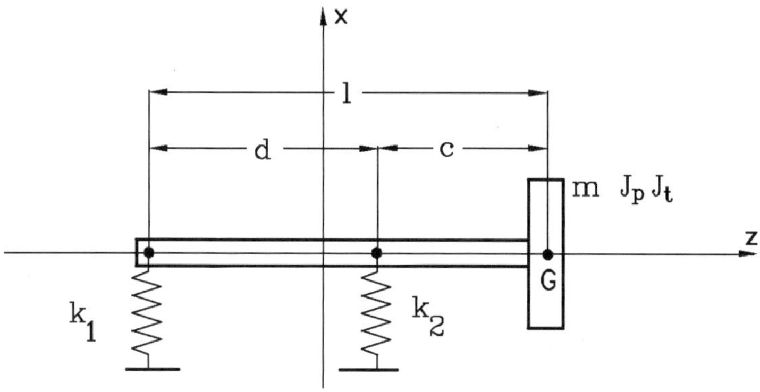

FIGURE 3.21. Sketch of a rigid rotor on compliant bearings. The bearings are on the same side with respect to the center of mass.

on the rotor in xz-plane are, for the two cases, respectively,

$$\begin{cases} F_1 = F\dfrac{c}{d} \\ F_2 = -F\dfrac{l}{d} \end{cases}, \quad \begin{cases} F_1 = \dfrac{M_y}{d} \\ F_2 = -\dfrac{M_y}{d} \end{cases}. \tag{3.132}$$

The corresponding displacements x_1 and x_2 are then

$$\begin{cases} x_1 = -\dfrac{F_1}{K_1} = -\dfrac{F}{K_1}\dfrac{c}{d} \\ x_2 = -\dfrac{F_2}{K_2} = \dfrac{F}{K_2}\dfrac{l}{d} \end{cases}, \quad \begin{cases} x_1 = -\dfrac{F_1}{K_1} = -\dfrac{M_y}{dK_1} \\ x_2 = -\dfrac{F_2}{K_2} = \dfrac{M_y}{dK_2} \end{cases}. \tag{3.133}$$

The displacement x of the center of mass and the rotation ϕ_y are, respectively,

$$\begin{cases} x = x_1 + \frac{l}{d}(x_2 - x_1) = \frac{F}{d^2}\left(\frac{l^2}{K_2} + \frac{c^2}{K_1}\right) = \frac{F}{K_1 K_2 d^2}\left(l^2 K_1 + c^2 K_2\right) , \\ \phi_y = \frac{x_2 - x_1}{d} = \frac{F}{d^2}\left(\frac{l}{K_2} + \frac{c}{K_1}\right) = \frac{F}{K_1 K_2 d^2}\left(l K_1 + c K_2\right) , \end{cases} \tag{3.134}$$

$$\begin{cases} x = x_1 + \frac{l}{d}(x_2 - x_1) = \frac{M_y}{d^2}\left(\frac{l}{K_2} + \frac{c}{K_1}\right) = \frac{M_y}{K_1 K_2 d^2}\left(l K_1 + c K_2\right) , \\ \phi_y = \frac{x_2 - x_1}{d} = \frac{M_y}{d^2}\left(\frac{1}{K_2} + \frac{1}{K_1}\right) = \frac{M_y}{K_1 K_2 d^2}\left(K_1 + K_2\right) . \end{cases} \tag{3.135}$$

The compliance matrix and the stiffness matrix in xz-plane are then

$$\mathbf{B} = \frac{1}{K_1 K_2 d^2}\begin{bmatrix} l^2 K_1 + c^2 K_2 & l K_1 + c K_2 \\ l K_1 + c K_2 & K_1 + K_2 \end{bmatrix}, \tag{3.136}$$

$$\mathbf{K} = \mathbf{B}^{-1} = \begin{bmatrix} K_1 + K_2 & -l K_1 - c K_2 \\ -l K_1 - c K_2 & l^2 K_1 + c^2 K_2 \end{bmatrix}. \tag{3.137}$$

In the case of the overhung rotor, the terms outside of the main diagonal of the stiffness matrix never vanish and translational motions never uncouple from rotational ones.

4
Discrete multi-degrees-of-freedom rotors

The models seen in the previous two chapters allow us to understand the basic features of the dynamic behavior of rotating systems and the main differences between rotordynamics and standard structural dynamics. Although suited for qualitative studies, they are unable to give accurate quantitative predictions for the complicated rotors encountered in real-world machinery, owing to their complexity.

At the macroscopic level typical of machine design, flexible bodies can be modeled as continuous systems, and linear elastic rotors can be studied by writing the relevant differential equations describing the behavior of linear elastic continua. Such equations are partial derivative differential equations, containing the derivatives with respect to time, usually up to the second derivative, as well as derivatives with respect to space coordinates. The computational difficulties developing both from the differential equations and, even more, from the boundary conditions, can, however, be tackled only in a few simple cases (see Chapter 5). The solution of most problems encountered in engineering practice requires dealing with complex rotors, and the use of continuous models is, consequently, ruled out; the only feasible approach is the discretization of the continuum to obtain a discretized model written in terms of ordinary differential equations, containing only derivatives with respect to time.

The substitution of a continuous model, characterized by an infinite number of degrees of freedom, with a discrete one, sometimes with a very large but finite number of degrees of freedom, is usually referred to as *discretization*. This step is of primary importance in the solution of practical problems, because the accuracy of the results depends largely on the adequacy of the discrete model to represent the actual system. The same discretization techniques widely used in structural dynamics can be applied in rotordynamics, provided that modifications aimed to introduce gyroscopic effects and other features typical of rotating systems are introduced.

After the discretization procedure has been applied, another problem, which is nowadays far less severe because of the growing power of computers, develops: The size of the discrete model, obtained through discretization, is usually very large, and in some cases, it may contain many hundreds, or even thousands, of degrees of freedom. Before the widespread use of computers, the solution of an eigenproblem containing matrices whose order was greater than a few units was very difficult and usually beyond actual possibilities. Many techniques were aimed at reducing to a minimum the size of the eigenproblem or transforming it into a form that could be solved using particular algorithms. The possibility of solving very large eigenproblems did actually change the basic approach to all problems in structural dynamics, and then also rotordynamics, making many popular methods obsolete.

The most common methods used in rotordynamics can be subdivided into two wide classes, the so-called *lumped-parameters methods* and the *finite element method* (FEM).

In the first case, the mass of the system (rotor and stator, if the latter is included in the system) is lumped into a certain number of rigid bodies (sometimes simply point masses) located at given stations in the deformable body. These lumped masses are then connected by massless fields that possess elastic and, sometimes, damping properties. Usually the properties of the fields are assumed to be uniform in space. Because the degrees of freedom of the lumped masses are used to describe the motion of the system, the model leads intuitively to a discrete system. Although the mass and gyroscopic matrices of such systems are easily obtained, it is often difficult to write the stiffness matrix, or, alternatively, the compliance matrix. To avoid such difficulty, together with that linked to the solution of large eigenproblems, an alternative approach can be followed. Instead of dealing with the system as a whole, the study can start at a certain station and proceed station by station using the so-called *transfer matrices*.

Remark 4.1 *Methods based on transfer matrices were very common in the past, because they could be worked out with tabular manual computations or implemented on very small computers. Their limitations are now making them yield to the finite element method.*

A separate class can be assigned to the FEM. As the FEM is widely used in many fields of engineering analysis (structural, thermal, magnetic, etc.), only a short account will be given here. In the FEM, the body is subdivided into a number of regions, called *finite elements*, as opposed to the vanishingly small regions used in writing the differential equations for continuous systems. The deformed shape of each finite element is assumed to be a linear combination of a set of functions of space coordinates through a certain number of parameters, considered as the degrees of freedom of the element. Usually such functions of space coordinates (called *shape functions*) are simple and the degrees of freedom have a direct physical meaning: Generalized displacements at selected points of the element, usually referred to as *nodes*. The analysis then proceeds to writing a set of differential equations of the same type as those obtained for discrete systems.

Actually, the use of the FEM can be limited to the writing of the stiffness matrix to be introduced into a lumped-parameters approach. Alternatively, the mass and gyroscopic matrices of the system can be obtained using the FEM approach. In this case, they are said to be *consistent*, because are obtained in a way that is consistent with that used for the computation of the stiffness matrix.

Although often considered as a separate approach, the dynamic stiffness method will be regarded here as a particular form of the FEM in which the shape functions are obtained from the actual deflected shape in free vibration.

Very often, rotors are assumed to be beam-like systems and are then modeled using beam elements, with inertial properties modeled using either the lumped or the consistent approach. Although there is no difficulty in using other types of elements, the relevant gyroscopic matrices are generally not available to users of standard finite element codes.

4.1 Transfer matrices approach: the Myklestadt-Prohl method

4.1.1 Undamped systems

The Myklestadt method was originally developed for the dynamic study of nonrotating beams, particularly cantilever aircraft wings [31]. It was then generalized by Prohl to include gyroscopic moments and since then has been the standard method for the dynamic study of rotors [32], particularly for the computation of critical speeds, for decades.

The system is lumped in a number of rigid bodies (often referred to as disks, particularly in the case of turbines) located at chosen points, often called *stations*, connected with each other by massless beams (*fields*) to which the elastic properties of the structure are ascribed. In many cases, the system lends itself very well to this discretization, as many turbine rotors are actually made of massive disks, connected together by lengths of shaft. In other cases, the discretization can be a more or less arbitrary procedure.

If the system possesses axial symmetry, an approach similar to that seen for complex coordinates in four degrees-of-freedom rotors can be used. Each end of a field has then four coordinates (displacements in x and y directions and rotations $\phi_{X'}$ and ϕ_y), which can be grouped into two complex coordinates (displacement r and rotation ϕ) and two generalized forces (shear force F and bending moment M) represented by using the complex notation. The state vectors are then of order four, and the transfer matrices are four by four. In nonisotropic systems, the displacements and rotations along x- and y-axes must be considered and the order of vectors and matrices is eight (real coordinates approach). However, in this case, the main advantage of the Myklestadt-Prohl method, i.e., its simplicity, is lost and it is better to use the FEM. There is no conceptual difficulty in also considering damping properties.

The Myklestạdt-Prohl method is used for the study of free whirling and the evaluation of the critical speeds but also for the computation of the response, if external forces are included in the model. Distributed forces are usually substituted by concentrated forces acting at the stations. The generalized displacements at the nodes are assumed to be generalized coordinates.

Remark 4.2 *As usual with transfer matrices methods, it is suited only for the study of in-line systems: The various fields must be connected to each other in series.As the structure is usually read like a written line, from left to right, each beam-like field has a left end, where it is connected to the preceding beam, and a right end, where the following beam starts.*

If the inertial properties of the system are lumped in n nodes, including the nodes at both ends, the fields are $n - 1$ (Figure 4.1). The computation starts at the first station, the *station at the left*, and ends at the last (right) station. It is clear that no branched systems or systems with multiple connections can be studied (unless by approximate approaches in which a branch is concentrated in the station where it stems from the main structure), and this is a serious drawback.

The aforementioned characteristics explain why the transfer-matrices method is now yielding to methods based on the study of the system as a whole, without having to go through it step by step from one station to the next.

The method is based on the definition of state vectors and transfer matrices. A *state vector* is a vector containing the generalized displacements and forces related to the degrees of freedom that characterize the ends of each field, considered as insulated from the rest of the structure. Consequently, each field has two state vectors, one at the left end and one at the right end. The state vectors $\mathbf{s}$ at the ends of a field are linked by the transfer matrix of the field

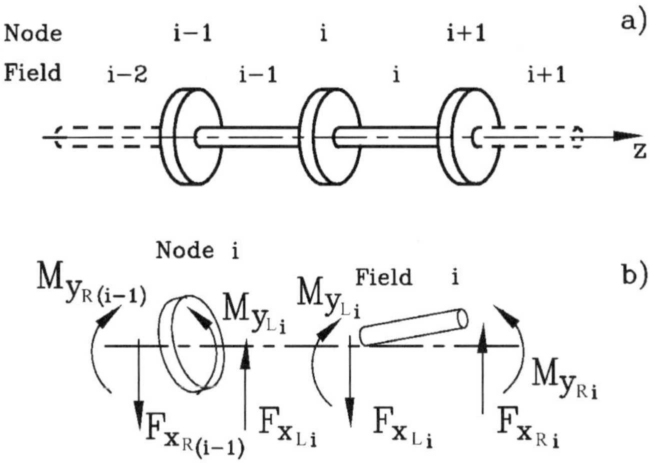

FIGURE 4.1. Mylestadt-Prohl method. (a) Sketch of the section included between the $(i-1)$th and the $(i+1)$th station of a system made of discs and beams; (b) forces and moments acting on the ith node and the ith field in the xz-plane.

$$\mathbf{s}_{R_i} = \mathbf{T}_{f_i}\mathbf{s}_{Li} , \tag{4.1}$$

where subscript i refers to the ith field and R and L designate the right and left ends, respectively, and $\mathbf{T}_{f_i}$ is the transfer matrix of the ith field. The left end of the ith field and the right end of the $(i-1)$th field are located at the ith station, and between them there is the ith lumped mass. The corresponding state vectors do not coincide because the mass exerts generalized inertia forces on the node. They are linked by the transfer matrix of the ith station (or node) $\mathbf{T}_{ni}$

$$\mathbf{s}_{Li} = \mathbf{T}_{ni}\mathbf{s}_{R_{i-1}} . \tag{4.2}$$

The station transfer matrix contains inertia forces caused by the lumped mass that, in harmonic free vibrations, are functions of the square of the frequency of vibration. On the contrary, the field is massless and the field transfer matrix is independent of the frequency.

If there is a linear elastic constraint at the ith node, its stiffness can be introduced into the expression of the station transfer matrix. In this way, it is possible to use the transfer-matrices method for systems in which the constraints are applied to nodes other than the first and last ones. The case of rigid constraints can be dealt with by introducing elastic constraints with very high stiffness.

Usually an element made of a node and a field is considered, and instead of writing Equations (4.1) and (4.2), a single relationship linking the state vector at the right of the ith field with that at the right of the $(i-1)$th field is stated

$$\mathbf{s}_{R_i} = \mathbf{T}_{f_i}\mathbf{T}_{n_i}\mathbf{s}_{R_{i-1}} = \mathbf{T}_i\mathbf{s}_{R_{i-1}} , \tag{4.3}$$

where the transfer matrix $\mathbf{T}_i$ is the product of the transfer matrix of a station and that of a field.

The state vector at the left of the first station $\mathbf{s}_0$ and that at the right of the last station $\mathbf{s}_n$ can be linked together by the equation

$$\mathbf{s}_n = \mathbf{T}_n\mathbf{T}_{n-1}\mathbf{T}_{n-2} \cdots \cdot \mathbf{T}_2\mathbf{T}_1\mathbf{s}_0 = \mathbf{T}_G\mathbf{s}_0 . \tag{4.4}$$

The overall transfer matrix

$$\mathbf{T}_G = \prod_{i=n}^{1} \mathbf{T}_i \tag{4.5}$$

is the product of all transfer matrices of all stations and fields from the last to the first, in the correct order. Note that the product of matrices depends on the order in which it is performed, so the overall transfer matrix must be computed by strictly following the aforementioned rule. If the system performs harmonic oscillations, the overall transfer matrix is a function of the frequency ω of the oscillations of the system.

To compute the transfer matrices, consider the field as a beam, bent in the xz-plane. The shear force at the right end is equal to the shear force at the left end, because no force acts on the field

$$F_{xR_i} = F_{xL_i} . \tag{4.6}$$

The bending moment at the right end is equal to the bending moment at the left end, increased by the moment caused by the shear force at the left end (for the signs, see Figure 4.1b)

$$M_{yR_i} = M_{yL_i} - l_i F_{xL_i} , \tag{4.7}$$

where l_i is the length of the ith field.

The displacement at the right end is equal to the displacement at the left end, increased by the displacements caused by the rigid rotation and to the deformation of the field

$$u_{xR_i} = u_{xL_i} + l_i\phi_{yL_i} + \Delta u_i . \tag{4.8}$$

The rotation at the right end is equal to the rotation at the left end increased by the rotation caused by bending deformation of the field

$$\phi_{yR_i} = \phi_{yL_i} + \Delta\phi_i . \tag{4.9}$$

The increments of displacement and rotation Δu_i and $\Delta\phi_i$ caused by the deformation of the field can be computed through the beam theory as the

Cross section			$\nu = 0.3$
Circular	$\frac{7+6\nu}{6(1+\nu)}$		1.128
Annular (thin)	$\frac{4+3\nu}{2(1+\nu)}$		1.88
Annular	$\frac{(7+6\nu)(1+m)^2+4m(5+3\nu)}{6(1+\nu)(1+m)^2}$	$\left(m = \frac{d_i}{d_o}\right)$	$1.128 + 3.026\frac{m}{(1+m)^2}$
Square	$\frac{12+11\nu}{10(1+\nu)}$		1.18
Square (thin)	$\frac{48+39\nu}{20(1+\nu)}$		2.30

TABLE 4.1. Values of the shear factor for beams with different cross section. The formula for an annular cross section with inner and outer diametrs d_i and d_o coincides with the formulas for solid circular and thin-walled annular sections when $m = 0$ and $m = 1$, respectively.

ith field can be considered as a prismatic beam clamped at the left end and loaded at the right end by a force F_{xR_i} and a moment M_{yR_i} [17]

$$\begin{cases} \Delta u_i = \left(-\frac{l_i^3}{6EI_y} + \frac{l_i\chi}{GA}\right) F_{xL_i} + \frac{l_i^2}{2EI_y} M_{yL_i} \ , \\ \Delta \phi_i = -\frac{l_i^2}{2EI_y} F_{xL_i} + \frac{l_i}{EI_y} M_{yL_i} \ , \end{cases} \tag{4.10}$$

where χ is the shear factor. Its value depends on the shape of the cross section, and for circular or annular beams, a value $\chi = 10/9$ is usually considered. More precise values (taken from [33]) referred to cross sections with two symmetry planes, i.e., whose stiffness is isotropic, are reported in Table 4.1 as functions of the Poisson's ratio of the material.

By writing the relationships obtained above in matrix form, the following expression for the field transfer matrix can be obtained:

$$\begin{Bmatrix} u_x \\ \phi_y \\ F_x \\ M_y \end{Bmatrix}_{R_i} = \begin{bmatrix} 1 & l_i & -\dfrac{l_i^3}{6EI_y} + \dfrac{l_i\chi}{GA} & \dfrac{l_i^2}{2EI_y} \\ 0 & 1 & -\dfrac{l_i^2}{2EI_y} & \dfrac{l_i}{EI_y} \\ 0 & 0 & 1 & 0 \\ 0 & 0 & -l_i & 1 \end{bmatrix} \begin{Bmatrix} u_x \\ \phi_y \\ F_x \\ M_y \end{Bmatrix}_{L_i} . \tag{4.11}$$

Now consider the ith node. The displacement and the rotation at the left of the node are equal to the displacement at the right

$$\begin{Bmatrix} u_x \\ \phi_y \end{Bmatrix}_{L_i} = \begin{Bmatrix} u_x \\ \phi_y \end{Bmatrix}_{R_{i-1}} . \tag{4.12}$$

The force at the right is equal to the force at the left increased by the inertia force caused by mass m_i, which in circular whirling conditions is $-\omega^2 m_i u_x$; if a constraint with stiffness k is located at the node, the force at the left of the node must be increased by ku_x

$$F_{xL_i} = F_{xR_{i-1}} + \left(-\omega^2 m_i + k\right) u_{xR_{i-1}} \, . \tag{4.13}$$

The moment at the right is equal to the moment at the left increased by the inertia torque added to the gyroscopic moment[1] $(-\omega^2 J_{t_i} + \omega\Omega J_{p_i})\phi_y$. If a constraint with angular stiffness κ is located at the node, the moment must be further increased by $\kappa\phi_y$

$$M_{y\,L_i} = M_{y\,R_{i-1}} + \left(-\omega^2 J_{t_i} + \omega\Omega J_{p_i} + \kappa\right)\phi_{y\,R_{i-1}}. \tag{4.14}$$

Using matrix notation, the following expression for the transfer matrix of the ith node is obtained:

$$\left\{\begin{array}{c} u_x \\ \phi_y \\ F_x \\ M_y \end{array}\right\}_{L_i} = \left[\begin{array}{cccc} 1 & 0 & 0 & 0 \\ 0 & 1 & 0 & 0 \\ -\omega^2 m_i + k & 0 & 1 & 0 \\ 0 & -\omega^2 J_{t_i} + \omega\Omega J_{p_i} + \kappa & 0 & 1 \end{array}\right] \left\{\begin{array}{c} u_x \\ \phi_y \\ F_x \\ M_y \end{array}\right\}_{R_{i-1}}. \tag{4.15}$$

Usually the transfer matrices method is used to compute the critical speeds: In this case, $\omega = \Omega = \Omega_{cr}$ and Equation (4.15) reduces to

$$\left\{\begin{array}{c} u_x \\ \phi_y \\ F_x \\ M_y \end{array}\right\}_{L_i} = \left[\begin{array}{cccc} 1 & 0 & 0 & 0 \\ 0 & 1 & 0 & 0 \\ -\Omega_{cr}^2 m_i + k & 0 & 1 & 0 \\ 0 & -\Omega_{cr}^2 (J_{t_i} - J_{p_i}) + \kappa & 0 & 1 \end{array}\right] \left\{\begin{array}{c} u_x \\ \phi_y \\ F_x \\ M_y \end{array}\right\}_{R_{i-1}}. \tag{4.16}$$

Once the transfer matrices of all nodes and all fields have been obtained, there is no difficulty in computing the global transfer matrix.

The boundary conditions must then be introduced. Consider, for instance, that both ends are simply supported: The displacement and the bending moment must vanish in both $\mathbf{s}_n$ and $\mathbf{s}_0$. Equation (4.4) then becomes

$$\left\{\begin{array}{c} 0 \\ \phi_y \\ F_x \\ 0 \end{array}\right\}_n = \left[\begin{array}{cccc} T_{11} & T_{12} & T_{13} & T_{14} \\ T_{21} & T_{22} & T_{23} & T_{24} \\ T_{31} & T_{32} & T_{33} & T_{34} \\ T_{41} & T_{42} & T_{43} & T_{44} \end{array}\right] \left\{\begin{array}{c} 0 \\ \phi_y \\ F_x \\ 0 \end{array}\right\}_0. \tag{4.17}$$

The first and fourth columns of the global transfer matrix can be canceled, because they multiply elements equal to zero in the state vector. The first and last equations (4.4) reduce to the homogenous equation

[1]This way of accounting for gyroscopic moments implies not only harmonic motion but also the use of complex coordinates. Gyroscopic moments are proportional to the velocities $\dot{\phi}_{X'}$ and $\dot{\phi}_y$, i.e., to the complex velocity $\dot{\phi}$, which in harmonic motion is proportional to $i\omega\phi$, and then the imaginary unit cancels with the other imaginary unit, which multiplies the gyroscopic term when complex coordinates are used.

$$\begin{bmatrix} T_{12} & T_{13} \\ T_{42} & T_{43} \end{bmatrix} \begin{Bmatrix} \phi_y \\ F_x \end{Bmatrix}_0 = \mathbf{0}\,, \tag{4.18}$$

which yields a solution different from the trivial one only if

$$\det \begin{bmatrix} T_{12} & T_{13} \\ T_{42} & T_{43} \end{bmatrix} = T_{12}T_{43} - T_{42}T_{13} = 0\,. \tag{4.19}$$

Because the elements of the global transfer matrix are functions of ω^2 (of ω if the gyroscopic moments are accounted for), this condition leads to an equation in ω, coinciding with the characteristic equation of the eigenproblem yielding the natural frequencies of the system.

The transfer-matrix approach is used to avoid solving an intricate eigenproblem, and an approach alternative to the solution of the characteristic equation in ω^2 is usually followed. A value of the frequency ω is assumed, and the transfer matrices are computed and multiplied to each other to obtain the overall transfer matrix. After canceling the rows and columns following the constraint conditions, its determinant is computed: If it vanishes, the frequency assumed is one of the natural frequencies. If this does not occur, a new frequency is assumed and the computation is repeated. By plotting the value of the determinant as a function of the frequency, it is easy to obtain as many natural frequencies as needed. Operating along these lines, no matrix of an order greater than that of the state vector must be dealt with. When the appropriate rows and columns of the overall transfer matrix have been canceled, the size of the determinant to be computed is usually not greater than 2. This explains why the method could be used without resorting to computers, even if long computations are usually involved.

The other boundary conditions can be dealt with in a similar way: If an end is clamped, both rotation and displacement vanish, whereas if it is free, the shear force and the bending moment are equal to zero. Six out of the many possible combinations of boundary conditions are reported here:

Right end	Left end	Equation
Free	Free	$T_{31}T_{42} - T_{41}T_{32} = 0$
Free	Supported	$T_{32}T_{43} - T_{42}T_{33} = 0$
Free	Clamped	$T_{33}T_{44} - T_{43}T_{34} = 0$
Supported	Supported	$T_{12}T_{43} - T_{42}T_{13} = 0$
Clamped	Free	$T_{11}T_{22} - T_{21}T_{12} = 0$
Clamped	Clamped	$T_{13}T_{24} - T_{23}T_{14} = 0$

As already stated, the value of the determinant of the matrix of the coefficients is plotted as a function of the frequency ω (or of the critical speed Ω_{cr}), and the values of the frequency for which it vanishes are obtained by searching the intersection of the curve with the ω- (or Ω_{cr}-) axis. Eigenfrequencies that are very close to each other may be difficult to find, as a very

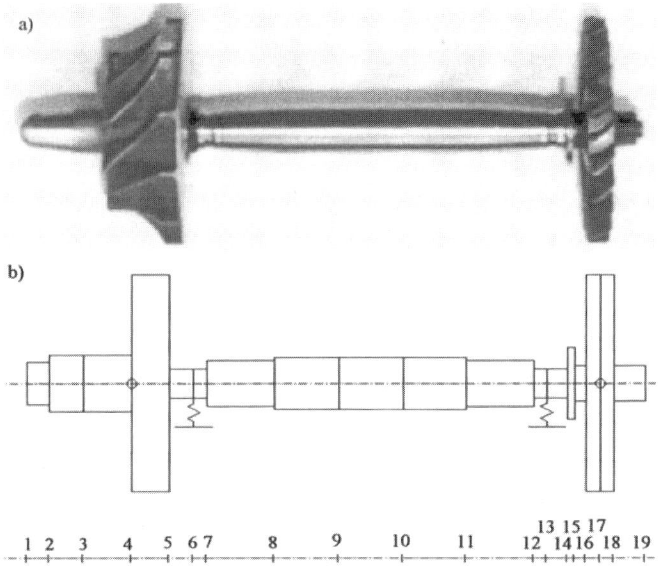

FIGURE 4.2. Rotor of a small turbojet engine. a): picture of the rotor; b) sketch of the FEM model.

small pitch may be required in the computation of the determinant. This disadvantage of transfer matrices methods is, however, not very serious.

It must be noted that the order in which the state variables are listed in the state vector can be different from the one shown here. Before using any formula from other books, it is necessary to verify the meaning of the subscripts of the elements of the transfer matrix.

Example 4.1 *Rotor of a very small turbojet.*

Consider the rotor of a small turbojet engine [Figure 4.2(a)]. A simple lumped parameters model is shown in Figure 4.2(b): It is made of 19 nodes or stations connected with 18 Timoshenko beams elements, two concentrated masses and two springs simulating the bearings.

The shaft between nodes 5 and 16 is modeled using beams with vanishing density and other properties coinciding the actual properties of the material (steel). The turbine wheel (between nodes 16 and 18) is modeled using two large beams, mostly for obtaining a realistic drawing, whose very large diameter coupled with the same Young's modulus of the material results in a very stiff part of the shaft. This models reasonably well the high stiffness of this part if the rotor, caused by the presence of the wheel on the shaft. The inertial properties are introduced with a concentrated mass in node 17, whereas the density of the beam elements is set to zero. The compressor wheel is modeled in the same way, using three massless (density equal to zero) beams, with a Young's modulus equal to that of aluminium, plus a concentrated mass. The two bearings are modeled using two springs. The

Field. #	1	2	3	4	5	6	7	8	9
ϕ [mm]	12	16	16	60	8	8	13	14.5	14.5
l [mm]	6	9	13	10	6.5	3.5	18	17	17
ρ [kg/m^2]	7810	0	0	0	7810	7810	7810	7810	7810
E [GN/m^2]	210	72	72	72	210	210	210	210	210

Field. #	10	11	12	13	14	15	16	17	18
ϕ [mm]	14.5	13	8	8	20	10	60	60	10
l [mm]	17	18	3.5	5.5	2	3	3.8	3.7	8.5
ρ [kg/m^2]	7810	7810	7810	7810	7810	7810	0	0	7810
E [GN/m^2]					210				

TABLE 4.2. Characteristics of the fields (ϕ: diameter; l: length; ρ density; E: Young's modulus).

Mass #	1	2	Spring. #	1	2
Node #	4	17	Node #	6	13
m [g]	72.1	65.8	k [MN/m]	1	1
J_p [kgm^2]	2.026×10^{-5}	2.099×10^{-5}	k_b [Nm/rad]	0	0
J_t [kgm^2]	1.219×10^{-5}	1.074×10^{-5}			

TABLE 4.3. Characteristics of concentrated masses and springs (m: mass; J_p: polar moment of inertia; J_t: diametral moment of inertia; k: stiffness; k_b bending stiffness).

front nut is modeled with a further beam. The properties of the beams, masses, and springs are listed in Tables 8.2 and 4.3.

The mass of each field is then computed and subdivided into two equal concentrated masses located at the end nodes of each field. The transfer matrices of the fields and of the stations are computed by assuming various values of the speed ranging from 0 to 16,000 rad/s, and then multiplied to obtain the global transfer matrix. The value of the determinant, which vanishes at the critical speeds, is then obtained. The result is reported in Figure 4.4. In the speed range considered, the determinant vanishes at three values of the speed, reported in Table 4.4.

The field transfer matrices were computed by taking into account also the deformation caused by shear; for the node transfer matrices, the mass of each field was lumped at its end and no moment of inertia of the fields was accounted for. The only moments of inertia introduced into the computation were then those of the turbine and compressor wheels, lumped in nodes 4 and 17.

Ω_{cr_I}	$\Omega_{cr_{II}}$	$\Omega_{cr_{III}}$
2,500	2,718	14,920

TABLE 4.4. Values of the first three critical speeds (in rad/s), computed using the Myklestadt-Prohl method.

FIGURE 4.3.

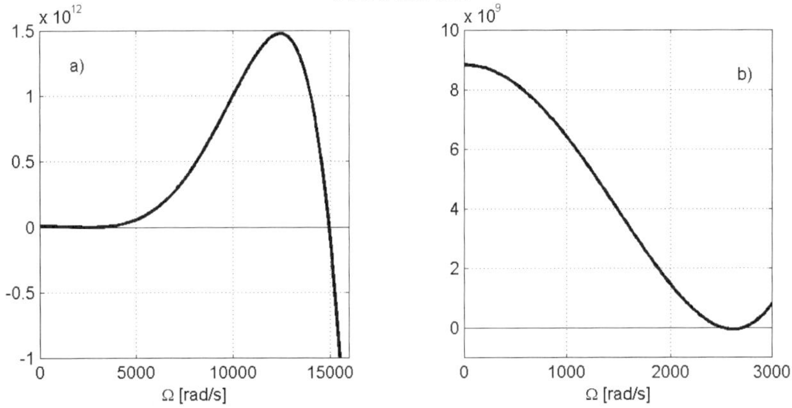

FIGURE 4.4. Value or the determinant of the global transfer matrix (after introducing the end conditions) as a function of the speed. a): range between 0 and 16,000 rad/s; b): range between 0 and 3,000 rad/s.

If real coordinates are used instead, the state vector has a size of 8 and the transfer matrices have 8 rows and columns. If the state vector is defined as

$$\mathbf{s} = \begin{bmatrix} u_x & \phi_y & F_x & M_y & u_y & -\phi_x & F_y & -M_x \end{bmatrix}^T , \tag{4.20}$$

the field transfer matrix of an axi-symmetrical beam is simply

$$\mathbf{T}_f = \begin{bmatrix} \mathbf{T}_f' & \mathbf{0} \\ \mathbf{0} & \mathbf{T}_f' \end{bmatrix} , \tag{4.21}$$

where $\mathbf{T}_f'$ is the transfer matrix of the field referred to the complex coordinates expressed by Equation (4.11). The two submatrices are equal because $-\phi_x$ instead of ϕ_x has been chosen as the degree of freedom for rotation in the yz-plane.

The transfer matrix of the node has an altogether different structure, because gyroscopic moments couple the two inflection planes. When the rotor spinning at speed Ω performs a harmonic oscillation in the two planes with frequency ω, the moment M_y at the left of a node is

$$M_{yL_i} = M_{yR_{i-1}} + \left(-\omega^2 J_{t_i} + \kappa\right) \phi_{yR_{i-1}} - i\omega\Omega J_{p_i} \phi_{xR_{i-1}} . \tag{4.22}$$

Ω_{cr_I}	$\Omega_{cr_{II}}$	$\Omega_{cr_{III}}$	$\Omega_{cr_{IV}}$	Ω_{cr_V}	$\Omega_{cr_{VI}}$	$\Omega_{cr_{VII}}$
$2,337^*$	$2,500$	$2,723^*$	$2,726$	$7,650^*$	$12,950^*$	$14,920$

TABLE 4.5. Values of the first seven critical speeds (in rad/s), computed using the Myklestadt-Prohl method, using real coordinates. * Values at the crossing of the $\omega = -\Omega$ line, which are not actually true critical speeds.

A similar expression holds also for moment M_{xL_i}. The node transfer matrix is then

$$\mathbf{T}_{n_i}=\begin{bmatrix} 1 & 0 & 0 & 0 & 0 & 0 & 0 & 0 \\ 0 & 1 & 0 & 0 & 0 & 0 & 0 & 0 \\ -\omega^2 m_i + k & 0 & 1 & 0 & 0 & 0 & 0 & 0 \\ 0 & -\omega^2 J_{t_i} + \kappa & 0 & 1 & 0 & i\omega\Omega J_{p_i} & 0 & 0 \\ 0 & 0 & 0 & 0 & 1 & 0 & 0 & 0 \\ 0 & 0 & 0 & 0 & 0 & 1 & 0 & 0 \\ 0 & 0 & 0 & 0 & -\omega^2 m_i + k & 0 & 1 & 0 \\ 0 & -i\omega\Omega J_{p_i} & 0 & 0 & 0 & -\omega^2 J_{t_i} + \kappa & 0 & 1 \end{bmatrix} . \tag{4.23}$$

For the computation of the critical speed, it is sufficient to substitute Ω_{cr} for both ω and Ω. However, as the use of real coordinates does not allow us to distinguish between forward and backward whirling, the intersections of the Campbell diagram with both the $\omega = \Omega$ and the $\omega = -\Omega$ lines are found. The latter are solutions of the mathematical problem but are not critical speeds in the actual sense and so must be discarded. The boundary conditions are imposed in the same way seen for the complex coordinates approach, the only difference being that now the determinant to be computed is that of a 4×4 and not of a 2×2 matrix.

Example 4.2 *Repeat the computations of the critical speeds of the turbojet studied in Example 4.1 using the real coordinates approach.*

The computation is repeated in the speed range from 0 to 16,000 rad/s, following the same lines seen in the previous example. The results are reported in Figure 4.5. In the speed range considered, the determinant vanishes at seven values of the speed, reported in Table 4.5. Three of them correspond with the intersections of the Campbell diagram with the $\omega = \Omega$ line and are the true critical speeds of the systems. The other four are the intersections with the $\omega = -\Omega$ line, i.e., represent the conditions of backward synchronous whirling.

4.1.2 Damped systems

If the system is damped, the computation of the free vibration frequencies and of the decay rates can proceed as already stated. Using a complex coordinates approach and assuming that the solution is of the type

$$\left\{ \begin{array}{c} r \\ \phi \end{array} \right\} = \left\{ \begin{array}{c} r_0 \\ \phi_0 \end{array} \right\} e^{st} , \tag{4.24}$$

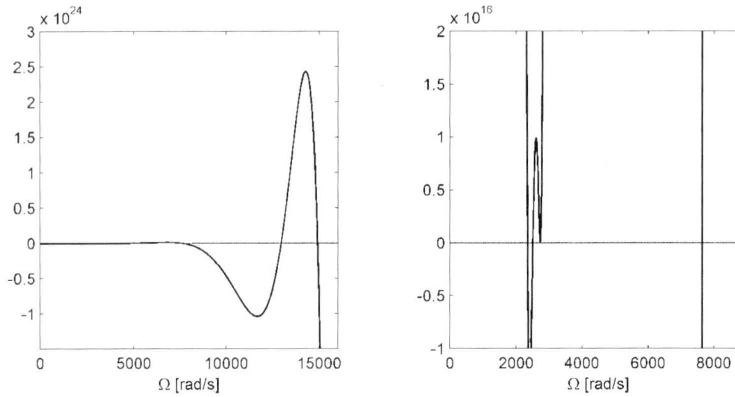

FIGURE 4.5. Value of the determinant of the global transfer matrix (after introducing the end conditions) as a function of the speed. (a) Range between 0 and 16,000 rad/s; (b) range between 0 and 9,000 rad/s.

the transfer matrices of the nodes and of the fields can be complex because of gyroscopic moments or to rotating damping. If, for instance, only the damping of the supports is added, the field transfer matrices are the same as above, whereas the transfer matrix of a node where a damped support with damping coefficient c_i for displacement and Γ_i for rotations is present, becomes

$$\mathbf{T}_{ni}=\left[\begin{array}{cccc} 1 & 0 & 0 & 0 \\ 0 & 1 & 0 & 0 \\ s^2 m_i + k_i + sc_i & 0 & 1 & 0 \\ 0 & s^2 J_{t_i} + is\Omega J_{p_i} + \kappa_i + s\Gamma_i & 0 & 1 \end{array}\right] . \tag{4.25}$$

In a similar way, it is possible to define the complex transfer matrices of other elements, remembering that nonrotating damping must be multiplied by $i\omega = s$ while rotating damping by $i(\omega - \Omega) = s - i\Omega)$.

There is no difficulty, in particular if a software able to deal with complex numbers is available, to multiply the various transfer matrices and then to compute the determinant. The point is that the unknown frequency to be computed is now complex; i.e., the unknowns are two and not a single one.

Several iterative schemes have been suggested, mostly based on Newton-Raphson technique. A simple numerical procedure is the following.

Let

$$\Delta = \det \left[\begin{array}{cc} T_{12} & T_{13} \\ T_{42} & T_{43} \end{array}\right] \tag{4.26}$$

in the case of a simply supported rotor. In the other cases, use the relevant determinant to define Δ. The complex equation $\Delta = 0$, allowing us to compute the complex frequency s can be split in its real and imaginary parts

$$\begin{cases} \operatorname{Re}(\Delta) = 0 \, , \\ \operatorname{Im}(\Delta) = 0 \, . \end{cases} \tag{4.27}$$

A starting value $s^{(0)}$ of s is assumed: In the case of a lightly damped system, it can be equal to i times the value of the natural frequency of the undamped system which has been previously computed. As Δ is a function of s, it can be expanded in series about the value it takes for $s = s^{(0)}$

$$\left\{ \begin{array}{c} \operatorname{Re}(\Delta(s)) \\ \operatorname{Im}(\Delta(s)) \end{array} \right\} = \left\{ \begin{array}{c} \operatorname{Re}\left(\Delta\left(s^{(0)}\right)\right) \\ \operatorname{Im}\left(\Delta\left(s^{(0)}\right)\right) \end{array} \right\} + \mathbf{S}\left(s^{(0)}\right) \left\{ \begin{array}{c} \operatorname{Re}(s) - \operatorname{Re}\left(s^{(0)}\right) \\ \operatorname{Im}(s) - \operatorname{Im}\left(s^{(0)}\right) \end{array} \right\} , \tag{4.28}$$

where the Jacobian matrix is

$$\mathbf{S}(s) = \begin{bmatrix} \frac{\partial \operatorname{Re}(\Delta)}{\partial \operatorname{Re}(s)} & \frac{\partial \operatorname{Re}(\Delta)}{\partial \operatorname{Im}(s)} \\ \frac{\partial \operatorname{Im}(\Delta)}{\partial \operatorname{Re}(s)} & \frac{\partial \operatorname{Im}(\Delta)}{\partial \operatorname{Im}(s)} \end{bmatrix} . \tag{4.29}$$

The algorithm to compute the value of s at the $(i+1)$th iteration from the known value at the ith iteration is then

$$\left\{ \begin{array}{c} \operatorname{Re}\left(s^{(i+1)}\right) \\ \operatorname{Im}\left(s^{(i+1)}\right) \end{array} \right\} = \left\{ \begin{array}{c} \operatorname{Re}\left(s^{(i)}\right) \\ \operatorname{Im}\left(s^{(i)}\right) \end{array} \right\} - \mathbf{S}\left(s^{(i)}\right)^{-1} \left\{ \begin{array}{c} \operatorname{Re}\left(\Delta\left(s^{(i)}\right)\right) \\ \operatorname{Im}\left(\Delta\left(s^{(i)}\right)\right) \end{array} \right\} . \tag{4.30}$$

At each iteration, the values of Δ and of the Jacobian matrix $\mathbf{S}$ are then needed. The first one can be computed by multiplying all of the transfer matrices and then computing the determinant; the first row of $\mathbf{S}$ can be obtained by increasing the real part of s by a small quantity, computing again Δ and approximating the derivatives with the incremental ratios. A further computation of Δ with the imaginary part of s incremented by a small quantity is needed to compute the second row of $\mathbf{S}$. The product of all transfer matrices must then be computed three times at each iteration.

As usual with the Newton-Raphson technique, if the solutions are many, there is no guarantee that the algorithm converges to the required one and the basin of attraction of the solutions may be very complex, often displaying a fractal geometry. Furthermore, the iteration can lock in a cycle, failing to converge to a solution.

Alternatively, it is possible to write the characteristic polynomial by multiplying the transfer matrices leaving s as an unknown, and then obtaining the complex roots of the polynomial, which has complex coefficients.

Example 4.3 *Compute the frequencies of free whirling and the corresponding decay rates for the turbojet studied in Example 4.1 at a speed Ω =10,000 rad/s, when a viscous damper with a damping coefficient $c = 100$ Ns/m is added at each support.*

The undamped system is studied first. The computation of the natural frequencies is performed using the complex coordinates approach, assuming a range of

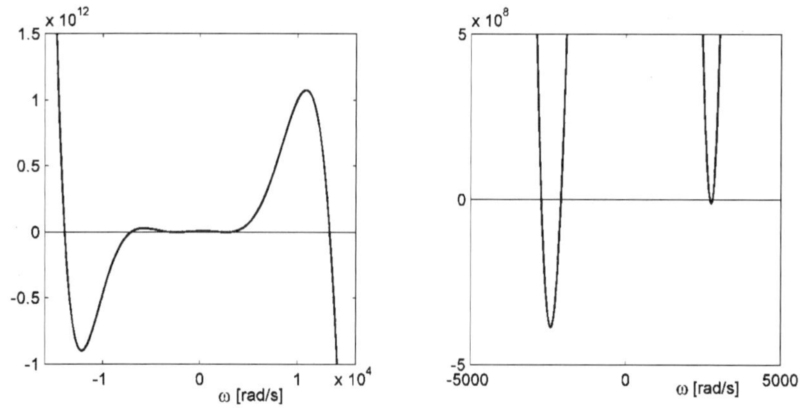

FIGURE 4.6. Value or the determinant of the global transfer matrix (after introducing the end conditions) at 10,000 rad/s as a function of the frequency. (a) Range between -16,000 and 16,000 rad/s; (b) range between -5,000 and 5,000 rad/s.

Mode #		1	2	3	4	5	6	7
Undam.	ω_n	-13,973	-7,063	-2,716	-2,078	2,698	2,780	13,300
Damped	ω	-13,948	-7,057	-2,695	-2,068	2,683	2,755	13,300
	σ	-480	-59	-360	-226	-347	-353	-66

TABLE 4.6. Values of the frequencies of free whirling (in rad/s) in the range between $-16,000$ and $16,000$ rad/s at a speed $\Omega = 10,000$ rad/s, for the undamped system and a system with dampers with $c = 100$ Ns/m at the support locations. In the latter case also, the decay rate σ (in 1/s) is reported.

frequencies spanning from $-16{,}000$ to 16,000 rad/s. The plot reporting the values of the determinant as a function of the frequency is reported in Figure 4.6. In the frequency range considered, the determinant vanishes at seven values of the frequency, four of which correspond to backward whirling modes. The results are reported in Table 4.6, second row.

The complex frequencies of the damped system are then computed using the Newton-Raphson method. The increment of the real and imaginary parts of s for the numerical computation of the Jacobian matrix is of 0.1 1/s. Each computation was started using the natural frequency of the undamped system. The results are reported in the last two rows of Table 4.6.

To investigate the convergency properties of the Newton-Raphson algorithm, the computation was repeated with many different starting values of the real and imaginary parts of s (201 values of the real part and 641 for the imaginary part, for a total of 128,841 computations). The map of the basins of attraction of the solutions is reported in Figure 4.7. The different shades of gray identify the basins of attraction, whereas the solutions are shown with the circles. The basins

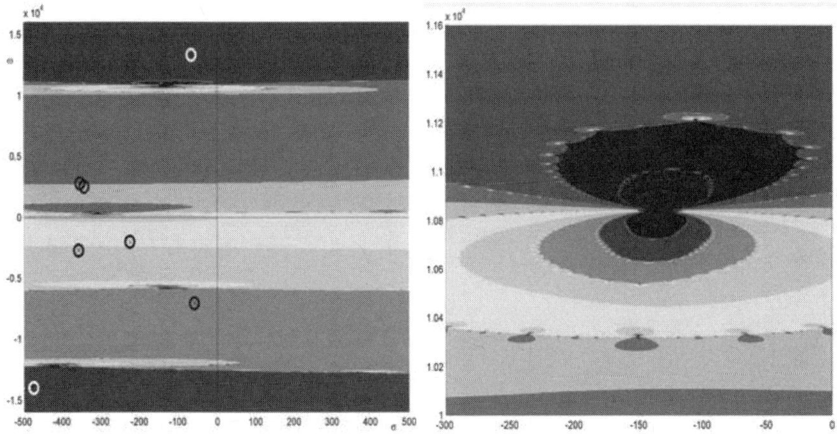

FIGURE 4.7. Basins of attraction of the solutions (shown by the circles) in the $\sigma\omega$-plane. The enlargement on the right shows a zone in which they have a marked fractal behavior.

of attraction are fairly well behaved, but zones in which they are interwoven and even zones in which they are fractal can be found, as shown by the enlargement on the right. There are even small zones in which no convergence was obtained after 100 iterations.

4.2 Lumped parameters stiffness method

The discretization is performed as seen for the Myklestadt-Prohl method, with the important difference that the method is not restricted to in-line systems. If complex coordinates are used, the relevant equations are either Equation (3.21) or (3.23), depending on whether damping is neglected. The size of all matrices and vectors is equal to the number of complex degrees of freedom, coinciding with the number of degrees of freedom related to bending behavior in the xz- and yz-planes. Mass, stiffness, and damping matrices $\mathbf{M}$, $\mathbf{K}$, and $\mathbf{C}$ are those related to the flexural behavior of the nonrotating system in the xz-plane, and the choice of the relevant plane depends on the way in which the complex coordinates were defined. This is obvious because the equation of motion must be the same describing the free behavior of the nonrotating system when the spin speed Ω tends to zero.

The stiffness matrix can be directly obtained using the FEM (see Section 4.3) or by inverting the compliance matrix, computed from the coefficients of influence [17].

If the complex coordinates are ordered as follows:

$$\mathbf{q} = \left[\begin{array}{cccccccc} r_1 & \phi_1 & r_2 & \phi_2 & \ldots & r_n & \phi_n \end{array} \right]^T , \tag{4.31}$$

matrices $\mathbf{M}$ and $\mathbf{G}$ and vector $\mathbf{f}$ of the unbalances are, respectively,

$$\mathbf{M} = \text{diag} \left[\begin{array}{cccccccc} m_1 & J_{t_1} & m_2 & J_{t_2} & \ldots & m_n & J_{t_n} \end{array} \right],$$

$$\mathbf{G} = \text{diag} \left[\begin{array}{cccccccc} 0 & J_{p_1} & 0 & J_{p_2} & \ldots & 0 & J_{p_n} \end{array} \right], \tag{4.32}$$

$$\mathbf{f} = \left[\begin{array}{ccccc} m_1\epsilon_1 e^{i\alpha_1} & (J_{t_1} - J_{p_1})\chi_1 e^{i\beta_1} & \ldots & m_n\epsilon_n e^{i\alpha_n} & (J_{t_n} - J_{p_n})\chi_n e^{i\beta_n} \end{array} \right]^T .$$

The phases α_i and β_i are needed to take into account the possible different orientations in space of the vectors expressing static and couple unbalances ϵ_i and χ_i. Rotating damping matrix $\mathbf{C}_r$ can be built in the same way as for general damping matrices, taking into account only the contribution to the overall energy dissipation caused by rotating parts of the machine.

In the case of nonisotropic systems, the complex coordinates approach can still be used, but the order of all matrices and vectors is at any rate doubled (see Chapter 6). The same occurs when real coordinates are used; also in this case, the number of degrees of freedom is doubled compared with that resulting from the complex coordinates approach.

4.3 The finite element method

The finite element method is a general discretization method for the solution of partial derivative differential equations, and consequently, it finds its application in many other fields beyond rotordynamic analysis. The aim of this section is not to provide a complete survey of the method, which can be dealt with only in a specialized text, [34, 35, 36] but to describe its main features to relate it to the other discretization techniques. The FEM is based on the subdivision of the structure into finite elements, i.e., into parts whose dimensions are not vanishingly small. Many different element formulations have been developed, depending on their shape and characteristics (beam elements, shell elements, plate elements, solid elements, and many others); however, the elements commonly used in elementary rotordynamics are just beam, mass, and spring elements.

Each element is essentially the model of a small deformable solid in which a limited number, usually quite small, of degrees of freedom is substituted for the infinity of degrees of freedom typical of continuous models. Inside each element, the displacement $\vec{u}(x, y, z)$ of the point of coordinates x, y, z is approximated by the linear combination of a number n of functions, the shape functions, which are assumed arbitrarily

$$\mathbf{u}(x, y, z, t) = \mathbf{N}(x, y, z)\mathbf{q}(t)\,, \tag{4.33}$$

where $\mathbf{u}$ is the displacement, written as a vector of order 3 in the three-dimensional space (sometimes of higher order, if rotations are also considered), $\mathbf{q}$ is a vector in which the n generalized coordinates of the element are listed, and $\mathbf{N}$ is the matrix containing the shape functions. There are as many rows in $\mathbf{N}$ as in $\mathbf{u}$ and as many columns as the number n of degrees of freedom.

Usually the degrees of freedom of the elements are the displacements at given points, which are referred to as *nodes*. In this case, Equation (4.33) is usually reduced to the simpler form

$$\left\{ \begin{array}{c} u_x(x, y, z, t) \\ u_y(x, y, z, t) \\ u_z(x, y, z, t) \end{array} \right\} = \left[\begin{array}{ccc} \mathbf{N}(x, y, z) & \mathbf{0} & \mathbf{0} \\ \mathbf{0} & \mathbf{N}(x, y, z) & \mathbf{0} \\ \mathbf{0} & \mathbf{0} & \mathbf{N}(x, y, z) \end{array} \right] \left\{ \begin{array}{c} \mathbf{q}_x(t) \\ \mathbf{q}_y(t) \\ \mathbf{q}_z(t) \end{array} \right\}, \tag{4.34}$$

where the displacements in each direction are functions of the nodal displacements in the same direction only. Matrix $\mathbf{N}$, in this case, has only one row and as many columns as the number of nodes of the element. Equation (4.34) has been written for a three-dimensional element; a similar formulation can also be easily obtained for one- or two-dimensional elements.

The shape functions are, as already stated, arbitrary. The freedom in the choice of such functions is, however, limited, because they must satisfy several conditions. A first requirement is a simple mathematical formulation, which is needed to lead to developments that are not too complex. Usually a set of polynomials in the space coordinates is assumed. To get results that are closer to the exact solution of the differential equations, which constitute the continuous model discretized by the FEM, while reducing the size of the elements, the shape functions must

- Be continuous and differentiable up to the required order, which depends on the type of element
- Be able to describe rigid-body motions of the element leading to vanishing elastic potential energy
- Lead to a constant strain field when the overall deformation of the element dictates so
- Lead to a deflected shape of each element that matches the shape of the neighboring elements. This means that when the nodes of two neighboring elements displace in a compatible way, all of the interface between the elements must displace in a compatible way.

Another condition, which is not always satisfied, is that the shape functions must be isotropic, i.e., must not show particular geometrical properties that depend on the orientation of the reference frame. Sometimes not all of these conditions are completely met; in particular, there are elements that fail to completely satisfy the matching of the deflected shapes of neighboring elements.

The nodes are usually located at the vertices or on the sides of the elements and are common to two or more of them, but points that are internal to an element can also be used.

The equation of motion of each element can be written through Lagrange equation, but there are plenty of alternative formulations, all leading to the same final expressions.

The strains can be expressed as functions of the derivatives of the displacements with respect to space coordinates. In general, it is possible to write a relationship of the type

$$\boldsymbol{\epsilon}(x,y,z,t) = \mathbf{B}(x,y,z)\mathbf{q}(t)\,, \tag{4.35}$$

where $\boldsymbol{\epsilon}$ is a column matrix in which the various elements of the strain tensor are listed (it is commonly referred to as a *strain vector*, but it is such only in the sense that it is a column matrix) and $\mathbf{B}$ is a matrix containing appropriate derivatives of the shape functions with respect to the x, y, z coordinates. $\mathbf{B}$ has as many rows as the number of components of the strain vector and as many columns as the number of degrees of freedom of the element.

If the element is free from initial stresses and strains and the behavior of the material is linear, the stresses can be directly expressed from the strains

$$\boldsymbol{\sigma}(x,y,z,t) = \mathbf{E}\boldsymbol{\epsilon} = \mathbf{E}(x,y,z)\mathbf{B}(x,y,z)\mathbf{q}(t)\,, \tag{4.36}$$

where $\mathbf{E}$ is the stiffness matrix of the material. It is a symmetric square matrix whose elements can theoretically be functions of the space coordinates but are usually constant within the element. The potential energy of the element can be easily expressed as

$$\mathcal{U} = \frac{1}{2}\int_V \boldsymbol{\epsilon}^T\boldsymbol{\sigma}dV = \frac{1}{2}\mathbf{q}^T\left(\int_V \mathbf{B}^T\mathbf{E}\mathbf{B}dV\right)\mathbf{q}\,. \tag{4.37}$$

The integral in Equation (4.37) is the stiffness matrix of the element

$$\mathbf{K} = \int_V \mathbf{B}^T\mathbf{E}\mathbf{B}dV\,. \tag{4.38}$$

Because the shape functions do not depend on time, the generalized velocities can be expressed as

$$\dot{\mathbf{u}}(x,y,z,t) = \mathbf{N}(x,y,z)\dot{\mathbf{q}}(t)\,. \tag{4.39}$$

If all generalized coordinates are related to displacements, the kinetic energy and the mass matrix of the element can be expressed as

$$\begin{aligned} \mathcal{T} &= \frac{1}{2}\dot{\mathbf{q}}^T \left(\int_V \rho \mathbf{N}^T \mathbf{N} dV \right) \dot{\mathbf{q}} , \\ \mathbf{M} &= \int_V \rho \mathbf{N}^T \mathbf{N} dV . \end{aligned} \tag{4.40}$$

When some generalized displacements are physically rotations, Equation (4.40) must be changed to introduce moments of inertia, but its basic structure remains the same. In the case of a non-natural (gyroscopic) system, the gyroscopic matrix can be obtained together with the mass matrix by taking also the spin speed into account when computing the kinetic energy.

As already stated, the FEM is often used just to compute the stiffness matrix to be used in the context of the lumped-parameters approach. In this case, the *consistent mass matrix* (4.40) is not computed and a diagonal matrix obtained by lumping the mass at the nodes is used. The advantage is that of dealing with a diagonal mass matrix, whose inversion is far simpler than that of the consistent mass matrix. The accuracy is, however, reduced or, better, a greater number of elements is needed to reach the same accuracy; as a consequence, the convenience between the two formulations must be assessed in each case. Generally speaking, the consistent approach leads to values of the natural frequencies that are higher than those computed using the elastic continuum model, whereas those obtained using the lumped-parameters approach are smaller.

If a force distribution $\mathbf{p}(x, y, z, t)$ acts on the body, the virtual work $\delta\mathcal{L}$ linked with the virtual displacement $\delta\mathbf{u} = \mathbf{N}\delta\mathbf{q}$ and the nodal force vector $\mathbf{f}$ can be expressed in the form

$$\begin{aligned} \delta\mathcal{L} &= \int_V \delta\mathbf{q}^T \mathbf{N}^T \mathbf{p}(x, y, z, t) dV , \\ \mathbf{f}(t) &= \int_V \mathbf{N}^T \mathbf{p}(x, y, z, t) dV . \end{aligned} \tag{4.41}$$

In a similar way, it is possible to obtain the nodal force vectors corresponding to surface force distributions or to concentrated forces acting on any point of the element.

The equation of motion of the nonrotating element is then the usual one for discrete undamped nonrotating systems

$$\mathbf{M}\ddot{\mathbf{q}} + \mathbf{K}\mathbf{q} = \mathbf{f}(t) , \tag{4.42}$$

where vector $\mathbf{f}$ contains all forces acting on the element.

4.3.1 Timoshenko beam element for rotordynamic analysis

As already stated, rotors are usually modeled as beam-like structures. Several beam formulations have been developed that differ from each other,

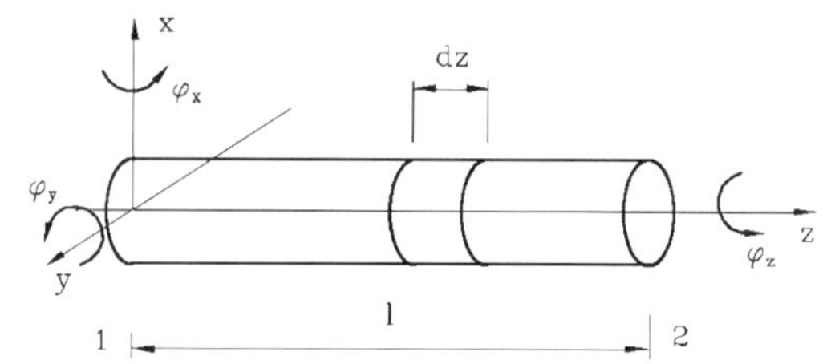

FIGURE 4.8. Beam element: geometrical definitions and reference frame.

owing to the number of nodes and degrees of freedom per node and to the theoretical formulation: Some of them are Euler-Bernoulli element, i.e., do not take into account shear deformation, whereas others include shear deformation following the simplified approach introduced by Timoshenko (Timoshenko beam elements). The element that will be studied here is often referred to as a *simple Timoshenko beam* [37, 38]. It has two nodes at the ends of the beam and six degrees of freedom per node, and it consists of a prismatic homogeneous beam with uncoupled axial, torsional, and flexural behavior. The relevant geometrical definitions and the reference frame used for the study are shown in Figure 4.8.

Each cross section has six degrees of freedom, three displacements, and three rotations, and the total number of degrees of freedom of the element is 12. The vector of the nodal displacements, i.e., of the generalized coordinates of the element, is

$$\mathbf{q} = [u_{x_1}, u_{y_1}, u_{z_1}, \phi_{x_1}, \phi_{y_1}, \phi_{z_1}, u_{x_2}, u_{y_2}, u_{z_2}, \phi_{x_2}, \phi_{y_2}, \phi_{z_2}]^T . \qquad (4.43)$$

As the beam has the properties needed to perform a complete uncoupling among axial, torsional, and flexural behavior in each of the coordinate planes, it is expedient to subdivide vector $\mathbf{q}$ into four smaller vectors

$$\begin{array}{ll} \mathbf{q}_A = [u_{z_1}, u_{z_2}]^T , & \mathbf{q}_{F1} = [u_{x_1}, \phi_{y_1}, u_{x_2}, \phi_{y_2}]^T , \\ \mathbf{q}_T = [\phi_{z_1}, \phi_{z_2}]^T , & \mathbf{q}_{F2} = [u_{y_1}, \phi_{x_1}, u_{y_2}, \phi_{x_2}]^T . \end{array} \qquad (4.44)$$

The flexural behavior of the beam element can thus be described using the complex coordinates vector

$$\mathbf{q}_F = [u_{x_1} + iu_{y_1}, \phi_{y_1} - i\phi_{x_1}, u_{x_2} + iu_{y_2}, \phi_{y_2} - i\phi_{x_2}]^T . \qquad (4.45)$$

A way of including shear deformation in the analysis without incurring in the so-called locking problem, a severe overestimate of the stiffness of the element, is to use the following shape functions:

$$N_{11} = \frac{1+\Phi(1-\zeta)-3\zeta^2+2\zeta^3}{1+\Phi}, \quad N_{12} = l\zeta\frac{1+\frac{1}{2}\Phi(1-\zeta)-2\zeta+\zeta^2}{1+\Phi},$$

$$N_{13} = \zeta\frac{\Phi+3\zeta-2\zeta^2}{1+\Phi}, \quad N_{14} = l\zeta\frac{-\frac{1}{2}\Phi(1-\zeta)-\zeta+\zeta^2}{1+\Phi},$$

$$N_{21} = 6\zeta\frac{\zeta-1}{l(1+\Phi)}, \quad N_{22} = \frac{1+\Phi(1-\zeta)-4\zeta+3\zeta^2}{1+\Phi},$$

$$N_{23} = -6\zeta\frac{\zeta-1}{l(1+\Phi)}, \quad N_{24} = \frac{\Phi\zeta-2\zeta+3\zeta^2}{1+\Phi}, \tag{4.46}$$

where $\zeta = z/l$ is a nondimensional coordinate,

$$\Phi = \frac{12EI_y\chi}{GAl^2}$$

and χ is the shear factor. When the slenderness of the beam increases, the value of Φ decreases, tending to zero for a Euler-Bernoulli beam. In the present chapter the beam is assumed to be axially symmetrical, i.e., $I_y = I_x$, and χ and Φ related to xz- and yz-planes are equal.

Equation (4.34) can thus be written in the form

$$\begin{Bmatrix} u_x \\ \phi_y \end{Bmatrix} = \begin{bmatrix} N_{11} & N_{12} & N_{13} & N_{14} \\ N_{21} & N_{22} & N_{23} & N_{24} \end{bmatrix} \begin{Bmatrix} u_{x_1} \\ \phi_{y_1} \\ u_{x_2} \\ \phi_{y_2} \end{Bmatrix},$$

$$\begin{Bmatrix} u_y \\ -\phi_x \end{Bmatrix} = \begin{bmatrix} N_{11} & N_{12} & N_{13} & N_{14} \\ N_{21} & N_{22} & N_{23} & N_{24} \end{bmatrix} \begin{Bmatrix} u_{y_1} \\ -\phi_{x_1} \\ u_{y_2} \\ -\phi_{x_2} \end{Bmatrix}. \tag{4.47}$$

Note that Equation (4.47) refers to the behavior in xy-plane, but holds also for complex coordinates, because the generalized coordinate for rotations about X'- axis is $-\phi_x$ instead of ϕ_x. In this way, the matrices referred to bending in xz-plane are the same as those referred to yz-plane (if the system is isotropic). Some of the generalized coordinates are related to rotations; as a consequence, Equations (4.38) and (4.40) cannot be used directly to express the stiffness and mass matrices. The potential energy obtained can be computed by adding the contributions caused by bending and shear deformations

$$d\mathcal{U} = \frac{1}{2}EI_y\left[\left(\frac{d\phi_y}{dz}\right)^2 + \left(\frac{d\phi_x}{dz}\right)^2\right]dz + \frac{1}{2}\frac{GA}{\chi}\left(\gamma_{xz}^2+\gamma_{yz}^2\right)dz\,. \tag{4.48}$$

The shear deformation γ is linked with the displacements by the relationships

$$\gamma_{xz} = \phi_y - \frac{du_x}{dz} \quad , \qquad \gamma_{yz} = -\phi_x - \frac{du_y}{dz} \; . \tag{4.49}$$

By using the symbols $\mathbf{N}_1$ and $\mathbf{N}_2$ to express the first and second rows of matrix $\mathbf{N}$ in Equation (4.47) and $\mathbf{q}_x$ and $\mathbf{q}_y$ for the vectors of the generalized coordinates in xz- and yz-planes, remembering Equations (4.47) and integrating over the whole beam, the expression for the potential energy of the element is

$$\begin{aligned}\mathcal{U} = & \frac{EI_y}{2l}\int_0^1 \mathbf{q}_x^T \frac{d}{dz}\mathbf{N}_2^T \frac{d}{dz}\mathbf{N}_2\mathbf{q}_x d\zeta + \frac{EI_y}{2l}\int_0^1 \mathbf{q}_y^T \frac{d}{dz}\mathbf{N}_2^T \frac{d}{dz}\mathbf{N}_2\mathbf{q}_y d\zeta + \\ & + \frac{6EI_y}{\Phi l}\int_0^1 \mathbf{q}_x^T\mathbf{N}_3^T\mathbf{N}_3\mathbf{q}_x d\zeta + \frac{6EI_y}{\Phi l}\int_0^1 \mathbf{q}_y^T\mathbf{N}_3^T\mathbf{N}_3\mathbf{q}_y d\zeta \; ,\end{aligned} \tag{4.50}$$

where $\mathbf{N}_3 = \mathbf{N}_2 - \frac{d}{dz}\mathbf{N}_1$. By introducing the stiffness matrix, the expression of the potential energy reduces to

$$\mathcal{U} = \frac{1}{2}\mathbf{q}_x^T\mathbf{K}\mathbf{q}_x + \frac{1}{2}\mathbf{q}_y^T\mathbf{K}\mathbf{q}_y \; . \tag{4.51}$$

As already stated, a $(-)$ sign is present before ϕ_x in vector $\mathbf{q}_y$ in Equation (4.50), as shown in Equation (4.47). In this way, the two stiffness matrices in Equation (4.51) are identical, which is essential for introducing complex coordinates. By introducing the expressions of the shape functions into that of the potential energy and performing the integrations, the bending stiffness matrix is obtained

$$\mathbf{K} = \frac{EI_y}{l^3(1+\Phi)}\begin{bmatrix} 12 & 6l & -12 & 6l \\ & (4+\Phi)l^2 & -6l & (2-\Phi)l^2 \\ & & 12 & -6l \\ & \text{symm.} & & (4+\Phi)l^2 \end{bmatrix} . \tag{4.52}$$

The length dz of the beam can be considered as a rigid body, and its kinetic energy can be computed using Equations (3.10) and (3.13)

$$d\mathcal{T} = \frac{1}{2}\rho A\left(\dot{u}_x{}^2 + \dot{u}_y{}^2\right) + \frac{1}{2}\rho\left[I_y(\dot{\phi}_x^2 + \dot{\phi}_y^2) + J_p(\Omega^2 + 2\Omega\dot{\phi}_x\phi_y)\right] . \tag{4.53}$$

In the case of an axi-symmetrical cross section, $J_p = 2J_y$. By introducing the shape functions into Equation (4.53), it yields

$$\begin{aligned}d\mathcal{T} = & \tfrac{1}{2}\rho A\left(\dot{\mathbf{q}}_x^T\mathbf{N}_1^T\mathbf{N}_1\dot{\mathbf{q}}_x + \dot{\mathbf{q}}_y^T\mathbf{N}_1^T\mathbf{N}_1\dot{\mathbf{q}}_y\right)dz + \tfrac{1}{2}\rho I_y\left(\dot{\mathbf{q}}_x^T\mathbf{N}_2^T\mathbf{N}_2\dot{\mathbf{q}}_x + \right. \\ & \left. + \dot{\mathbf{q}}_y^T\mathbf{N}_2^T\mathbf{N}_2\dot{\mathbf{q}}_y\right) + \rho J_y(\Omega^2 - 2\Omega\dot{\mathbf{q}}_y^T\mathbf{N}_2^T\mathbf{N}_2\mathbf{q}_x)dz \; .\end{aligned} \tag{4.54}$$

The kinetic energy of the element is then

$$\mathcal{T} = \frac{1}{2}\rho A l \int_0^1 \left(\dot{\mathbf{q}}_x^T \mathbf{N}_1^T \mathbf{N}_1 \dot{\mathbf{q}}_x + \dot{\mathbf{q}}_y^T \mathbf{N}_1^T \mathbf{N}_1 \dot{\mathbf{q}}_y\right) dz + \frac{1}{2}\rho I_y l \int_0^1 \left(\dot{\mathbf{q}}_x^T \mathbf{N}_2^T \mathbf{N}_2 \dot{\mathbf{q}}_x + \right.$$

$$\left. + \dot{\mathbf{q}}_y^T \mathbf{N}_2^T \mathbf{N}_2 \dot{\mathbf{q}}_y\right) + \rho J_y l \Omega^2 - 2\rho J_y l \Omega \int_0^1 \mathbf{q}_y^T \mathbf{N}_2^T \mathbf{N}_2 \mathbf{q}_x dz \, , \tag{4.55}$$

i.e.,

$$\begin{aligned} \mathcal{T} = {} & \tfrac{1}{2}\dot{\mathbf{q}}_x^T \mathbf{M}_T \dot{\mathbf{q}}_x + \tfrac{1}{2}\dot{\mathbf{q}}_y^T \mathbf{M}_T \dot{\mathbf{q}}_y + \tfrac{1}{2}\dot{\mathbf{q}}_x^T \mathbf{M}_R \dot{\mathbf{q}}_x + \tfrac{1}{2}\dot{\mathbf{q}}_y^T \mathbf{M}_R \dot{\mathbf{q}}_y + \\ & + \rho J_y l \Omega^2 - 2\Omega \dot{\mathbf{q}}_y^T \mathbf{M}_R \mathbf{q}_x \, . \end{aligned} \tag{4.56}$$

Matrices $\mathbf{M}_T$ and $\mathbf{M}_R$ are the mass matrices linked with translational and rotational inertia

$$\mathbf{M}_T = \frac{\rho A l}{420(1+\Phi)^2} \begin{bmatrix} m_1 & l m_2 & m_3 & -l m_4 \\ & l^2 m_5 & l m_4 & -l^2 m_6 \\ & & m_1 & -l m_2 \\ & \text{symm.} & & l^2 m_5 \end{bmatrix} , \tag{4.57}$$

$$\mathbf{M}_R = \frac{\rho I_y}{30 l(1+\Phi)^2} \begin{bmatrix} m_7 & l m_8 & -m_7 & l m_8 \\ & l^2 m_9 & -l m_8 & -l^2 m_{10} \\ & & m_7 & -l m_8 \\ & \text{symm.} & & l^2 m_9 \end{bmatrix} , \tag{4.58}$$

where

$$\begin{array}{ll} m_1 = 156 + 294\Phi + 140\Phi^2 \, , & m_2 = 22 + 38.5\Phi + 17.5\Phi^2 \, , \\ m_3 = 54 + 126\Phi + 70\Phi^2 \, , & m_4 = 13 + 31.5\Phi + 17.5\Phi^2 \, , \\ m_5 = 4 + 7\Phi + 3.5\Phi^2 \, , & m_6 = 3 + 7\Phi + 3.5\Phi^2 \, , \\ m_7 = 36 \, , & m_8 = 3 - 15\Phi \, , \\ m_9 = 4 + 5\Phi + 10\Phi^2 \, , & m_{10} = 1 + 5\Phi - 5\Phi^2 \, . \end{array}$$

The consistent mass and gyroscopic matrices of the element are thus

$$\mathbf{M} = \mathbf{M}_T + \mathbf{M}_R \, , \qquad \mathbf{G} = 2\mathbf{M}_R \, . \tag{4.59}$$

A linear distribution of static unbalance can be defined by assuming that the centers of mass of the various cross sections lie on a straight line connecting the eccentricities ϵ_{ξ_1} and ϵ_{η_1} in the directions of ξ and η axes at node 1 and ϵ_{ξ_2} and ϵ_{η_2}at node 2. A couple unbalance can be defined by stating an angle χ (or, better, angles χ_η and χ_ξ to account for the phasing of the unbalance) at nodes 1 and 2. The consistent unbalance vector to be introduced into the equation of motion of the element can thus be computed through Equation (4.41)

$$\mathbf{f} = \mathbf{a}_1 \left\{ \begin{array}{c} \epsilon_{\xi_1} + i\epsilon_{\eta_1} \\ \epsilon_{\xi_2} + i\epsilon_{\eta_2} \end{array} \right\} + \mathbf{a}_2 \left\{ \begin{array}{c} \chi_{\eta_1} - i\chi_{\xi_1} \\ \chi_{\eta_2} - i\chi_{\xi_2} \end{array} \right\}, \tag{4.60}$$

where

$$\mathbf{a}_1 = \frac{\rho A l}{120(1-\Phi)} \left[\begin{array}{cc} 42+40\Phi & 18+20\Phi \\ l(6+5\Phi) & l(4+5\Phi) \\ 18+20\Phi & 42+40\Phi \\ -l(4+5\Phi) & -l(6+5\Phi) \end{array} \right],$$

$$\mathbf{a}_2 = \frac{\rho I_y}{12(1-\Phi)} \left[\begin{array}{cc} -6 & -6 \\ l(1+4\Phi) & l(-1+2\Phi) \\ 6 & 6 \\ l(-1+2\Phi) & l(1+4\Phi) \end{array} \right].$$

If the beam element is loaded by an axial force, it behaves as if its stiffness was increased (tensile axial forces) or decreased (compressive axial forces). Within the frame of the FEM, this effect can be easily accounted for by using the geometric matrix.

Consider the flexural vibration in the xz-plane of a Timoshenko beam element, on which a constant and known axial force F_a is acting. If force F_a is not known but results from the bending loads acting on the element, an iterative procedure must be followed to solve the resulting nonlinear problem: The unknown axial force F_z caused by to the loads must be computed first, and then the same force can be introduced in a further dynamic computation as F_a.

Taking into account the lateral inflection of the beam, the axial strain in correspondence of the neutral axis of the beam, which is usually expressed as $\epsilon_z = \partial u_z/\partial z$, becomes

$$\epsilon_z = \frac{\partial u_z}{\partial z} + \frac{1}{2}\left(\frac{\partial u_y}{\partial z}\right)^2. \tag{4.61}$$

If the small term linked with lateral deflection is also considered, the elastic potential energy caused by the axial strain becomes

$$\mathcal{U} = \frac{1}{2}\int_0^l EA\epsilon_z^2 dz = \frac{1}{2}\int_0^l EA\left[\frac{\partial u_z}{\partial z} + \frac{1}{2}\left(\frac{\partial u_y}{\partial z}\right)^2\right]^2 dz \approx$$

$$\approx \frac{1}{2}\int_0^l EA\left(\frac{\partial u_z}{\partial z}\right)^2 dz + \frac{1}{2}\int_0^l EA\frac{\partial u_z}{\partial z}\left(\frac{\partial u_y}{\partial z}\right)^2 dz. \tag{4.62}$$

The last expression has been obtained by neglecting the term in $(\partial u_y/\partial z)^4$. The first term of the potential energy has already been taken into account in the computation of the axial stiffness matrix of the element. The second one causes an increase of the potential energy, which can be considered by adding a suitable matrix, the so-called geometric stiffness matrix, or simply geometric matrix, to the stiffness matrix of the element for lateral deformations. Assuming that the axial force F_a is constant, the increment of potential energy can be written in terms of matrix $\mathbf{K}_{g_1}$ for the flexural behavior in the xz-plane in the form

$$\Delta\mathcal{U} = \frac{1}{2}F_a \int_0^l EA \left(\frac{\partial u_y}{\partial z}\right)^2 dz = \frac{1}{2}F_a \mathbf{q}_x^T \mathbf{K}_{g_1} \mathbf{q}_x \, , \tag{4.63}$$

where

$$\mathbf{K}_{g_1} = \frac{F_a}{30l(1+\Phi)^2} \begin{bmatrix} k_1 & lk_2 & -k_1 & lk_2 \\ & l^2k_3 & -lk_2 & -l^2k_4 \\ & & k_1 & -lk_2 \\ & \text{symm.} & & l^2k_3 \end{bmatrix} ,$$

$$k_1 = 36 + 60\Phi + 3\Phi^2 \, , \quad k_2 = 3 \, ,$$
$$k_3 = 4 + 5\Phi + 2.5\Phi^2 \, , \quad k_4 = 1 + 5\Phi + 2.5\Phi^2 \, .$$

The geometric matrix can thus be added to the stiffness matrix, being immaterial whether the complex coordinates or the real coordinates approach is used. Note that if the beam is axially symmetrical, the values of parameter Φ in the two inflection planes coincide and the geometric matrices in the two planes are equal (if $-\phi_x$ is used instead of ϕ_x as generalized coordinate; otherwise, the signs of some elements of the matrices are different).

4.3.2 Mass element

Consider a concentrated mass or, better, a rigid body located at the ith node. The mass and the gyroscopic matrices are the same seen for the rotor with four degrees of freedom and introduced into Equation (3.21). The size of the relevant matrices is 2×2 if complex coordinates are used or 4×4 when using real coordinates.

4.3.3 Spring element

Consider a spring element, i.e., an element that introduces a concentrated stiffness between two nodes, say node 1 and node 2, of the structure. When the nodes have a two complex degrees of freedom, the generalized coordinates of the element are

$$\mathbf{q} = [u_{x_1} + iu_{y_1}, \phi_{y_1} - i\phi_{x_1}, u_{x_2} + iu_{y_2}, \phi_{y_2} - i\phi_{x_2}]^T ,$$

and the stiffness matrix is

$$\mathbf{K} = \begin{bmatrix} k & 0 & -k & 0 \\ 0 & \kappa & 0 & -\kappa \\ -k & 0 & k & 0 \\ 0 & -\kappa & 0 & \kappa \end{bmatrix}, \tag{4.64}$$

where k and κ are the stiffness for displacements and rotations, respectively.

When using real coordinates, the stiffness matrix has eight rows and columns. Spring elements are very useful in rotordynamics as they can be used to model joints between shafts or linearized bearings located between a rotor and a stator, which are both modeled. If the stator is not included in the model, a similar element with a single node can be used to model a compliant constraint of the shaft.

4.3.4 Assembling the structure

The equations of motion of the element are written with reference to a *local* or *element reference frame* that has an orientation determined by the features of the element. In a beam element, for example, the z-axis usually coincides with the axis of the beam, whereas x- and y-axes are principal axes of inertia of the cross section. In rotordynamics, usually the various local reference frames of the elements have the same z-axis, which coincides with the axis of rotation. If the rotor is axially symmetrical the position of x- and y-axes in xy-plane is immaterial and it is possible to assume a *global reference frame* for the whole system that coincides with the reference frames of all elements.

If this is not the case, the orientation in space of any local frame can be expressed, with reference to the orientation of the global frame, by a suitable rotation matrix

$$\mathbf{R} = \begin{bmatrix} l_x & m_x & n_x \\ l_y & m_y & n_y \\ l_z & m_z & n_z \end{bmatrix}, \tag{4.65}$$

where l_i, m_i, and n_i are the direction cosines of the axes of the former in the global frame. The expressions $\mathbf{q}_{il}$ and $\mathbf{q}_{ig}$ of the displacement vector $\mathbf{q}_i$ of the ith node in the local and global reference frames are linked by the usual coordinate transformation

$$\mathbf{q}_{il} = \mathbf{R}\mathbf{q}_{ig} \,. \tag{4.66}$$

Remark 4.3 *The mentioned coordinate transformation holds only if the generalized coordinates are defined in a way consistent with the reference frames. No three-dimensional rotation can be performed by premultiplying by the rotation matrix if the complex coordinates defined above are used or if the generalized coordinate for rotation about x-axis is $-\phi_x$. Complex* *coordinates are suited only for systems in which all of the elements share the common z-axis and rotations are only in the xy-plane.*

The generalized coordinates in the displacement vector of the element can be transformed from the local to the global reference frame using a similar relationship in which an expanded rotation matrix $\mathbf{R}'$ is used to deal with all of the relevant generalized coordinates. It is essentially made by a number of matrices of the type of Equation (4.65) suitably assembled together. In the case of a beam element where the generalized coordinates are ordered as in Equation (4.43), the expanded rotation matrix has 12 rows and columns

$$\mathbf{R}' = \begin{bmatrix} \mathbf{R} & \mathbf{0} & \mathbf{0} & \mathbf{0} \\ \mathbf{0} & \mathbf{R} & \mathbf{0} & \mathbf{0} \\ \mathbf{0} & \mathbf{0} & \mathbf{R} & \mathbf{0} \\ \mathbf{0} & \mathbf{0} & \mathbf{0} & \mathbf{R} \end{bmatrix} . \tag{4.67}$$

The assumption of small displacements and rotations allows consideration of the rotations about the axes as the components of a vector, which can be rotated in the same way as displacements.

Because the inverse of a rotation matrix is coincident with its transpose, the expressions of the mass, gyroscopic and stiffness matrices of the element rotated from the local to the global frame are

$$\mathbf{M}_g = \mathbf{R}'^T \mathbf{M}_l \mathbf{R}' , \quad \mathbf{G}_g = \mathbf{R}'^T \mathbf{G}_l \mathbf{R}' , \quad \text{and} \quad \mathbf{K}_g = \mathbf{R}'^T \mathbf{K}_l \mathbf{R}' .$$

Similarly, the nodal load vector can be rotated using the relationship

$$\mathbf{f}_g = \mathbf{R}'^T \mathbf{f}_l .$$

Once the mass, gyroscopic, and stiffness matrices of the various elements, have been computed with reference to the global frame, it is possible to easily obtain the matrices of the whole structure. The n generalized coordinates of the structure can be ordered in a single vector $\mathbf{q}_g$. The matrices of the various elements can be rewritten in the form of matrices of order $n \times n$, containing all elements equal to zero except those that are in the rows and columns corresponding to the generalized coordinates of the relevant element. Because the kinetic and potential energies of the structure can be obtained simply by adding the energies of the various elements, it follows that

$$\mathbf{M} = \sum_{\forall i} \mathbf{M}_i , \qquad \mathbf{G} = \sum_{\forall i} \mathbf{G}_i , \qquad \mathbf{K} = \sum_{\forall i} \mathbf{K}_i . \tag{4.68}$$

In practice, the various matrices of size $n \times n$ for the elements are never written: Each term of the matrices of all elements is just added into the global mass and stiffness matrices in the correct place. If the generalized coordinates are taken into a suitable order, the assembled matrices have a band structure; many general-purpose computer codes have a routine that reorders the coordinates in such a way that the bandwidth is the smallest possible.

The nodal force vector can be assembled in a similar way

$$\mathbf{f} = \sum_{\forall i} \mathbf{f}_i .$$

The forces exchanged between the elements at the nodes cancel each other in this assembling procedure, and the force vectors that must be inserted into the global equation of motion of the structure are only those related to the external forces.

4.3.5 Constraining the structure

One of the advantages of the FEM is the ease with which the constraints can be defined. If the ith degree of freedom is rigidly constrained, the corresponding generalized displacement vanishes, and as a consequence, the ith column of the stiffness and mass matrices can be neglected, because they multiply a displacement and an acceleration, respectively, that are equal to zero. As one of the generalized displacements is known, one of the equations of motion can be neglected when solving for the deformed configuration of the system. The ith equation can thus be separated from the rest of the set of equations, which amounts to canceling the ith row of all matrices and of the force vector. Note that the ith equation could, in this case, be used after all displacements have been computed to obtain the value of the ith generalized nodal force, which, in this case, is the unknown reaction of the constraint.

To rigidly constrain a degree of freedom, it is sufficient to cancel the corresponding row and column in all matrices and vectors. This approach allows simplification of the formulation of the problem, which is particularly useful in dynamic problems, but this simplification is often marginal, as the number of constrained degrees of freedom is very small, compared with the total number of degrees of freedom. To avoid restructuring the whole model and rewriting all the matrices, rigid constraints can be transformed into very stiff elastic constraints.

If the ith degree of freedom is constrained through a linear spring with stiffness k_i, the potential energy of the structure is increased by the potential energy of the spring

$$\mathcal{U} = \frac{1}{2} k_i q_i^2 . \tag{4.69}$$

To take the presence of the constraint into account, it is sufficient to add the stiffness k_i to the element in the ith row and ith column (i.e., the ith element on the main diagonal) of the global stiffness matrix. This procedure is very simple, which explains why a very stiff elastic constraint is often added instead of canceling a degree of freedom in the case of rigid constraints. An additional advantage is that the reaction of the constraint can be obtained simply by multiplying the very high generalized stiffness k_i by the corresponding very small generalized displacement q_i.

4.3.6 Damping matrices

It is possible to take into account the damping of the structure in a way that closely follows what has been said for the stiffness. If elements that can be modeled as viscous dampers are introduced into the structure between two nodes or between a node and the ground, a viscous damping matrix can be obtained using the same procedures used for the stiffness matrix of spring elements or elastic constraints. Actually, the relevant equations are equal, once the damping coefficient is substituted for the stiffness and the velocity for the displacements. If the damping of some of the elements can be modeled as hysteretic damping, within the limits of validity of the complex stiffness model, an imaginary part of the element stiffness matrix can be obtained by simply multiplying the real part by the loss factor. Note that if there is a geometric stiffness matrix, the imaginary part of the stiffness matrix must be computed before adding the geometric matrix to the stiffness matrix.

Viscous or structural damping matrices are then assembled following the same rules seen for mass and stiffness matrices. The real and imaginary parts of the stiffness matrices must be assembled separately, because, when the loss factor is not constant along the structure, they are not proportional to each other.

In rotordynamics, the damping matrices of rotating elements must be assembled separately from those of the nonrotating ones, and the result is that a rotating damping matrix and a nonrotating damping matrix are obtained. Their role is that already seen in the case of the Jeffcott rotor and of the rotor with four degrees of freedom.

4.3.7 Transfer matrices methods and the FEM

Although the FEM and the transfer matrices methods work in different ways, they are essentially equivalent. Consider for instance the relationship between the forces and the displacements through the stiffness matrix in xz-plane for a beam element [equation (4.52)] and partition it as

$$\left\{ \begin{Bmatrix} F_{x_1} \\ M_{y_1} \end{Bmatrix} \\ \begin{Bmatrix} F_{x_2} \\ M_{y2} \end{Bmatrix} \right\} = \begin{bmatrix} \mathbf{K}_{11} & \mathbf{K}_{12} \\ \mathbf{K}_{12}^T & \mathbf{K}_{22} \end{bmatrix} \left\{ \begin{Bmatrix} u_{x_1} \\ \phi_{y_1} \end{Bmatrix} \\ \begin{Bmatrix} u_{x_2} \\ \phi_{y2} \end{Bmatrix} \right\} . \tag{4.70}$$

The first two equations can be solved as

$$\left\{ \begin{array}{c} u_{x2} \\ \phi_{y2} \end{array} \right\} = \mathbf{K}_{12}^{-1} \left\{ \begin{array}{c} F_{x_1} \\ M_{y_1} \end{array} \right\} - \mathbf{K}_{12}^{-1}\mathbf{K}_{11} \left\{ \begin{array}{c} u_{x1} \\ \phi_{y_1} \end{array} \right\} . \qquad (4.71)$$

Equation (4.71), together with the last two Equation (4.70), yield

$$\left\{ \begin{array}{c} \left\{ \begin{array}{c} u_{x2} \\ \phi_{y2} \end{array} \right\} \\ \left\{ \begin{array}{c} F_{x_2} \\ M_{y2} \end{array} \right\} \end{array} \right\} = \left[\begin{array}{cc} -\mathbf{K}_{12}^{-1}\mathbf{K}_{11} & -\mathbf{K}_{12}^{-1} \\ \mathbf{K}_{12}^{T} - \mathbf{K}_{22}\mathbf{K}_{12}^{-1}\mathbf{K}_{11} & -\mathbf{K}_{22}\mathbf{K}_{12}^{-1} \end{array} \right] \left\{ \begin{array}{c} \left\{ \begin{array}{c} u_{x1} \\ \phi_{y_1} \end{array} \right\} \\ \left\{ \begin{array}{c} -F_{x_1} \\ -M_{y_1} \end{array} \right\} \end{array} \right\} . \qquad (4.72)$$

Equation (4.72) is the transfer matrix formulation for a field having the characteristics of a Timoshenko beam element and is identical to Equation (4.11), but for the signs of the last two elements of the state vector at node 1. This is because of the different sign conventions used for generalized forces in the two methods. By introducing the values of the elements of the stiffness matrix obtained from Equation (4.52), the transfer matrix of Equation (4.11) can be readily found.

The two methods are then equivalent, but for the fact that the Myklestadt-Prohl method is based on the lumped approach, and the FEM is based on the consistent approach, and each one of them can be obtained starting from the other. If lumped mass matrices are used in the context of the FEM, the same results that can be obtained from the Myklestadt-Prohl method are reached.

4.4 Real versus complex coordinates

The equations of motion of axi-symmetrical rotors have been obtained in the preceding sections with reference to a set of complex coordinates of the type defined by Equation (3.19). The general equation for an axi-symmetrical rotor is then

$$\mathbf{M}\ddot{\mathbf{q}}+(\mathbf{C}_n+\mathbf{C}_r-i\Omega\mathbf{G})\dot{\mathbf{q}}+(\mathbf{K}+\mathbf{K}_\Omega\Omega^2-i\Omega\mathbf{C}_r)\mathbf{q}=\Omega^2\mathbf{f}_r e^{i\Omega t}+\mathbf{f}_n(t) , \qquad (4.73)$$

where the term $\mathbf{K}_\Omega\Omega^2$ has been explicitly added to account for centrifugal stiffening, which can be present when also the compliance of disks and blades is considered. $\mathbf{f}_r$ and $\mathbf{f}_n(t)$ are, respectively, the force caused by unbalance and a generic force nonrotating force vector, function of time.

The use of complex coordinates is very expedient in the case of axi-symmetrical systems, particularly if damping is neglected, because it allows the study of the system using a model whose size is just half the size of the same problem expressed in real coordinates. Note that even if the coordinates are complex, all relevant matrices are real and the computation of the critical speeds, the undamped Campbell diagram, and the undamped unbalance response does not involve actual working with complex numbers, as it will be shown in detail in the following sections.

Another advantage of the use of complex coordinates is that of obtaining the response in terms of rotating vectors that rotate in the physical space, directly giving complete information about the orbits of the various nodes and the direction of the whirling motions. Using real coordinates, however, the rotating vectors are defined with reference to the Argand plane and the whirling motion is obtained as the compositions of harmonic vibrations in the directions of x- and y-axes. The main limitation of the complex coordinate approach is that it deals with difficulty with elliptical orbits, as those occurring in the case of non-axi-symmetrical machines, as will be seen in Chapter 6.

Equation (4.73) can be rewritten by separating the real part of each equation from the imaginary part, obtaining

$$\begin{bmatrix} \mathbf{M} & \mathbf{0} \\ \mathbf{0} & \mathbf{M} \end{bmatrix} \ddot{\mathbf{x}} + \left(\Omega \begin{bmatrix} \mathbf{0} & \mathbf{G} \\ -\mathbf{G} & \mathbf{0} \end{bmatrix} + \begin{bmatrix} \mathbf{C} & \mathbf{0} \\ \mathbf{0} & \mathbf{C} \end{bmatrix} \right) \dot{\mathbf{x}} + \tag{4.74}$$

$$\left(\begin{bmatrix} \mathbf{K} + \mathbf{K}_\Omega \Omega^2 & \mathbf{0} \\ \mathbf{0} & \mathbf{K} + \mathbf{K}_\Omega \Omega^2 \end{bmatrix} + \Omega \begin{bmatrix} \mathbf{0} & \mathbf{C}_r \\ -\mathbf{C}_r & \mathbf{0} \end{bmatrix} \right) \mathbf{x} =$$

$$= \Omega^2 \left[\mathbf{f}_{rc} \cos(\Omega t) + \mathbf{f}_{rs} \sin(\Omega t) \right] + \mathbf{f}_n(t) \, ,$$

where

$$\mathbf{x} = [\mathrm{Re}(\mathbf{q})^T, \mathrm{Im}(\mathbf{q})^T]^T \; , \quad \mathbf{C} = \mathbf{C}_n + \mathbf{C}_r \; ,$$

and

$$\mathbf{f}_{rc} = [\mathrm{Re}(\mathbf{f}_r)^T, \mathrm{Im}(\mathbf{f}_r)^T]^T \; , \quad \mathbf{f}_{rs} = [-\mathrm{Im}(\mathbf{f}_r)^T, \mathrm{Re}(\mathbf{f}_r)^T]^T$$

are the cosine and sine components of the unbalance. Vector $\mathbf{f}_n(t)$ is similar to the vector with the same name in the complex coordinates approach, but it contains the x- and y-components of the forces instead being based on the complex representation.

Note that in the case of complex coordinates, all matrices were symmetrical, but the use of real coordinates results in skew-symmetrical gyroscopic and rotating damping (circulatory) matrices. The real coordinates $\mathbf{x}$ defined earlier differ from the standard real coordinates used for the study of nonrotating systems for what the sign of the rotational degrees of freedom related to rotation about the x-axis is concerned. In many cases, a generalized form of Equation (4.74) is used, in which the sign conventions for rotational coordinates are the standard ones used in the FEM. The equation of motion of the damped system is then

$$\begin{bmatrix} \mathbf{M}_x & \mathbf{0} \\ \mathbf{0} & \mathbf{M}_y \end{bmatrix} \ddot{\mathbf{x}}^* + \left(\Omega \begin{bmatrix} \mathbf{0} & \mathbf{G}_{xy} \\ -\mathbf{G}_{yx} & \mathbf{0} \end{bmatrix} + \begin{bmatrix} \mathbf{C}_x & \mathbf{0} \\ \mathbf{0} & \mathbf{C}_y \end{bmatrix} \right) \dot{\mathbf{x}}^* +$$

$$+\left(\begin{bmatrix} \left(\mathbf{K}+\mathbf{K}_\Omega\Omega^2\right)_x & \mathbf{0} \\ \mathbf{0} & \left(\mathbf{K}+\mathbf{K}_\Omega\Omega^2\right)_y \end{bmatrix} + \Omega\begin{bmatrix} \mathbf{0} & \mathbf{C}_{r_{xy}} \\ -\mathbf{C}_{r_{yx}} & \mathbf{0} \end{bmatrix}\right)\mathbf{x}^* = \tag{4.75}$$

$$= \Omega^2 \left\{ \begin{array}{c} \mathbf{f}_x \cos(\Omega t) - \mathbf{f}_y \sin(\Omega t) \\ \mathbf{f}_x \sin(\Omega t) + \mathbf{f}_y \cos(\Omega t) \end{array} \right\},$$

where matrices with subscript x are the same as for complex coordinates, and matrices with subscript y are similar, except for the sign of elements with subscripts made by two numbers whose sum is an odd number. Matrices with subscripts xy and yx have signs that differ from those of the corresponding matrices in Equation (4.73) and are such that the global gyroscopic and rotating damping matrices are skew-symmetrical. Vector $\mathbf{x}^*$ is a vector containing the generalized coordinates and is of the type of the vector of the generalized coordinates of Equation (4.74).

The order of the degrees of freedom shown in Equations (4.74) and (4.75), with all coordinates related to the xz-plane listed one after the other, followed by those related to the other plane, is just an indication because it would lead to matrices with a very large bandwidth. Actually, the degrees of freedom are mixed at the element level and the structure of the matrices shown holds for the matrices of the elements and not for those related to the whole structure.

In the following section, the real coordinates equation will be written in the form

$$\mathbf{M}'\ddot{\mathbf{x}} + \left(\mathbf{C}' + \Omega\mathbf{G}'\right)\dot{\mathbf{x}} + \left(\mathbf{K}' + \mathbf{K}'_\Omega\Omega^2 + \Omega\mathbf{C}'_r\right)\mathbf{x} = \tag{4.76}$$

$$= \Omega^2\left[\mathbf{f}_{rc}\cos(\Omega t) + \mathbf{f}_{rs}\sin(\Omega t)\right] + \mathbf{f}_n(t)\,.$$

4.5 Fixed versus rotating coordinates

Although the flexural behavior of rotating axi-symmetrical systems can be studied using an inertial reference frame, as seen in the preceding sections, it is possible to use a set of coordinates based on the rotating frame O ΞHZ. A set of complex coordinates $\mathbf{r}$, related to the coordinates $\mathbf{q}$, can thus be obtained by the relationship

$$\mathbf{r} = \mathbf{q}e^{-i\Omega t}\,, \tag{4.77}$$

and the equation of motion (4.73) transforms into the following equation:

$$\mathbf{M}\ddot{\mathbf{r}} + \left[\mathbf{C}_n + \mathbf{C}_r + i\Omega(2\mathbf{M} - \mathbf{G})\right]\dot{\mathbf{r}} + \tag{4.78}$$

$$+\left[\mathbf{K} - \Omega^2(\mathbf{M} - \mathbf{G} - \mathbf{K}_\Omega) + i\Omega\mathbf{C}_n\right]\mathbf{r} = \Omega^2\mathbf{f}\,.$$

The same considerations seen for Equation (2.169), obtained for the Jeffcott model, also hold for Equation (4.78). Also in the case of rotating coordinates, it is possible to choose between the complex coordinates approach seen earlier and an alternative formulation of the problem based on real coordinates. The relevant equation can be easily obtained by separating the real and imaginary parts of Equation (4.78). The gyroscopic, Coriolis, and nonrotating damping terms give way to skew-symmetrical matrices. The use of rotating coordinates, which is not very convenient in the case of axi-symmetrical systems, becomes advisable when dealing with machines including a nonisotropic rotor (see Chapter 6).

4.6 Complex state-space equations

All preceding equations can be written with reference to the state space, yielding the usual state-space equation

$$\dot{\mathbf{z}} = \mathbf{A}\mathbf{z} + \mathbf{B}\mathbf{u}\,. \tag{4.79}$$

If complex coordinates are used, the state vector referred to an inertial reference frame is

$$\mathbf{z} = \left\{ \begin{array}{c} \dot{\mathbf{q}} \\ \mathbf{q} \end{array} \right\}. \tag{4.80}$$

The corresponding state matrix and product **Bu** caused by unbalance are, respectively,

$$\mathbf{A} = \left[\begin{array}{cc} -\mathbf{M}^{-1}\left(\mathbf{C}_n + \mathbf{C}_r - i\Omega\mathbf{G}\right) & -\mathbf{M}^{-1}\left(\mathbf{K} + \Omega^2\mathbf{K}_\Omega - i\Omega\mathbf{C}_r\right) \\ \mathbf{I} & \mathbf{0} \end{array} \right], \tag{4.81}$$

$$\mathbf{B}\mathbf{u} = \Omega^2 e^{i\Omega t} \left\{ \begin{array}{c} \mathbf{f} \\ \mathbf{0} \end{array} \right\}. \tag{4.82}$$

Note that the dynamic matrix is complex, because of the presence of the gyroscopic and rotating damping (circulatory) terms, and as a consequence, its eigenvalues are nonconjugated. When the roots locus, with the spin speed used as a parameter, is plotted, it is not symmetrical with respect to the real axis. This was also the case for the Jeffcott rotor, as shown in Figure 2.12, but in the current case, it also occurs for the undamped system owing to the gyroscopic term.

State-space equations can also be written with reference to real coordinates, in which case a real dynamic matrix is always obtained, or rotating coordinates.

4.7 Static solution

In this section, the term *static solution* will be referred to as the computation of the deformed shape and of the stress and strain fields of a rotor under the effects of constant forces. The relevant equation yielding the static displacement is easily obtained by adding a vector of static or nonrotating forces $\mathbf{f}_n$ to the right-hand side of Equation (4.73) and neglecting all terms containing the time derivatives of the generalized coordinates. There is no difficulty in introducing the hysteretic damping matrix $\mathbf{K}''_r$, i.e., the imaginary part of the complex stiffness matrix related to the rotating part of the system into the equation of motion. Following the same procedures as in Section 2.4.5, and remembering that in the case of a static solution the whirl speed ω vanishes, the relevant equation, which takes into account also centrifugal stiffening, is

$$\left(\mathbf{K} + \Omega^2\mathbf{K}_\Omega - i\Omega\mathbf{C}_r - i\mathbf{K}''_r\right)\mathbf{q} = \mathbf{f}_n. \tag{4.83}$$

Using the real coordinates approach and neglecting hysteretic damping, Equation (4.83) becomes

$$\left(\mathbf{K}' + \mathbf{K}'_\Omega\Omega^2 + \Omega\mathbf{C}'_r\right)\mathbf{x} = \mathbf{f}_n \ , \tag{4.84}$$

where matrix $\mathbf{C}'_r$ is skew symmetric. If hysteretic damping also was included, matrix $(\mathbf{K}''_r)'$ would have been skew symmetric.

Neglecting centrifugal stiffening, from Equation (4.83), it is clear that, in the case of an undamped rotor, the static solution of the rotating system coincides with that of the corresponding nonrotating structure, but when rotating damping is present, the deflected shape is influenced by the spin speed. This effect is easily understood if the force vector is assumed to be real, i.e., if all forces and moments act in the same plane, e.g., in the xz-plane, as in the case of a rotor loaded by self-weight. The presence of imaginary terms in the matrix of the coefficients of Equation (4.83) results in a complex displacement vector $\mathbf{q}$. The presence of damping then causes a lateral deviation of the deformed shape of the rotor from the plane in which all loads are acting. In the case of viscous damping, this deviation depends on the spin speed. Its measurement in a rotating-bending fatigue test has been suggested several times and was actually used to measure the internal damping of materials.

Using the real coordinates approach, the coupling between the two inflection planes is caused by the skew-symmetric rotating damping matrix.

4.8 Critical-speed computation

Critical speeds are usually computed with reference to the undamped system. They can be easily computed from the homogeneous equation of motion (4.73) in which all damping matrices have been neglected, yielding free whirling. By assuming a solution of the type $\mathbf{q} = \mathbf{q}_0 e^{i\Omega t}$ for the free circular whirling, the following algebraic homogeneous equation is obtained:

$$\left(-\Omega^2(\mathbf{M} - \mathbf{G} - \mathbf{K}_\Omega) + \mathbf{K} \right)\mathbf{q}_0 = \mathbf{0}\,. \tag{4.85}$$

The critical speeds can thus be computed by solving the eigenproblem

$$\det \left(-\Omega^2(\mathbf{M} - \mathbf{G} - \mathbf{K}_\Omega) + \mathbf{K} \right) = 0. \tag{4.86}$$

Note that the size of the eigenproblem, in which only real quantities are involved, is equal to the number of complex degrees of freedom. Mathematically, the problem is the same as for the computations of the natural frequencies of an undamped vibrating system whose mass matrix is $\mathbf{M} - \mathbf{G} - \mathbf{K}_\Omega$. The only difference is that in the current case, the mass matrix can be nonpositive definite. The eigenvectors in this case describe the shapes the system takes at the critical speeds. All of their properties, typical of vibrating systems, and in particular their orthogonality with respect to the mass and stiffness matrices, still hold, because they are related only to the symmetry of the relevant matrices. It is also possible to use the eigenvectors, which can be obtained from Equation (4.86) to perform a modal analysis and to obtain a diagonal stiffness and mass matrix, as described in Section 3.6 for systems with two complex degrees of freedom. However, if matrix $\mathbf{M} - \mathbf{G} - \mathbf{K}_\Omega$ is nonpositive definite, some of the modal masses are negative and the corresponding eigenvalues are imaginary.

Remark 4.4 *The effects of gyroscopic moments and centrifugal stiffening on the critical speeds are similar: Both increase their values, and if they are high enough, they make some of them disappear.*

Remark 4.5 *As rotating damping is destabilizing only in the supercritical field, they reduce the destabilizing effect of rotating damping or, better, increase the speed at which instability ranges eventually start. As it will be shown in Chapter 13, this effect is particularly strong in the case of bladed discs, and it prevents the damping of blades from causing instability.*

There are cases where forcing functions whose frequency is a multiple of the spin speed are present: As will be seen later, very often the excitation provided by constant forces on slightly nonsymmetrical rotors is accounted for by intersecting the Campbell diagram with the straight line $\omega = 2\Omega$, whereas the excitation caused by lubricated journal bearings is studied using the straight line $\omega = \Omega/2$. The secondary critical speeds occurring at the intersection of the generic straight line $\omega = h\Omega$ can be computed from the eigenproblem

$$\left(-\Omega^2(h^2\mathbf{M} - h\mathbf{G} - \mathbf{K}_\Omega) + \mathbf{K} \right)\mathbf{q}_0 = \mathbf{0} \,. \tag{4.87}$$

Also, in this case, the use of complex coordinates allows us to perform the modal analysis of the system using mode shapes that are k- and m-orthogonal.

If real coordinates are used, the equation yielding the critical speeds is

$$\det\left(-\Omega^2(\mathbf{M}' - i\mathbf{G}' - \mathbf{K}'_\Omega) + \mathbf{K}' \right) = \mathbf{0} \,, \tag{4.88}$$

or, for secondary critical speeds

$$\det\left(-\Omega^2(h^2\mathbf{M}' - ih\mathbf{G}' - \mathbf{K}'_\Omega) + \mathbf{K}' \right) = \mathbf{0} \,. \tag{4.89}$$

Remark 4.6 *Equations (4.86) and (4.88) are not exactly equivalent. The first one states the conditions for resonant forward synchronous whirling, whereas the second for resonant synchronous whirling, being immaterial whether it is forward or backward. Together with the true critical speeds, also resonant conditions for backward whirling are obtained.*

4.9 Computation of the unbalance response

The response to a generic unbalance distribution can be easily obtained by computing a particular integral of Equation (4.73), in the form $\mathbf{q} = \mathbf{q}_0 e^{i\Omega t}$. The algebraic equation yielding the response of the damped system is

$$\left(-\Omega^2(\mathbf{M} - \mathbf{G} - \mathbf{K}_\Omega) + i\Omega\mathbf{C}_n + \mathbf{K} \right)\mathbf{q}_0 = \Omega^2\mathbf{f}_r. \tag{4.90}$$

In the case of undamped systems, Equation (4.90) has no solution when the spin speed coincides with a critical speed and the matrix of the coefficients is singular. If vectors ϵ_i and angles χ_i are not all contained in the same plane, i.e., phases α_i and β_i are not all equal, vector $\mathbf{f}_r$ is complex and the solution $\mathbf{q}$ is also complex: Physically, this means that the deflected shape is a skew line. However, the matrix of the coefficients of Equation (4.90) remains real, if damping is not accounted for, and no actual working with complex numbers is needed: The real and imaginary parts of the solution can be obtained separately, solving two sets of real linear equations with the same matrix of the coefficients.

The computation of the response to an arbitrary unbalance distribution is formally similar to the computation of the response of a vibrating system excited by a harmonic forcing function, the differences being that the rotational speed Ω is used in the equation instead of the frequency ω, the excitation vector $\Omega^2\mathbf{f}_r$ is proportional to Ω^2, and the mass matrix of the system is $\mathbf{M} - \mathbf{G} - \mathbf{K}_\Omega$, which may be nonpositive definite. Some of its eigenvalues may be negative, in terms of Ω^2, and then some of the critical speed may be imaginary, as was seen in detail in Sections 3.3 and 3.6. The response in terms of displacement is more similar to an inertance than to a dynamic compliance.

When damping is not neglected, the response to unbalance is very similar to the response to harmonic excitation of a damped vibrating system, but only nonrotating damping enters the equation of motion. The matrix of the coefficients never becomes singular, because it has no real eigenvalues. It is, however, complex, and, if software for use with complex numbers is not available, the solution of Equation (4.90) involves the solution of a set of equations whose size must be doubled to work with real numbers: The number of equations is then equal to the number of real coordinates, and there is little computational advantage to using complex coordinates in this case.

When using real coordinates, the solution for the unbalance response can be assumed to be

$$\mathbf{x} = \mathbf{x}_{0c}\cos(\Omega t) + \mathbf{x}_{0s}\sin(\Omega t)\ . \tag{4.91}$$

By introducing this solution into Equation (4.76) and separating the sine and cosine terms, the following algebraic equation yielding the response is obtained:

$$\begin{bmatrix} \mathbf{K}'+\Omega\mathbf{C}'_r - \Omega_x^2\left(\mathbf{M}' - \mathbf{K}'_\Omega\right) & \Omega\left(\mathbf{C}' + \Omega\mathbf{G}'\right) \\ -\Omega\left(\mathbf{C}' + \Omega\mathbf{G}'\right) & \mathbf{K}' + \Omega\mathbf{C}'_r - \Omega_x^2\left(\mathbf{M}' - \mathbf{K}'_\Omega\right) \end{bmatrix} \left\{ \begin{array}{c} \mathbf{x}_{0c} \\ \mathbf{x}_{0s} \end{array} \right\} = \\ = \Omega^2 \left\{ \begin{array}{c} \mathbf{f}_{rc} \\ \mathbf{f}_{rs} \end{array} \right\}\ . \tag{4.92}$$

The response to unbalance of an axi-symmetrical system could also be studied using rotating coordinates. In this case, the response is nothing other than the static response to forces that are constant in both modulus and direction, and the critical speeds are a sort of elastic instability condition, in which the stiffness matrix of the system becomes singular

4.10 Plotting the Campbell diagram and the roots locus

To plot the Campbell diagram, it is necessary to compute the whirl frequencies of the system as functions of the spin speed. If the stability of the system has to be studied, damping must also be taken into account. This does not make the problem conceptually more complicated, but it makes actual computation much longer. In the case of viscous damping, the basic equation is the homogeneous equation associated with Equation (4.73)

$$\mathbf{M}\ddot{\mathbf{q}} + (\mathbf{C}_n + \mathbf{C}_r - i\Omega\mathbf{G})\dot{\mathbf{q}} + (\mathbf{K} + \Omega^2\mathbf{K}_\Omega - i\Omega\mathbf{C}_r)\mathbf{q} = \mathbf{0}\,. \tag{4.93}$$

The roots locus is directly obtained by computing the eigenvalues of the dynamic matrix (4.81); to take into account structural damping, it is sufficient to add the expression $\pm i\mathbf{K}''_n \pm i\mathbf{K}''_r$, where the double signs have the same meaning seen in Equation (2.116) and in Figure 2.20, to the stiffness matrix $\mathbf{K}$.

However, the Campbell diagram is often computed using a solution of the type $\mathbf{q} = \mathbf{q}_0 e^{i\omega t}$ instead of $\mathbf{q} = \mathbf{q}_0 e^{st}$. In this case, it is possible to find the complex whirl frequencies ω as eigenvalues of the matrix

$$\begin{bmatrix} \mathbf{M}^{-1}(\Omega\mathbf{G} + i\mathbf{C}_n + i\mathbf{C}_r) & \mathbf{M}^{-1}(\mathbf{K} + \Omega^2\mathbf{K}_\Omega \pm i\mathbf{K}''_n \pm i\mathbf{K}''_r - i\Omega\mathbf{C}_r) \\ \mathbf{I} & \mathbf{0} \end{bmatrix}. \tag{4.94}$$

Operating in this way, a real eigenproblem must be solved in the case of undamped systems.

If the Campbell diagram has to be plotted by scanning the $\omega\Omega$-plane using m values of the spin speed Ω, an eigenproblem of order $2n$, where n is the number of complex degrees of freedom, must be solved m times. The computation is then very time-consuming, and large-scale condensation may be necessary to keep computer time within reasonable limits.

In the case of damped systems, the solution of a complex eigenproblem requires a further doubling of the size of the relevant matrices, if software for use with complex algebra is not available. Moreover, if structural damping is present, several eigenproblems with the different signs included in Equation (4.94) must be solved for each value of the spin speed. A maximum of four eigenproblems must be solved, in the case where all forms of damping are present. After obtaining the solutions, there is no difficulty in checking where the real part of the various solutions is located in the Campbell diagram and then in choosing which eigenvalues are to be discarded, following the scheme of Figure 2.20. The computation of the damped Campbell diagram is then much heavier than that of the undamped diagram. In most cases, however, the presence of damping has little effect on the values of the whirl frequencies, while deeply affecting the decay rate. A good strategy is that of first studying the undamped system to locate the critical speeds and the general pattern of the whirl frequencies and then computing the complex whirl frequencies at some selected values of the speed, mainly in the supercritical range, where the occurrence of instability can be suspected.

In the case of undamped systems, a solution of the type

$$\mathbf{q} = \mathbf{q}_0 e^{i\omega t}$$

allows an eigenproblem with real matrices to be obtained while assuming that

$$\mathbf{q} = \mathbf{q}_0 e^{st}$$

the dynamic matrix turns out to be complex. On the contrary, if damping is not neglected, the two notations lead to similar complexities and the second solution can be used with advantage.

If the real-coordinates approach is used, the size of the problem is doubled and the information on the direction of the whirl motion included in the eigenvalues is lost: To distinguish between forward and backward modes, the eigenvectors must be studied.

Assuming a solution of the type

$$\mathbf{x} = \mathbf{x}_0 e^{st} \ ,$$

the eigenvalues of matrix

$$\begin{bmatrix} -\mathbf{M}'^{-1}\left(\mathbf{C}'+\Omega\mathbf{G}'\right) & -\mathbf{M}'^{-1}\left(\mathbf{K}' + \Omega^2\mathbf{K}'_\Omega + \Omega\mathbf{C}'_r\right) \\ \mathbf{I} & \mathbf{0} \end{bmatrix} \tag{4.95}$$

must be found.

As it is easily seen from Equation (4.94), the reasons for which the Campbell diagram of an axi-symmetrical rotor may be not *flat* (i.e., not made of straight lines parallel to the Ω-axis) are three: The presence of gyroscopic effect, of centrifugal stiffening, and of rotating damping. Centrifugal stiffening has the same effect for backward and forward whirling: an increase of the whirl speed. Gyroscopic effect causes the forward whirl speed to increase and the backward whirl speed to decrease (in absolute value). Rotating damping (within the small damping assumption) has usually a negligible effect on the whirl speed, and its effect on the decay rate is large.

Example 4.4 *Compute the critical speeds, the Campbell diagram, and the unbalance response of the rotor of Example 4.1 using the FEM.*

The rotor is modeled using 18 beam, 2 mass, and 2 spring elements, for a total of 36 degrees of freedom. An eccentricity of 1 μm of the compressor wheel has been assumed for the computation of the unbalance response.

Critical speeds. *The values of the first three critical speeds are reported in Table 4.7; the values earlier obtained using the Myklestadt-Prohl method are also reported for comparison. The difference between the results obtained using the Myklestadt method and the FEM is small, and it is mainly because in the latter, the consistent mass matrix has been used. The mode shapes corresponding to the first three critical speeds are reported in Figure 4.9. The first two are mainly rigid-body modes, a conical and a cylindrical one, whereas the third mode involves larger deformations of the shaft.*

Campbell diagram *The Campbell diagram of the system is reported in Figure 4.10. The rigid-body modes are not much affected by gyroscopic effects, because*

	Ω_{cr_I}	$\Omega_{cr_{II}}$	$\Omega_{cr_{III}}$
FEM	2,511	2,726	14,866
Myklestadt	2,500	2,718	14,920

TABLE 4.7. Values of the first three critical speeds, computed using the FEM and the Myklestadt-Prohl method.

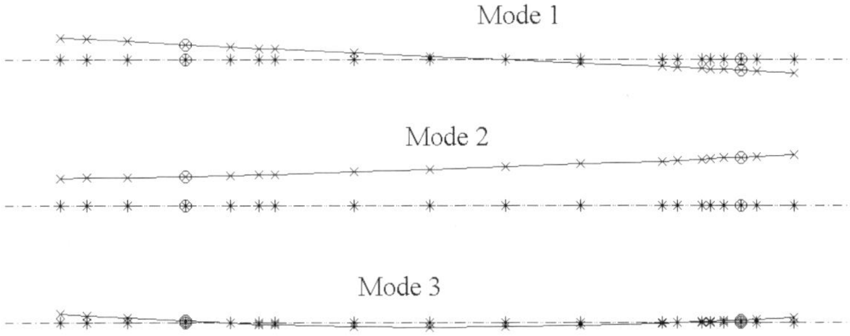

FIGURE 4.9. Mode shapes corresponding to the first three critical speeds.

the polar moment of inertia of the whole rotor is small if compared with the transversal moment of inertia; the third mode on the contrary is much more speed-dependent. This can be ascribed to the fact that the compressor and, above all, the turbine discs are much more similar to thin discs.

Unbalance response. *The undamped unbalance response is plotted, in amplitude and phase, in Figure 4.11. As damping was not taken into account, the values corresponding to the crossing of the critical speeds are larger than the actual ones. Note the large plateau extending for the whole working range of the machine (between 50,000 and 120,000 rpm), followed by an antiresonance.*

Example 4.5 *Consider the very simple model of a twin spool turbine [39] shown in Figure 4.12(a). The two rotors are shown separately in Figure 4.12(b).*

The model consists of 13 nodes (26 complex degrees of freedom), 11 timoshenko beam elements, 4 concentrated mass elements, and 4 spring elements to model the bearings, which are assumed to be isotropic.

The characteristics of the beam elements are shown in Table 4.8. The material is the same for both rotors and is characterized by $E = 206.9$ GN/m^2 *and* $\rho =$ 8304 kg/m^3. *The characteristics of the mass and spring elements are shown in Table 4.9. The mass elements are considered as thin discs, with* $J_p = 2J_t$.

The spin speed of the hig-pressure rotor is assumed to be 1.5 times that of the low-pressure rotor, which is assumed as a reference.

The critical speeds are readily computed, obtaining the following values 864 rad/s (8254 rpm), 1,602 rad/s (15,298 rpm), and 2,287 rad/s (21,839 rpm). These are are the critical speeds caused to the unbalance of the low-presssure

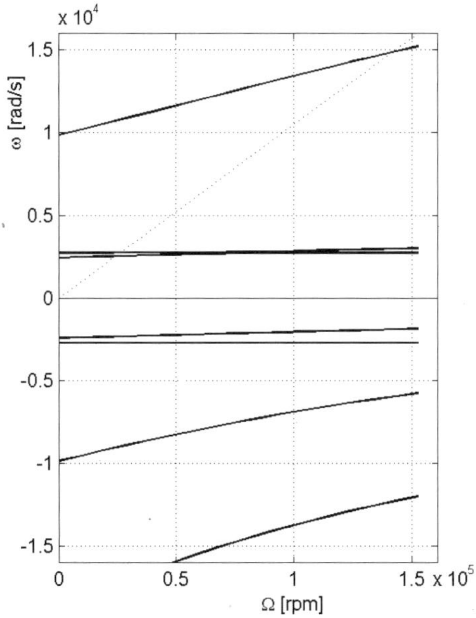

FIGURE 4.10. Campbell diagram.

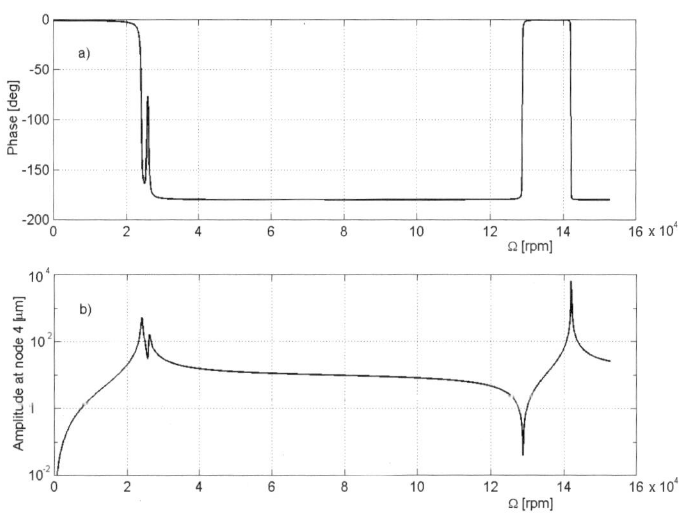

FIGURE 4.11. Response to a static unbalance caused by an eccentricity of 1 μm of the compressor wheel. The amplitude and phase of the orbit of the center of the compressor are plotted.

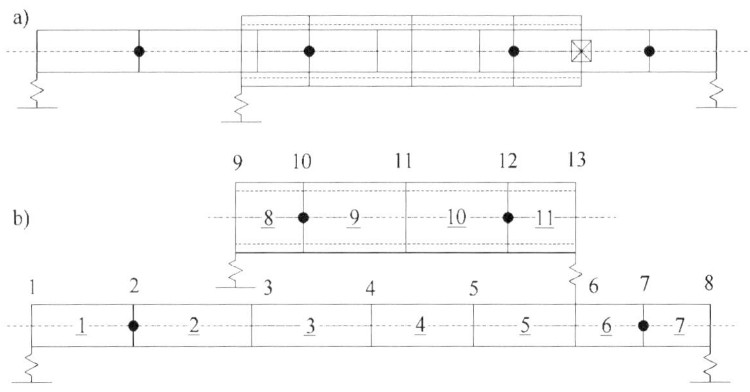

FIGURE 4.12. Simple FEM model of a twin spool turbine. (a) Assembled model; (b) models for the high-pressure and low-pressure rotors (underlined numbers designate elements).

Element #	1	2-3	4-5	6-7	8-11	9-10
ϕ_i [mm]	0	0	0	0	38.1	38.1
ϕ_0 [mm]	30.4	30.4	30.4	30.4	50.8	50.8
l [mm]	76.2	88.9	76.4	50.8	50.8	76.2

TABLE 4.8. Characteristics of the beam elements (ϕ_i: inner diameter; ϕ_o: outer diameter; l: length;).

rotor. The unbalance of the high-pressure rotor causes other critical speeds, which can be found by intersecting the Campbell diagram with the line $\omega = 1.5\Omega$: *548 rad/s (5,232 rpm), 1,054 rad/s (10,067 rpm), 1,517 rad/s (14,490 rpm) and 2,408 rad/s (23,000 rpm), again in terms of the spin speed of the low-pressure rotor. Note that only the critical speeds lower than 3,000 rad/s have been reported.*

The Campbell diagram is reported in Figure 4.13.

Mass #	1	2	3	4
Node #	2	7	10	12
m [g]	4.904	4.203	3.327	2.227
J_p [kgm^2]	0.02712	0.02034	0.01469	0.00972

Spring. #	1	2-3	4
Node #	1	8-9	from 6 to 13
k [MN/m]	26.2795	17.519	8.7598

TABLE 4.9. Characteristics of concentrated masses and springs (m: mass; J_t: diametral moment of inertia; k: stiffness).

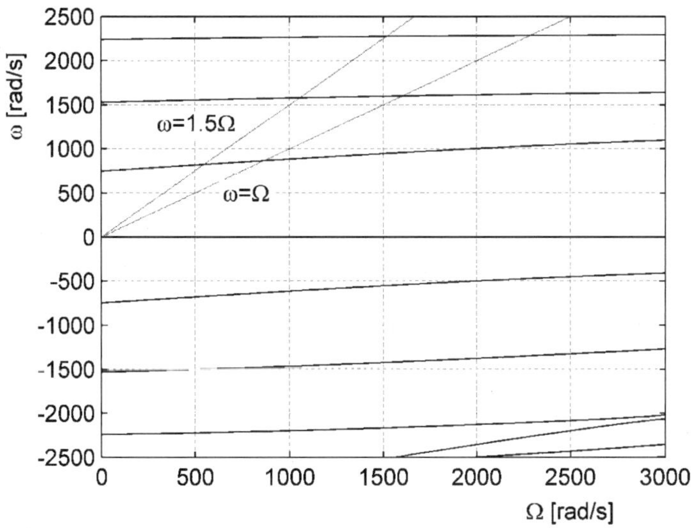

FIGURE 4.13. Campbell diagram of a twin spool turbine. Ω is the speed of the low-pressure rotor.

4.11 Reduction of the number of degrees of freedom

The FEM usually yields models of very large size: It is not uncommon to use models with thousands or tens of thousands of degrees of freedom. Although that does not constitute a problem for modern computers, when studying static problems, in dynamic analysis, the solution of an eigenproblem of that size can still be a formidable problem. This is particularly true in rotordynamics, when the Campbell diagram of a complex rotor is to be plotted and a large eigenproblem must be solved for many values of the spin speed Ω.

As the FEM is a displacement method, i.e., first solves the displacements and then computes stresses and strains as derivatives of the displacements, the precision with which displacements, and all other quantities directly linked with them, including mode shapes and natural frequencies, are obtained is far greater, for a given mesh, than that achievable for stresses and strains. Conversely, this means that the mesh needs to be much finer when solving the stress field, which is typical of static problems, than when searching for natural frequencies and mode shapes. Because it is often expedient to use the same mesh for both static and dynamic analysis, a reduction of the number of degrees of freedom for dynamic solution is useful, particularly when only a limited number of natural frequencies is required.

Two approaches can be used: reducing the size of the model or leaving the model as it is and using algorithms that search only the lowest natural frequencies. Although the two are more or less equivalent, the first leaves the choice of which degrees of freedom to retain to the user and the second operates automatically. As a consequence, a skilled operator can use advantageously reduction techniques, which allow us to obtain very good results with very few degrees of freedom. A general-purpose code for routine computations, which is sometimes used by analysts who are not very well trained, on the contrary, can advantageously use the second approach.

Remark 4.7 *Before computers were available, remarkable results were obtained using models with very few (often a single) degrees of freedom, but this required great computational ability and physical insight.*

4.11.1 Nodal reduction techniques

Static reduction

The term *static reduction* is used for a procedure to reduce the size of the stiffness matrix. It is based on the subdivision of the generalized coordinates $\mathbf{q}$ of the model into two types: master degrees of freedom $\mathbf{q}_1$ and slave degrees of freedom $\mathbf{q}_2$. When used to solve a static problem (or, better, a problem in which only the stiffness properties of the system are involved, because in the computation of the static deformation of a damped rotor also its damping properties enter the analysis), it yields exact results, i.e., the same results that would be obtained from the complete model.

The stiffness matrix and the nodal force vector can be partitioned following the distinction between master and slave coordinates, and the equation expressing the static problem becomes

$$\begin{bmatrix} \mathbf{K}_{11} & \mathbf{K}_{12} \\ \mathbf{K}_{21} & \mathbf{K}_{22} \end{bmatrix} \left\{ \begin{array}{c} \mathbf{q}_1 \\ \mathbf{q}_2 \end{array} \right\} = \left\{ \begin{array}{c} \mathbf{f}_1 \\ \mathbf{f}_2 \end{array} \right\} . \tag{4.96}$$

Note that matrices $\mathbf{K}_{11}$ and $\mathbf{K}_{22}$ are symmetrical, and $\mathbf{K}_{12} = \mathbf{K}_{21}^T$ are neither symmetrical nor square. Solving the second set of Equations (4.96) in $\mathbf{q}_2$, the following relationship linking the slave to the master coordinates is obtained:

$$\mathbf{q}_2 = -\mathbf{K}_{22}^{-1}\mathbf{K}_{21}\mathbf{q}_1 + \mathbf{K}_{22}^{-1}\mathbf{f}_2 . \tag{4.97}$$

Introducing Equation (4.97) into Equation (4.96), the latter yields

$$\mathbf{K}_{cond}\mathbf{q}_1 = \mathbf{f}_{cond} , \tag{4.98}$$

where

$$\begin{cases} \mathbf{K}_{cond} = \mathbf{K}_{11} - \mathbf{K}_{12}\mathbf{K}_{22}^{-1}\mathbf{K}_{12}^T , \\ \mathbf{f}_{cond} = \mathbf{f}_1 - \mathbf{K}_{12}\mathbf{K}_{22}^{-1}\mathbf{f}_2 . \end{cases}$$

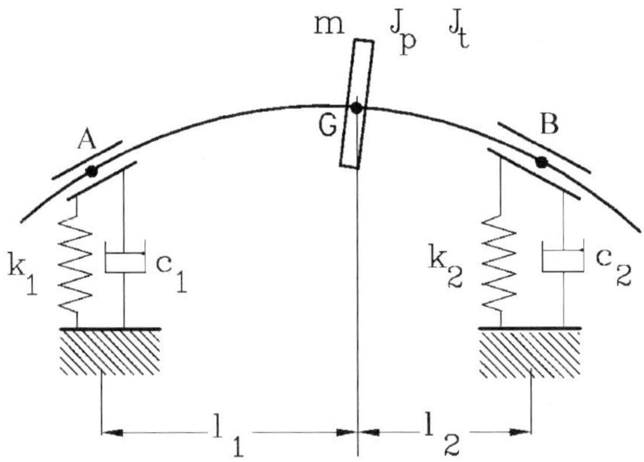

FIGURE 4.14. Rotor with four real degrees of freedom on compliant bearings.

Equation (4.98) yields the master generalized displacements $\mathbf{q}_1$. The slave displacements can be obtained directly from Equation (4.97) simply by multiplying some matrices.

The subdivision of the degrees of freedom between vectors $\mathbf{q}_1$ and $\mathbf{q}_2$ can be based on different criteria. The master degrees of freedom can simply be those in which the user is directly interested. Another type of choice can be that of physically subdividing the structure into two parts. The second practice, which can be generalized by subdividing the generalized coordinates into many subsets, is generally known as *solution by substructures* or *substructuring*.

Static reduction does not introduce any approximation into the model in the case of static analysis. It can be used also in dynamic analysis without introducing approximations only if no generalized inertia is associated with the slave degrees of freedom. In this case, static reduction is advisable because the mass matrix of the original system is singular and the condensation procedure allows removal of the singularity. Generally speaking, however, the mass matrix is not singular and it is not possible to just neglect the inertia linked with some degrees of freedom.

Example 4.6 *Rotor with four degrees of freedom on compliant bearings.*

Consider a rotor of the same type of that studied in Section 3.6.3, but with a compliant shaft and bearings that may include also damping (Figure 4.14). The system can be modeled using 12 real degrees of freedom or 6 complex ones, namely, the displacements and rotations of points A, G, and B. The generalized coordinates, expressed using the complex coordinates approach, are

$$\mathbf{q} = \begin{bmatrix} r_A & \phi_A & r_G & \phi_G & r_B & \phi_B \end{bmatrix}^T . \tag{4.99}$$

Undamped system. *The shaft can be modeled using two Euler-Bernoulli beam elements. The stiffness matrix of the first one is*

$$\mathbf{K}_1 = \frac{E_1 I_{y1}}{l_1^3}\begin{bmatrix} 12 & 6l_1 & -12 & 6l_1 \\ & 4l_1^2 & -6l_1 & 2l_1^2 \\ & & 12 & -6l_1 \\ & symm. & & 4l_1^2 \end{bmatrix} = \tag{4.100}$$

$$= \kappa_1 \begin{bmatrix} 4 & 2l_1 & -4 & 2l_1 \\ & \frac{4}{3}l_1^2 & -2l_1 & \frac{2}{3}l_1^2 \\ & & 4 & -2l_1 \\ & symm. & & \frac{4}{3}l_1^2 \end{bmatrix},$$

where $\kappa_1 = 3E_1 I_{y1}/l_1^3$*. A similar expression, with subscript 2 instead of 1, can be used for the second element. By assembling the two matrices and adding the constraints, the stiffness matrix of the whole system is*

$$\mathbf{K} = \begin{bmatrix} 4\kappa_1 + k_1 & 2\kappa_1 l_1 & -4\kappa_1 & 2\kappa_1 l_1 & 0 & 0 \\ & \frac{4}{3}\kappa_1 l_1^2 & -2\kappa_1 l_1 & \frac{2}{3}\kappa_1 l_1^2 & 0 & 0 \\ & & 4(\kappa_1 + \kappa_2) & 2(-l\kappa_1 + l\kappa_2) & -4\kappa_2 & 2\kappa_2 l_2 \\ & & & \frac{4}{3}(\kappa_1 l_1^2 + \kappa_2 l_2^2) & -2\kappa_2 l_2 & \frac{2}{3}\kappa_2 l_2^2 \\ & & & & 4\kappa_2 + k_2 & -2\kappa_2 l_2 \\ & symm. & & & & \frac{4}{3}\kappa_2 l_2^2 \end{bmatrix}. \tag{4.101}$$

As the shaft and the bearings are assumed to be massless, the mass and gyroscopic matrices are

$$\mathbf{M} = \begin{bmatrix} 0 & 0 & 0 & 0 & 0 & 0 \\ & 0 & 0 & 0 & 0 & 0 \\ & & m & 0 & 0 & 0 \\ & & & J_t & 0 & 0 \\ & & & & 0 & 0 \\ & symm. & & & & 0 \end{bmatrix}, \quad \mathbf{G} = \begin{bmatrix} 0 & 0 & 0 & 0 & 0 & 0 \\ & 0 & 0 & 0 & 0 & 0 \\ & & 0 & 0 & 0 & 0 \\ & & & J_p & 0 & 0 \\ & & & & 0 & 0 \\ & symm. & & & & 0 \end{bmatrix}. \tag{4.102}$$

A first reduction of the number of degrees of freedom can be performed by eliminating the rotational degrees of freedom at the supporting points ϕ_A *and* ϕ_B*. As the mass and gyroscopic matrices have vanishing rows and columns corresponding to them, static reduction can be used. It does not introduce any error, and only the condensation of the stiffness matrix is required. By subdividing the coordinates into master and slave*

$$\mathbf{q}_1 = \begin{bmatrix} r_A & r_G & \phi_G & r_B \end{bmatrix}^T, \quad \mathbf{q}_2 = \begin{bmatrix} \phi_A & \phi_B \end{bmatrix}^T, \tag{4.103}$$

the stiffness matrix can be partitioned

$$\mathbf{K}_{11} = \begin{bmatrix} 4\kappa_1 + k_1 & -4\kappa_1 & 2\kappa_1 l_1 & 0 \\ & 4(\kappa_1 + \kappa_2) & 2(-\kappa_1 l_1 + \kappa_2 l_2) & -4\kappa_2 \\ & & \frac{4}{3}(\kappa_1 l_1^2 + \kappa_2 l_2^2) & -2\kappa_2 l_2 \\ & symm. & & 4\kappa_2 + k_2 \end{bmatrix}, \tag{4.104}$$

$$\mathbf{K}_{21} = \mathbf{K}_{12}^T = \begin{bmatrix} 2\kappa_1 l_1 & -2\kappa_1 l_1 & \frac{2}{3}\kappa_1 l_1^2 & 0 \\ 0 & 2\kappa_2 l_2 & \frac{2}{3}\kappa_2 l_2^2 & -2\kappa_2 l_2 \end{bmatrix}, \tag{4.105}$$

$$\mathbf{K}_{22} = \begin{bmatrix} \frac{4}{3}\kappa_1 l_1^2 & 0 \\ 0 & \frac{4}{3}\kappa_2 l_2^2 \end{bmatrix} . \tag{4.106}$$

The reduced stiffness matrix is then

$$\mathbf{K}_{cond} = \mathbf{K}_{11} - \mathbf{K}_{12}\mathbf{K}_{22}^{-1}\mathbf{K}_{21} = \begin{bmatrix} \kappa_1 + k_1 & -\kappa_1 & \kappa_1 l_1 & 0 \\ & \kappa_1 + \kappa_2 & -l\kappa_1 + l\kappa_2 & -\kappa_2 \\ & & \kappa_1 l_1^2 + \kappa_2 l_2^2 & -\kappa_2 l_2 \\ & symm. & & \kappa_2 + k_2 \end{bmatrix} , \tag{4.107}$$

and the matrices expressing the inertia properties are

$$\mathbf{M}_{cond} = diag \left\{ 0 \quad m \quad J_t \quad 0 \right\} , \quad \mathbf{G}_{cond} = diag \left\{ 0 \quad 0 \quad J_p \quad 0 \right\} \tag{4.108}$$

As the system is undamped, no generalized forces caused by damping or inertia act on points A and B: In this case, a further static reduction procedure, aimed at eliminating the generalized coordinates r_A and r_B, can be performed without introducing any approximation into the model. Now master and slave coordinates are

$$\mathbf{q}'_1 = \left[r_G \quad \phi_G \right]^T , \quad \mathbf{q}'_2 = \left[r_A \quad r_B \right]^T . \tag{4.109}$$

The stiffness matrix can be partitioned

$$\mathbf{K}'_{11} = \begin{bmatrix} 4(\kappa_1 + \kappa_2) & 2(-l\kappa_1 + l\kappa_2) \\ symm. & \frac{4}{3}\left(\kappa_1 l_1^2 + \kappa_2 l_2^2\right) \end{bmatrix} , \tag{4.110}$$

$$\mathbf{K}'_{21} = \mathbf{K}'^T_{12} = \begin{bmatrix} -\kappa_1 & \kappa_1 l_1 \\ -\kappa_2 & -\kappa_2 l_2 \end{bmatrix} , \quad \mathbf{K}'_{22} = \begin{bmatrix} \kappa_1 + k_1 & 0 \\ 0 & \kappa_2 + k_2 \end{bmatrix} . \tag{4.111}$$

The reduced stiffness matrix is then

$$\mathbf{K}'_{cond} = \mathbf{K}'_{11} - \mathbf{K}'_{12}\mathbf{K}'^{-1}_{22}\mathbf{K}'_{21} = \begin{bmatrix} \frac{\kappa_1 k_1}{\kappa_1 + k_1} + \frac{\kappa_2 k_2}{\kappa_2 + k_2} & -\frac{l_1 \kappa_1 k_1}{\kappa_1 + k_1} + \frac{l_2 \kappa_2 k_2}{\kappa_2 + k_2} \\ symm. & \frac{l_1^2 \kappa_1 k_1}{\kappa_1 + k_1} + \frac{l_2^2 \kappa_2 k_2}{\kappa_2 + k_2} \end{bmatrix} , \tag{4.112}$$

and the matrices expressing the inertia properties are

$$\mathbf{M}_{cond} = diag \left\{ m \quad J_t \right\} , \quad \mathbf{G}_{cond} = diag \left\{ 0 \quad J_p \right\} . \tag{4.113}$$

Note that in this case the model reduces to a rotor with four (real) degrees of freedom, of the type studied in Chapter 3.

Symmetrical system. *In this case, the system is symmetrical, i.e., the two shafts and the two supports are equal, and the equations of motion uncouple*

$$\mathbf{K}'_{cond} = 2 \begin{bmatrix} \frac{\kappa k}{\kappa + k} & 0 \\ 0 & \frac{l^2 \kappa k}{\kappa + k} \end{bmatrix} . \tag{4.114}$$

For the translational degree of freedom (cylindrical mode), the shaft and the bearings behave like two springs, with stiffness κ and k in series, whereas for the *rotational degree of freedom (conical mode), they are like two torsional springs, with stiffness κl^2 and kl^2 in series.*

Damped system. *The first reduction to four complex (eight real) degrees of freedom can be performed in the usual way, whereas the second one, although being still possible as an approximation, introduces errors that may be large if the system is much damped. In particular, the damping at the bearings may be large enough to prevent from performing this second reduction. For this reason, reference will be made to the four complex degrees of freedom model expressed by Equations (4.107) and (4.108).*

The damping of the system must be subdivided into nonrotating and rotating damping. The former is easily modeled using a viscous damping matrix

$$\mathbf{C}_n = \begin{bmatrix} c_1 & 0 & 0 & 0 \\ 0 & 0 & 0 & 0 \\ 0 & 0 & 0 & 0 \\ 0 & 0 & 0 & c_2 \end{bmatrix}. \tag{4.115}$$

Rotating damping is best modeled in this case by using the hysteretic damping approach. If the loss factor η is constant, the rotating damping matrix is simply given by the loss factor multiplied by the stiffness matrix of Equation (4.107) from which the stiffness of the bearings has been removed

$$\mathbf{K}"_r = \eta \begin{bmatrix} \kappa_1 & -\kappa_1 & \kappa_1 l_1 & 0 \\ & \kappa_1 + \kappa_2 & -l\kappa_1 + l\kappa_2 & -\kappa_2 \\ & & \kappa_1 l_1^2 + \kappa_2 l_2^2 & -\kappa_2 l_2 \\ & symm. & & \kappa_2 \end{bmatrix}. \tag{4.116}$$

Guyan reduction

The so-called *Guyan reduction* is based on the assumption that the slave generalized displacements $\mathbf{q}_2$ can be computed directly from master displacements $\mathbf{q}_1$, neglecting inertia forces and external forces $\mathbf{f}_2$. In this case, Equation (4.97), without the last term, can also be used for dynamic solution. By partitioning the mass matrix in the same way seen for the stiffness matrix, the part of the kinetic energy of the rotor that depends only on the generalized velocities can be expressed as

$$\mathcal{T}_1 = \frac{1}{2} \left\{ \begin{array}{c} \dot{\mathbf{q}}_1 \\ -\mathbf{K}_{22}^{-1}\mathbf{K}_{21}\dot{\mathbf{q}}_1 \end{array} \right\}^T \begin{bmatrix} \mathbf{M}_{11} & \mathbf{M}_{12} \\ \mathbf{M}_{21} & \mathbf{M}_{22} \end{bmatrix} \left\{ \begin{array}{c} \dot{\mathbf{q}}_1 \\ -\mathbf{K}_{22}^{-1}\mathbf{K}_{21}\dot{\mathbf{q}}_1 \end{array} \right\}. \tag{4.117}$$

The kinetic energy is then

$$\mathcal{T} = \frac{1}{2}\dot{\mathbf{q}}_1^T \mathbf{M}_{cond}\dot{\mathbf{q}}_1, \tag{4.118}$$

where the condensed mass matrix is

$$\begin{aligned}\mathbf{M}_{cond} = \mathbf{M}_{11} - \mathbf{M}_{12}\mathbf{K}_{22}^{-1}\mathbf{K}_{12}^T - \left[\mathbf{M}_{12}\mathbf{K}_{22}^{-1}\mathbf{K}_{12}^T\right]^T + \\ +\mathbf{K}_{12}\mathbf{K}_{22}^{-1}\mathbf{M}_{22}\mathbf{K}_{22}^{-1}\mathbf{K}_{12}^T .\end{aligned} \tag{4.119}$$

By operating in the same way for the part of the kinetic energy of the rotor in which the products of generalized velocities and displacements are present, a reduced gyroscopic matrix can be defined

$$\begin{aligned}\mathbf{G}_{cond} = \mathbf{G}_{11} - \mathbf{G}_{12}\mathbf{K}_{22}^{-1}\mathbf{K}_{12}^{T} - \left[\mathbf{G}_{12}\mathbf{K}_{22}^{-1}\mathbf{K}_{12}^{T}\right]^{T} + \\ +\mathbf{K}_{12}\mathbf{K}_{22}^{-1}\mathbf{G}_{22}\mathbf{K}_{22}^{-1}\mathbf{K}_{12}^{T} .\end{aligned} \tag{4.120}$$

In a similar way, also the damping matrices can be reduced. By writing the expressions of the Rayleigh dissipation function, the condensed damping matrices can be obtained

$$\begin{aligned}\mathbf{C}_{cond} = \mathbf{C}_{11} - \mathbf{C}_{12}\mathbf{K}_{22}^{-1}\mathbf{K}_{12}^{T} - \left[\mathbf{C}_{12}\mathbf{K}_{22}^{-1}\mathbf{K}_{12}^{T}\right]^{T} + \\ +\mathbf{K}_{12}\mathbf{K}_{22}^{-1}\mathbf{C}_{22}\mathbf{K}_{22}^{-1}\mathbf{K}_{12}^{T} .\end{aligned} \tag{4.121}$$

Equation (4.121) holds separately both for the rotating and the nonrotating damping matrices; also, structural damping matrices can be reduced using Equation (4.119), in which $\mathbf{M}$ has been substituted with $\mathbf{K}''$.

Guyan reduction is not much more demanding from a computational viewpoint than static reduction because the only matrix inversion is that of $\mathbf{K}_{22}$, which has already been performed for the computation of the condensed stiffness matrix. In addition, if matrices $\mathbf{M}$ and $\mathbf{G}$ are diagonal, two of the terms of Equations (4.119) and (4.120) vanish.[2]

Although approximate, it introduces errors that are usually very small, at least if the choice of the slave degrees of freedom is appropriate. Inertia forces related to slave degrees of freedom are actually not neglected, but their contribution to the kinetic energy is computed from a deformed configuration obtained on the basis of the master degrees of freedom alone. If the relevant mode shapes are only slightly influenced by the presence of some of the generalized masses or if some parts of the structure are so stiff that their deflected shape can be determined by a few coordinates, the results can be very good, even when very few master degrees of freedom are used.

Also, the reduction of damping matrices introduces errors that depend on the choice of the slave degrees of freedom but are usually very small when the degrees of freedom in which viscous dampers are applied or, in the case of hysteretic damping, where the loss factor of the material changes, are not eliminated. Alternatively, these degrees of freedom can be neglected when their displacement is well determined by some master displacement, as in the case of very stiff parts of the structure.

[2]Here implicit reference is made to the complex coordinates approach: In real coordinates, $\mathbf{G}$ cannot be diagonal, because it is skew-symmetric.

	FEM				Myklestadt
Master degrees of freedom	2	10	19	All	
Ω_{cr_I} [rad/s]	2,512	2,511	2,511	2,511	2,500
$\Omega_{cr_{II}}$ [rad/s]	2,759	2,726	2,726	2,726	2,718
$\Omega_{cr_{III}}$ [rad/s]	–	14,942	14,872	14,866	14,940

TABLE 4.10. Values of the first three critical speeds, computed using the FEM (different condensation schemes) and the Myklestadt-Prohl method.

Example 4.7 *Repeat the computation of the critical speeds of the same turbojet rotor studied in Example 4.1, using different reduction schemes. Three condensation schemes for Guyan reduction are considered; a large-scale reduction in which only two degrees of freedom are retained (translations at nodes 4 and 17), and two more detailed schemes, in which 10 and 19 master degrees of freedom are considered. In the latter case, all translational degrees of freedom were retained. The results are reported in Table 4.10.*

Note that if a condensation scheme with only two master degrees of freedom is used, the model reduces to that of a rotor with four real (two complex) degrees of freedom. The corresponding reduced matrices are

$$\mathbf{K}_{cond} = \begin{bmatrix} 7.208 & 1.907 \\ 1.907 & 7.628 \end{bmatrix} \times 10^5, \quad \mathbf{M}_{cond} = \begin{bmatrix} 0.1091 & 0.0151 \\ 0.0151 & 0.1080 \end{bmatrix}, \tag{4.122}$$

$$\mathbf{G}_{cond} = \begin{bmatrix} 0.0036 & -0.0032 \\ -0.0032 & 0.0036 \end{bmatrix}.$$

These reduced matrices are referred to the displacements at nodes 4 and 17, i.e., at the locations of the compressor and turbine wheels. As the center of mass of the rotor is at 90.34 mm from the front end, the distances of nodes 4 and 17 from the center of mass are $a = 62.34$ *mm and* $b = 62.46$ *mm. If the rotor was a rigid body, the transformation matrix to pass from the coordinates of the center of mass (displacement and rotation) to those of nodes 4 and 17 was*

$$\mathbf{T} = \begin{bmatrix} 1 & -62.34 \\ 1 & 62.46 \end{bmatrix}. \tag{4.123}$$

The stiffness, mass, and gyroscopic matrices transformed using matrix $\mathbf{T}$ *are*

$$\mathbf{K}^* = \begin{bmatrix} 1865 & 2.727 \\ 2.727 & 4.292 \end{bmatrix} \times 10^3, \quad \mathbf{M}^* = \begin{bmatrix} 0.247 & -5.566 \times 10^{-5} \\ -5.566 \times 10^{-5} & 7.276 \times 10^{-4} \end{bmatrix}, \tag{4.124}$$

$$\mathbf{G}^* = \begin{bmatrix} 0.881 & -0.000370 \\ -0.000370 & 0.0525 \end{bmatrix} \times 10^{-3}.$$

The matrices obtained assuming that the rotor is rigid are

$$\mathbf{K}^* = \begin{bmatrix} 2000 & 2.318 \\ 2.318 & 4.421 \end{bmatrix} \times 10^3, \quad \mathbf{M}^* = \begin{bmatrix} 0.266 & 0 \\ 0 & 7.256 \times 10^{-4} \end{bmatrix}, \tag{4.125}$$

$$\mathbf{G}^* = \begin{bmatrix} 0 & 0 \\ 0 & 4.426 \times 10^{-3} \end{bmatrix} .$$

The fact that the matrices obtained using a rigid rotor model, at least for what the stiffness and mass matrices are concerned, are close to those expressed by Equation (4.124) shows that the first two modes are close to be rigid-body modes for the rotor, in which all of the flexibility is concentrated in the bearings.

Dynamic reduction

The reduction can be operated directly on the dynamic stiffness matrix (dynamic reduction). This procedure does not introduce approximations, but the frequency must be stated before performing it. It can then be useful for the computation of the response, but not for computing the natural frequencies or plotting the Campbell diagram, except if an iterative scheme, similar to that described when dealing with the Myklestadt-Prohl method, is used.

Dynamic reduction can be a powerful tool in the study of systems that are partly nonlinear, because it allows us to reduce the number of equations, eliminating all degrees of freedom not directly involved in the nonlinearities.

4.11.2 Modal reduction

Configuration space approach

In structural dynamics, a common way to reduce the order of the problem is that of resorting to modal coordinates and then considering only a reduced number of modes. Modal reduction is always approximated, but its accuracy is usually very good, even when only a reduced number of modes are considered, particularly when the modes are completely uncoupled, as it occurs when undamped systems are studied.

As already stated in Section 3.7, the eigenvectors of the undamped natural systems (i.e., the eigenvectors of matrix $\mathbf{M}^{-1}\mathbf{K}$) can be used to define the modal coordinates of any gyroscopic, damped (and circulatory) system. However, the modal equations are not uncoupled, because the gyroscopic and the damping matrices (and the centrifugal stiffening matrix, if present) are not diagonalized by the eigenvectors of the undamped natural systems. As a result of this coupling, the number of modes required to obtain a good result can be large. Note that if all modes are used, the results are exact, but the mathematical model obtained through modal reduction is as complex as the original one.

Example 4.8 *Repeat the computation of the critical speeds and of the Campbell diagram of the rotor of Example 4.1 by resorting to modal reduction with three modes.*

The FEM model of the rotor has 36 degrees of freedom. By computing the eigenvector matrix of the natural system, normalizing it in such a way that the

	Ω_{cr_I}	$\Omega_{cr_{II}}$	$\Omega_{cr_{III}}$
Modal (3 modes)	2,513	2,726	16,638
Modal (3 modes, uncoupled)	2,514	2,726	16,624
Non modal (36 d.o.f.)	2,511	2,726	14,866

TABLE 4.11. Values of the first three critical speeds, computed using the modal approach (three modes) and the complete FEM model.

modal masses have a unit value, and then using only the first three modes, the following modal matrices are obtained

$$\overline{\mathbf{M}} = \begin{bmatrix} 1 & 0 & 0 \\ 0 & 1 & 0 \\ 0 & 0 & 1 \end{bmatrix}, \quad \overline{\mathbf{K}} = \begin{bmatrix} 0.5866 & 0 & 0 \\ 0 & 0.7414 & 0 \\ 0 & 0 & 9.7145 \end{bmatrix} \times 10^7, \quad (4.126)$$

$$\overline{\mathbf{G}} = \begin{bmatrix} 0.0717 & 0.0093 & -0.0071 \\ 0.0093 & 0.0021 & 0.0225 \\ -0.0071 & 0.0225 & 0.6485 \end{bmatrix}.$$

An uncoupled model can be obtained by neglecting the out of diagonal elements of the gyroscopic matrix. In this case, the uncoupled model can be predicted to yield results that are not much different from those of the coupled model, because the system is not strongly gyroscopic.

Critical speeds. *The values of the first three critical speeds obtained from both the coupled and uncoupled models are reported in Table 4.11; the values earlier obtained from the complete model are also reported for comparison. Note that a very good approximation has been obtained for the first two critical speeds and a fair one for the third.*

Campbell diagram. *The Campbell diagram, limited to the first three modes, is reported in Figure 4.15. The lines related to the coupled and uncoupled models are completely superimposed. The results of the complete FEM model are also reported for comparison: It is clear that the frequencies of the first two modes are computed with good precision through the modal approach, whereas the approximation on the third forward mode is worse.*

State-space approach; complex coordinates

The same modal transformation based on a state-space complex coordinates representation seen in Section 3.7 for the system with four real degrees of freedom can be applied also to multi-degrees-of-freedom systems. The system with n complex degrees of freedom can be transformed into n uncoupled single-degree-of-freedom gyroscopic systems, each one with its modal mass, stiffness, gyroscopic term, and rotating and nonrotating damping. However, all of these modal parameters are speed dependent: This means that the modal transformation must be performed for all values of the speed the analyst wants to study. The relevant equations are reported in Section 3.7 and will not be repeated here.

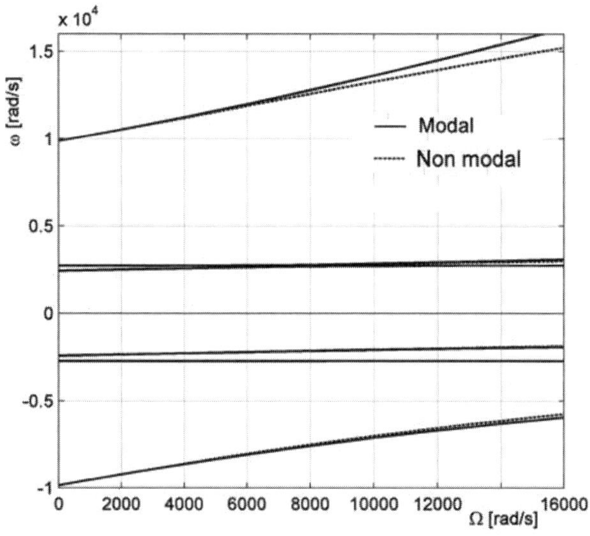

FIGURE 4.15. Campbell diagram, limited to the first three modes, computed using the modal and the nonmodal approach.

	M	G	K
Modal system #1	1	0.0546	5.156×10^6
Modal system #2	1	0.0168	8.026×10^6
Modal system #3	1	0.5970	8.885×10^7

TABLE 4.12. Values of the modal mass, stiffness and gyroscopic term for the first three modal systems at a speeds $\Omega = 15,000$ rad/s.

Example 4.9 *Find the modal parameters for the first three modes as functions of the speed for the rotor of Example 4.1.*

Owing to the size of the matrices, no intermediate result can be shown. The modal stiffness and gyroscopic terms for the first three modal systems are reported in Figure 4.16 (the modal masses have unit values). The values at a speed $\Omega = 15,000$ *rad/s are reported in Table 4.12; The Campbell diagram for the first three forward and backward modes is not reported, because it coincides with that of the complete system.*

State-space approach; real coordinates

If the system is undamped, the modal transformation based on a state-space real coordinates representation seen in Section 3.7 can also be applied. If the system has n complex coordinates, it has $2n$ real coordinates and the space state has dimension $4n$. At each speed, a modal analysis yielding the $4n$ modal masses (which can be set at unit value by normalizing the eigen-

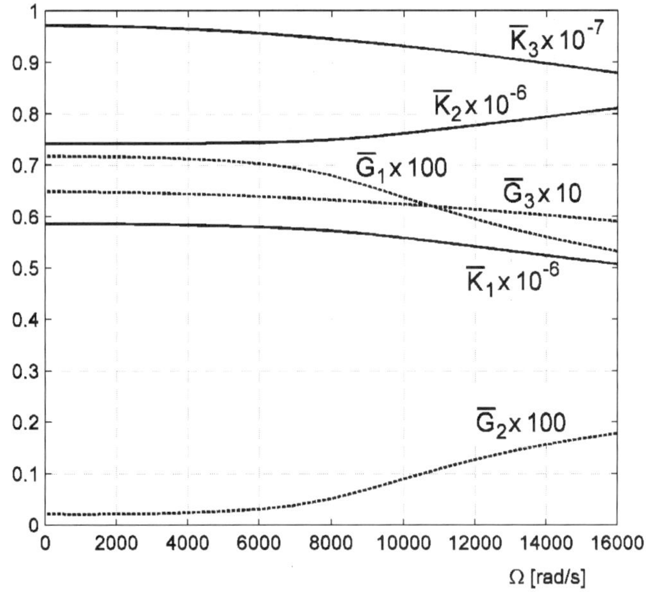

FIGURE 4.16. Modal stiffness and gyroscopic terms for the first three modal systems as functions of the speed.

vectors) and modal stiffness must be performed. Note that if the system is axially symmetrical, the modal systems are identical in pairs, i.e. for each mode is repeated with reference to xz and yz plane. Moreover, as stated in section 3.7, there is no possibility of distinguishing between forward and backward modes from the eigenvalues or the modal parameters. The relevant equations are identical to those reported in section 3.7 (except for the size of the matrices) and will not be reported here.

Example 4.10 *Find the modal parameters for the first three modes as functions of the speed for the rotor of Example 4.1, using the state-space real coordinates approach.*

The size of the matrices is fairly large: The system has 38 degrees of freedom in each plane, leading to a total of 76 real degrees of freedom. Matrices $\mathbf{M}'$, $\mathbf{K}'$, *and* $\mathbf{G}'$ *have then 76 rows and columns, whereas matrices* $\mathbf{M}^*$ *and* $\mathbf{K}^*$ *in Equation (3.100) have 152 rows and columns.*

By performing a modal analysis on matrices $\mathbf{M}^*$ *and* $\mathbf{K}^*$, *the speed-dependent modal parameters of the system can be found. The modal stiffness for the first 14 modal systems are reported in Figure 4.17 (the modal masses have unit values). Fourteen modal systems are needed to find the lowest seven natural frequencies (they come in pairs) corresponding to the first three forward and backward modes, plus the fourth backward mode whose frequency is lower than that of the third forward mode at high speed and must be included not to lose the latter. The values*

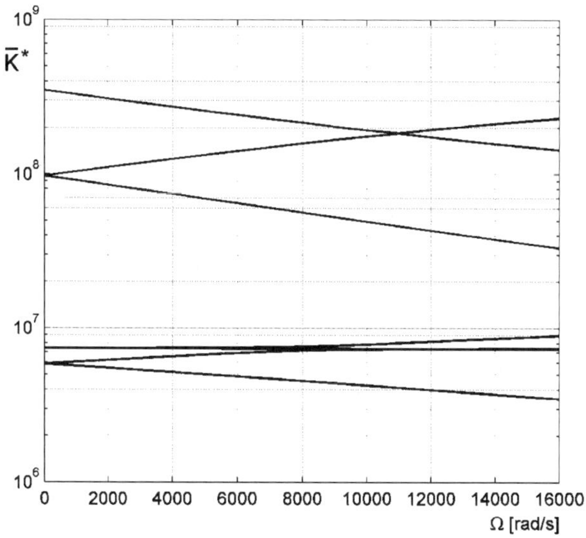

FIGURE 4.17. Modal stiffness $\overline{\mathbf{K}}^*$ for the first 14 modal sytems.

at a speed $\Omega = 15,000$ *rad/s are* 3.60×10^6, 7.34×10^6, 7.38×10^6, 8.77×10^6, 3.549×10^6, 1.505×10^8, *and* 2.224×10^8, *each one with multiplicity 2. They correspond to the squares of the whirl frequencies.*

4.11.3 Component mode synthesis

When substructuring is used, the degrees of freedom of each structure can be divided into two sets: internal degrees of freedom and boundary degrees of freedom. The latter are all degrees of freedom that the substructure has in common with other parts of the structure. They are often referred to as *constraint degrees of freedom* because they express how the substructure is constrained to the rest of the system. *Internal degrees of freedom* are those belonging only to the relevant substructure. The largest possible reduction scheme is that in which all internal degrees of freedom are considered as slave coordinates and all boundary degrees of freedom are considered as master coordinates. In this way, however, all modes in which the motion of the internal points of the substructure is important with respect to the motion of its boundary are approximated in a rough way.

A simple way to avoid this drawback is to also consider as master coordinates, together with the boundary degrees of freedom, some of the modal coordinates of the substructure constrained at its boundary. This procedure would obviously lead to exact results if all modes were retained, but because the total number of modes is equal to the number of internal degrees of freedom, the model obtained has as many degrees of freedom as the original one. As usual with modal practices, the computational advantages grow with the number of modes that can be neglected. The relevant matrices are partitioned as seen for reduction techniques, with subscript 1 referring to the boundary degrees of freedom and subscript 2 to the internal degrees of freedom. The displacement vector $\mathbf{q}_2$ can be assumed to be equal to the sum of the constrained modes $\mathbf{q}_2'$, i.e., the deformation pattern caused by the displacements $\mathbf{q}_1$ when no force acts on the substructure, plus the constrained normal modes $\mathbf{q}_2''$, i.e., the natural modes of free vibration of the substructure when the boundary generalized displacements $\mathbf{q}_1$ are equal to zero.

The constrained modes $\mathbf{q}_2'$ can be expressed by Equation (4.97) once the force vector $\mathbf{f}_2$ is set equal to zero.

When the present reduction technique is applied to nonrotating structures, there is no problem in computing the natural modes of vibration of the relevant substructures. In rotordynamics, the modes computed at the relevant velocity should be introduced, but this causes further difficulties. The reduction should be repeated for each value of the spin speed considered in the computations, and the modes are more difficult to define (see Section 4.4.3.). As a result, a further approximation is introduced, by using the modes of the undamped substructures computed at standstill, i.e., the modes of the undamped, nongyroscopic system. The constrained normal modes can thus be computed by solving the eigenproblem

$$\left(-\omega^2\mathbf{M}_{22} + \mathbf{K}_{22}\right)\mathbf{q}_2'' = \mathbf{0}\,. \tag{4.127}$$

A similar consideration holds also in the case in which the stiffness matrix is itself a function of the speed, as it occurs when centrifugal stiffening is present.

This approximation is acceptable, provided that the gyroscopic moments acting on the substructure do not influence greatly its behavior at the speed of interest or that a larger number of constrained normal modes are used. Clearly, an exact solution is at any rate reached if all of them are retained, but the usefulness of the reduction technique vanishes.

Once the eigenproblem has been solved, the matrix of the eigenvectors $\mathbf{\Phi}$ can be used to perform the modal transformation $\mathbf{q}_2'' = \mathbf{\Phi}\boldsymbol{\eta}_2$. The generalized coordinates of the substructure can thus be expressed as

$$\begin{gathered}\left\{\begin{array}{c}\mathbf{q}_1\\ \mathbf{q}_2\end{array}\right\} = \left\{\begin{array}{c}\mathbf{q}_1\\ -\mathbf{K}_{22}^{-1}\mathbf{K}_{21}\mathbf{q}_1 + \mathbf{\Phi}\boldsymbol{\eta}_2\end{array}\right\} = \\ = \left[\begin{array}{cc}\mathbf{I} & \mathbf{0}\\ -\mathbf{K}_{22}^{-1}\mathbf{K}_{21} & \mathbf{\Phi}\end{array}\right]\left\{\begin{array}{c}\mathbf{q}_1\\ \boldsymbol{\eta}_2\end{array}\right\} = \mathbf{\Psi}\left\{\begin{array}{c}\mathbf{q}_1\\ \boldsymbol{\eta}_2\end{array}\right\}.\end{gathered} \tag{4.128}$$

Equation (4.128) defines a coordinate transformation, allowing the expression of the deformation of the internal part of the substructure in terms of constrained and normal modes. Matrix $\mathbf{\Psi}$ expressing this transformation can be used to compute the new mass, gyroscopic, stiffness, and, where needed, damping matrices and force vector

$$\begin{array}{c}\mathbf{M}^* = \mathbf{\Psi}^T\mathbf{M}\mathbf{\Psi}\,, \qquad \mathbf{G}^* = \mathbf{\Psi}^T\mathbf{G}\mathbf{\Psi}\,, \qquad \mathbf{K}^* = \mathbf{\Psi}^T\mathbf{K}\mathbf{\Psi}\,, \\ \mathbf{C}^* = \mathbf{\Psi}^T\mathbf{C}\mathbf{\Psi}\,, \qquad \mathbf{f}^* = \mathbf{\Psi}^T\mathbf{f}\,. \end{array} \tag{4.129}$$

If there are m constrained coordinates and n internal coordinates and if only k constrained normal modes are considered ($k < n$), then the size of the original matrices $\mathbf{M}$, $\mathbf{G}$, ... is $m+n$, whereas that of matrices $\mathbf{M}^*$, $\mathbf{G}^*$, ... is $m + k$.

Remark 4.8 *It is easy to verify that if $k = 0$, i.e., if no constrained normal modes are considered, this formulation reduces to Guyan reduction.*

Once the coordinate transformation of the substructures has been performed, they can be assembled in the same way already seen for elements: The boundary coordinates belong to a number of substructures and are actually assembled whereas the modal coordinates are typical of only one substructure at a time, in the same way as the coordinates of internal nodes, when elements with such nodes are used. Actually each substructure can be regarded as a large element, sometimes referred to as a *superelement*, and the relevant procedures do not differ from those that are standard in the FEM.

The main advantage of substructuring is that of allowing the construction of the model and the analysis of the various parts of a large structure in an independent way. In particular, in rotordynamics, it can be very expedient to study separately the stator and the rotor of the machine, and then to connect them at the bearing locations or, in the case of multishaft machines, to study the various rotors separately. In a sense, this way of reducing the size of the model is ideally suited for rotordynamic analysis as rotating machinery are usually made of different parts connected with each other in a very limited number of locations, the bearings and the dampers, if present.

The results can then be assembled, and the behavior of the structure can be assessed from that of its parts. If this is done, however, the connecting nodes must be defined in such a way that the same boundary degrees of freedom are considered in the analysis of the various parts. It is, however, possible to use algorithms allowing us to connect otherwise incompatible meshes.

All of the methods discussed in this section, which are closely related to each other and are found in the literature in a variety of versions, are general for discrete systems and can be used outside the FEM even if they became popular only with the use of the latter, owing to the large number of degrees of freedom.

	Ω_{cr_I}	$\Omega_{cr_{II}}$	$\Omega_{cr_{III}}$
Comp. mode synth. (1 mode)	2,527	2,737	17,797
Comp. mode synth. (2 modes)	2,512	2,734	17,520
Comp. mode synth. (3 modes)	2,512	2,727	15,014
Non modal (36 d.o.f.)	2,511	2,726	14,866

TABLE 4.13. Values of the first three critical speeds, computed using the component modes synthesis approach with one, two, and three modes (three, four, and five degrees of freedom) and the complete FEM model.

Example 4.11 *Consider the rotor of Example 4.1 as made by two substructures, the rotor and the bearings.*

The stiffness, mass, and gyroscopic matrices are built in the same way as for the previous examples, with the difference being that now the stiffness matrix is computed without taking into account the bearings. The stiffness matrix is then singular. The bearings are located at nodes 6 and 13, so the displacements at these nodes are taken as master degrees of freedom. All matrices are accordingly partitioned: Matrices with subscript 11 have two rows and two columns, those with subscript 12 have two rows and 36 columns, and matrices with subscript 22 have 36 rows and columns.

The eigenproblem related to matrices $\mathbf{M}_{22}$ *and* $\mathbf{K}_{22}$ *is then solved, yielding the 36 natural frequencies and mode shapes of the rotor with locked bearings. The transformation matrix* $\mathbf{\Psi}$ *is then built.*

Consider first the case in which only the first modal coordinate is used as master degree of freedom, in addition to the displacements at the bearings. Matrix $\mathbf{\Psi}$ *has then 38 rows and 3 columns. The reduced matrices of the rotor are then*

$$\mathbf{K}^* = \begin{bmatrix} 0 & 0 & 0 \\ 0 & 0 & 0 \\ 0 & 0 & 7.4584 \end{bmatrix} \times 10^5, \quad \mathbf{M}^* = \begin{bmatrix} 0.1519 & -0.0157 & 0.0192 \\ -0.0157 & 0.1453 & -0.0014 \\ 0.0192 & -0.0014 & 0.0096 \end{bmatrix},$$

$$\mathbf{G}^* = \begin{bmatrix} 0.005009 & -0.005009 & 0.001650 \\ -0.005009 & 0.005009 & -0.001650 \\ 0.001650 & -0.001650 & 0.005078 \end{bmatrix}.$$

To assemble the bearing to the structure, a stiffness of 10^6 *N/m is added to the terms* $\mathbf{K}^*_{11}$ *and* $\mathbf{K}^*_{22}$. *The values of the three critical speeds obtained from the reduced model are reported together with those obtained from the complete model in Table 4.13. The Campbell diagram obtained from the same model is shown in Figure 4.18. Clearly, this approximation is fair, but the approximation in the computation of the third mode is hardly acceptable.*

The computation was repeated keeping two and thee modal coordinates of the rotor (a total of four and five degrees of freedom). The results have been reported in Table 4.13 and Figure 4.18.

Example 4.12 *Compute the Campbell diagram and the critical speeds of the twin spool turbine of Example 4.5 using the component mode synthesis approach.*

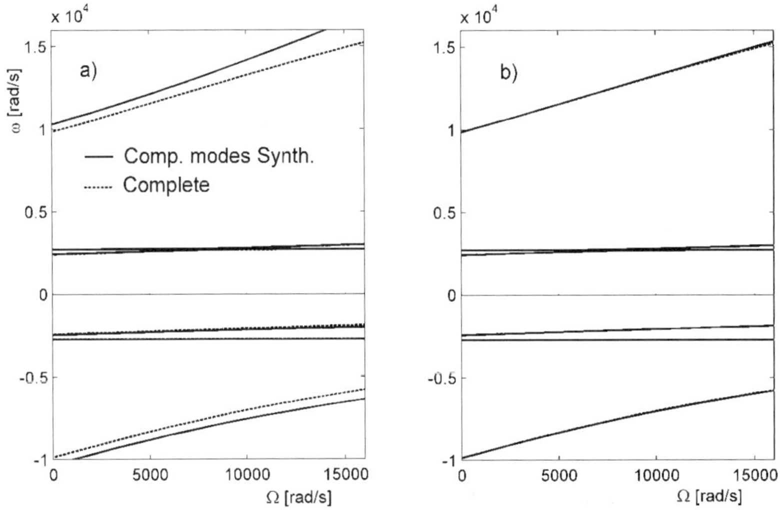

FIGURE 4.18. Campbell diagram obtained using the component mode synthesis approach with (a) one mode and (b) three modes. Also, the plot obtained using the complete model has been reported.

Assume the low-pressure rotor as substructure 1 and the high-pressure rotor as substructure 2. The first has a total of 16 degrees of freedom, 3 of which are taken as master coordinates (displacements at nodes 1, 6, and 8) in the modal reduction. The second substructure has a total of 10 degrees of freedom, 2 of which are taken as master coordinates (displacements at nodes 9 and 13).

Assuming the modal coordinate of the first deformation mode of each substructure as a further degree of freedom, the model has a total of 7 (4 + 3) degrees of freedom. The computations of the gyroscopic matrices of the substructures are performed by referring to the spin speed of each one of them.

The matrices related to the two substructures (4×4 and 3×3 matrices) are assembled to obtain 7×7 matrices. At this point, the gyroscopic matrix of the second substructure is multiplied by 1.5.

The springs simulating the bearings are then added. Those attached at nodes 2, 8, and 9 are simply accounted for by adding the relevant stiffness on the main diagonal of the stiffness matrix. The spring simulating the intershaft bearing is

	Ω_{cr_1}	Ω_{cr_2}	Ω_{cr_3}	Ω_{cr_4}	Ω_{cr_5}	Ω_{cr_6}	Ω_{cr_7}
Comp. mode synth.	549	865	1,057	1,539	1.603	2.287	2,323
Non modal	548	864	1,054	1,517	1,602	2,287	2,408

TABLE 4.14. Values of the first seven critical speeds (in rad/s), computed using the component modes synthesis approach with one mode for each substructure (seven degrees of freedom) and the complete FEM model (26 degrees of freedom).

then added between the two substructures (nodes 6 and 13). This element is the only element that supplies a coupling between the two parts of the stiffness matrix, following the physical configuration of the system. The mass and gyroscopic matrices are made by two uncoupled parts.

The results in terms of the first seven critical speeds are compared with those obtained in Example 4.5 from the complete model in Table 4.14. Both the critical speeds caused by resonance with the low-pressure and the high-pressure rotors are listed.

5

Continuous systems: Transmission shafts

Transmission shafts are often easily modeled as beams rotating about their axis. If such beams are prismatic and homogenous, i.e., their geometric and material properties are constant along their length, it is possible to resort to continuous models, based on the beam theory. Moreover, if they are slender, it is possible to neglect both shear deformation and the moments of inertia of their cross section. In this case, usually referred to as Euler-Bernoulli beam, there is no gyroscopic effect and the Campbell diagram is made of straight lines. The behavior of the rotating system can be deduced from that at standstill, and the usual results typical of vibrating slender beams are used. In most catalogs of tubular transmission shafts with Hook joints, for instance, a formula for the computation of the critical speed is reported that coincides with the well-known expression yielding the first natural frequency of a simply supported Euler-Bernoulli beam.

The dynamic behavior of a slender nonrotating beam is analyzed in the following sections, together with a more complete analysis of the behavior of a beam rotating about its axis.

5.1 The Euler-Bernoulli vibrating beam

The study of the elastic behavior of beams dates back to Galileo, with important contributions by Daniel Bernoulli, Euler, De Saint Venant, and many others. A *beam* is essentially an elastic solid in which one dimension is prevalent over the others. Often the beam is prismatic (i.e., the cross

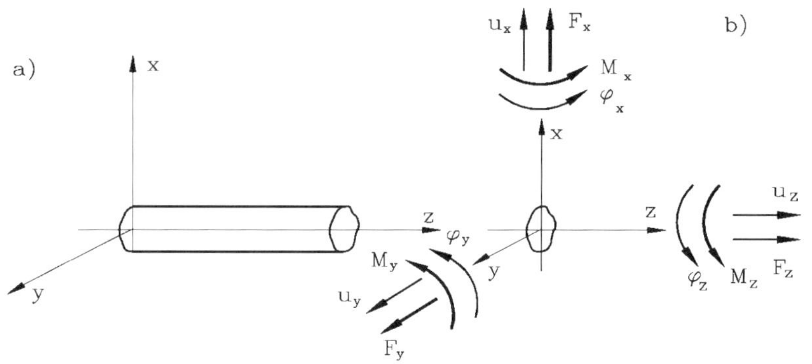

FIGURE 5.1. Straight, prysmatic and untwisted beam. a): Reference frame; b): generalized coordinates and generalized forces for the generic cross section.

sections are all equal), homogeneous (i.e., with constant material characteristics), straight (i.e., its axis is a part of a straight line), and untwisted (i.e., the principal axes of elasticity of all sections are equally directed in space). The unidimensional nature of beams allows simplification of the study: Each cross section is considered as a rigid body whose thickness in the axial direction is vanishingly small; it has six degrees of freedom, three translational and three rotational. The problem is then reduced to a unidimensional problem, in the sense that a single coordinate, namely, the axial coordinate, is required.

Setting the z-axis of the reference frame along the axis of the beam (Figure 5.1), the six generalized coordinates of each cross section are the axial displacement u_z, the lateral displacements u_x and u_y, the torsional rotation ϕ_z about the z-axis, and the flexural rotations ϕ_x and ϕ_y about axes x and y. Displacements and rotations are assumed to be small, so that rotations can be regarded as vector quantities, which simplifies all rotation matrices by linearizing trigonometric functions. The three rotations will then be considered as components of a vector in the same way as the three displacements are components of vector $\mathbf{u}$. The generalized forces acting on each cross section and corresponding to the six degrees of freedom defined earlier are the axial force F_z, shear forces F_x and F_y, the torsional moment M_z about the z-axis, and bending moments M_x and M_y about x- and y-axes.

From the aforementioned assumptions, it follows that all normal stresses in directions other than z (σ_x and σ_y) are assumed to be small enough to be neglected. When geometric and material parameters are not constant along the axis they must change at a sufficiently slow rate in order not to induce stresses σ_x and σ_y, which cannot be considered in this model. If the axis of the beam is assumed to be straight, the axial translation is uncoupled from the other degrees of freedom, at least in first approximation. If the area

Type of behavior	Degrees of freedom	Generalized forces
Axial	Displacement u_z	Axial force F_z
Torsional	Rotation ϕ_z	Torsional moment M_z
Flexural (xz-plane)	Displacement u_x Rotation ϕ_y	Shearing force F_x Bending moment M_y
Flexural (yz-plane)	Displacement u_y Rotation ϕ_x	Shearing force F_y Bending moment M_x

TABLE 5.1. Generalized coordinates and forces of a generic cross section of a beam. Uncoupling axial, torsional, and flexural behavior.

center of all cross sections coincides with their shear center, which happens if all cross sections have two perpendicular planes of symmetry, also the torsional-rotation degree of freedom is uncoupled from the others. If the planes of symmetry of all sections are equally oriented (the beam is not twisted) and x- and y-axes are perpendicular to such planes, the flexural behavior in xz-plane is uncoupled from that in yz-plane (Table 5.1).

Consider the flexural behavior in xz-plane and assume that both shear deformation and rotational inertia of the cross sections are negligible compared with bending deformation and translational inertia, respectively. The latter assumption allows us to consider each cross section as a point mass instead of a rigid body whose axial thickness is vanishingly small. This assumption leads to a good approximation if the beam is very slender, i.e., if the thickness in the x-direction is much smaller than length l. Note that, at any rate, the thickness must be small enough to use beam theory.

The relevant degrees of freedom are displacement u_x and rotation ϕ_y.

The inertia force acting on the length dz of the beam is (Figure 5.2)

$$dm\frac{d^2u_x}{dt^2} = \rho A\frac{d^2u_x}{dt^2}dz ,$$

and the resultant of the forces caused by the other parts of the beam is

$$\left(F_x + \frac{1}{2}\frac{\partial F_x}{\partial z}dz\right) - \left(F_x - \frac{1}{2}\frac{\partial F_x}{\partial z}dz\right) = \frac{\partial F_x}{\partial z}dz .$$

The equilibrium equation for translations in x-direction of the length dz of the beam is then

$$\rho A\frac{d^2u_x}{dt^2}dz = \frac{\partial F_x}{\partial z}dz + f_x(z,t)dz , \tag{5.1}$$

where $f_x(z,t)dz$ is a generic external force acting in x-direction.

If the rotational inertia of the length dz of the beam is neglected, and no distributed bending moment acts on the beam, the equilibrium equation for rotations about y-axis of the length dz of the beam is

$$F_xdz + \frac{\partial M_y}{\partial z}dz = 0 . \tag{5.2}$$

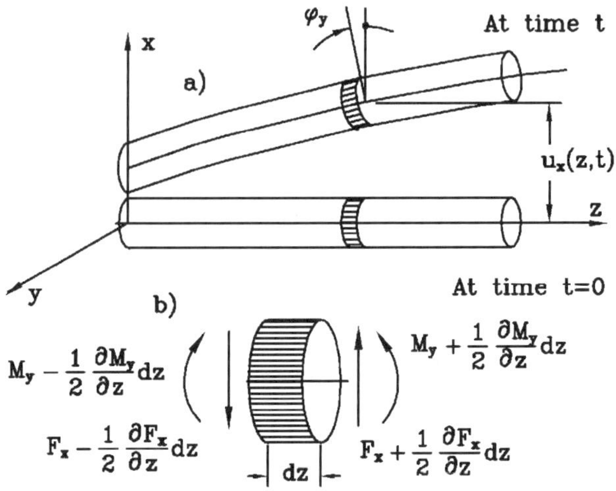

FIGURE 5.2. (a) Sketch of a beam deflected in xz-plane; (b) forces and moments acting on the length dz of the beam.

By introducing Equation (5.2) into Equation (5.1), it follows that

$$\rho A \frac{d^2 u_x}{dt^2} = -\frac{\partial^2 M_y}{\partial z^2} + f_x(z,t)\,. \tag{5.3}$$

Neglecting shear deformation and using elementary beam theory, the bending moment is proportional to the curvature of the inflected shape of the beam; the latter is approximated as the second derivative of the displacement u_x

$$M_y = EI_y \frac{\partial^2 u_x}{\partial z^2}\,, \tag{5.4}$$

where I_y is the area moment of inertia of the cross section of the beam about y-axis. The following equilibrium equation can thus be obtained:

$$m(z)\frac{d^2 u_x}{dt^2} + \frac{\partial^2}{\partial z^2}\left[k(z)\frac{\partial^2 u_z}{\partial z^2}\right] = f_x(z,t)\,, \tag{5.5}$$

where the mass per unit length and the bending stiffness are, respectively,

$$m(z) = \rho(z)A(z)\,, \qquad k(z) = E(z)I_y(z)\,. \tag{5.6}$$

In the case of a prismatic homogeneous beam, Equation (5.5) reduces to

$$\rho A \frac{d^2 u_x}{dt^2} + EI_y \frac{\partial^4 u_x}{\partial z^4} = f_x(z,t)\,. \tag{5.7}$$

The solution of the homogeneous equation associated with Equation (5.5) can be expressed as the product of a function of time $\eta(t)$ and a function of the space coordinate $q(z)$

$$u_x(z,t) = q(z)\eta(t)\,. \tag{5.8}$$

Introducing Equation (5.8) into the homogeneous equation associated with Equation (5.5) and separating the variables, it follows that

$$\frac{1}{\eta(t)}\frac{d^2\eta(t)}{dt^2} = \frac{1}{m(z)q(z)}\frac{d^2}{dz^2}\left[k(z)\frac{d^2q(z)}{dz^2}\right]. \tag{5.9}$$

The function on the left-hand side depends on time but not on the space coordinate z. Conversely, the function on the right-hand side is a function of z but not of t. The only possibility of satisfying Equation (5.9) for all values of time and of coordinate z is to state that both sides are constant and that the two constants are equal. This constant can be indicated as $-\omega^2$. The condition on the function of time on the left-hand side is

$$\frac{1}{\eta(t)}\frac{d^2\eta(t)}{dt^2} = \text{constant} = -\omega^2\,, \tag{5.10}$$

i.e.,

$$\frac{d^2\eta(t)}{dt^2} + \omega^2\eta(t) = 0. \tag{5.11}$$

Neglecting a proportionality constant that will be introduced later into function $q(z)$, Equation (5.11) yields a harmonic oscillation with frequency ω

$$\eta(t) = \sin(\omega t + \phi)\,. \tag{5.12}$$

The solution of the equation of motion for the free bending oscillations of the bar is then

$$u_x(z,t) = q(z)\sin(\omega t + \phi)\,, \tag{5.13}$$

i.e., each point of the bar performs a harmonic motion with frequency ω, phase ϕ and amplitude $q(z)$.

Function $q(z)$ is said to be the *principal function*. The resultant motion is then a standing wave, with all points of the bar vibrating in phase. By introducing Equation (5.13) into Equation (5.9), it follows that

$$\omega^2 m(z)q(z) = \frac{d^2}{dz^2}\left[k(z)\frac{d^2q(z)}{dz^2}\right], \tag{5.14}$$

or in the case of constant parameters,

$$\omega^2 q(z) = \frac{EI_y}{\rho A}\frac{d^4q(z)}{dz^4}\,. \tag{5.15}$$

The general solution of Equation (5.15) is

$$q(z) = C_1 \sin(az) + C_2 \cos(az) + C_3 \sinh(az) + C_4 \cosh(az) \ , \qquad (5.16)$$

where

$$a^2 = \omega \sqrt{\frac{\rho A}{E I_y}} \ . \qquad (5.17)$$

Equation (5.16) expresses the mode shapes of bending vibration of the beam, after constants C_i have been computed from the boundary conditions.

On each end of the beam, two boundary conditions must be stated, for a total of four boundary conditions allowing us to compute constants C_i. If the translational degree of freedom is locked, the function $q(z)$ in the relevant point must be equated to zero. If it is free, the shear force must vanish; i.e., the third derivative

$$\frac{d^3 q(z)}{dz^3}$$

must be equated to zero. If the rotational degree of freedom is locked, the rotation of the cross section vanishes, i.e.,

$$\frac{dq(z)}{dz} = 0,$$

whereas if it is free, what vanishes is the bending moment:

$$\frac{d^2 q(z)}{dz^2} = 0.$$

The case usually considered in rotordynamics is that with both ends simply supported, i.e., with vanishing displacements and bending moments at both ends. This is the case, for instance, of a transmission shaft supported by hook joints at both ends, if the inertia of the joints is negligible.

Assuming that the origin is at the left end of the beam of length l, the first condition yields

$$q(0) = q(l) = 0 \ , \qquad (5.18)$$

i.e.,

$$\begin{cases} C_2 + C_4 = 0 \ , \\ C_1 \sin(al) + C_2 \cos(al) + C_3 \sinh(al) + C_4 \cosh(al) = 0 \ . \end{cases} \qquad (5.19)$$

The second condition implies that the curvature of the beam, i.e., with the usual approximations, d^2q/dz^2, vanishes at both ends

$$\left[\frac{d^2 q(z)}{dz^2}\right]_{z=0} = \left[\frac{dq^2(z)}{dz^2}\right]_{z=l} = 0 \ . \qquad (5.20)$$

With simple computations, Equation (5.20) yields

$$\left\{ \begin{array}{l} C_2 - C_4 = 0 \; , \\ C_1 \sin(al) + C_2 \cos(al) - C_3 \sinh(al) - C_4 \cosh(al) = 0 \; . \end{array} \right. \tag{5.21}$$

The first Equation (5.19) and the first Equation (5.21) yield $C_2 = C_4 = 0$. The second Equation (5.19) and the second Equation (5.21) yield

$$C_1 \sin(al) = 0 \; , \qquad C_3 \sinh(al) = 0 \; . \tag{5.22}$$

The second condition leads to $C_3 = 0$. The first one yields a solution different from the trivial solution, with all $C_i = 0$, only if $al = i\pi$, i.e., if

$$\omega_i = \frac{i^2\pi^2}{l^2}\sqrt{\frac{EI_y}{\rho A}} \; , \qquad i = 1, 2, 3, \ldots \; . \tag{5.23}$$

Equation (8.2) yields the natural frequencies of the bending vibrations of the beam. Constant C_1 is not determined, which is obvious because it is the amplitude of the mode of free vibration. The mode shapes are

$$q_i(z) = q_{i_0} \sin(i\pi\zeta) \; , \tag{5.24}$$

where

$$\zeta = \frac{z}{l} \; .$$

If the beam is clamped on one side (e.g., at the left end) and free at the other one (like the Stodola-Green rotor, but without the mass at the end), the boundary conditions are

$$\left\{ \begin{array}{l} q(0) = C_2 + C_4 = 0 \; , \\ \left[\frac{dq(z)}{dz}\right]_{z=0} = a\left(C_1 + C_3\right) = 0 \; , \\ \left[\frac{d^2q(z)}{dz^2}\right]_{z=l} = a^2\left[-C_1 \sin(al) - C_2 \cos(al) + C_3 \sinh(al) + \right. \\ \qquad\qquad \left. + C_4 \cosh(al)\right] = 0 \; , \\ \left[\frac{d^3q(z)}{dz^3}\right]_{z=l} = a^3\left[-C_1 \cos(al) + C_2 \sin(al) + C_3 \cosh(al) + \right. \\ \qquad\qquad \left. + C_4 \sinh(al)\right] = 0 \; . \end{array} \right. \tag{5.25}$$

The first and second equations state that the displacement and the rotation vanish at the left end, whereas the third and the fourth ones state that at the right end, the bending moment and the shear force are nil. By solving the first and second equations in C_4 and C_3 and substituting in the following two equations, it follows that

$$\begin{bmatrix} \sin(al) + \sinh(al) & \cos(al) + \cosh(al) \\ \cos(al) + \cosh(al) & -\sin(al) + \sinh(al) \end{bmatrix} \left\{ \begin{array}{c} C_1 \\ C_2 \end{array} \right\} = \left\{ \begin{array}{c} 0 \\ 0 \end{array} \right\} \; . \tag{5.26}$$

Boundary condition; $i =$	0	1	2	3	4	$i > 4$
Free-free	0	4.730	7.853	10.996	14.137	$\approx (i+1/2)\pi$
Supported-free	0	1.25π	2.25π	3.25π	4.25π	$(i+1/4)\pi$
Clamped-free	–	1.875	4.694	7.855	10.996	$\approx (i-1/2)\pi$
Supported-supported	–	π	2π	3π	4π	$i\pi$
Supported-clamped	–	3.926	7.069	10.210	13.352	$\approx (i+1/4)\pi$
Clamped-clamped	–	4.730	7.853	10.996	14.137	$\approx (i+1/2)\pi$

TABLE 5.2. Values of constants $\beta_i = a_i l$ for the various modes with different boundary conditions.

To obtain a solution different from the trivial solution with all $C_i = 0$, the determinant of the coefficients of the homogeneous Equation (5.26) must vanish, i.e.,

$$\cos(al)\cosh(al) + 1 = 0 \ . \tag{5.27}$$

This equation cannot be solved in closed form; however, its numerical solution is straightforward, yielding $al = 1.875$, $al = 4.694$, $al = 7.855$, $al = 10.996$, ... For large values of al, Equation (5.27) reduces to

$$\cos(al) \approx 0 \ , \tag{5.28}$$

yielding $al \approx (i - 1/2)\,\pi$.

Operating in the same way for the other boundary conditions, the following general expression for the natural frequencies can be obtained

$$\omega_i = \frac{\beta_i^2}{l^2}\sqrt{\frac{EI_y}{\rho A}} \ , \tag{5.29}$$

where the values of constants $\beta_i = a_i l$ depend on the boundary conditions (Table 5.2).

The mode shapes corresponding to the different boundary conditions are plotted in Figure 5.3.

Remark 5.1 *The classic formulas found in vibration handbooks for the computation of natural frequencies and critical speeds of transmission shafts are based on Equation (5.29).*

The situation in the yz-plane is clearly very similar to that already studied with reference to the xz-plane. The physical problem is clearly the same; however, there are some differences in the equations because in the xz-plane, the positive direction for rotations is from z-axis to x-axis, whereas in the yz-plane, a rotation is positive if it goes from y-axis to z-axis. Actually, the difficulties of notation arising from the differences existing in the two planes could be easily circumvented by assuming $-\phi_x$ instead of ϕ_x as the generalized coordinate for rotation. If the corresponding generalized force is taken as $-M_x$ instead of M_x, the situation in the two planes is exactly the same.

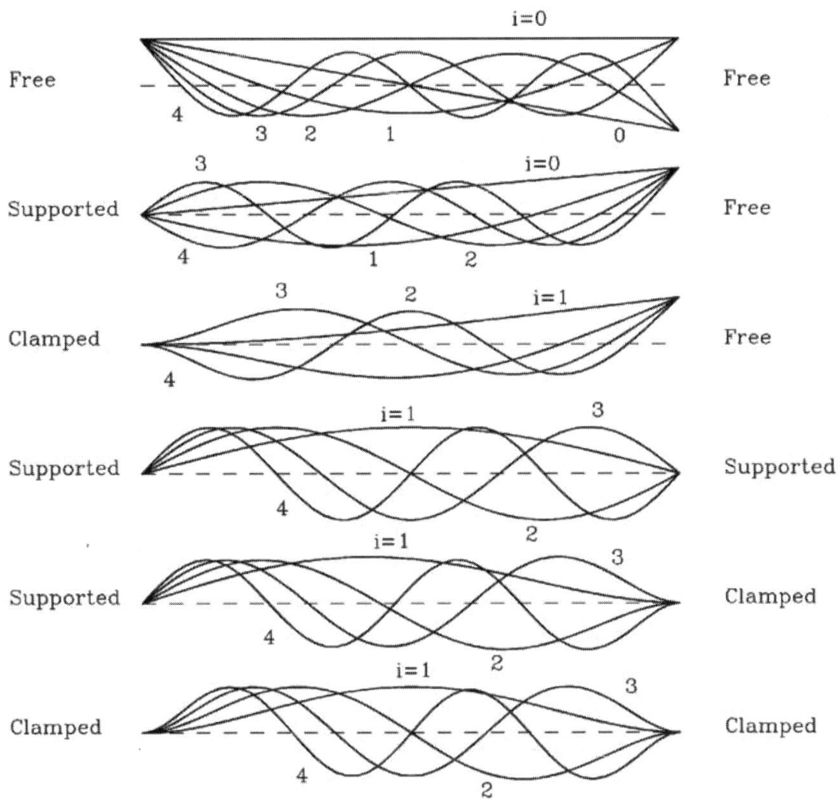

FIGURE 5.3. First four mode shapes (plus the rigid body modes, when they exist) for an Euler-Bernoulli beam with different boundary conditions.

Also in the present case, it is possible to resort to complex coordinates, defining them in the same way as seen for discrete systems.

5.2 Other boundary conditions

The boundary conditions in which the two degrees of freedom at each end are either locked or free are just two particular cases of a more general situation, in which the boundary conditions may be different. For instance, the end of the beam may be supported by an elastic constraint, supplying a force proportional to the displacement through a constant K, the stiffness of the spring. As in the beam the shear force is the derivative with respect to z of the bending moment; the condition stating that the shear force is equal to the reaction of the constraint can be written in the form:

$$EI_y \left[\frac{d^3q(z)}{dz^3}\right]_{z=z_c} = K\left[q(z)\right]_{z=z_c} \ , \tag{5.30}$$

i.e.

$$\begin{aligned} &EI_y a^3 \left[-C_1 \cos(az_c) + C_2 \sin(az_c) + C_3 \cosh(az_c) + C_4 \sinh(az_c)\right] = \\ &= K\left[C_1 \sin(az_c) + C_2 \cos(az_c) + C_3 \sinh(az_c) + C_4 \cosh(az_c)\right] \ , \end{aligned} \tag{5.31}$$

where z_c is the z coordinate of the constraint.

In a similar way, if the elastic constraint is a spring that exerts a bending moment proportional to the rotation through the rotational stiffness κ, the boundary condition is

$$EI_y \left[\frac{d^2q(z)}{dz^2}\right]_{z=z_c} = \kappa \left[\frac{dq(z)}{dz}\right]_{z=z_c} \ . \tag{5.32}$$

If at the end of the beam there is a concentrated mass M, the boundary condition can be written by stating that the shear force equals the inertia force, which in harmonic motion is proportional to the displacement and the square of the frequency

$$EI_y \left[\frac{d^3q(z)}{dz^3}\right]_{z=z_c} = -M\omega^2 \left[q(z)\right]_{z=z_c} \ . \tag{5.33}$$

The case of a moment of inertia J_y can be dealt with by equating the bending moment with the inertia torque

$$EI_y \left[\frac{d^2q(z)}{dz^2}\right]_{z=z_c} = -J_y\omega^2 \left[\frac{dq(z)}{dz}\right]_{z=z_c} \ . \tag{5.34}$$

Consider for instance the case of a beam supported at both ends by elastic constraints with stiffness K and negligible bending stiffness [Figure 5.4(a)]. The boundary conditions are

$$\begin{cases} EI_y a^3 \left(-C_1 + C_3\right) = k\left(C_2 + C_4\right) \ , \\ a^2\left(-C_2 + C_4\right) = 0 \ , \\ EI_y a^3 \left[-C_1 \cos(al) + C_2 \sin(al) + C_3 \cosh(al) + C_4 \sinh(al)\right] = \\ \qquad = K\left[C_1 \sin(al) + C_2 \cos(al) + C_3 \sinh(al) + C_4 \cosh(al)\right] \ , \\ a^2\left[-C_1 \sin(al) - C_2 \cos(al) + C_3 \sinh(al) + C_4 \cosh(al)\right] = 0 \ . \end{cases} \tag{5.35}$$

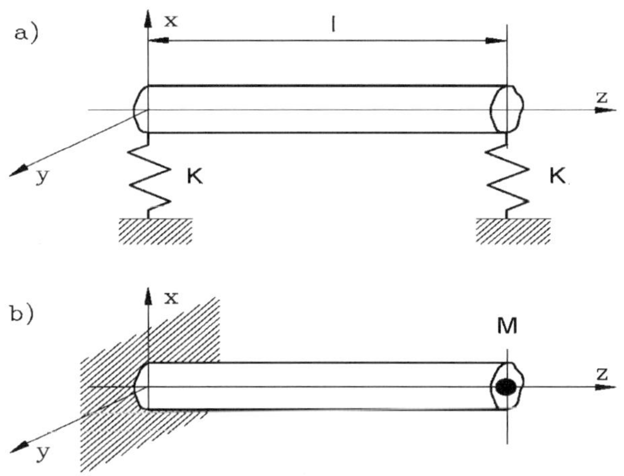

FIGURE 5.4. Prismatic Euler-Bernoulli beam. (a) Supported at both ends by elastic constraints. (b) Clamped at one end and with a mass M attached at the other end.

By introducing the nondimensional stiffness ratio

$$K^* = \frac{Kl^3}{EI_y} ,$$

the characteristic equation in the nondimensional unknown al can be written in the form

$$\det \begin{bmatrix} -(al)^3 & -K^* & (al)^3 & -K^* \\ 0 & -1 & 0 & 1 \\ -(al)^3 c - K^* s & (al)^3 s - K^* c & (al)^3 C - K^* S & (al)^3 S - K^* C \\ -s & -c & S & C \end{bmatrix} = 0 , \tag{5.36}$$

where

$$c = \cos(al) , \quad s = \sin(al) , \quad C = \cosh(al) , \quad S = \sinh(al) .$$

Equation (5.36) can be solved numerically, obtaining the values of $\beta_i = a_i l$ to be inserted into Equation (5.29) to obtain the natural frequencies.

Example 5.1 *Study the effect of the stiffness of the bearings on the natural frequencies of a shaft simply supported at its ends.*

The problem can be easily solved by introducing different values of the stiffness of the supports or, better, of the nondimensional parameter K^* into Equation (5.36). The plot reported in Figure 5.5 shows the first four natural frequencies of the beam as functions of K^*. Note that for $K^* = 0$, the values for the free-free conditions are found, whereas for $K^* \to \infty$, the beam is supported on rigid constraints at both sides.

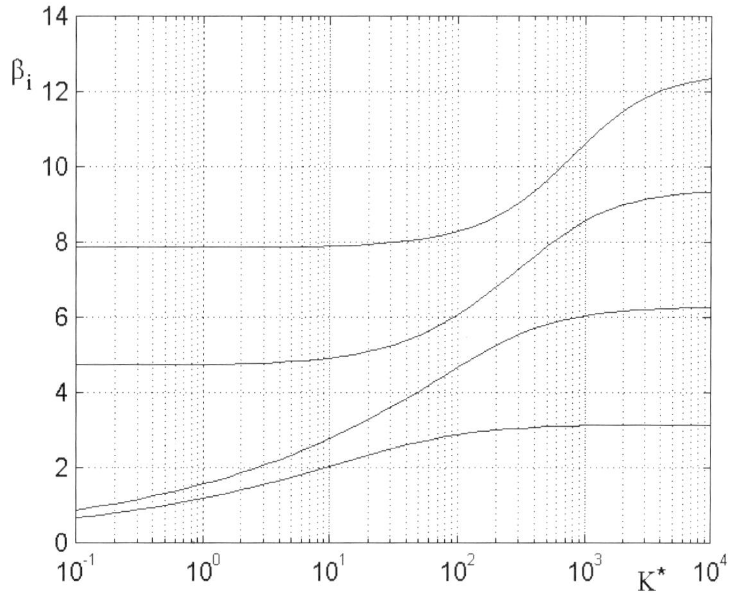

FIGURE 5.5. First four natural frequencies, in terms of parameter $\beta_i = a_i l$, for a beam on elastic supports as functions of the nondimensional stiffness K^*.

Another interesting case is that of Figure 5.4(b): a beam clamped on one side (e.g., on the left) with a concentrated mass at the other end.

The boundary conditions in this case are

$$\begin{cases} C_2 + C_4 = 0 \ , \\ a\left(C_1 + C_3\right) = 0 \ , \\ EI_y a^3\left(-C_1 c + C_2 s + C_3 C + C_4 S\right) = -M\omega^2\left(C_1 s + C_2 c + C_3 S + C_4 C\right) \ , \\ a^2\left[-C_1 s - C_2 c + C_3 S + C_4 C\right] = 0 \ . \end{cases} \tag{5.37}$$

By introducing into Equation (5.37) the value of the frequency given by Equation (5.29) and the nondimensional mass ratio

$$M^* = \frac{M}{\rho A l}$$

and solving the first two equations in C_3 and C_4, the characteristic equation reduces to

$$\det\begin{bmatrix} -s + S - M^* al\left(c - C\right) & c + C \\ c + C - M^* al\left(s - S\right) & s + S \end{bmatrix}^T = 0 \ . \tag{5.38}$$

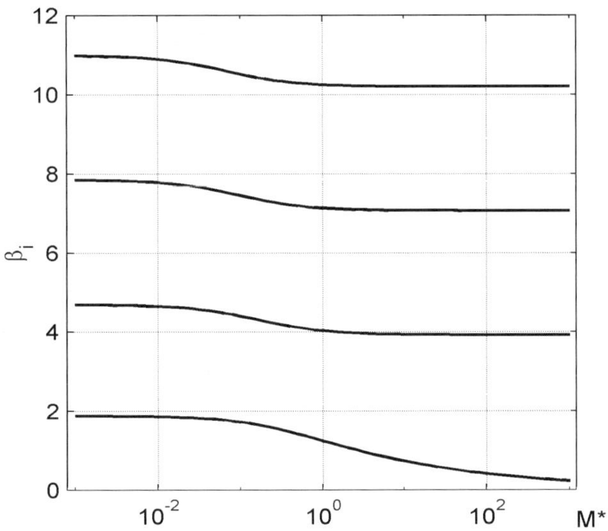

FIGURE 5.6. First four natural frequencies, in terms of parameter $\beta_i = a_i l$, for a beam clamped at one end and with a mass at the other one as functions of the nondimensional mass M^*.

It is easy to verify that for $M^* = 0$, the results coincide with those for a clamped-free beam, whereas for $M^* \to \infty$, the natural frequencies for a clamped-supported beam are obtained. For $M^* = 1$, for instance, the results are $\beta_1 = 1.248$, $\beta_2 = 4.031$, $\beta_3 = 7.134$, $\beta_4 = 10.257$, ..., which are lower than those for the clamped free beam (the presence of the added mass lowers the natural frequencies, as expected). The results are reported in Figure 5.6.

5.3 Effect of the moments of inertia: Timoshenko beam

In the case of the Euler-Bernoulli beam, the rotational inertia of the cross section and shear deformation were not taken into account. Moreover, gyroscopic effect caused by rotation of the beam about z-axis causes the behavior of the beam in xz- and yz-planes to be coupled.

Neglecting external distributed forces that may act on the beam, the equilibrium equation for translations is still Equation (5.1), without the term in f, together with a similar equation for the behavior in the yz-plane

$$\begin{cases} \rho A \dfrac{d^2 u_x}{dt^2} dz = \dfrac{\partial F_x}{\partial z} dz \ , \\ \\ \rho A \dfrac{d^2 u_y}{dt^2} dz = \dfrac{\partial F_y}{\partial z} dz \ . \end{cases} \tag{5.39}$$

If the beam is axially symmetrical and no distributed bending moments act on the beam, the equilibrium equations for rotations about x- and y-axes of the length dz of the beam can be computed starting from Equation (5.2), in which also inertia torques and gyroscopic moments are accounted for

$$\begin{cases} F_y dz - \dfrac{\partial M_x}{\partial z} dz = -\rho I_y \dfrac{d^2 \phi_x}{dt^2} dz - 2\rho I_y \Omega \dfrac{d\phi_y}{dt} dz \ , \\ \\ F_x dz + \dfrac{\partial M_y}{\partial z} dz = \rho I_y \dfrac{d^2 \phi_y}{dt^2} dz - 2\rho I_y \Omega \dfrac{d\phi_x}{dt} dz \ , \end{cases} \tag{5.40}$$

where Ω is the spin speed of the shaft and the polar moment of inertia of the cross section is $I_p = 2I_y$.

By differentiating Equations (5.40) with respect to z and introducing them into Equations (5.39), it follows that

$$\begin{cases} \rho A \dfrac{d^2 u_x}{dt^2} + \dfrac{\partial^2 M_y}{\partial z^2} - \rho I_y \dfrac{\partial}{\partial z} \dfrac{d^2 \phi_y}{dt^2} + 2\rho I_y \Omega \dfrac{\partial}{\partial z} \dfrac{d\phi_x}{dt} = 0 \ , \\ \\ \rho A \dfrac{d^2 u_y}{dt^2} - \dfrac{\partial^2 M_x}{\partial z^2} + \rho I_y \dfrac{\partial}{\partial z} \dfrac{d^2 \phi_x}{dt^2} + 2\rho I_y \Omega \dfrac{\partial}{\partial z} \dfrac{d\phi_y}{dt} = 0 \ . \end{cases} \tag{5.41}$$

If shear deformation is neglected, the rotation of the cross section is simply

$$\phi_y = \frac{\partial u_x}{\partial z} \ , \qquad \phi_x = -\frac{\partial u_y}{\partial z} \ . \tag{5.42}$$

The bending moment can be linked to the inflected shape of the beam by the relationships

$$M_y = EI_y \frac{\partial^2 u_x}{\partial z^2} \ , \qquad M_x = -EI_y \frac{\partial^2 u_y}{\partial z^2} \ . \tag{5.43}$$

Equations (5.41), can thus be written in the form

$$\begin{cases} \rho A \dfrac{d^2 u_x}{dt^2} + EI_y \dfrac{\partial^4 u_x}{\partial z^4} - \rho I_y \dfrac{\partial^2}{\partial z^2} \left(\dfrac{d^2 u_x}{dt^2} \right) - 2\rho I_y \Omega \dfrac{\partial^2}{\partial z^2} \left(\dfrac{du_y}{dt} \right) = 0 , \\ \\ \rho A \dfrac{d^2 u_y}{dt^2} + EI_y \dfrac{\partial^4 u_y}{\partial z^4} - \rho I_y \dfrac{\partial^2}{\partial z^2} \left(\dfrac{d^2 u_y}{dt^2} \right) + 2\rho I_y \Omega \dfrac{\partial^2}{\partial z^2} \left(\dfrac{du_x}{dt} \right) = 0 . \end{cases} \tag{5.44}$$

By introducing the complex coordinate $u = u_x + iu_y$, Equations (5.44) reduce to

$$\rho A \frac{d^2 u}{dt^2} + EI_y \frac{\partial^4 u_x}{\partial z^4} - \rho I_y \frac{\partial^2}{\partial z^2} \left(\frac{d^2 u}{dt^2} \right) + 2i\rho I_y \Omega \frac{\partial^2}{\partial z^2} \left(\frac{du}{dt} \right) = 0 \, . \quad (5.45)$$

Equation (5.45) can be used for the computation of the Campbell diagram and of the critical speeds, but it has a disadvantage, linked with the assumptions used to obtain it: When the slenderness of the beam is low enough and the rotational inertia of the cross sections becomes important in determining its dynamic behavior, also shear deformations are large enough to require their inclusion in the model. A beam based on the *Timoshenko beam* model needs to be considered, and shear deformation can be taken into account as a deviation of the direction of the deflected shape of the beam not accompanied by a rotation of the cross section. The latter is then no more perpendicular to the deformed shape of the beam, and the rotation of the cross section in the two inflection planes can be expressed as

$$\phi_y = \frac{\partial u_x}{\partial z} - \gamma_x \, , \qquad \phi_x = -\frac{\partial u_y}{\partial z} - \gamma_y \, . \quad (5.46)$$

The shear strains γ_x and γ_y are linked to the shear forces by the relationship

$$\gamma_x = \frac{\chi F_x}{GA} \, , \qquad \gamma_y = -\frac{\chi F_y}{GA} \, , \quad (5.47)$$

where the already mentioned shear factor χ depends on the shape of the cross section, even if there is not a complete accord on its value. For a circular beam, a value of 10/9 is usually assumed; for other shapes, the expressions reported in Table 4.1 can be used.

Remark 5.2 *In rotordynamics, the so-called Euler-Bernoulli and Timoshenko beam differ from each other because the latter includes (a) shear deformation, (b) inertia of the cross section for rotations about x- and y-axes, and (c) gyroscopic effect.*

Equation (5.46) can thus be written in the form

$$\phi_y = \frac{\partial u_x}{\partial z} - \frac{\chi F_x}{GA} \, , \qquad \phi_x = -\frac{\partial u_y}{\partial z} + \frac{\chi F_y}{GA} \, . \quad (5.48)$$

The bending moment has no effect on the shear deformation: If the latter is accounted for, the relationship linking the bending moment to the inflected shape of the beam becomes

$$M_y = EI_y \frac{\partial \phi_y}{\partial z} \, , \qquad M_x = EI_y \frac{\partial \phi_x}{\partial z} . \quad (5.49)$$

By solving Equation (5.48) in F_x and F_y and introducing it into Equation (5.39), it follows that

$$\begin{cases} \rho A \dfrac{d^2 u_x}{dt^2} = \dfrac{GA}{\chi}\left(\dfrac{\partial^2 u_x}{\partial z^2} - \dfrac{\partial \phi_y}{\partial z}\right) , \\ \\ \rho A \dfrac{d^2 u_y}{dt^2} = \dfrac{GA}{\chi}\left(\dfrac{\partial^2 u_y}{\partial z^2} + \dfrac{\partial \phi_x}{\partial z}\right) . \end{cases} \tag{5.50}$$

Operating in the same way as with Equation (5.40), it follows that

$$\begin{cases} \rho I_y \dfrac{d^2 \phi_x}{dt^2} + 2\rho I_y \Omega \dfrac{d\phi_y}{dt} = -\dfrac{GA}{\chi}\left(\dfrac{\partial u_y}{\partial z} + \phi_x\right) + EI_y \dfrac{\partial^2 \phi_x}{\partial z^2} , \\ \\ \rho I_y \dfrac{d^2 \phi_y}{dt^2} - 2\rho I_y \Omega \dfrac{d\phi_x}{dt} = \dfrac{GA}{\chi}\left(\dfrac{\partial u_x}{\partial z} - \phi_y\right) + EI_y \dfrac{\partial^2 \phi_y}{\partial z^2} . \end{cases} \tag{5.51}$$

By differentiating equations (5.50) with respect to z and eliminating ϕ_x and ϕ_y, the following equations can be obtained:

$$\begin{cases} EI_x \dfrac{\partial^4 u_y}{\partial z^4} - \rho I_y \left(1 + \dfrac{E\chi}{G}\right) \dfrac{\partial^4 u_y}{\partial z^2 \partial t^2} + \dfrac{\rho^2 I_y \chi}{G} \dfrac{d^4 u_y}{dt^4} + \\ +2\rho I_y \Omega \left[\dfrac{\partial^2}{\partial z^2}\left(\dfrac{du_x}{dt}\right) - \dfrac{\rho\chi}{G}\dfrac{d^3 u_x}{dt^3}\right] + \rho A \dfrac{d^2 u_x}{dt^2} = 0 , \\ \\ EI_y \dfrac{\partial^4 u_x}{\partial z^4} - \rho I_y \left(1 + \dfrac{E\chi}{G}\right) \dfrac{\partial^4 u_x}{\partial z^2 \partial t^2} + \dfrac{\rho^2 I_y \chi}{G} \dfrac{d^4 u_x}{dt^4} \\ -2\rho I_y \Omega \left[\dfrac{\partial^2}{\partial z^2}\left(\dfrac{du_y}{dt}\right) - \dfrac{\rho\chi}{G}\dfrac{d^3 u_y}{dt^3}\right] + \rho A \dfrac{d^2 u_x}{dt^2} = 0 . \end{cases} \tag{5.52}$$

If the complex coordinate $u = u_x + iu_y$ is introduced, Equations (5.52) reduce to

$$\begin{aligned} &EI_y \frac{\partial^4 u}{\partial z^4} - \rho I_y \left(1 + \frac{E\chi}{G}\right) \frac{\partial^4 u}{\partial z^2 \partial t^2} + \frac{\rho^2 I_y \chi}{G} \frac{d^4 u}{dt^4} + \\ &+2i\rho I_y \Omega \left[\frac{\partial^2}{\partial z^2}\left(\frac{du}{dt}\right) - \frac{\rho\chi}{G}\frac{d^3 u}{dt^3}\right] + \rho A \frac{d^2 u}{dt^2} = 0 . \end{aligned} \tag{5.53}$$

The solution of Equation (5.53) is

$$u_x(z,t) = q(z) e^{i\omega t} , \tag{5.54}$$

which yields

$$EI_y \frac{d^4q(z)}{dz^4} + \rho I_y \left[\omega^2\left(1+\frac{E\chi}{G}\right) - 2\omega\Omega\right]\frac{d^2q(z)}{dz^2} +$$
$$+\rho\left[\omega^4\frac{\rho I_y \chi}{G} - 2\frac{\rho I_y\chi}{G}\omega^3\Omega - A\omega^2\right]q(z) = 0\ . \tag{5.55}$$

The same considerations regarding the form of the eigenfunctions seen in the preceding section also hold in this case. If the beam is simply supported at both ends, the same eigenfunction seen in the case of the Euler-Bernoulli beam $q = q_{i_0}\sin(i\pi\zeta)$ still holds. From Equation (5.55), it is thus possible to obtain the characteristic equation

$$\omega^4\frac{\rho I_y\chi}{GA} - 2\frac{\rho I_y\chi}{GA}\omega^3\Omega - \left[1 + I_y\frac{i^2\pi^2}{Al^2}\left(1+\frac{E\chi}{G}\right)\right]\omega^2 +$$
$$+2I_y\frac{i^2\pi^2}{Al^2}\omega\Omega + \frac{EI_y}{\rho A}\frac{i^4\pi^4}{l^4} = 0\ . \tag{5.56}$$

By introducing the nondimensional frequency and speed

$$\omega^* = \frac{\omega}{\frac{\pi^2}{l^2}\sqrt{\frac{EI_y}{\rho A}}}\ , \qquad \Omega^* = \frac{\Omega}{\frac{\pi^2}{l^2}\sqrt{\frac{EI_y}{\rho A}}}\ , \tag{5.57}$$

obtained by dividing them by the value of the first natural frequency of the Euler-Bernoulli beam, the characteristic equation can be expressed in nondimensional form as

$$\omega^{*4} - 2\Omega^*\omega^{*3} - \omega^{*2}\frac{i^2\alpha^2}{\pi^2\chi^*}\left(1+\chi^*+\frac{\alpha^2}{i^2\pi^2}\right) + 2\Omega^*\omega^*\frac{i^2\alpha^2}{\pi^2\chi^*} + \frac{i^4\alpha^4}{\pi^4\chi^*} = 0\ , \tag{5.58}$$

where the slenderness α of the beam and χ^* are

$$\alpha = l\sqrt{A/I_y} = l/r\ , \quad \chi^* = \chi E/G$$

and r is the radius of inertia of the cross section.

The results obtained from Equation (5.58) for a beam with circular cross section ($\chi = 10/9$) and material with $\nu = 0.3$ are reported in the form of a nondimensional Campbell diagram for two different values of the slenderness in Figure 5.7. Note that the Campbell diagrams have been plotted showing the first and second quadrants, to expand the ω^* scale more than the Ω^* scale. The effects of both shear deformation and rotational inertia tend to lower the value of the natural frequencies, as seen by the intersections with the ω^*-axis (for an Euler-Bernoulli beam, they should be at the values 1, 4, 9, and 16).

A particular transformation can be devised to plot figures like Figure 5.7, in which some branches of the plot extend to very high values of the

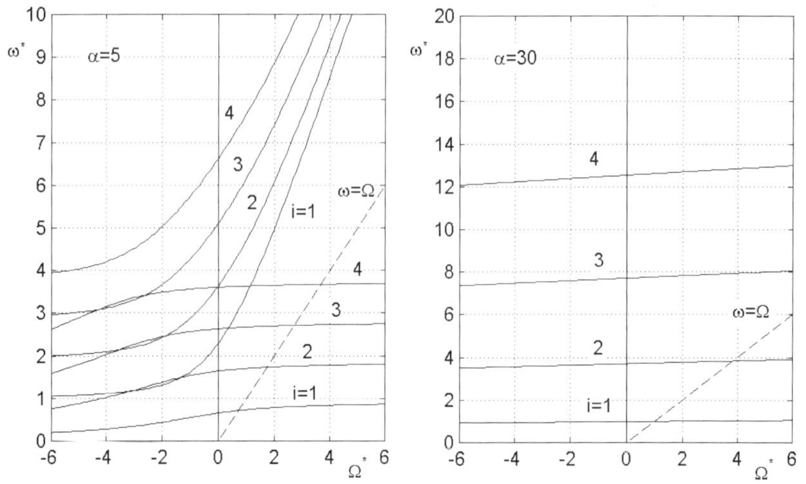

FIGURE 5.7. Nondimensional Campbell diagram of a Timoshenko beam simply supported at the ends with two different values of the slenderness ($\alpha = 5$ and $\alpha = 30$). The first four natural frequencies have been plotted.

abscissas or the ordinates, whereas other stay about the origin, without amplifying too much the zone about the origin as it occurs with logarithmic scales. The transformation is very simple: The quantity x spanning from 0 to infinity is transformed into the quantity $\bar{x}$ laying in the interval [0, 1] using the relationship

$$\bar{x} = \frac{x}{1-x} \, . \tag{5.59}$$

Figure 5.8 has been obtained by transforming Figure 5.7 by using Equation (5.59) for both the Ω^*- and ω^*-axes.

From the figure, it is clear that even with a fairly low value of the slenderness ($\alpha = 30$), the Campbell diagram is reasonably flat and only very high-frequency modes are affected by the gyroscopic effect. In Figure 5.7 they are out of the plot, because their nondimensional frequency is higher than 50. As a consequence, the critical speeds and the whirling frequencies can be computed with sufficient precision, even neglecting gyroscopic effect and computing the natural frequencies at standstill. The effect of shear deformation is, however, not negligible.

The plot computed for a slenderness of five shows a greater gyroscopic effect. However, the higher branches of the Campbell diagram, which have an asymptote whose equation is $\omega = 2\Omega$, do not intersect the line $\omega = \Omega$, i.e., do not give way to critical speeds. The critical speeds are, even in this case, not very far from the natural frequencies at standstill, as the relevant branches of the Campbell diagram are still close to horizontal straight lines. A slenderness equal to five is at any rate too low: The Timoshenko beam

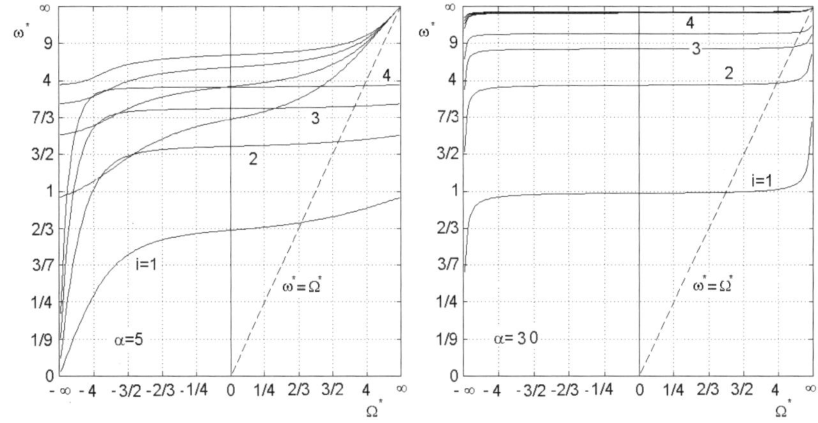

FIGURE 5.8. Same as Figure 5.7 but with the scales transformed using Equation (5.59). The whole Campbell diagram spanning from -∞ to ∞ is reported.

model is an approximation, because it is based on the beam theory, and the very model of a one-dimensional solid is no more satisfactory for very low values of α.

5.4 Dynamic stiffness matrix

The finite element method yields results that, even in the case of beam elements, are approximated (although they can be very close to the correct ones) because the shape functions cannot yield the exact inflected shape. To overcome this limitation, which is more apparent than real, an exact method has been developed for beam elements, in which the expression of the deflected shape obtained through the continuous model is assumed.

In the case of a homogeneous, prismatic Euler-Bernoulli beam performing harmonic oscillations with frequency ω in the xz-plane, for example, the general solution of Equation (5.7) yielding the deflected shape with any boundary condition is

$$q(z) = C_1 \sin(az) + C_2 \cos(az) + C_3 \sinh(az) + C_4 \cosh(az) , \qquad (5.60)$$

where $a = \sqrt{\omega}\sqrt[4]{\rho A/EI_y}$.

Constants C_i can be obtained from the boundary conditions, i.e., from the nodal generalized displacements. Note that in this case, the shape functions that can be obtained from Equation (5.60) depend on the frequency ω of the harmonic motion. They can be used to compute the mass and stiffness matrices of the element in the usual way. In this case, these matrices contain the frequency ω of the harmonic motion, and it is possible to directly write the dynamic stiffness matrix

$$\mathbf{K}_{dyn} = \mathbf{K} - \omega^2\mathbf{M} ,$$

which for the Euler-Bernoulli beam, is [40]

$$\mathbf{K}_{dyn} = \frac{EI_y}{l^3\left[1-\cos(b)\cosh(b)\right]}\begin{bmatrix} k_1 & lk_2 & -k_3 & -lk_4 \\ & l^2k_5 & lk_4 & l^2k_6 \\ & & k_1 & -lk_2 \\ & \text{symm.} & & l^2k_5 \end{bmatrix} , \tag{5.61}$$

where

$$\begin{array}{ll} k_1 = b^3\left[\sin(b)\cosh(b)+\cos(b)\sinh(b)\right] , & k_2 = b^2\sin(b)\sinh(b) , \\ k_3 = b^3\left[\sin(b)+\sinh(b)\right] , & k_4 = b^2\left[\cos(al)-\cosh(b)\right] , \\ k_5 = b\left[\sin(b)\cosh(b)-\cos(b)\sinh(b)\right] , & k_6 = b\left[\sinh(b)-\sin(b)\right] , \end{array}$$

and

$$b = al = \sqrt{\omega}\sqrt[4]{\frac{\rho A l^4}{EI_y}} .$$

A similar approach can be used, at least in principle, for any element, provided that an expression of the type of Equation (5.60) can be obtained from a continuous model. The dynamic matrices of the various elements can be assembled in the same way seen for mass and stiffness matrices. When the response to harmonic excitation has to be computed, the frequency of the motion is known and the computation is straightforward. On the contrary, when computing the natural frequencies of the system, the solution of the eigenproblem is complicated because it is impossible to express it in standard form and then apply the usual numerical methods. The solution follows the lines seen for methods based on transfer matrices: A value of the frequency is assumed, and the determinant of the dynamic stiffness matrix is computed. The determinant is not equal to zero, unless a natural frequency has been chosen, and new values of the frequency are assumed. A plot of the value of the determinant as a function of the frequency is then drawn. From the plot, the values of the frequency causing the determinant to vanish can be obtained. The problem here is that the determinant of a fairly large matrix has to be computed many times and this is, from a computational point of view, far more involving than solving an eigenproblem in standard form using standard techniques.

The dynamic stiffness matrix, which is now a function of the frequency, can be expressed as a power series in the frequency ω. By performing the series expansion of the trigonometric and hyperbolic functions included into Equation (5.60), it is possible to show that only even powers of ω are present:

$$\mathbf{K}_{dyn} = \mathbf{K}_{d_0} + \omega^2\mathbf{K}_{d_2} + \omega^4\mathbf{K}_{d_4} + \ldots .$$

The first term $\mathbf{K}_{d_0}$ is

$$\mathbf{K}_{d_0} = \frac{EI_y}{l^3} \begin{bmatrix} 12 & 6l & -12 & 6l \\ & 4l^2 & -6l & 2l^2 \\ & & 12 & -6l \\ & \text{symm.} & & 4l^2 \end{bmatrix} . \tag{5.62}$$

It depends only on the elastic properties and coincides with the stiffness matrix of the Euler Bernoulli beam.

The second term $\mathbf{K}_{d_2}$

$$\mathbf{K}_{d_2} = -\frac{\rho Al}{420} \begin{bmatrix} 156 & 22l & 54 & -13l \\ & 4l^2 & 13l & -3l^2 \\ & & 156 & -22l \\ & \text{symm.} & & 4l^2 \end{bmatrix} \tag{5.63}$$

depends only on the inertial properties and coincides, except for the sign, which is obviously changed, with the mass matrix.

The third term $\mathbf{K}_{d_4}$

$$\mathbf{K}_{d_4} = -\frac{\rho^2 A^2 l^5}{69854400 EI_y} \begin{bmatrix} 25488 & 5352l & 23022 & -5043l \\ & 1136^2 & 5043l & -1097l^2 \\ & & 25488 & -5352l \\ & \text{symm.} & & 1136l^2 \end{bmatrix} \tag{5.64}$$

and all other terms of the series depend on both inertial and stiffness properties and represent a correction of the potential and kinetic energy of the element caused by a more precise formulation of the displacement field. The standard approach can thus be thought of as a truncation at the second term of the series for the exact dynamic stiffness matrix.

The dynamic stiffness matrix approach was sometimes used in rotordynamics; however, its drawbacks, mainly those of leading to long and costly computations and of being applicable only to harmonic motion, are greater, in the opinion of the author, than its advantage of giving a better approximation, which can be obtained simply by using a finer mesh, with far less computational difficulty. Note that the term *exact* used in this context means simply that it leads to the same results of the beam theory, i.e., of the continuous model. All of the approximations linked to that model are, however, present.

Example 5.2 *Compute the natural frequencies of a beam on two identical elastic supports with* $K^* = Kl^3/EI_y = 100$ *located at its end using the dynamic stiffness approach, and compare the results obtained with those computed through the FEM.*

A sketch of the system is shown in Figure 5.4(a). If the beam is modeled with a single element, the dynamic stiffness matrix is, after assembling the stiffness of

		b_1	b_2	b_3	b_4
Dynamic stiffness	1 el.	2.877	4.663	6.076	8.276
FEM (consistent)	1 el.	2.974	4.751	6.551	9.973
	2 el.	2.881	4.711	6.101	8.758
	4 el.	2.877	4.667	6.086	8.307
FEM (lumped)	1 el.	3.761	3.761	-	-
	2 el.	2.935	4.472	4.770	-
	4 el.	2.892	4.646	5.641	7.167
	8 el.	2.881	4.661	5.964	7.939

TABLE 5.3. Values of the first four natural frequencies (or better, of nondimensional parameter $b = al = \sqrt{\omega}\sqrt[4]{\rho A l^4/EI_y}$) for a beam on two elastic supports computed using the dynamic stiffness matrix approach and the FEM.

the end supports

$$\mathbf{K}_{dyn} = \frac{EI_y}{cl^3}\begin{bmatrix} k_1 + K^* & lk_2 & -k_3 & -lk_4 \\ & l^2k_5 & lk_4 & l^2k_6 \\ & & k_1 + K^* & -lk_2 \\ & symm. & & l^2k_5 \end{bmatrix},$$

where

$$c = 1 - \cos(b)\cosh(b)\ , \qquad K^* = c\frac{Kl^3}{EI_y}\ .$$

To study the general case of the beam with any length, instead of using the rotations at the end nodes ϕ_1 and ϕ_2 as degrees of freedom, the products ϕl are used. The vector of the generalized coordinates is then

$$\mathbf{q} = \begin{bmatrix} r_1 & l\phi_1 & r_2 & l\phi_2 \end{bmatrix}^T .$$

The equation yielding the natural frequencies is then

$$\det\begin{bmatrix} k_1 + K^* & k_2 & -k_3 & -k_4 \\ & k_5 & k_4 & k_6 \\ & & k_1 + K^* & -k_2 \\ & symm. & & k_5 \end{bmatrix} = 0\ .$$

No closed-form solutions can be found, but there is little difficulty in plotting the determinant as a function of b and then finding the values of b, which cause the determinant to vanish. Such a plot is reported in Figure 5.9. The first four values of b are reported in Table 5.3. The values obtained using the dynamic matrices approach coincide with those computed using the beam theory. In the same table, the results computed using the standard FEM approach and a different number of elements are reported.

By using the consistent mass matrices, fairly good results are obtained by approximating the structure with just two elements, particularly for what the first two natural frequencies are regarded. A larger number of elements are required for higher order modes. When using the lumped mass matrix, a large number of elements is required to obtain a comparable accuracy.

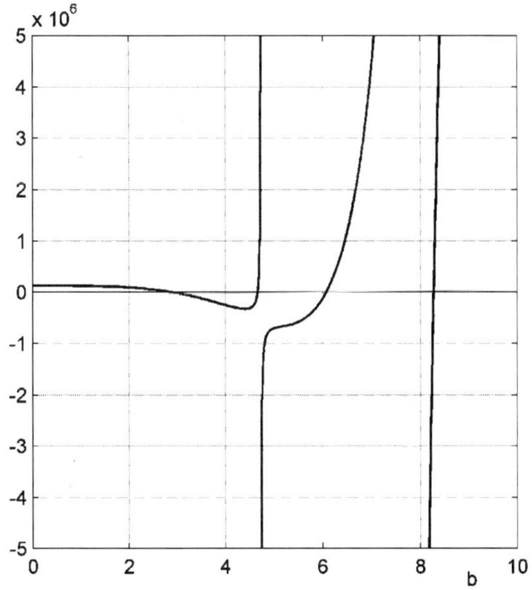

FIGURE 5.9. Determinant of the dynamic stiffness marix as a function of the frequency.

Example 5.3 *Propeller shaft for a front engine, rear-wheel drive, motor car.*

Consider the propeller shaft of a motor car with front engine and gearbox and rear-wheel drive. The shaft is made in two sections, with a hook joint between them and a central elastic support, attached to the vehicle body. The shaft is connected to the gearbox and the final drive through elastic joints. A model of the system is shown in Figure 5.10.

Also, the output shaft of the gearbox and the input shaft of the final drive have been included in the model. The various shafts have been modeled using 14 Timoshenko beam elements (because the shafts are very slender, also Euler-Bernoulli beams could have been used). Three spring elements plus six concentrated mass elements were used to model the joints, whereas a further spring element was used for the central support. The bearings (two on the input shaft of the final drive and three on the output shaft of the gearbox) were modeled as rigid constraints, by eliminating the relevant degrees of freedom.

The first three values of the critical speeds are reported in Table 5.4, together with the first three natural frequencies at standstill. The modes corresponding to the first three critical speeds are shown in Figure 5.11.

All computations were performed using a reduction scheme with ten master degrees of freedom. The first mode is a rigid-body mode, in which the two parts of the shaft vibrate on the central support. This mode could be obtained simply by reducing the whole system to a single-degree-of-freedom system, by condensing everything on the translational degree of freedom of node 9 or 10. The result is

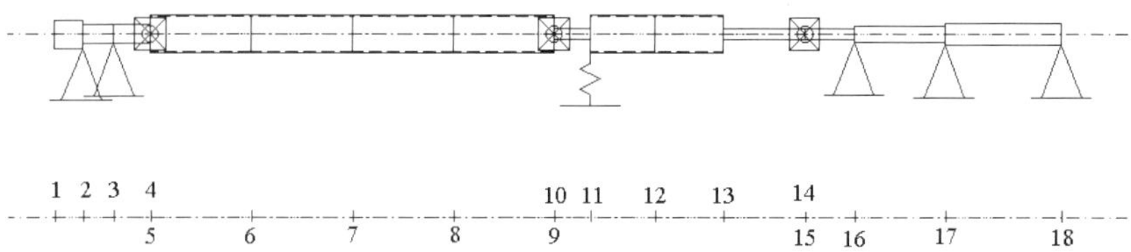

FIGURE 5.10. FEM model of a transmission shaft. The output shaft of the gearbox and the input shaft of the final drive have been included in the model.

	Ω_{cr}[rad/s]		ω_n[rad/s]	
Master degrees of freedom	1	10	1	10
I	118.2	117.9	117.1	117.0
II	–	1236	–	1061
III	–	1891	–	1507

TABLE 5.4. Values of the first three critical speeds together with the first three natural frequencies at standstill

a single-degree-of-freedom system with a mass of 5.29 kg and a stiffness of 72.6 *kN/m. The fact that the critical speed obtained using this model is close to that obtained through more complex models shows that the first mode is actually a rigid-body mode.*

The Campbell diagram is reported in Figure 5.12. Note that although the first mode is almost independent from the speed, as expected for a transmission shaft, the other modes show a stronger dependence. This is mostly because of the polar moments of inertia of the joints.

For evaluating the unbalance response, an unbalance grade G40 (see Appendix B), typical of automotive transmission shafts, is assumed. If the maximum speed of the drive shaft is 1,000 rad/s, such unbalance corresponds to an eccentricity of about 40 μm. The response is reported in Figure 5.13, in terms of amplitude and phase of the orbit of node 10. For the computation of the amplitude when crossing the critical speed, a hysteretic damping with loss factor $\eta = 0.05$ *has been assumed for the central support. The amplitude of the orbit and the modulus of the force exerted on the body of the car during the critical speed crossing are, respectively, 310 μm and 26 N. Note that after the critical speed, the system self-centers almost completely; however, the value of the amplitude of the orbit is smaller than 40 μm, because the eccentric mass is only that of the joint, and not all the mass associated with node 10.*

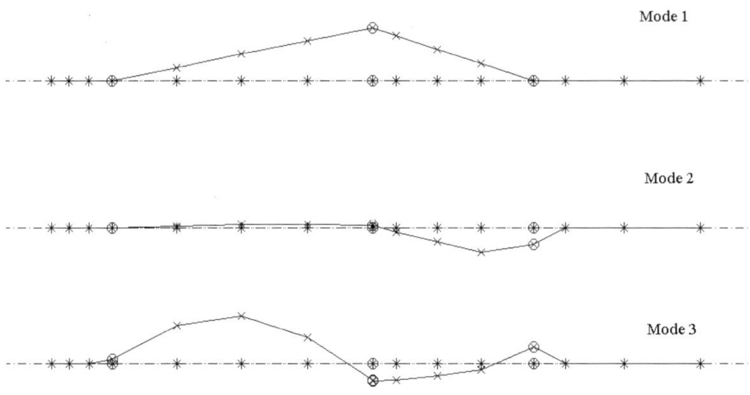

FIGURE 5.11. Modes corresponding to the first three critical speeds of the transmission shaft.

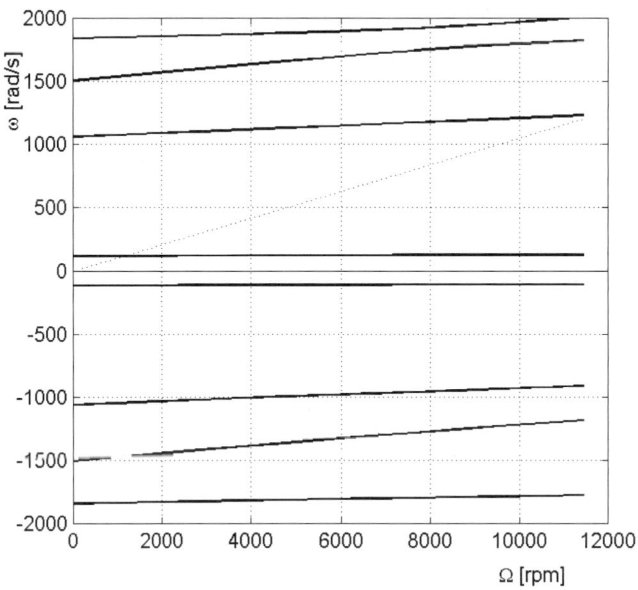

FIGURE 5.12. Campbell diagram of the transmission shaft, computed using ten master degrees of freedom.

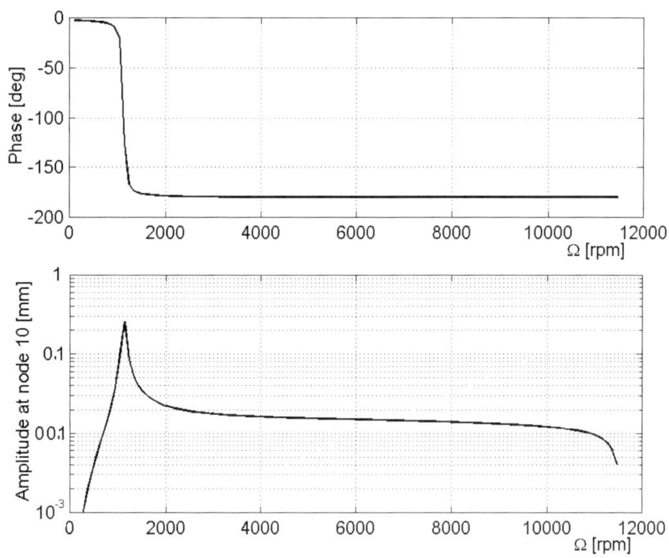

FIGURE 5.13. Response to a static unbalance (grade G40) of the central joint.

6

Anisotropy of rotors or supports

All of the models studied in the preceding sections are based on the assumptions of axial symmetry of the whole system. There are many cases, however, in which some of the parts of the machine that are included in the model do not possess axial symmetry. If either the stator or the rotor are isotropic about the axis of rotation, the rotordynamic analysis can still be performed in closed form without great difficulties. If on the contrary both lack axial symmetry, the computation of the natural frequencies, critical speed, and unbalance response becomes far more difficult and closed-form solutions are not available.

It must be noted that there are cases in which the system displays a nonisotropic behavior, even if all of the parts of the machine are geometrically axi-symmetrical. This occurs particularly when the rotor runs on lubricated journal bearings: Under the effect of external forces, the journal takes an eccentric position within the bearing and reacts in different ways to the forces in the various planes through the rotation axis.

In the following sections, the dynamic behavior of both an axially symmetrical rotor running on anisotropic supports and an asymmetrical rotor supported by isotropic bearings will be dealt with. A very simple configuration, based on the Jeffcott rotor, will be studied first. After the relevant phenomena have been qualitatively understood using this simplified model, a more complete study, allowing quantitative results to be obtained in the case of very complex systems, will be expounded.

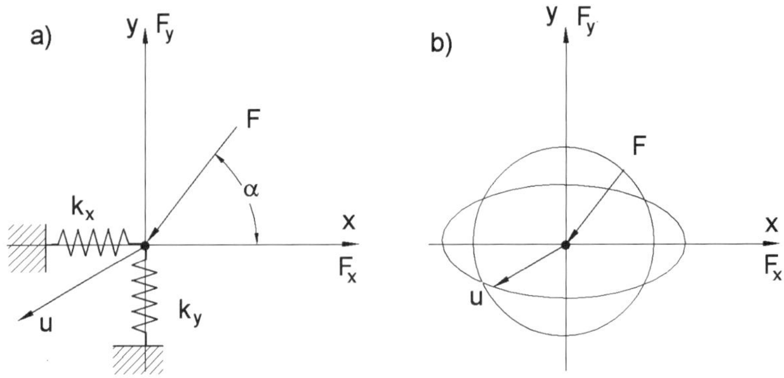

FIGURE 6.1. Force-displacement relationship for an anisotropic support. (a) Force F and displacement u; (b) polar plot of the displacements caused by a force with constant modulus and varying direction.

6.1 Isotropic rotors on anisotropic supports

6.1.1 *Jeffcott rotor on nonisotropic supports*

Consider the case of the rotor in Figure 2.1(b) but assume now that the stiffness of the supports is not isotropic in the xy-plane. All other assumptions made in Section 4.5, in particular, the linearity of the system and the assimilation of the rotor to a point mass, will be retained. The motion will be studied in the xy-plane.

The anisotropic nature of the supports makes the displacement to occur in a direction that is in general different from that of the applied force. Consider the case of Figure 6.1(a), in which the support is made by two springs acting in the direction of x- and y-axis with stiffness k_x and k_y, respectively. If the components of the force are

$$\begin{cases} F_x = F\cos(\alpha) \ , \\ F_y = F\sin(\alpha) \ , \end{cases}$$

the components of the displacement are

$$\begin{cases} u_x = \dfrac{F_x}{k_x} = \dfrac{F}{k_x}\cos(\alpha) \ , \\ \\ u_y = \dfrac{F_y}{k_y} = \dfrac{F}{k_y}\sin(\alpha) \ . \end{cases} \tag{6.1}$$

If the force is kept constant in modulus while its direction α changes, the displacement u describes an ellipse in xy-plane [Figure 6.1(b)], whereas if it is the modulus of the displacement that is kept constant, the force F

describes an ellipse in the F_xF_y-plane. The modulus of the stiffness is then

$$|k| = \frac{|F|}{|u|} = \frac{k_x k_y}{\sqrt{k_x^2 \sin^2(\alpha) + k_y^2 \cos^2(\alpha)}} \; . \tag{6.2}$$

The polar plot of the stiffness, defined as

$$\begin{cases} k_x^* = |k| \cos(\alpha) \; , \\ k_y^* = |k| \sin(\alpha) \; , \end{cases} \tag{6.3}$$

is an ellipse, the so-called *ellipse of elasticity*, provided that $|k|$ is reported in the direction of the force and not in the direction of the displacement.

If axes x and y are directed along the axes of the ellipse of elasticity, i.e., in the direction of the springs of Figure 6.1(a), the force and the displacement have the same direction

$$\begin{cases} F_x = -k_x x \; , \\ F_y = -k_y y \; . \end{cases} \tag{6.4}$$

Note that here the forces are the reactions of the support and not the forces acting on it, as shown by the negative sign.

If axes x and y are not the axes of the ellipse of elasticity, the force-displacement relationship cannot be expressed by Equation (6.4), but becomes

$$\left\{ \begin{array}{c} F_x \\ F_y \end{array} \right\} = - \left[\begin{array}{cc} k_x & k_{xy} \\ k_{yx} & k_y \end{array} \right] \left\{ \begin{array}{c} x \\ y \end{array} \right\} \tag{6.5}$$

with $k_{xy} = k_{yx}$.

In the case of the Jeffcott rotor, axes x and y will be assumed to coincide with the axes of the ellipse of elasticity, i.e., to be the principal axes of elasticity of the supporting structure, without loss of generality.

Remark 6.1 *In the case of rotors with a larger number of degrees of freedom, the ellipses of elasticity at the various nodes can be directed in a different way and no reference frame in which Equation (6.4) holds may exist.*

Assume that the stiffness along the x-direction is lower than that along the y-direction. By introducing the two different values of the stiffness into the equation of motion (2.6), the latter transforms into

$$\begin{cases} m\ddot{x} + k_x x = m\epsilon\Omega^2 \cos(\Omega t) + F_x \, , \\ m\ddot{y} + k_y y = m\epsilon\Omega^2 \sin(\Omega t) + F_y \, . \end{cases} \tag{6.6}$$

From the homogeneous equation, it is clear that there are two natural frequencies, one (the lower) related to the motion in the xz-plane and the other related to the motion in the yz-plane:

$$\omega_{n_1} = \sqrt{\frac{k_x}{m}} \, , \qquad \omega_{n_2} = \sqrt{\frac{k_y}{m}} \, . \tag{6.7}$$

They are not influenced by the spin speed, and then the Campbell diagram is made by two horizontal straight lines. Note that in this case, the free motions in the two planes occur at different frequencies, so that the two harmonic motions cannot combine to make circles or ellipses.

The fact that the two natural frequencies are independent of the spin speed causes the two critical speeds to coincide with them:

$$\Omega_{cr_1} = \omega_{n_1} = \sqrt{\frac{k_x}{m}}, \qquad \Omega_{cr_2} = \omega_{n_2} = \sqrt{\frac{k_y}{m}}. \tag{6.8}$$

At the first critical speed, the motion reduces to a straight vibration along the x-axis, and at the other critical speed, it reduces to a straight motion along the y-axis.

The unbalance response can be obtained directly from Equation (6.6), assuming a solution of the type

$$\begin{cases} x = x_0 \cos(\Omega t), \\ y = y_0 \sin(\Omega t). \end{cases} \tag{6.9}$$

The response in each plane is then equal to the response of the Jeffcott rotor, computed using the stiffness related to that plane. The amplitudes of the two harmonic motions are

$$x_0 = \frac{m\epsilon\Omega^2}{k_x - m\Omega^2}, \qquad y_0 = \frac{m\epsilon\Omega^2}{k_y - m\Omega^2}, \tag{6.10}$$

and the orbit in xy-plane is an ellipse whose axes are directed along x- and y-axes.

The unbalance response can be subdivided into three speed ranges (Figure 6.2):

- From standstill to the first critical speed [Figure 6.3(a) and (b)]: The responses in the two planes have the same sign and are out of phase from each other by 90°, as clearly seen from Equations (6.9) and (6.10), where the response is expressed by a sine and a cosine function. The orbit grows mainly along the x-axis and has the shape of an elongated ellipse. Approaching the first critical speed, the axis of the orbit along the x-axis tends to infinity.

- From the first to the second critical speed [Figure 6.3(c), (d), and (e)]: The response along the x-axis is negative, having already crossed the critical speed; that along the y-axis is positive. When they combine, they give way to an elliptical motion in the backward direction. Near the first critical speed, the ellipse is elongated along the x-axis (c), and near the second one, it is elongated in the other direction (e). There is an intermediate speed at which the amplitudes in the two planes are equal and the orbit is circular (d). Unbalance, an excitation that by definition is applied in the forward direction, can thus excite a backward synchronous whirling of the rotor.

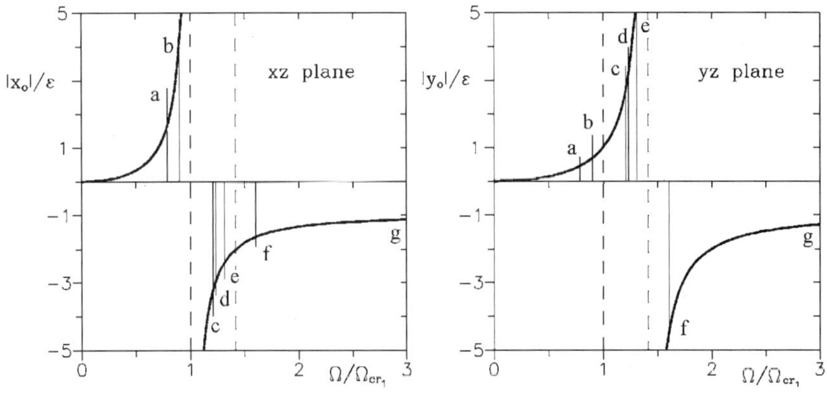

FIGURE 6.2. Jeffcott rotor on anisotropic supports. Amplitude of the motion along x-axis (where the stiffness is lower) and along y-axis.

- Above the second critical speed [Figure 6.3(f) and (g)]: The signs of the amplitudes in both planes are the same, and they both tend to the same value, namely, to $-\epsilon$. An elliptic forward whirling that tends to become circular with increasing speed is so obtained; self-centering takes place normally. It could easily be expected that in the high supercritical field, the elastic anisotropy has little influence on the dynamic behavior of the rotor: The behavior of the system is dominated by inertia forces that are clearly isotropic.

The behavior of the Jeffcott rotor on anisotropic supports was studied using real coordinates. Although not very common, it is also possible to use complex coordinates to study nonsymmetric systems. As usual, it is possible to add the first Equation (6.6) to the second equation multiplied by the imaginary unit i. By introducing the mean stiffness

$$k_m = (k_x + k_y)/2 \tag{6.11}$$

and the deviatoric stiffness

$$k_d = (k_x - k_y)/2 \ , \tag{6.12}$$

the equation for the unbalance response in terms of complex coordinates is

$$m\ddot{r} + k_m r + k_d \overline{r} = m\epsilon\Omega^2 e^{i\Omega t} \ , \tag{6.13}$$

where $\overline{r}$ is the conjugate of the complex number r. The solution of the homogeneous equation is of the type

$$r = r_1 e^{i\omega t} + r_2 e^{-i\overline{\omega} t} \ ,$$

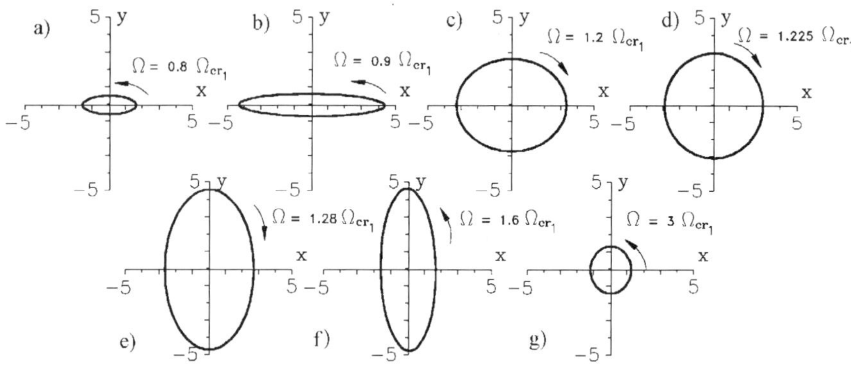

FIGURE 6.3. Jeffcott rotor on anisotropic supports. Orbits at different values of the speed. (a) and (b) subcritical; (c), (d), and (e) between the critical speeds; (f) and (g) supercritical (the letters refer to Figure 6.2).

which gives way to elliptical orbits. It can be introduced into the homogeneous equation of motion, yielding

$$\left(-\omega^2\begin{bmatrix} m & 0 \\ 0 & m \end{bmatrix}+\begin{bmatrix} k_m & k_d \\ k_d & k_m \end{bmatrix}\right)\left\{\begin{array}{c} r_1 \\ \overline{r}_2 \end{array}\right\}=\left\{\begin{array}{c} 0 \\ 0 \end{array}\right\}. \tag{6.14}$$

Equation (6.14) yields an eigenproblem in ω, yielding the values of the whirl frequencies coinciding with those expressed by Equation (6.7).

The particular integral allowing us to compute the unbalance response is

$$r = r_1 e^{i\Omega t} + r_2 e^{-i\Omega t}.$$

By introducing it into the equation of motion and solving for the amplitudes of the forward and backward components r_1 and r_2, the following unbalance response is obtained:

$$r = \frac{m\epsilon\Omega^2}{(k_x - m\Omega^2)(k_y - m\Omega^2)}\left[(k_m - m\Omega^2)e^{i\Omega t} - k_d e^{-i\Omega t}\right]. \tag{6.15}$$

It is easy to demonstrate that the orbits expressed by Equation (6.15) coincide with those expressed by Equation (6.10). The possibility of backward whirling to be excited by unbalance is, however, more obvious when complex coordinates are used: At the speed $\Omega_b = \sqrt{k_m/m}$, the amplitude of the forward component vanishes, and the orbit is a circular backward whirl with amplitude $z_{0_b} = \epsilon k_d/k_m$.

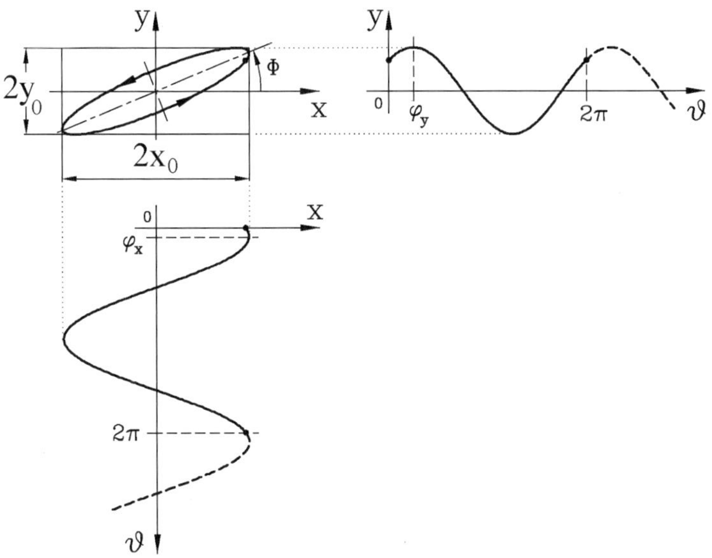

FIGURE 6.4. Composition of the harmonic motions in the direction of x- and y-axes into an ellipse in xy-plane (peak-to-peak amplitudes $xpp = 2x_0$ and.$ypp = 2y_0$).

6.1.2 Effect of damping

If the system is damped, the equation of motion can be written in the form

$$\begin{bmatrix} m & 0 \\ 0 & m \end{bmatrix}\begin{Bmatrix} \ddot{x} \\ \ddot{y} \end{Bmatrix}+\begin{bmatrix} c_x + c_r & 0 \\ 0 & c_y + c_r \end{bmatrix}\begin{Bmatrix} \dot{x} \\ \dot{y} \end{Bmatrix}+ \\ +\begin{bmatrix} k_x & \Omega c_r \\ -\Omega c_r & k_y \end{bmatrix}\begin{Bmatrix} x \\ y \end{Bmatrix}=m\epsilon\Omega^2\begin{Bmatrix} \cos(\Omega t) \\ \sin(\Omega t) \end{Bmatrix}. \tag{6.16}$$

Although in the case of undamped systems [Equation (6.6)] the unbalance response takes the form of an ellipse with both axes aligned with the principal axes of elasticity of the supports (x- and y-axes), now the ellipse is rotated; i.e., the phases of the two harmonic motions are no more equal to 0° and 90°(Figure 6.4). The unbalance response can thus be written in the form

$$\begin{cases} x = x_0 \cos(\Omega t - \varphi_x) \,, \\ y = y_0 \cos(\Omega t - \varphi_y) \,, \end{cases} \tag{6.17}$$

where y_0 and x_0 are the peak amplitudes of the displacements along the axes and φ_x and φ_y are the phase delays with respect to the common phase reference. By introducing the solution (6.17), the average parameters of the system

$$\Omega_0 = \sqrt{\frac{k_x + k_y}{2m}} = \sqrt{\frac{k_m}{m}}, \tag{6.18}$$

$$\zeta_n = \frac{c_{n_x} + c_{n_y}}{2\sqrt{2m(k_x + k_y)}}, \qquad \zeta_r = \frac{c_r}{2\sqrt{2m(k_x + k_y)}}, \tag{6.19}$$

the anisotropy parameters

$$\alpha = \frac{k_x - k_y}{k_x + k_y}, \qquad \beta = \frac{c_{n_x} - c_{n_y}}{c_{n_x} + c_{n_y}}, \tag{6.20}$$

and the nondimensional speed

$$\Omega^* = \frac{\Omega}{\Omega_0}, \tag{6.21}$$

into Equation (6.16), the equation yielding the nondimensional amplitudes x_0/ϵ and y_0/ϵ and the phases of the harmonic motions along the two axes becomes

$$\begin{bmatrix} 1+\alpha-\Omega^{*2} & 2\Omega^* R & 2\Omega^*\xi_r & 0 \\ 2\Omega^* R & 1+\alpha-\Omega^{*2} & 0 & 2\Omega^*\xi_r \\ -2\Omega^*\xi_r & 0 & 1-\alpha-\Omega^{*2} & 2\Omega^* S \\ 0 & -2\Omega^*\xi_r & 2\Omega^* S & 1-\alpha-\Omega^{*2} \end{bmatrix} \begin{Bmatrix} x_0\cos\varphi_x \\ x_0\sin\varphi_x \\ y_0\cos\varphi_y \\ y_0\sin\varphi_y \end{Bmatrix} = \epsilon\Omega^{*2} \begin{Bmatrix} 1 \\ 0 \\ 0 \\ 1 \end{Bmatrix}, \tag{6.22}$$

where

$$R = \xi_n(1+\beta) + \xi_r \qquad \text{and} \qquad S = \xi_n(1-\beta) + \xi_r\,.$$

Also in the case of damped systems, a range in which backward whirling occurs may exist between the two critical speeds of the undamped system, located at speeds equal to $\sqrt{k_x/m}$ and $\sqrt{k_y/m}$. This range is smaller than that characterizing the undamped system, which spans from one critical speed to the other, and reduces with increasing nonrotating damping. If the latter is large enough, it disappears altogether.

To assess the direction of the whirling motion, it is expedient to introduce the difference of the two phases

$$\varphi = \varphi_y - \varphi_x\,. \tag{6.23}$$

Defining angle $\vartheta = \Omega t - \varphi_y$, the equations describing the elliptical orbit (6.17) become

$$\begin{cases} x = x_0\cos(\vartheta + \varphi)\,, \\ y = y_0\cos\vartheta\,. \end{cases} \tag{6.24}$$

Note that during a whole revolution of the shaft, $\vartheta \in (0, 2\pi)$.

To demonstrate the dependence of the whirling direction with φ, define γ as the angle of vector $\overline{C-O}$ with x-axis

$$\gamma = \arctan\left(\frac{y}{x}\right) = \arctan\left[\frac{y_0 \cos\vartheta}{x_0 \cos(\vartheta + \varphi)}\right] . \tag{6.25}$$

The velocity at which such vector rotates is

$$\frac{d\gamma}{dt} = \Omega \frac{d\gamma(\Omega)}{d\vartheta} = \frac{\Omega\, x_0 y_0 \sin\varphi}{x_0^2 \cos^2(\vartheta + \varphi) + y_0^2 \cos^2\vartheta} . \tag{6.26}$$

For $\vartheta = 0$, it becomes

$$\frac{d\gamma}{dt} = \Omega \frac{x_0 y_0 \sin\varphi}{x_0^2 \cos^2\varphi + y_0^2} . \tag{6.27}$$

The sign of $d\gamma/dt$ is determined by the sign of its numerator, i.e., by the sign of $\sin\varphi$, positive for $\varphi \in (0°, 180°)$ (whirling direction coincident with the rotational speed, forward whirling), negative for $\varphi \in (180°, 360°)$ (backward whirling). The speeds at which the reversal of the whirl speed occurs are thus identified by the condition $\varphi = 180°$.

The nondimensional amplitudes and angle φ for a rotor with $k_y/k_x = 1.5$, ($\alpha = -0.2$), $\zeta_n = 0.05$, $\zeta_r = 0$, and $\beta = 0$ are reported as functions of the nondimensional spin speed Ω^* in Figure 6.5(a).

At low speed, the orbit grows as an ellipse with larger axis in the direction in which the supports are softer, and at the critical speed, its amplitude is limited and its direction is slightly misaligned with the principal direction of the supports [Figure 6.5(b)]. The orbit then starts to rotate in the xy-plane and to become thinner: If at a certain speed it reduces to a line, a reversal of the direction of whirling occurs. The backward orbit continues to rotate, getting more circular, and at a certain speed, circular backward whirling may or may not take place. Then the ellipse becomes very thin again, reduces to a line, returns to forward whirling, and after a rotation of almost 90°, the crossing of the second critical speed takes place. After the second critical speed, the orbit tends again to circularize, until self-centering occurs at high speed.

This behavior is clearly shown in the orbital tube representation and its related views of Figures 6.6 and 6.7.

Remark 6.2 *As the orbit is not circular, the unbalance response is influenced by rotating damping as well as nonrotating damping.*

Remark 6.3 *The main feature of rotors on anisotropic supports is the eventual presence of backward whirling in the unbalance response: Because acting in forward direction (unbalance) can thus produce a backward response.*

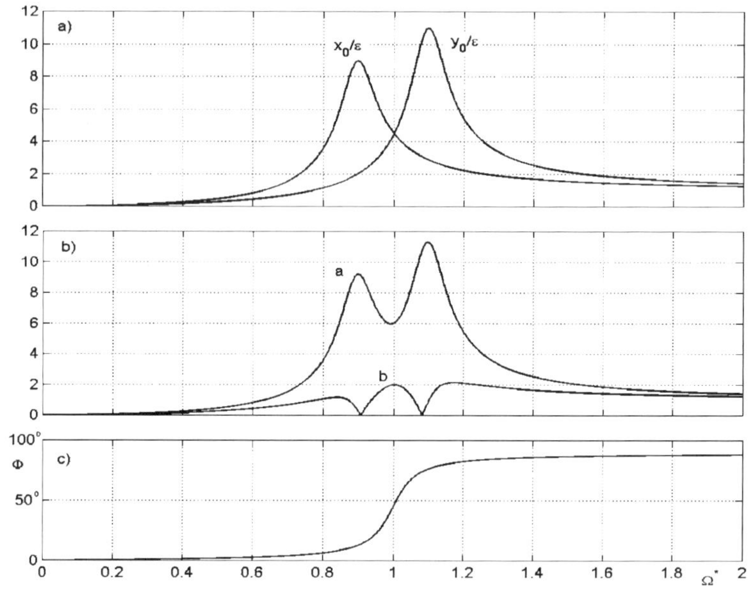

FIGURE 6.5. (a) Nondimensional amplitudes and angle φ for a rotor with $k_y/k_x = 1.5$ ($\alpha = -0.2$), $\zeta_n = 0.05$, $\zeta_r = 0$, and $\beta = 0$ as functions of the nondimensional spin speed Ω^*. (b) Nondimensional values of the major axis a and the minor axis b of the elliptical orbit. (c) Inclination Φ of the orbit (defined in Figure 6.4) as functions of the nondimensional speed. The minor axis vanishes at the speed in which the reversal of the whirling direction occurs.

Example 6.1 *First vibrational mode of the motor car transmission shaft studied as Example 5.3, but on anisotropic supports.*

Consider the propeller shaft of a motor car with front engine and gearbox and rear-wheel drive already studied in the mentioned example (Figure 5.10) and assume that the central support is not isotropic, with a stiffness of 120 kN/m in vertical direction and of 80 kN/m in horizontal direction, instead of being isotropic with a stiffness of 100 kN/m.

By proceeding in the same way already seen in Example 5.3, a Jeffcott rotor model with mass $m = 5.292$ *kg and two values of the stiffness equal to 87.12 kN/m and 58.08 kN/m in vertical and horizontal direction, respectively, are obtained. This leads to a mean stiffness* $k_m = 72.58$ *kN/m and a deviatoric stiffness* $k_d =$ 14.43 *kN/m.*

The natural frequencies are 128.3 rad/s and 104.8 rad/s instead of 117.1 rad/s.

6.1.3 System with many degrees of freedom

The equation of motion of a general multi-degrees-of-freedom axi-symmetrical system is Equation (4.75), if real coordinates with the standard definition of

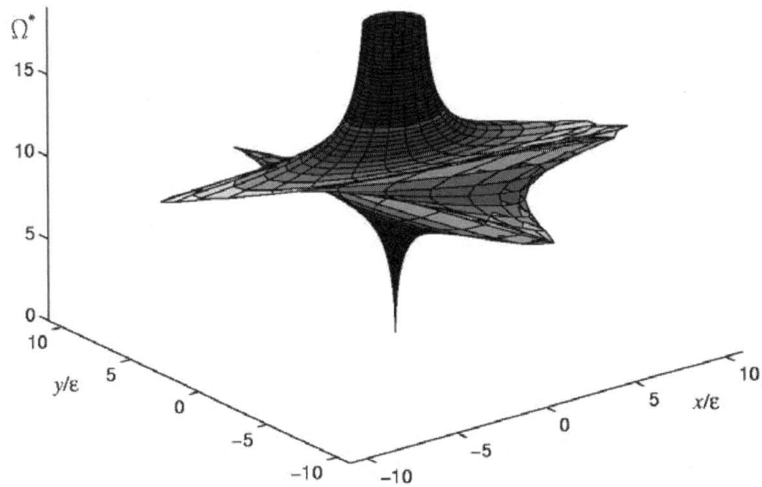

FIGURE 6.6. Nondimensional orbital tube for the Jeffcott rotor on anisotropic supports studied in the previous figures. The color scale (in the black and white representation the gray scale) is used to stress the information on the amplitude of the orbit.

the rotational degrees of freedom are used. For convenience, that equation is repeated here

$$\begin{bmatrix} \mathbf{M}_x & \mathbf{0} \\ \mathbf{0} & \mathbf{M}_y \end{bmatrix} \ddot{\mathbf{x}}^* + \left(\begin{bmatrix} (\mathbf{C}_n + \mathbf{C}_r)_x & \mathbf{0} \\ \mathbf{0} & (\mathbf{C}_n + \mathbf{C}_r)_y \end{bmatrix} + \right.$$
$$\left. + \Omega \begin{bmatrix} \mathbf{0} & \mathbf{G}_{xy} \\ -\mathbf{G}_{yx} & \mathbf{0} \end{bmatrix} \right) \dot{\mathbf{x}}^* + \left(\begin{bmatrix} \left(\mathbf{K} + \mathbf{K}_\Omega \Omega^2\right)_x & \mathbf{0} \\ \mathbf{0} & \left(\mathbf{K} + \mathbf{K}_\Omega \Omega^2\right)_y \end{bmatrix} + \right.$$
$$\left. + \Omega \begin{bmatrix} \mathbf{0} & \mathbf{C}_{r_{xy}} \\ -\mathbf{C}_{r_{yx}} & \mathbf{0} \end{bmatrix} \right) \mathbf{x}^* = \Omega^2 \left\{ \begin{array}{c} \mathbf{f}_x \cos(\Omega t) - \mathbf{f}_y \sin(\Omega t) \\ \mathbf{f}_x \sin(\Omega t) + \mathbf{f}_y \cos(\Omega t) \end{array} \right\} .$$

Matrices with subscripts x and y differ from each other only because of the usual sign conventions used for xz- and yz-planes: If the degrees of freedom are listed in such a way that the elements of vector $\mathbf{x}^*$ with odd subscript are related to translations and those with even subscript to rotations, all elements of the relevant matrices with subscript x or xy are identical to those of the matrices with subscripts y or yx if their subscripts add to give an even number. They have equal absolute value but opposite sign otherwise.

The same equation can be written in the form of Equation (4.74)

$$\begin{bmatrix} \mathbf{M} & \mathbf{0} \\ \mathbf{0} & \mathbf{M} \end{bmatrix} \ddot{\mathbf{x}} + \left(\Omega \begin{bmatrix} \mathbf{0} & \mathbf{G} \\ -\mathbf{G} & \mathbf{0} \end{bmatrix} + \begin{bmatrix} \mathbf{C}_n + \mathbf{C}_r & \mathbf{0} \\ \mathbf{0} & \mathbf{C}_n + \mathbf{C}_r \end{bmatrix} \right) \dot{\mathbf{x}} +$$

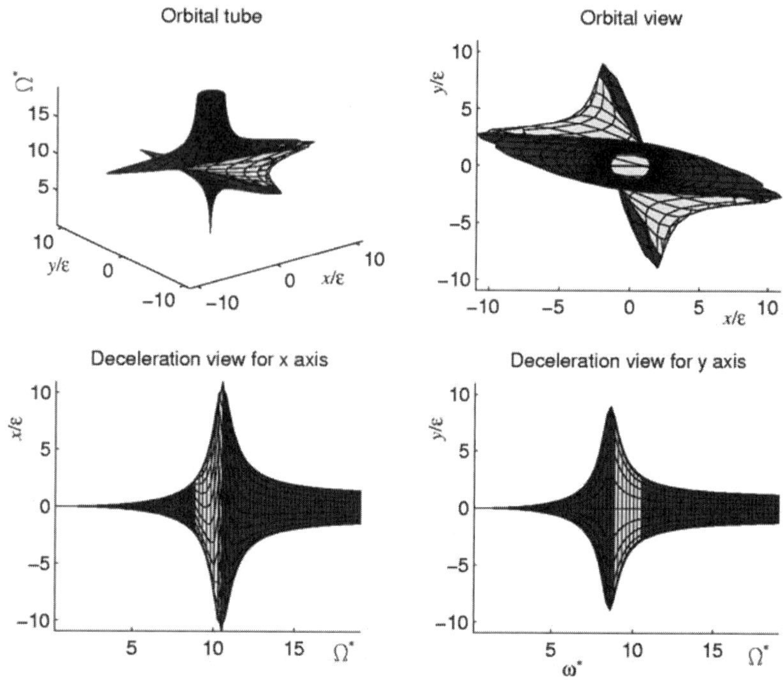

FIGURE 6.7. Nondimensional unbalance response of the Jeffcott rotor on anisotropic supports studied in the previous figures. (a) Orbital tube; (b) orbital view; (c) and (d) projections on the Ωx- and Ωy-planes, respectively. The light zone identifies the speed range in which backward whirling occurs.

$$\left(\begin{bmatrix} \mathbf{K}+\mathbf{K}_\Omega\Omega^2 & \mathbf{0} \\ \mathbf{0} & \mathbf{K}+\mathbf{K}_\Omega\Omega^2 \end{bmatrix} + \Omega\begin{bmatrix} \mathbf{0} & \mathbf{C}_r \\ -\mathbf{C}_r & \mathbf{0} \end{bmatrix}\right)\mathbf{x} = \Omega^2\left\{\begin{array}{c} \mathrm{Re}(\mathbf{f}e^{i\Omega t}) \\ \mathrm{Im}(\mathbf{f}e^{i\Omega t}) \end{array}\right\},$$

if the coordinates related to rotations about x-axis, and their related generalized forces (moments) are taken with the $(-)$ sign.

If complex coordinates are used, the relevant equation of motion is Equation (4.73)

$$\mathbf{M}\ddot{\mathbf{q}} + (\mathbf{C}_n + \mathbf{C}_r - i\Omega\mathbf{G})\dot{\mathbf{q}} + (\mathbf{K} + \mathbf{K}_\Omega\Omega^2 - i\Omega\mathbf{C}_r)\mathbf{q} = \Omega^2\mathbf{f}e^{i\Omega t}.$$

The stiffness matrix for the flexural behavior in xz- and yz-planes of a general nonisotropic system can be written in the form

$$\mathbf{K} = \begin{bmatrix} \mathbf{K}_x & \mathbf{K}_{xy} \\ \mathbf{K}_{yx} & \mathbf{K}_y \end{bmatrix}, \tag{6.28}$$

where submatrices $\mathbf{K}_x$ and $\mathbf{K}_y$ are different from each other and matrices $\mathbf{K}_{xy}$ and $\mathbf{K}_{yx}$ are nonvanishing except in the case in which axes x and y are principal axes of elasticity for the whole system. However, they are usually linked by the relationship $\mathbf{K}_{xy} = \mathbf{K}_{yx}^T$, which comes from the consideration that the stiffness matrix must be symmetrical if it can be obtained from the potential energy of the system. An important exception to this rule is the case of the elements used to model lubricated bearings: When linearizing their behavior, the stiffness matrix has components that should be more properly included into the circulatory matrix, because they are skew symmetric.

The same considerations hold also for the mass and damping matrices. The matrices of the stator take then a form similar to Equation (6.28), whereas those related to the rotor are those typical of isotropic system. The overall matrices of the system (except for the rotating damping matrix) have the form of Equation (6.28), and the equation of motion of the rotor is

$$\begin{bmatrix} \mathbf{M}_x & \mathbf{M}_{xy} \\ \mathbf{M}_{yx} & \mathbf{M}_y \end{bmatrix} \ddot{\mathbf{x}} + \left(\Omega \begin{bmatrix} \mathbf{0} & \mathbf{G} \\ -\mathbf{G} & \mathbf{0} \end{bmatrix} + \begin{bmatrix} \mathbf{C}_x & \mathbf{C}_{xy} \\ \mathbf{C}_{yx} & \mathbf{C}_y \end{bmatrix} \right) \dot{\mathbf{x}} +$$
$$+ \left(\begin{bmatrix} \mathbf{K}_x & \mathbf{K}_{xy} \\ \mathbf{K}_{yx} & \mathbf{K}_y \end{bmatrix} + +\Omega \begin{bmatrix} \mathbf{0} & \mathbf{C}_r \\ -\mathbf{C}_r & \mathbf{0} \end{bmatrix} \right) \mathbf{x} = \tag{6.29}$$
$$= \Omega^2 \left\{ \begin{array}{c} \mathrm{Re}(\mathbf{f}_r e^{i\Omega t}) \\ \mathrm{Im}(\mathbf{f}_r e^{i\Omega t}) \end{array} \right\} + \left\{ \begin{array}{c} \mathrm{Re}(\mathbf{f}_n) \\ \mathrm{Im}(\mathbf{f}_n) \end{array} \right\} ,$$

where also a vector of nonrotating forces $\mathbf{f}_n$ has been added to the forces caused by unbalance $\mathbf{f}_r$ and the real-coordinates vector is defined as in Equation (4.74).

Also in the case of a nonisotropic system, it is possible to resort to complex coordinates, by introducing the mean and deviatoric matrices. For the stiffness matrix, they are defined as

$$\begin{cases} \mathbf{K}_m = \frac{1}{2}(\mathbf{K}_x + \mathbf{K}_y) + i\frac{1}{2}(\mathbf{K}_{yx} - \mathbf{K}_{xy}) \ , \\ \mathbf{K}_d = \frac{1}{2}(\mathbf{K}_x - \mathbf{K}_y) + i\frac{1}{2}(\mathbf{K}_{yx} + \mathbf{K}_{xy}) \ . \end{cases} \tag{6.30}$$

Note that, except in the mentioned case of elements used for the linearized modeling of hydrodynamic bearings, the matrices with subscripts xy and yx are equal and the mean matrices are real. On the contrary, deviatoric matrices are, in general, complex. Similarly, it is possible to define also mean and deviatoric matrices for the mass and nonrotating damping properties of the system. Note that, owing to the isotropy of the rotor, no rotating damping deviatoric matrix exists. Using the complex-coordinates approach and the definitions of mean and deviatoric matrices given by Equation (6.30), the equation of motion describing the flexural behavior of a general system with an isotropic rotor and a nonisotropic stator can be shown to be [41]

$$\mathbf{M}_m\ddot{\mathbf{q}} + (\mathbf{C}_m - i\Omega\mathbf{G})\dot{\mathbf{q}} + (\mathbf{K}_m - i\Omega\mathbf{C}_{r_m})\mathbf{q} + \mathbf{M}_{n_d}\ddot{\overline{\mathbf{q}}} +$$
$$+\mathbf{C}_{n_d}\dot{\overline{\mathbf{q}}} + \mathbf{K}_d\overline{\mathbf{q}} = \mathbf{F}_n + \Omega^2\mathbf{F}_r e^{i\Omega t}\ , \tag{6.31}$$

where $\overline{\mathbf{q}}$ is the conjugate of $\mathbf{q}$.

Equation (6.31) involves actually working with complex coordinates, because deviatoric matrices are generally complex, and consequently, the advantage of resorting to complex coordinates depends on the time and cost-effectiveness of the available subroutines for computations involving complex numbers.

The solution for static loading is similar to the corresponding solution for axi-symmetrical systems, i.e., a constant vector $\mathbf{x} = \mathbf{x}_0$ (or $\mathbf{q} = \mathbf{q}_0$ if complex coordinates are used), leading to the equation

$$\left(\begin{bmatrix} \mathbf{K}_x & \mathbf{K}_{xy} \\ \mathbf{K}_{yx} & \mathbf{K}_y \end{bmatrix} + \Omega\begin{bmatrix} \mathbf{0} & \mathbf{C}_r \\ -\mathbf{C}_r & \mathbf{0} \end{bmatrix}\right)\mathbf{x}_0 = \left\{\begin{array}{c} \mathrm{Re}(\mathbf{f}_n) \\ \mathrm{Im}(\mathbf{f}_n) \end{array}\right\}. \tag{6.32}$$

or, if complex coordinates are used,

$$(\mathbf{K}_m - i\Omega\mathbf{C}_{r_m})\mathbf{q}_0 + \mathbf{K}_d\overline{\mathbf{q}}_0 = \mathbf{f}_n\ . \tag{6.33}$$

The inflected shape is a line (generally a skew line) fixed in space. Rotating damping couples the behavior in the xz- and yz-planes, even if the coordinate planes are planes of symmetry for the stator.

The unbalance response is a synchronous elliptical whirling. The solution of the equation of motion can be expressed in the form

$$\mathbf{q} = \mathbf{q}_1 e^{i\Omega t} + \mathbf{q}_2 e^{-i\Omega t}\ ,$$

i.e., as the sum of two circular whirling motions taking place at speed Ω in opposite directions. Both $\mathbf{q}_1$ and $\mathbf{q}_2$ are, generally speaking, complex vectors that physically correspond to elliptical orbits not having axes x and y as axes of symmetry. The unknowns of the problem are then $4n$ in number, i.e., the imaginary and real parts of two vectors of size n. By introducing this solution into the equation of motion (6.31), the latter yields

$$\begin{bmatrix} \mathbf{A}_{11} & \mathbf{A}_{12} \\ \overline{\mathbf{A}}_{21} & \overline{\mathbf{A}}_{22} \end{bmatrix}\left\{\begin{array}{c} \mathbf{q}_1 \\ \overline{\mathbf{q}}_2 \end{array}\right\} = \Omega^2\left\{\begin{array}{c} \mathbf{f}_r \\ \mathbf{0} \end{array}\right\}, \tag{6.34}$$

where

$$\begin{aligned} \mathbf{A}_{11} &= -\Omega^2(\mathbf{M}_m - \mathbf{G}) + i\Omega\mathbf{C}_{n_m} + \mathbf{K}_m\ , \\ \mathbf{A}_{12} &= -\Omega^2\mathbf{M}_d + i\Omega\mathbf{C}_{n_d} + \mathbf{K}_d\ , \\ \mathbf{A}_{21} &= -\Omega^2\mathbf{M}_d - i\Omega\mathbf{C}_{n_d} + \mathbf{K}_d\ , \\ \mathbf{A}_{22} &= -\Omega^2(\mathbf{M}_m + \mathbf{G}) - i\Omega(\mathbf{C}_{n_m} + 2\mathbf{C}_r) + \mathbf{K}_m\ . \end{aligned} \tag{6.35}$$

Remark 6.4 *As already noted for the Jeffcott rotor on nonisotropic supports, rotating damping now enters the equation yielding the unbalance response: The shaft no longer rotates in the deformed configuration but actually vibrates, in the sense that each part of it experiences stresses that vary with time.*

In the case of undamped systems, Equation (6.34) reduces to

$$\left(\begin{bmatrix} \mathbf{K}_m & \mathbf{K}_{n_d} \\ \overline{\mathbf{K}}_{n_d} & \overline{\mathbf{K}}_m \end{bmatrix} - \Omega^2 \begin{bmatrix} \mathbf{M}_m - \mathbf{G} & \mathbf{M}_{n_d} \\ \overline{\mathbf{M}}_{n_d} & \overline{\mathbf{M}}_m + \mathbf{G} \end{bmatrix}\right)\left\{\begin{array}{c} \mathbf{q}_1 \\ \overline{\mathbf{q}}_2 \end{array}\right\} = $$
$$= \Omega^2 \left\{\begin{array}{c} \mathbf{f}_r \\ \mathbf{0} \end{array}\right\}, \tag{6.36}$$

which is real if the stator is symmetrical with respect to the coordinate planes. Note that the mean mass matrix is always real and coincides with its conjugate.

By equating to zero the determinant of the matrix of the coefficients of Equation (6.36), an eigenproblem in Ω^2 is obtained, which allows the critical speeds to be computed. At certain speeds, vector $\mathbf{q}_1$ vanishes; this physically corresponds to a circular backward whirling motion caused by unbalance, as was shown in Section 6.1 for the Jeffcott rotor.

The equation corresponding to Equation (6.36), but obtained using real coordinates, is very similar, and the complexity of the actual computations to be performed is also similar.

In the case of free whirling, the orbits of the system are elliptical. The relevant solution of the homogeneous equation of motion is of the type

$$\mathbf{q} = \mathbf{q}_1 e^{i\omega t} + \mathbf{q}_2 e^{-i\overline{\omega} t},$$

which leads to the following algebraic equation:

$$\left(-\omega^2 \begin{bmatrix} \mathbf{M}_m & \mathbf{M}_d \\ \overline{\mathbf{M}}_d & \mathbf{M}_m \end{bmatrix} + \omega\Omega \begin{bmatrix} \mathbf{G} & \mathbf{0} \\ \mathbf{0} & -\mathbf{G} \end{bmatrix} + i\omega \begin{bmatrix} \mathbf{C}_{n_m} & \mathbf{C}_{n_d} \\ \overline{\mathbf{C}}_{n_d} & \overline{\mathbf{C}}_{n_m} \end{bmatrix} + \right.$$
$$\left. + \begin{bmatrix} \mathbf{K}_m & \mathbf{K}_d \\ \overline{\mathbf{K}}_d & \overline{\mathbf{K}}_m \end{bmatrix} - i\Omega \begin{bmatrix} \mathbf{C}_r & \mathbf{0} \\ \mathbf{0} & -\mathbf{C}_r \end{bmatrix}\right)\left\{\begin{array}{c} \mathbf{q}_1 \\ \overline{\mathbf{q}}_2 \end{array}\right\} = \mathbf{0}\,. \tag{6.37}$$

The corresponding solution in terms of real coordinates is $\mathbf{x} = \mathrm{Re}(\mathbf{x}_0 e^{st})$, which yields the following algebraic equation:

$$\left(s^2 \begin{bmatrix} \mathbf{M}_x & \mathbf{M}_{xy} \\ \mathbf{M}_{yx} & \mathbf{M}_y \end{bmatrix} + s\Omega \begin{bmatrix} \mathbf{0} & \mathbf{G} \\ -\mathbf{G} & \mathbf{0} \end{bmatrix} + s \begin{bmatrix} \mathbf{C}_{n_x} + \mathbf{C}_r & \mathbf{C}_{n_{xy}} \\ \mathbf{C}_{n_{yx}} & \mathbf{C}_{n_y} + \mathbf{C}_r \end{bmatrix} + \right.$$
$$\left. + \begin{bmatrix} \mathbf{K}_x & \mathbf{K}_{xy} \\ \mathbf{K}_{yx} & \mathbf{K}_y \end{bmatrix} + \Omega \begin{bmatrix} \mathbf{0} & \mathbf{C}_r \\ -\mathbf{C}_r & \mathbf{0} \end{bmatrix}\right)\mathbf{x}_0 = \mathbf{0}\,. \tag{6.38}$$

	ω_n[rad/s]	
	Isotropic	Non isotropic
I	117	104.7 128.1
II	1061	1061.1 1061.3
III	1507	1507.0 1507.2

TABLE 6.1. Values of the first six natural frequencies at standstill for the transmission shaft on nonisotropic supports.

In the case of an undamped system whose stator is symmetrical with respect to the coordinate planes, the two approaches are exactly equivalent, because both lead to a set of $2n$ real algebraic equations. The eigenvalues from Equation (6.37) are real, whereas those from Equation (6.38) are imaginary. The eigenvectors of the former are real, and those of the latter are made up of real and imaginary terms, depending on the phasing of the various motions added to give the various orbits. In the most general case, Equation (6.37) yields a set of $2n$ complex equations, and Equations (6.38) are always real. In the author's opinion, however, the physical interpretation is more straightforward in the case of the former equation, even if it is not sufficient to find out the sign of the eigenvalue ω to assess whether the whirl motion occurs in the forward or backward direction. In fact, if ω is a solution of the eigenproblem, $-\overline{\omega}$ is also a solution, and consequently, each mode is found twice, with opposite signs of $\text{Re}(\omega)$.

Remark 6.5 *There are modes, sometimes referred to as* mixed modes, *in which the whirling occurs in the forward direction at some points of the rotor and in the backward direction at other points.*

Although only the study of the eigenvectors can make clear which mode occurs in the forward or backward direction, the authors feel that the physical interpretation of the solution is somehow more clear when using complex coordinates.

Example 6.2 *Transmission shaft on anisotropic supports.*

Consider the propeller shaft of a motor car with front engine and gearbox and rear-wheel drive already studied as Example 5.3 (Figure 5.10). Assume that the central support is not isotropic, with a stiffness of 120 kN/m in vertical direction and of 80 kN/m in horizontal direction, instead of the isotropic value of 100 kN/m assumed in the previous example.

By proceeding in the same way already seen in Example 5.5.1, the first six values of the natural frequencies, computed using ten master degrees of freedom, at standstill are reported in Table 6.1.

The lack of symmetry of the central support has a large effect on the first mode, which were computed with very good accuracy by using the Jeffcott rotor

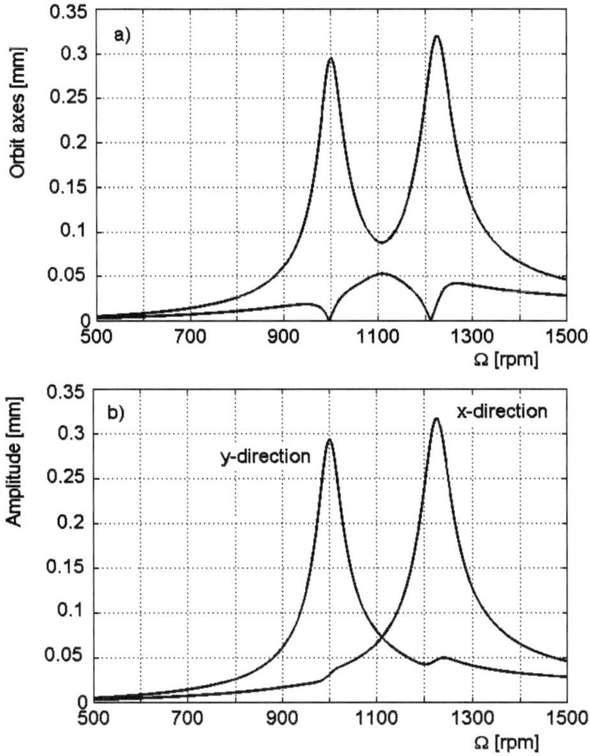

FIGURE 6.8. Unbalance response at node 10 in a speed range close to the first critical speed. (a) Size of the two semi-axes of the elliptical orbit; (b)amplitudes in vertical (x) and horizontal (y) directions.

model obtained condensing the whole structure on node 10 (Example 6.1). The other modes, which are mostly affected by the characteristics of the rotor, which is isotropic, are essentially the same as those of Example 5.3.

The unbalance response at node 10 at the crossing of the first two critical speeds is plotted in Figure 6.8.

Note that there are two values of the speed at which the small axis of the elliptical orbit vanishes: The speed range between these two values is characterized by backward whirling; however, there is no speed at which the orbit reduces to a circle. The orbital tube is reported in Figure 6.9.

Example 6.3 *Consider the small turbojet studied in Example 4.1, but assume that the vertical stiffness of the supports is $k_y = 4$ MN/m, whereas the horizontal stiffness is $k_x = 1$ MN/m Compute the critical speeds by using the Myklestadt-Prohl method. As the stator is not axially symmetrical, the real coordinates approach must be followed and the transfer matrices have eight rows and columns. The transfer matrix of the nodes in which the supports are located must take into*

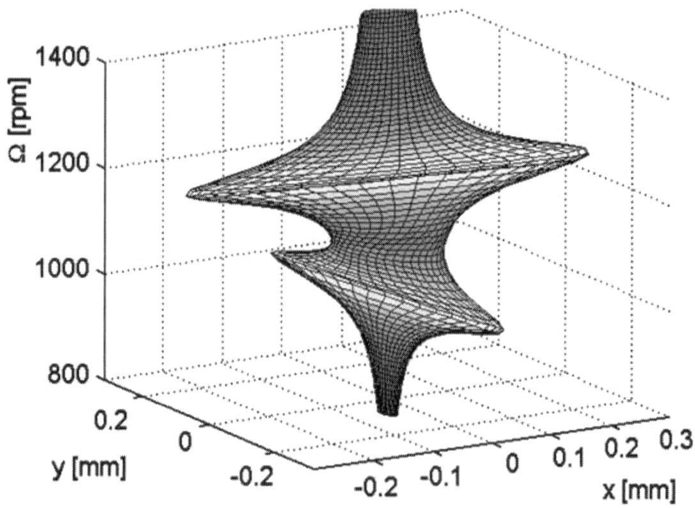

FIGURE 6.9. Orbital tube of the system of Figure 6.8.

Ω_{cr_I}	$\Omega_{cr_{II}}$	$\Omega_{cr_{III}}$	$\Omega_{cr_{IV}}$	Ω_{cr_V}	$\Omega_{cr_{VI}}$	$\Omega_{cr_{VII}}$
2,415	2,725	4,590	5,275	7,800	13,383	15,018

TABLE 6.2. Values of the first seven critical speeds (in rad/s), computed using the Myklestadt-Prohl method, using real coordinates.

account the anisotropy of the latter, and Equation (4.16) becomes

$$\mathbf{T}_{n_i}=\begin{bmatrix} 1 & 0 & 0 & 0 & 0 & 0 & 0 & 0 \\ 0 & 1 & 0 & 0 & 0 & 0 & 0 & 0 \\ -\omega^2 m_i + k_x & 0 & 1 & 0 & 0 & 0 & 0 & 0 \\ 0 & -\omega^2 J_{t_i} + \kappa_x & 0 & 1 & 0 & i\omega\Omega J_{p_i} & 0 & 0 \\ 0 & 0 & 0 & 0 & 1 & 0 & 0 & 0 \\ 0 & 0 & 0 & 0 & 0 & 1 & 0 & 0 \\ 0 & 0 & 0 & 0 & -\omega^2 m_i + k_y & 0 & 1 & 0 \\ 0 & -i\omega\Omega J_{p_i} & 0 & 0 & 0 & -\omega^2 J_{t_i} + \kappa_y & 0 & 1 \end{bmatrix} .$$

The determinant of the global stiffness matrix, computed in the speed range from 0 to 16,000 rad/s after introducing the free-free end conditions, is reported in Figure 6.10. In the speed range considered, the determinant vanishes at seven values of the speed, reported in Table 6.2.

Example 6.4 *Repeat the computations performed in the previous example by using the component mode synthesis by considering the rotor as made by two substructures, the rotor, and the bearings. Plot the Campbell diagram of the rotor.*

The rotor is axially symmetrical, and as a consequence, its modal characteristics are identical in the two planes. The stiffness, mass, and gyroscopic matrices

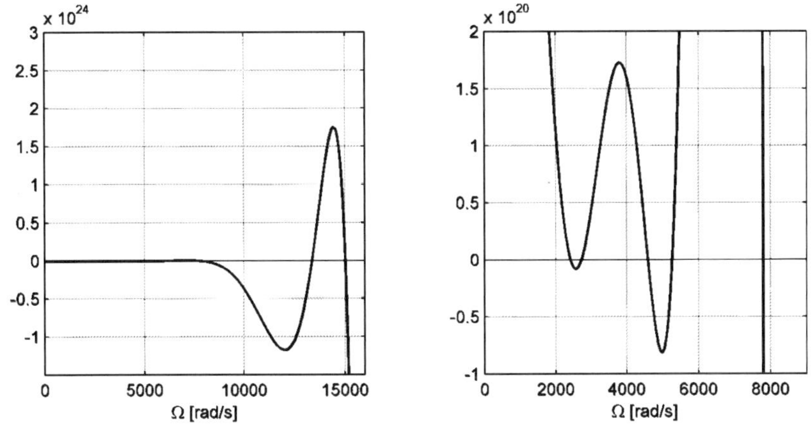

FIGURE 6.10. Value or the determinant of the global transfer matrix (after introducing the end conditions) as a function of the speed. (a) Range between 0 and 16,000 rad/s; (b) range between 0 and 9,000 rad/s.

of the rotor can be built in the same way seen in Example 4.10. The bearings are located at nodes 6 and 13, so the displacements at these nodes are taken as master degrees of freedom. All matrices are accordingly partitioned: Matrices with subscript 11 have two rows and two column, those with subscript 12 have two rows and 36 columns, and matrices with subscript 22 have 36 rows and columns. The reduced matrices for the rotor are then computed: They have $n+2$ *rows and columns, if* n *is the number of pairs of identical modes (in the two planes) considered. The expressions for* $\mathbf{K}^*$, $\mathbf{M}^*$, *and* $\mathbf{G}^*$ *are those seen in Example 4.10.*

As the stator is not axially symmetrical, either the deviatoric matrices of the stator are written or the computation is carried on using real coordinates. Although in this case the first alternative is straightforward, here the real coordinates approach is followed. The reduced stiffness, mass, and gyroscopic matrices are then computed

$$\mathbf{K}' = \begin{bmatrix} \mathbf{K}^* & \mathbf{0} \\ \mathbf{0} & \mathbf{K}^* \end{bmatrix}, \quad \mathbf{M}' = \begin{bmatrix} \mathbf{M}^* & \mathbf{0} \\ \mathbf{0} & \mathbf{M}^* \end{bmatrix}, \quad \mathbf{G}' = \begin{bmatrix} \mathbf{0} & \mathbf{G}^* \\ -\mathbf{G}^* & \mathbf{0} \end{bmatrix}.$$

Their size is $2(n+2)$. *To assemble the bearings to the structure, a stiffness of* 10^6 *N/m is added to the terms* K'_{11} *and* K'_{22}, *whereas a stiffness of* 4×10^6 *N/m is added to the elements on the diagonal of the stiffness matrix in position* $n+3$ *and* $n+4$.

The values of the first seven critical speeds obtained from a model in which three mode shapes of the rotor are retained and reported in Table 6.3.

The Campbell diagram obtained from the same model is shown in Figure 6.11. As the real coordinates approach has been used, only the part with positive ω *of the plot has been reported.*

Ω_{cr_I}	$\Omega_{cr_{II}}$	$\Omega_{cr_{III}}$	$\Omega_{cr_{IV}}$	Ω_{cr_V}	$\Omega_{cr_{VI}}$	$\Omega_{cr_{VII}}$
2,417	2,723	4,602	5,272	7,778	13,570	15,106

TABLE 6.3. Values of the first seven critical speeds (in rad/s), computed using the component mode synthesis approach (three modes).

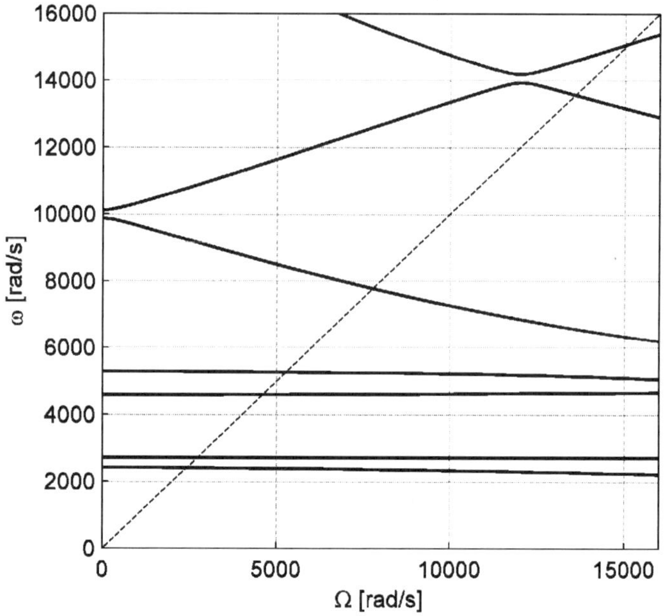

FIGURE 6.11. Campbell diagram of the small tyurbojet on anisotropic supports computed through component mode synthesis. Three modes of the rotor have been retained.

6.2 Nonisotropic rotors on isotropic supports

If the rotor is not axially symmetrical, a fixed observer sees the elastic, damping and in some cases inertial properties of the system to vary periodically in time. As such properties vary following an elliptical pattern with the angle of rotation, the same situation repeats identically every half a revolution, i.e., with a frequency equal to 2Ω. The equation of motion written in the inertial reference frame is thus a linear differential equation with periodic coefficients, whose period is π/Ω.

However, if the nonrotating parts of the machine are axially symmetrical, an observer that rotates together with the rotor sees all system properties as constants. The equation of motion is then a linear differential equation with constant coefficients in a rotating reference frame. Note that such reference frame is not inertial, and the equation of motion includes the usual noninertial terms (e.g., Coriolis and centrifugal acceleration, etc.).

6.2.1 *Nonisotropic Jeffcott rotor*

Consider a Jeffcott rotor of the type shown in Figure 2.1(a), in which the stiffness of the shaft is not isotropic. The polar diagram of the stiffness is now an ellipse whose axes can be assumed, without loss of generality, to lie along the ξ- and η-axes. It is then possible to write an equation similar to Equation (6.4) but referred to the $O\xi\eta z$-frame:

$$F_\xi = -k_\xi \xi\,, \qquad F_\eta = -k_\eta \eta\,.$$

The equation of motion of an isotropic Jeffcott rotor written with reference to the rotating frame is Equation (2.166)

$$\begin{bmatrix} m & 0 \\ 0 & m \end{bmatrix} \begin{Bmatrix} \ddot{\xi} \\ \ddot{\eta} \end{Bmatrix} + \left(\begin{bmatrix} c_n + c_r & 0 \\ 0 & c_n + c_r \end{bmatrix} + 2\Omega \begin{bmatrix} 0 & -m \\ m & 0 \end{bmatrix} \right) \times$$

$$\times \begin{Bmatrix} \dot{\xi} \\ \dot{\eta} \end{Bmatrix} + \left(\begin{bmatrix} k & 0 \\ 0 & k \end{bmatrix} - \Omega^2 \begin{bmatrix} m & 0 \\ 0 & m \end{bmatrix} + \Omega \begin{bmatrix} 0 & -c_n \\ c_n & 0 \end{bmatrix} \right) \times$$

$$\times \begin{Bmatrix} \xi \\ \eta \end{Bmatrix} = \begin{Bmatrix} m\epsilon\Omega^2 \\ 0 \end{Bmatrix} + \begin{Bmatrix} F_x \cos(\Omega t) + F_y \sin(\Omega t) \\ F_y \cos(\Omega t) - F_x \sin(\Omega t) \end{Bmatrix} .$$

By introducing two different values of the stiffness in ξ- and η-direction, neglecting damping and introducing a phase angle α between ξ-axis and the unbalance vector ϵ, the equation of motion of an anisotropic undamped Jeffcott rotor is

$$\begin{bmatrix} m & 0 \\ 0 & m \end{bmatrix} \begin{Bmatrix} \ddot{\xi} \\ \ddot{\eta} \end{Bmatrix} + \begin{bmatrix} 0 & -2m\Omega \\ 2m\Omega & 0 \end{bmatrix} \begin{Bmatrix} \dot{\xi} \\ \dot{\eta} \end{Bmatrix} +$$

$$+ \begin{bmatrix} k_\xi - m\Omega^2 & 0 \\ 0 & k_\eta - m\Omega^2 \end{bmatrix} \begin{Bmatrix} \xi \\ \eta \end{Bmatrix} = \tag{6.39}$$

$$= m\epsilon\Omega^2 \begin{Bmatrix} \cos(\alpha) \\ \sin(\alpha) \end{Bmatrix} + \begin{Bmatrix} F_x \cos(\Omega t) + F_y \sin(\Omega t) \\ F_y \cos(\Omega t) - F_x \sin(\Omega t) \end{Bmatrix} .$$

The unbalance response is easily obtained as a steady-state solution of Equation (6.39)

$$\xi = \frac{m\epsilon\Omega^2 \cos(\alpha)}{k_\xi - m\Omega^2}\,, \qquad \eta = \frac{m\epsilon\Omega^2 \sin(\alpha)}{k_\eta - m\Omega^2}\,, \tag{6.40}$$

which represents a circular whirling in a fixed reference frame.

The denominators of Equation (6.40) vanish for two values of the spin speed, which are the critical speeds of the system

$$\Omega_{cr_I} = \sqrt{\frac{k_\xi}{m}}, \qquad \Omega_{cr_{II}} = \sqrt{\frac{k_\eta}{m}}. \tag{6.41}$$

By introducing the mean and deviatoric stiffness in the rotating frame

$$k_m = \frac{k_\xi + k_\eta}{2}, \qquad k_d = \frac{k_\xi - k_\eta}{2}, \tag{6.42}$$

the same equation can be written with reference to the complex coordinate $\rho = \xi + i\eta$

$$m\ddot{\rho} + 2i\Omega m\dot{\rho} - m\Omega^2\rho + k_m\rho + k_d\overline{\rho} = m\epsilon\Omega^2 e^{i\alpha}. \tag{6.43}$$

The free whirling of the system can easily be obtained from the homogeneous Equation (6.43) or (6.39), being immaterial whether real or complex coordinates are used. In the first case, the solution of the homogeneous Equation (6.39) is

$$\xi = \xi_0 e^{s't}, \qquad \eta = \eta_0 e^{s't},$$

where s' is the Laplace variable in the rotating frame

$$s' = \sigma + i\omega' \tag{6.44}$$

and ω' is the whirl speed in the $\xi\eta$-plane; it does not coincide with the whirl speed ω in the xy-plane, but it is linked to it by the relationship

$$\omega' = \omega - \Omega. \tag{6.45}$$

The homogeneous Equation (6.39) then yields an eigenproblem in s'

$$\begin{bmatrix} \Omega_{cr_I}^2 + s'^2 - \Omega^2 & -2\Omega s' \\ 2\Omega s' & \Omega_{cr_{II}}^2 + s'^2 - \Omega^2 \end{bmatrix} \begin{Bmatrix} \xi_0 \\ \eta_0 \end{Bmatrix} = \mathbf{0}. \tag{6.46}$$

By introducing the anisotropy parameter

$$\alpha^* = \frac{k_d}{k_m} = \frac{k_\xi - k_\eta}{k_\xi + k_\eta}, \tag{6.47}$$

and the nondimensional whirl and spin speeds

$$\omega'^* = \frac{\omega'}{\omega_0}, \qquad \Omega^* = \frac{\Omega}{\omega_0}, \qquad \omega^* = \frac{\omega}{\omega_0}, \qquad s'^* = \frac{s'}{\omega_0} \tag{6.48}$$

computed with reference to the natural frequency of the average system

$$\omega_0 = \sqrt{\frac{k_\xi + k_\eta}{2m}} = \sqrt{\frac{k_m}{m}}, \tag{6.49}$$

the characteristic equation can be written in nondimensional form

$$s'^{*^4} + 2s'^{*^2}(1 + \Omega^{*^2}) + (1 - \Omega^{*^2})^2 - \alpha^{*2} = 0 \,. \tag{6.50}$$

The anisotropy parameter α^* spans in the range -1, 1, with the central value 0 for isotropic systems. If ξ axis is chosen in such a way that $k_\xi < k_\eta$, α^* is negative. By first solving Equation (6.50) in s'^{*^2}, it follows that

$$s'^{*^2} = -\left(1 + \Omega^{*^2}\right) \pm \sqrt{4\Omega^{*2} + \alpha^{*2}} \,. \tag{6.51}$$

The expression under the radical sign in Equation (6.51) is always positive: The two solutions for s'^{*^2} are then always real. The one with the lower sign ($-$) is always negative and yields two imaginary solutions in s'^*, one positive and one negative, i.e., two oscillatory solutions in the $\xi\eta$-plane, corresponding to circular whirling. The solution with the upper sign (+) is negative only if

$$\Omega^{*^4} - 2\Omega^{*^2} + 1 - \alpha^{*2} > 0.$$

As α^* was assumed to be negative and with absolute value smaller than unity, the last condition can be written in the form

$$\Omega^* < \sqrt{1 + \alpha^*} \,, \qquad \sqrt{1 - \alpha^*} < \Omega^* \,; \tag{6.52}$$

i.e.,

$$\Omega < \sqrt{\frac{k_\xi}{m}} \,, \qquad \sqrt{\frac{k_\eta}{m}} < \Omega \,. \tag{6.53}$$

If Condition (6.53) is satisfied, the characteristic Equation (6.50) has four imaginary roots – two positive and two negative. The whirling of the system is then an undamped circular whirling, and no instability is present. However, as could be expected from an undamped system, the amplitude of the whirl orbit does not decay in time.

If the value of the spin speed Ω lies between the two critical speeds of the system, as shown by Condition (6.52), one of the solutions for s'^{*^2} is positive, and together with the two imaginary solutions, two real roots, with opposite signs, are found. A positive real solution in s'^* then exists, which corresponds to an unstable behavior of the system, with amplitude growing indefinitely with exponential law.

Remark 6.6 *The presence of an elastic anisotropy of the rotating parts of the system causes the occurrence of an instability range spanning from the lowest to the highest critical speed.*

The Campbell diagram $\omega(\Omega)$ and the decay rate plot $\sigma(\Omega)$ for a system with $k_\eta/k_\xi = 2$ ($\alpha^* = -1/3$) obtained from Equation (6.51) is plotted in

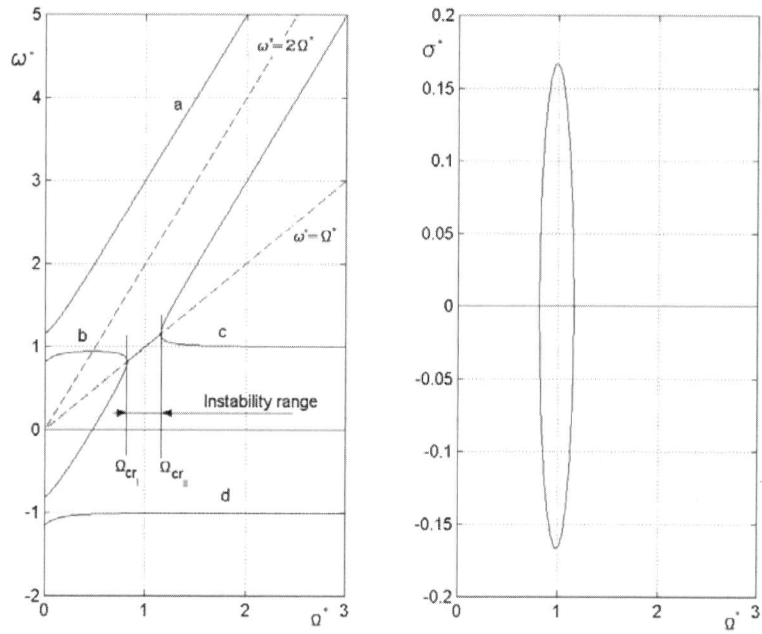

FIGURE 6.12. Nondimensional Campbell diagram and plot of the decay rate against the speed for a nonisotropic Jeffcott rotor on isotropic supports with $k_\eta/k_\xi = 2$ ($\alpha^* = -1/3$).

Figure 6.12. Note that $\omega(\Omega)$ has been obtained as

$$\omega = \omega' + \Omega = \mathrm{Im}(s') + \Omega\,.$$

The same conclusions already seen can be drawn from the figure. At low speed, up to the first critical speed, there are four imaginary solutions. Those on branches a and d of the curve come from the Solution (6.51) with the lower sign ($-$) and are a forward and a backward whirl. Those on branch b come from the solution with the upper sign ($+$) and are one backward and one forward or both forward, depending on the value of Ω. At high speed, above the second critical speed, the situation is similar, the difference being that the solutions coming from the expression with the upper sign ($+$) are both forward motions and lie on branch c of the curve. If the value of the speed lies in the instability range spanning between the critical speeds, there are two imaginary solutions, on branches a and d of the curve, and two real ones ($\omega' = 0$, and then in correspondence to the real roots $\omega = \Omega$), reported in the plot $\sigma(\Omega)$. The positive real solution causes the behavior of the system to become unstable.

The system of Figure 2.16 can be considered as a limiting case of an asymmetrical rotor: The stiffness along the η-axis is infinitely high, causing a second critical speed that tends to infinity. The field of instability then extends for all values of Ω that are above the critical speed.

6.2.2 *Effect of damping*

The equation of motion of a nonisotropic Jeffcott rotor written with reference to the rotating frame can be obtained from Equation (2.166)

$$\begin{bmatrix} m & 0 \\ 0 & m \end{bmatrix}\begin{Bmatrix} \ddot{\xi} \\ \ddot{\eta} \end{Bmatrix}+\begin{bmatrix} c_n + c_{r_\xi} & -2\Omega m \\ 2\Omega m & c_n + c_{r_\eta} \end{bmatrix}\begin{Bmatrix} \dot{\xi} \\ \dot{\eta} \end{Bmatrix}+$$

$$+\begin{bmatrix} k_\xi - m\Omega^2 & -\Omega c_n \\ \Omega c_n & k_\eta - m\Omega^2 \end{bmatrix}\begin{Bmatrix} \xi \\ \eta \end{Bmatrix}= \tag{6.54}$$

$$= m\epsilon\Omega^2\begin{Bmatrix} \cos(\alpha) \\ \sin(\alpha) \end{Bmatrix}+\begin{Bmatrix} F_x\cos(\Omega t)+F_y\sin(\Omega t) \\ F_y\cos(\Omega t)-F_x\sin(\Omega t) \end{Bmatrix}.$$

By introducing the solution for free whirling

$$\xi = \xi_0 e^{s't}, \qquad \eta = \eta_0 e^{s't}$$

into the homogenous Equation (6.54), the following characteristic equation is obtained:

$$\det\begin{bmatrix} ms'^2 + s'(c_n + c_r) - m\Omega^2 + k_\xi & -\Omega(2ms' + c_n) \\ \Omega(2ms' + c_n) & ms'^2 + s'(c_n + c_r) - m\Omega^2 + k_\eta \end{bmatrix} = 0\,. \tag{6.55}$$

Equation (6.55) can be written in nondimensional form by introducing the same parameters seen for the undamped system, plus the damping ratios

$$\zeta_n = \frac{c_n}{\sqrt{2m(k_\xi + k_\eta)}}\,, \qquad \zeta_r = \frac{c_{r_\xi} + c_{r_\eta}}{2\sqrt{2m(k_\xi + k_\eta)}} \tag{6.56}$$

and the anisotropy parameter for damping

$$\beta^* = \frac{c_d}{c_m} = \frac{c_{r_\xi} - c_{r_\eta}}{c_{r_\xi} + c_{r_\eta}}\,. \tag{6.57}$$

The characteristic equation in nondimensional form is then

$$\det\begin{bmatrix} f_1(s'^*) & -2\Omega^*[s'^* + \xi_n] \\ 2\Omega^*[s'^* + \xi_n] & f_2(s'^*) \end{bmatrix} = 0\,, \tag{6.58}$$

where

$$\begin{aligned} f_1(s'^*) &= 1 + \alpha^* - \Omega^{*2} + s'^{*2} + 2s'^*[\xi_r(1+\beta^*) + \xi_n]\,, \\ f_2(s'^*) &= 1 - \alpha^* - \Omega^{*2} + s'^{*2} + 2s'^*[\xi_r(1-\beta^*) + \xi_n]\,. \end{aligned} \tag{6.59}$$

By solving the determinant, it follows

$$\begin{gathered} s'^{*4} + 4s'^{*3}\left(\xi_r+\xi_n\right) + 2s'^{*2}\left[\left(1+\Omega^{*2}\right)+2\left(\xi_r+\xi_n\right)^2-2\beta^{*2}\xi_r^2\right]+ \\ +4s'^{*}\left[\left(\xi_r+\xi_n\right)\left(1-\Omega^{*2}\right)-\beta^*\alpha^*\xi_r+2\Omega^{*2}\xi_n\right]+ \\ +(1-\Omega^{*2})^2-\alpha^{*2}+4\Omega^{*2}\xi_n^2=0\,. \end{gathered} \tag{6.60}$$

Also in the case of damped systems an instability range may exist between the two critical speeds of the undamped system, located at speeds equal to $\sqrt{k_\xi/m}$ and $\sqrt{k_\eta/m}$. This range is smaller than that characterizing the undamped system, which spans from one critical speed to the other, and reduces with increasing damping. If the latter is large enough, the instability range may disappear altogether. The nondimensional Campbell diagram $\omega^*(\Omega^*)$ and the decay rate plot $\sigma^*(\Omega^*)$ for a system with $k_\eta/k_\xi=2$ $(\alpha^*=-1/3)$, $\beta=-1/3$, $\zeta_n=0.2$, and $\zeta_r=0.1$ obtained from Equation (6.60) is plotted in Figure 6.13. The roots locus is reported in Figure 6.14.

Remark 6.7 *On the decay rate plot, the effect of damping is that of moving down the loop, which in undamped systems is centered on the Ω-axis, until it lies all in the negative half-plane, showing that no instability is present.*

Remark 6.8 *Although rotating damping has a stabilizing effect, together with nonrotating damping, for what this instability range is concerned, it can cause the system to become unstable at high speed.*

6.2.3 Response to a static force

All conditions in which there is resonance between one of the natural frequencies of the system and an exciting force different from that caused by unbalance will be referred to as *secondary critical speeds*. It is well known that when constant bending forces, such as the self-weight of a rotor whose axis is horizontal, act on the rotor, the critical speeds and the Campbell diagram are not influenced by the presence of such forces. Whirling takes place about the deflected configuration of the rotor, but because of linearity, the two effects, namely, static bending and whirling, do not interact. It is, however, well known that the weight of a rotor with a horizontal axis can cause the occurrence of secondary critical speeds, whose values are about half those of primary critical speeds or, more exactly, are located at the intersections on the Campbell diagram of the curves for free whirling with the straight line $\omega=2\Omega$. The presence of these secondary critical speeds is linked to the deviations from a perfect axial symmetry of the rotor.

In the case of the Jeffcott rotor studied in the preceding section, these secondary critical speeds are easily deduced from Figure 6.12. In the figure, an intersection of branch b of the curve with the Ω^*-axis is clearly visible. The system has, at a well-determined speed, a natural frequency that is

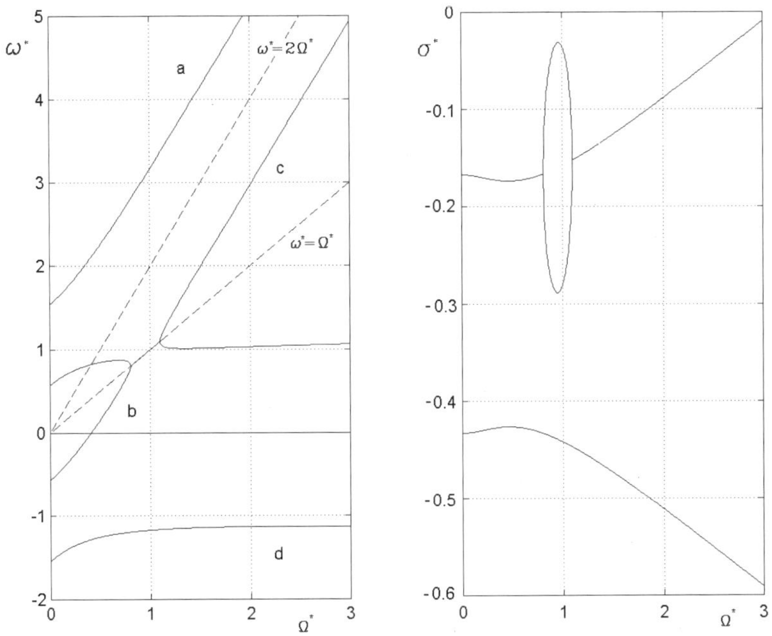

FIGURE 6.13. Nondimensional Campbell diagram and plot of the decay rate against the speed for a nonisotropic damped Jeffcott rotor on isotropic supports with $k_\eta/k_\xi = 2$ ($\alpha^* = -1/3$), $\beta^* = -1/3$, $\zeta_n = 0.2$, and $\zeta_r = 0.1$. Note that the system is always stable in the speed range shown in the figure.

vanishingly small and then a sort of resonance with a static force, i.e., with a force constant in modulus and direction, is possible at that speed.

The phenomenon may be easier to understand with reference to the rotating frame $O\xi\eta z$, where condition $\omega = 0$ is seen as $\omega' = -\Omega$. A constant force in the xy-plane, as self-weight of a horizontal rotor, is seen in the $\xi\eta$-plane as a force rotating with speed $-\Omega$, which can cause resonance when the natural frequency of the system has the same frequency. The same phenomenon can also be seen in a different way. When the stiffness of the shaft is not isotropic in the $\xi\eta$-plane, its polar diagram is an ellipse, the ellipse of elasticity. The functions of time expressing the stiffness $k_x(t)$ and $k_y(t)$ are periodic in time, if the angular speed Ω is constant, and their period is equal to half a revolution (frequency equal to 2Ω). The conditions for resonance occur when the curves on the Campbell diagram intersect the straight line $\omega = 2\Omega$.

The two ways of seeing the same phenomenon are equivalent: From Figure 6.12, it is clear that at the same value of the speed at which a branch of

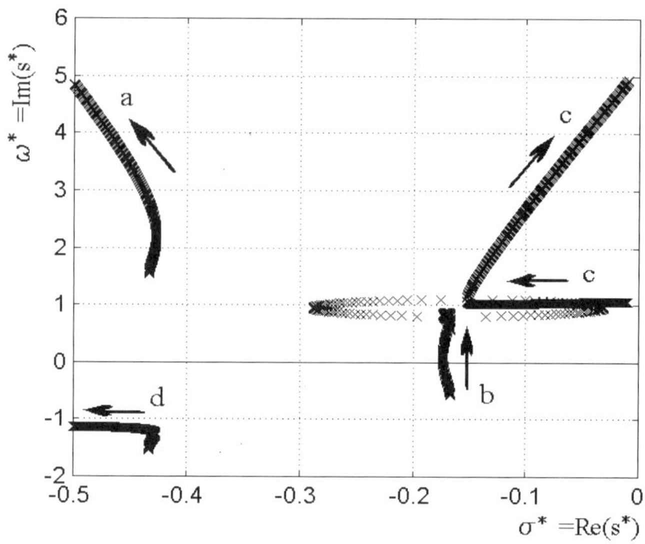

FIGURE 6.14. Nondimensional roots locus for the same system of Figure 6.13.

the curve intersects the Ω^*-axis, another one intersects the line $\omega^* = 2\Omega^*$. The value of the secondary critical speed for the Jeffcott rotor of Figure 6.12 can be obtained by looking for a solution

$$s = i\omega = 2i\Omega \tag{6.61}$$

in the inertial reference frame. The same solution in the rotating frame (and in nondimensional terms) is

$$s' = s - i\Omega = i\Omega \; . \tag{6.62}$$

By introducing Solution (6.62) into Equation (6.50) and solving in Ω^*, it follows that

$$\Omega^* = \frac{\sqrt{1-\alpha^*}}{2} = \frac{\sqrt{k_\xi k_\eta}}{k_\xi + k_\eta} \, . \tag{6.63}$$

If k_ξ and k_η tend to a single value k, $\alpha^* \rightarrow 0$ and the value of the secondary critical speed tends to

$$\Omega_{cr_s} = \frac{1}{2}\sqrt{\frac{k}{m}} = \frac{1}{2}\Omega_{cr} \, . \tag{6.64}$$

Remark 6.9 *Equation (6.64) holds only in the case of a Jeffcott rotor; in all other cases, the secondary critical speeds can be found at the intersections of the curve $\omega(\Omega)$ with the line $\omega = 2\Omega$.*

Remark 6.10 *Also, inertial anisotropy, i.e., difference in the moments of inertia about transversal baricentrical axes, has effects similar to those seen for elastic anisotropy.*

All secondary critical speeds characterized by the condition $\omega > \Omega$, as is the case for those excited by self-weight, occur in the subcritical region and usually cannot trigger unstable behavior. The internal damping of the rotor is in these conditions stabilizing, and generally speaking, no unstable behavior is expected in the subcritical region.

The response to a static force can be studied directly using Equation (6.39) (Equation (6.54) if the system is damped). Assume that the force is applied along the x-axis (this does not detract from the generality of the solution, as a static force is fixed to the stator, which is isotropic): In the rotating frame, it is seen as a force rotating backward at the speed Ω. For the study of the response of the undamped system, a solution of the type

$$\begin{cases} \xi = \xi_1 \cos(\Omega t) + \xi_2 \sin(\Omega t) \ , \\ \eta = \eta_1 \cos(\Omega t) + \eta_2 \sin(\Omega t) \ , \end{cases} \tag{6.65}$$

can then be introduced into Equation (6.39), in which the unbalance term has been neglected, obtaining

$$\begin{bmatrix} k_\xi - 2m\Omega^2 & 0 & 0 & -2m\Omega^2 \\ 0 & k_\xi - 2m\Omega^2 & 2m\Omega^2 & 0 \\ 0 & 2m\Omega^2 & k_\eta - 2m\Omega^2 & 0 \\ -2m\Omega^2 & 0 & 0 & k_\eta - 2m\Omega^2 \end{bmatrix} \begin{Bmatrix} \xi_1 \\ \xi_2 \\ \eta_1 \\ \eta_2 \end{Bmatrix} = \\ = F \begin{Bmatrix} 1 \\ 0 \\ 0 \\ -1 \end{Bmatrix} . \tag{6.66}$$

By adding the first equation to the fourth and the third one to the second, it follows that

$$\begin{cases} \left(k_\xi - 4m\Omega^2\right) \xi_1 + \left(k_\eta - 4m\Omega^2\right) \eta_2 = 0 \ , \\ k_\xi \xi_2 + k_\eta \eta_1 = 0 \ . \end{cases} \tag{6.67}$$

The first and second Equations (6.66) yield

$$\begin{array}{ll} \xi_1 = F\frac{k_\eta - 4m\Omega^2}{k_\xi k_\eta - 2m\Omega^2(k_\xi + k_\eta)} \ , & \xi_2 = 0 \ , \\ \eta_1 = 0 \ , & \eta_2 = -F\frac{k_\xi - 4m\Omega^2}{k_\xi k_\eta - 2m\Omega^2(k_\xi + k_\eta)} \ . \end{array} \tag{6.68}$$

The motion of point C in the rotor fixed frame is then a backward elliptical whirling. By transforming it in the fixed reference frame, it follows that

$$\begin{cases} x = \xi \cos(\Omega t) - \eta \sin(\Omega t) \ , \\ y = \xi \sin(\Omega t) + \eta \cos(\Omega t) \ , \end{cases} \tag{6.69}$$

i.e., remembering Equation (6.65),

$$\begin{cases} x = \xi_1 \cos^2(\Omega t) - \eta_2 \sin^2(\Omega t) , \\ y = (\xi_1 + \eta_2) \sin(\Omega t) \cos(\Omega t) , \end{cases} \tag{6.70}$$

or

$$\begin{cases} x = \frac{1}{2} (\xi_1 - \eta_2) + \frac{1}{2} (\xi_1 + \eta_2) \cos(2\Omega t) , \\ y = \frac{1}{2} (\xi_1 + \eta_2) \sin(2\Omega t) . \end{cases} \tag{6.71}$$

The motion of point C is then the combination of a static displacement in x direction, i.e., in the direction of the static applied force F, with amplitude

$$\frac{1}{2} (\xi_1 - \eta_2) = F \frac{k_\xi + k_\eta - 8m\Omega^2}{2k_\xi k_\eta - 4m\Omega^2 (k_\xi + k_\eta)} , \tag{6.72}$$

plus a circular whirling motion, occurring at frequency 2Ω, with amplitude

$$\frac{1}{2} (\xi_1 + \eta_2) = F \frac{k_\eta - k_\xi}{2k_\xi k_\eta - 4m\Omega^2 (k_\xi + k_\eta)} . \tag{6.73}$$

The whirl amplitude decreases with increasing speed, tending to zero when the speed tends to infinity, whereas the static displacement tends to $2F/(k_\xi + k_\eta)$.

If $k_\xi \to k_\eta$, i.e., the rotor tends to be isotropic, the whirling component vanishes, whereas the static component tends to F/k, as expected.

When

$$\Omega^* = \frac{\sqrt{k_\xi k_\eta}}{k_\xi + k_\eta} ,$$

i.e., when the Campbell diagram crosses the $\omega = 2\Omega$ line, $\xi_1 - \eta_2 \to \infty$ and $\xi_1 + \eta_2 \to \infty$.

The presence of damping couples the behavior in xz- and yz-planes, and it makes the static component no more aligned with the load.

6.2.4 Anisotropic rotors with many degrees of freedom

As already stated, the study of a system in which the rotor has no axial symmetry while the stator is isotropic is best performed with reference to a rotating frame, in which the equation of motion is a differential equation with constant coefficients.

The equation of motion, written in rotating coordinates for an isotropic system is Equation (4.78). If real coordinates are used instead of complex ones, it becomes

$$\begin{bmatrix} \mathbf{M} & \mathbf{0} \\ \mathbf{0} & \mathbf{M} \end{bmatrix} \ddot{\mathbf{x}} + \left(-\Omega \begin{bmatrix} \mathbf{0} & 2\mathbf{M} - \mathbf{G} \\ -2\mathbf{M} + \mathbf{G} & \mathbf{0} \end{bmatrix} + \begin{bmatrix} \mathbf{C} & \mathbf{0} \\ \mathbf{0} & \mathbf{C} \end{bmatrix} \right) \dot{\mathbf{x}} +$$

$$+ \left(\begin{bmatrix} \mathbf{K} & \mathbf{0} \\ \mathbf{0} & \mathbf{K} \end{bmatrix} - \Omega^2 \begin{bmatrix} \mathbf{M} - \mathbf{G} - \mathbf{K}_\Omega & \mathbf{0} \\ \mathbf{0} & \mathbf{M} - \mathbf{G} - \mathbf{K}_\Omega \end{bmatrix} + \right.$$

$$+\Omega\begin{bmatrix}\mathbf{0} & -\mathbf{C}_n\\ \mathbf{C}_n & \mathbf{0}\end{bmatrix}\Bigg)\mathbf{x}=\left\{\begin{array}{c}\mathrm{Re}(\mathbf{f}_n e^{-i\Omega t})\\ \mathrm{Im}(\mathbf{f}_n e^{-i\Omega t})\end{array}\right\}+\Omega^2\left\{\begin{array}{c}\mathrm{Re}(\mathbf{f}_r)\\ \mathrm{Im}(\mathbf{f}_r)\end{array}\right\},$$

where $\mathbf{C}=\mathbf{C}_r+\mathbf{C}_n$, the real-coordinates vector is

$$\mathbf{x}=[\mathrm{Re}(\mathbf{r})^T,\mathrm{Im}(\mathbf{r})^T]^T$$

and the complex coordinates vector is simply

$$\mathbf{r}=\mathbf{q}e^{-i\Omega t}\ .$$

Note that in the rotating frame, the unbalance forces (vector $\mathbf{f}_r$) are seen as forces constant in time, whereas static forces (vector $\mathbf{f}_n$) are seen as rotating backward at the spin speed Ω. The circulatory and gyroscopic matrices are skew-symmetric, as expected.

If the rotor is nonisotropic, the equation of motion in real coordinates transforms into[1]

$$\begin{bmatrix}\mathbf{M}_\xi & \mathbf{M}_{\xi\eta}\\ \mathbf{M}_{\eta\xi} & \mathbf{M}_\eta\end{bmatrix}\ddot{\mathbf{x}}+\Bigg(-\Omega\begin{bmatrix}\mathbf{M}_{\eta\xi}-\mathbf{M}_{\xi\eta} & \mathbf{M}_\xi+\mathbf{M}_\eta-\mathbf{G}\\ -\mathbf{M}_\xi-\mathbf{M}_\eta+\mathbf{G} & \mathbf{M}_{\eta\xi}-\mathbf{M}_{\xi\eta}\end{bmatrix}+ \qquad (6.74)$$

$$+\begin{bmatrix}\mathbf{C}_\xi & \mathbf{C}_{r\xi\eta}\\ \mathbf{C}_{r\xi\eta} & \mathbf{C}_\eta\end{bmatrix}\Bigg)\dot{\mathbf{x}}+\Bigg(\begin{bmatrix}\mathbf{K}_\eta & \mathbf{K}_{\xi\eta}\\ \mathbf{K}_{\eta\xi} & \mathbf{K}_\eta\end{bmatrix}-\Omega^2\begin{bmatrix}\mathbf{M}_\eta-\mathbf{G} & -\mathbf{M}_{\eta\xi}\\ -\mathbf{M}_{\xi\eta} & \mathbf{M}_\eta-\mathbf{G}\end{bmatrix}+$$

$$+\Omega\begin{bmatrix}\mathbf{0} & -\mathbf{C}_n\\ \mathbf{C}_n & \mathbf{0}\end{bmatrix}\Bigg)\mathbf{x}=\left\{\begin{array}{c}\mathrm{Re}(\mathbf{f}_n e^{-i\Omega t})\\ \mathrm{Im}(\mathbf{f}_n e^{-i\Omega t})\end{array}\right\}+\Omega^2\left\{\begin{array}{c}\mathrm{Re}(\mathbf{f}_r)\\ \mathrm{Im}(\mathbf{f}_r)\end{array}\right\}.$$

Note that although in the Jeffcott rotor the anisotropy was caused by the elastic and damping properties of the material, in multi-degrees-of-freedom rotors, the anisotropy can also be caused by the inertia properties, i.e., to different values of the moments of inertia in the $\xi\eta$-plane.

By introducing the mean and deviatoric matrices, it is possible to obtain an equation written in term of complex coordinates

$$\mathbf{M}_m\ddot{\mathbf{r}}+\Big[\mathbf{C}_m+i\Omega(2\mathbf{M}_m-\mathbf{G})\Big]\dot{\mathbf{r}}+\left[-\Omega^2(\mathbf{M}_m-\mathbf{G})+\mathbf{K}_m+i\Omega\mathbf{C}_n)\right]\mathbf{r}+ \qquad (6.75)$$

$$+\mathbf{M}_{r_d}\ddot{\bar{\mathbf{r}}}+\mathbf{C}_{r_d}\dot{\bar{\mathbf{r}}}+\left(\Omega^2\mathbf{M}_{r_d}+\mathbf{K}_{r_d}\right)\bar{\mathbf{r}}=\mathbf{F}_n e^{-i\Omega t}+\Omega^2\mathbf{F}_r\,.$$

The deviatoric matrices related to the stator vanish, owing to its isotropy. The solution for the free whirling of the system is of the type

$$\mathbf{r}=\mathbf{r}_1 e^{s't}+\mathbf{r}_2 e^{\bar{s}'t}\,,$$

[1] Centrifugal stiffening (matrix $\mathbf{K}_\Omega$) is not included in the following equations to simplify the formulation. There is no difficulty in including the relevant terms into the equations of motion when needed.

i.e., an elliptical whirling with reference to the rotating frame $\xi\eta z$. By introducing it into the equation of motion (6.75), the following algebraic equation is obtained:

$$\left(s'^2 \begin{bmatrix} \mathbf{M}_m & \mathbf{M}_d \\ \overline{\mathbf{M}}_d & \overline{\mathbf{M}}_m \end{bmatrix} - is'\Omega \begin{bmatrix} 2\mathbf{M}_m + \mathbf{G} & \mathbf{0} \\ \mathbf{0} & 2\overline{\mathbf{M}}_m - \overline{\mathbf{G}} \end{bmatrix} + \right.$$

$$+ s' \begin{bmatrix} \mathbf{C}_m + \mathbf{C}_{r_m} & \mathbf{C}_{r_d} \\ \overline{\mathbf{C}}_{r_d} & \overline{\mathbf{C}}_n + \overline{\mathbf{C}}_{r_m} \end{bmatrix} + i\Omega \begin{bmatrix} \mathbf{C}_n & \mathbf{0} \\ \mathbf{0} & -\overline{\mathbf{C}}_n \end{bmatrix} +$$

$$\left. + \begin{bmatrix} -\mathbf{M}_m + \mathbf{G} & \mathbf{M}_d \\ \overline{\mathbf{M}}_d & -\overline{\mathbf{M}}_m + \overline{\mathbf{G}} \end{bmatrix} + \begin{bmatrix} \mathbf{K}_m & \mathbf{K}_{n_d} \\ \overline{\mathbf{K}}_{n_d} & \overline{\mathbf{K}}_m \end{bmatrix} \right) \begin{Bmatrix} \mathbf{r}_1 \\ \overline{\mathbf{r}}_2 \end{Bmatrix} = \mathbf{0}. \quad (6.76)$$

By expressing the same solution in the fixed frame, it yields

$$\mathbf{q} = \mathbf{r}_1 e^{st} + \mathbf{r}_2 e^{(2i\Omega + \overline{s})t}.$$

The orbits are then elliptical when seen in the rotating frame, but become Lissajous curves when seen in the fixed frame. The equation of motion could be written directly in terms of fixed coordinates, obtaining an equation in which s, referred to the inertial frame, is present instead of s', referred to the rotating frame

$$\left(s^2 \begin{bmatrix} \mathbf{M}_m & \mathbf{M}_d \\ \overline{\mathbf{M}}_d & \overline{\mathbf{M}}_m \end{bmatrix} - is\Omega \begin{bmatrix} \mathbf{G} & 2\mathbf{M}_d \\ 2\overline{\mathbf{M}}_d & 4\overline{\mathbf{M}}_m - \overline{\mathbf{G}} \end{bmatrix} + \right.$$

$$+\Omega^2 \begin{bmatrix} \mathbf{0} & \mathbf{0} \\ \mathbf{0} & -4\overline{\mathbf{M}}_m + 2\overline{\mathbf{G}} \end{bmatrix} + s \begin{bmatrix} \mathbf{C}_m + \mathbf{C}_{r_m} & \mathbf{C}_{r_d} \\ \overline{\mathbf{C}}_{r_d} & \overline{\mathbf{C}}_n + \overline{\mathbf{C}}_{r_m} \end{bmatrix} +$$

$$\left. -i\Omega \begin{bmatrix} \mathbf{C}_{r_m} & \mathbf{C}_{r_d} \\ \overline{\mathbf{C}}_{rd} & 2\overline{\mathbf{C}}_n + \overline{\mathbf{C}}_{r_m} \end{bmatrix} + \begin{bmatrix} \mathbf{K}_m & \mathbf{K}_{n_d} \\ \overline{\mathbf{K}}_{n_d} & \overline{\mathbf{K}}_m \end{bmatrix} \right) \begin{Bmatrix} \mathbf{r}_1 \\ \overline{\mathbf{r}}_2 \end{Bmatrix} = \mathbf{0}. \quad (6.77)$$

In the current case, the study of the unbalance response is easier than that of the response to a static loading because the first gives way to a deformed configuration that is stationary with respect to the system of reference.

The solution of Equation (6.75) for static loading is of the type

$$\mathbf{r} = \mathbf{r}_1 e^{-i\Omega t} + \mathbf{r}_2 e^{i\Omega t},$$

leading to the equation

$$\begin{bmatrix} \mathbf{A}_{11} & \mathbf{A}_{12} \\ \mathbf{A}_{21} & \mathbf{A}_{22} \end{bmatrix} \begin{Bmatrix} \mathbf{q}_1 \\ \overline{\mathbf{q}}_2 \end{Bmatrix} = \begin{Bmatrix} \mathbf{f}_n \\ \mathbf{0} \end{Bmatrix}, \quad (6.78)$$

where

$$\begin{aligned} \mathbf{A}_{11} &= -i\Omega \mathbf{C}_{r_m} + \mathbf{K}_m, \\ \mathbf{A}_{12} &= -i\Omega \mathbf{C}_{r_d} + \mathbf{K}_d, \\ \mathbf{A}_{21} &= -i\Omega \overline{\mathbf{C}}_{r_d} + \overline{\mathbf{K}}_d, \\ \mathbf{A}_{22} &= -2\Omega^2(2\overline{\mathbf{M}}_m - \overline{\mathbf{G}}) - i\Omega(\overline{\mathbf{C}}_{r_m} + 2\overline{\mathbf{C}}_n) + \overline{\mathbf{K}}_m. \end{aligned}$$

The same solution can be written with reference to the fixed frame as

$$\mathbf{q} = \mathbf{r}e^{i\Omega t} = \mathbf{r}_1 + \mathbf{r}_2 e^{2i\Omega t}.$$

The obvious meaning of $\mathbf{r}_1$ is then the mean inflected shape, which is fixed in space, whereas that of $\mathbf{r}_2$ is a component of the deflected shape, which rotates at a speed equal to 2Ω. A static loading then causes the onset of vibrations, which are seen by the stator as occurring with a frequency 2Ω and by the rotor with a frequency Ω.

By equating to zero the matrix of the coefficients of Equation (6.78), an eigenproblem in Ω is obtained. It yields the values of the secondary critical speeds caused by a constant load distribution, as seen in the previous section for the Jeffcott rotor.

The solution of the problem related to a given unbalance distribution is straightforward, leading to a synchronous circular whirling. The solution of the equation of motion is constant, $\mathbf{r} = \mathbf{r}_0$, leading to the equation

$$\left[-\Omega^2(\mathbf{M}_m - \mathbf{G}) + \mathbf{K}_m + i\Omega\mathbf{C}_n\right]\mathbf{r}_0 + \left(\Omega^2\mathbf{M}_{r_d} + \mathbf{K}_{r_d}\right)\overline{\mathbf{r}}_0 = \Omega^2\mathbf{f}_r\,. \quad (6.79)$$

A constant solution in the rotating frame, once written in the fixed reference frame, is of the type

$$\mathbf{q} = \mathbf{r}_0 e^{i\Omega t}.$$

Remark 6.11 *The response to unbalance is a pure circular synchronous whirling, and rotating damping has no effect on the behavior of the system because the rotor does not vibrate but merely rotates in the deflected configuration.*

Remark 6.12 *The main feature of the behavior of anisotropic rotors on isotropic supports is the presence of one or more fields of instability (which can be reduced by adding damping to the system), and the presence of a vibration occurring at a frequency equal to 2Ω, together with the usual static component, in the response to a static force.*

Example 6.5 *Nonisotropic high-speed rotor.*

Consider a very unconventional rotating machine: a rotating gravitational quadrupole built with the aim of calibrating the detector used for demonstrating experimentally the existence of gravitational waves. The aim of the machine is that of spinning a rotor with the largest possible gravitational quadrupole moment (which coincides with the difference between the mass principal moments of inertia about two axes perpendicular to the spin axis) at a rotational frequency equal to half the resonance frequency of the detector. In the case under examination, this means a spin speed of 27,450 rpm = 457.5 Hz, leading to a very high peripheral velocity, namely, 575 m/s. A schematic drawing of the rotor is reported in Figure 6.15.

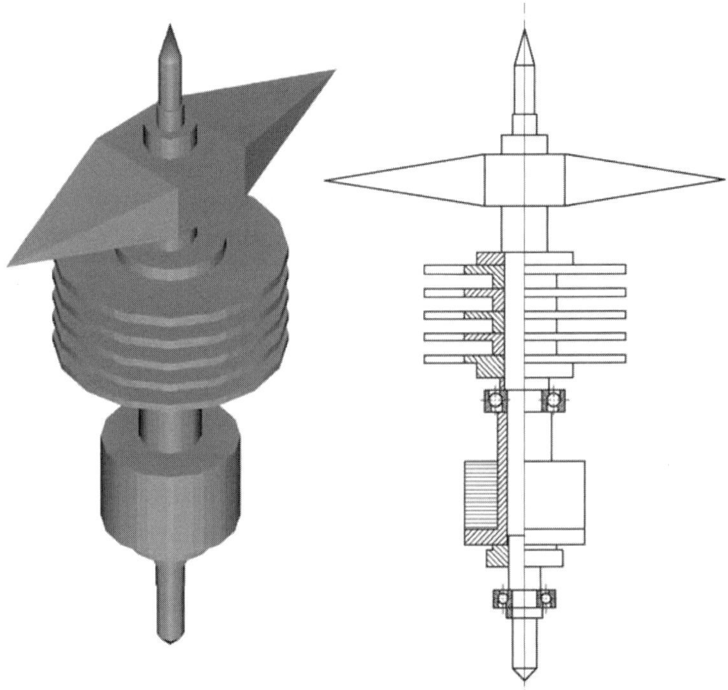

FIGURE 6.15. Schematic drawing of the rotor of a high speed machine used for calibrating a gravitational waves detector.

The bar must spin in a vacuum. To keep the pressure in the chamber in which the bar is housed lower than that in the bearings' case, five stages of a turbomolecular pump are fitted to the shaft (note that in the 3D plot, the blades are not represented). The whole machine is derived from a turbomolecular pump, with the gravitational quadrupole and a point used to measure the distance between the spin axis and the detector, added on top. The angular contact bearings are soft mounted using a conventional elastomeric damped support, as in the case of the machine from which this device derives.

A FEM model is shown in Figure 6.16. The bar and the stages of the pump are modeled as rigid bodies. However, although the former is modeled using an asymmetrical mass element, the latter are modeled using beam elements, just to avoid computing separately their masses and moments of inertia. The same is done for the rotor of the electric motor.

A plot of the first six critical speeds as functions of the stiffness of the bearings is reported in Figure 6.17.

The first four critical speeds, i.e., those related to the two rigid-body modes in each one of the two coordinate planes containing the spin axis, must be as low as possible, to be easily passed, whereas the fifth one must be above the working speed range. The latter condition is always verified, and a stiffness of 1×10^6 *N/m can*

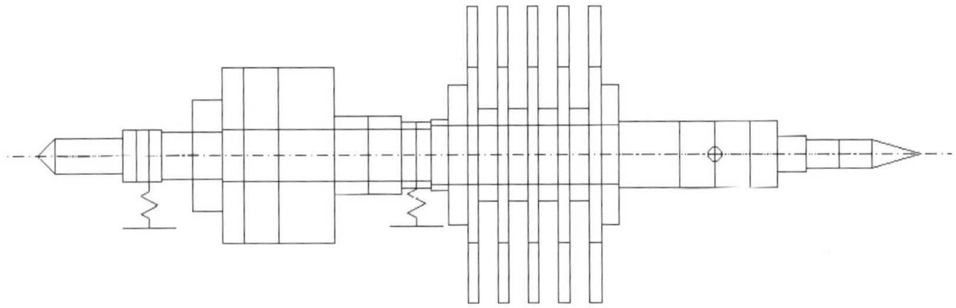

FIGURE 6.16. FEM model of the rotating gravitational quadrupole.

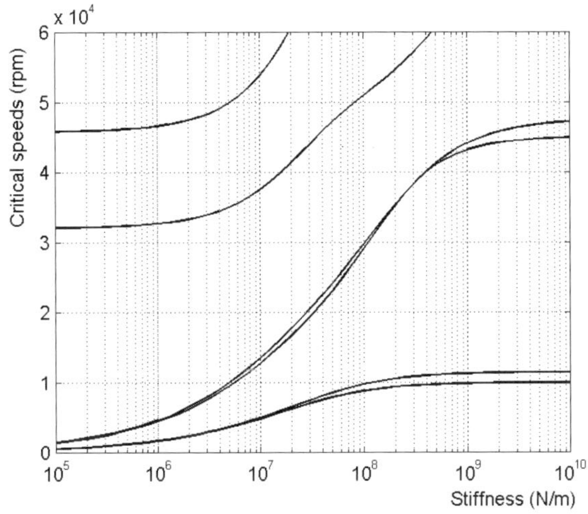

FIGURE 6.17. First six critical speeds as functions of the stiffness of the bearings.

be chosen. The corresponding values of the critical speeds are reported in Table 6.4

The fifth critical speed is then at 32,699 rpm, i.e., 120% of the maximum operating speed. Such safety margin is usually considered sufficient.

The lack of symmetry of the rotor causes the critical speeds to split and some fields of instability to be present, at least in the case of the undamped system. The Campbell diagram and the decay rate plot are reported in Figure 6.18. It is clear that a very strong instability range is present above the fifth critical speed, but other, weaker, instabilities are present within the working range of the machine.

To stabilize the system, damping must be added. If the supports have a structural damping with $\eta = 0.8$, *no instability is encountered within the working range of the machine, as shown by the decay rate plot of Figure 6.19. From the plot, it is clear that during spin-up, a frequency field at about 18,000 rpm in which*

Ω_{cr}[rad/s]					
178.0	181.9	470.9	491.7	3,424	4,885

TABLE 6.4. Values of the first six critical speeds.

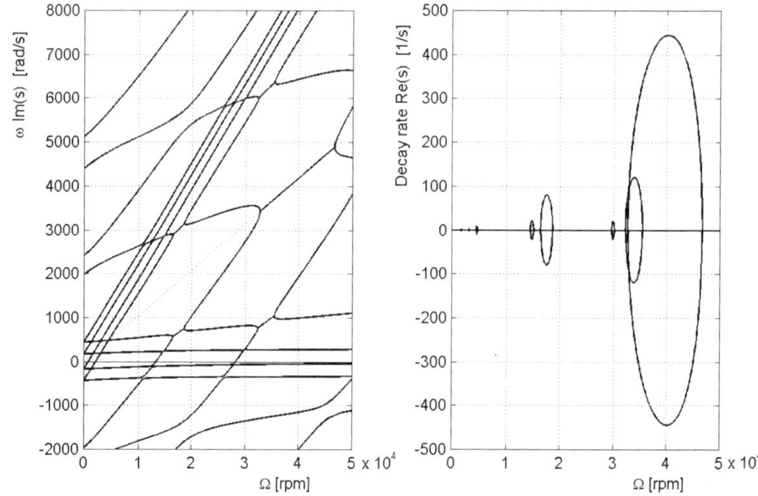

FIGURE 6.18. Campbell diagram and decay rate plot for the undamped system.

the machine has a low margin of stability is encountered, but the threshold of instability is at about 33,000 rpm.

The values of the first six natural frequencies, with the corresponding decay rates, at the operating speed of 27,450 rpm are reported in Table 6.5.

Backward		Forward	
Im(s) [rad/s]	Re(s) [1/s]	Im(s) [rad/s]	Re(s) [1/s]
71.2	−38.8	281.3	−79.7
131.1	−45.7	807.0	−122.0
377.5	−137.3	2283	−44.1
1417	−60.5	3467	−44.1
3537	−44.5	4943	−122.0
4337	−47.5	5469	−79.7

TABLE 6.5. Values of the first six natural frequencies at the operating speed of 27,450 rpm.

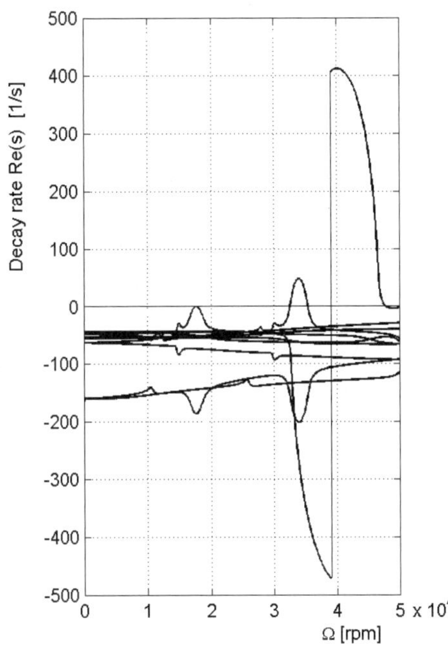

FIGURE 6.19. Decay rate plot of the damped system.

7
Torsional and axial dynamics

The lateral dynamics of rotating systems was studied in the previous chapters under the assumption of complete uncoupling. Actually, even if in elementary rotordynamics such uncoupling is always assumed to hold, many phenomena encountered in actual rotors can be explained only by torsional-lateral or axial-lateral coupling. The physical mechanisms causing such coupling can be of different type, from the geometry of the shaft, which can be different from a simple straight beam (an important case is that of crankshafts), to the presence of twisted blades. Although these effects are present also in the case of linearized models, usually coupling becomes important only when nonlinear effects are taken into account. Also, unbalance can cause such coupling: As this effect is a second-order effect, it needs to be taken into account only in case of rotors with very large unbalance.

Only uncoupled torsional and axial vibrations will be considered in the present chapter. Moreover, no parts of the machine performing reciprocating motion will be considered.

7.1 Torsional free vibration

7.1.1 Lumped parameters approach

If the rotor is assumed to be a beamlike structure and its torsional behavior is studied using a lumped paramcters approach, each node has a single degree of freedom, namely, torsional rotation, and a single generalized force, the torsional moment, acts on it. Moreover, the rotor is free to rotate and, as a consequence, in the torsional study it can be considered as an insulated system, with both ends free. This is true only when the whole rotating system is considered, including all transmission shafts, gearing etc. In the study of a turbo-alternator, for instance, both the rotor of the turbine and of the electrical machine must be included in the model, together with the shaft, which couples them, and the gearing, which may be interposed in between.

There are, however, cases in which only a part of the rotating system is considered, to simplify the problem. This can occur when a part of the machine has a very large moment of inertia, so large that the cross section of the shaft in which it is located can be considered as constrained. This is the case of marine propellers, and a propulsion system of a ship driven by a turbine is sometimes studied considering the turbine rotor and the propeller shaft as constrained at the propeller location. On the opposite side, when a coupling that is torsionally very soft is interposed between two parts of a rotor, it is possible to assume that the coupling does not react with a torque to the relative rotation and the two parts can be considered as uncoupled from each other.

Transfer matrices approach

In all cases, when the system reduces to a single shaft on which a number of moments of inertia are located, it is possible to deal with such an in-line system using the transfer-matrices method. The Holzer method has been for decades the most common tool for dealing with torsional vibrations of engines and drivelines. It is very similar to the Myklestadt method seen in Section 4.1, for beamlike structures vibrating in bending.

Remark 7.1 *Nowadays it is however possible, and more computationally efficient, to resort to a stiffness approach writing the stiffness matrix of each span of the shaft, modeled as a beam element, and assembling them into a global stiffness matrix. This allows us to study also branched systems.*

Consider a lumped torsional system, i.e., a system consisting of rigid discs connected by straight shafts possessing all the properties needed for the uncoupling of the torsional modes from flexural ones [Figure 7.1(a)]. The model can result from the lumping of a more or less uniform continuous system or from a system that is actually made of concentrated rotors and lightweight shafts. Let the torsional generalized coordinate of each station be the rotation ϕ_z and the torsional moment be M_z. The state vectors is then

$$\mathbf{s} = \left\{ \begin{array}{c} \phi_z \\ M_z \end{array} \right\} , \tag{7.1}$$

and its order is 2.

If the stations are located at the positions of the lumped moments of inertia, the field transfer matrices are easily computed by considering that

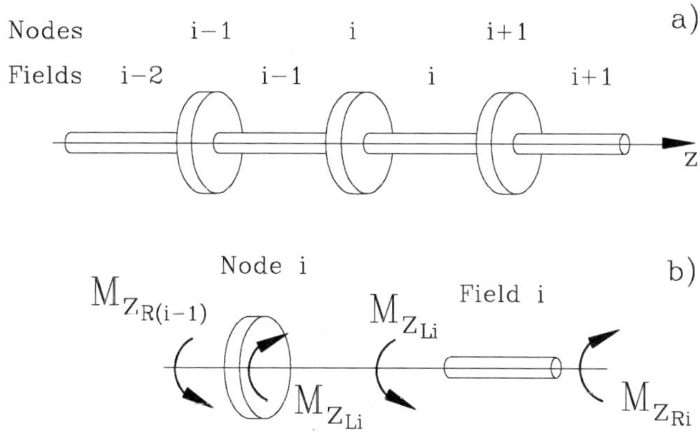

FIGURE 7.1. Holzer's method. (a) Sketch of a lumped parameter torsional system; (b) torques acting on the ith node and field.

b/a	1	1.5	2	3	4	6	10	∞
Ψ	7.14	5.10	4.37	3.80	3.56	3.34	3.19	3.00

TABLE 7.1. Values of coefficient Ψ for the computation of I'_{p} of rectangular cross sections.

the moment at the left is equal to the moment at the right because no moment acts along the field, whereas the rotation at the right end is equal to the rotation at the left end increased by the twisting of the field,

$$\phi_{z_{R_i}} = \phi_{z_{L_i}} + \frac{l_i}{G_i I'_{P_i}} M_{z_{L_i}} \ , \tag{7.2}$$

where the torsional moment of inertia I'_p coincides with the polar moment of inertia I_z if the cross section is circular or annular, whereas in other cases the two quantities are different. The values for I'_p for the cases of greater practical interest can be found in the literature. If the cross section is elliptical with axes a and b, for instance, the following formula can be used:

$$I'_p - \frac{a^3 b^3}{5.1\,(a^2 + b^2)} \ , \tag{7.3}$$

whereas in the case of a rectangular cross section with sides a and b ($a \leq b$)

$$I'_p = \frac{a^3 b}{\Psi} \ , \tag{7.4}$$

where Ψ is reported as a function of ratio a/b in Table 7.1.

The station *transfer matrix* is obtained by considering that the rotation at the left of the ith station (i.e., the rotation at the right of the $(i-1)$th field) is equal to the rotation at the right, whereas the moment at the right of the station is equal to that at the left of the station, increased by the concentrated moment acting on it. If no external moment is included in the model, the latter is only caused by the inertia reaction, which, in harmonic motion, is proportional to the square of the frequency and to the rotation $\phi_{z_{L_i}}$. If the station is constrained by a torsional spring of stiffness K_t, the moment $K_t\phi_{z_{L_i}}$ also acts on the node.

The equations defining the field and station transfer matrices are, respectively,

$$\left\{ \begin{array}{c} \phi_z \\ M_z \end{array} \right\}_{R_i} = \left[\begin{array}{cc} 1 & \dfrac{l_i}{G_i I'_{P_i}} \\ 0 & 1 \end{array} \right] \left\{ \begin{array}{c} \phi_z \\ M_z \end{array} \right\}_{L_i} ,$$
$$\left\{ \begin{array}{c} \phi_z \\ M_z \end{array} \right\}_{L_i} = \left[\begin{array}{cc} 1 & 0 \\ -\omega^2 J_{z_i} + K_t & 1 \end{array} \right] \left\{ \begin{array}{c} \phi_z \\ M_z \end{array} \right\}_{R_{i-1}} . \tag{7.5}$$

Once the transfer matrices have been obtained, it is easy to multiply them to obtain the overall transfer matrix. Equation (4.4) takes the simple form

$$\left\{ \begin{array}{c} \phi_z \\ M_z \end{array} \right\}_{n} = \left[\begin{array}{cc} T_{11} & T_{12} \\ T_{21} & T_{22} \end{array} \right] \left\{ \begin{array}{c} \phi_z \\ M_z \end{array} \right\}_{0} . \tag{7.6}$$

The boundary conditions are easily assessed. If the end at the left is free, moment M_{z_0} must vanish. The second column of the matrix is of no interest, because it multiplies a vanishing moment. If the end at the right is also free, as often happens with torsional systems, the second equation (7.6) reduces to

$$M_{z_n} = T_{21}\phi_{z_0} = 0 , \tag{7.7}$$

yielding a solution different from the trivial one $\phi_{z_0} = 0$ only if $T_{21} = 0$.

Working in the same way for the other end conditions, the following equations yielding the natural frequencies can be obtained:

Right end	Left end	Equation
Free	Free	$T_{21} = 0$
Free	Clamped	$T_{22} = 0$
Clamped	Free	$T_{11} = 0$
Clamped	Clamped	$T_{12} = 0$

As already said for transfer matrices methods, the solution is usually obtained numerically, plotting the appropriate element T_{ij} as a function of ω, and looking for the values of the frequency at which the curve crosses the frequency axis

In the case of the Holzer method, the structure of the transfer matrix is so simple that it is possible to write the relevant equations in a more explicit way. The inflected shape can be obtained by setting an arbitrary value (usually a unit value) for ϕ_{z_0}, if the left end is clamped, or for M_{z_0} if it is free, and computing the state vectors after having introduced a value of the frequency. Assume that the left end is free. The equations can be reformulated by eliminating the moments and writing the rotation at the $(i+1)$th station as function of the rotations at the ith and $(i-1)$th station

$$\phi_{z_{i+1}} = \left(1 - \frac{\omega^2 J_{z_{i-1}}}{k_i} + \frac{k_{i-1}}{k_i}\right)\phi_{z_i} - \frac{k_{i-1}}{k_i}\phi_{z_{i-1}} , \tag{7.8}$$

where the rotations are referred to the nodes. From the assumed value at the first node (node 0), the rotation at node 1 can be computed from Equation (7.8)

$$\phi_{z_1} = \left(1 - \frac{\omega^2 J_{z0}}{k_1}\right)\phi_{z0} . \tag{7.9}$$

In a similar way, the rotations at all nodes can be computed. If also the last node is free, the constraint condition can be written in the form

$$\sum_{i=0}^{n} \omega^2 J_{z_i}\phi_{z_i} = 0 , \tag{7.10}$$

which has an immediate meaning: Because both ends are free, the sum of all the inertia torques $-\omega^2 J_{z_i}\phi_{z_i}$ must vanish. Equation (7.10) is generally not satisfied, as the arbitrary value of the frequency ω does not coincide with one of the natural frequencies of the system. By changing the values of the frequency and by plotting $\sum_{i=0}^{n} \omega^2 J_{z_i}\phi_{z_i}$ as a function of ω, it is straightforward to find the values of the frequency at which the total inertia torque vanishes, i.e., the natural frequencies of the system. Note that the value of the first natural frequency is 0, and the corresponding mode is a rigid rotation.

The mode shapes are so obtained. The arbitrarity because the rotation at the first node was chosen arbitrarily is not inconvenient, because only the shape of the modes is defined; and can be removed by normalizing the eigenvectors.

FEM approach

Instead of using the transfer matrices method, it is possible and, when computers are used, more convenient, to resort to the stiffness approach. The mass matrix is a diagonal matrix, containing the moments of inertia, whereas, in the case of in-line systems, a tridiagonal stiffness matrix is obtained. In the case of a system without constraints, the stiffness matrix is singular and rigid rotations are possible. It cannot be inverted, and the compliance matrix is not available.

The equation of motion for the study of the free behavior of the system is

$$\begin{bmatrix} J_1 & 0 & \cdots & 0 & 0 \\ 0 & J_2 & \cdots & 0 & 0 \\ \cdots & \cdots & \cdots & \cdots & \cdots \\ 0 & 0 & \cdots & J_{n-1} & 0 \\ 0 & 0 & \cdots & 0 & J_n \end{bmatrix} \begin{Bmatrix} \ddot{\phi}_1 \\ \ddot{\phi}_2 \\ \cdots \\ \ddot{\phi}_{n-1} \\ \ddot{\phi}_n \end{Bmatrix} + \\ + \begin{bmatrix} k_1 & -k_1 & \cdots & 0 & 0 \\ -k_1 & k_1 + k_2 & \cdots & 0 & 0 \\ \cdots & \cdots & \cdots & \cdots & \cdots \\ 0 & 0 & \cdots & k_{n-1} + k_n & -k_n \\ 0 & 0 & \cdots & -k_n & k_n \end{bmatrix} \begin{Bmatrix} \phi_1 \\ \phi_2 \\ \cdots \\ \phi_{n-1} \\ \phi_n \end{Bmatrix} = \mathbf{0} . \tag{7.11}$$

The stiffness matrix of a branched or multiple-connected system can be built using the assembly procedures in Sections 4.3.4 and 4.3.5. It is no longer tridiagonal, even if it usually has a band structure, and it is possible to resort to the algorithms commonly used to reorder the list of the generalized coordinates to reduce the bandwidth to a minimum. The mass matrix, on the contrary, is always a diagonal matrix if the model is based on the lumped-parameters approach.

7.1.2 Consistent parameters approach

The inertia of the shafts, which in the lumped parameters approach can be considered by concentrating it in the nodes at the ends of the various elements, can be accounted for in a more detailed way by resorting to the consistent parameters approach, typical of the finite elements method.

Consider the two-nodes beam element whose lateral behavior was studied in Section 4.3.1 (Figure 4.8). The degree of freedom of each cross section involved in the torsional behavior is the torsional rotation, and hence, the element has two degrees of freedom: The rotations about z-axis at the nodes

$$\mathbf{q} = \begin{Bmatrix} \phi_{z_1} \\ \phi_{z_2} \end{Bmatrix} . \tag{7.12}$$

Because each point of the element has a single degree of freedom, the vector $\mathbf{u}$ of the displacements of the internal points of the element (defined by Equation 4.33) has a single component ϕ_z and matrix $\mathbf{N}$ has one row and two columns (the element has two degrees of freedom). ϕ_z can be expressed as a polynomial in z, or, better, in the nondimensional axial coordinate $\zeta = z/l$

$$\phi_z = a_0 + a_1\zeta + a_2\zeta^2 + a_3\zeta^3 + \ldots . \tag{7.13}$$

The polynomial must yield the values of the rotations ϕ_{z_1} and ϕ_{z_2}, respectively, at the left end (node 1, $\zeta = 0$) and at the right end (node 2,

$\zeta = 1$). These two conditions allow computing only two coefficients a_i and then the polynomial expression of the displacement must include only two terms, i.e., the constant and the linear terms. With simple computations [17], the matrix of the shape functions is obtained:

$$\mathbf{N} = [1 - \zeta,\ \zeta] . \tag{7.14}$$

The shear strain γ_{xy} can be expressed as $\gamma_{xy} = d\phi_z/dz$, or, using vector $\boldsymbol{\epsilon}$, which in this case has only one element,

$$\boldsymbol{\epsilon} = \left[\frac{d}{dz}(1-\zeta)\,,\ \frac{d}{dz}\zeta\right]\left\{\begin{array}{c}\phi_{z_1}\\ \phi_{z_2}\end{array}\right\} . \tag{7.15}$$

Matrix

$$\mathbf{B} = \left[\frac{d}{dz}(1-\zeta)\,,\ \frac{d}{dz}\zeta\right] = \frac{1}{l}\left[-1\,,\ 1\right] \tag{7.16}$$

has one row and two columns.

The torsional moment M_z is linked to the rotation of the cross section by the usual formula

$$M_z = GI'_p \frac{\partial \phi_z}{\partial z} . \tag{7.17}$$

The potential energy of a length dz of beam is

$$d\mathcal{U} = \frac{1}{2} M_z \gamma dz = \frac{1}{2} GI'_p \mathbf{q}^T \mathbf{B}^T \mathbf{B} \mathbf{q} dz . \tag{7.18}$$

The stiffness matrix can be obtained by integrating Equation (7.18)

$$\mathbf{K} = \int_0^l I'_p \mathbf{B}^T G \mathbf{B} dz = \frac{GI'_p}{l} \int_0^1 \left[\begin{array}{cc} 1 & -1 \\ -1 & 1 \end{array}\right] d\zeta = \frac{GI'_p}{l} \left[\begin{array}{cc} 1 & -1 \\ -1 & 1 \end{array}\right] , \tag{7.19}$$

The kinetic energy of a length dz of beam is

$$d\mathcal{T} = \frac{1}{2} \rho I_p \dot{\phi}_z^2 dz = \frac{1}{2} \rho I_p \dot{\mathbf{q}}^{\mathbf{T}} \mathbf{N}^T \mathbf{N} \dot{\mathbf{q}} dz . \tag{7.20}$$

The mass matrix is then

$$\mathbf{M} = \int_0^l \rho I_p \mathbf{N}^T \mathbf{N} dz = \rho I_p l \int_0^1 \left[\begin{array}{cc} (1-\zeta)^2 & \zeta(1-\zeta) \\ \zeta(1-\zeta) & \zeta^2 \end{array}\right] d\zeta =$$

$$= \frac{\rho I_p l}{6} \left[\begin{array}{cc} 2 & 1 \\ 1 & 2 \end{array}\right] . \tag{7.21}$$

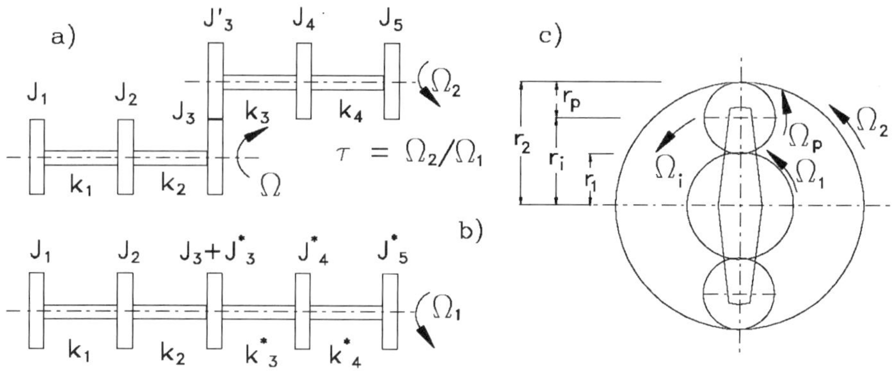

FIGURE 7.2. Geared system: Sketch of the (a) actual system and of the (b) equivalent system. (c) Planetary gear train: sketch of the system and notation.

7.1.3 Geared systems

Consider the system sketched in Figure 7.2(a) in which the two shafts are linked by a pair of gear wheels, with transmission ratio τ. For the study of the torsional vibrations of the system, it is possible to substitute the system with a suitable equivalent system, in which one of the two shafts is substituted by an expansion of the other [Figure 7.2(b)]. This substitution can be performed only if no allowance is taken for backlash, which would introduce nonlinearities. Assuming also that the deformation of gear wheels is negligible, the equivalent rotations ϕ_i^* can be obtained from the actual rotations ϕ_i simply by dividing the latter by the transmission ratio $\tau = \Omega_2/\Omega_1$:

$$\phi_i^* = \phi_i/\tau. \tag{7.22}$$

The kinetic energy of the ith flywheel, whose moment of inertia is J_i, and the elastic potential energy of the ith span of the shaft are, respectively,

$$\begin{aligned} \mathcal{T} &= \tfrac{1}{2} J_i \dot{\phi}_i^2 = \tfrac{1}{2} J_i^* \dot{\phi}_i^{*^2} , \\ \mathcal{U} &= \tfrac{1}{2} k_i \left(\phi_{i+1}^2 - \phi_i^2\right) = \tfrac{1}{2} k_i^* \left(\phi_{i+1}^{*^2} - \phi_i^{*^2}\right) , \end{aligned} \tag{7.23}$$

where the equivalent moment of inertia and stiffness are, respectively,

$$J_i^* = \tau^2 J_i \ , \qquad k_i^* = \tau^2 k_i \ . \tag{7.24}$$

If the system includes a planetary gear train, the computation can be performed without difficulties. The equivalent stiffness can be computed simply from the overall transmission ratio, whereas in the computation of the equivalent inertia, the total kinetic energy of the rotating parts must be taken into account. The angular velocities of the central gear Ω_1, of the ring gear Ω_2, of the revolving carrier Ω_i, and of the intermediate pinions Ω_p of the planetary gear shown in Figure 7.2(c) are linked by the equation

$$\frac{\Omega_1 - \Omega_i}{\Omega_2 - \Omega_i} = -\frac{r_2}{r_1}, \qquad \Omega_p = (\Omega_1 - \Omega_i)\frac{r_1}{r_p} - \Omega_i \,. \tag{7.25}$$

The kinetic energy of the system made of the internal gear, with moment of inertia J_1, the ring gear, with moment of inertia J_2, the revolving carrier, with moment of inertia J_i, and n intermediate pinions, each with mass m_p and moment of inertia J_p, is

$$\mathcal{T} = \frac{1}{2}\left[J_1\Omega_1^2 + J_2\Omega_2^2 + (J_i + nm_p r_i^2)\Omega_i^2 + nJ_p\Omega_p^2\right]. \tag{7.26}$$

The equivalent moment of inertia, computed with reference to the shaft of the internal gear spinning at speed Ω_1, is

$$J_{eq} = \frac{2\mathcal{T}}{\Omega_1^2} = J_1 + J_2\left(\frac{\Omega_2}{\Omega_1}\right)^2 + (J_i + nm_p r_i^2)\left(\frac{\Omega_i}{\Omega_1}\right)^2 + nJ_p\left(\frac{\Omega_p}{\Omega_1}\right)^2 . \tag{7.27}$$

If the deformation of the meshing teeth must be accounted for, it is possible to introduce into the model two separate degrees of freedom for the two meshing gear wheels, modeled as two different inertias, and to introduce between them a shaft whose compliance simulates the compliance of the transmission. This is particularly important when a belt or flexible transmission of some kind is used instead of the stiffer gear wheels. In a machine, there may be several shafts connected to each other, in series or in parallel, by gear wheels with different transmission ratios. The equivalent system is defined with reference to one of the shafts, and the equivalent inertia and stiffness of the elements of the other ones are all computed using the ratios between the speeds of the relevant element and the reference one. The equivalent system will then be made of a set of elements, in series or in parallel, following the scheme of the actual system, but with rotations that are all consistent. If the compliance of the gears is to be taken into account in detail, the nonlinearities caused by the contacts between the meshing teeth and backlash must be considered.

Example 7.1 *Multi-engine marine propulsion system.*

Consider a marine unit made by four in-line, six cylinder engines driving two propellers through gear wheels. The engines work in pairs, the two engines being assembled with the flywheels one close to the other, connected through a shaft. The two pairs are located side by side, driving a common shaft through gear wheels with a transmission ratio equal to 0.9. Two propeller shafts are driven with a transmission ratio of 0.4 from the central intermediate shaft.

A detailed sketch of the system is shown in Figure 7.3.

The engines are modeled with six equivalent flywheels for the crank-connecting rod-piston systems, a moment of inertia for the flywheel and six equivalent shafts

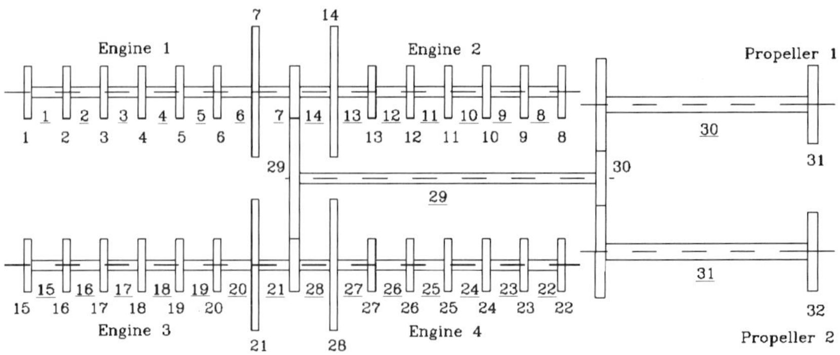

FIGURE 7.3. Sketch of a marine unit with 4 in-line, six cylinder engines driving 2 propellers (the numbers of the elements are underlined).

Node #	J [kg/m^2]	Element #	k [kN/m]
1, 8, 15, 22	5.51×10^{-3}	1→5, 8→12	392
2, 5, 9, 12, 16, 19, 23, 26	4.89×10^{-3}	15→19, 22→26	392
3, 4, 10, 11, 17, 18, 24, 25	6.02×10^{-3}	6, 13, 20, 27	455
6, 12, 20, 27	6.25×10^{-3}	7, 14, 21, 28	67.63
7, 14, 21, 28	6.753×10^{-2}	29	28.66
29	0.1416	30, 31	4.1122
30	0.4025		
31,32	1.5552		

TABLE 7.2. Data for the nodes and the elements.

that model the torsional stiffness of the various parts of the crankshaft. The compliance of the gearings is neglected, and consequently, the inertia of the first three gear wheels is reduced to the engine shaft and concentrated in node 29. The inertia of the other three gear wheels is reduced to the same shaft and concentrated in node 30. The data for a lumped parameters model, all reduced to the engines' shafts, are reported in Table 7.2.

The mass matrix is diagonal, and because the system is not in-line, the stiffness matrix is not tridiagonal. By assembling the mass and stiffness matrices and computing the eigenvalues of the matrix $\mathbf{M}^{-1}\mathbf{K}$*, the natural frequencies are readily obtained. Note that, as usual with torsional systems, the stiffness matrix is singular and it is not possible to compute matrix* $\mathbf{K}^{-1}\mathbf{M}$*. The first eigenvalue is vanishing, because a rigid-body mode is present. The following three eigenfrequencies are as follows:*

$$\omega_1 = 51.42 \quad rad/s, \quad \omega_2 = 100.94 \quad rad/s, \quad \omega_3 = 359.3 \quad rad/s.$$

The corresponding mode shapes are reported in Figure 7.4.

The first mode is characterized by an oscillation of the propellers in opposite directions while the whole system from node 1 to node 30 remains still. The frequency of the propellers on their shafts is 51.42 rad/s, the same value obtained

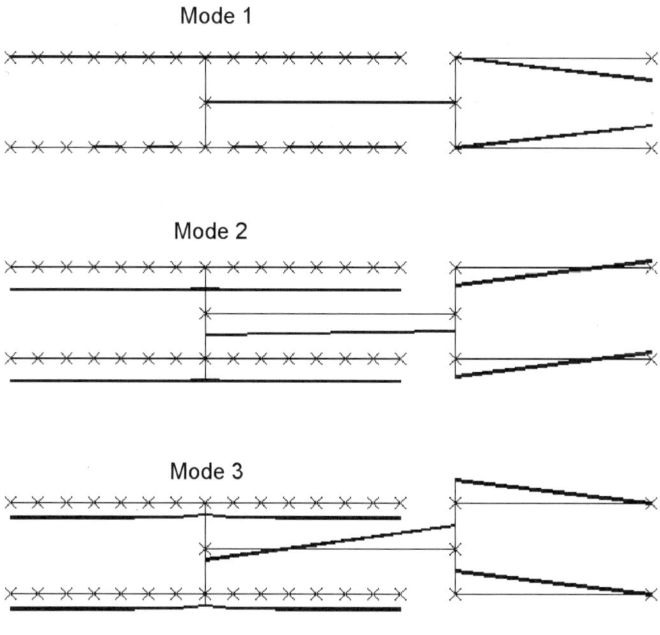

FIGURE 7.4. First three mode shapes of the marine unit. The rigid-body mode has not been plotted.

for the first natural frequency of the system. In the second mode, the propellers oscillate together in one direction, and the four engines oscillate in the opposite one. In the third mode, the propellers are almost stationary, and the gear wheels of the final drive oscillate in the opposite direction than the engines.

7.2 Forced vibrations

The torques acting on rotating machines are very often periodic, with a period coincident with the period of rotation or with one of its multiples or submultiples. In such cases, it is possible to express the driving or resisting torques acting on the nodes of the lumped parameter system used to model the actual rotor in the form of a Fourier series and to study the response to each one of the harmonic terms separately. The total response will then be the sum of the response to the different harmonics. The same approach holds also in the case in which the consistent approach is used to compute the nodal forces corresponding to a torsional moment distribution.

Once the forcing functions have been obtained, there is no difficulty in computing the response of the system at various speeds. Note that torsional systems are usually lightly damped and the evaluation of the response to the harmonics of the forcing function that are not in resonance with the torsional natural frequencies of the rotor can be performed by resorting to an undamped model. However, in the case in which a forcing function has a frequency close to a natural frequency, no useful result can be obtained without introducing damping into the model.

A vector of the driving functions, whose terms are expressed by trigonometric series, can be added on the right-hand side of the equation of motion (7.11), to which a damping matrix has also been added

$$\mathbf{M}\ddot{\boldsymbol{\phi}}_z + \mathbf{C}\dot{\boldsymbol{\phi}}_z + \mathbf{K}\boldsymbol{\phi}_z = \sum_{k=1}^{m} \mathbf{m}_{s_k} \sin(k\omega_0 t) + \sum_{k=1}^{m} \mathbf{m}_{c_k} \cos(k\omega_0 t) \ , \qquad (7.28)$$

where $\mathbf{m}_c$ and $\mathbf{m}_s$ indicates the vectors in which the coefficients of the terms in cosine and sine of the moments are listed, ω_0 is the fundamental frequency, usually a multiple or a submultiple of the rotation frequency, and m is the number of harmonics that is considered. Note that the static component of the torque has been neglected, but it could be considered without difficulties.

The solution of Equation (7.28) can be obtained directly by adding the responses to the various harmonic components:

$$\boldsymbol{\phi} = \sum_{k=1}^{m} \boldsymbol{\phi}_{z_k} = \sum_{k=1}^{m} \boldsymbol{\phi}_{z_{s\,k}} \sin(k\omega_0 t) + \sum_{k=1}^{m} \boldsymbol{\phi}_{z_{c\,k}} \cos(k\omega_0 t) \ . \qquad (7.29)$$

The amplitudes of the components in sine and cosine of the response can be obtained by solving the following linear set of equations:

$$\begin{bmatrix} \mathbf{K} - k^2{\omega_0}^2\mathbf{M} & -k\omega_0\mathbf{C} \\ k\omega_0\mathbf{C} & \mathbf{K} - k^2{\omega_0}^2\mathbf{M} \end{bmatrix} \left\{ \begin{array}{c} \boldsymbol{\phi}_{z_{s\,k}} \\ \boldsymbol{\phi}_{z_{c\,k}} \end{array} \right\} = \left\{ \begin{array}{c} \mathbf{m}_{s_k} \\ \mathbf{m}_{c_k} \end{array} \right\} . \qquad (7.30)$$

If damping is neglected, as is customary when the response in conditions that are far from resonance is to be obtained, Equation (7.30) uncouples into two separate linear sets of equations. Equation (7.30) must be solved at each rotational speed and for each harmonic of the forcing function; however, the computation may be fairly simple, because the order of the matrices, equal to the number of the degrees of freedom of the system, is usually small.

The amplitude of the oscillations at each node of the system, caused by each harmonic component of the forcing function, can be easily computed. Once the amplitudes are known, there is no difficulty in obtaining the dynamic stressing of each span of the shaft. The maximum value of the shear stress τ in the shaft spanning from the jth to the $(j+1)$th node is

$$\tau_{max_j}(t) = \frac{k_j(\phi_{z_{j+1}} - \phi_{z_j})}{W_j}, \tag{7.31}$$

where W_j is the torsional section modulus of the relevant shaft element [42]. The torsional displacements at the nodes are not in phase, and then it is impossible to obtain the twist angle of each span as the difference of amplitude between the end sections. The components in phase and in quadrature must be accounted for separately, and the amplitude of the twist angle, which is needed to compute the stress, must be obtained from the two components. The shear stress so computed is variable in time with polyharmonic time history; the time histories of the stresses caused by the various harmonics, each with its amplitude, phase, and frequency, should then be added to each other and to the static stressing, and the fact that their consequences on the overall fatigue of the shaft are different should be considered.

In practice, a much simpler approach is followed, not because of the long computations involved, which with a modern computer could be dealt with without problem, but because the phasing of the harmonics can be difficult to evaluate (near the resonance the phase is quickly variable) and predicting the fatigue life of a machine element subject to polyharmonic stressing is still difficult. The amplitudes of the stress cycles caused by the various harmonics are computed and added together, often limiting the sum to the few most important harmonics. This procedure leads to overestimating the amplitude of the variable component of the stress and then is conservative. The shear stresses so computed must be added to those from other causes and, using a suitable failure criterion, to stresses caused by bending, axial forces, shrink fitting, surface forces, and so on.

Although the response of the system in conditions far from resonance can be computed from an undamped model, the response at resonance can be obtained only after the damping of the system has been evaluated, and the precision of the results is strictly dependent on the precision with which damping is known. Because the resonant conditions are usually the most dangerous, this part of the dynamic analysis is very important, and the difficulty in achieving a good estimate of the damping is one of the factors limiting the usefulness of the dynamic analysis in the current case and compels one to resort to extensive experimentation. When damping is mostly caused by the internal damping of the material constituting the various shaft elements, there is no difficulty in introducing a proportional damping with modal damping ratio equal for all modes: $\zeta_j = \eta/2$, where η is the loss factor of the material of the crankshaft.

In many torsional systems, however, damping is from many causes, among which friction between moving parts, electromagnetic forces, and the presence of fluid in which some rotating parts move can be important. Neglecting them would lead to a large underestimate of damping, and it is usually necessary to resort to experimental results, obtained from machines similar to the one under study and to empirical or semiempirical formulas and numerical values reported in the literature [43, 44].

Generally, the damping caused by the various elements connected to the shaft (propellers, brakes, electrical machines) can be evaluated by assuming that they provide a braking torque on the relevant node that is proportional to the instantaneous angular velocity at the power p through coefficient k''

$$M_{d_j}(t) = k'' \left[\Omega + \sum_{k=1}^{m} \dot{\phi}_{z_{j_k}}(t)\right]^p , \tag{7.32}$$

where subscript j refers to the jth node of the model and k denotes the kth harmonic. Note that all harmonics are assumed to act in phase, which is clearly an approximation but does not affect the results obtained later. This type of damping is clearly nonlinear, but it can be linearized by introducing in the computation a viscous damping for each harmonic, which is equivalent from the viewpoint of energy dissipation. By introducing the harmonic time history of each component of the velocity $\dot{\phi}_{z_{j_k}}(t)$, the energy dissipated in a cycle is

$$\mathcal{L} = \int_0^T k'' \left[\Omega + \sum_{k=1}^{m} k\omega_0 \phi_{z_{j_{k_0}}} \cos(k\omega_0 t)\right]^{p+1} dt \, . \tag{7.33}$$

By computing the $(p{+}1)$th power of the binomial within the integral sign, it follows that

$$\begin{aligned}\mathcal{L} = &\int_0^T k''\Omega^{p+1} dt + \binom{p+1}{1} \int_0^T k''\Omega^p \omega_0 \sum_{k=1}^{m} k\phi_{z_{j_{k_0}}} \cos(k\omega_0 t) dt + \\ &+ \binom{p+1}{2} \int_0^T k''\Omega^{p-1}\omega_0^2 \left[\sum_{k=1}^{m} k\phi_{z_{j_{k_0}}} \cos(k\omega_0 t)\right]^2 dt + \dots \, .\end{aligned} \tag{7.34}$$

The first integral of Equation (7.34) is constant and corresponds to the average power $P = k''\Omega^{p+1}$ applied in the relevant node. It can be used to compute coefficient k'', because the average power is generally known.

The second integral vanishes, and the third contains terms in the squares of cosine function and terms in which there are products of cosines with different arguments; the latter vanish once they are integrated over a whole period. The fourth term also vanishes. If the series (7.34) is truncated after the fourth term, the braking torque, excluding the contribution caused by the constant term, is

$$\mathcal{L} = \binom{p+1}{2} \pi k''\Omega^{p-1}\omega_0 \sum_{k=1}^{m} k\phi^2_{z_{j_{k_0}}} , \tag{7.35}$$

corresponding to that caused by a viscous damper with equivalent damping coefficient

$$c_{eq} = \binom{p+1}{2} k'' \Omega^{p-1} . \tag{7.36}$$

If $p = 1$, there is a true viscous damper and c_{eq} coincides with k''. If $p = 2$, as is often assumed in the case of propellers, k'' is expressed in S.I. units as Nms^2 and $c_{eq} = 3k''\Omega$. If $p = 3$, k'' is expressed in S.I. units in Nms^3 and $c_{eq} = 6k''\Omega^2$.

The damping matrix of the system can be obtained by applying in the various nodes the viscous dampers whose damping coefficients have been computed above and is diagonal. This way of evaluating the damping of the system is clearly an approximation that can, in many cases, be very rough. However, it is often used in practice because there are no simple alternatives.

7.3 Torsional critical speeds

As stated in the previous section, often the torsional moments acting on the nodes of the system are periodic and their fundamental frequency is a multiple or a submultiple of the rotation frequency. In the case of two-stroke-cycle internal-combustion engines or other reciprocating machines in which the duration of the working cycle corresponds to a single revolution of the crankshaft, for instance, the frequency of the various harmonics is equal to whole multiples of the rotational speed $\omega_k = k\Omega$, whereas in the case of four-stroke-cycle internal-combustion engines, the frequency of the fundamental harmonic is equal to half of the rotational speed Ω and the frequency of the ith harmonic is $\omega_k = k\Omega/2$. Note that there are cases in which a large number of harmonics with nonvanishing amplitude are present, particularly in the case in which the torques vary very quickly, in comparison with the time needed to perform a revolution. Again an example is that of internal combustion engines, in which more than 20 harmonics usually need to be taken into account.

In most cases the torsional natural frequencies of the system are independent from the rotational speed. The resonance conditions, defining the torsional critical speeds, can then be studied using a Campbell diagram of the type shown in Figure 7.5. There are many resonance conditions, and it is very difficult, usually impossible, to avoid some of them being located within the working range of the machine. Not all resonance conditions are equally dangerous, and the dynamic stressing of the rotor at the various critical speeds must be evaluated to understand their severity.

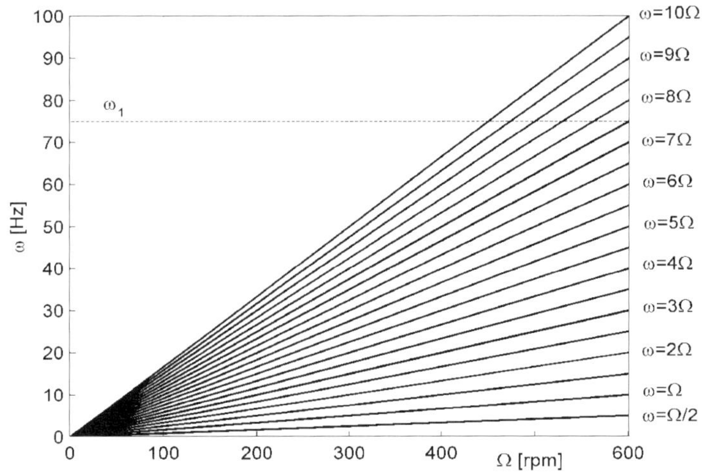

FIGURE 7.5. Campbell diagram for a six-cylinder, four-stroke in-line internal-combustion engine for the computation of the resonance conditions: The first 20 harmonics of the forcing function and the first torsional natural frequency ω_1 are plotted.

7.4 Axial vibration

Axial vibration of straight rotors are usually uncoupled with torsional and lateral vibrations and can be studied separately. Moreover, their natural frequencies are higher than those of other vibration modes and very little excitation acts on axial modes. As a consequence, the study of the axial vibration of rotors is usually omitted.

The simplified models used for the study of axial vibrations are very similar to those seen in the context of torsional vibrations: In lumped parameters or consistent models, each node has only a single degree of freedom, as it was the case for torsional vibrations. A difference is that although the models used for the study of torsional vibrations are underconstrained, rotors are axially kept in position by the thrust bearing, usually modeled as an elastic constraint. All other bearings are often not considered; they allow axial displacements of the order of magnitude of those encountered in axial vibration. Once the model has been obtained, there is no difficulty in studying free vibrations. Usually, an in-line system is obtained and transfer-matrices procedures, like the Holzer method, can be used even if more modern approaches based on the FEM are more common now.

8

Rotor-bearings interaction

Almost all rotors are supported by some sort of bearings, i.e., cylindrical hinges that keep the axis of rotation in a given position in space. There are, however, cases of rotors that are completely free in space, like spinning spacecraft or celestial bodies. They are usually defined as *bearingless* or *free rotors*, and they will be studied in detail later. In the present chapter, the influence of the characteristics of the bearing on the behavior of *fixed rotors* will be studied in detail.

Like any material body, bearings cannot be infinitely stiff and hence cannot constrain the axis of rotation to remain exactly in the required position. Depending on whether the stiffness of the bearings is low or high if compared with the stiffness of the rotor, the latter can be said to be *soft mounted* or *hard mounted.* Hard-mounted rotors are used when the characteristics of the machine require that the rotor has a well-determined position in space and that forces applied to it do not cause deformations exceeding the allowable limits. An example of rotors that have to be hard mounted are the spindles of machine tools. In other cases, on the contrary, it is not so important to keep the position of the rotor under strict tolerances and larger displacements are acceptable, although there is an advantage in having the critical speed of the machine low enough to allow an easy passage into the supercritical regime to exploit self-centering. This is the case of many turbomachines and centrifugal pumps, in which, however, the displacements of the rotor cannot be too high to avoid the need of leaving a large gap between rotor and stator, which would degrade the fluid dynamics performance. A case in which large displacements are allowed and a particularly soft suspension is used is that of the domestic washing machines, where the possibility of exploiting self-centering is very important.

With reference to the Jeffcott rotor, and particularly to Figure 2.22, it can be said that a hard-mounted rotor tends to rotate about its geometric center C (clearly if the bow of the shaft itself is neglected), whereas a soft-mounted rotor tends to rotate about its mass center P.

In the case of hard-mounted rotors, it is sometimes possible to neglect the compliance of the bearings and to study the dynamic behavior assuming that the bearings are stiff.

On the contrary, in the case of soft-mounted rotors (and often for not-so-hard-mounted ones), the compliance of the bearings must be accounted for. There are also cases in which not only the bearings, but also the whole supporting structure introduces a compliance that cannot be neglected.

The simplest alternative to considering the bearings as rigid bodies is that of modeling them as linear springs and dampers. If the compliance of the supporting structure is neglected, they are just elastic and damped constraints applied to the rotor, whereas if also the nonrotating parts of the machine are modeled, the bearings are considered as springs and dampers connecting the rotor with the stator. However, all of the bearings used in real-world machines (rolling elements, hydrodynamic, magnetic, ...) display a more or less pronounced nonlinear behavior, and linearized models are always approximations that in most cases are acceptable only if the amplitude of the vibration is small.

Bearings have also another role, apart from keeping the axis of rotation in a position close to the nominal one: They provide the nonrotating damping needed for operation in the supercritical range. If the bearings are not able to supply enough damping, purposely designed dampers can be added to achieve the required stability. As bearings exert forces on the rotor, it is possible to use them to supply also control forces to perform an active control of the vibrations of the system. This is the case of magnetic bearings, but other types of bearings can be used to transfer the forces generated by actuators of different types. Bearings may also be supplied with sensors to measure the displacements of the rotor, both to close the loop in case the latter is actively controlled or simply to monitor the working conditions of the machine.

8.1 Rigid-body and flexural modes

In the case of soft-mounted rotors, the lowest natural frequencies are usually related to *rigid-body modes* of the rotor, i.e., modes in which the rotor behaves as a rigid body on soft supports. Strictly speaking, perfectly rigid-body modes are possible only in the case of free rotors, i.e., if the stiffness of the bearings is vanishingly small. The natural frequencies of the rigid-body modes is equal to zero. In actual cases in which the stiffness of the bearings is small, but not zero, no exactly rigid-body mode can exist; however, if the rotor is much stiffer than the bearings, some modes exist in which the deformation of the rotor is very small if compared with that of the bearings and the natural frequencies, although not being zero, are small. These are still referred to as rigid-body modes, even if this is an approximation. If the system is axially symmetrical (both the rotor and the stator), there are two of such modes, linked with mainly conical and cylindrical whirling, even if only particular symmetry conditions allow us to have really conical and really cylindrical modes. The higher order modes involve the deformation of both the rotor and the stator, and they are often defined as *flexural modes* or *deformation modes.*

In rigid-body modes, there is, by definition, no deformation of the rotating parts of the machine and rotating damping has no effect on its behavior. In the supercritical range, no instability is usually related with modes of this type, at least for what the destabilizing effect of rotating damping is concerned. The damping of the bearings, on the contrary, is very effective in stabilizing the motion. However, as it will be seen later, the destabilizing effects of lubricated bearings can destabilize rigid-body modes.

Remark 8.1 *As already stated, no true rigid-body mode exists. However, when the natural frequency related to a mode is much higher than the highest values of the forcing frequency actually present, that mode is considered as a rigid-body mode. In particular, because in rotors synchronous excitation caused by unbalance is usually the most important excitation, a rotor whose critical speeds linked with deformation modes are well above the working range, is considered as a rigid rotor.*

Flexural modes, on the contrary, involve actual deformation of the shaft and hence are affected by rotating damping. If the relevant mode shape involves also some deformation of the bearings or, in general, of nonrotating parts of the machine, stability in the supercritical range may be granted if the stabilizing effect of nonrotating damping overcomes the destabilizing one of rotating damping. If, on the contrary, the mode shape is such that the deformation of the bearings (of the stator in general) is very small, the mode is easily unstable in supercritical conditions. If the rotor has some sort of active control of the vibrations, when the displacement in correspondence to the sensors is vanishing small, the mode is not observable. When the displacement at the actuator location is vanishing, the mode is not controllable.

For stability reasons, often soft-mounted rotors work in supercritical conditions with respect to rigid-body modes (to exploit self-centering) and in subcritical conditions with respect to flexural modes to assure stability.

When the stiffness of the bearings is not much smaller than that of the rotor, it is no more possible to speak of rigid-body modes not even in an approximated way, because both the rotor and the supporting structure inflect in a comparable way. Finally, in the case of hard-mounted rotors, the deformation of the bearings is usually negligible with respect to that of the rotor.

To state whether a mode can be considered as a rigid-body mode or a deformation mode, the ratio between the potential energy (caused by the modal deformation) stored in the bearings and that stored in the whole system can be defined. Such *modal bearing deformation index* of the ith mode $\mathbf{q}_i$ can be defined as

$$I_{di} = \frac{\sum_{k=1}^{n_b} K_k q_{k_i}}{\mathbf{q}_i^T \mathbf{K} \mathbf{q}_i}, \tag{8.1}$$

where $\mathbf{K}$ is the stiffness matrix of the rotor-bearing system and K_k is the stiffness of the kth bearing. The values of I_{di} are between 0 and 1. It tends to zero in case of stiff bearings, to increase with increasing bearing flexibility. However, when the stiffness of the bearings vanishes, rigid body modes are characterized by $I_{di} = 1$ (both energies of the rotor and the bearings are vanishingly small).

These considerations are summarized in Figure 8.1 for the same system studied in Figure 5.5: a constant diameter shaft supported at both ends by two identical springs. Instead of plotting the product $\beta_i = al$ as a function of the nondimensional stiffness of the supports K^* as in Figure 5.5, the first four nondimensional natural frequency

$$\omega_i^* = (a_i l)^2 = \omega_i l^2 \sqrt{\frac{\rho A}{E I_y}}, \qquad i = 1, 2, 3, 4 \tag{8.2}$$

are reported, together with the mode shapes for some values of K^*.

Also the values of I_{di} (computed through a discretized FEM model) are reported.

The figure deals with very slender beam modeled using the Euler Bernoulli scheme, and consequently, the natural frequencies are not affected by the spin speed. The critical speeds then coincide with the natural frequencies at standstill. In other cases, the gyroscopic effect may be non-negligible, and separate computations for the natural frequencies at different speeds and for the critical speeds must be performed, obtaining plots similar to that of Figure 8.1.

8.2 Linearization of the characteristics of the bearings

Consider a bearing whose behavior is both nonlinear and axially symmetrical. The radial load F acting on the bearing is linked with the radial

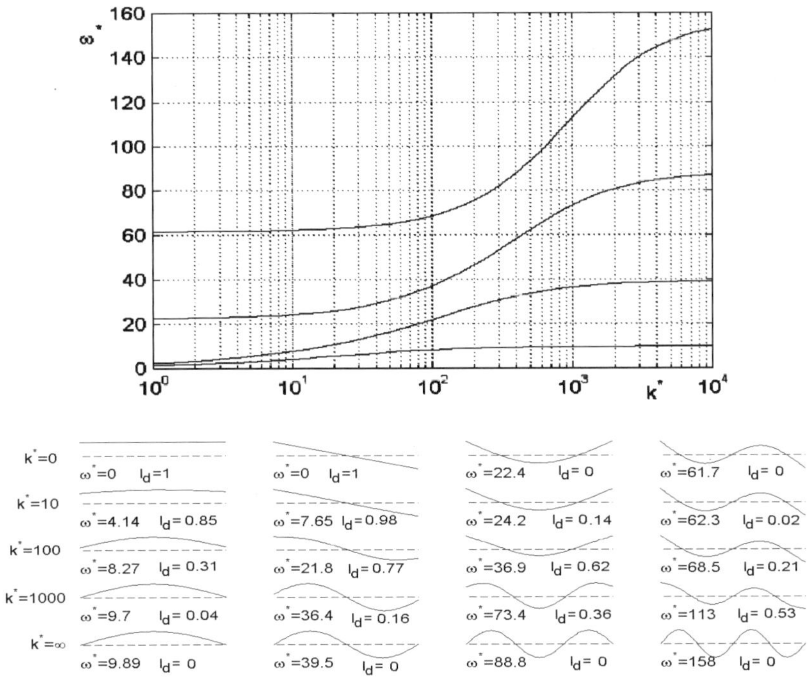

FIGURE 8.1. Nondimensional natural frequency at standstill ω^* as a function of the ratio between the stiffness of the supports and the stiffness of the beam k^* for a constant diameter beam supported by two identical bearings at the ends. The first four mode shapes for different values of k^* are also shown.

displacement r by the nonlinear relationship $F = F(r)$. If the bearing is axially symmetrical, the force-displacement characteristics are independent of the direction in the radial plane and the displacement occurs in the same direction of the applied force. Using real coordinates, the radial force can be written in the form

$$\begin{cases} F_x = -xf\left(|r|^2\right) , \\ F_y = -yf\left(|r|^2\right) , \end{cases} \tag{8.3}$$

where $|r| = \sqrt{x^2 + y^2}$ is the modulus of the displacement of the bearing and $f\left(|r|^2\right)$ is a function that depends on the characteristics of the bearing. The $(-)$ sign denotes a restoring force. By using the complex coordinate $r = x + iy$, Equation (8.3) becomes

$$F = F_x + iF_y = -rf\left(|r|^2\right) . \tag{8.4}$$

To study the dynamic behavior of the rotor, it is possible to compute the point of the characteristics $F(r)$ in which the bearing operates under the effect of the static forces and then to linearize the characteristics about that point. Let the static force be $F_0 = F_{x_0} + iF_{y_0}$ and the static displacement be $r_0 = x_0 + iy_0$. The Jacobian matrix containing the derivatives of the components of the force with respect to the displacements is

$$\mathbf{S}_{r=r_0} = \begin{bmatrix} \frac{\partial F_x}{\partial x} & \frac{\partial F_x}{\partial y} \\ \frac{\partial F_y}{\partial x} & \frac{\partial F_y}{\partial x} \end{bmatrix}_{r=r_0} =$$

$$= \begin{bmatrix} f\left(|r_0|^2\right) + 2x_0^2 \left(\frac{\partial f(|r|^2)}{\partial(|r|^2)}\right)_{r=r_0} & 2x_0 y_0 \left(\frac{\partial f(|r|^2)}{\partial(|r|^2)}\right)_{r=r_0} \\ 2x_0 y_0 \left(\frac{\partial f(|r|^2)}{\partial(|r|^2)}\right)_{r=r_0} & f\left(|r_0|^2\right) + 2y_0^2 \left(\frac{\partial f(|r|^2)}{\partial(|r|^2)}\right)_{r=r_0} \end{bmatrix} . \tag{8.5}$$

The stiffness matrix of the bearing for small motions about the static equilibrium position coincides with the Jacobian matrix $S_{r=r_0}$. Note that the linearized characteristics of the bearings are not isotropic: Not only the element with subscript 11 is in general different from the element with subscript 22, but also the elements with subscripts 12 and 21 do not vanish. The axes of the ellipse of elasticity of the linearized bearing do not coincide with the x- and y-axes, except for the case in which the static force, and hence the static displacement, are directed along the coordinate axes.

Consider, for instance, a case in which function $f\left(|r|^2\right)$ is given by the equation

$$f\left(|r|^2\right) = k\left(1 + \mu |r|^2\right) . \tag{8.6}$$

Equation (8.6) can be considered the expression of generic nonlinear function $f\left(|r|^2\right)$ truncated at the second term of its series expansion. Parameter μ has the dimensions of a length at the power -2. Constant k is usually positive, a condition necessary to lead to a stable static equilibrium position at $x = 0$. If μ is positive, the behavior of the bearing is said to be of the *hardening* type, because its stiffness increases with the displacement. On the contrary, if μ is negative, the bearing is said to be of the *softening* type. Equation (8.6) is reported in nondimensional form in Figure 8.2. The restoring force obtained using Expression (8.6) for function $f\left(|r|^2\right)$ is usually defined as a *Duffing type* force, because nonlinear systems with forces of this type were studied extensively by G. Duffing in 1918 [45].

By introducing Equation (8.6) into the Expression (8.5), the following stiffness matrix of the linearized bearing is obtained:

$$\mathbf{K}_{r=r_0} = k \begin{bmatrix} 1 + \mu\left(3x_0^2 + y_0^2\right) & 2\mu x_0 y_0 \\ 2\mu x_0 y_0 & 1 + \mu\left(x_0^2 + 3y_0^2\right) \end{bmatrix} . \tag{8.7}$$

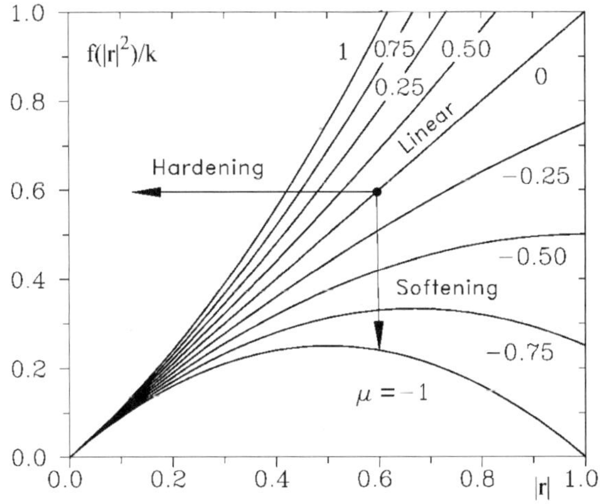

FIGURE 8.2. Function $f\left(|r|^2\right)$ expressed by Equation (8.6) for different values of μ. The values of r are normalized in the range $0-1$.

If the rotor is statically determined, as it is the case of a rotor on two bearings, there is no difficulty in computing the static loads on the bearings and, once the forces are known, to compute the static displacements at the bearing locations. This involves the solution of a set of two nonlinear equations (8.3) in which the forces F_x and F_y are known and the displacements x and y are the unknowns. As the force and the displacement have the same direction, the simplest way is to compute the modulus of the resultant force on the bearing $|F|=\sqrt{F_x^2+F_y^2}$; from the force, the displacement is obtained from the equation

$$|F|=|r|\,f\left(|r|^2\right)\,, \tag{8.8}$$

which in some cases can be solved in closed form [for instance, if $f\left(|r|^2\right)$ is expressed by Equation (8.6)] or in general can be solved using an iterative procedure like the Newton-Raphson algorithm. Once the modulus of the displacement $|r|$ has been obtained, its components are computed as

$$x=|r|\frac{F_x}{|F|}\,,\qquad y=|r|\frac{F_y}{|F|}\,. \tag{8.9}$$

If the compliance of the stator of the machine is accounted for, these displacements must be considered as relative displacements of the shaft with respect to the corresponding points on the stator. The components x_0 and y_0 of the displacement of each bearing allow us to compute through Equation (8.5) the stiffness matrix of that bearing. The stiffness matrices of the bearings may then be assembled into the model of the system. Note that identical bearings may lead to different stiffness matrices owing to the fact that different loads cause the bearings to work in a different point of the force-displacement characteristics. Moreover, nonlinearity causes the bearing to behave in an anisotropic way, even if it is geometrically axially symmetrical.

If the system is not statically determinate, like a rotor on three or more bearings, the loads on the bearings depend on the deformation of both the stator and the rotor and on possible misalignments. A coupled problem that is far more complex must be solved. The equation allowing the static deflected configuration of the rotor to be studied is Equation (4.83), written with reference to a set of complex coordinates. By writing explicitly the forces caused by weight, it yields

$$\left(\mathbf{K} + \Omega^2\mathbf{K}_\omega - i\Omega\mathbf{C}_r - i\mathbf{K}''_r\right)\mathbf{q} = \mathbf{f}_n + g\mathbf{M}(\boldsymbol{\delta}_x + i\boldsymbol{\delta}_y)\,, \qquad (8.10)$$

where vector $\mathbf{f}_n$ contains static forces, not including weight, whereas the elements of vector $\boldsymbol{\delta}_x + i\boldsymbol{\delta}_y$ vanish in correspondence with rotational degrees of freedom and are equal to the cosines of the angle between the vertical direction and the x- and y-axes for translational coordinates. Note that the deformed equilibrium position depends on the angular velocity only if there is viscous damping. The internal damping of the rotor will be assumed to be of the structural (hysteretic) type, and centrifugal stiffening will be neglected to allow the computations referred to the rotor to be performed only once. However, there is no difficulty modifying the equations to also take into account viscous damping

As when the Newton-Raphson technique is used for the solution of a nonlinear set of equations, there is some advantage to resorting to real coordinates, Equation (8.10) can be rewritten in the form

$$\mathbf{K}^*\left\{\begin{array}{c}\mathbf{x}_r\\ \mathbf{y}_r\end{array}\right\} = \mathbf{f}_r^*\,, \qquad (8.11)$$

where

$$\mathbf{K}^* = \left[\begin{array}{cc}\mathbf{K} & \mathbf{K}''\\ -\mathbf{K}'' & \mathbf{K}\end{array}\right], \qquad \mathbf{f}_r^* = \left\{\begin{array}{c}\mathbf{f}_{\mathbf{x_n}}\\ \mathbf{f}_{\mathbf{y_n}}\end{array}\right\} + g\left\{\begin{array}{c}\mathbf{M}\boldsymbol{\delta}_x\\ \mathbf{M}\boldsymbol{\delta}_y\end{array}\right\},$$

and vectors $\mathbf{x}_r$ and $\mathbf{y}_r$ contain the real and imaginary parts, respectively, of the complex coordinates of the rotor. The stiffness matrix is singular because the rotor has been considered unsupported, and this part of the system is underconstrained.

It is possible to separate the vectors of the generalized coordinates into two subsets: the first, labeled with subscript 1, containing the displacements at the supporting points (i.e., the nodes of the rotor at which the bearings are connected), and the second, with subscript 2, containing all other generalized displacements. The bearings will be assumed to react only to translations of the rotor, so the first set contains a number of elements equal to twice the number of the bearings. If the model of the bearing is modified to also include the moment caused by to angular displacements, the inclusion of the rotational degrees of freedom in this set of displacements is straightforward. By partitioning accordingly the relevant matrices and vectors, it follows that

$$\begin{bmatrix} \mathbf{K}^*_{11} & \mathbf{K}^*_{12} \\ \mathbf{K}^*_{21} & \mathbf{K}^*_{22} \end{bmatrix} \left\{ \begin{array}{c} \left\{ \begin{array}{c} \mathbf{x} \\ \mathbf{y} \end{array} \right\}_1 \\ \left\{ \begin{array}{c} \mathbf{x} \\ \mathbf{y} \end{array} \right\}_2 \end{array} \right\} = \left\{ \begin{array}{c} \mathbf{f}^*_1 \\ \mathbf{f}^*_2 \end{array} \right\}. \tag{8.12}$$

By applying the usual techniques of static reduction and assuming the generalized coordinates of the first group as master degrees of freedom, Equation (8.10) reduces to

$$\mathbf{K}_{cond_r} \left\{ \begin{array}{c} \mathbf{x} \\ \mathbf{y} \end{array} \right\}_1 = \mathbf{f}_{cond_r} , \tag{8.13}$$

where the expressions of the condensed matrices are the usual ones and the internal generalized coordinates of the rotor are expressed by Equation (4.98).

The model of the stator can be built in a way similar to that seen for the rotor, with two important differences. The stator does not need to be axially symmetrical, and because it is stationary, its static deformation is not affected by its damping. The equilibrium equation for the stator is Equation (8.11) where subscript r has been substituted with s. Matrices $\mathbf{M}^*$ and $\mathbf{K}^*$ are, in general, symmetrical and contain the coupling terms between the behavior in the xz- and yz-planes. Also, in this case, it is possible to separate the vectors of the generalized coordinates into two subsets: The first, labeled with subscript 3, containing the displacements at the supporting points (i.e., the nodes of the stator at which the bearings are connected), and the second, with subscript 4, containing all other generalized displacements, and then to resort to static reduction techniques.

The equilibrium equation referred to the displacements of the bearings is then Equation (8.13) where subscripts r and 1 have been substituted with s and 3, respectively. It can be reduced with the usual algorithm.

Note that because the interface between stator and rotor is represented by the bearings, which react only to translations, the sets of generalized coordinates with subscripts 1 and 3 contain only displacements and no rotations. This allows the use of conventions for the rotations of the stator and the rotor that are not consistent and the use of any standard FEM code to build the model of the stator even if the conventions for rotations about the x-axis are different from those used in the model of the rotor. Any type of element can be used in both models, provided that the displacements at the interface are measured with reference to the same axes used for the rotor.

Consider the ith bearing; the displacements of the node of the rotor in which the bearing is located can thus be indicated as x_{i_1} and y_{i_1}, whereas the displacements of the corresponding point of the stator can be indicated as x_{i_3} and y_{i_3} (where subscripts 1 and 3 refer to the above-mentioned partitioning of the generalized coordinates).

The relative displacements x_i and y_i of the rotor with respect to the stator at the bearing location are then

$$\left\{ \begin{array}{c} x_i \\ y_i \end{array} \right\} = \left\{ \begin{array}{c} x_{i_1} - x_{i_3} - \Delta_{i_x} \\ y_{i_1} - y_{i_3} - \Delta_{i_y} \end{array} \right\} . \tag{8.14}$$

Equation (8.14) takes into account the possibility that the center of the ith bearing may be displaced by the quantities Δ_{i_x} and Δ_{i_y}, with respect to the nominal position when no force acts on it. Equation (8.14) allows the computation of the relative displacement of the bearing and then of the forces F_x and F_y that the rotor receives from the stator, by using the nonlinear bearing characteristics. The interaction between stator and rotor can then be expressed by applying the forces on the rotor and the stator, which the two exchange through the bearing (remembering that they are equal in magnitude and opposite in sign)

$$\left\{ \begin{array}{l} \mathbf{K}_{cond_r} \left\{ \begin{array}{c} x \\ y \end{array} \right\}_1 + \left\{ \begin{array}{c} \mathbf{f}_x \\ \mathbf{f}_y \end{array} \right\} = \mathbf{f}_{cond_r} , \\ \mathbf{K}_{cond_s} \left\{ \begin{array}{c} x \\ y \end{array} \right\}_3 - \left\{ \begin{array}{c} \mathbf{f}_x \\ \mathbf{f}_y \end{array} \right\} = \mathbf{f}_{cond_s} . \end{array} \right. \tag{8.15}$$

If the number of bearings is m, Equation (8.15) is a set of $4m$ nonlinear equations with the $4m$ unknowns representing the displacements of the stator and of the rotor in the x- and y-directions. The number of nonlinear equations can, however, be reduced because the actual unknowns of the nonlinear part of the equation are the differences between the displacements of the rotor and the stator, which are only $2m$ in number. Because the reduced stiffness matrix of the rotor is generally singular and cannot be inverted, the second set of Equations (8.15) can be multiplied by $\mathbf{K}_{cond_r}\mathbf{K}_{cond_s}^{-1}$ and added to the first, obtaining the following nonlinear equation, which can be solved by resorting to the Newton-Raphson iterative technique:

$$\mathbf{K}_{cond_r}\mathbf{x} + \mathbf{A} \left\{ \begin{array}{c} \mathbf{f}_x(\{x - \Delta\}) \\ \mathbf{f}_y(\{x - \Delta\}) \end{array} \right\} = \mathbf{f}^{**} , \tag{8.16}$$

where

$$\mathbf{x} = \left\{ \begin{array}{c} \mathbf{x}_1 - \mathbf{x}_3 \\ \mathbf{y}_1 - \mathbf{y}_3 \end{array} \right\} , \qquad \mathbf{A} = \mathbf{I} + \mathbf{K}_{cond_r}\mathbf{K}_{cond_s}^{-1} ,$$

$$\mathbf{f}^{**} = \mathbf{f}_{cond_r} - \mathbf{K}_{cond_r}\mathbf{K}_{cond_s}^{-1}\mathbf{f}_{cond_s} .$$

Once the values of the relative displacements $\mathbf{x}$ of the bearings in the static equilibrium position have been computed, their stiffness matrices can be easily obtained. Note that if the bearing characteristics depend on the speed or if the damping of the rotor is modeled as viscous damping, the solution of the static problem must be repeated for each value of the speed at which the natural frequencies are to be computed.

The dynamic study in the small can thus be performed using a linear model of the whole system. As the presence of nonlinear bearings makes the nonrotating parts of the machine to behave in an anisotropic way, even if everything does possess geometrical axial symmetry, both real or complex coordinates may be used. Both free whirling and unbalance response may be studied, but the latter only if the amplitude of the response is small enough not to exceed the limits of linearization. As a last consideration, it must be noted that if the static forces acting on the bearings are vanishingly small, as in the case of a vertical bearing without static loading, the linearized model may in some cases supply only a very rough approximation of the actual behavior of the system: If a clearance is present in the bearing or if the stiffness in the central position vanishes, the equilibrium position is not well defined. This, however, is usually a sign of a poor design, which is usually solved by adding some preload.

To study the motion in the large and to find the limit cycles that can occur in the cases in which the motion in the small is unstable, it is necessary to study the complete nonlinear problem by resorting to numerical integration of the equations of motion, but this goes beyond the scope of the present chapter.

8.3 Rolling elements bearings

The stiffness of rolling element bearings depend on both the load and the preload, showing a pronounced nonlinear behavior. In general, it may be said that roller bearings are stiffer than ball bearings and generally more linear. Bearings used in high-speed machines have a light preload or, particularly in the case of roller and needle bearings, may have no preload at all. In this case, a certain play within the bearing can be expected.

Generally speaking, a bearing is considered as a constraint that reacts only in radial and axial direction, while leaving the shaft completely free for what the rotational degrees of freedom are concerned. This is exactly verified only in the case of particular bearing geometries, like oscillating ball or roller bearings, and it is just an approximation for other geometries, like deep groove or angular contract bearings. The rotational stiffness is, however, very low and this justifies the practice of neglecting it with any type of rolling elements bearings.

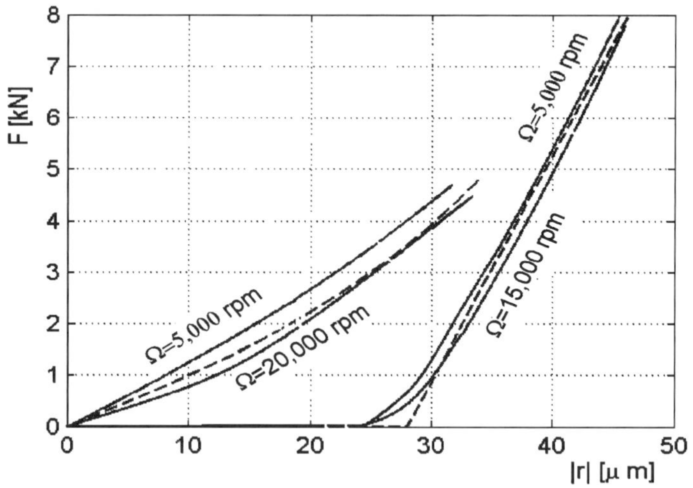

FIGURE 8.3. Force-displacement characteristics of two rolling element bearings. (a) Preloaded pair (X layout) of angular contact ball bearings (SKF 71907 c/PG) with a preload of 160 N. (b) Cylindrical roller bearing (SKF BC1B 649247B) with no preload. For each bearing, the characteristics at two different speeds are reported, together with a simple analytical approximation (dashed lines).

By applying the Hertz theory regarding the contact stresses and strain in the rolling elements and the races, the radial displacement x (or y) is proportional to $F^{2/3}$, $F^{3/4}$, $F^{0.9}$ (F indicates the radial force), respectively, for ball bearings, roller bearings with linear contact on one race and a point contact on the other one, and roller bearings with linear contact on both races. Similar relationships hold also for the case of axial displacements and forces [46].

The force-displacement characteristic $F(r)$ is the of the *hardening* type, i.e., the force increases more than linearly with the displacement. The actual characteristics of rolling elements bearings are, however, strongly affected by many features, like the deformation of the races, clearances, and so on, and more complex formulations must be used.

The force-displacements characteristics of a roller bearing and an angular contact preloaded pair are reported in Figure 8.3. Note that the characteristic depends slightly on the speed (the curves obtained at two different values of the speed are reported). Also, two simple analytical approximations (dashed lines) are plotted in the figure. For the preloaded pair, a cubic expression of the type of Equation (8.6) with

$$k = 0.95 \times 10^8 , \qquad \mu = 4.25 \times 10^8 ,$$

with the force F expressed in N and the displacement in m, can be used.

For the roller bearing, a linear characteristic with clearance is suitable. The analytical expression of function $f\left(|r|^2\right)$ is given by the equation

$$\begin{array}{ll} f=0 & \text{for } |r|<c\,, \\ |f|=k\frac{(|r|-c)}{|r|} & \text{for } |r|\geq c\,. \end{array} \tag{8.17}$$

The clearance and the stiffness of the bearing in the figure are, respectively,

$$c=2.8\times 10^{-5}\text{m}\,,\qquad k=4.44\times 10^{8}\text{N/m}\,.$$

The stiffness in the deflected position x_0, y_0 can be computed through Equation (8.5), which yields

$$\mathbf{K}_{r=r_0}=k\begin{bmatrix} 1 & 0 \\ 0 & 1 \end{bmatrix}+\frac{kc}{\left(x_0^2+y_0^2\right)^{3/2}}\begin{bmatrix} -y_0^2 & x_0y_0 \\ x_0y_0 & -x_0^2 \end{bmatrix}. \tag{8.18}$$

If the force acts along one of the coordinate axes, say x-axis, the stiffness matrix simplifies as

$$\mathbf{K}_{r=r_0}=k\begin{bmatrix} 1 & 0 \\ 0 & 1-\frac{c}{x_0} \end{bmatrix}, \tag{8.19}$$

the ellipse of elasticity is less and less elongated with increasing loading, and the stiffness in y-direction vanishes with vanishing load. However, if the load tends to zero, the contact becomes unilateral and no linearized model can be used. The axes of the ellipse are directed along x- and y-axes.

Example 8.1 *Turbojet rotor on rolling elements bearings. Rigid-body model.*

Consider the rotor of the very simple turbojet engine whose simplified drawing is shown in Figure 8.4. The rotor, which includes a single-stage centrifugal compressor and a single-stage axial turbine, is supported on a deep groove ball bearing and a roller bearing. In the subsequent analysis, the stator of the machine is considered as a rigid body; in this first study, also the rotor is assumed to be rigid.

The stiffness of the bearings depends on the load. The mass of the whole rotor has been evaluated in 17.040 kg, with its center of mass located at 188.6 mm from the front (compressor) end. The distances of the front and rear bearings from the center of mass are 175.6 mm and 42.4 mm, respectively. In these conditions, the radial loads caused by weight are 32.49 N and 134.67 N for the ball and the roller bearings, respectively.

Following the Hertz theory for the deformation of bodies in contact, the displacements at the bearings are assumed to be proportional to $F^{2/3}$ and $F^{0.9}$. The force-displacement characteristic of the ball bearing is then

$$\begin{cases} F_x=ax\left(x^2+y^2\right)^{1/4}\,, \\ F_y=ay\left(x^2+y^2\right)^{1/4}\,, \end{cases}$$

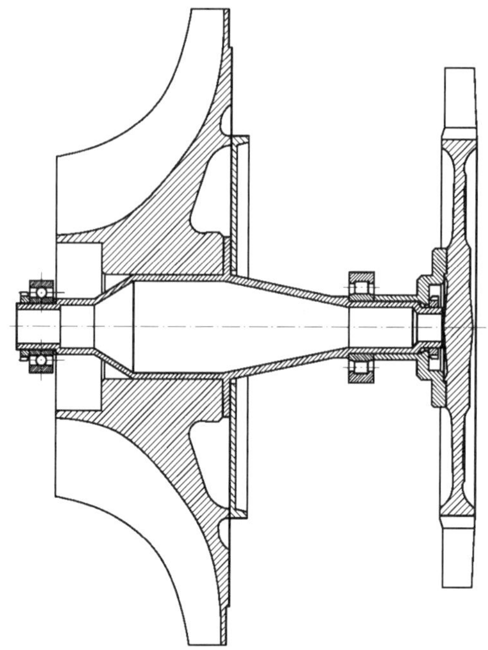

FIGURE 8.4. Simplified drawing of the rotor of a turbojet engine.

where the constant a is evaluated, following the data reported in the book by Palmgren [46], as $a = 10^{10}$ *in S.I. units. Following Equation (8.5), the Jacobian matrix about the static inflected position is*

$$\mathbf{S}_{r=r_0} = \begin{bmatrix} \frac{\partial F_x}{\partial x} & \frac{\partial F_x}{\partial y} \\ \frac{\partial F_y}{\partial x} & \frac{\partial F_y}{\partial x} \end{bmatrix}_{r=r_0} =$$

$$= a \begin{bmatrix} |r_0|^{1/2} + \frac{1}{2}x_0^2 |r_0|^{-3/2} & \frac{1}{2}x_0 y_0 |r_0|^{-3/2} \\ \frac{1}{2}x_0 y_0 |r_0|^{-3/2} & |r_0|^{1/2} + \frac{1}{2}y_0^2 |r_0|^{-3/2} \end{bmatrix} ,$$

where $|r_0| = \sqrt{x_0^2 + y_0^2}$.

If the load acts in the direction of x-axis, the static displacement in y-direction vanishes and the stiffness of the bearing is simply

$$\mathbf{S}_{r=r_0} = \begin{bmatrix} \frac{\partial F_x}{\partial x} & \frac{\partial F_x}{\partial y} \\ \frac{\partial F_y}{\partial x} & \frac{\partial F_y}{\partial x} \end{bmatrix}_{r=r_0} = a \begin{bmatrix} \frac{3}{2}\sqrt{x_0} & 0 \\ 0 & \sqrt{x_0} \end{bmatrix} .$$

By computing the static displacement under load, the stiffness of the ball bearing (in N/m) can be assumed to be

$$\begin{cases} k_x = 6.96 \times 10^6 \sqrt[3]{F_x} = 2.261 \times 10^8 , \\ k_y = 4.64 \times 10^6 \sqrt[3]{F_x} = 1.510 \times 10^8 . \end{cases}$$

	$\omega_n(\Omega=0)$[rad/s]		Ω_{cr}[rad/s]	
	Isotropic	Nonisotropic	Isotropic	Nonisotropic
I	2,027	870	2,301	866
II	6,025	2,725	10,046	2,751
III	-	5,375	-	5,048
IV	-	6,613	-	10,107

TABLE 8.1. Values of the first four natural frequencies at standstill and of the first four critical speeds. Rigid-body model.

For the roller bearing, a characteristic of the type $F = a\,|r_O|^{1.1}$ *with a value for constant a equal to* 4.8×10^8 *can be assumed, to which a clearance must be added. The characteristics is close to a straight line, so a linear characteristic with clearance can be used. The value of the stiffness for a clearance of some micrometers is* $k = 1.2 \times 10^8$. *By assuming a clearance of 10* μ*m, and using Equation (8.5), the linearized stiffness of the roller bearing is*

$$\begin{cases} k_x = 1.20 \times 10^8 \ , \\ k_y = 1.21 \times 10^7 \ . \end{cases}$$

Note that the roller bearing is less stiff than the ball bearing; this is because the latter is very lightly loaded, the load being much smaller than the load capacity.

The other relevant data are the moments of inertia of the rigid rotor, $J_t = 0.233$ *kg m*^{2} *and* $J_p = 0.168$ *kg m*^{2}. *The matrices referred to the xz- and yz- planes are*

$$\mathbf{M}_x = \mathbf{M}_y = \begin{bmatrix} 17.040 & 0 \\ 0 & 0.233 \end{bmatrix} , \qquad \mathbf{G} = \begin{bmatrix} 0 & 0 \\ 0 & 0.168 \end{bmatrix} ,$$

$$\mathbf{K}_x = \begin{bmatrix} 3.461 & 0.346 \\ 0.346 & 0.0719 \end{bmatrix} \times 10^8 \ , \qquad \mathbf{K}_y = \begin{bmatrix} 1.631 & 0.260 \\ 0.260 & 0.0468 \end{bmatrix} \times 10^8 \ .$$

The mean stiffness matrix, used to study the dynamics of an averaged system, is

$$\mathbf{K}_m = \begin{bmatrix} 2.546 & 0.303 \\ 0.303 & 0.0593 \end{bmatrix} \times 10^8 \ .$$

The first four natural frequencies at standstill and the first four critical speeds, computed using both the anisotropic model and the 'averaged' model with isotropic characteristics are reported in Table 8.1.

Example 8.2 *Turbojet rotor on rolling elements bearings. FEM.*

Now consider also the flexibility of the rotor, using the finite element model shown in Figure 8.5. The model consists of 21 nodes, 20 beam elements to model the shaft, 2 mass elements to model the compressor and turbine wheels, and 2 spring elements to model the bearings. The inner races of the bearings are included into the model, which may result in overestimating the stiffness, because they add probably more stiffness in the model than in the real object. The same is

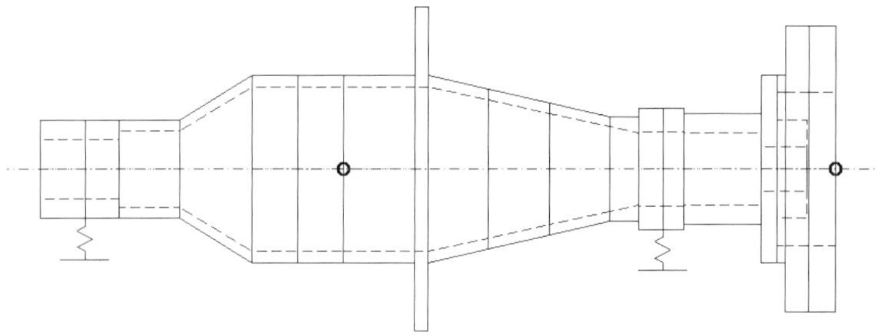

FIGURE 8.5. Sketch of the finite element model of the turbojet of Figure 8.4.

	$\omega_n(\Omega = 0)$[rad/s]		Ω_{cr}[rad/s]	
	Isotropic	Nonisotropic	Isotropic	Nonisotropic
I	1,539	831	2,103	809
II	2,832	1,707	3,975	1,492
III	4,825	2,365	38,524	2,700
IV	7,540	3,262	43,477	3,487

TABLE 8.2. Values of the first 4 natural frequencies at standstill and of the first 4 critical speeds. FEM model.

for the connection between the turbine wheel and the shaft, which in the model is one piece with the latter, whereas in the actual machine is fit on it. On the contrary, the stiffening effect of the compressor wheel on the shaft has been neglected, which may result in underestimating the stiffness. The latter problem may be alleviated by adding some massless beam elements to model the stiffening effect of the compressor wheel, as it was done in Example 4.1; However, in that case, the compressor wheel was locked in place using a nut on the shaft, whereas here it is bolted on a single section (the disc at midshaft) and its stiffening effect is much lower and depends strictly on the fit between the shaft and the wheel. Note that the fit changes with the speed and the temperature (in both cases, the wheel expands more than the shaft) and it may be wise to neglect the stiffening altogether, unless experimental data are available. A short massless beam element has been used to connect the mass element modeling the turbine wheel to the shaft.

The first four natural frequencies at standstill and the first four critical speeds, computed using both an anisotropic model and an averaged model with isotropic characteristics are reported in Table 8.2.

Note that the first mode, particularly when the nonisotropic model is used, was computed with a fair accuracy also using the rigid-body model: This suggest that the deformations occur mainly in the bearings, as it will be shown later when plotting the mode shape. The other modes are mainly deformation modes, and the rigid-rotor model yields values of the frequencies that are much higher than the correct ones.

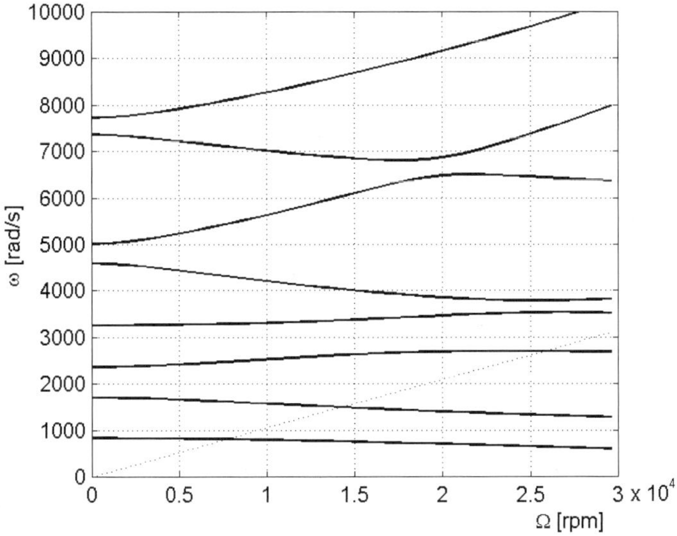

FIGURE 8.6. Campbell diagram of the turbojet of Figure 8.5.

The Campbell diagram is reported in Figure 8.6. The first four modes are neither forward nor backward modes, because some of the nodes of the shaft whirl in one direction and others in the opposite one. However, the fact that the first two are characterized by a whirl frequency that decreases with the speed, whereas that of the other two increases, shows that the first two modes are mainly backward modes and the other ones are mainly forward modes.

The mode shapes corresponding to the critical speeds are shown in Figure 8.7.

From the shape of the orbits, it is clear that the first two modes occur mostly in the horizontal (yz) plane, with the turbine wheel (node 21, at the right end) whirling along a circular orbit owing to its own gyroscopic effect in the second mode. The third mode is mainly a mode in the vertical plane. As shown in the sketch of the axes, the forward direction for whirling (forward whirling) is counterclockwise when looking at the rotor from the origin. The turbine wheel whirls in forward direction in the second and third mode, and in backward direction in the first mode, whereas the compressor whirls in forward direction in the first and in backward direction in the others.

An unbalance corresponding to grade 2.5 at 20,000 rpm has been given to the turbine wheel. This corresponds to an eccentricity of 1.2 μm and to a static unbalance $m\epsilon = 7.6$ *gmm. The response computed at the turbine wheel (node 21) and at the rear bearing (node 14) is reported in Figure 8.8.*

Note that although the damping of the bearings has been assumed small ($\eta = 0.01$)*, the orbits, which are elliptical, are very small.*

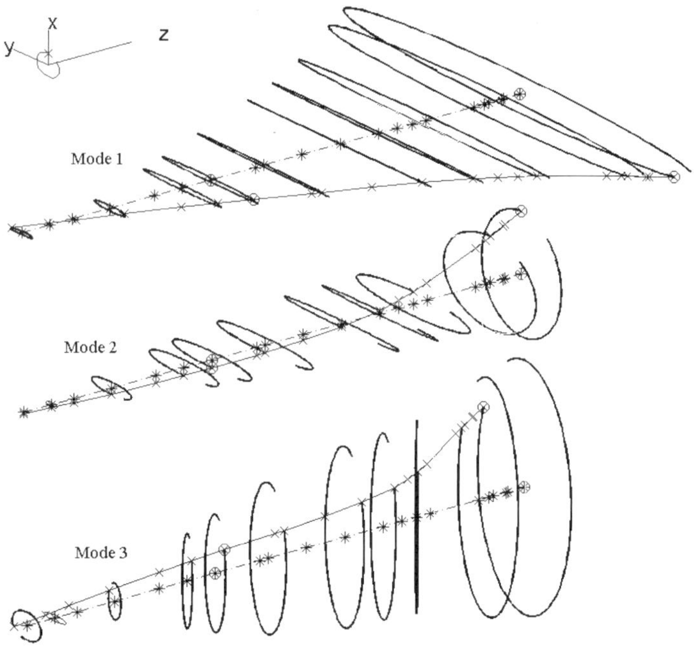

FIGURE 8.7. First three mode shapes at the critical speeds for the turbojet of Figure 8.5. Note that the orbits are elliptical and that their direction is forward or backward depending on the node considered (mixed modes).

8.4 Fluid film bearings

8.4.1 Forces exerted by the oil film on the journal in stationary conditions

The study of journal bearing is much more complex than that of rolling elements bearings: In this case, the forces that support the rotor are supplied by a fluid film whose behavior depends on many parameters in a complex way and above all on the spin speed. It is then impossible to supply a single plot like the one of Figure 8.3 from which a single characteristic of the bearing can be obtained.

Above all, the force supporting the shaft is in this case not directed in the same way as the displacement, so that Equation (8.3) does not hold any more. In the simplest case, as it will be shown later, the direction of the force is even perpendicular to that of the displacement. In the present section, the behavior of journal bearings will be dealt with in a simplified way: The reader can find a more detailed analysis in many specialized texts, like the book *Journal-Bearing Databook* by Someya [47].

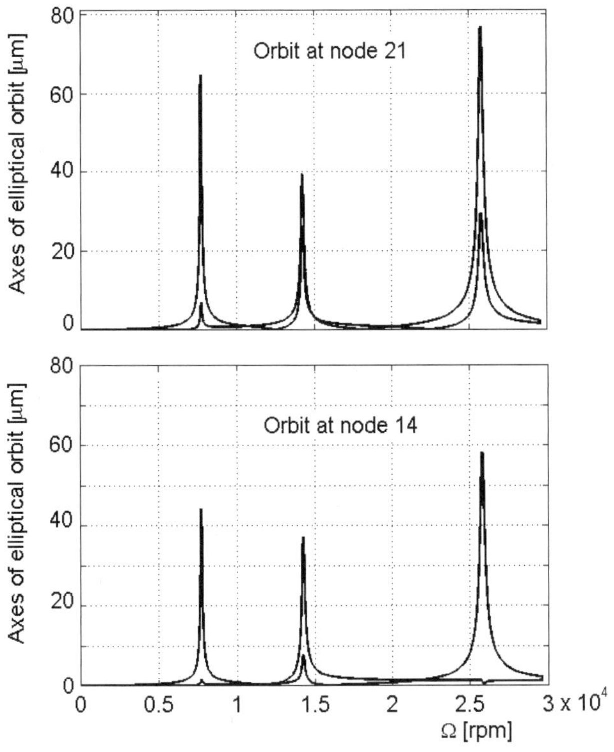

FIGURE 8.8. Unbalance response of the turbojet of Figure 8.5. Semi-axes of the elliptical orbits at nodes 14 and 21.

Consider the journal bearing sketched in Figure 8.9(a). Assume that the bearing is perfectly aligned, i.e., that the axes of the bearing and the journal are parallel. The nonrotating reference frame $Oxyz$ is centered in the center of the bearing and has the directions of its axes fixed in space, and the directions of the axes of reference frame $Ox'y'z$, whose axis x' contains the center of the journal C, are not fixed in space if point C moves about point O. In this section, the way of working of the bearing will be defined as *stationary* if the coordinates x and y of the center of the journal are independent from time t, and consequently, reference frame $Ox'y'z$ is also fixed in space and the *attitude angle* β is constant.

Assume that the pressure is linked to the thickness of the fluid film by the well-known Reynolds equation [48]

$$\frac{1}{6}\left[\frac{1}{R_j^2}\frac{\partial}{\partial\theta}\left(\frac{h^3}{\mu}\frac{\partial p}{\partial\theta}\right)+\frac{\partial}{\partial z}\left(\frac{h^3}{\mu}\frac{\partial p}{\partial z}\right)\right]=\Omega\frac{\partial h}{\partial\theta}+2\frac{\partial h}{\partial t}\,, \tag{8.20}$$

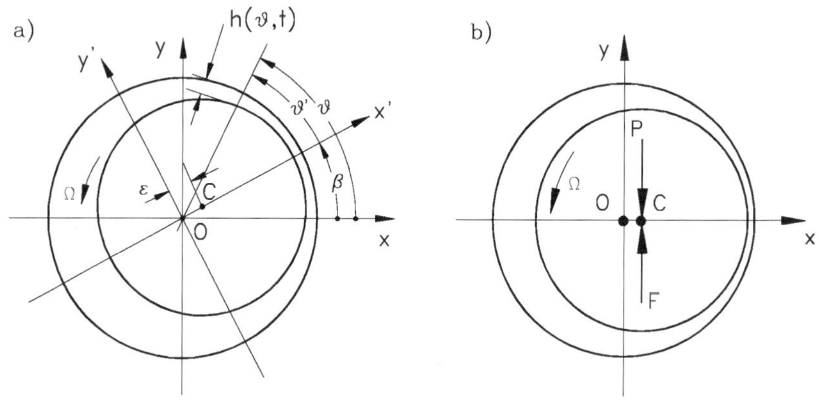

FIGURE 8.9. Lubricated journal bearing; (a) geometrical definitions, (b) position of the journal when a load P acts along the y-axis; solution obtained for stationary conditions from Equation (8.27).

where μ is the viscosity of the lubricant and $\Omega = \Omega_j - \Omega_b$, subscripts b and j referring to the bearing and the journal, respectively. Usually, the bearing does not rotate; hereafter, it will be assumed that $\Omega_b = 0$ and $\Omega_j = \Omega$.

The film thickness h is easily expressed as a function of the coordinates of the center of the journal with respect to the center of the bearing

$$h = c[1 - x^* \cos(\theta) - y^* \sin(\theta)] = c[1 - \epsilon^* \cos(\theta')] , \tag{8.21}$$

where the clearance c is simply given by the difference of the radii $c = R_b - R_j$ and the nondimensional eccentricity and coordinates are $\epsilon^* = \epsilon/c$, $x^* = x/c$, and $y^* = y/c$.

Equation (8.20) can be used to obtain the pressure distribution in the fluid film. However, if a numerical solution is considered as not general enough, some simplifications must be introduced to allow the pressure to be computed in closed form. If the bearing is assumed to be very long, it is possible to neglect the fluid flow and pressure gradient in axial direction, obtaining the so-called long-bearing approximation, often associated with the name of Sommerfeld. Equation (8.20) reduces to

$$\frac{1}{6R_j^2}\frac{\partial}{\partial\theta}\left(\frac{h^3}{\mu}\frac{\partial p}{\partial\theta}\right) = (\Omega x - 2\dot{y})\sin(\theta) - (\Omega y + 2\dot{x})\cos(\theta) . \tag{8.22}$$

If the center of the journal does not move, i.e., in stationary conditions, Equation (8.22) reduces to

$$\frac{1}{6R_j^2}\frac{d}{d\theta}\left(\frac{h^3}{\mu}\frac{dp}{d\theta}\right) = \Omega\left[x\sin(\theta) - y\cos(\theta)\right] .$$

By introducing the variable θ' and remembering that the viscosity μ can be considered as constant, it follows that

$$\frac{d}{d\theta'}\left(h^3\frac{dp}{d\theta'}\right) = 6\mu\Omega R_j^2 c\epsilon^* \sin\theta' . \tag{8.23}$$

By integrating Equation (8.23), it yields:

$$\frac{dp}{d\theta'} = 6\mu\Omega\left(\frac{R_j^2}{c}\right)^2 \frac{-\epsilon^*\cos\theta' + C}{\left(1-\epsilon^*\cos\theta'\right)^2} ,$$

where C is a constant of integration. By introducing the thickness h_0 (still unknown), corresponding to the point where $dp/d\theta'$ vanishes, it follows that

$$\frac{dp}{d\theta'} = 6\mu\Omega\left(\frac{R_j^2}{c}\right)^2 \left[\frac{1}{\left(1-\epsilon^*\cos\theta'\right)^2} - \frac{h_0}{c\left(1-\epsilon^*\cos\theta'\right)^3}\right] . \tag{8.24}$$

The integration of Equation (8.24) is not straightforward. Sommerfeld [49] resorted to the substitution:

$$\left(1+\epsilon^*\cos\theta'\right) = \frac{1-\epsilon^{*2}}{1-\epsilon^*\cos\alpha} \tag{8.25}$$

and, stating the condition $p(0) = p(2\pi)$, obtained the following pressure distribution:

$$p - p_0 = -6\mu\Omega\left(\frac{R_j}{c}\right)^2 \frac{\epsilon^*}{2+\epsilon^{*2}} \frac{2-\epsilon^*\cos(\theta')}{[1-\epsilon^*\cos(\theta')]^3}\sin(\theta') . \tag{8.26}$$

The pressure $p - p_0$ obtained from Equation (8.26) is plotted in nondimensional form as a function of θ' for different values of ϵ^* in Figure 8.10.

Pressure p_0 is the pressure at $\theta' = 0$, i.e., in the point at which the oil film is at its minimum thickness. It can be easily computed by assuming that the pressure attains a known value at the locations in which the oil supply is located. The components of the force the journal receives from the oil film in the x'- and y'-directions can be obtained simply by integrating the pressure profile on the journal surface

$$\begin{cases} F_{x'} = \displaystyle\int_0^{2\pi} (p-p_0)\cos(\theta')d\theta = 0 , \\ F_{y'} = \displaystyle\int_0^{2\pi} (p-p_0)\sin(\theta')d\theta = 12\pi\mu\Omega R_j l \left(\frac{R_j}{c}\right)^2 \frac{\epsilon^*}{(2+\epsilon^{*2})\sqrt{1-\epsilon^{*2}}} , \end{cases} \tag{8.27}$$

where l is the length of the bearing in the axial direction.

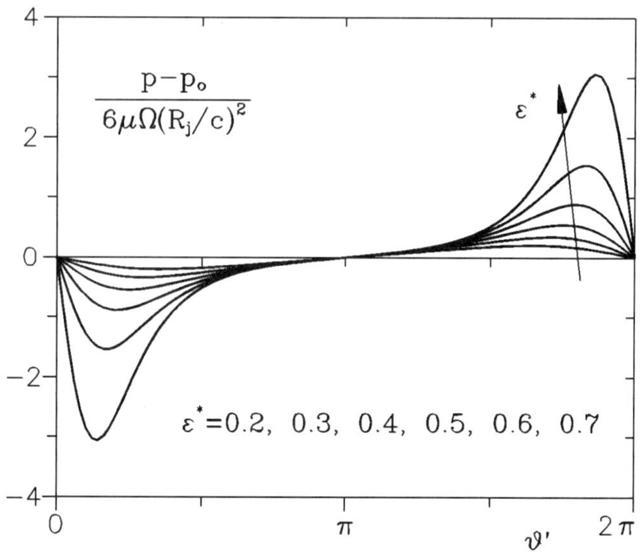

FIGURE 8.10. Pressure distribution on the journal along angle θ' for different values of the nondimensional eccentricity ϵ^*. Long bearing assumption; Equation (8.26).

The force exerted by the oil film on the journal is then directed along y'-axis, and the displacement is directed along x'-axis. The journal is displaced in a direction perpendicular to the direction of the load, as shown in Figure 8.9(b), in which the load P is assumed to act in the vertical direction. By linearizing the expression of the forces about the central position (vanishing ϵ^*), the following expression is obtained:

$$\left\{ \begin{array}{c} F_x \\ F_y \end{array} \right\} = -\mathbf{K} \left\{ \begin{array}{c} x \\ y \end{array} \right\} = -6\pi\mu\Omega l \left(\frac{R_j}{c} \right)^3 \left[\begin{array}{cc} 0 & 1 \\ -1 & 0 \end{array} \right] \left\{ \begin{array}{c} x \\ y \end{array} \right\} . \quad (8.28)$$

Note that the stiffness matrix obtained from the linearization of the bearing has vanishing elements on the main diagonal and is skew symmetrical. It is then more a circulatory matrix than a stiffness matrix.

Remark 8.2 *The fact that the stiffness matrix is skew symmetrical is linked to the particular oversimplified formulation used. In general, it is a nonsymmetric matrix, which can be decomposed into a symmetric and a skew symmetric matrix, i.e., into stiffness and circulatory matrices.*

From Figure 8.9, it is clear that when the eccentricity is high and the inlet pressure is low, very low values of the absolute pressure can be reached in some parts of the oil film, or even negative absolute pressures, which is physically without meaning. When the pressure becomes lower than the vapor pressure of the lubricant at the relevant temperature, cavitation occurs and the oil film ruptures. A usual approach for the study of bearings with partially cavitated oil film is assuming that the pressure in the noncavitated part of the bearing is equal to that which can be computed by assuming a complete oil film. Because the pressure in the cavitated part of the bearing can be neglected, the same formulas already seen for the computation of the forces can be used, provided that the integration is performed between angles θ_1 and θ_2, which define the region on which the oil film extends.

A simple approach, usually referred to as *fully cavitated bearing*, is assuming that the oil film extends between $\theta' = \pi$ and $\theta' = 2\pi$, i.e., in the region in which the value of the pressure is higher than p_0. Once angles θ_1 and θ_2 have been defined, there is little difficulty computing the static forces corresponding to a given displacement of the journal, i.e., to a pair of values x_{st} and y_{st} of the coordinates of point C

$$\mathbf{f} = \frac{6\mu R_j^3 l \Omega}{c^2} \left\{ \begin{array}{c} \int_{\theta_1}^{\theta_2} \Lambda \cos(\theta) d\theta \\ \int_{\theta_1}^{\theta_2} \Lambda \sin(\theta) d\theta \end{array} \right\} , \tag{8.29}$$

where

$$\Lambda = \frac{x_{st}^* \sin(\theta) - y_{st}^* \cos(\theta)}{h_{st}^{*3}} , \quad h_{st}^* = 1 - x_{st}^* \cos(\theta) + y_{st}^* \sin(\theta) .$$

The integrals in Equation (8.29) must be solved numerically, but this does not imply long and costly computations.

If the bearing is relatively short and the long-bearing approach does not seem to be applicable, the flow in the circumferential direction may be neglected, obtaining the so-called short-bearing approximation. By neglecting the term linked with the circumferential pressure gradients in Reynolds Equation (8.20) and introducing into the latter the expression of the film thickness, it follows that

$$\frac{1}{6} \frac{\partial}{\partial z} \left(\frac{h^3}{\mu} \frac{\partial p}{\partial z} \right) = (\Omega x - 2\dot{y}) \sin(\theta) - (\Omega y + 2\dot{x}) \cos(\theta) . \tag{8.30}$$

By operating in the same way as for the long-bearing Sommerfeld approximation, the following expressions for the forces in static conditions and the stiffness and damping matrices are obtained

$$\mathbf{f} = \frac{\mu R_j l^3 \Omega}{2c^2} \left\{ \begin{array}{c} \int_{\theta_1}^{\theta_2} \Lambda \cos(\theta) d\theta \\ \int_{\theta_1}^{\theta_2} \Lambda \sin(\theta) d\theta \end{array} \right\} , \tag{8.31}$$

where Λ and h_{st}^* have the same expression as in Equation (8.29)[50].

Apart from the plain lubricated bearing, there are many other bearing types (cylindrical grooves, multilobe, tilting pads, etc.), mostly evolved with the aim of reducing the inherent instability (see below) of plain bearings. In the case of complex geometries, there is no chance of obtaining a closed-form solution, or even a solution computed with simple numerical integrations. The possible approaches either use experimental results or perform a more complex numerical modeling of the fluid film. When the results obtained experimentally or by numerical modeling are presented in the form of general charts applicable to a given family of bearings, nondimensional parameters, related to the average pressure

$$p_m = \frac{F}{2R_j l} \tag{8.32}$$

are commonly used. One of them is the Sommerfeld number, defined as

$$S = \frac{\mu\Omega}{2\pi p_m}\left(\frac{R_j}{c}\right)^2 = \frac{\mu\Omega R_j l}{\pi F}\left(\frac{R_j}{c}\right)^2 . \tag{8.33}$$

A low Sommerfeld number corresponds to low-speed, high-load working conditions. The Sommerfeld number is well suited to the study of long bearings. For the short-bearing model, the load factor

$$O = \frac{2p_m}{\mu\Omega}\left(\frac{c}{R_j}\right)^2\left(\frac{2R_j}{l}\right)^2 = \frac{F}{\mu\Omega l R_j}\left(\frac{c}{R_j}\right)^2\left(\frac{2R_j}{l}\right)^2 \tag{8.34}$$

is commonly used. Barwell [51] in 1956 proposed calling the load factor the Ocvirk number, and here it will be referred to with the symbol O. The two nondimensional parameters are linked by the relationship

$$O = \frac{1}{\pi S}\left(\frac{2R_j}{l}\right)^2 . \tag{8.35}$$

Note that a low value of the Sommerfeld number corresponds to a high value of the load factor and vice versa.

The position of the center of the journal in stationary conditions is univocally determined, given a certain type of bearing, once the Sommerfeld number (or any other relevant nondimensional parameter) is stated. Charts giving the nondimensional coordinates of the center of the bearing x^* and y^*, or better, the eccentricity ϵ^* and the attitude angle β, as functions of the Sommerfeld number, summarize the static behavior of the bearing. The attitude angle defines the direction of the displacement of the center of the journal with respect to the direction of the force F, as shown in Figure 8.11(a). In the case of an uncavitated long bearing, the attitude angle is then equal to 90° for any value of the Sommerfeld number.

In many cases, the minimum thickness of the oil film $h^*_{min} = 1 - \epsilon^*$ is given instead of the eccentricity. The graph giving the eccentricity and the

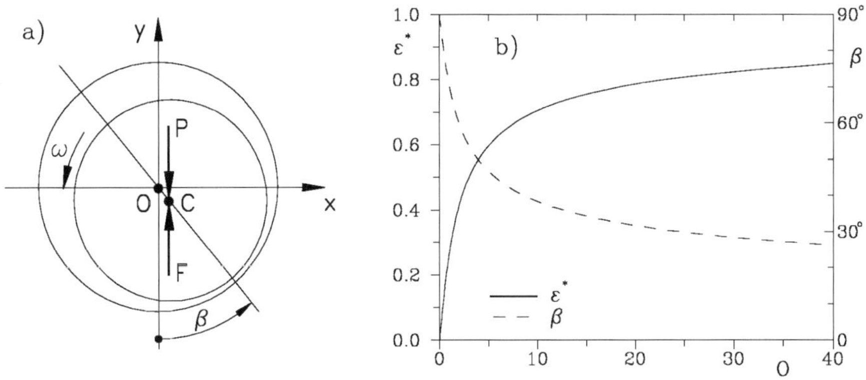

FIGURE 8.11. Lubricated journal bearing; (a) definition of the attitude angle β (load P acting along y-axis); (b) eccentricity and attitude angle as functions of the load factor for a fully cavitated short bearing.

attitude angle as functions of the load factor for a fully cavitated short bearing is reported in Figure 8.11(b).

In the case of short bearings, to pass from the load factor to the Sommerfeld number the ratio R/l must be stated. The same plot of Figure 8.11 is reported in Figure 8.12 as a function of the Sommerfeld number, for a short bearing with $R/l = 1$, together with a similar plot for a two-lobes bearing reported as Computation 12 on the already mentioned book edited by T. Someya [47]. Note that in the latter case, ϵ^* may take values larger than 1: The clearance is not uniform around the journal, so when the attitude angle approaches 90°, a far larger clearance than that used to define ϵ^* is available.

A lubricated journal bearing does not work at standstill, i.e., with a vanishingly low Sommerfeld number. In this case, the eccentricity is at its maximum value (for cylindrical bearings $\epsilon^* = 1$) and the attitude angle is equal to 0: The journal touches the bearing at the bottom. When a minimum speed is reached, the oil film is formed and the bearing works properly. With increasing speed (increasing Sommerfeld number), the eccentricity reduces, i.e., the journal tends to work more centered in the bearing, and the attitude angle increases. At very high speed (or low load, i.e., at high values of the Sommerfeld number), the eccentricity further reduces, and the attitude angle tends, for cylindrical bearings, to 90°, approaching the theoretical situation for a noncavitated long bearing.

8.4.2 Linearized dynamics of the bearing

Once the stationary conditions have been defined, there is no difficulty in computing the linearized dynamics of the system. If the bearing is dis-

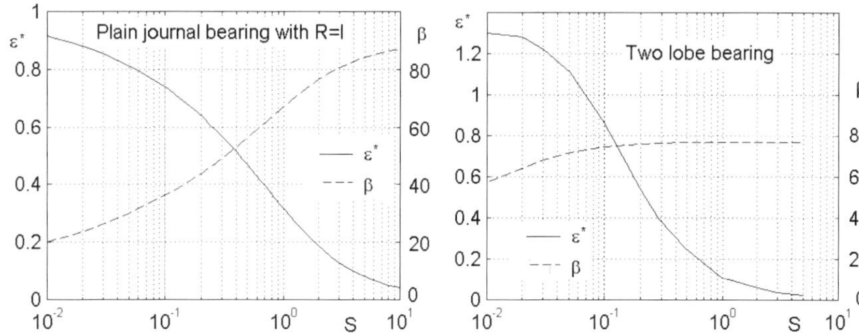

FIGURE 8.12. Eccentricity and attitude angle as functions of the Sommerfeld number for a fully cavitated short bearing with $R/l = 1$ and for a two-lobes bearing (see T. Someya (editor), *Journal-Bearing Databook*, Springer, Berlin, 1989, calculation n. 12).

placed of a small quantity Δx, Δy from the static equilibrium position (corresponding to the static force f_{st}) and moves with a speed $\dot{x}$, $\dot{y}$, the force the journal receives from the fluid film can be expressed as

$$\mathbf{f} = \mathbf{f}_{st} - \begin{bmatrix} \frac{\partial F_x}{\partial \dot{x}} & \frac{\partial F_x}{\partial \dot{y}} \\ \frac{\partial F_y}{\partial \dot{x}} & \frac{\partial F_y}{\partial \dot{y}} \end{bmatrix}_{x_{st}, y_{st}} \begin{Bmatrix} \dot{x} \\ \dot{y} \end{Bmatrix} - \begin{bmatrix} \frac{\partial F_x}{\partial x} & \frac{\partial F_x}{\partial y} \\ \frac{\partial F_y}{\partial x} & \frac{\partial F_y}{\partial y} \end{bmatrix}_{x_{st}, y_{st}} \begin{Bmatrix} \Delta x \\ \Delta y \end{Bmatrix} . \tag{8.36}$$

Equation (8.36) holds if the inertia of the oil film is neglected. The linearized behavior of the bearing is then expressed by a set of coefficients: the eight derivatives of the components of the force with respect to the displacements (stiffness coefficients) and the velocities (damping coefficients). This way of linearizing the behavior of lubricated bearings is usually referred to as the eight-coefficients model.

In the case of the long-bearing assumption, the expressions of the eight coefficients are

$$\mathbf{K} = -\left[\left(\frac{\partial F_i}{\partial x_j}\right)\right]_{x_{st}, y_{st}} = -\frac{6\mu R_j^3 l \Omega}{(2+n^2)c^3} \begin{bmatrix} I_1 + I_2 & -I_4 + I_5 \\ I_6 + I_5 & -I_1 + I_3 \end{bmatrix},$$

$$\mathbf{C} = -\left[\left(\frac{\partial F_i}{\partial \dot{x}_j}\right)\right]_{x_{st}, y_{st}} = \frac{6\mu R_j^3 l}{n^2 c^3} \begin{bmatrix} I_7 & I_8 \\ I_9 & I_{10} \end{bmatrix}, \tag{8.37}$$

where

$$
\begin{aligned}
I_1 &= \int_{\theta_1}^{\theta_2} \frac{1+h_{st}^*}{h_{st}^{*3}} \sin(\theta)\cos(\theta)d\theta\,, & I_2 &= \int_{\theta_1}^{\theta_2} (2+h_{st}^*)\frac{\Lambda}{h_{st}^*} \cos^2(\theta)d\theta\,,\\
I_3 &= \int_{\theta_1}^{\theta_2} (2+h_{st}^*)\frac{\Lambda}{h_{st}^*} \sin^2(\theta)d\theta\,, & I_4 &= \int_{\theta_1}^{\theta_2} \frac{1+h_{st}^*}{h_{st}^{*3}} \cos^2(\theta)d\theta\,,\\
I_5 &= \int_{\theta_1}^{\theta_2} (2+h_{st}^*)\frac{\Lambda}{h_{st}^*} \sin(\theta)\cos(\theta)d\theta\,, & I_6 &= \int_{\theta_1}^{\theta_2} \frac{1+h_{st}^*}{h_{st}^{*3}} \sin^2(\theta)d\theta\,,\\
I_7 &= \int_{\theta_1}^{\theta_2} (Ry_{st}^* + Sx_{st}^*)\cos(\theta)d\theta\,, & I_8 &= \int_{\theta_1}^{\theta_2} (-Rx_{st}^* + Sy_{st}^*)\cos(\theta)d\theta\,,
\end{aligned}
$$

$$
\begin{aligned}
I_9 &= \int_{\theta_1}^{\theta_2} (Ry_{st}^* + Sx_{st}^*)\sin(\theta)d\theta\,, & I_{10} &= \int_{\theta_1}^{\theta_2} (-Rx_{st}^* + Sy_{st}^*)\sin(\theta)d\theta\,,\\
h_{st}^* &= 1 - x_{st}^*\cos(\theta) + y_{st}^*\sin(\theta)\,, & \Lambda &= \frac{x_{st}^*\sin(\theta) - y_{st}^*\cos(\theta)}{h_{st}^{*3}}\,,\\
R &= 2\Lambda h_{st}^* \frac{1+h_{st}^*}{2+\epsilon^{*2}}\,, & S &= \frac{1}{(1+\epsilon^*)^2} - \frac{1}{h_{st}^{*2}}\,.
\end{aligned}
$$

Again, the integrals in Equations (8.37) must be solved numerically, which is straightforward.

By using the short bearing assumption, the following expressions for the stiffness and damping matrices are obtained:

$$
\mathbf{K} = -\left[\left(\frac{\partial F_i}{\partial x_j}\right)\right]_{x_{st},y_{st}} = -\frac{\mu R_j l^3 \Omega}{2c^3}\begin{bmatrix} I_1+I_2 & -I_4+I_5 \\ I_6+I_5 & -I_1+I_3 \end{bmatrix},
$$

$$
C = -\left[\left(\frac{\partial F_i}{\partial \dot{x}_j}\right)\right]_{x_{st},y_{st}} = \frac{\mu R_j l^3}{c^3}\begin{bmatrix} I_4 & I_1 \\ I_1 & I_6 \end{bmatrix}, \tag{8.38}
$$

where

$$
\begin{aligned}
I_1 &= \int_{\theta_1}^{\theta_2} \frac{\sin(\theta)\cos(\theta)}{h_{st}^{*3}} d\theta\,, & I_2 &= 3\int_{\theta_1}^{\theta_2} \frac{\Lambda}{h_{st}^*}\cos^2(\theta)d\theta\,,\\
I_3 &= 3\int_{\theta_1}^{\theta_2} \frac{\Lambda}{h_{st}^*}\sin^2(\theta)d\theta\,, & I_4 &= \int_{\theta_1}^{\theta_2} \frac{\cos^2(\theta)}{h_{st}^{*3}} d\theta\,,\\
I_5 &= 3\int_{\theta_1}^{\theta_2} \frac{\Lambda}{h_{st}^*}\sin(\theta)\cos(\theta)d\theta\,, & I_6 &= \int_{\theta_1}^{\theta_2} \frac{\sin^2(\theta)}{h_{st}^{*3}} d\theta\,,\\
h_{st}^* &= 1 - x_{st}^*\cos(\theta) - y_{st}^*\sin(\theta)\,, & \Lambda &= \frac{x_{st}^*\sin(\theta) - y_{st}^*\cos(\theta)}{h_{st}^{*3}}\,.
\end{aligned}
$$

The eight coefficients summarizing the linearized behavior of the different bearings are reported in nondimensional form as functions of the Sommerfeld number in many handbooks. The plots or tables may be computed using numerical methods (e.g., the FEM) or measured experimentally. As an example, the stiffness and damping parameters of the same bearings of Figure 8.12 (a plain cylindrical bearing, modeled using the short bearing model, and a two-lobe bearing) are reported in Figure 8.13. The coefficients are made nondimensional by dividing them by the constant

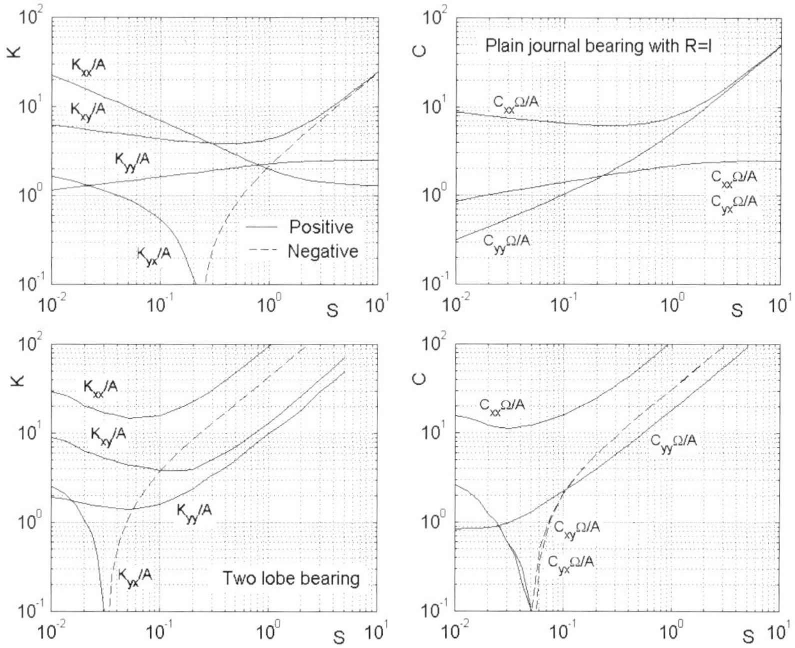

FIGURE 8.13. Elements of the stiffness and damping matrices as functions of the Sommerfeld number for a fully cavitated short bearing with $R/l = 1$ and for a two-lobe bearing (see T. Someya (editor), *Journal-Bearing Databook*, Springer, Berlin, 1989, calculation n. 12). The elements are made nondimensional through constant $A = c/F_{st}$.

$$A = \frac{c}{F_{st}} \ .$$

In both cases, the out-of-diagonal terms of the stiffness matrix are different, showing that the stiffness matrix is nonsymmetric. In other terms, the presence of lubricated bearings introduce a circulatory matrix, and hence a destabilizing effect, into the equation of motion. This is typical of lubricated bearings, except for some types, which are purposely designed to avoid instability: Tilting pad bearings have the out-of-diagonal elements vanishingly small. The damping matrix of the first bearing is symmetrical, whereas that of the second one is almost such.

If also the inertia of the oil film must be accounted for, it is possible to resort to more complex models, like the 12-coeffcient model in which 4 further coefficients are used to define a mass matrix. The data for its implementation are, however, not readily available.

8.4.3 Stability problems linked with the use of lubricated bearings

It is very well known that rotors running on journal bearings show particular dynamic problems linked intrinsically with the behavior of the fluid film within the bearings. These problems are usually referred to as *oil whirl* and *oil whip*.

The first phenomenon is a whirling of the rotor, which takes place at a frequency of about half the speed of rotation (sometimes referred to as *half-frequency whirl*) and is superimposed on the other whirling motions, particularly the synchronous whirling caused by unbalance. Its amplitude is usually not large and does not constitute an actual problem. At a speed that is usually not far from twice the first critical speed, the motion becomes more severe and can rapidly degenerate in a very violent, often destructive, whirling that takes place at a frequency almost independent of the speed and coincident with the first natural whirl frequency of the rotor at the relevant speed. This motion is usually referred to as *oil whip*.

A very simple heuristic explanation of the phenomenon accounts for the fact that the whirling takes place at about half the rotational frequency by noting that the oil in a plain journal bearing moves around at a speed that is about half of the peripheral velocity of the journal, providing a sort of rotating damping whose speed of rotation is about half the spin speed. In the region of the Campbell diagram, which lies below the straight line whose equation is $\omega = \Omega/2$, the behavior of the system is unstable, and the threshold of instability can be found easily by intersecting the lowest branch of the Campbell diagram with the mentioned line of equation $\omega = \Omega/2$ (Figure 8.14). If the lowest branch of the Campbell diagram is a horizontal straight line, as is the case of the Jeffcott rotor, the threshold of instability occurs at twice the critical speed of the rotor.

Note that this way of reasoning assumes that the presence of the bearings does not change deeply the natural frequencies of the rotor, i.e., that the bearing are much stiffer than the latter. If this does not occur, the Campbell diagram is changed and it has little meaning to refer to the diagram of the rotor on stiff bearings. Moreover, if the bearings' dynamics affects deeply the behavior of the rotor, their anisotropy (because of the load) makes the distinction between forward and backward modes to lose importance, and the intersection of the $\omega = \Omega/2$ line with all branches of the Campbell diagram must be considered.

The mentioned heuristic explanation also accounts for the fact that the frequency of the oil whirl is slightly lower than half the rotational speed, usually in the range of 0.45 to 0.48 Ω. The average velocity of the oil film is slightly lower than half the peripheral speed of the journal, depending on the clearance and the exact velocity profile.

However, the phenomenon is much more complex, and the actual behavior of the lubricant film must be modeled in some detail. In particular, the

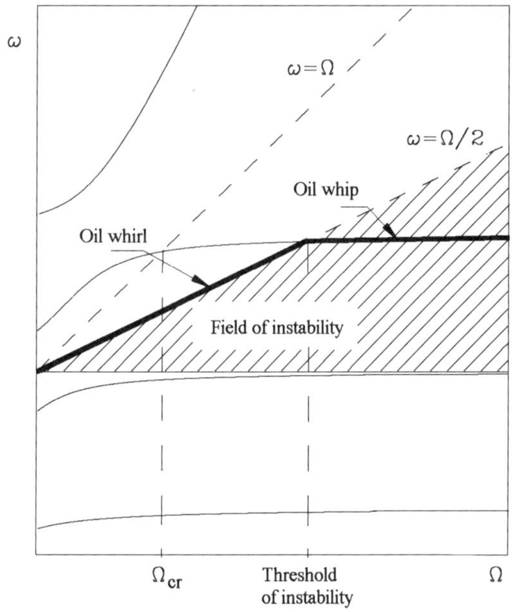

FIGURE 8.14. Oil whip and oil whirl on the Campbell diagram.

intrinsic nonlinear nature of the bearing cannot be neglected, which makes the study of the dynamic behavior of the system much more complex than the simple linear rotor dynamics study that is often sufficient when no allowance is taken for the compliance of journal bearings. In particular, the whirl-whip transition is not as abrupt as shown in Figure 8.14, and because the system is nonlinear, the free and self-excited vibrations are not really independent from the forced vibrations, like the unbalance response: It is well known that in many cases, an increase of the unbalance, while increasing the forced vibrations of the rotor, decreases the tendency to instability and consequently rises the threshold of instability and lowers the amplitude of the self-excited vibration, when present.

A detailed study of the stability in the small can, however, be performed by using the eight-coefficients model, i.e., by remaining within the simplification of linearized models. No interaction between free and forced behavior can be obtained in this way, but the study of the motion in the small allows us to obtain much information on the dynamic behavior of the rotor in a straightforward way.

The inherent instability of plain journal bearings makes them unsuitable for high-speed supercritical machinery, and many other bearing configurations have been developed with the aim of overcoming this difficulty. In particular, tilting pad bearings allow the instability problem to be solved completely at the cost of a reduction of damping at low speed and of added overall complexity. Among the many papers and books existing on this subject, those by Tondl [52] and Muszynska [21] are worth mentioning.

Example 8.3 *Study the dynamic behavior of the rotor described in Example 4.5 (Figure 4.14), running on a pair of plain lubricated bearings.*

Consider the following data:

- *Inertial properties:* $m = 20$ *kg*, $J_t = 0.2$ *kg* m^2, $J_p = 0.25$ *kg* m^2
- *Shaft geometric and material data: diameter* $\phi = 30$ *mm*, $l_1 = 100$ *mm*, $l_2 = 200$ *mm*, $E = 2.1 \times 10^{11}$ N/m^2
- *Bearings data:* $R = 12$ *mm*, $l = 12$ *mm*, $c = 35$ μm, $\mu = 0.02$ N/sm^2

The rotor is axially symmetrical, so the complex coordinates approach will be first followed. Assuming as generalized coordinates the displacements at the bearings and at the disc and the rotation of the latter, the 4×4 stiffness matrix is [Equation (4.107)]

$$\mathbf{K} = \begin{bmatrix} 1.0020 & -1.0020 & 0.0501 & 0 \\ -1.0020 & 1.1272 & -0.0376 & -0.1252 \\ 0.0501 & -0.0376 & 0.0050 & -0.0125 \\ 0 & -0.1252 & -0.0125 & 0.1252 \end{bmatrix} \times 10^8 \; .$$

Note that the stiffness of the bearings has not been included into the computations. To avoid a singular mass matrix, a concentrated mass of 0.001 kg is added at the bearing locations. The mass and gyroscopic matrices are [Equation (4.108)]

$$\mathbf{M} = diag \left\{ \begin{array}{cccc} 0.001 & 20 & 0.2 & 0.001 \end{array} \right\} , \quad \mathbf{G} = diag \left\{ \begin{array}{cccc} 0 & 0 & 0.25 & 0 \end{array} \right\} .$$

As the bearings are not isotropic, the study is performed using the real coordinates approach. The relevant matrices are then

$$\mathbf{K}' = \begin{bmatrix} \mathbf{K} & \mathbf{0} \\ \mathbf{0} & \mathbf{K} \end{bmatrix} , \quad \mathbf{M}' = \begin{bmatrix} \mathbf{M} & \mathbf{0} \\ \mathbf{0} & \mathbf{M} \end{bmatrix} , \quad \mathbf{G}' = \begin{bmatrix} \mathbf{0} & \mathbf{G} \\ -\mathbf{G} & \mathbf{0} \end{bmatrix} .$$

The forces on the bearings are easily computed:

$$F_1 = mg\frac{l_2}{l_1 + l_2} = 130.8 \; N \; , \qquad F_2 = mg\frac{l_1}{l_1 + l_2} = 65.4 \; N \; .$$

The Sommerfeld numbers are then

$$S_1 = \frac{\mu \Omega R l}{\pi F_1} \left(\frac{R}{c} \right)^2 = 8.24 \times 10^4 \; \Omega \; , \qquad S_2 = \frac{\mu \Omega R l}{\pi F_2} \left(\frac{R}{c} \right)^2 = 16.48 \times 10^4 \; \Omega \; .$$

For instance, at a speed of 500 rad/s, their values are

$$S_1 = 0.412 \ , \qquad S_2 = 0.824 \ ,$$

Assuming two fully cavitated plain cylindrical bearings, at that speed, the eccentricity ϵ *and the attitude angles are (Figure 8.12)*

$$\epsilon_1 = 0.51 \ , \qquad \beta_1 = 53.44^\circ \ , \qquad \epsilon_2 = 0.36 \ , \qquad \beta_2 = 64.03^\circ \ .$$

The eight coefficients of the two bearings can be obtained from Figure 8.13. At 500 rad/s, they are

$$\mathbf{K}_1 = \frac{F_1}{c}\begin{bmatrix} 3.162 & 3.826 \\ -0.580 & 2.002 \end{bmatrix} = \begin{bmatrix} 1.182 & 1.430 \\ -0.217 & 0.748 \end{bmatrix} \times 10^7 \ ,$$

$$\mathbf{K}_2 = \frac{F_2}{c}\begin{bmatrix} 2.180 & 4.091 \\ -1.680 & 2.212 \end{bmatrix} = \begin{bmatrix} 4.073 & 7.645 \\ -3.139 & 4.134 \end{bmatrix} \times 10^6 \ ,$$

$$\mathbf{C}_1 = \frac{F_1}{c\Omega}\begin{bmatrix} 6.297 & 1.864 \\ 1.864 & 2.515 \end{bmatrix} = \begin{bmatrix} 4.707 & 1.393 \\ 1.393 & 1.880 \end{bmatrix} \times 10^4 \ ,$$

$$\mathbf{C}_2 = \frac{F_2}{c\Omega}\begin{bmatrix} 7.216 & 2.106 \\ 2.106 & 4.327 \end{bmatrix} = \begin{bmatrix} 2.697 & 0.787 \\ 0.787 & 1.617 \end{bmatrix} \times 10^4 \ .$$

The four elements of matrices $\mathbf{K}_1$ *and* $\mathbf{C}_1$ *must be assembled in position 11, 15, 51, and 55 of the stiffness matrix* $\mathbf{K}'$ *and of the global damping matrix, whereas the elements of matrices* $\mathbf{K}_2$ *and* $\mathbf{C}_2$ *must be assembled in position 44, 48, 84, and 88.*

The model of the system is then completed, and the whirling frequencies can be computed. At 500 rad/s, there are six real values of s

$$s_1 = -841 \ , \quad s_2 = -7096 \ , \quad s_3 = -1.202 \times 10^7 \ ,$$

$$s_4 = -1.308 \times 10^7 \ , \quad s_5 = -3.111 \times 10^7 \ , \quad s_6 = -5.278 \times 10^7 \ ,$$

representing six nonoscillatory (overdamped) modes, plus 5 pairs of complex conjugate solutions

$$s_{7,8} = -199 \pm 355\, i \ , \quad s_{9,10} = -176 \pm 666\, i \ , \quad s_{11,12} = -642 \pm 816\, i \ ,$$

$$s_{13,14} = -136 \pm 1504\, i \ , \quad s_{15,16} = -864 \pm 2265\, i \ ,$$

related to oscillatory modes.

The real part of all solutions is negative: The system is stable at 500 rad/s.

The computations can be repeated for different values of the speed, obtaining the Campbell diagram and the decay rate plot (Figure 8.15). The Campbell diagram of the rotor on rigid supports is reported on the same figure: As it was easily predictable from the observation that the elements of the stiffness matrix of the shaft are of the same order of magnitude of those of the bearings, the latter have a deep influence on the dynamic behavior of the system. The static working conditions of the bearings are reported in Figure 8.16

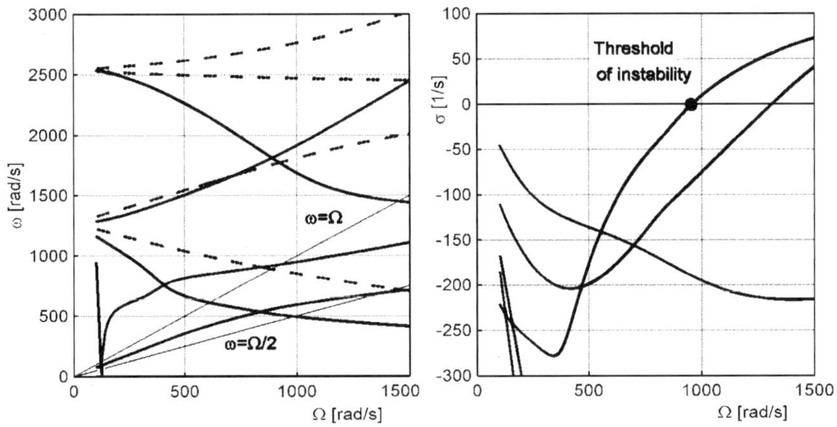

FIGURE 8.15. Campbell diagram and decay rate of a rotor on plain lubricated bearings (full lines). The Campbell diagram is compared with that of the rotor on stiff bearings (dashed lines).

From the decay rate plot, it is clear that a threshold of instability is present. Its value is $\Omega_{th} = 959$ *rad/s. Note that, even if the bearing is fully cavitated, the threshold of instability is not far from the intersection of the line related to the first mode with the line* $\omega = \Omega/2$.

Example 8.4 *Repeat the previous example, substituting a pair of tilting pad bearings to the plain lubricated bearings.*

The values of the eight coefficients for a five tilting pad bearing with the load between the pads are reported in Figure 8.17; the data are taken from [47], calculation n. 40. Note that coefficients K_{xy}, K_{yx}, C_{xy}, *and* C_{yx} *are equal to zero, as typical of tilting pads bearings.The rotordynamic study is repeated and the Campbell diagram and the decay rate plot (Figure 8.18), together with the Campbell diagram of the rotor on rigid supports.*

No instability is present, even at high speed. Note that the whirling frequencies are lowered with respect to the previous case, showing that the bearings are less stiff.

8.4.4 *Effect of seals, clearances, and dampers*

Problems similar to those linked with lubricated journal bearings are also encountered in all other cases in which a fluid is interposed between the stator and the rotor, like, for example, in labyrinth and liquid seals. In all cases, one of the most effective measures aimed at reducing the instability problems is to decrease the peripheral velocity of the fluid around the shaft. This can be done using antiswirl vanes, by roughening the stator walls, or by injecting the fluid in the tangential, backward direction.

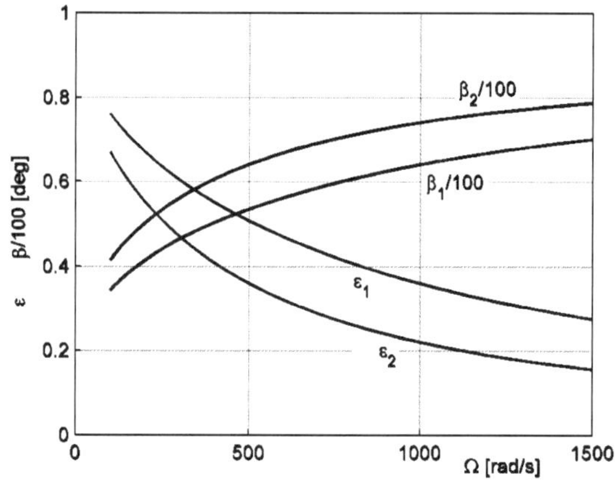

FIGURE 8.16. Eccentricity and attitude angle of the two bearings as functions of the speed.

A linearized analysis of the effect of such elements can be performed using the eight-coefficients model: Charts similar to those seen for the bearings can be obtained also for the various types of seals. In the simplest cases, it is even possible to use the short- or the long-bearing model; otherwise, the relevant coefficients may be computed using numerical methods (e.g., the FEM) or measured on purposely designed test rigs. In many cases, the clearance in seals is much larger than that in bearings, and this causes the forces exerted by the first ones to be far smaller than those caused by the latter. In this case, the static equilibrium position of the shaft can be computed by neglecting the presence of the seals, which are then introduced in the following dynamic analysis.

The instability of bearings and seals is because the fluid entrained by the rotor moves around, and thus its damping acts as a sort of rotating damping, although its velocity is smaller than the spin speed (usually close to half of it, as already said). It is then clear that no destabilizing effect can be produced by a nonrotating fluid film that, thanks to its viscosity, supplies a nonrotating, and hence stabilizing, damping.

A damper can thus be made by a journal, connected with the shaft through a bearing (usually a rolling elements bearing) and prevented from rotating by a pin, which can whirl in a bearing. Such an arrangement is usually referred to as a squeeze-film damper (Figure 8.19). The oil between the nonrotating journal and bearing is prevented from moving axially by seals at the ends. Radial movements of the journal cause the oil to move circumferentially, and this movement provides the required damping.

By using the models seen in the previous section, it is easy to state that the stiffness of the oil film is vanishingly small, because no relative angular

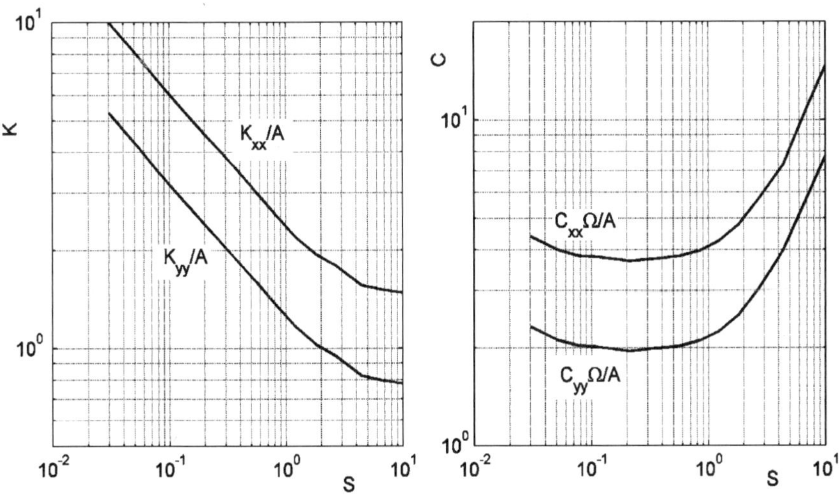

FIGURE 8.17. Stiffness and damping coefficients for a five tilting pad bearing with the load between the pads as function of the Sommerfeld number (T. Someya (editor), *Journal-Bearing Databook*, Springer, Berlin, 1989, calculation n. 40).

velocity between journal and bearing exists. The damper does not exert any steady-state force on the shaft and can be neglected in the computation of the static equilibrium position. If the bearing is cylindrical and the oil is prevented from moving axially, the long-bearing assumption can be used, and the eight-coefficient model is expressed by Equations (8.37). From the first equation, it is clear that the stiffness matrix **K** vanishes (owing to the vanishing value of the spin speed Ω), and that the damper can supply only damping forces. If the damper works in the central position, the forces received by the journal can be computed in closed form as

$$\left\{ \begin{array}{c} F_x \\ F_y \end{array} \right\} = -\mathbf{C} \left\{ \begin{array}{c} \dot{x} \\ \dot{y} \end{array} \right\} = -12\pi\mu l \left(\frac{R_j}{c} \right)^3 \left[\begin{array}{cc} 1 & 0 \\ 0 & 1 \end{array} \right] \left\{ \begin{array}{c} \dot{x} \\ \dot{y} \end{array} \right\} . \qquad (8.39)$$

The damping matrix is not only symmetrical, but also even diagonal and the effect is stabilizing.

In the damper of Figure 8.19(b), the journal is connected to the bearing by a spring and the device can provide both a damping and a restoring force. More than a damper, it can then be defined as a damped support.

Squeeze-film dampers are very common, particularly in the case of rotors running in supercritical conditions, where nonrotating damping must be added to achieve the required stability. In machines containing coaxial rotors spinning at different speeds, like multishaft turbines, it is also possible to use intershaft dampers, i.e., dampers located between two shafts rotating at different speeds. They must, however, be designed with great

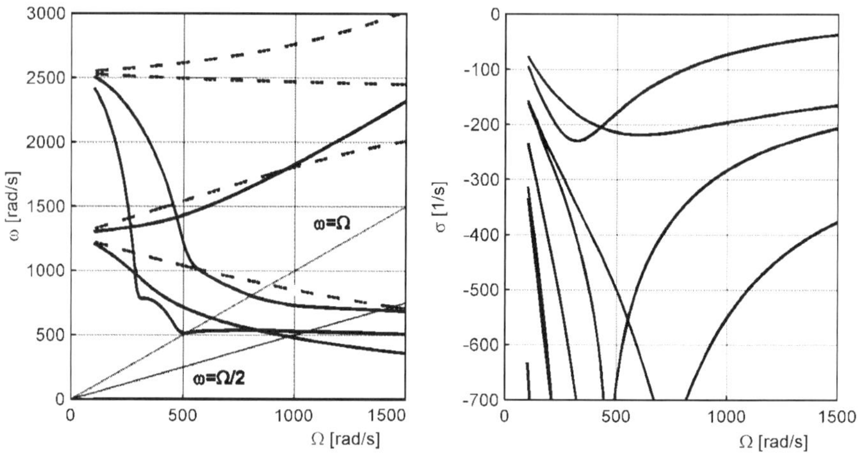

FIGURE 8.18. Campbell diagram and decay rate of a rotor on tilting pad bearings (full lines). The Campbell diagram is compared with that of the rotor on stiff bearings (dashed lines).

care because the element that dissipates energy in this case spins with the velocity of one of the two rotors, usually the slower one, and then they can have a destabilizing effect in particular working conditions.

Remark 8.3 *Squeeze film dampers are intrinsically nonlinear devices, and although their behavior in the small may be linearized through the eight-coefficients approach, they introduce nonlinearities into the system. The aforementioned linearization holds only for motions with very small amplitude.*

8.5 Magnetic bearings

Magnetic bearings use electromagnetic forces to keep the rotor in place, without any material contact between stator and rotor. They can drastically reduce bearing drag, while completely avoiding the presence of lubricant and wear. The field in which the advantages of magnetic bearings are most important is, however, that of the dynamic behavior of the rotor: The stiffness and damping of the bearing system can be tailored for the application and can be adjusted following the operating conditions. In the case of active bearings, the control system can be used not only to maintain the rotor in the required position as a rigid body, minimizing the effects of unbalance and reducing the needs of strict balancing tolerances, but also to control the deformation modes. As a magnetic suspension must keep the suspended body in the required position, constraining rigid-body motion, a

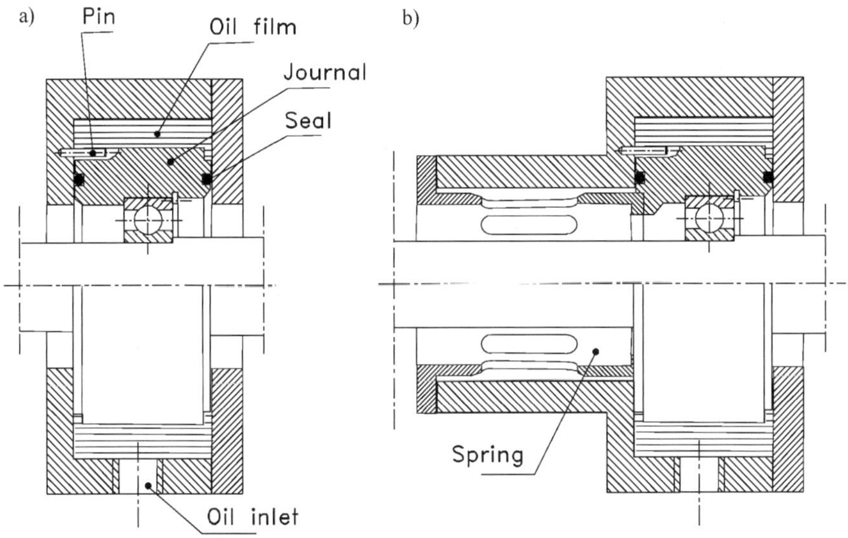

FIGURE 8.19. Squeeze-film dampers: (a) Damper and (b) damped support.

complete suspension must constrain the six modes of a rigid body in space. In the case of a rotor, one of the rigid-body modes (namely rotation about the axis) must be left unconstrained and a five-axis suspension is needed.[1] If the axial suspension can be considered as uncoupled, the lateral suspension must deal with the four rigid-body modes, to which the flexural modes of the rotor must be added.

Electromagnetic forces can be exerted by a passive device, based on permanent magnets or on uncontrolled electromagnets, or by an active device outfitted by a suitable control system. Strictly speaking, the two types of magnetic suspension systems should be defined as uncontrolled or controlled, because a device based on an uncontrolled electromagnet is active in the sense that it receives energy from the outside. However, the use of the terms *active* and *passive* with this meaning is well established in magnetic-bearing technology and will be followed in this section.

A magnetic suspension based on a five-passive-axes layout is unstable (the so-called Earnshaw theorem [53, 54]), except if diamagnetic or superconducting materials [55] are used.[2] The choice is then between the use of a hybrid suspension system, in which at least one of the degrees of freedom

[1] Actually the electric motor (if present) can be considered as a part of the suspension system, and because it takes care of the sixth axis, also a rotor on active magnetic bearings driven by an electrical machine can be considered as restrained by a six-axes suspension.

[2] It is possible to use electrodynamic phenomena or even gyroscopic moments (see Section 12.1.3) to stabilize a fully passive magnetic suspension. As such types of sus-

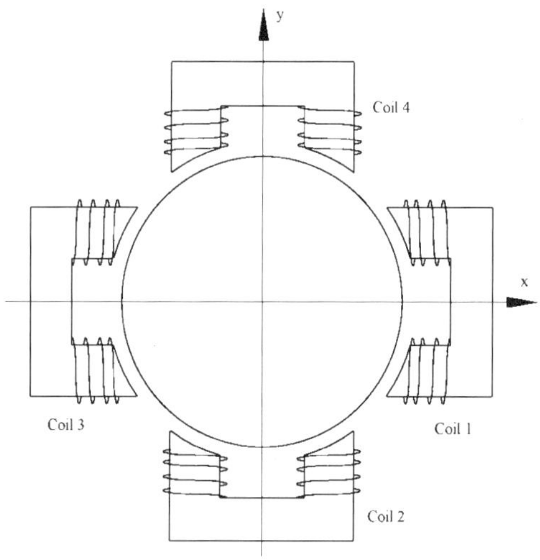

FIGURE 8.20. Sketch of an electromagnetic actuator of a magnetic bearing.

of the rotor is constrained by a mechanical bearing and that of a device that is at least partially active. The solutions span from one-active-axis suspensions, in which two passive radial bearings control the four degrees of freedom linked with the lateral behavior of the rotor, and one axial active bearing restrains axial motion to fully active five-axis suspensions, with active radial and axial bearings.

In this section, only plain radial heteropolar electromagnetic actuators will be dealt with; for other layouts, the reader may refer to the specialized literature [56]. Each actuator of a magnetic bearing, controlling two of the rigid-body lateral degrees of freedom of the rotor, can be assumed to be made by four electromagnets as shown in Figure 8.20: Coils 1 and 3 are responsible for the force in the x-direction and coils 2 and 4 for the force in the y-direction. Other arrangements, in which the pole pieces are aligned in axial direction (homopolar bearings), are also possible, but their working is essentially similar.

As a first approximation, the force exerted by a single coil depends on the current i and the air gap t between stator and rotor following the law

$$F = K \left(\frac{i}{t} \right)^2 , \tag{8.40}$$

where K is a constant that includes all design parameters of the actuator.

---

pensions are seldom considered for rotating machinery, they will not be dealt with here. Also diamagnetic and superconducting suspension will not be dealt with.

Remark 8.4 *Another cause of approximations is the assumption that the force exerted by the actuator in one direction is independent from that exerted in other directions. Coupling effects depend on the exact geometry of the actuator.*

Usually, to linearize the overall behavior of the bearing, the coils are fed with a constant current, referred to as *bias current*, equal for the two opposite electromagnets of the same axis, on which the control current is superimposed. A so-called static compensation current i'_0 (of opposite sign on the two coils of the same axis) is also superimposed to the bias current i_0 of the two coils acting in the same direction to withstand any static load.

The sum of the bias and compensation current in coils 1 and 3 acting in the x-direction can be written as $i_0+i'_0$ and $i_0-i'_0$. The total x-component of the force F_c exerted by the actuator is

$$F_x = K\left[\left(\frac{i_0+i'_0+i_x}{c-d-u_x}\right)^2 - \left(\frac{i_0-i'_0-i_x}{c+d+u_x}\right)^2\right] , \tag{8.41}$$

where i_x, c, d and u_x are, respectively, the control current, the radial clearance, the static offset, and the radial displacement, and the constants K of the two coils are assumed to be equal. Equation (8.41) holds only if $i_0 - i'_0 - i_x > 0$: In the opposite case, the controller should switch off the coil for which this relationship is not verified, and the system works in nonlinear conditions.

An actuator that works with a bias current high enough to allow all coils to be always energized is said to work as a "class A" actuator whereas a "class B" actuator works with at least one coil switched on and off.

The force expressed by Equation (8.41) can be linearized about the condition with $i_x = 0$, $u_x = 0$ as

$$F_x = F_0 + K_i i_x + K_u u_x , \tag{8.42}$$

where

$$F_0 = K\left[\left(\frac{i_0+i'_0}{c-d}\right)^2 - \left(\frac{i_0-i'_0}{c+d}\right)^2\right] ,$$

$$K_i = \left(\frac{\partial F_x}{\partial i_x}\right)_{\substack{i_x=0\\u_x=0}} = 2K\left[\frac{i_0+i'_0}{(c-d)^2} + \frac{i_0-i'_0}{(c+d)^2}\right] , \tag{8.43}$$

$$K_u = \left(\frac{\partial F_x}{\partial u_x}\right)_{\substack{i_x=0\\u_x=0}} = 2K\left[\frac{(i_0+i'_0)^2}{(c-d)^3} + \frac{(i_0-i'_0)^2}{(c+d)^3}\right] .$$

If the actuator operates in a central position, i.e., if $d = 0$, Equation (8.43) reduces to

$$F_0 = 4K\frac{i_0 i_0'}{c^2}\,, \qquad K_i = 4K\frac{i_0}{c^2}\,, \qquad K_u = 4K\frac{i_0^2 + {i_0'}^2}{c^3}. \tag{8.44}$$

The force F_0 acting on the bearing determines the value of the current i_0' and hence affects the stiffness of the bearing. As a consequence, if the static forces acting in the x- and y-directions are not equal, as in the case of horizontal rotors, the behavior is anisotropic even if the geometrical and electrical characteristics are equal in the two directions. This effect is particularly strong in the case of bearings operating with low bias currents.

In the case in which the sensors and the actuators related to a bearing are located in the same cross section of the shaft (colocated bearing), each actuator is controlled following the data supplied by a single sensor (decentralized control) and the control current is proportional to the displacement and the velocity measured by the sensor (ideal PD controller), and the linearized model of the magnetic bearing is equivalent to a mechanical bearing with a stiffness k and damping c given by the relationship

$$\begin{aligned} k &= -K_u + K_i K_{cp} K_a K_s\,, \\ c &= K_i K_{cd} K_a K_s\,, \end{aligned} \tag{8.45}$$

where the gains of the control chain are K_s, gain of the sensor; K_a, gain of the power amplifier; K_{cp}, proportional gain of the controller; and K_{cd}, derivative gain of the controller. Note that K_u, is the negative stiffness, expressing the instability of the bearing in open-loop conditions.

This seldom applies, because usually it is impossible to perfectly colocate the sensor and the actuator, the controller (but also the sensors and the power amplifiers) has its own dynamics and often a more complex control strategy is used. In particular, sensor-actuator non-colocation may induce instabilities in the system and must be dealt with in detail at the design stage.

The dynamics of a rotor running on magnetic bearings can still be studied using a linearized model (provided that all relevant elements may be linearized) through a multi-degrees-of-freedom model in the state space, introducing a suitable number of states of the control loop together with the mechanical states, positions, and velocities. Non-colocation can thus be accounted for very easily, by simply introducing suitable nodes in the sensor and actuator locations, and so its effects on stability can be assessed. A sketch and a block diagram of a rotor on two magnetic bearings are reported in Figure 8.21; some details on dynamics of the control loop will be dealt with in Chapter 16.

Magnetic bearings are very well suited for soft-mounted rotors, owing to the possibility of tailoring stiffness and damping in the required way. However, in many cases, it may be difficult to control adequately high-frequency deformation modes, so in many cases rotors on magnetic bearings

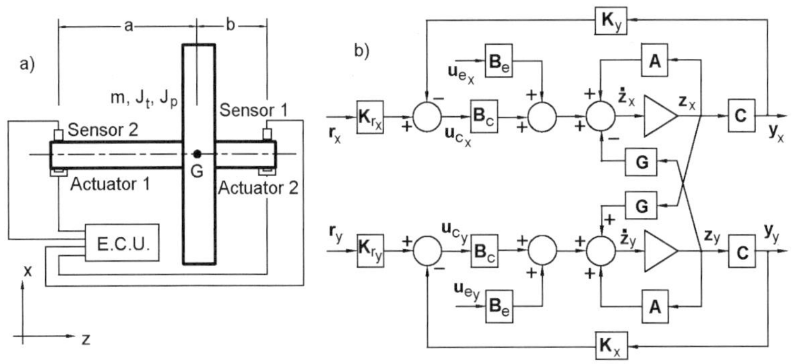

FIGURE 8.21. Rigid rotor on a four-axis active magnetic suspension: (a) Sketch of the system; (b) block diagram, evidencing gyroscopic coupling between the xz- and yz-planes.

work in a speed range located between the rigid-body critical speeds and the critical speeds related to deflection modes of the rotor.

Example 8.5 *Jeffcott rotor on an ideal magnetic bearing with a proportional-derivative (PI) controller.*

Assume the following data: mass $m = 3$ *kg, actuator constant* $K = 1.5246 \times 10^{-6}$ *Nm*^{2}*/A*^{2}*, nominal air gap* $c = 0.75$ *mm, bias current* $i_0 = 3$ *A, proportional gain of the controller* $k_c = 1.25$ *V/m, derivative gain* $K_{cd} = 0.001129$ *Vs/m, gain of the sensor* $K_s = 7500$ *V/m, gain of the power amplifier* $K_a = 1.027$ *A/V.*

Using complex coordinates, the equation of motion of the rotor is

$$m\ddot{r} = f_c + m\epsilon\Omega^2 e^{i\Omega t} + f_n \ .$$

The static force f_n *is equal to the weight of the rotor* mg *acting in vertical direction. If the direction of* x*-axis is vertical, it is expressed by a real number. The control force* f_c *is made by two components, namely a proportional and a derivative one*

$$f_c = -kr - c\dot{r} \ ,$$

where k *and* c *are expressed by Equation (8.45). The final equation of motion of the Jeffcott rotor is thus*

$$m\ddot{r} + (K_i K_{cd} K_a K_s)\dot{r} + (-K_u + K_i K_{cp} K_a K_s) r = m\epsilon\Omega^2 e^{i\Omega t} + f_n \ .$$

The linearized characteristics of the bearings are linked to the bias and compensation currents through the relationships (8.44)

$$F_0 = 4K\frac{i_0 i'_0}{c^2} \ , \qquad K_i = 4K\frac{i_0}{c^2} \ , \qquad K_u = 4K\frac{i_0^2 + {i'_0}^2}{c^3} .$$

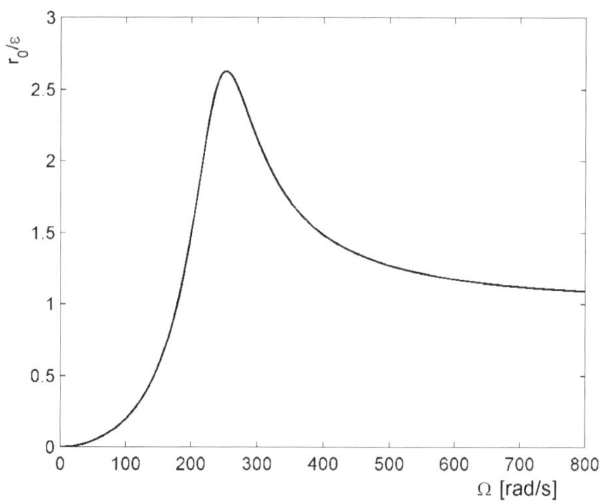

FIGURE 8.22. Nondimensional unbalance response of the Jeffcott rotor on magnetic bearings.

The static force is equal to the weight in vertical direction (x-axis) and 0 in horizontal direction (y-axis). The corresponding compensation currents are

$$i'_{0x} = F_0 \frac{c^2}{4Ki_0} = 0.9048\ A\,, \qquad i'_{0y} = 0.$$

The current and displacement bearing stiffness is

$$K_{ix} = K_{iy} = 32.525\ \ N/A, \quad K_{ux} = 1.419 \times 10^5\ \ N/m, \quad K_{uy} = 1.301 \times 10^5\ \ N/m.$$

As the bearings are almost isotropic, they will be assumed to be such and an average value of $K_u = 1.36 \times 10^5\ \ N/m$.

By introducing the values of the various gains into the homogeneous equation of motion, the following characteristic equation is obtained:

$$3s^2 + 282.83s + 1.7715 \times 10^5 = 0\ ,$$

whose poles are

$$s = -47.14 \pm 238.4\ i\ .$$

The two complex pairs correspond to the two damped whirl frequencies, one forward and one backward, as is typical for a damped Jeffcott rotor.

The unbalance response is easily computed and coincides with that of a standard isotropic Jeffcott rotor. It is reported in nondimensional form in Figure 8.22.

Example 8.6 *Study the radial suspension of a rigid, isotropic rotor supported by a five-active-axis magnetic suspension with perfectly colocated bearings and an* *ideal PD controller (Figure 8.21). The main inertial and geometric data are* $m = 3$ *kg;* $J_p = 0.02$ *kg m*^{2}*;* $J_t = 0.015$ *kg m*^{2}*;* $a = 200$ *mm;* $b = 100$ *mm.*

Neglecting the effect of weight on the position of the rotor and on the gains of the bearings, compute the gain matrix of the control system assuming that each bearing is controlled separately and that a complete axial symmetry is required. Plot the Campbell diagram, and compute the control forces at 20,000 rpm if the rotor has a residual static unbalance following the ISO quality grade G 2.5. The control system should provide uncoupling between translational and rotational modes and locate the first critical speed in the vicinity of 2,500 rpm = 261.8 rad/s.

Using complex coordinates and considering only static unbalance, the equation of motion of the rotor is

$$\begin{bmatrix} m & 0 \\ 0 & J_t \end{bmatrix} \begin{Bmatrix} \ddot{r} \\ \ddot{\phi} \end{Bmatrix} - i\Omega \begin{bmatrix} 0 & 0 \\ 0 & J_p \end{bmatrix} \begin{Bmatrix} \dot{r} \\ \dot{\phi} \end{Bmatrix} = \mathbf{f}_c + m\epsilon\Omega^2 \begin{Bmatrix} 1 \\ 0 \end{Bmatrix} e^{i\Omega t} + \mathbf{f}_n ,$$

where the control force f_c *is made by two components, namely a proportional and a derivative one*

$$\mathbf{f}_c = -\mathbf{K}\mathbf{q} - \mathbf{C}\dot{\mathbf{q}} .$$

Indicating the displacements at the bearing locations (sensor and actuator, owing to the colocation assumption) with r_1 *and* r_2*, the forces acting on the rotor are*

$$\begin{Bmatrix} f_{c1} \\ f_{c2} \end{Bmatrix} = - \begin{bmatrix} k_{d1} & 0 \\ 0 & k_{d2} \end{bmatrix} \begin{Bmatrix} \dot{r}_1 \\ \dot{r}_2 \end{Bmatrix} - \begin{bmatrix} k_{p1} & 0 \\ 0 & k_{p2} \end{bmatrix} \begin{Bmatrix} r_1 \\ r_2 \end{Bmatrix} ,$$

where the derivative and proportional overall gain (the latter including the softening open-loop effect of thc actuator K_u*) are expressed by Equation (8.45). The relationship linking the displacements at the bearing locations and the displacement and rotation at the mass center of the rotor is*

$$\begin{Bmatrix} r_1 \\ r_2 \end{Bmatrix} = \begin{bmatrix} 1 & -a \\ 1 & b \end{bmatrix} \begin{Bmatrix} r \\ \phi \end{Bmatrix} = \mathbf{T} \begin{Bmatrix} r \\ \phi \end{Bmatrix} .$$

A similar relationship can be written for the control forces. The relationship yielding the control forces at the center of mass of the system is then

$$\mathbf{f}_c = -\mathbf{T}^T \begin{bmatrix} k_{d1} & 0 \\ 0 & k_{d2} \end{bmatrix} \mathbf{T} \begin{Bmatrix} \dot{r} \\ \dot{\phi} \end{Bmatrix} - \mathbf{T}^T \begin{bmatrix} k_{p1} & 0 \\ 0 & k_{p2} \end{bmatrix} \mathbf{T} \begin{Bmatrix} r_1 \\ r_2 \end{Bmatrix} .$$

The closed-loop dynamic matrix A_{cl} *is then*

$$\mathbf{A}_{cl} = \begin{bmatrix} -\dfrac{k_1}{m} & -\dfrac{k_2}{m} & -\dfrac{k_4}{m} & -\dfrac{k_5}{m} \\ -\dfrac{k_2}{J_t} & -\dfrac{k_3 - i\Omega J_p}{J_t} & -\dfrac{k_5}{J_t} & -\dfrac{k_6}{J_t} \\ 1 & 0 & 0 & 0 \\ 0 & 1 & 0 & 0 \end{bmatrix} ,$$

where $k_1 = k_{d_1} + k_{d_2}$*,* $k_2 = -ak_{d_1} + bk_{d_2}$*,* $k_3 = a^2 k_{d_1} + b^2 k_{d_2}$*,* $k_4 = k_{p_1} + k_{p_2}$*,* $k_5 = -ak_{p_1} + bk_{p_2}$*, and* $k_6 = a^2 k_{p_1} + b^2 k_{p_2}$*.*

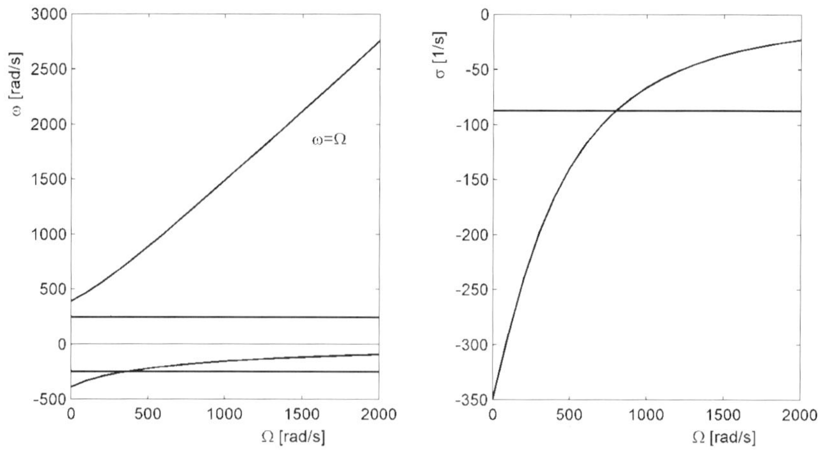

FIGURE 8.23. Campbell diagram and decay rate for the rigid rotor on magnetic bearings.

Uncoupling between translational and rotational modes implies that $ak_{d_1} = bk_{d_2}$, *and* $ak_{p_1} = bk_{p_2}$, *and the condition on the critical speed of the corresponding undamped system yields* $\sqrt{(k_{p_1} + k_{p_2})/m} = \Omega_{cr} = 261.8$. *The values of the stiffness gains thus obtained are* $k_{p_1} = 6.85 \times 10^4$ *Nm, and* $k_{p_2} = 1.37 \times 10^5$ *Nm and yield the following terms to be included in the closed-loop gain matrix:* $k_4 = k_{p_1} + k_{p_2} = 2.056 \times 10^5$ N/m *and* $k_6 = a^2 k_{p_1} + b^2 k_{p_2} = 4{,}110$ Nm/rad.

The damping gains can be obtained assuming a given value for the damping ratio of a particular mode, say, the translational mode. Assuming a damping ratio of 1/3, the following values are obtained: $k_{d_1} = 349.1$ Ns/m, $k_{d_2} = 174.5$ Ns/m, $k_1 = k_{v_1} + k_{v_2} = 523.6$ Ns/m, *and* $k_3 = a^2 k_{v_1} + b^2 k_{v_2} = 10.47$ Nsm/rad. *The Campbell diagram and the decay rate plot are reported in Figure 8.23.*

The static deflection of the system is immediately obtained: $x_{st} = -mg/k_{d_t} = 0.14$ *mm. Note that the control system can be set to compensate for the static deflection and to maintain the rotor in the center of the bearing notwithstanding the weight.*

Because of uncoupling between translational and rotational rigid-body modes, the response to static unbalance can be plotted by resorting to the equation for translational motion alone

$$\left(-m\Omega^2 + i\Omega k_1 + k_4\right)(x + iy) = m\epsilon\Omega^2 e^{i\omega t} .$$

The amplitude of the response to static unbalance is plotted in nondimensional form in Figure 8.24. Note that the system is very much damped, and the rotor self-centers very quickly in the supercritical range. Unbalance quality grade G 2.5 at 20,000 rpm = 2,094 rad/s corresponds to an eccentricity $\epsilon = 1.19\mu m$, *i.e., with a mass of 3 kg, to a residual unbalance* $m\epsilon = 3.58$ g mm. *The response of the system at 20,000 rpm is* $(x + iy)_0 = -1.204 - 0.102i$ μm.

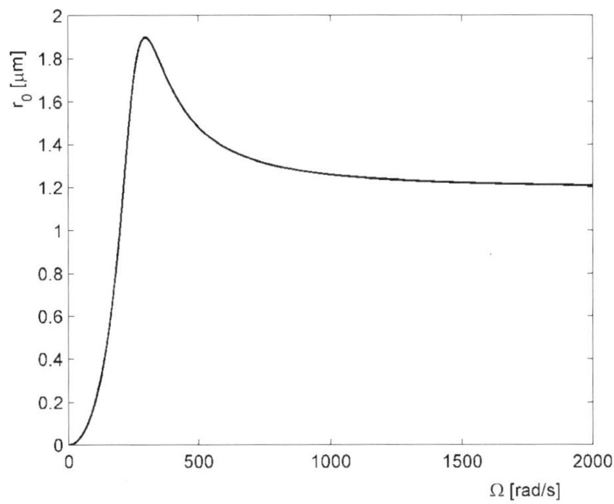

FIGURE 8.24. Unbalance response of a rigid rotor on magnetic bearings.

The total control force acting on the rotor is simply

$$F_c = (i\Omega k_1 + k_4)(x + iy) = m\Omega^2(x + iy) + m\epsilon\Omega^2 e^{i\omega t} .$$

The modulus of the control force is then $|F_c| = 1.347$ *N, and it is shared between the two bearings following their stiffness (2/3 on bearing 1 and 1/3 on bearing 2). Note that the control forces are very small, even if the balancing quality grade is not very high. Actually, very smooth running is obtained with even less strict balancing tolerances.*

Example 8.7 *Consider a rotor running on a one-active-axis magnetic suspension. The radial bearings are of the passive type, based on permanent magnets.*

The axial suspension must be active, because the passive radial bearings have a negative (destabilizing) axial stiffness. The mass of the rotor is $m = 0.8$ *kg, and the linearized axial stiffness of the radial suspension is* $k_z = -5 \times 10^4$ *N/m. The axial actuator is made by a couple of electromagnets whose force-current characteristic is* $F_c = \mu_0 S N^2 I^2 / 4d^2$*, where* I *is the current,* $\mu_0 = 4\pi \times 10^{-7}$ *is the vacuum permeability,* $S = 1.2 \times 10^{-3} m^2$ *is the surface of the pole pieces,* $N = 160$ *is the number of turns, and* $d = 0.5$ *mm is the nominal air gap.*

The axial position is measured by a sensor that outputs a voltage proportional to the axial displacement u *from the nominal position (air gap of 2.5 mm) through the law* $V_s = V_0 + au$*, where constants* V_0 *and* a *take the values 0.85 V and 300 V/m, respectively, when the input voltage to the sensor is 12 V.*

The quadratic law linking the magnetic force to the current in the coil compels a choice between two alternatives: to use a control system that supplies the coil with a current proportional to the square root of the displacement in such a way that a restoring force proportional to the displacement is exerted or to add a constant

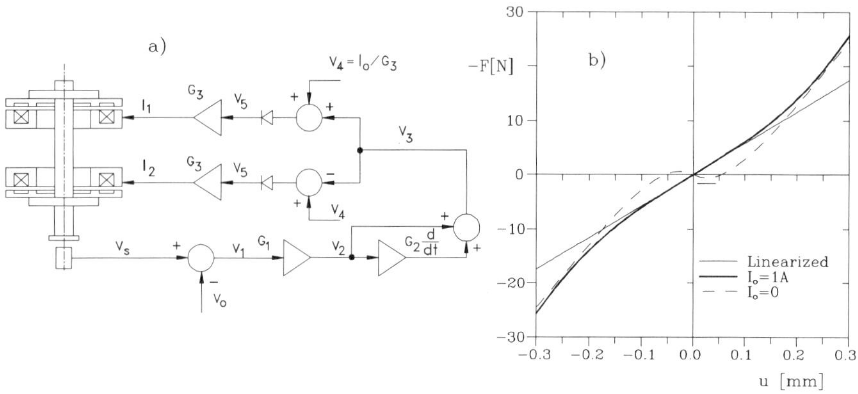

FIGURE 8.25. Active axial magnetic bearing. (a) Scheme of the control system; (b) force-displacement characteristics.

current (bias current) to the current proportional to the displacement supplied by the control system. In the latter case, the coils exert two forces, which in the equilibrium position are equal and opposite; when the shaft is displaced axially, the force caused by one coil is reduced and the other increases and the restoring effect is obtained. A detailed scheme of the control system, following the second of the two approaches, is shown in Figure 8.25(a).

Note that the voltage signal $V_2 = V_1G_1$ *is differentiated to supply an input to the current generators that is proportional to both the axial displacement and velocity. A damping effect is thus obtained. The voltage supplied to the current generators driving the coils is then*

$$V_5 = V_4 \mp G_1V_1 \mp G_1G_2\frac{dV}{dt} = V_4 \mp aG_1(u + G_2\dot{u}) \,,$$

where the upper signs hold for the coil nearer the end at which the sensor is located. The current supplied to the coils is then $I_i = G_3V_5 = I_0 \pm aG_1G_3(u + G_2\dot{u})$*, the upper sign being referred to coil 1. When both coils are operating, the force they exert (positive if directed upward, i.e., in a direction to cause an increase of the displacement u) is*

$$F = \frac{\mu_0 SN^2}{4} I_0 \left[\frac{I_2^2}{(d-u)^2} - \frac{I_1^2}{(d+u)^2}\right] .$$

The expression of the force can be linearized, obtaining

$$F = \frac{\mu_0 SN^2}{4} \left[aG_1G_3(u + G_2\dot{u}) - \frac{I_0}{d}u\right] .$$

The total stiffness and damping of the axial suspension are, respectively,

$$K_z = \frac{\mu_0 SN^2}{4} I_0 \left[aG_1G_3 - \frac{I_0}{d}\right] - k_z \,, \qquad C_z = \frac{\mu_0 SN^2}{4} I_0 aG_1G_3G_3 \,.$$

The second term in brackets is the open-loop stiffness, which is negative. It can cause the axial instability of the bearing, which, if the current is not controlled, is unstable.

To achieve a stable behavior in the axial direction, the stiffness must be positive. Assuming the following values for the current I_0 *and the gains,* $I_0 = 1\ A$, $G_1 = 10$, $G_2 = 3 \times 10^{-4}\ s$, *and* $G_3 = 0.9\ A/V$, *the values of the axial stiffness and damping are* $K_z = 58,100\ N/m$ *and* $C_z = 125\ Ns/m$.

The system is then axially stable. The axial natural frequency and the damping ratio are $\lambda_n = 269\ rad/s$ *and* $\zeta = 0.29$. *The force-displacement characteristic of the axial suspension is reported in Figure 8.25(b). The curve obtained from the linearized equation is compared with those obtained from the nonlinearized equation and the curve obtained when no constant current is supplied* ($I_0 = 0$).

8.6 Bearing alignment in multibearing rotors

If the rotor is supported on more than two bearings, the loads on the latter cannot be found without considering the deformations of both stator and rotor. Moreover, as in all statically undermined structures, the stress distribution and the loads in the supports depends on their exact placement.

A solution to this difficulty is to introduce concentrated flexibility in the structure to make it statically determined: In the case of a rotor, this can be easily implemented by splitting it in several spans and introducing flexible joints between them. This is easily done in lightly loaded long transmission shafts, which need to be supported in several points to raise (or lower, if needed, by soft mounting some supports) the critical speed. Examples are the propeller shafts of front-engine, rear-wheel drive motor cars and trucks and the transmission powering the tail rotor in helicopters. As an added bonus, the stressing of the shaft becomes independent from the deformations of the supporting structure.

However, this solution cannot be used in general. There are cases in which the very high torque to be transmitted rules out the use of flexible joints. A case is that of the connection between large steam turbines and alternators, where the shafts of the various machines are rigidly coupled to one another. The alignment of the various parts of the rotor must be done very accurately, and the supports must be set with great accuracy. This is, however, not enough: Even if the alignment was perfect, the stresses caused by self-weight may be large, and because they are seen as cyclic stresses with a frequency equal to the rotation frequency of the rotor, they may cause fatigue problems.

In these cases, it is a common practice to set the supports in such a way that the loads on the bearings and the stresses in the rotor are the same as in the case in which the various rotors are disconnected. This can be easily done by inserting shims under the supports of all rotors except one

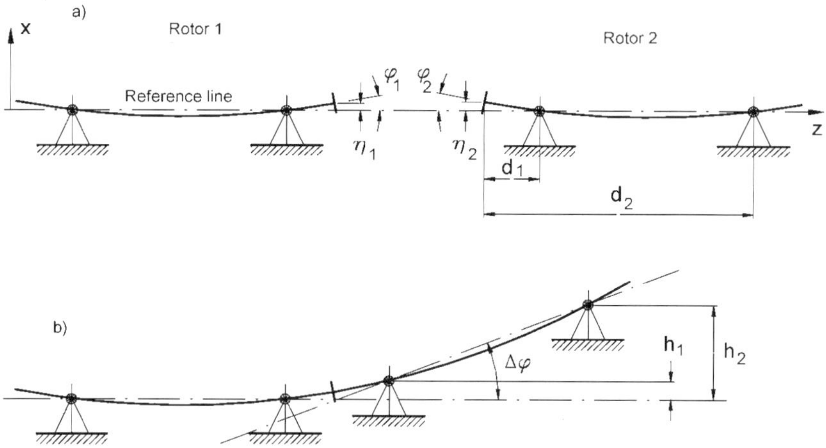

FIGURE 8.26. Aligning the supports of a rotor to account for static deformation. (a) Inflected shape of the two spans when the joint is disconnected; (b) inflection of the rotor after the supports of the second span have been raised. Note that all deformations are grossly exaggerated.

(which is taken as a reference) in such a way that the displacements and the rotations at the mating ends are the same, so as not to be disturbed by connecting the joints.

Consider the situation shown in Figure 8.26(a). The two rotors inflect under their own weight, and at the interface, the first one has a displacement η_1 and a rotation ϕ_1. The corresponding displacement and rotation of the second rotor are η_2 and ϕ_2. Note that in the figure, both displacements are positive, and ϕ_1 is positive and ϕ_2 is negative.

To compensate for the difference in displacement at the connection, the second rotor must be raised of

$$\Delta\eta = \eta_1 - \eta_2 , \tag{8.46}$$

and to realign the inflected configurations of the two parts, the second rotor must be rotated of an angle

$$\Delta\phi = \phi_1 - \phi_2 . \tag{8.47}$$

As usually the two rotations ϕ_1 and ϕ_2 have opposite signs, $\Delta\phi$ is the sum of the absolute values of the two angles.

Assuming that angles ϕ_i are small, the two supports of the second rotor must then be raised of

$$h_i = \Delta\eta + d_i\Delta\phi . \tag{8.48}$$

Part II

Advanced topics

9
Anisotropy of rotors and supports

One of the assumptions on which the analysis presented in the preceding chapters was based is the axial symmetry of either the stator or the rotor. Under this assumption, the equations of motion could be reduced to a set of differential equations with constant coefficients.

If on the contrary both the stator and the rotor lack axial symmetry, it is impossible to find a reference frame in which the equation of motion has constant coefficients. If the assumption of linearity is retained, a linear differential equation with periodic coefficients is obtained, which cannot in general be solved in closed form. The alternatives are then the numerical solution in the time domain or the approximation of the solution using a truncated series; however, linearity guarantees the existence of a unique solution. Moreover, a general solution, which can be found as the superimposition of the general solution of the homogenous equation and a particular integral of the complete equation, still exists.

9.1 Nonisotropic Jeffcott rotor

Consider a nonisotropic Jeffcott rotor running on nonisotropic bearings, and let k_x and k_y be the values of the stiffness of the bearing in the x- and y-directions and k_ξ and k_η the values of the stiffness of the rotor in directions of the rotating ζ- and η-axes (Figure 9.1).

Assuming as generalized coordinates the displacements in x- and y-direction of the bearing (x_1 and y_1) and those of the geometrical center

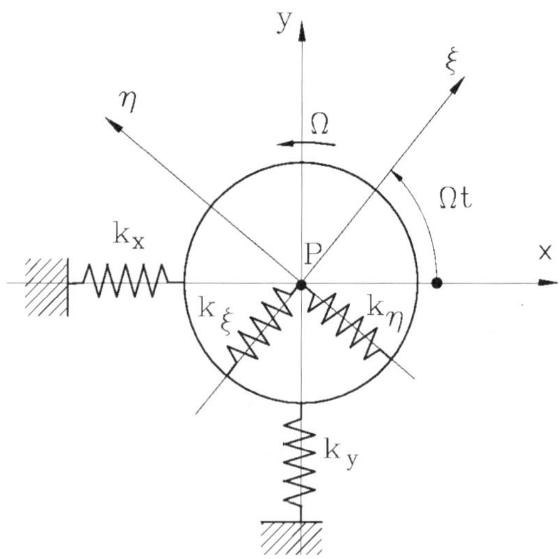

FIGURE 9.1. Sketch of a nonisotropic Jeffcott rotor running on nonisotropic bearings.

of the rotor (point C, x_2 and y_2), the equation of motion of the system is

$$\begin{bmatrix} 0 & 0 & 0 & 0 \\ 0 & 0 & 0 & 0 \\ 0 & 0 & m & 0 \\ 0 & 0 & 0 & m \end{bmatrix} \begin{Bmatrix} \ddot{x}_1 \\ \ddot{y}_1 \\ \ddot{x}_2 \\ \ddot{y}_2 \end{Bmatrix} + (\mathbf{K}_n + \mathbf{K}_r) \begin{Bmatrix} x_1 \\ y_1 \\ x_2 \\ y_2 \end{Bmatrix} = \\ = \begin{Bmatrix} 0 \\ 0 \\ f_{n_x} \\ f_{n_y} \end{Bmatrix} + m\epsilon\Omega^2 \begin{Bmatrix} 0 \\ 0 \\ \cos(\Omega t) \\ \sin(\Omega t) \end{Bmatrix}, \tag{9.1}$$

where

$$\mathbf{K}_n = \begin{bmatrix} k_x & 0 & 0 & 0 \\ 0 & k_y & 0 & 0 \\ 0 & 0 & 0 & 0 \\ 0 & 0 & 0 & 0 \end{bmatrix},$$

$$\mathbf{K}_r = \begin{bmatrix} \mathbf{R}^T \begin{bmatrix} k_\xi & 0 \\ 0 & k_\eta \end{bmatrix} \mathbf{R} & -\mathbf{R}^T \begin{bmatrix} k_\xi & 0 \\ 0 & k_\eta \end{bmatrix} \mathbf{R} \\ -\mathbf{R}^T \begin{bmatrix} k_\xi & 0 \\ 0 & k_\eta \end{bmatrix} \mathbf{R} & \mathbf{R}^T \begin{bmatrix} k_\xi & 0 \\ 0 & k_\eta \end{bmatrix} \mathbf{R} \end{bmatrix},$$

and the rotation matrix is

$$\mathbf{R} = \begin{bmatrix} \cos(\Omega t) & \sin(\Omega t) \\ -\sin(\Omega t) & \cos(\Omega t) \end{bmatrix}.$$

It is possible to introduce the mean and deviatoric stuffiness of both the rotor and the stator

$$\begin{array}{ll} k_n = \frac{1}{2}(k_x + k_y) \ , & k_{d_n} = \frac{1}{2}(k_x - k_y) \ , \\ k_r = \frac{1}{2}(k_\xi + k_\eta) \ , & k_{d_r} = \frac{1}{2}(k_\xi - k_\eta) \ , \end{array} \tag{9.2}$$

and the nondimensional ratios

$$\alpha_n = \frac{k_{d_n}}{k_n} = \frac{(k_x - k_y)}{(k_x + k_y)} \ , \qquad \alpha_r = \frac{k_{d_r}}{k_r} = \frac{(k_\xi - k_\eta)}{(k_\xi + k_\eta)} \ , \qquad \beta = \frac{k_n}{k_r} \ . \tag{9.3}$$

The stiffness matrix can then be easily computed, obtaining

$$\mathbf{K}_n + \mathbf{K}_r = k_r \begin{bmatrix} R & \alpha_r s & -1 - \alpha_r c & -\alpha_r s \\ \alpha_r s & S & -\alpha_r s & -1 + \alpha_r c \\ -1 - \alpha_r c & -\alpha_r s & 1 + \alpha_r c & \alpha_r s \\ -\alpha_r s & -1 + \alpha_r c & \alpha_r s & 1 - \alpha_r c \end{bmatrix} , \tag{9.4}$$

where

$$\begin{array}{ll} c = \cos(2\Omega t) \ , & R = \beta(1 + \alpha_n) + 1 + \alpha_r c \ , \\ s = \sin(2\Omega t) \ , & S = \beta(1 - \alpha_n) + 1 - \alpha_r c \ . \end{array}$$

The first two Equations (9.1) can be solved separately, yielding the following values for x_1 and y_1:

$$x_1 = \frac{\left[1 - \alpha_r^2 + \beta(1 - \alpha_n) + \beta\alpha_r(1 - \alpha_n)\cos(2\Omega t)\right] x_2 + \beta\alpha_r(1 - \alpha_n)\sin(2\Omega t)\, y_2}{(1 + \beta)^2 - \beta^2\alpha_n^2 - \alpha_r^2 - 2\beta\alpha_r\alpha_n\cos(2\Omega t)} ,$$

$$y_1 = \frac{\beta\alpha_r(1 + \alpha_n)\sin(2\Omega t)\, x_2 + \left[1 - \alpha_r^2 + \beta(1 + \alpha_n) - \beta\alpha_r(1 + \alpha_n)\cos(2\Omega t)\right] y_2}{(1 + \beta)^2 - \beta^2\alpha_n^2 - \alpha_r^2 - 2\beta\alpha_r\alpha_n\cos(2\Omega t)} . \tag{9.5}$$

By introducing Equation (9.5) into Equation (9.1), the following equation of motion containing only the coordinates of point C can be obtained:

$$\begin{aligned} &[1 + U\cos(2\Omega t)] \begin{bmatrix} m & 0 \\ 0 & m \end{bmatrix} \begin{Bmatrix} \ddot{x}_2 \\ \ddot{y}_2 \end{Bmatrix} + \\ &+ k_r \begin{bmatrix} R + S + T\cos(2\Omega t) & T\sin(2\Omega t) \\ T\sin(2\Omega t) & R - S - T\cos(2\Omega t) \end{bmatrix} \begin{Bmatrix} x_2 \\ y_2 \end{Bmatrix} = \\ &= [1 + U\cos(2\Omega t)] \left[\begin{Bmatrix} f_{n_x} \\ f_{n_y} \end{Bmatrix} + m\epsilon\Omega^2 \begin{Bmatrix} \cos(\Omega t) \\ \sin(\Omega t) \end{Bmatrix} \right] , \end{aligned} \tag{9.6}$$

where

$$R=\beta\,\frac{1+\beta\left(1-\alpha_n^2\right)-\alpha_r^2}{\left(1+\beta\right)^2-\beta^2\alpha_n^2-\alpha_r^2}\,,$$

$$S=\beta\,\frac{\alpha_n\left(1-\alpha_r^2\right)}{\left(1+\beta\right)^2-\beta^2\alpha_n^2-\alpha_r^2}\,,$$

$$T=\beta^2\,\frac{\alpha_r\left(1-\alpha_n^2\right)}{\left(1+\beta\right)^2-\beta^2\alpha_n^2-\alpha_r^2}\,,$$

$$U=-\frac{\beta\alpha_r\alpha_n}{\left(1+\beta\right)^2-\beta^2\alpha_n^2-\alpha_r^2}\,.$$

By multiplying the second Equation (9.6) by the imaginary unit i and adding it to the first one, and introducing the complex coordinate $r = x_2 + iy_2$ and its conjugate $\overline{r}$, the following equation is readily obtained:

$$\left(1+Ue^{2i\Omega t}+Ue^{-2i\Omega t}\right)\ddot{r}+\frac{k_r}{m}\left(Rr+S\overline{r}+Te^{2i\Omega t}\overline{r}\right)= \tag{9.7}$$

$$=\left(1+\frac{U}{2}e^{2i\Omega t}+\frac{U}{2}e^{-2i\Omega t}\right)\left(\frac{f_n}{m}+\epsilon\Omega^2e^{i\Omega t}\right)\,.$$

Equation (9.7) is a linear differential equation with periodic coefficients, with period equal to half of the rotation period of the rotor (i.e., equal to π/Ω), and it cannot be reduced to an equation with constant coefficients simply by resorting to a change of reference frame.

Linear equations with variable coefficients are studied by the Floquet theory: As in the case of linear equations with constant coefficients, they can be solved by adding a particular integral of the complete equation to the general solution of the homogeneous equation. However, no general method exists to reach such a solution. Because the coefficients are periodic functions of time with period π/Ω, the solution of the homogeneous equation associated with Equation (9.7) is of the type

$$r=r_1(t)e^{i\omega t}+r_2(t)e^{-i\overline{\omega}t},$$

where both vectors r_1 and r_2 are periodic functions of time with the same period as the coefficients, i.e., with fundamental frequency equal to 2Ω. The general solution is then the sum of a number of terms of the type mentioned earlier, each with its value of ω, plus a solution of the complete equation.

The study of the stability of the system can then be performed in the same way as in the case with constant r_1 and r_2.

Unknown functions r_1 and r_2 can be expressed by trigonometric series

$$r=\sum_{j=-\infty}^{\infty}\left(r_{1_j}e^{i(\omega+2j\Omega)t}+r_{2_j}e^{-i(\overline{\omega}+2j\Omega)t}\right). \tag{9.8}$$

By introducing Equation (9.5) into the homogeneous equation associated with Equation (9.7), the following algebraic equation is readily obtained:

$$\begin{gathered}\sum_{j=-\infty}^{\infty}\left(A_j r_{1_j}e^{i(\omega+2j\Omega)t}+A_j r_{2_j}e^{-i(\overline{\omega}+2j\Omega)t}+B_j r_{1_j}e^{i[\omega+2(j+1)\Omega]t}+\right.\\ +B_j r_{2_j}e^{-i[\overline{\omega}+2(j-1)\Omega]t}+B_j r_{1_j}e^{i[\omega+2(j-1)\Omega]t}+B_j r_{2_j}e^{-i[\overline{\omega}+2(j+1)\Omega]t}+\\ +C\overline{r}_{1_j}e^{-i(\overline{\omega}+2j\Omega)t}+C\overline{r}_{2_j}e^{i(\omega+2j\Omega)t}+\\ \left.+D\overline{r}_{1_j}e^{-i[\overline{\omega}+2(j-1)\Omega]t}+D\overline{r}_{2_j}e^{i[\omega+2(j+1)\Omega]t}\right)=0\ .\end{gathered} \tag{9.9}$$

By separately equating to zero the various terms of Equation (9.9), the following infinite set of algebraic equations is readily obtained:

$$\left[\begin{array}{ccccccccc} \dots & \dots & \dots & & & & & & \\ D & A_{-1} & C & B_0 & & & & & \\ B_{-2} & C & A_{-1} & D & B_0 & & & & \\ & B_{-1} & D & A_0 & C & B_1 & & & \\ & & B_{-1} & C & A_0 & D & B_1 & & \\ & & & B_0 & D & A_1 & C & B_2 & \\ & & & & B_0 & C & A_1 & D & B_2 \\ & & & & & B_1 & D & A_2 & C \\ & & & & & & \dots & \dots & \dots \end{array}\right]\times\left\{\begin{array}{c} \dots \\ r_{1_{-1}} \\ \overline{r}_{2_{-1}} \\ r_{1_0} \\ \overline{r}_{2_0} \\ r_{1_1} \\ \overline{r}_{2_1} \\ r_{1_2} \\ \dots \end{array}\right\}=\mathbf{0}\ , \tag{9.10}$$

where

$$\begin{array}{ll} A_j=-(\omega+2j\Omega)^2+\frac{k_r}{m}R\ , & C=\frac{k_r}{m}S\ , \\ B_j=-(\omega+2j\Omega)^2U\ , & D=\frac{k_r}{m}T\ . \end{array}$$

The homogeneous Equation (9.10) has solutions different from the trivial one only if the matrix of the coefficients is singular. An eigenproblem, similar to the one linked with the well-known Hill's infinite determinant, is obtained. The bandwidth of the determinant is five in the present case, and it is three in the latter. Approximate solutions for the eigenproblem can then be obtained by considering a limited array of m rows by m columns, centered on the matrices with subscript 0. Let these limited matrices be denoted by $\mathbf{Z}_m$, and let the number A_0 be $\mathbf{Z}_0$.

The first-order approximation can, for example, be obtained by considering matrix

$$\mathbf{Z}_1=\left[\begin{array}{cc} A_0 & C \\ C & A_0 \end{array}\right].$$

$m=0$	$m=1$	$m=2$	$m=3$	$m=4$	$m=5$	$m=6$
±0.6374	±0.5621	±0.5349	±0.5246	±0.5215	±0.5196	±0.5187
	±0.7046	±0.5941	±0.5484	±0.5368	±0.5292	±0.5258
		±0.6567	±0.5824	±0.5622	±0.5448	±0.5375
		±0.7467	±0.6557	±0.6058	±0.5659	±0.5540
			±0.7142	±0.6339	±0.5885	±0.5752
			±0.7670	±0.6870	±0.6424	±0.6103
				±0.7364	±0.6778	±0.6258
				±0.7743	±0.7171	±0.6628
					±0.7532	±0.6981
					±0.7793	±0.7322
						±0.7615
						±0.7816

TABLE 9.1. Nondimensional frequencies at standstill for various values of m.

Matrices $\mathbf{Z}_m$ contain terms in ω^2 and ω that are inside the terms in A_j and B_j: The relevant eigenproblem can this be written in the form

$$\det\left[-\mathbf{M}\omega^2+\mathbf{G}\omega+\mathbf{K}\right]=0\ , \tag{9.11}$$

where matrices **M**, **G**, and **K** have the same expression of the matrix in Equation (9.10), once the constants A_j, B_j, C, and D are given the values

$$\begin{array}{ll} A_j=1\ , & C=0\ , \\ B_j=U\ , & D=0\ , \end{array}$$

for **M**,

$$\begin{array}{ll} A_j=-4j\Omega\ , & C=0\ , \\ B_j=-4j\Omega U\ , & D=0\ , \end{array}$$

for **G**, and

$$\begin{array}{ll} A_j=-4j^2\Omega^2+\frac{k_r}{m}R\ , & C=\frac{k_r}{m}S\ , \\ B_j=-4j^2\Omega^2U\ , & D=\frac{k_r}{m}T\ , \end{array}$$

for **K**.

Note that in general matrices **G** and **K** are in this case neither positive definite nor symmetrical. As the eigenproblem is not in standard form, the size of the matrices must be doubled and the values of ω are the eigenvalues of the matrix

$$\begin{bmatrix} \mathbf{M}^{-1}\mathbf{G} & \mathbf{M}^{-1}\mathbf{K} \\ \mathbf{I} & \mathbf{0} \end{bmatrix} .$$

Example 9.1 *Consider a Jeffcott rotor with $\alpha_n=0.4$, $\alpha_r=0.4$, and $\beta=0.8$. By dividing the speeds and the frequencies by $\sqrt{k_r/m}$, and solving the eigenproblem with different values of m, the nondimensional frequencies at standstill reported in Table 9.1 can be found.*

$m=0$	$m=1$	$m=2$	$m=3$	$m=4$	$m=5$	$m=6$
±0.6374	±0.5621	±0.5609	±0.5616	±0.5616	±0.5616	±0.5616
	±0.7046	±0.7103	±0.7113	±0.7114	±0.7114	±0.7114
		±1.3588	±1.2909	±1.2890	±1.2886	±1.2886
		±2.6387	±1.4378	±1.4387	±1.4384	±1.4384
			±2.5615	±2.5612	±2.5616	±2.5616
			±2.7069	±2.7107	±2.7113	±2.7114
				±3.3588	±3.2909	±3.2890
				±4.6387	±3.4378	±3.4387
					±4.5615	±4.5612
					±4.7069	±4.7107
						±5.3588
						±6.6387

TABLE 9.2. Nondimensional frequencies at a nondimensional speed $\Omega/\sqrt{k_r/m} = 1$ for various values of m.

Note that the value computed with $m = 0$ does not coincide exactly with the natural frequency obtained using the average value of the stiffness $\omega/\sqrt{k_r/m} = 0.667$. When the speed vanishes, the various values correspond to the natural frequencies in the various positions of the rotor: Obviously there should be an infinity of natural frequencies, spanning from a minimum value (the natural frequency in the "softest" direction when the springs with minimum stiffness are aligned; in the example, $\omega/\sqrt{k_r/m} = 0.5164$) and a maximum value (in the example, $\omega/\sqrt{k_r/m} = 0.7888$).

The same computation performed at a nondimensional speed $\Omega/\sqrt{k_r/m} = 1$ yields the results reported in Table 9.2.

Note that in practice, there are only four distinct values of the natural frequency (in the example, $\omega/\sqrt{k_r/m} = \pm0.5616$ and ±0.7114), whereas the other values can be obtained by adding to them $\pm j\Omega$, with $j = 1, 2, 3, \ldots$. This explains the apparent inconsistency of the number of the eigenvalues, which tend to infinity when increasing the size of the matrix $\mathbf{Z}_m$: While obtaining a better precision, solutions whose real parts are equal to those already obtained plus a multiple of 2Ω are obtained.

The Campbell diagram and the decay rate plot computed using $m = 3$ and $m = 15$ are shown in Figure 9.2. In the figure labeled $m = 3$, only four branches of the plot are shown; from them all other branches can be obtained by simply adding multiples of the speed 2Ω.

Remark 9.1 *Three fields of instability can be detected from the decay rate plot. They are caused by the rotating anisotropy and can be avoided only by adding damping to the system.*

The response to a static force can be obtained by introducing a solution of the type

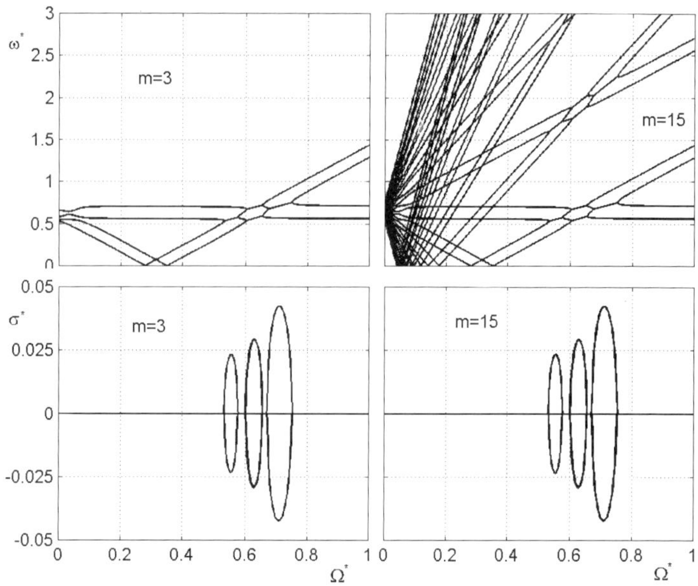

FIGURE 9.2. Campbell diagram and the decay rate plot for a Jeffcott rotor with $\alpha_n = 0.4$, $\alpha_r = 0.4$, and $\beta = 0.8$, computed using $m = 3$ and $m = 15$. Nondimensional frequency, decay rate, and speed $\omega^* = \mathrm{Re}(\sqrt{k_r/m}\omega)$, $\sigma^* = \mathrm{Im}(\sqrt{k_r/m}\omega)$, $\Omega^* = \sqrt{k_r/m}$.

$$r = \sum_{j=-\infty}^{\infty} r_j e^{2ij\Omega t} , \tag{9.12}$$

i.e.,

$$r = r_0 + \sum_{j=1}^{\infty} r_{1_j} e^{2ij\Omega t} + \sum_{j=1}^{\infty} r_{2_j} e^{-2ij\Omega t} , \tag{9.13}$$

into the complete Equation (9.7). In the latter equation, the mean response, the forward components, and the backward components are written separately.

In a similar way, the unbalance response can be obtained by introducing a solution of the type

$$r = \sum_{j=1}^{\infty} r_j e^{i(2j+1)\Omega t} ; \tag{9.14}$$

i.e.,

$$r = \sum_{j=1}^{\infty} r_{1_j} e^{i(2j+1)\Omega t} + \sum_{j=1}^{\infty} r_{2_j} e^{-i(2j-1)\Omega t} . \tag{9.15}$$

A number of harmonics (in theory, an infinity of them) is then present in the response. However, the present model does not include damping, and there is little interest in obtaining the undamped response, as the amplitudes of all harmonics have infinitely high peaks. If damping is introduced into the model, a further degree of freedom must be considered, because it is no more possible to solve the equation of motion for x_1 and y_1 as it was the case for the first two Equations (9.1) and the system has to be dealt with as shown below for multi-degrees-of-freedom systems.

9.2 Equation of motion for an anisotropic machine with many degrees of freedom

The mass and stiffness matrices, related to flexural behavior, of an anisotropic beam or spring element whose principal axes of inertia and elasticity lie in the xz- and yz-planes (hereafter designated by subscripts x and y), can be written in the form

$$\mathbf{M} = \begin{bmatrix} \mathbf{M}_x & \mathbf{0} \\ \mathbf{0} & \mathbf{M}_y \end{bmatrix}, \qquad \mathbf{K} = \begin{bmatrix} \mathbf{K}_x & \mathbf{0} \\ \mathbf{0} & \mathbf{K}_y \end{bmatrix}. \tag{9.16}$$

As usual, if the generalized coordinate for rotation in the yz-plane is $-\phi_x$ instead of ϕ_x, the matrices related to the xz- and yz-planes of an isotropic element are equal. In this way, the introduction of complex coordinates is straightforward.

When assembling the structure, assume that the global reference frame has the same z-axis as those of each element, but that the x-axes of the elements are rotated of an angle α with respect to the global reference frame. The rotation matrix is

$$\mathbf{R}' = \begin{bmatrix} \cos(\alpha)\,\mathbf{I} & \sin(\alpha)\,\mathbf{I} \\ -\sin(\alpha)\,\mathbf{I} & \cos(\alpha)\,\mathbf{I} \end{bmatrix}, \tag{9.17}$$

and the stiffness matrix in the global reference frame is

$$\mathbf{K}_g = \begin{bmatrix} \cos^2(\alpha)\mathbf{K}_x + \sin^2(\alpha)\mathbf{K}_y & \sin(\alpha)\cos(\alpha)(\mathbf{K}_x - \mathbf{K}_y) \\ \sin(\alpha)\cos(\alpha)(\mathbf{K}_x - \mathbf{K}_y) & \sin^2(\alpha)\mathbf{K}_x + \cos^2(\alpha)\mathbf{K}_y \end{bmatrix}. \tag{9.18}$$

By introducing the mean and deviatoric stiffness matrices of the elements

$$\mathbf{K}_m = \frac{1}{2}(\mathbf{K}_x + \mathbf{K}_y), \qquad \mathbf{K}_d = \frac{1}{2}(\mathbf{K}_x - \mathbf{K}_y),$$

Equation (9.18) can be written as

$$\mathbf{K}_g = \begin{bmatrix} \mathbf{K}_m + \mathbf{K}_d\cos(2\alpha) & \mathbf{K}_d\sin(2\alpha) \\ \mathbf{K}_d\sin(2\alpha) & \mathbf{K}_m - \mathbf{K}_d\cos(2\alpha) \end{bmatrix}. \tag{9.19}$$

If the element belongs to the rotor, angle α must be substituted by $\alpha+\theta$, or, in the case of constant spin speed equal to Ω, by $\alpha+\Omega t$.

All of the aforementioned considerations hold for mass and damping matrices and for elements other than beam or spring elements. Note that, in general, all mean and deviatoric matrices of structural elements are symmetrical, with the exception of the stiffness and damping matrices of the elements used to model lubricated journal bearings in linearized theories (see Section 8.4.2). Once all matrices of the elements have been obtained and expressed in the global reference frame, it is possible to assemble the various elements to obtain the matrices related to the whole structure. Obviously, the rotating elements must be assembled separately from the nonrotating elements.

Because of the presence of the deviatoric matrices (it is sufficient that a single element has a nonvanishing deviatoric matrix), the structure of the assembled matrices is more complex than that of Equation (9.16). For the stiffness matrix, it follows that

$$\mathbf{K}=\begin{bmatrix}\mathbf{K}_x & \mathbf{K}_{xy}\\ \mathbf{K}_{yx} & \mathbf{K}_y\end{bmatrix}. \tag{9.20}$$

Because of the presence of the coupling terms with xy- and yx-subscripts, a new definition of the mean and deviatoric matrices for the whole structure is needed

$$\begin{cases}\mathbf{K}_m=\dfrac{1}{2}(\mathbf{K}_x+\mathbf{K}_y)+i\dfrac{1}{2}(\mathbf{K}_{yx}-\mathbf{K}_{xy})\,,\\[2ex] \mathbf{K}_d=\dfrac{1}{2}(\mathbf{K}_x-\mathbf{K}_y)+i\dfrac{1}{2}(\mathbf{K}_{yx}+\mathbf{K}_{xy})\,.\end{cases} \tag{9.21}$$

Note that, except in the mentioned case of elements used for the linearized modeling of hydrodynamic bearings, the matrices with subscripts xy and yx are equal and the mean matrices are real. On the contrary, deviatoric matrices are, in general, complex. Using the complex-coordinate approach and the definitions of mean and deviatoric matrices given by Equation (9.21), the equation of motion describing the flexural behavior of a general system containing stationary elements and elements rotating at constant spin speed Ω can be shown to be [41]

$$\begin{aligned}&\mathbf{M}_m\ddot{\mathbf{q}}+(\mathbf{C}_m-i\Omega\mathbf{G})\dot{\mathbf{q}}+(\mathbf{K}_m-i\Omega\mathbf{C}_{r_m})\mathbf{q}+\mathbf{M}_{n_d}\ddot{\overline{\mathbf{q}}}+\\ &+\mathbf{M}_{r_d}e^{2i\Omega t}(\ddot{\overline{\mathbf{q}}}+2i\Omega\dot{\overline{\mathbf{q}}})+\mathbf{C}_{n_d}\dot{\overline{\mathbf{q}}}+\mathbf{C}_{r_d}e^{2i\Omega t}\dot{\overline{\mathbf{q}}}+\\ &+\mathbf{K}_{n_d}\overline{\mathbf{q}}+(\mathbf{K}_{r_d}-i\Omega\mathbf{C}_{r_d})e^{2i\Omega t}\overline{\mathbf{q}}=\mathbf{F}_n+\Omega^2\mathbf{F}_r e^{i\Omega t}\,.\end{aligned} \tag{9.22}$$

Matrices and vectors with subscript r are related to the rotating elements, and those with subscript n are related to the stator of the machine. The mean matrices without r or n subscripts are related to the whole system and are obviously the sum of the corresponding nonrotating and rotating mean matrices. The nonrotating force vector is related to static forces, and the rotating vector is related to forces that are stationary in a reference frame rotating at the spin speed Ω. Because the latter are usually unbalance forces, their magnitude is proportional to the square of the spin speed. Note that the complex conjugate of the vector of the generalized coordinates is present in the terms related to deviatoric matrices, and that all terms containing the deviatoric matrices of rotating elements have coefficients that are periodic functions of time, with periods equal to half the period of rotation.

Like the equation of motion for the nonisotropic Jeffcott rotor, studied in Section 5.1.1., Equation (9.22) is a linear differential equation with periodic coefficients. If all deviatoric matrices vanish, as with axisymmetrical systems, it reduces to the usual constant coefficient equation typical of isotropic rotors.

As already stated, linear equations with variable coefficients can be solved by adding a particular integral of the complete equation to the general solution of the homogeneous equation. However, no general method exists to reach such a solution, particularly for large sets of equations. Because the coefficients are periodic functions of time with period π/Ω, the solution of the homogeneous equation associated with Equation (9.22) is of the type

$$\mathbf{q} = \mathbf{q}_1(t)e^{i\omega t} + \mathbf{q}_2(t)e^{-i\overline{\omega}t},$$

where both vectors $\mathbf{q}_1$ and $\mathbf{q}_2$ are periodic functions of time with the same period, i.e., with fundamental frequency equal to 2Ω. The general solution is then the sum of a number of terms of the type mentioned earlier, each with its value of ω, plus a solution of the complete equation. The study of the stability of the system can then be performed in the same way as in the case with constant $\mathbf{q}_1$ and $\mathbf{q}_2$.

Unknown functions $\mathbf{q}_1$ and $\mathbf{q}_2$ can be expressed by trigonometric series

$$\mathbf{q} = \sum_{j=-\infty}^{\infty} \left(\mathbf{q}_{1_j} e^{i(\omega+2j\Omega)t} + \mathbf{q}_{2_j} e^{-i(\overline{\omega}+2j\Omega)t} \right). \tag{9.23}$$

By introducing Equation (9.23) into the homogeneous equation associated with Equation (9.22), the following algebraic equation is readily obtained:

$$\sum_{j=-\infty}^{\infty} \left(\mathbf{A}_j \mathbf{q}_{1_j} e^{i(\omega+2j\Omega)t} + \overline{\mathbf{B}}_j \mathbf{q}_{2_j} e^{-i(\overline{\omega}+2j\Omega)t} + \overline{\mathbf{C}}_j \overline{\mathbf{q}}_{1_j} e^{-i(\overline{\omega}+2j\Omega)t} + \right.$$
$$\left. + \mathbf{D}_j \overline{\mathbf{q}}_{2_j} e^{i(\omega+2j\Omega)t} + \overline{\mathbf{E}}_j \overline{\mathbf{q}}_{1_j} e^{-i[\overline{\omega}+2(j-1)\Omega]t} + \mathbf{E}_j \overline{\mathbf{q}}_{2_j} e^{i[\omega+2(j+1)\Omega]t} \right) = 0. \tag{9.24}$$

By separately equating to zero the various terms of Equation (9.24), the following infinite set of algebraic equations is readily obtained:

$$\begin{bmatrix} \dots & \dots & \dots & & & & & & \\ \mathbf{F}_{-2} & \mathbf{A}_{-1} & \mathbf{D}_{-1} & & & & & & \\ & \mathbf{C}_{-1} & \mathbf{B}_{-1} & \mathbf{E}_0 & & & & & \\ & & \mathbf{F}_{-1} & \mathbf{A}_0 & \mathbf{D}_0 & & & & \\ & & & \mathbf{C}_0 & \mathbf{B}_0 & \mathbf{E}_1 & & & \\ & & & & \mathbf{F}_0 & \mathbf{A}_1 & \mathbf{D}_1 & & \\ & & & & & \mathbf{C}_1 & \mathbf{B}_1 & \mathbf{E}_2 & \\ & & & & & & \mathbf{F}_1 & \mathbf{A}_2 & \mathbf{D}_2 \\ & & & & & & \dots & \dots & \dots \end{bmatrix} \times \begin{Bmatrix} \dots \\ \mathbf{q1}_{-1} \\ \overline{\mathbf{q}}\mathbf{2}_{-1} \\ \mathbf{q1}_0 \\ \overline{\mathbf{q}}\mathbf{2}_0 \\ \mathbf{q1}_1 \\ \overline{\mathbf{q}}\mathbf{2}_1 \\ \mathbf{q1}_2 \\ \dots \end{Bmatrix} = \{0\}, \tag{9.25}$$

where

$$\begin{aligned}
\mathbf{A}_j &= -(\omega + 2j\Omega)^2\mathbf{M}_m + \Omega(\omega + 2j\Omega)\mathbf{G} + i(\omega + 2j\Omega)\mathbf{C}_m + \mathbf{K}_m - i\Omega\mathbf{C}_{r_m}, \\
\mathbf{B}_j &= -(\omega + 2j\Omega)^2\mathbf{M}_m - \Omega(\omega + 2j\Omega)\mathbf{G} + i(\omega + 2j\Omega)\overline{\mathbf{C}}_m + \overline{\mathbf{K}}_m + i\Omega\overline{\mathbf{C}}_{r_m}, \\
\mathbf{C}_j &= -(\omega + 2j\Omega)^2\overline{\mathbf{M}}_{n_d} + i(\omega + 2j\Omega)\overline{\mathbf{C}}_{n_d} + \overline{\mathbf{K}}_{n_d}, \\
\mathbf{D}_j &= -(\omega + 2j\Omega)^2\mathbf{M}_{n_d} + i(\omega + 2j\Omega)\mathbf{C}_{n_d} + \mathbf{K}_{n_d}, \\
\mathbf{E}_j &= [-\omega^2 - 2\omega\Omega(2j-1) - 4j\Omega^2(j-1)]\overline{\mathbf{M}}_{r_d} + i(\omega + 2j\Omega)\overline{\mathbf{C}}_{r_d} + \\
&\qquad + \overline{\mathbf{K}}_{r_d} - i\Omega\overline{\mathbf{C}}_{r_d}, \\
\mathbf{F}_j &= [-\omega^2 - 2\omega\Omega(2j+1) - 4j\Omega^2(j+1)]\mathbf{M}_{r_d} + i(\omega + 2j\Omega)\mathbf{C}_{r_d} + \\
&\qquad + \mathbf{K}_{r_d} + i\Omega\mathbf{C}_{r_d}.
\end{aligned} \tag{9.26}$$

The homogeneous Equation (9.25) has solutions different from the trivial one only if the matrix of the coefficients is singular. The situation is very similar to that seen for the Jeffcott rotor, but here the bandwidth of the determinant is $3n$ instead of 5. Approximate solutions for the eigenproblem can then be obtained by considering a limited array of $2m$ rows by $2m$ columns, in terms of matrices (the actual number of rows and columns is $2\ m \times n$), centered on the matrices with subscript 0. Let these limited matrices be denoted by $\mathbf{Z}_m$, and let matrix $\mathbf{A}_0$ be $\mathbf{Z}_0$. It can readily be seen that the solution of the zero-order approximation involving matrix $\mathbf{Z}_0$ coincides with the solution of a symmetrical system having the mean properties of the actual system.

By inspecting Equations (9.26), it is clear that matrices $\mathbf{A}_j$ and $\mathbf{B}_j$ contain the mean properties of the system, matrices $\mathbf{C}_j$ and $\mathbf{D}_j$ are related to the unsymmetrical characteristics of the stator (hence, they vanish when the stator is symmetrical), and matrices $\mathbf{E}_j$ and $\mathbf{F}_j$ are related to the unsymmetrical properties of the rotor.

The first-order approximation obtained by considering that matrix

$$\mathbf{Z}_1 = \begin{bmatrix} \mathbf{A}_0 & \mathbf{B}_0 \\ \mathbf{C}_0 & \mathbf{D}_0 \end{bmatrix}$$

coincides with the solution of a system with isotropic rotor (with mean properties) running on an asymmetrical stator (with the actual properties).

The equation that allows the study of a system with a symmetrical stator and an asymmetrical rotor can be written as

$$\begin{bmatrix} \mathbf{B}_0 & \mathbf{E}_1 \\ \mathbf{F}_0 & \mathbf{A}_1 \end{bmatrix} \begin{Bmatrix} \overline{\mathbf{q}}_{2_0} \\ \mathbf{q}_{1_1} \end{Bmatrix} = \mathbf{0}. \tag{9.27}$$

The infinite set of Equations (9.25) and the corresponding nonhomogeneous set that includes the forcing functions represent a general model for rotating machinery. The solution of the eigenproblem, however, is not easy. Thc equations must be rearranged to obtain an eigenproblem in standard form, which results in a further doubling of the size of the problem, becoming $4mn$, and then a problem of a very large order must be faced. It must be noted that if $\text{Re}(\omega)$ is the real part of the solution of the eigenproblem, then $\text{Re}(\omega) + 2j\Omega$ and $\text{Re}(\omega) - 2j\Omega$ (for any value of j) is the real part of a solution. This explains the apparent inconsistency of the number of the eigenvalues, which tend to infinity when increasing the size of the matrix $\mathbf{Z}_m$: While obtaining a better precision, solutions whose real parts are equal to those already obtained plus a multiple of 2Ω are obtained. Vectors $\mathbf{q}_j$ with j greater than 1 usually add only a small ripple on the basic solution, which is given by the vectors with $j \leq 1$. Third-order approximations should, consequently, give results that accurately simulate the behavior of the actual system.

Solutions for nonhomogeneous problems, like those related to the response to static loading and unbalance, can be obtained in a similar way. In the former case, the solution can be expressed by the series

$$\mathbf{q} = \mathbf{q}_0 + \sum_{j=1}^{\infty} \mathbf{q}_{1_j} e^{2ij\Omega t} + \sum_{j=1}^{\infty} \mathbf{q}_{2_j} e^{-2ij\Omega t}. \tag{9.28}$$

The mean response, the forward components, and the backward components are written separately in Equation (9.28) . By using this expression of the deflected shape, the following infinite set of equations can be obtained from the equation of motion (9.25):

$$\begin{cases} \mathbf{A}_0\mathbf{q}_0 + \mathbf{C}_0\overline{\mathbf{q}}_0 + \mathbf{E}_1\overline{\mathbf{q}}_{1_1} = \mathbf{f}_n \ , \\ \mathbf{F}_0\overline{\mathbf{q}}_0 + \mathbf{A}_1\mathbf{q}_{1_1} + \mathbf{D}_1\overline{\mathbf{q}}_{2_1} = \{0\} \ , \\ \mathbf{C}_1\mathbf{q}_{1_1} + \mathbf{B}_1\overline{\mathbf{q}}_{2_1} + \mathbf{E}_2\mathbf{q}_{1_2} = \{0\} \ , \\ \mathbf{F}_1\overline{\mathbf{q}}_{2_1} + \mathbf{A}_2\mathbf{q}_{1_2} + \mathbf{D}_1\overline{\mathbf{q}}_{2_2} = \{0\} \ , \\ \cdots\cdots\cdots\cdots\cdots\cdots\cdots\cdots\cdots , \end{cases} \tag{9.29}$$

where the relevant matrices can be obtained from Equation (9.26) by setting $\omega = 0$. Here, again, a good approximation can be obtained by considering only the components $\mathbf{q}_0$, $\mathbf{q}_{11}$, and $\mathbf{q}_{21}$, i.e., by resorting to a reduced set formed by the first three equations (9.29) in which the term in $\mathbf{E}_2$ has been neglected. The size of the problem is thus reduced to $3n$ equations.

The response to an arbitrary unbalance distribution is similarly obtained by assuming a solution of the type

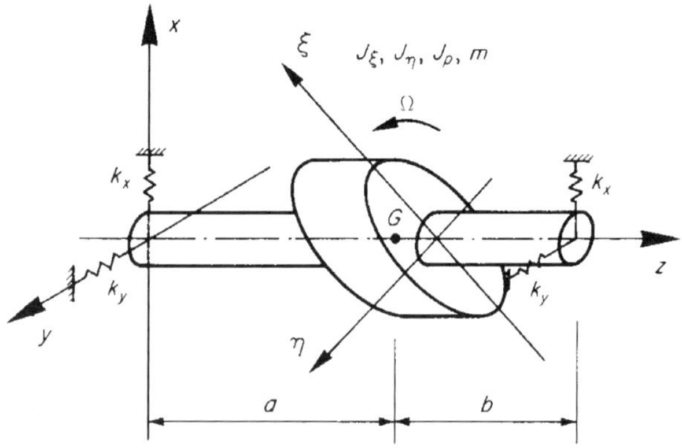

FIGURE 9.3. Sketch of an asymmetric rigid rotor on asymmetric elastic supports.

$$\mathbf{q} = \sum_{j=1}^{\infty} \mathbf{q}_{1_j} e^{i(2j+1)\Omega t} + \sum_{j=1}^{\infty} \mathbf{q}_{2_j} e^{-i(2j-1)\Omega t} . \tag{9.30}$$

By using the Expressions (9.30) for the deflected shape, the following infinite set of equations can be obtained from the equation of motion (9.25):

$$\begin{bmatrix} \mathbf{A}_0 & \mathbf{D}_0 & & & \\ \mathbf{C}_0 & \mathbf{B}_0 & \mathbf{E}_1 & & \\ & \mathbf{F}_0 & \mathbf{A}_1 & \mathbf{D}_1 & \\ & & \mathbf{C}_1 & \mathbf{B}_1 & \mathbf{E}_2 \\ & & \dots & \dots & \dots \end{bmatrix} \begin{Bmatrix} \mathbf{q}_{1_0} \\ \overline{\mathbf{q}}_{2_0} \\ \mathbf{q}_{1_1} \\ \overline{\mathbf{q}}_{2_1} \\ \dots \end{Bmatrix} = \Omega^2 \begin{Bmatrix} \mathbf{f}_r \\ \{0\} \\ \{0\} \\ \{0\} \\ \dots \end{Bmatrix} . \tag{9.31}$$

Here the relevant equations can be obtained from Equation (9.26) by setting $\omega = \Omega$. In this case, a good approximation can be obtained by considering only the first four equations in which the term in $\mathbf{E}_2$ has been neglected. The size of the problem thus reduces to $4n$ equations.

Example 9.2 *Anisotropic system with four degrees of freedom.*

Consider a rigid unsymmetrical rotor running on two identical anisotropic bearings (Figure 9.3). It can be modeled as a four degrees-of-freedom (two complex) system of the same type studied as Example 3.1 but with anisotropic properties added. Assume that the principal directions of both supports are the same, so that matrices $\mathbf{K}_{xy}$ and $\mathbf{K}_{yx}$ vanish. The only mean and deviatoric matrices that are not equal to zero are

$$\mathbf{M}_m = m\,\mathbf{M}^* = m \begin{bmatrix} 1 & 0 \\ 0 & J/ml^2 \end{bmatrix} , \quad \mathbf{M}_{rd} = m\,\Delta\mathbf{M}^* = m \begin{bmatrix} 1 & 0 \\ 0 & \Delta J/ml^2 \end{bmatrix} ,$$

$$\mathbf{K}_m = k\,\mathbf{K}^* = k \begin{bmatrix} 2 & \frac{a}{l} - \frac{b}{l} \\ \frac{a}{l} - \frac{b}{l} & \left(\frac{a}{l}\right)^2 + \left(\frac{b}{l}\right)^2 \end{bmatrix} , \quad \mathbf{K}_{nd} = \Delta k\,\mathbf{K}^* ,$$

$$\mathbf{G} = m\ \mathbf{G}^* = m \begin{bmatrix} 0 & 0 \\ 0 & J_p/ml^2 \end{bmatrix} ,$$

where l is the length of the shaft ($l = a + b$) and

$$J = \frac{1}{2}(J_\xi + J_\eta) , \quad \Delta J = \frac{1}{2}(J_\xi - J_\eta) ,$$

$$k = \frac{1}{2}(k_x + k_y) , \quad \Delta k = \frac{1}{2}(k_x - k_y) .$$

Equations (9.26) can be written in nondimensional form

$$\mathbf{A}_j = -\omega^{*2}\mathbf{M}^* + \Omega^*\omega^*(\mathbf{G}^* - 4j\mathbf{M}^*) + \Omega^{*2}(2j\mathbf{G}^* - 4j^2\mathbf{M}^*) + \mathbf{K}^*,$$

$$\mathbf{B}_j = -\omega^{*2}\mathbf{M}^* - \Omega^*\omega^*(\mathbf{G}^* + 4j\mathbf{M}^*) - \Omega^{*2}(2j\mathbf{G}^* + 4j^2\mathbf{M}^*) + \mathbf{K}^*,$$

$$\mathbf{C}_j = \mathbf{D}_j = \frac{\Delta k}{k}\mathbf{K}^*,$$

$$\mathbf{E}_j = [-\omega^{*2} - 2\omega^*\Omega^*(2j-1) - 4j\Omega^{*2}(j-1)]\Delta\mathbf{M}^*,$$

$$\mathbf{F}_j = [-\omega^{*2} - 2\omega^*\Omega^*(2j+1) - 4j\Omega^{*2}(j+1)]\Delta\mathbf{M}^*,$$

where the nondimensional spin and whirl speeds are $\Omega^* = \Omega/\sqrt{k/m}$ *and* $\omega^* = \omega/\sqrt{k/m}$.

The behavior of the system is thus controlled by five nondimensional parameters:

$$J/ml^2, \quad J_P/ml^2, \quad \Delta J/J, \quad \Delta k/k, \quad a/l \text{ (or } b/l).$$

Note that there are physical limits to some of these parameters: $0 < J_P/J < 2$, $-1 < \Delta J/J < 1$, $-1 < \Delta k/k < 1$, *and* $-1 < a/l < 1$. *Even more strict limitations are present because when* J_p/J *approaches its lower or upper limit, the system no more behaves like a rigid body.*

Consider a case with

$$J/ml^2 = 0.3, \quad J_p/ml^2 = 0.35, \quad \Delta J/J = 0.35, \quad \Delta k/k = 0.2, \quad a/l = 0.25.$$

The Campbell diagram and the decay rate plot computed using three harmonics in the series are reported in Figure 9.4. The presence of instability ranges, also caused by the lack of damping, is clear.

The orbits may be far from being circular or elliptical, as shown in the three-dimensional plot of the fourth mode shape at a nondimensional speed $\Omega^* = 1$ *reported in Figure 9.5. The orbits are reported for a time equal to one period of the fundamental harmonic. Not only the orbits are noncircular, but they are qualitatively much different from station to station of the rotor. In the case shown, the left bearing and the center of mass whirl on a less complicated pattern than the right bearing.*

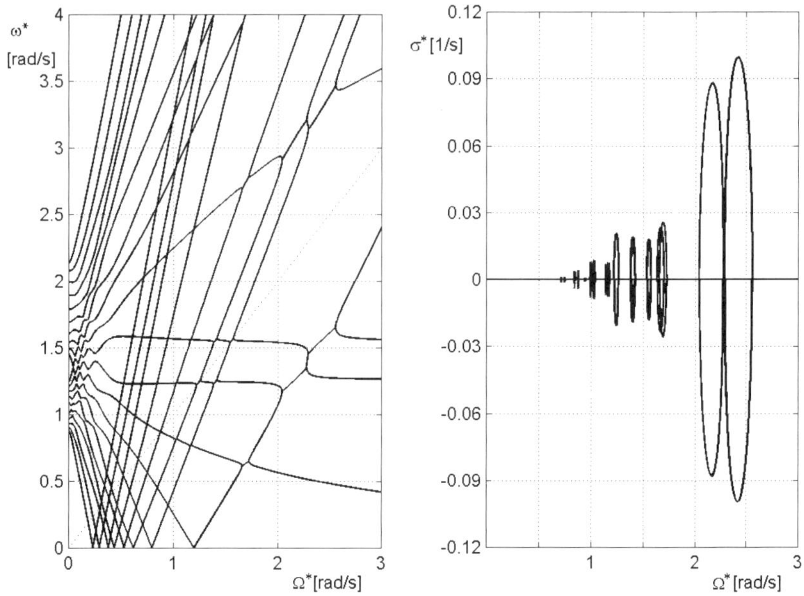

FIGURE 9.4. Campbell diagram and decay rate plot computed using three harmonics.

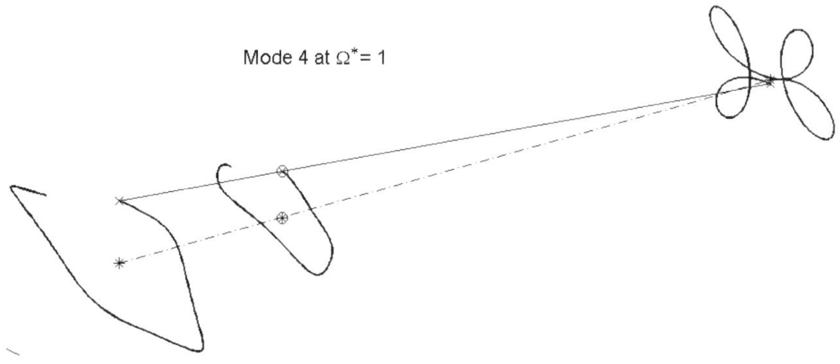

FIGURE 9.5. Orbits and mode shape of the fourth mode at a nondimensional speed $\Omega^* = 1$.

10
Nonlinear rotordynamics

Another of the assumptions on which the analysis seen in Part 1 was based is that of linearity. If this assumption is dropped, no general solution of the equations of motion can be achieved and the alternatives are the numerical solution by step-by-step integration in the time domain or the search for approximate solutions, using the classic approximation techniques typical on nonlinear dynamics.

The difference between the behavior of axi-symmetrical and nonisotropic systems is even larger in the case of nonlinear rotors than that already seen for linear ones. If the system is axially symmetrical, circular whirling is an exact solution for the unbalance response, although the nonlinearity of the system makes it possible for other solutions to exist. This differentiates the behavior of a nonlinear rotating system from that of a nonlinear oscillator, where no closed-form solutions can usually be found. The closed-form solutions (in some cases, a single solution, but in other cases, the existence of multiple solutions has been confirmed) constitute attractors, whose basins of attraction share the phase space with other possible solutions of different type, if and when they exist at all. The impossibility of demonstrating that a solution exists and that it is unique leads to the possible existence, even in those cases, of multiple solutions, with the jump phenomenon typical of nonlinear dynamics, polyharmonic, and even chaotic solutions.

On the contrary, if no axial symmetry exists, particularly if the nonrotating parts of the machine are nonisotropic, a situation similar to that characterizing nonlinear vibrating systems occurs: No closed-form solution of the equations of motion can be found. As already seen for the case of the linearized behavior of bearings, introducing a constant force acting in radial direction into the model of a geometrically isotropic system may cause a departure from isotropy.

Note that nonlinearity makes it impossible to define critical speeds, at least in the usual terms, and to plot the Campbell diagram; the study is then generally limited to the computation of the response of the system to given forcing causes, usually unbalance distributions. It is still possible to define critical speeds as those speeds at which the response of the system becomes very large, but the fact that these speeds depend on the particular unbalance distribution that has been assumed makes their definition of limited use. The nonlinearity of the problem makes it impossible to study separately the effects of static and dynamic loads and the free whirling of the system, and the dynamic behavior can be strongly influenced by the presence of static loads.

10.1 Nonlinear isotropic Jeffcott rotor

10.1.1 Equation of motion

The simplest model that can be used to study the flexural behavior of nonlinear rotors is a Jeffcott rotor, consisting of a point mass m connected to a massless shaft and running on massless bearings. The shaft and the nonrotating parts of the machine act on mass m with forces that are in general nonlinear functions of displacement and velocity, expressed in the nonrotating $Oxyz$ and rotating $O\xi\eta z$ reference frames, respectively (Figure 10.1). The angular velocity Ω is assumed to be constant.

Such a system has complete axial symmetry about the nominal rotation axis. The center P of mass m is not located exactly in the center C of the cross section of the shaft, and the eccentricity ϵ causes a static unbalance $m\epsilon$, which affects strongly the behavior of the system. The nonlinearity of the problem makes it impossible to study separately the response to rotating forces, as those caused by unbalance, to nonrotating forces, as those caused by self-weight, and the free whirling of the system.

The motion in the xy- and $\xi\eta$-planes can be studied using the complex coordinates defined by Equations (2.40) and (2.167)

$$r = x + iy \; , \quad \rho = \xi + i\eta = re^{-i\Omega t} \; .$$

The forces exerted on point C by both the rotating and the nonrotating parts of the machine can be subdivided into a restoring force $\vec{F}_r$, which has the direction of the displacement, and a damping force $\vec{F}_d$, which has the direction of the velocity. Owing to the axial symmetry, the absolute value of these forces depends in general only on the absolute values of the displacement and the velocity. As the expression of the velocity in the rotating frame is different from that in the fixed frame, the forces caused

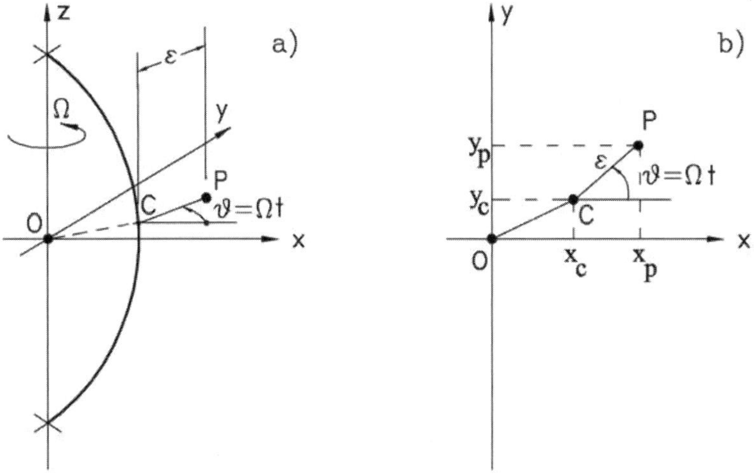

FIGURE 10.1. Isotropic nonlinear Jeffcott rotor. The system is the same as shown in Figure 2.2, but here the restoring and the damping forces are nonlinear functions of the displacement and the velocity, respectively.

by the nonrotating and rotating parts of the system can be expressed as

$$\begin{cases} \vec{F}_{rn} = -f_n(|r|, |\dot{r}|)\, r \ , \\ \vec{F}_{rr} = -f_r(|r|, |\dot{r} - i\Omega r|)\, r \ , \\ \vec{F}_{dn} = -g_n(|r|, |\dot{r}|)\, \dot{r} \ , \\ \vec{F}_{dr} = -g_r(|r|, |\dot{r} - i\Omega r|)\ (\dot{r} - i\Omega r) \ , \end{cases} \tag{10.1}$$

where f_n, f_r, g_n, and g_r are scalar functions of the amplitudes of the displacement and the velocity; the directions of the forces are given by vectors r, $\dot{r}$, and $\dot{r} - i\Omega r$. In the following, the restoring forces expressed by the first two Equations (10.1) are assumed to depend only on the displacement whereas the damping forces expressed by the other two equations are assumed to depend only on the speed. In these conditions, the former can be obtained from a potential energy, and the damping forces imply the existence of a Rayleigh-type dissipation function. The potential energy from which function $f_n(|r|)$ is obtained and the dissipation function for $g_n(|\dot{r}|)$ must be written in the fixed frame, and those used to obtain $f_r(|r|)$ and $g_r(|\dot{r} - i\Omega r|)$ must be written in the rotating one. A consequence of axial symmetry is that the elastic restoring forces caused by the shaft or the supporting structure play the same role, and a function $f(|r|) = f_n(|r|) + f_r(|r|)$ can be defined:

The above assumptions lead to the following nonlinear equation of motion written in the inertial coordinate frame

$$m\ddot{r} + g_n(|\dot{r}|)\dot{r} + g_r(|\dot{r} - i\Omega r|)(\dot{r} - i\Omega r) + f(|r|)r = m\epsilon\Omega^2 e^{i\Omega t} + F_n \ , \tag{10.2}$$

where F_n is an external nonrotating force. Without any loss of generality, ϵ can be assumed to be real, which amounts to assuming that at time $t = 0$, ξ-axis coincides with x-axis.

For a qualitative study of the effects of nonlinearity, functions f and g can be expanded in Taylor series and truncated. Owing to the larger importance of the nonlinearity of the restoring force than that of the damping force, $f(|r|)$ will be truncated at the first nonlinear term, whereas only the constant terms of $g_n(|\dot{r}|)$ and $g_r(|\dot{r} - i\Omega r|)$ will be considered. As the restoring force is an odd function of the displacement, the term linear in $|r|$ will not be considered and an expression of the restoring force of the same type of that included in the well-known Duffing's equation is assumed [see Equation (8.6)]

$$\vec{F}_r = -k(1 + \mu|r|^2)\, r\ , \tag{10.3}$$

or in terms of real coordinates,

$$\begin{cases} F_x = -xk\left[1 + \mu\left(x^2 + y^2\right)\right]\ , \\ F_y = -yk\left[1 + \mu\left(x^2 + y^2\right)\right]\ . \end{cases} \tag{10.4}$$

The stiffness k is usually positive, and in this case, the force may be said to be attractive, but there are cases in which it takes a negative value (repulsive force). The parameter of nonlinearity μ may be either positive or negative. The possible cases are reported in the following table:

	Attractive	Repulsive
Hardening	$k > 0$; $\mu > 0$	$k < 0$; $\mu < 0$
Softening	$k > 0$; $\mu < 0$	$k < 0$; $\mu > 0$

For attractive systems, see Figure 8.2.

The equation of motion (10.2) reduces to

$$m\ddot{r} + c\dot{r} + k(1 + \mu|r|^2)\, r - i\Omega c_r r = m\epsilon\Omega^2 e^{i\Omega t} + F_n\ , \tag{10.5}$$

where $c = c_n + c_r$.

Equation (10.2) can be written in nondimensional form by introducing the nondimensional coordinate and time

$$r^* = r/\epsilon\ , \quad \tau = t\sqrt{|k|/m}$$

and parameters

$$\mu^* = \mu\epsilon^2\ , \quad \Omega^* = \frac{\Omega}{\sqrt{|k|/m}}\ , \quad c_n^* = \frac{c_n}{2\sqrt{|k|m}}\ , \quad c_r^* = \frac{c_r}{2\sqrt{|k|m}}\ . \tag{10.6}$$

Ω^* is the ratio between the spin speed and the critical speed of the linearized system (in the case of attractive restoring force; for repulsive restoring force, the linearized system is unstable). The nondimensional amplitude

r^* has been obtained by dividing the amplitude by the eccentricity ϵ, as it is common for linear systems. If the eccentricity is nil, any other parameter having the dimensions of a length may be used.

The following nondimensional equation of motion is thus obtained:

$$r^{*\prime\prime} + 2c^* r^{*\prime} + \left(\pm 1 + \mu^* |r^*|^2\right) r^* - 2i\Omega^* c_r^* r^* = \Omega^{*2} e^{i\Omega^* \tau} + \frac{F_n}{|k|\epsilon} , \qquad (10.7)$$

where prime indicates differentiation with respect to the nondimensional time τ and the ($\pm$) sign becomes (+) for attractive systems and ($-$) for repulsive ones.

For the study of the stability of the circular whirling, it is useful to rewrite Equation (10.7) in the rotating coordinate frame $O\xi\eta z$ using the complex coordinate ρ (or, better, its nondimensional equivalent $\rho^* = \rho/\epsilon$)

$$\rho^{*\prime\prime} + 2\left[i\Omega^* + c^*\right] \rho^{*\prime} + (\pm 1 + \mu^* |r^*|^2)\rho^* + \left(-\Omega^{*2} + 2i\Omega^* c_n^*\right) \rho^* =$$
$$= \Omega^{*2} + \tfrac{F_n}{|k|\epsilon} e^{-i\Omega^* \tau} . \qquad (10.8)$$

Equations (10.2) and the following ones can be used to study the free behavior of the system (free circular whirling), the effect of a static load F_n (of weight, for example), or that of an eccentricity of mass m. The presence of nonlinear terms, however, makes it impossible to perform a general study by superimposing the various solutions and to obtain noncircular whirling by adding forward and backward whirling of different amplitudes.

10.1.2 Unbalance response – circular whirling

Amplitude and phase

Consider the case in which no nonrotating force F_n acts on the system. A possible solution to the equation of motion (10.2) is

$$r = r_0 e^{i(\Omega t + \Phi)} ; \qquad (10.9)$$

i.e.,

$$\rho = \rho_0 e^{i\Phi} , \qquad (10.10)$$

where $r_0 = \rho_0$ is a real, positive number and the phase Φ is defined in the range $[-\pi, 0]$. r_0 could have been assumed to be a complex number, instead of introducing the phase Φ.

By introducing Solution (10.9) into the equation of motion, the following nonlinear algebraic equation is readily obtained:

$$\left[-m\Omega^2 + i g_n(|\Omega r_0|)\Omega + f(|r_0|)\right] r_0 e^{i\Phi} = m\epsilon\Omega^2 . \qquad (10.11)$$

Note that Equation (10.11) is an exact solution of the nonlinear equation of motion and not just an approximation of the fundamental harmonics of the response. As a consequence, there are no higher order harmonics in the response. This does not mean that there cannot be higher order harmonics, because circular whirling is only one of the possible solutions; however, it will be demonstrated to be stable and has been found both experimentally and by numerical experimentation.

Remark 10.1 *Also in the nonlinear case, when the unbalance response is a circular synchronous whirling, rotating damping has no influence on the steady state response of the system.*

By separating the real and imaginary parts of Equation (10.11), the latter yields

$$\begin{cases} \left[-m\Omega^2 + f(|r_0|)\right] r_0 = m\epsilon\Omega^2 \cos(\Phi) \ , \\ g_n(|\Omega r_0|) r_0 = -m\epsilon\Omega^2 \sin(\Phi) \ . \end{cases} \tag{10.12}$$

The amplitude and the phase of the response can then be computed from the equations

$$\begin{cases} \left\{\left[-m\Omega^2 + f(|r_0|)\right]^2 + \left[g_n(|\Omega r_0|)\right]^2\right\} r_0^2 = m^2\epsilon^2\Omega^4 \ , \\ \Phi = \text{artg}\left\{\dfrac{g_n(|\Omega r_0|)}{\left[-m\Omega^2 + f(|r_0|)\right]}\right\} \ . \end{cases} \tag{10.13}$$

By introducing the simplified expressions for function f (quadratic) and g_n (constant) and writing the equations in nondimensional form, Equations (10.13) reduce to

$$\begin{cases} \mu^{*2} r_0^{*6} + 2\mu^* \left(\pm 1 - m\Omega^{*2}\right) {r_0}^4 + (1 + \Omega^{*4} + 4c_n^{*2}\Omega^{*2} \mp 2\Omega^{*2}) r_0^2 - \Omega^{*4} = 0 \ , \\ \Phi = \text{artg}\left\{\dfrac{2c_n^*\Omega^*}{-\Omega^2 + \left(\pm 1 - \mu^* r_0^{*2}\right)}\right\} \ , \end{cases} \tag{10.14}$$

where the upper signs are for attractive systems and the lower ones for repulsive systems.

The first Equation (10.14) is a cubic equation in ${r_0}^2$ and can be solved in closed-form.

Note that the equation yielding the amplitude of the orbit is nonlinear and, like all nonlinear equations may have multiple solution, or even no real solution at all. In the case of the cubic Equation (10.14), the real solutions are either 1 or 3. When more than one solution is possible, the system may settle in any of them, depending both on the operating parameters and on its history, and in some cases, it may shift from one to the other. This is a phenomenon usually referred to as the *jump*: In the case of nonlinear rotors, for instance, the amplitude of the orbit can increase while slowly increasing speed up to a point, when a sudden decrease of the amplitude is experienced. This is a downward jump, which is in many cases essential to achieving self-centering at high speed. On the contrary, when slowing down, the amplitude may suddenly jump up. The speed at which this occurs may be different (usually lower) than that at which a jump occurs during the run-up phase.

Backbone and limit envelope

The backbone can be obtained as the unbalance response of a nonlinear system with vanishing eccentricity and damping from the equation of motion referred to the rotating coordinate frame or, simply, by equating to zero the left-hand side of the first Equation (10.12)

$$\left[-m\Omega^2 + f(|r_0|)\right] r_0 = 0 \,, \tag{10.15}$$

or in nondimensional form, for a Duffing-type nonlinearity,

$$r_0^* = \sqrt{\frac{\Omega^{*2} \mp 1}{\mu^*}} \,. \tag{10.16}$$

The backbone defines, in the case of vibrating systems, the conditions for a sort of nonlinear resonance; here the situation on the backbone is similar to that occurring at a critical speed.

Remark 10.2 *A result that is similar to the classic solution of Duffing's equation for vibrating systems is so obtained. Some important differences, however, must be mentioned: Ω here is the spin speed and not a circular frequency, and the excitation is proportional to Ω^2. In the case of Duffing's equation, such a solution is only a first approximation of the fundamental harmonic of the response, whereas here the solution gives the whole response. This can be related to the fact that in the present case, the nonlinear element does not oscillate along the force-displacement characteristic but it rotates, maintaining a given deformation. Most considerations regarding multiple solutions and the jump phenomenon, however, hold. These considerations hold in general, regardless of the particular law $f(|z|)$.*

Remark 10.3 *In the case of linear systems, the solution in terms of r^* is independent from the eccentricity, although this clearly does not hold in the present case. When ϵ tends to zero, also μ^* vanishes and r^* tends to the solution of the linearized system, if it exists at all.*

The limit envelope can be obtained simply by introducing $\Phi = -\pi/2$ into the second Equation (10.12)

$$g_n(|\Omega r_0|) r_0 = m\epsilon\Omega^2 \,; \tag{10.17}$$

i.e., in nondimensional form for the case with g_n constant,

$$r_0 = \frac{\Omega^*}{2c_n^*} \tag{10.18}$$

The intersection point between backbone and limit envelope gives a good estimate of the point at which the jump phenomenon occurs.

Stability of the circular synchronous whirling

The equation of motion (10.7) (without the external nonrotating force F_n, or the equation corresponding to Equation (10.2), but written in rotating coordinates) can be written in the form of a set of four first-order differential equations in the state space

$$\mathbf{z}' = \mathbf{f}(\mathbf{z}) , \tag{10.19}$$

where $\mathbf{z} = \begin{bmatrix} \xi^{*\prime} & \eta^{*\prime} & \xi^* & \eta^* \end{bmatrix}^T$ is the state vector, ξ^* and η^* are the nondimensional rotating coordinates, and $\mathbf{f}$ is a vector function of $\mathbf{z}$

$$\mathbf{f} = \begin{bmatrix} -2c^* & 2\Omega^* & \Omega^* \mp 1 & 2\Omega^* c_n^* \\ -2\Omega^* & -2c^* & -2\Omega^* c_n^* & \Omega^* \mp 1 \\ 1 & 0 & 0 & 0 \\ 0 & 1 & 0 & 0 \end{bmatrix} \mathbf{z} + \tag{10.20}$$

$$+ \begin{Bmatrix} -\mu^* \xi^* \left(\xi^{*2} + \eta^{*2}\right) \\ -\mu^* \eta^* \left(\xi^{*2} + \eta^{*2}\right) \\ 0 \\ 0 \end{Bmatrix} + \begin{Bmatrix} \Omega^{*2} \\ 0 \\ 0 \\ 0 \end{Bmatrix} .$$

The study of the stability in the neighborhood of an equilibrium condition (stability in the small), characterized by the state variables $\mathbf{z} = \mathbf{z}_E$ (by definition of equilibrium point $\mathbf{f}(\mathbf{z}_E) = 0$), can be performed by expressing the state vector as

$$\mathbf{z} = \boldsymbol{\delta} + \mathbf{z}_E , \tag{10.21}$$

where $\boldsymbol{\delta}$ is a small displacement from the equilibrium state (in the state space). The state equation

$$(\boldsymbol{\delta} + \mathbf{z}_E)' = \mathbf{f}(\boldsymbol{\delta} + \mathbf{z}_E) \tag{10.22}$$

can then be linearized by expanding function $\mathbf{f}(\boldsymbol{\delta} + \mathbf{z}_E)$ in Taylor series in the neighborhood of $\mathbf{z}_E$ and truncating the series after the second term, yielding

$$\boldsymbol{\delta}' = \mathbf{J}(\mathbf{z}_E)\boldsymbol{\delta} , \tag{10.23}$$

where $\mathbf{J}(\mathbf{z}_E)$, the Jacobian matrix of function $\mathbf{f}(\mathbf{z})$ computed in the equilibrium state $\mathbf{z}_E$, acts as the dynamic matrix of the linearized system.

The explicit expression of the Jacobian matrix $\mathbf{J}(\mathbf{z})$ is

$$\begin{bmatrix} -2c^* & 2\Omega^* & \Omega^* \mp 1 - \mu^* \left(3\xi^{*2} + \eta^{*2}\right) & 2\Omega^* c_n^* - 2\mu^* \xi^* \eta^* \\ -2\Omega^* & -2c^* & -2\Omega^* c_n^* - 2\mu^* \xi^* \eta^* & \Omega^* \mp 1 - \mu^* \left(\xi^{*2} + 3\eta^{*2}\right) \\ 1 & 0 & 0 & 0 \\ 0 & 1 & 0 & 0 \end{bmatrix} . \tag{10.24}$$

The motion *in the small* can be expressed in the form

$$\boldsymbol{\delta} = \boldsymbol{\delta}_0 e^{s^*\tau} , \tag{10.25}$$

where s^* are the Liapunov exponents. They can be found by solving the eigenproblem

$$[\mathbf{J}(\mathbf{z}_E) - s^*\mathbf{I}]\, \boldsymbol{\delta}_0 = \mathbf{0} . \tag{10.26}$$

The characteristic equation can be expressed in nondimensional form as

$$\begin{gathered} s^{*4} - 2J_{11}s^{*3} + (J_{11} + J_{12} - J_{13} - J_{24})\, s^{*2} + \\ + [J_{11}(J_{13} + J_{24}) + J_{12}(J_{14} + J_{23})]\, s^* + J_{13}J_{24} - J_{14}J_{23} = 0 . \end{gathered} \tag{10.27}$$

Equation (10.27) can be solved in closed-form, and the nondimensional eigenvalues s^* can be obtained. If any one of the real parts of the eigenvalues is positive, the system is unstable following Liapunov criterion; if all real parts of s^* are negative the system is asymptotically stable. When the real part of at least one s^* is equal to zero, the system undergoes undamped oscillations.

Remark 10.4 *Although rotating damping does not affect the amplitude of the unbalance response, it enters into the equation yielding the Liapunov exponents and can be a critical factor in assessing the stability. This feature was already present in the linear system.*

Attractive systems

The linear part of the restoring force is actually restoring, because it tends to pull point C (see Figure 10.1) toward the axis of rotation, and the upper signs shown in the previous equations apply. Depending on the sign of μ^*, the system can be hardening (μ^* positive) or softening (μ^* negative).

As the unbalance response of the system depends on four nondimensional parameters, it is impossible to summarize all possible cases in a small number of plots. However, typical results can be obtained for different parameter combinations and general considerations can be drawn.

The results obtained for two hardening attractive and two softening attractive systems are reported in Figure 10.2. The nondimensional amplitude r_0^* of the orbit of point C is plotted against the nondimensional spin speed Ω^*; solid lines denote stable solutions, whereas dashed lines imply instability. Also, the backbone of the response and the limit envelope have been plotted. The results are typical and qualitatively very similar to those usually linked with Duffing oscillators, with the obvious and already mentioned difference that in the latter case, they are only first approximations of the response. Increasing the nonlinear parameter μ^*, the slope of the backbone decreases, as usual.

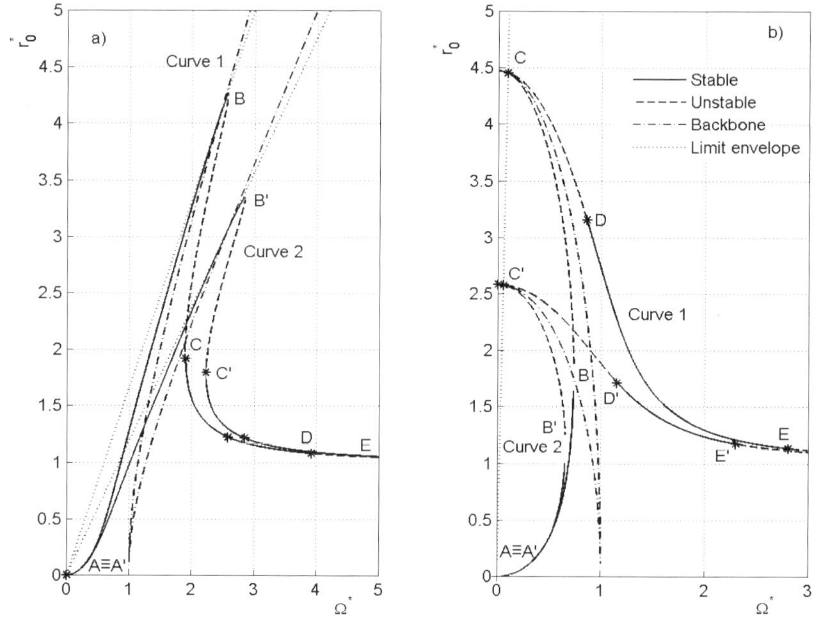

FIGURE 10.2. Unbalance response of attractive nonlinear Jeffcott rotors of the (a) hardening and (b) softening type. (a) Curve 1: $c_n^* = 0.3$, $c_r^* = 0.15$, $\mu^* = 0.3$; Curve 2: $c_n^* = 0.42$, $c_r^* = 0.15$, $\mu^* = 0.6$. The dashed line DE is superimposed to the solid line C'E. (b) Curve 1: $c_n^* = 0.01$, $c_r^* = 0.005$, $\mu^* = -0.05$; Curve 2: $c_n^* = 0.01$, $c_r^* = 0.005$, $\mu^* = -0.15$.

Consider an acceleration of a hardening attractive rotor from standstill, which occurs at a rate small enough to allow the use of the model based on steady-state whirling (the angular acceleration is so small that at any instant, the motion can be approximated with a steady-state motion with the corresponding parameters). The amplitude grows along the path AB in figure 10.2(a). When the conditions of point B are reached, any further speed increase will cause the amplitude to *jump down* to conditions on the line CE. Then the rotor self-centers going on toward E until instability is reached and no circular whirling is possible anymore. When spinning down, the rotor follows line EC up to point C. Any further decrease of speed causes the amplitude to *jump up* to a point on line AB, to proceed then toward point A. Note that the maximum amplitude reached during speeding up is larger than that obtained during spinning down.

In the case of hardening systems, to obtain the jump, the nonrotating damping ratio c_n^* must be larger than a minimum value, which increases with increasing nonlinearity parameter μ^*; otherwise, the backbone and the limit envelope do not intersect each other. This effect is shown in Figure 10.3. Branches AB and A'B' of the curves are stable. The self-centered solution, when it exists, is stable only up to a certain speed (branch CD). Above this threshold of instability, the circular whirling is unstable in the small (branch DE). This does not imply necessarily instability in the large for the system: Attractors different from equilibrium points may develop in this region. The unstable solutions belonging to the interval BC, B'C' are obviously responsible for the jump.

The differences with the usual results obtained for damped Duffing's equation are that the peak amplitude occurs now at the right of the backbone instead of at the left (the same thing happens in linear systems), and for low values of the damping, the curves do not close. This suggests a greater difficulty in experiencing the jump phenomenon and, consequently, self-centering.

It is easily predictable that self-centering is much more difficult in the case of nonlinear rotors than in the linear case, and that higher nonrotating damping is needed to work in the supercritical range, because the jump, which is needed to obtain the self-centered configuration, takes place when the energy supplied by the forcing function is not sufficient to sustain the motion with higher amplitude. In rotating systems, the forcing function is caused by unbalance, and the amount of energy supplied by the centrifugal field is large. This can be seen from the shape of the limit envelope, which may not cross the backbone.

Also in the case of the two softening attractive systems with different nonlinear parameters μ^*, the response is similar to that of a Duffing oscillator of the corresponding type; in that case, however, the approximations in some parts of the response curves of the latter are unacceptable whereas in the present case, an exact response is again obtained. Also, in this case, an increase of μ^* leads to a decrease in the backbone slope.

The small values adopted for damping parameters c_n^* and c_r^* cause the instability of both the branches at the left and the right of the nonlinear resonance peak (BC, CD, and B'C', C'D'), up to the crossing of the backbone. In the corresponding frequency ranges BD and B'D', no standard jump phenomenon may occur and the basin of attraction of stable circular whirling solutions vanishes. In all of the mentioned unstable ranges, as well as at speeds beyond the instability threshold corresponding to points E and E', only numerical integration of the equation of motion can give information about the possibility of stability in the large and on the nature of the attractors.

Remark 10.5 *The values for μ^* used to plot the responses are large, which can mean either a very large unbalance or a very large nonlinearity. The figures reported here are aimed to show the response of a Jeffcott rotor, which behaves in a much nonlinear way, either because of its very nonlinear nature or because its large unbalance causes it to operate in a very nonlinear part of its characteristic.*

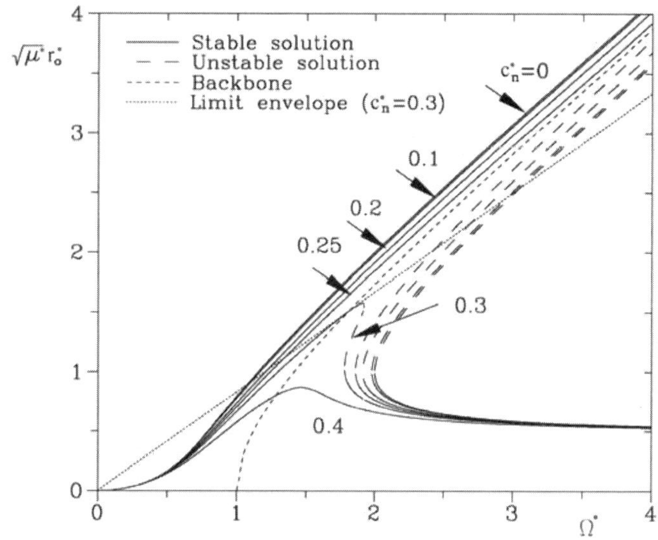

FIGURE 10.3. Nondimensional response of a nonlinear hardening attractive Jeffcott rotor, with various values of nonrotating damping. If damping is too low, the backbone and the limit envelope do not intersect, and no jump is possible. The plot has been obtained for a nondimensional nonlinear parameter $\mu^* = 0.25$ and various values of the nondimensional damping c_n^*.

Repulsive systems

Nonlinear rotors with repulsive behavior have been rarely considered in the literature; however, the total restoring force acting on the rotor of an electric motor running on lightly preloaded ball bearings is of the type here described as repulsive because it results from the combination of an unstabilizing effect, the so-called unbalanced magnetic pull of the electromagnetic field, and a stabilizing reaction of the bearings. If the latter are lightly preloaded, the linear term caused by the first effect overcomes the linear term of the bearing force. In such systems in fact, the centered static equilibrium configuration $r_0^* = 0$ is unstable in the small and no linearized analysis is possible. In the case of hardening repulsive systems, there is another equilibrium condition that, at least at standstill, is stable while softening repulsive systems are unstable also in the large: Because the latter are of little practical interest, they will not be studied any longer.

The unbalance response and stability in the small of some different damped and undamped hardening repulsive Jeffcott rotors with different values of μ^* are studied in Figure 10.4. Figure 10.4(a) is representative of systems with $\mu^* < 1$; in this case, if the system is undamped, three equilibrium positions can always be found. On the contrary, if $\mu^* > 1$ [Figure

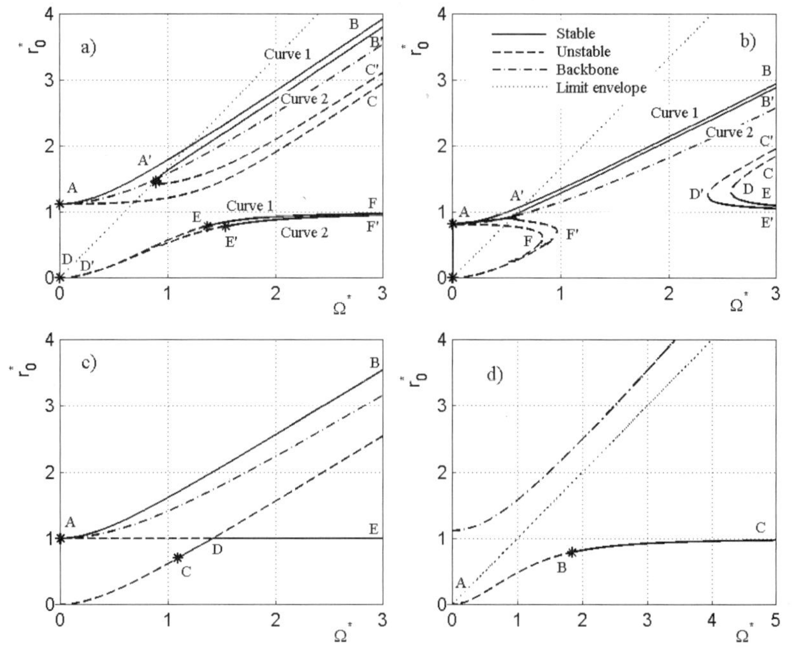

FIGURE 10.4. Unbalance response of repulsive nonlinear Jeffcott rotors of the hardening type. (a) Curve 1: undamped ($c_n^* = c_r^* = 0$) system with $\mu^* = 0.8$; Curve 2: $c_n^* = 0.3$, $c_r^* = 0.01$, $\mu^* = 0.8$. (b) Curve 1: undamped ($c_n^* = c_r^* = 0$) system with $\mu^* = 1.5$; Curve 2: $c_n^* = 0.3$, $c_r^* = 0.01$, $\mu^* = 1.5$. (c) undamped system with $\mu^* = 1$. (d) $c_n^* = 0.5$, $c_r^* = 0.01$, $\mu^* = 0.8$.

(10.4b)] there is a spin-speed range where the undamped system has only one equilibrium configuration (frequency range FD).

For small values of Ω^*, the number of solutions is modified by the introduction of damping, which leads to a spin-speed range where only an unstable configuration exists (Figures 10.4(a) and (b), $\Omega^* < \Omega_A^*$). If the nonrotating damping is small, the intersection between the backbone and the limit envelope occurs in A'; this excludes the possibility of a connection between the curves A'B' and A'C' [Figure 10.4(a)] or A'B' and D'C' [Figure 10.4(b)] for speed values $\Omega^* > \Omega_A^*$. The system cannot experience a jump during a slow acceleration phase and achieve spontaneously the self-centered configuration.

Figure 10.4(c) shows the case of an undamped hardening repulsive system with nonlinear parameter $\mu^* = 1$. It can be considered as a limiting case between the undamped systems of Figure 10.4(a) and Figure 10.4(b). If damping was present, a path similar to that of Figure 10.4(c) would have been obtained with $\mu^* > 1$. In this case, an increase of c_n^* from 0 leads to a decrease of the distance between points F' and D' of Figure 10.4(b). If the nonrotating damping is high enough to avoid an intersection between backbone and limit envelope [Figure 10.4(d)], only one solution can be found for every value of Ω^*. In this case, the self-centered solution represents the only possible synchronous motion.

Noncircular whirling

When circular synchronous whirling is possible, point C moves along a circumference in xy-plane, a solution that in the rotating $\xi\eta$ frame corresponds to a fixed point. By resorting to a state-space approach, circular synchronous whirling corresponds, in the phase space related to the rotating coordinates, to a fixed point attractor. As already stated, such attractor, when exists, is only one of the possible attractors of the system. When no point attractor exists, the motion may be stable in the large and other types of attractors can be possible. Steady-state solutions that can be represented in the rotating frame by limit cycles and *strange*, i.e., *chaotic*, attractors are then possible.

As no general solution of the equation of motion can be found, the study of noncircular whirling can be performed only through numerical experimentation by integrating numerically in time the equation of motion. Even in the simplest case of a nonlinear Jeffcott rotor modeled using Equation (10.7), a large number of numerical experiments needs to be performed since the time history of the system is defined by four parameters c_n^*, c_r^*, μ^*, and Ω^* and by the four initial conditions ξ_0^*, η_0^*, $\xi_0'^*$, and $\eta_0'^*$. Only a few typical results will be reported in the following pages: three values of each one of parameters c_n^*, c_r^* and μ^*, corresponding to a hardening attractive, a softening attractive, and a hardening repulsive system, were chosen, and each case was studied at various values of the speed Ω^*. In all cases, the system has been considered to be initially at rest in the origin of the state space ($\xi_0^* = \eta_0^* = \xi_0'^* = \eta_0'^* = 0$).

Hardening attractive system. The numerical integration has been performed using the same values of the parameters of Figure 10.2(a), Curve 1. The trajectories in the $\xi^*\eta^*$-plane for six values of the speed are reported in Figure 10.5. All trajectories have been plotted in the same scale except that in Figure 10.5(f), where a smaller scale is used. Three speed ranges in which the trajectories tend to different attractors can be identified.

- For $0 < \Omega^* \leq 1.966$, after the transient motion, the system falls on the stable attractor, a stable focus, which corresponds to the higher amplitude solutions of Figure 10.2(a) [$\Omega^* = 1$, Figure 10.5(a)]. Increasing Ω^* the trajectory initially tends, during the transient motion, to the unstable solution located on line BC of Figure 10.2(a) (a saddle point), to reach finally the stable focus [$\Omega^* = 1.966$, Figure 10.5(b)]. The origin of the phase space lies within the domain of attraction of the non self-centered solution.

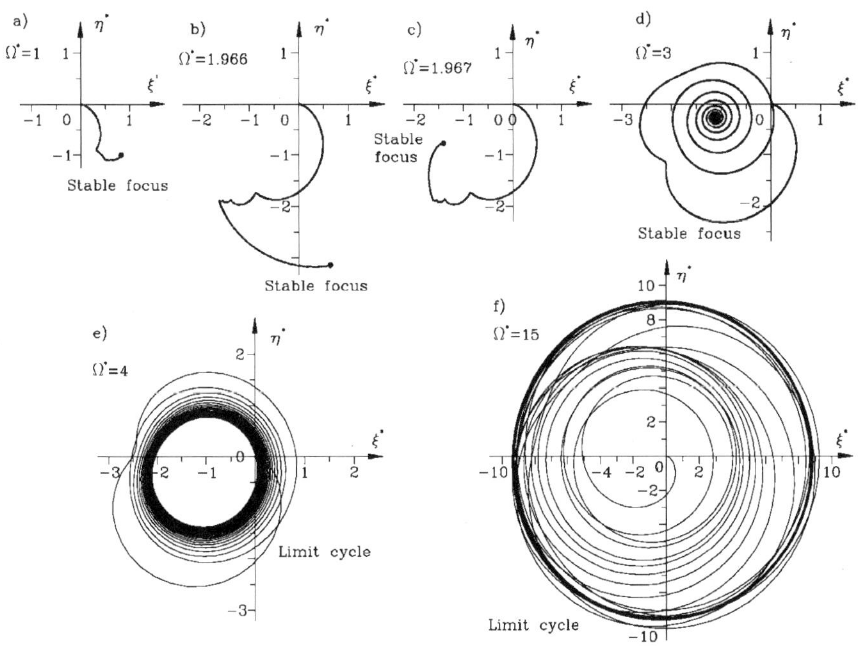

FIGURE 10.5. Trajectories in the $\xi^*\eta^*$-plane of an attractive nonlinear Jeffcott rotor of the hardening type with $c_n^* = 0.3$, $c_r^* = 0.15$, $\mu^* = 0.3$. In all cases the initial conditions are $\xi_0^* = \eta_0^* = \xi_0'^* = \eta_0'^* = 0$. The numerical integration was performed over a time at least sufficient to perform 10,000 revolutions of the rotor; the duration of the numerical experiment is always very long in comparison with the time needed to extinguish all transient motion.

- For $1.967 \leq \Omega^* < 4$, the system reaches the self-centered solution [line CD Figure 10.2(a)]. For the lower values of Ω^*, the self-centered condition is reached without oscillation, although the system tends initially to the saddle point [$\Omega^* = 1.967$, Figure 10.5(c)]. For higher values of Ω^* the system self-centers with an oscillatory motion [$\Omega^* = 3$, Figure 10.5(d)], which is decreasingly damped as the speed increases and nears the threshold of instability. The origin of the phase space lies within the domain of attraction of the self-centered solution.

- For $\Omega^* \geq 4$, the speed is higher than the threshold of instability and no point attractor exists. The trajectory tends to a limit cycle, whose amplitude increases with Ω^* [Figures 10.5(e)] and 10.5(f)].

When multiple solutions exist, the solution that is actually obtained depends on the initial conditions. In the case of a Duffing-type hardening attractive restoring force, when more than one solution is present, they are three: A larger one, with a phase laying in the field $0 > \Phi > -90°$, which

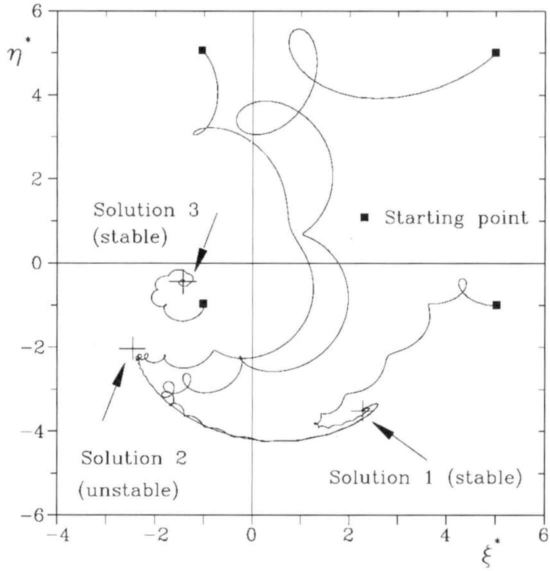

FIGURE 10.6. Trajectories in the rotating plane starting from four different points, one belonging to the domain of attraction of the lowest solution and three to the other domain. Values of parameters: $\Omega^* = 2$, $\mu^* = 0.2$, $c_n^* = 0.2$, $c_r^* = 0.03$.

is stable, an intermediate, unstable one (in the state space it is a saddle point) and a smaller, stable solution with phase $-90° > \Phi > -180°$. As already shown, the smallest solution (the self-centered one) is stable only up to a certain speed, which depends essentially on the ratio between the nonrotating and rotating damping.

An example of trajectories starting from different points in the $\xi^*\eta^*$-plane (different values of ξ_0^* and η_0^*) but all with $\xi_0'^* = \eta_0'^* = 0$ are shown in Figure 10.6. The figure has been obtained in conditions yielding three stationary solutions, two of which are stable.

The trajectories tend to the stable solutions, even if a certain attraction is also felt toward the saddle point, which is unstable. By integrating the equations of motion of the system for different values of the speed, using as starting positions different pairs of values of coordinates ξ^*, η^*, the domains of attraction of the various solutions at different speeds can be obtained. The results for some values of the speed are reported in Figure 10.7. The figure was obtained using the same parameter values as for Figure 10.3 and integrating numerically the equations of motion 81,000 times for each value of the speed until a stable equilibrium position was found. In some cases, a few hundred steps were sufficient; in others, the motion had to be followed for thousands of steps.

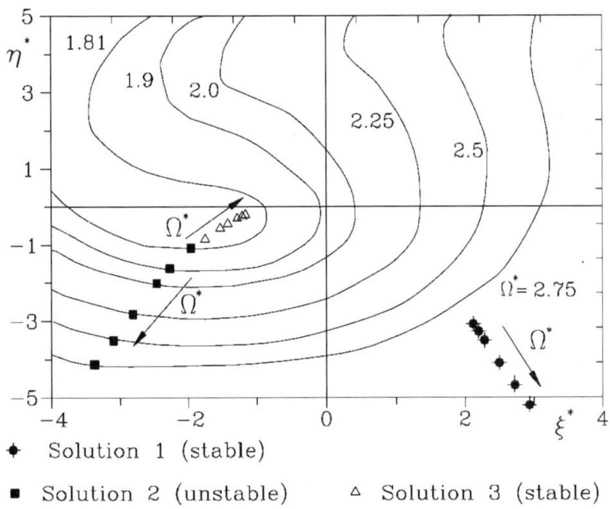

FIGURE 10.7. Basins of attraction in the $\xi^*\eta^*$-plane of the equilibrium positions at different values of the speed computed by numerically integrating the equations of motion; the same values of the parameters as in Figure 10.3.

The two domains of attraction are clearly well defined, and the unstable solution lies on the separatrix of the two attraction domains. At low speed, when the self-centered solution just starts existing, its domain of attraction is very small. By increasing the speed, the domain of attraction of the self-centered solution grows, and at the speed at which the jump occurs, it extends for the whole plane. The physics of the phenomenon consequently does not show any critical dependence on the initial conditions, and chaotic behavior was never encountered.

Softening attractive system. The parameters of the system are the same used for plotting curve 2 in Figure 10.2(b). The trajectories in the $\xi^*\eta^*$ plane for six values of the speed are reported in Figure 10.8. Two velocity ranges in which the system behaves in a completely different way have been found.

- For $0 < \Omega^* \leq 0.6104$, after the transient motion dies out, the system falls on the stable attractor corresponding to line A'B' of Figure 10.2(b). Some examples of transient motion are reported in Figures 10.8, from (a) to (d). The trajectories show a lobed pattern, and the number of lobes increases with Ω^*. The overall trajectory has a polygonal shape, until the fixed point attractor is reached.

- For $\Omega^* \geq 0.6105$, the trajectories of the system always diverge [Figures 10.8(e) and 10.8(f)]: The system is unstable in the large as well as in the small. Note that for $1.15 < \Omega^* \leq 2.32$, there is a stable so-

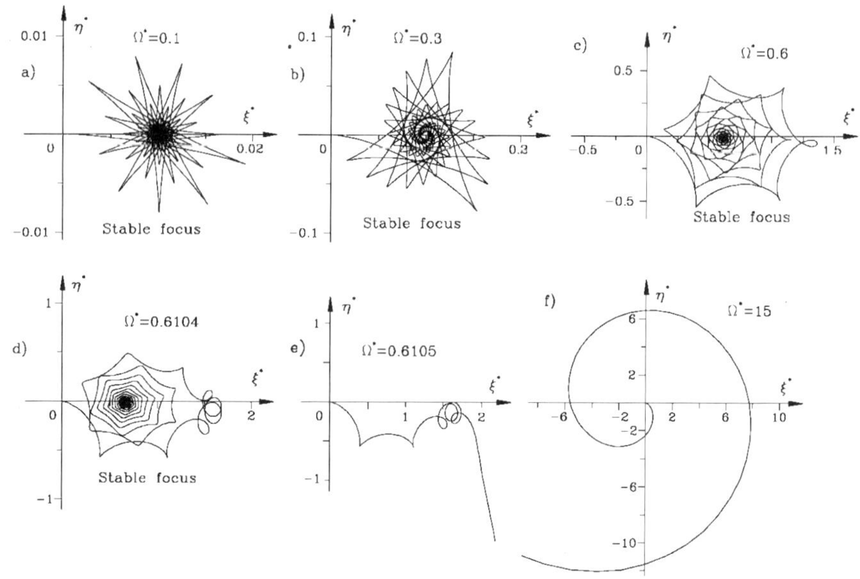

FIGURE 10.8. Trajectories in the $\xi^*\eta^*$-plane of an attractive nonlinear Jeffcott rotor of the softening type with $c_n^* = 0.01$, $c_r^* = 0.005$, $\mu^* = -0.15$. In all cases, the initial conditions are $\xi_0^* = \eta_0^* = \xi_0'^* = \eta_0'^* = 0$.

lution corresponding to branch D'E' in Figure 10.2(b). However, all numerical simulations reported in Figure 10.8 were performed starting from the origin of the state space, and evidently such point is outside the domain of attraction of the stable solution. A section of the domain of attraction in the $\xi^*\eta^*$-plane has been obtained for a value of the speed $\Omega^* = 1.8$. The domain of attraction of the stable circular whirling was seen to be a regular zone about the stable solution, but its boundaries are uncertain as there is an ample zone in which the convergence or the divergence are extremely slow. In some trajectories, even after 12,000 revolutions, there is no clue to understand whether the point attractor will eventually be reached or the trajectory will end diverging.

Hardening repulsive system. The parameters of the system are the same used for plotting Curve 2 in Figure 10.4(b). The trajectories in the $\xi^*\eta^*$-plane for nine values of the speed are reported in Figure 10.9. Seven velocity ranges in which the system behaves in a different way have been found.

- For $0 < \Omega^* \leq 0.555$, the trajectory tends to a limit cycle that is circular at very low speed to become increasingly deformed with increasing Ω^* [Figures 10.9(a) and 10.9(b)]. Increasing Ω^* the system

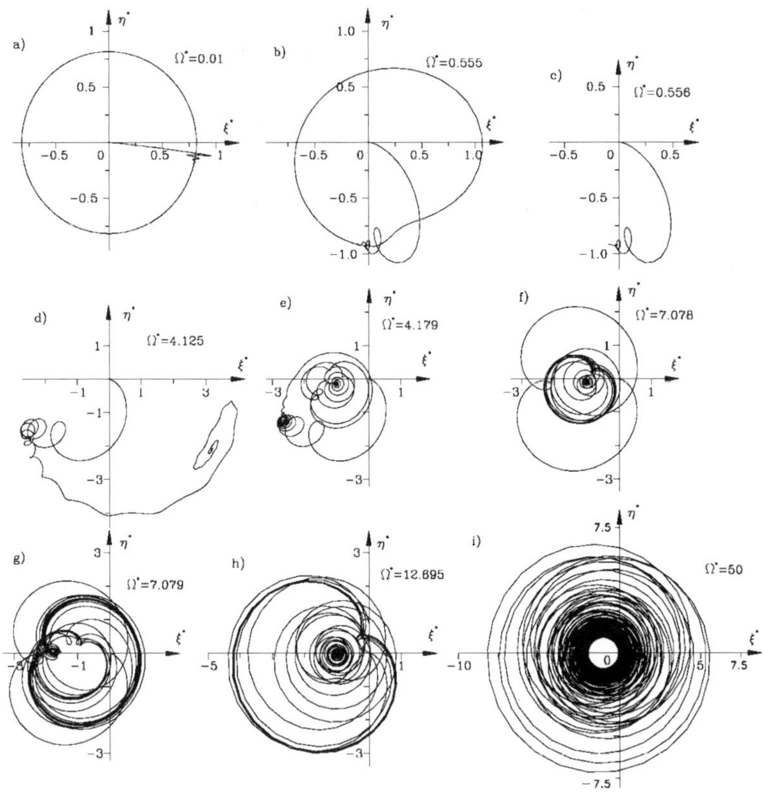

FIGURE 10.9. Trajectories in the $\xi^*\eta^*$-plane of a repulsive nonlinear Jeffcott rotor of the hardening type with $c_n^* = 0.3$, $c_r^* = 0.01$, $\mu^* = 1.5$. In all cases, the initial conditions are $\xi_0^* = \eta_0^* = \xi_0'^* = \eta_0'^* = 0$.

tends to remain for a longer time near a point that will become the equilibrium position for higher values of Ω^*.

- For $0.556 \leq \Omega^* \leq 4.125$, the attractor coincides with the stable equilibrium point corresponding to the higher amplitude solutions of Figure 10.4(b), Curve 2. The length and the complexity of the transient motion increase with increasing Ω^* [Figures 10.9(c) and 10.9(d)].

- For $4.125 < \Omega^* \leq 4.179$, the trajectory keeps on falling on the same stable attractor, but at the beginning of the transient motion, the system tends to a saddle-point [Figure 10.9(d)].

- For $\Omega^* = 4.179$, the domain of attraction of the self-centered solution includes the origin of the phase space: The phase projection passes through the saddle-point and falls on that stable point attractor on the lower branch of Curve 2, Figure 10.4(b) [Figure 10.9(e)].

- For $4.179 \leq \Omega^* \leq 7.078$, the trajectory at the end falls on the attractor corresponding to the self-centered equilibrium position, but for $\Omega^* \approx 7.078$, initially the system tends to a limit cycle, which is unstable [Figure 10.9(f)].
- For $7.078 \leq \Omega^* < 12.695$, the system, after some revolutions on the same unstable limit cycle mentioned above, reaches a new limit cycle that is stable [Figure 10.9(g)].
- For $12.695 \leq \Omega^* < 43.9$, after the transient motion, the system reaches the attractor corresponding to the self-centered equilibrium point [Figure 10.9(h)].
- For $\Omega^* > 43.9$, the system tends to a limit cycle [Figure 10.9(i)].

The system shows a complex behavior, characterized by attractors of both the fixed point and the limit cycle type. In some cases, the transient to reach the limit cycle is long, and the attractor has an intricate shape. However, never strange attractors were found.

Free circular whirling

The free response of the system can be computed by solving the homogeneous Equation (10.2), modeling a perfectly balanced Jeffcott rotor. A possible solution is also in this case a circular whirling, although not synchronous. It can be expressed in the form

$$r = r_0 e^{st} \,. \tag{10.28}$$

Remark 10.6 *This solution is again an exact solution of the equation of motion.*

By introducing the solution for circular free whirling into the homogeneous equation of motion, the following algebraic equation is obtained:

$$ms^2 + g_n(|sr_0|) + g_r(|sr_0 - i\Omega r_0|)(r_0 - i\Omega r_0) + f(|r_0|)r_0 = 0 \,. \tag{10.29}$$

In the case of linear damping, it reduces to

$$\left\{ms^2 + f(|r_0|) + sc_n + (s - i\Omega)c_r\right\} r_0 = 0 \,. \tag{10.30}$$

Equation (10.30) has a solution different from the trivial solution $r_0 = 0$ only if the expression in braces is equal to zero. An equation in s is so obtained, which yields the frequency and the decay rate of the free whirling

$$\begin{aligned} \sigma &= \mathrm{Re}\,(s) = -\frac{c}{2m} \pm \frac{1}{\sqrt{2}}\sqrt{-\Gamma + \sqrt{\Gamma^2 + \left(\frac{\Omega c_r}{m}\right)^2}}\,, \\ \omega &= \mathrm{Im}\,(s) = \pm\frac{1}{\sqrt{2}}\sqrt{\Gamma + \sqrt{\Gamma^2 + \left(\frac{\Omega c_r}{m}\right)^2}}\,, \end{aligned} \tag{10.31}$$

where

$$\Gamma = \frac{f(|r_0|)}{m} - \frac{c^2}{4m^2},$$

or for a Duffing-type restoring force,

$$\Gamma = \frac{k(\pm 1 + \mu |r_0|^2)}{m} - \frac{c^2}{4m^2}.$$

Equation (10.31) is very similar to the corresponding expression for the linear case, with the difference that here the whirl speed is a function of the amplitude of the motion. As is obvious for a nonlinear system, the Campbell diagram loses any meaning. A three-dimensional plot in which the amplitude $|r_0|$ is reported as a function of Ω and ω can, however, be introduced. The Ω,ω–plane of the tridimensional plot coincides with the Campbell diagram of the linearized system. The second Equation (10.31) defines a surface expressing all of the possible conditions of free whirling; it can be considered as the backbone of the system at varying spin speed Ω. The intersection of the surface with the $|r_0|, \omega$–plane expresses the relationship linking the frequency of free circular vibration with the amplitude at standstill: It is then the backbone of the nonrotating system for a whirling mode in the xy-plane.

One of the mentioned plots has been reported in nondimensional form in Figure 10.10. The figure has been plotted for a system with a Duffing-type hardening attractive restoring force with $c_n^* = c_r^* = 0.3$. The Campbell diagram of the linearized system coincides with that plotted in Figure 10.3 (curve for the corresponding value of the damping ratio). The stability threshold of the linearized system is at $\Omega^* = 2$.

The intersection of the surface with the plane of equation $\omega = \Omega$ gives the conditions for free synchronous whirling: It then coincides with the backbone of the unbalance response shown in Figure 10.3.

From the first Equation (10.31), it is possible to obtain the condition for stability

$$\Omega < \sqrt{\frac{f(|z_0|)}{m}} \left(1 + \frac{c_n}{c_r}\right). \tag{10.32}$$

The threshold of stability depends on the amplitude, as shown in Figure 10.10, where the unstable part of the surface is dashed. If the spin speed Ω is lower than the threshold of instability of the linearized system, the motion is always stable and the amplitude of free whirling decays to zero. On the contrary, when the linearized analysis shows an unstable behavior in the small, the nonlinear effects reduce the instability, and the result is a motion with growing amplitude until the border separating the full lines from the dotted lines is reached. The amplitude corresponding to these conditions is that of a sort of limit cycle that constitutes an attractor for all free whirl motions. The amplitude of the limit cycle is a function of the speed and grows with increasing Ω.

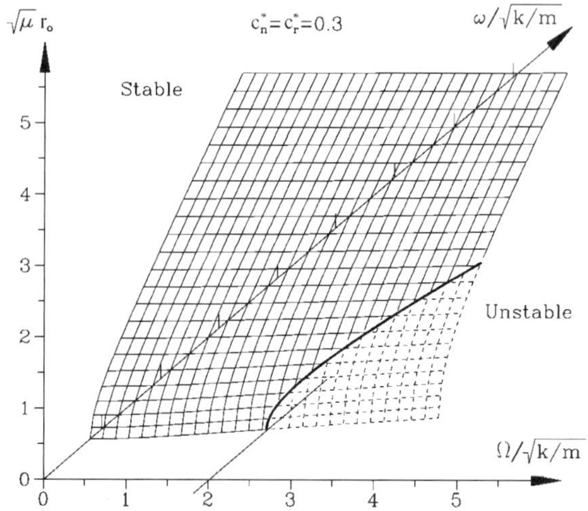

FIGURE 10.10. Tridimensional plot for the study of the free whirling of the nonlinear Jeffcott rotor with a Duffing-type hardening attractive nonlinearity studied in Figure 10.3.

Note that the nonlinearity of the system ensures that the stability condition for the unbalance response will not coincide with the stability condition of the free whirling, i.e., of the perfectly balanced system.

Some solutions found in the literature in which chaotic behavior of an axi-symmetrical rotor has been found are available. However, a key factor that can trigger chaotic behavior of rotating systems seems to be the lack of axial symmetry, either because of geometric or material anisotropy or to the presence of mechanisms such as bearing clearance or static loading, which make an isotropic system operate in an offset position. At any rate, it must be stressed that the domains of attraction should be plotted in phase space, which in the current case has four dimensions (two positions and two velocities) and not in the space of the configurations as in Figure 10.7.

10.2 Nonlinear isotropic Jeffcott rotor running on nonsymmetric supports

Consider the Jeffcott rotor studied in the previous section, but add a nonlinear restoring force caused by two nonlinear spring systems acting, respectively, in xz- and yz-planes. Note that although in the case of linear supports, the stiffness varies in the xy-plane following an elliptical pattern, in the case of nonlinear supports, more complicated patterns are obtained and the fact that the elastic characteristics of the supports in xz- and yz-

planes are equal does not mean that the supports are isotropic. Owing to the lack of axial symmetry of the system, the restoring and the nonrotating damping forces do not have the same directions as the displacement and the velocity. Rotating forces are in general still expressed by the second and fourth Equations (10.1), whereas for nonrotating forces, the following equations hold:

$$\left\{\begin{array}{c} F_x \\ F_y \end{array}\right\}_{rn} = -\left\{\begin{array}{c} f_x(x,\dot{x})x \\ f_y(y,\dot{y})y \end{array}\right\} , \quad \left\{\begin{array}{c} F_x \\ F_y \end{array}\right\}_{dn} = -\left\{\begin{array}{c} g_x(x,\dot{x})\dot{x} \\ g_y(y,\dot{y})\dot{y} \end{array}\right\} , \tag{10.33}$$

where subscripts r and d refer to the restoring and damping force, respectively. Moreover, the forces in xz- and yz-planes are assumed to act independently from each other.

The above assumptions lead to the following nonlinear equation of motion written in the inertial coordinate frame $Oxyz$:

$$\begin{cases} m\ddot{x} + g_x(\dot{x})\dot{x} + g_r(|\dot{r} - i\Omega r|)(\dot{x} + \Omega y) + f_r(|r|)x + f_x(x)x = \\ \qquad = m\epsilon\Omega^2 \cos(\Omega t) + F_{nx} \ , \\ m\ddot{y} + g_y(\dot{y})\dot{y} + g_r(|\dot{r} - i\Omega r|)(\dot{y} - \Omega x) + f_r(|r|)y + f_y(y)y = \\ \qquad = m\epsilon\Omega^2 \sin(\Omega t) + F_{ny} \ . \end{cases} \tag{10.34}$$

Note that, in order to simplify the notation, restoring forces are explicitly written as functions of the displacements only and damping forces of the velocity; there is, however, no difficulty in introducing a more complex dependence into the equation. Note also that rotating forces are functions of the modulus of the displacement $|r| = \sqrt{x^2 + y^2}$ and of the velocity, whereas nonrotating forces are functions of their components in x- and y-directions.

In a way that is similar to what has been done above for axi-symmetrical systems, functions f will be assumed to be of the Duffing type (they depend only on the displacements and contain only a constant and a quadratic term), whereas functions g are assumed to be constant.

Equation (10.34) reduces then consequently to

$$\begin{cases} m\ddot{x} + c_x\dot{x} + c_r\Omega y + k_r(\pm 1 + \mu_r|r|^2)x + k_x(\pm 1 + \mu_x x^2)x = \\ \qquad = m\epsilon\Omega^2 \cos(\Omega t) + F_{nx} \ , \\ m\ddot{y} + c_y\dot{y} - c_r\Omega x + k_r(\pm 1 + \mu_r|r|^2)y + k_y(\pm 1 + \mu_y y^2)y = \\ \qquad = m\epsilon\Omega^2 \sin(\Omega t) + F_{ny} \ , \end{cases} \tag{10.35}$$

where $c_x = c_{nx} + c_r$ and $c_y = c_{ny} + c_r$. The values of the stiffness k are assumed to be positive, and the double signs take care of the possibility of having attractive or repulsive systems, whereas μ is positive or negative for hardening and softening systems, respectively.

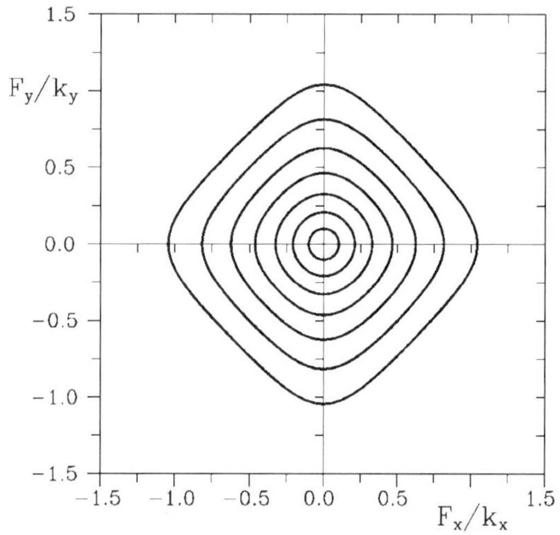

FIGURE 10.11. Polar plot of the restoring force. Ratios F_x/k_x and F_y/k_y have been reported on the axes, and the curves refer to different values of the nondimensional displacements $x/\sqrt{\mu_x}$ and $y/\sqrt{\mu_y}$.

By neglecting the presence of a rotating restoring force and introducing the nondimensional coordinates $x^* = x/\epsilon$, $y^* = y/\epsilon$ and time $\tau = t\sqrt{k_x/m}$ and parameters

$$\alpha = \frac{k_y}{k_x}\,, \quad \mu_i^* = \mu_i \epsilon^2\,, \quad \Omega^* = \frac{\Omega}{\sqrt{k_x/m}}\,, \quad c_i^* = \frac{c_i}{2\sqrt{k_x m}} \quad (i = x, y, r)\,, \tag{10.36}$$

the following nondimensional equation of motion is obtained:

$$\begin{cases} x^{*\prime\prime} + 2c_x^* x^{*\prime} + 2c_r^* \Omega^* y^* + (\pm 1 + \mu_x^* x^{*2})x^* = \Omega^{*2}\cos(\Omega^* \tau) + \frac{F_{nx}}{|k|\epsilon}\,, \\ y^{*\prime\prime} + 2c_y^* y^{*\prime} - 2c_r^* \Omega^* x^* + \alpha(\pm 1 + \mu_y^* y^{*2})y^* = \Omega^{*2}\sin(\Omega^* \tau) + \frac{F_{ny}}{|k|\epsilon}\,, \end{cases} \tag{10.37}$$

where prime denotes differentiation with respect to τ. The behavior of the system depends then on six parameters (α, μ_x^*, μ_y^*, c_x^*, c_y^*, c_r^*), and on the nondimensional spin speed Ω^*.

Note that the system is not isotropic even if the parameters related to x- and y-directions are equal, owing to nonlinearities. The polar plot of Figure 10.11 shows clearly that the restoring force tends to be isotropic for small displacements, whereas for large displacements, when nonlinearities are significant, the curve tends to a square instead of a circle. The figure has been obtained for a hardening case, but similar results would have been obtained in the case of a softening system.

The solution of the equation of motion (10.37) is easily performed by numerical integration. However, it can also be approximated by the polyharmonic expression

$$\begin{cases} x^* = \sum_{i=0}^{m} a_{ci} \cos(i\Omega^*\tau) + \sum_{i=0}^{m} a_{si} \sin(i\Omega^*\tau) \ , \\ y^* = \sum_{i=0}^{m} b_{ci} \cos(i\Omega^*\tau) + \sum_{i=0}^{m} b_{si} \sin(i\Omega^*\tau) \ , \end{cases} \tag{10.38}$$

which has been truncated after m terms. By introducing the truncated series (10.38) into the equation of motion and balancing the harmonics, a set of nonlinear algebraic equations is obtained. Each harmonics introduces four unknowns and four equations, and the computations may be involving even if only few harmonics are considered.

If only the fundamental harmonic of the unbalance response is retained, the solution is an elliptical whirling. Note that if the system is linear, the elliptical solution (in the fixed frame) is the exact solution of the equation of motion. In the rotating frame, the solution is a stationary solution, superimposed on a vibration component with frequency 2Ω, as is well known.

If rotating damping is vanishingly small, the two pairs of equations describing the behavior in xz- and yz-planes uncouple, and in each plane, the same equations commonly used to compute a first approximation of the response of a Duffing oscillator are found. The nonlinear set of equations can be easily solved using the Newton-Raphson algorithm. Problems can, however, be encountered owing to the presence of multiple solutions whose basins of attraction may have very complicated shapes. The solution so obtained is, however, just an approximation, and no actual elliptical whirling can take place. The backbone of the response can be computed in a very simple way, obtaining, in each coordinate plane, the same backbone typical of a Duffing oscillator.

The response of a hardening attractive system is shown in Figure 10.12. Note that fields in which multiple solutions exist can be found. The method used to obtain the response does not allow one to know whether other solutions exist or whether the solutions found are stable. The velocity range considered can be divided into three parts:

- Up to $\Omega^* = 1.75$, there is only one solution, a forward elliptical whirling.
- For $1.75 < \Omega^* < 2.50$, there are two solutions, a forward and a backward elliptical whirling.
- For $1.75 > 2.50$, there are two solutions, one of which is backward elliptical whirling and the other is a forward whirling, which is increasingly circular at increasing speed. The latter can be considered as the self-centered solution. Note that in the whole frequency range,

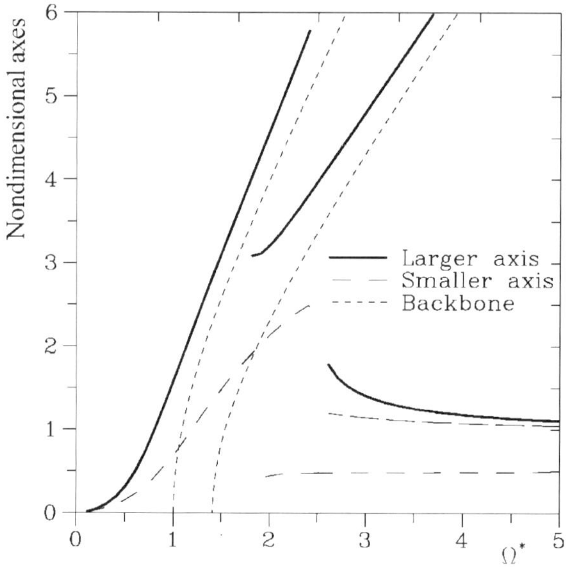

FIGURE 10.12. Semiaxes of the elliptical response as functions of the nondimensional speed. Hardening attractive system. Data: $\alpha = 2$; $\mu^* = 0.25$; $c^*_{nx} = c^*_{ny} = 0.25$; $c^*_r = 0.034$. Fundamental harmonic of the response.

a high solution exists and no downward jump is expected during a run-up of the rotor.

The orbits at some selected values of the speed ($\Omega^* = 1, 2, 3$) are plotted in Figure 10.13. The first approximation (fundamental harmonic only) and the results obtained through numerical integration of the equation of motion are reported. Note that all solutions reported in Figure 10.12 have been found using harmonic balance and numerical integration: The last results indicate that all solutions (also those causing backward whirling) correspond to stable whirling. Note that in the cases shown, the fundamental harmonics yields already a very close approximation and that by including also a further harmonics (the third), a result that is completely superimposed to the numerical solution is obtained.

The results obtained for the symmetrical nonlinear Jeffcott rotor running on nonisotropic bearings show a close similarity with those typical of the linear case. In the latter case, if damping is neglected, three velocity ranges can be defined. In the lowest, spanning up to the first critical speed, unbalance causes forward elliptical whirling with larger axis situated in the direction of lowest stiffness. In the highest, starting at the highest critical speed, forward elliptical whirling with larger axis situated in the direction of larger stiffness is obtained. However, the orbit tends to become circular at a very high speed. Between the critical speeds, forward unbalance causes

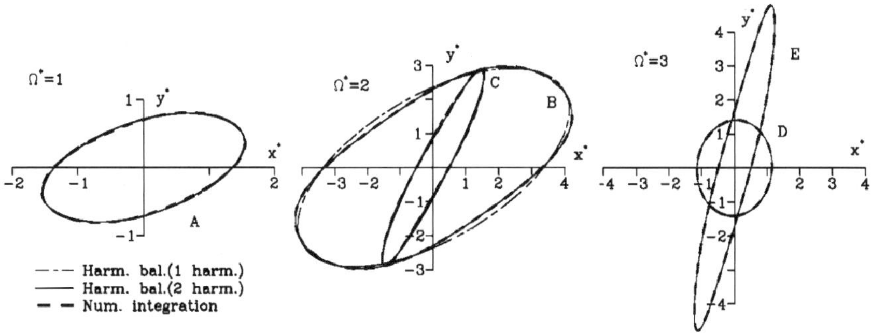

FIGURE 10.13. Orbits at three different values of the speed for the system of Figure 10.12. Note that different scales have been used.

elliptical backward whirling, which at an intermediate speed becomes circular. The presence of damping restricts the intermediate field, making the speeds at which the reversal of whirling direction occurs, closer to each other. At these speeds, whirling degenerates to a vibration along a straight line in a direction, making an angle with the principal axes of elasticity of the supports. There is a value of the damping above which backward whirling is no longer possible and forward whirling occurs at all speeds.

The nonlinearity of the systems modifies these results in the following way:

- No exact solution of the equation of motion is possible, and the shape of the orbit is no longer elliptical. Numerical simulation has, however, shown that in most cases, the orbit obtained is not distinguishable from an elliptical orbit, and even in the cases in which higher order harmonics are important, it is sufficient to take into account only a further one (the third) to obtain very good results. This result has been obtained for supports with cubic stiffness characteristics and holds only for supports with symmetrical characteristics. Moreover, being based on numerical experiments, this conclusion does not exclude the possibility of very complicated whirling patterns, periodic or even chaotic, which, however, might not be very common.

- At speeds higher than the lowest critical speed of the linearized system, multiple solutions can exist. Some of them consist in forward whirling and others in backward whirling. As a general, empirical rule, when a solution can be considered as a combination of vibrations in x- and y-directions, occurring both with the larger or the smaller amplitude, whirling takes place in forward direction and backward whirling occurs when a large vibration combines with a small vibration.

- The presence of damping complicates this picture, and very high damping makes backward whirling impossible. Rotating damping couples the behavior in xz- and yz-planes and destabilizes the system, as in the case of isotropic rotors.

The above-mentioned conclusions, drawn from the Jeffcott rotor model, are believed to hold qualitatively for real-life rotors, which are much more complex. However, the presence of gyroscopic effects will result in a stronger coupling between the inflection planes, even in cases of very low damping.

10.3 Nonlinear anisotropic Jeffcott rotor running on symmetric supports

Consider the same model studied in the previous section, but now assume that the elastic nonlinear restoring force is caused by two nonlinear spring systems acting, respectively, in ξz- and ηz-planes. By operating in the same way as for the previous section and assuming a Duffing-type restoring force and linear damping, defining nondimensional coordinates and parameters as in Equation (10.37) and neglecting the nonrotating external force, the above assumptions lead to the following nonlinear equation of motion written in the rotating coordinate frame:

$$\begin{cases} \xi^{*\prime\prime} + 2\left(c_n^* + c_{r\xi}^*\right)\xi^{*\prime} - 2\Omega^*\eta^{*\prime} + (\pm 1 - \Omega^{*2} + \mu_\xi^* \xi^{*2})\xi^* + \\ \qquad -2c_n^*\Omega^*\eta^* = \Omega^{*2}\cos(\beta) \ , \\ \eta^{*\prime\prime} + 2\left(c_n^* + c_{r\eta}^*\right)\eta^{*\prime} + 2\Omega^*\xi^{*\prime} + (\pm\alpha - \Omega^{*2} + \alpha\mu_\eta^*\eta^{*2})\eta^* + \\ \qquad +2c_n^*\Omega^*\xi^* = \Omega^{*2}\sin(\beta) \ , \end{cases} \tag{10.39}$$

where β is the angle between the direction of static unbalance and ξ-axis. An exact solution of Equation (10.39) is circular whirling defined by constant values of $\xi^* = \xi_0^*$ and $\eta^* = \eta_0^*$. By introducing such a solution into the equation of motion (10.39), it reduces to the algebraic equation

$$\begin{bmatrix} \pm 1 - \Omega^{*2} + \mu_\xi^*\xi^{*2} & 2c_n^*\Omega^* \\ 2c_n^*\Omega^* & \pm\alpha - \Omega^{*2} + \alpha\mu_\eta^*\eta^{*2} \end{bmatrix} \begin{Bmatrix} \xi_0^* \\ \eta_0^* \end{Bmatrix} = \begin{Bmatrix} \Omega^{*2}\cos(\beta) \\ \Omega^{*2}\sin(\beta) \end{Bmatrix} , \tag{10.40}$$

which yields the *in phase* and *in quadrature* components of the amplitude of the orbit. In this case, the situation is very close to that characterizing the unbalance response of an axi-symmetrical Jeffcott rotor. If no nonrotating damping is present, the two equations are uncoupled and the system responds to the components of unbalance in each coordinate plane as an isotropic rotor with the properties related to that plane. The presence of damping introduces a coupling that modifies the quantitative results but not the type of the response.

The typical feature of linear nonisotropic rotors is the presence of fields of instability at speeds laying between the critical speeds, which are narrowed by the presence of damping, both rotating and nonrotating, until they disappear for large values of the damping coefficients. The study of the stability in the small of the solution expressed by Equation (10.40) can be performed in the same way seen for isotropic systems (Section 5.5.1). The equilibrium condition in the space state is characterized by $\mathbf{z} = \mathbf{z}_E$, and consequently, the state vector can be expressed as

$$\mathbf{z} = \boldsymbol{\delta} + \mathbf{z}_E \ , \tag{10.41}$$

where $\mathbf{z} = \begin{bmatrix} \xi^{*\prime} & \eta^{*\prime} & \xi^* & \eta^* \end{bmatrix}^T$ is the state vector and $\boldsymbol{\delta}$ is a small displacement from the equilibrium position (in the state space).

The equation of motion can be expressed as

$$(\boldsymbol{\delta} + \mathbf{z}_E)' = \mathbf{f}(\boldsymbol{\delta} + \mathbf{z}_E) \ , \tag{10.42}$$

where the function $\mathbf{f}(\mathbf{z})$ obtainable from Equation (10.40) is

$$\mathbf{f} = \begin{bmatrix} -2\left(c_n^* + c_{r\xi}^*\right) & 2\Omega^* & \Omega^* \mp 1 & 2\Omega^* c_n^* \\ -2\Omega^* & -2\left(c_n^* + c_{r\eta}^*\right) & -2\Omega^* c_n^* & \Omega^* \mp \alpha \\ 1 & 0 & 0 & 0 \\ 0 & 1 & 0 & 0 \end{bmatrix} \mathbf{z} + \tag{10.43}$$

$$+ \begin{Bmatrix} -\mu_\xi^* \xi^{*3} \\ -\mu_\eta^* \eta^{*3} \\ 0 \\ 0 \end{Bmatrix} + \begin{Bmatrix} \Omega^{*2} \cos(\beta) \\ \Omega^{*2} \sin(\beta) \\ 0 \\ 0 \end{Bmatrix} .$$

The Liapunov exponents allowing the study of the stability in the small are then the eigenvalues of the Jacobian matrix $\mathbf{J}(\mathbf{z})$ of function $\mathbf{f}(\mathbf{z})$, whose explicit expression is

$$\begin{bmatrix} -2\left(c_n^* + c_{r\xi}^*\right) & 2\Omega^* & \Omega^* \mp 1 - 3\mu_\xi^* \xi^{*2} & 2\Omega^* c_n^* \\ -2\Omega^* & -2\left(c_n^* + c_{r\eta}^*\right) & -2\Omega^* c_n^* & \Omega^* \mp \alpha - 3\mu_\eta^* \eta^{*2} \\ 1 & 0 & 0 & 0 \\ 0 & 1 & 0 & 0 \end{bmatrix} . \tag{10.44}$$

As usual, if any one of the real parts of the eigenvalues s^* of $\mathbf{J}(\mathbf{z})$ is positive, the system is unstable following Liapunov criterion; if all real parts of s^* are negative, the system is asymptotically stable. When the real part of at least one s^* is equal to zero, the system behaves as an ubdamped system.

The response of a hardening attractive system is reported in Figure 10.14. Note that no speed range in which no stable solution exists can be found.

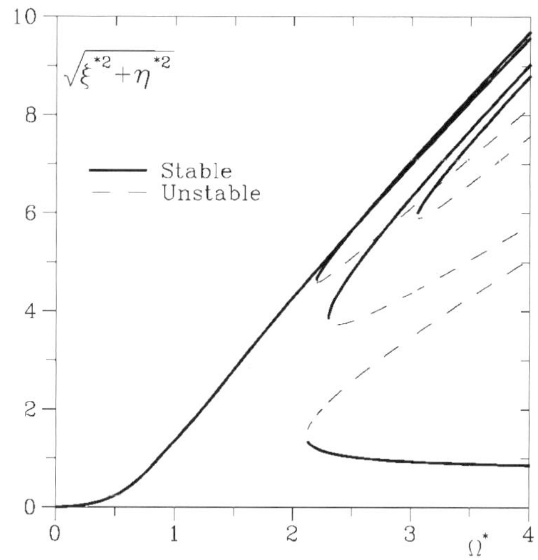

FIGURE 10.14. Amplitude of the circular response as a function of the nondimensional speed. Hardening attractive system, with $\alpha = 2$; $\mu^*_\xi = \mu^*_\eta = 0.25$; $c^*_{r\xi} = c^*_{r\eta} = 0.03$; $c^*_n = 0.22$; $\beta = 45°$.

This result, which is conflicting with those obtained for linear systems, can be ascribed to the relatively high values of damping and nonlinearity coefficients used. However, several computer simulations performed with decreasing values of both showed that even with vanishingly small damping and nonlinearity coefficients, at least one stable solution exists at any speed. This inconsistency with the results obtained for linear systems is only apparent: The presence of a vanishingly small nonlinearity causes a stable solution to be present where one should expect instability, but the amplitude of this solution is very large, tending to infinity when the nonlinearity tends to zero: A stable solution with unbounded amplitude is then found instead of an unstable solution.

Obviously circular whirling, although being an exact solution, is not necessarily unique; alternative solutions, linked with noncircular orbits must be found through numerical integration of the equations of motion (10.40).

The Jeffcott rotor model is inadequate for the study of a system in which both the stator and the rotor are nonisotropic. The complexity of this problem suggests to resort to a numerical approach, using the equations that will be introduced when dealing with multi-degrees-of-freedom systems.

10.4 Systems with many degrees of freedom

A very general mathematical model of a nonlinear rotor can be obtained from Equation (9.22) by adding a generic vector function $\mathbf{f}(q_i, \dot{q}_i, \theta, t)$ to take into account the behavior of the nonlinear part of the system

$$\begin{aligned}&\mathbf{M}_m\ddot{\mathbf{q}} + \left(\mathbf{C}_m - i\Omega\mathbf{G}\right)\dot{\mathbf{q}} + \left(\mathbf{K}_m - i\Omega\mathbf{C}_{r_m}\right)\mathbf{q} + \mathbf{M}_{n_d}\ddot{\overline{\mathbf{q}}}+\\&+\mathbf{M}_{r_d}e^{2i\Omega t}\ddot{\overline{\mathbf{q}}} + \mathbf{C}_{n_d}\dot{\overline{\mathbf{q}}} + \left(\mathbf{C}_{r_d} + 2i\Omega\mathbf{M}_{r_d}\right)e^{2i\Omega t}\dot{\overline{\mathbf{q}}} + \mathbf{K}_{n_d}\overline{\mathbf{q}}+\\&+\left(\mathbf{K}_{r_d} + i\Omega\mathbf{C}_{r_d}\right)e^{2i\Omega t}\overline{\mathbf{q}} + \mathbf{f}(q_i, \dot{q}_i, \theta, t) = \Omega^2\mathbf{f}_r e^{i\Omega t} + \mathbf{f}_n\,.\end{aligned} \tag{10.45}$$

Equation (10.45) has been obtained with the only assumption of uncoupling among flexural, axial, and torsional behavior. If the system is axially symmetrical, it reduces to

$$\begin{aligned}\mathbf{M}\ddot{\mathbf{q}} + \left(\mathbf{C}_n + \mathbf{C}_r - i\Omega\mathbf{G}\right)\dot{\mathbf{q}} + \left(\mathbf{K} - i\Omega\mathbf{C}_r\right)\mathbf{q}+&\\+\mathbf{f}(q_{0_i}, \dot{q}_{0_i}) = \Omega^2\mathbf{f}_r e^{i\Omega t} + \mathbf{f}_n\,.&\end{aligned} \tag{10.46}$$

If no nonrotating force acts on the structure, a possible solution for the unbalance response is a circular whirling that can be expressed as

$$\mathbf{q} = \mathbf{q}_0 e^{i\Omega t}\,. \tag{10.47}$$

Remark 10.7 *Also, in the case of multi-degrees-of-freedom systems, circular whirling is an exact solution of the equation of motion, not just an approximation of the fundamental harmonic of the response, as is customary in nonlinear vibrating systems.*

Introducing this solution into Equation (10.46) and considering the presence of a nonrotating structural damping matrix $\mathbf{K}''_n$, the unbalance response of a nonlinear rotor with many degrees of freedom is readily obtained

$$\left(\mathbf{K} - \Omega^2(\mathbf{M} - \mathbf{G}) + i(\Omega\mathbf{C}_n + \mathbf{K}''_n)\right)\mathbf{q}_0 + \mathbf{f}_i(q_{0_i}) = \Omega^2\mathbf{f}_r\,. \tag{10.48}$$

To reduce the size of the problem, a combined Guyan and static reduction technique can be used, particularly when fewer generalized coordinates are directly linked with nonlinearities, as when dealing with a rotor that is linear in itself but runs on nonlinear bearings. A first reduction of the number of degrees of freedom is performed by eliminating those not directly involved in nonlinearities, with a small fraction of the total mass associated with them. This procedure, basically a Guyan reduction, introduces approximations, which are usually very small if the degrees of freedom to be dropped are chosen with care. The equations of motion are then divided into two groups: a linear and a nonlinear set of equations. A second dynamic condensation procedure, to be applied at each value of the spin speed at which the unbalance response is to be obtained, can then follow to eliminate all linear degrees of freedom.

In some particular cases, the solution of the nonlinear problem can easily be performed, because it reduces to that of a single nonlinear equation. This includes the case of an undamped linear rotor supported by two nonlinear bearings, one of which has a behavior that can be approximated by a cubic characteristic. For the solution of the general problem, in which the nonlinear functions have complicated expressions and there are many degrees of freedom, the Newton-Raphson method seems to be the most appropriate choice. In this case, it is advisable to write the equation of motion using real coordinates instead of complex ones.

The iterative algorithm allowing the computation of the solution of Equation (10.48), after introducing a set of real coordinates, at the $(i+1)$th iteration from that at the ith iteration is

$$\mathbf{x}_{i+1} = \mathbf{x}_i - h\mathbf{S}(x_i)^{-1}\mathbf{p}(x_i)\,, \tag{10.49}$$

where h is a relaxation constant and the vector of the unknown $\mathbf{x}$, the elements of Jacobian matrix $\mathbf{S}(x)$, and functions $\mathbf{p}(x)$ are defined as

$$\mathbf{x} = \left\{ \begin{array}{c} \mathrm{Re}(\mathbf{q})_0 \\ \mathrm{Im}(\mathbf{q})_0 \end{array} \right\} = \left\{ \begin{array}{c} \mathbf{x}_\xi \\ \mathbf{x}_\eta \end{array} \right\}, \qquad S_{ij} = \frac{\partial p_i(x)}{\partial x_j},$$

$$\mathbf{p}(\mathbf{x}) = \left[\begin{array}{cc} \mathbf{K} - \Omega^2(\mathbf{M} - \mathbf{G}) & -\Omega(\mathbf{C}_n + \mathbf{K}''_n) \\ \Omega(\mathbf{C}_n + \mathbf{K}''_n) & \mathbf{K} - \Omega^2(\mathbf{M} - \mathbf{G}) \end{array} \right] \mathbf{x} +$$

$$+\{f(\mathbf{x})\} - \Omega^2 \left\{ \begin{array}{c} \mathrm{Re}(\mathbf{f})_r \\ \mathrm{Im}(\mathbf{f})_r \end{array} \right\}. \tag{10.50}$$

If there is no damping in the system and the unbalance distribution is contained in a plane (i.e., vector $\mathbf{f}_r$ is real), the computation of the response is much simpler because vector $\mathbf{q}_0$ is also real. All aforementioned equations still hold, but they can be written directly using the unknowns $\mathbf{q}_0$. The number of equations is, consequently, halved, and, in some particular cases, the solution of the nonlinear part can be reduced to the solution of a single nonlinear equation.

Because the backbone curves are related to the undamped system, they can be computed using the same approach as for the latter, simply by neglecting the forcing vector $\mathbf{f}_r$ in the definition of functions $\mathbf{p}(\mathbf{q}_0)$.

The convergence of the generalized Newton-Raphson method can be, in some cases, a source of potential problems. The basins of attraction of the solutions may take very complicated shapes, or the computation can lock itself into a cycle without reaching any actual solution of the basic

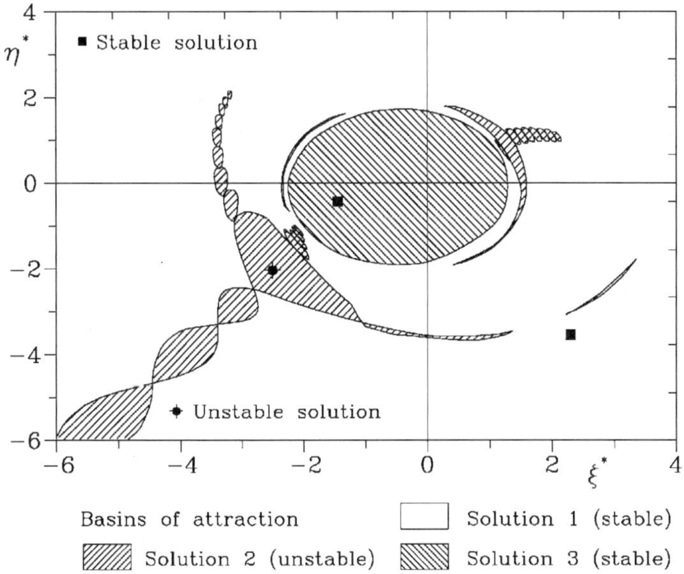

FIGURE 10.15. Basins of attraction of the solutions for the equilibrium positions of the system of Figure 10.7 at a speed $\Omega^* = 2$, computed through the Newton-Raphson iterative technique.

equation. For example, the domains of attraction of the various solutions plotted for the Jeffcott rotor already studied in Figure 10.7 are reported in Figure 10.15. The map has been plotted for a value of the nondimensional speed $\Omega^* = 2$, at which three solutions (two stable and an unstable one) exist.

The structure of the map is fractal, as can be seen by enlarging selected zones, and the unstable solution also has a nonvanishing attraction domain. A consequence of the fractal structure of the map is that there are zones in which very small changes of the initial values assumed for the computation cause large differences on the solution obtained. When comparing the maps of Figures 10.7 and 10.15, it is clear that the fractal nature of the second is strictly linked with the mathematical procedure used for the solution of the Equation (i.e., the Newton-Raphson technique) and has nothing to do with the actual physical behavior of the system. No improvement of the mathematical behavior of the equations has been obtained by introducing a relaxation factor. The map of the domains of attraction changes but retains its fractal structure and the characteristics of showing a domain of attraction of the unstable solution; however, the number of iterations increases.

The results obtained for the Jeffcott rotor show fairly good convergence characteristics of the Newton-Raphson technique. Obviously this does not guarantee an equally well-behaved nature of the general mathematical model for systems with many degrees of freedom. The following considerations, however, can be extrapolated and used as guidelines in the solution of more complex problems:

- It seems that numerical damping is not of great use in avoiding non-convergence.
- Attractive stable cycles can be found, particularly in the fields just above a critical speed of the linearized system. After each iteration, it is necessary to check not only whether convergence has been obtained, but also whether the computation has been locked in a stable cycle.
- When this occurs, the computation can be started again using a different set of initial values. When operating in a self-centered branch, a good guess could be trying to start from a vector $\mathbf{x}$ obtained by multiplying the one of the preceding attempt by a constant smaller than one; in a high branch of the response, a constant greater than 1 can be used.
- The fractal nature of the domains of attraction can cause large differences in the results to be produced by small changes in the trial vector $\mathbf{x}$. The solution may also converge on an unstable branch of the response.
- The stable branches should be obtained using as a trial vector the result obtained in the previous computation.
- The aforementioned considerations also hold for the computation of the various branches of the backbone. Here, however, a trivial solution with all the elements of $\mathbf{x}$ vanishingly small exists, and therefore, the procedure must be started outside the basin of attraction of the trivial solution. A suggestion is to start at a speed just above a critical speed of the linearized system (for hardening systems) with a trial vector proportional to the eigenvector of the linearized system corresponding to the mentioned critical speed.

Example 10.1 *System with four degrees of freedom on nonlinear bearings.*

Consider a rigid, vertical, symmetrical rotor running on two identical nonlinear isotropic bearings. It can be modeled as a system with four real (two complex) degrees of freedom of the same type studied in Section 3.6.1 but with nonlinear properties. However, the fact that the axis of rotation is vertical and no external forces act in radial direction allows us to consider the rotor as an isotropic system and ensures that circular whirling is a possible solution for the unbalance response. The rotor, a flywheel, has a working range between 8,500 and 17,000 rpm, a mass $m = 125$ *kg and moments of inertia* $J_p = 2.272$ *kg m*^{2} *and* $J_t = 1.477$ *kg m*^{2}*. It is supported by a pair of ball bearings whose nonlinear characteristics can be* *approximated by the expression* $F = -2\times10^7(1+10^{10}(|r|)^2)r$ *and whose damping properties can be modeled as hysteretic damping applied only to the linear part of the stiffness with loss factor* η *= 0.08. Knowing that the span between the bearings is 400 mm and that the center of gravity is at 30% of the span, study the response to a static unbalance* $m\epsilon = 438\times10^{-6}$ *kg m and state whether self-centering is possible. Compute the forces exerted on the stator.*

Because the distances of the center of gravity of the rotor from the bearings are a = 120 mm and b = 280 mm, respectively, the matrices entering into the mathematical model of the linearized system are

$$\mathbf{M} = \begin{bmatrix} 125 & 0 \\ 0 & 1.477 \end{bmatrix}, \qquad \mathbf{G} = \begin{bmatrix} 0 & 0 \\ 0 & 2.272 \end{bmatrix},$$

$$\mathbf{K} = k\begin{bmatrix} 2 & a-b \\ a-b & a^2+b^2 \end{bmatrix} = \begin{bmatrix} 40 & -3.2 \\ -3.2 & 1.856 \end{bmatrix}\times10^6 .$$

In this case, it is expedient to use as generalized coordinates the displacements at the bearing locations. The complex coordinates r and Φ *can be obtained from the complex coordinates* r_1 *and* r_2 *using a transformation matrix* $\mathbf{T}$

$$\left\{ \begin{array}{c} r \\ \Phi \end{array} \right\} = \mathbf{T}\left\{ \begin{array}{c} r_1 \\ r_2 \end{array} \right\} = \frac{1}{l}\begin{bmatrix} b & a \\ 1 & -1 \end{bmatrix}\left\{ \begin{array}{c} r_1 \\ r_2 \end{array} \right\}.$$

The equation of motion of the undamped system is then

$$\mathbf{M}^*\left\{ \begin{array}{c} \ddot{r}_1 \\ \ddot{r}_2 \end{array} \right\} - i\Omega\mathbf{G}^*\left\{ \begin{array}{c} \dot{r}_1 \\ \dot{r}_2 \end{array} \right\} + \mathbf{K}^*\left\{ \begin{array}{c} r_1 \\ r_2 \end{array} \right\} + k\mu\left\{ \begin{array}{c} |r_1|^2 r_1 \\ |r_2|^2 r_2 \end{array} \right\} = \Omega^2\mathbf{f}^* e^{i\Omega t},$$

where

$$\mathbf{M}^* = \mathbf{T}^T\mathbf{M}\mathbf{T}, \qquad \mathbf{G}^* = \mathbf{T}^T\mathbf{G}\mathbf{T}$$

$$\mathbf{K}^* = \mathbf{T}^T\mathbf{K}\mathbf{T} = \begin{bmatrix} k & 0 \\ 0 & k \end{bmatrix}, \qquad \mathbf{f}^* = \mathbf{T}^T\mathbf{f} = \frac{m\epsilon}{l}\left\{ \begin{array}{c} b \\ a \end{array} \right\}.$$

The solution for circular whirling can be obtained from Equation (10.48):

$$\left(-\Omega^2(\mathbf{M}^* - \mathbf{G}^*) + \mathbf{K}^*(1+i\eta)\right)\left\{ \begin{array}{c} r_{1_0} \\ r_{2_0} \end{array} \right\} + k\mu\left\{ \begin{array}{c} |r_{1_0}|^2 r_{1_0} \\ |r_{2_0}|^2 r_{2_0} \end{array} \right\} = \Omega^2\mathbf{f}^* .$$

The solution can be performed using the Newton-Raphson algorithm. The expression of the Jacobian matrix is simply

$$\mathbf{S} = \left(-\Omega^2(\mathbf{M}^* - \mathbf{G}^*) + \mathbf{K}^*(1+i\eta)\right) + 3k\mu\begin{bmatrix} |r_{1_0}|^2 & 0 \\ 0 & |r_{2_0}|^2 \end{bmatrix}.$$

Note that all matrices and vectors are complex, and if software allowing the use of complex numbers is available, there is no need to double the number of equations to work with real numbers.

The solution is performed twice. By increasing the speed from zero, the higher branch is obtained, and by reducing the speed from the maximum value the lower,

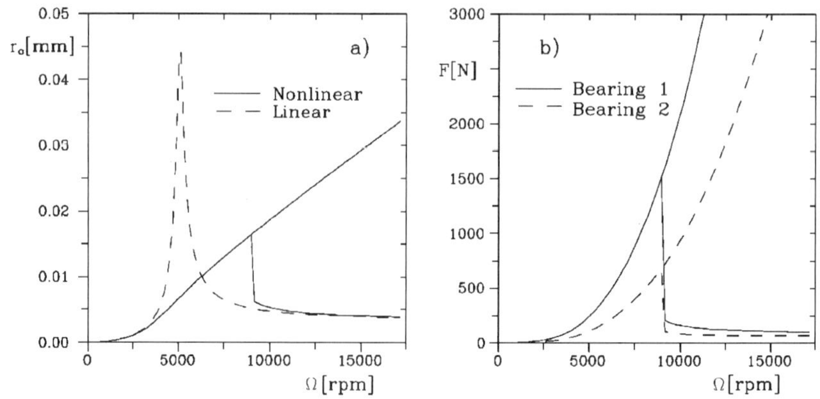

FIGURE 10.16. (a) Amplitude of the orbit of the center of gravity of the nonlinear and linearized rotors. (b) Bearing forces as functions of the speed.

the self-centered branch is computed. At each speed, the solution obtained in the previous computation is chosen as the starting solution. The results are plotted in Figure 10.16. From the figure, it is clear that self-centering is not possible, because of the low value of nonrotating damping. Correspondingly, the forces on the bearings are very high.

Example 10.2 *Chaotic motion of a Jeffcott rotor on plain Journal bearings.*

Consider a Jeffcott rotor running on plain journal bearings, and apply the short bearing model. The equation of motion, written using real coordinates, under the effect of unbalance and weight is

$$\begin{cases} m\ddot{x} = f_x(x, y, \dot{x}, \dot{y}) + m\epsilon\Omega^2 \cos(\Omega t) \ , \\ m\ddot{y} = f_y(x, y, \dot{x}, \dot{y}) + m\epsilon\Omega^2 \sin(\Omega t) - mg \ , \end{cases}$$

where f_x and f_y are the forces exerted on the bearings on the rotor.

By, introducing the nondimensional coordinates and time x^*, y^*, and τ and parameter ρ

$$x^* = \frac{x}{c} \ , \quad x^* = \frac{x}{c} \ , \quad \tau = \frac{t}{\Omega} \ , \quad \rho = \frac{\epsilon}{c} \ ,$$

where c is the clearance of the bearings, and indicating as $'$ and $''$ the derivatives with respect to τ, the following nondimensional equation is obtained:

$$\begin{cases} x^{*''} = \frac{1}{cm\Omega^2} f_x\left(x^*, y^*, x^{*'}, y^{*'}\right) + \rho\cos(\tau) \ , \\ y^{*''} = \frac{1}{cm\Omega^2} f_y\left(x^*, y^*, x^{*'}, y^{*'}\right) + \rho\sin(\tau) - \frac{g}{c\Omega^2} \ . \end{cases}$$

The forces exerted on the journal can be expressed as functions of the load factor [Ocwirk number, Equation (8.34)]

$$f_{x,y} = \frac{O f^*_{x,y}}{mg} \ ,$$

where

$$O = \frac{mg}{\mu\Omega RL}\left(\frac{c}{R}\right)^2\left(\frac{2R}{L}\right)^2$$

and L, R, and μ are the length and radius of the bearing and the viscosity of the oil.

Introducing a further nondimensional parameter

$$\gamma = O\frac{c\Omega^2}{g} \; ,$$

the equations can be written in a very simple form

$$\begin{cases} x^{*''} = \frac{1}{\gamma}f_x^*\left(x^*, y^*, x^{*'}, y^{*'}\right) + \rho\cos(\tau) \; , \\ y^{*''} = \frac{1}{\gamma}f_y^*\left(x^*, y^*, x^{*'}, y^{*'}\right) + \rho\sin(\tau) - \frac{O}{\gamma} \; , \end{cases}$$

which depend only on three parameters, namely, γ, σ, and ρ.

To easily integrate in time the equation of motion, it is expedient to resort to the state space approach

$$\begin{cases} x_1' = x_3 \; , \\ x_2' = x_4 \; , \\ x_3' = \frac{1}{\gamma}f_x^*(x_1, \; x_2, \; x_3, \; x_4) + \rho\cos(\tau) \; , \\ x_4' = \frac{1}{\gamma}f_y^*(x_1, \; x_2, \; x_3, \; x_4) + \rho\sin(\tau) - \frac{O}{\gamma} \; . \end{cases}$$

where

$$x_1 = x^*, \quad x_2 = y^*, \quad x_3 = x^{*'}, \quad x_4 = y^{*'} .$$

Following the suggestion in [57, 58], functions f_x^* and f_y^* can be approximated by the equation

$$\begin{Bmatrix} f_x^* \\ f_y^* \end{Bmatrix} = -\frac{\sqrt{(x^* - 2y^{*'})^2 + (y^* + 2x^{*'})^2}}{1 - x^{*2} - y^{*2}}\begin{Bmatrix} 3x^*V - \sin(\alpha)G - 2\cos(\alpha)S \\ 3y^*V + \cos(\alpha)G - 2\sin(\alpha)S \end{Bmatrix} ,$$

where angle α defining the arc $(\alpha, \; \alpha+\pi)$ in which the oil film is cavitated is

$$\alpha = artg\left(\frac{y^* + 2x^{*'}}{x^* - 2y^{*'}}\right) - \frac{\pi}{2}sign\left(\frac{y^* + 2x^{*'}}{x^* - 2y^{*'}}\right) - \frac{\pi}{2}sign\left(y^* + 2x^{*'}\right)$$

and functions V, G, and S are

$$\begin{aligned} V(x^*, \; y^*, \; \alpha) &= \frac{2 + G\left[y^*\cos(\alpha) - x^*\sin(\alpha)\right]}{1 - x^{*2} - y^{*2}} \; , \\ G(x^*, \; y^*, \; \alpha) &= \frac{2}{\sqrt{1 - x^{*2} - y^{*2}}}\left\{\frac{\pi}{2} + artg\left[\frac{y^*\cos(\alpha) - x^*\sin(\alpha)}{\sqrt{1 - x^{*2} - y^{*2}}}\right]\right\} \; , \\ S(x^*, \; y^*, \; \alpha) &= \frac{x^*\cos(\alpha) + y^*\sin(\alpha)}{1 - \left[x^*\cos(\alpha) + y^*\sin(\alpha)\right]^2} \; . \end{aligned}$$

Following [57], a set of values of parameters O, ρ, and γ giving way to chaotic motion is

$$O = 10, \quad \rho = 0.3, \quad \gamma = 70.$$

The steady-state values of the attitude angle β and of the eccentricity ϵ^* for a fully cavitated short bearing working at $O = 10$ are

$$\epsilon^* = 0.706, \quad \beta = 38.6^\circ.$$

The equations of motion were integrated numerically in time using an adaptive Runge-Kutta algorithm, starting from an initial state with initial conditions $[0.0001, \ -.1, \ 0.0001, \ 0.0001]$*. Some of the results, recorded after 900 periods* $(\tau = 5,655)$ *of the forcing function have passed to let the system reach steady-state chaotic conditions, are reported in Figure 10.17.*

The time histories $x^*(\tau)$ *and* $y^*(\tau)$ *of the journal center are reported in (a) and (b). The projections of the state trajectories in* $x^*x^{*'}$*- and* $y^*y^{*'}$*-planes are reported in (c) and (d), and the trajectory in the configuration space* x^*y^* *is reported in (g). The integration was performed for 30 periods of the forcing function, from* $\tau = 5,655$ *to* $\tau = 5,843$.

The Poincaré sections in $x^*x^{*'}$*-,* $y^*y^{*'}$*-, and* x^*y^**-planes are reported in (e), (f) and (h). To plot the Poincaré sections, the integration was performed for a much longer time, 4,200 periods after the initial 900. The fractal nature of the Poincaré sections is well visible and together with the computation of the Liapunov exponents and the spectra reported in [57] substantiates the claim that the behavior is chaotic.*

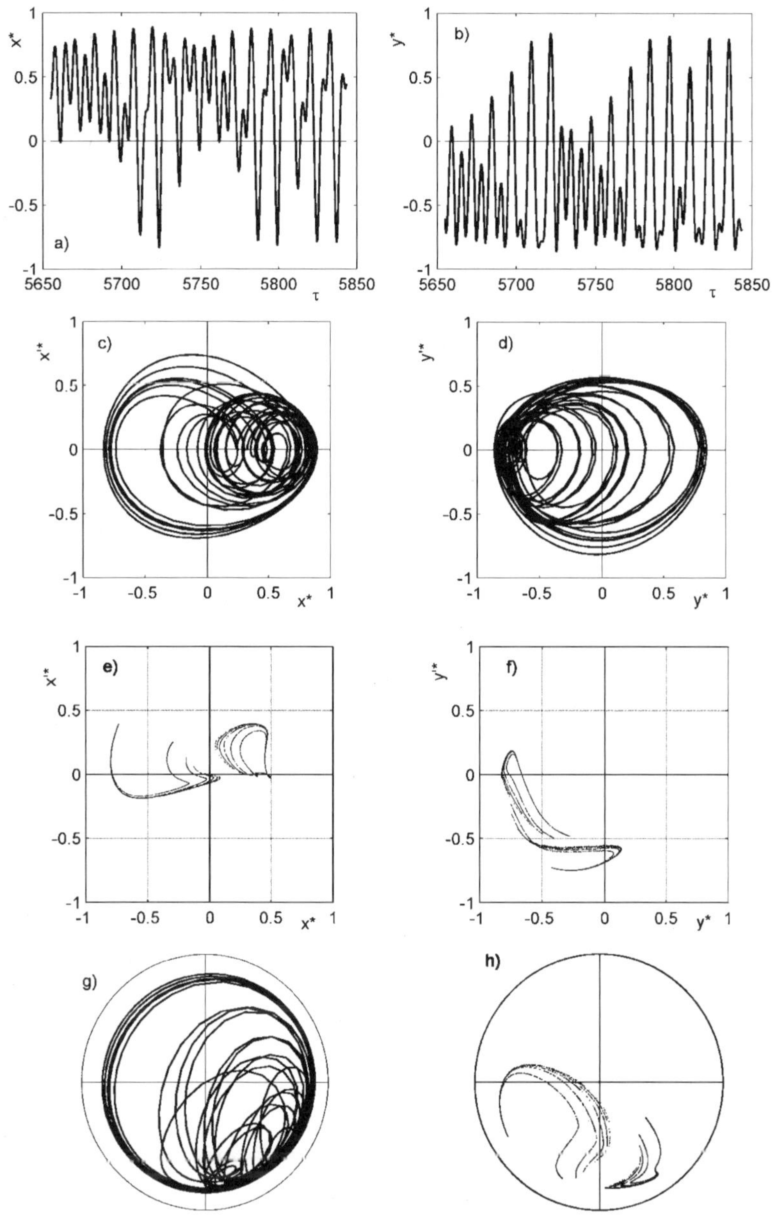

FIGURE 10.17. Chaotic response of a Jefcott rotor on plain journal bearings, modeled as fully cavitated short bearings, with $O = 10$, $\rho = 0.3$, $\gamma = 70$. (a) and (b) Time history of the journal center; (c) and (d) evolution of the state in $x^*x^{*'}$- and $y^*y^{*'}$-planes; (e) and (f) Poincaré sections in $x^*x^{*'}$- and $y^*y^{*'}$-planes; (g) trajectory in x^*y^*-plane; (h) Poincaré section in x^*y^*-plane.

11

Nonstationary rotordynamics

The dynamic study of rotors is usually performed assuming that the spin speed is maintained constant by a suitable driving device. As seen in Chapter 2, the torque exerted by the driving system must overcome not only the bearing drag, but also a peculiar form of drag caused by rotordynamic reasons or, to use a different phrasing, must supply a power equal to that dissipated by the rotating and nonrotating damping.

When studying the unbalance response, it was clearly stated that the plot of the amplitude as a function of speed had to be interpreted as a representation of the amplitude of the synchronous whirling occurring during steady-state working at different values of the spin sped and not as the amplitude during an acceleration or a spinning down of the rotor. In many cases, this is, however, just a theoretical statement, because the angular accelerations encountered in operation are so low that their effects on the overall dynamic behaviour are very weak.

In the present chapter, on the contrary, the assumption of constant speed is removed and the effects of nonstationary operation on the dynamic behavior of rotors are studied.

11.1 Nonstationary linear Jeffcott rotor

11.1.1 Equations of motion

As usual, the study is started using the simplest rotor model: the Jeffcott rotor. The scheme of the system is again that sketched in Figure 2.2(a)

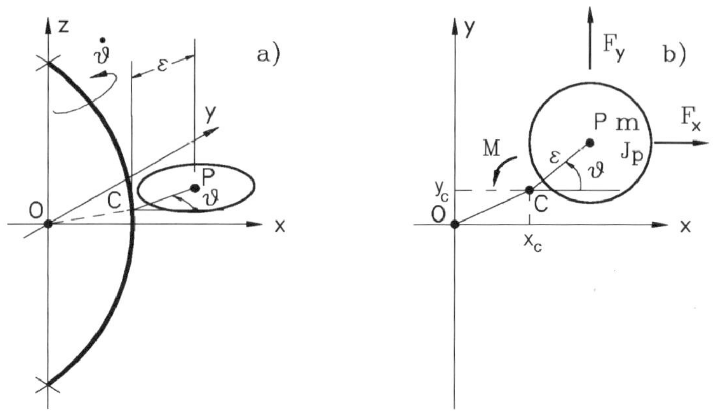

FIGURE 11.1. Sketch of an accelerating Jeffcott rotor.

and (b), with the difference that now angle θ is no more equal to Ωt but must be considered as a variable, together with the coordinates x_C and y_C of point C or coordinates x_P and y_P of point P (Figure 11.1). Note that a driving torque M is applied to the rotor by the shaft and that, although the Jeffcott rotor is considered to be a point mass, a polar moment of inertia J_p is associated with it. This is needed to study the acceleration of the rotor once the external torque is imposed. No assumption has been made on the torsional behavior of the shaft: It may be stiff, and then the torque M applied to the disc is equal to the torque applied to the end of the shaft, or compliant, in which case additional degrees of freedom can be added to study the torsional vibration that may occur.

By following the usual practice of choosing as generalized coordinates those of point C, the position and velocity of point P can be expressed by Equations (2.1) and (2.2), modified as

$$\overline{\text{P-O}} = \left\{ \begin{array}{c} x_P \\ y_P \end{array} \right\} = \left\{ \begin{array}{c} x_C + \epsilon\cos(\theta) \\ y_C + \epsilon\sin(\theta) \end{array} \right\} , \tag{11.1}$$

$$V_P = \left\{ \begin{array}{c} \dot{x}_P \\ \dot{y}_P \end{array} \right\} = \left\{ \begin{array}{c} \dot{x}_C - \epsilon\dot{\theta}\sin(\theta) \\ \dot{y}_C + \epsilon\dot{\theta}\cos(\theta) \end{array} \right\} . \tag{11.2}$$

The kinetic and potential energy are still expressed by Equations (2.3) with slight modifications

$$\mathcal{T} = \frac{1}{2}m\left\{\dot{x}_C^2 + \dot{y}_C^2 + \epsilon^2\dot{\theta}^2 + 2\epsilon\dot{\theta}\left[-\dot{x}_C\sin(\theta t) + \dot{y}_C\cos(\theta t)\right]\right\} + \frac{1}{2}J_p\dot{\theta}^2 , \tag{11.3}$$

$$\mathcal{U} = \frac{1}{2}k\left(x_C^2 + y_C^2\right) . \tag{11.4}$$

Assuming that an external force acts on point P in xy-plane (e.g., the weight of the rotor in case the axis of rotation is horizontal), to obtain the generalized forces Q_i to be introduced into the equation of motion, the virtual displacement of point P must be written

$$\left\{ \begin{array}{c} \delta x_{\rm P} \\ \delta y_{\rm P} \end{array} \right\} = \left\{ \begin{array}{c} \delta x_{\rm C} - \epsilon \sin(\theta)\delta\theta \\ \delta y_{\rm C} + \epsilon \cos(\theta)\delta\theta \end{array} \right\} . \tag{11.5}$$

The virtual work of force F with components F_x and F_y acting on point P and of the moment M_d is

$$\delta\mathcal{L} = F_x \delta x_{\rm C} + F_y \delta y_{\rm C} + \left[M_d - F_x \epsilon \sin(\theta) + F_y \epsilon \cos(\theta)\right] \delta\theta . \tag{11.6}$$

The equations of motion can then be obtained from the Lagrange equation, written in the form

$$\frac{d}{dt}\left(\frac{\partial (\mathcal{T} - \mathcal{U})}{\partial \dot{q}_i}\right) - \frac{\partial (\mathcal{T} - \mathcal{U})}{\partial q_i} = Q_i .$$

By performing the relevant derivatives and remembering that the spin speed $\Omega = \dot{\theta}$ has no more been assumed to be constant, the following equations of motion are obtained:

$$\left\{ \begin{array}{l} m\ddot{x}_{\rm C} - m\epsilon \left[\dot{\theta}^2 \cos(\theta) + \ddot{\theta}\sin(\theta)\right] + kx_{\rm C} = F_x , \\ m\ddot{y}_{\rm C} - m\epsilon \left[\dot{\theta}^2 \sin(\theta) - \ddot{\theta}\cos(\theta)\right] + ky_{\rm C} = F_y , \\ \left(J_p + m\epsilon^2\right)\ddot{\theta} + m\epsilon\left[-\ddot{x}\sin(\theta) + \ddot{y}\cos(\theta)\right] = M_d - \epsilon F_x \sin(\theta) + \epsilon F_y \cos(\theta) . \end{array} \right. \tag{11.7}$$

By introducing the complex coordinate $r_{\rm C} = x_{\rm C} + iy_{\rm C}$ and the nonrotating force $F_n = F_x + iF_y$, the equations of motion become

$$\left\{ \begin{array}{l} m\ddot{r}_{\rm C} + kr_{\rm C} - m\epsilon\left(\dot{\theta}^2 - i\ddot{\theta}\right)e^{i\theta} = F_n , \\ \left(J_p + m\epsilon^2\right)\ddot{\theta} + m\epsilon \,{\rm Im}\left(\ddot{r}e^{-i\theta}\right) = M_d + \epsilon \,{\rm Im}\left(F_n e^{-i\theta}\right) , \end{array} \right. \tag{11.8}$$

By introducing the terms related to rotating and nonrotating damping, the first Equation (11.8) becomes

$$m\ddot{r}_{\rm C} + c\dot{r}_{\rm C} + (k - i\dot{\theta}c_r)r_{\rm C} - m\epsilon\left(\dot{\theta}^2 - i\ddot{\theta}\right)e^{i\theta} = F_n , \tag{11.9}$$

where $c = c_r + c_n$ is the total damping. The equations of motion can be written in a rotating reference frame. By remembering the transformation allowing us to express the inertial complex coordinate r and its derivatives in terms of the rotating coordinate ρ [Equation (2.168)], the equation of motion in the rotating frame can be obtained

$$\left\{\begin{array}{l} m\ddot{p}+(c+2im\dot{\theta})\dot{p}+(k-m\dot{\theta}^2+i\dot{\theta}c_n+im\ddot{\theta})\rho-m\epsilon(\dot{\theta}^2-i\ddot{\theta})=F_ne^{-i\theta}, \\ \left(J_p+m\epsilon^2\right)\ddot{\theta}+m\epsilon\,\mathrm{Im}\left(\ddot{\rho}+2i\dot{\theta}\dot{\rho}+i\ddot{\theta}\rho-\dot{\theta}^2\rho\right)=M_d+\epsilon\,\mathrm{Im}\left(F_ne^{-i\theta}\right), \end{array}\right. \tag{11.10}$$

or, using real coordinates,

$$\begin{bmatrix} m & 0 \\ 0 & m \end{bmatrix}\left\{\begin{array}{c}\ddot{\xi}\\ \ddot{\eta}\end{array}\right\}+\begin{bmatrix} c & -2m\dot{\theta} \\ 2m\dot{\theta} & c \end{bmatrix}\left\{\begin{array}{c}\dot{\xi}\\ \dot{\eta}\end{array}\right\}+\begin{bmatrix} k-m\dot{\theta}^2 & -\dot{\theta}c_n-m\ddot{\theta} \\ \dot{\theta}c_n+m\ddot{\theta} & k-m\dot{\theta}^2 \end{bmatrix}\cdot$$

$$\cdot\left\{\begin{array}{c}\xi\\ \eta\end{array}\right\}-m\epsilon\left\{\begin{array}{c}\dot{\theta}^2\\ -\ddot{\theta}\end{array}\right\}=\left\{\begin{array}{c}F_x\cos(\theta)+F_y\sin(\theta)\\ -F_x\sin(\theta)+F_y\cos(\theta)\end{array}\right\}, \tag{11.11}$$

$$(J_p+m\epsilon^2)\ddot{\theta}+m\epsilon\left(\ddot{\eta}+2\dot{\theta}\dot{\xi}+\ddot{\theta}\xi-\dot{\theta}^2\eta\right)=M_d-\epsilon F_x\sin(\theta)+\epsilon F_y\cos(\theta).$$

By solving the second Equation (11.11) in $\ddot{\eta}$ and substituting it into the third one, the latter simplifies as

$$J_p\ddot{\theta}-\epsilon\left(c\dot{\eta}+\dot{\theta}c_n\xi+k\eta\right)=M_d. \tag{11.12}$$

Finally, by solving Equation (11.12) in $\ddot{\theta}$ and substituting it into the first two equations (11.11) and writing the equations of motion in the space state, it follows that

$$\left\{\begin{array}{l} \dot{z}_1=-\frac{c}{m}z_1+2z_2z_3-\left(\frac{k}{m}-z_3^2\right)z_4+\frac{c_n}{m}z_3z_5+\epsilon z_3^2+\frac{M_d}{J_p}z_5+\frac{\epsilon c}{J_p}z_2z_5+ \\ \quad +\frac{\epsilon c_n}{J_p}z_3z_4z_5+\frac{\epsilon k}{J_p}z_5^2+\frac{F_x}{m}\cos(z_6)+\frac{F_y}{m}\sin(z_6), \\ \dot{z}_2=-\frac{c}{m}z_2-2z_1z_3-\left(\frac{k}{m}-z_3^2\right)z_5-\frac{c_n}{m}z_3z_4+z_3^2z_5-\frac{M_d}{J_p}(z_4+\epsilon)+ \\ \quad -\frac{\epsilon c}{J_p}z_2(z_4+\epsilon)-\frac{\epsilon c_n}{J_p}z_3z_4(z_4+\epsilon)-\frac{\epsilon k}{J_p}z_5(z_4+\epsilon)-\frac{F_x}{m}\sin(z_6)+ \\ \quad +\frac{F_y}{m}\cos(z_6), \\ \dot{z}_3=\frac{M_d}{J_p}+\frac{\epsilon}{J_p}(cz_2+c_nz_3z_4+kz_5), \\ \dot{z}_4=z_1, \\ \dot{z}_5=z_2, \\ \dot{z}_6=z_3, \end{array}\right. \tag{11.13}$$

where $z_1=\dot{\xi}$, $z_2=\dot{\eta}$, $z_3=\dot{\theta}=\Omega$, $z_4=\xi$, $z_5=\eta$, and $z_6=\theta$.

11.1.2 Torsionally stiff rotor with imposed acceleration

The equations of motion (11.7) and following are nonlinear in the variable θ, and their general solution is impossible. Although there is no difficulty in performing a numerical integration in the time domain, some simplified solution can yield interesting results.

Assume that the rotor is torsionally stiff, and the driving device imposes a given law $\theta(t)$. As a consequence also, the spin speed $\Omega(t) = \dot{\theta}(t)$ and the angular acceleration $a(t) = \ddot{\theta}(t)$ are known, the simplest case being a constant acceleration. The first two Equations (11.7) uncouple from the third and yield the lateral behavior of the rotor. Once the first two have been solved yielding the laws $x_C(t)$ and $y_C(t)$, the third one can be solved in the torque M and yields the driving (or braking) torque that must be supplied to obtain the prescribed law $\theta(t)$.

By neglecting also nonrotating forces, the equations of motion can thus be written in the form

$$\begin{cases} m\ddot{x}_C + kx_C = m\epsilon\left[\dot{\theta}^2\cos(\theta) + \ddot{\theta}\sin(\theta)\right] , \\ m\ddot{y}_C + ky_C = m\epsilon\left[\dot{\theta}^2\sin(\theta) - \ddot{\theta}\cos(\theta)\right] , \end{cases} \tag{11.14}$$

$$M_d = \left(J_p + m\epsilon^2\right)\ddot{\theta} + m\epsilon\left[-\ddot{x}\sin(\theta) + \ddot{y}\cos(\theta)\right] .$$

Even if in the literature some solutions for the constant acceleration case can be found [27], they are so complicated that today it is easier to perform the numerical integration of the equations of motion. If the angular acceleration is constant

$$\ddot{\theta} = a = \text{constant},$$

it is possible to introduce the following nondimensional time and coordinates

$$\xi^* = \frac{\xi}{\epsilon} , \quad \eta^* = \frac{\eta}{\epsilon} , \quad \tau = \frac{t}{t_1} = t\frac{a}{\Omega_{cr}} ,$$

where $\Omega_{cr} = \sqrt{k/m}$ is the critical speed of the Jeffcott rotor and $t_1 = \Omega_{cr}/a$ is the time needed to reach the critical speed from standstill. The other nondimensional parameters are

$$c_n^* = \frac{c_n}{2\sqrt{km}} , \quad c_r^* = \frac{c_r}{2\sqrt{km}} , \quad \Omega^* = \frac{\Omega}{\Omega_{cr}} , \quad a^* = \frac{a}{\Omega_{cr}^2} .$$

The nondimensional equation in real, rotating coordinates is thus

$$\left\{ \begin{array}{c} \xi^{*''} \\ \eta^{*''} \end{array} \right\} + \frac{2}{a^*}\left[\begin{array}{cc} c^* & -\tau \\ \tau & c^* \end{array} \right]\left\{ \begin{array}{c} \xi^{*'} \\ \eta^{*'} \end{array} \right\} + \frac{1}{a^{*2}}\left[\begin{array}{cc} 1-\tau^2 & -2\tau c_n^* - a^* \\ 2\tau c_n^* + a^* & 1-\tau^2 \end{array} \right] \times$$

$$\times \left\{ \begin{array}{c} \xi^* \\ \eta^* \end{array} \right\} = \frac{1}{a^{*2}}\left\{ \begin{array}{c} \tau^2 \\ -a^* \end{array} \right\} , \tag{11.15}$$

$$\frac{M_d - J_p a}{m\epsilon^2\Omega_{cr}^2} = -2a^*c^*\eta^{*'} - 2\tau c_n^*\xi^* - \eta^* .$$

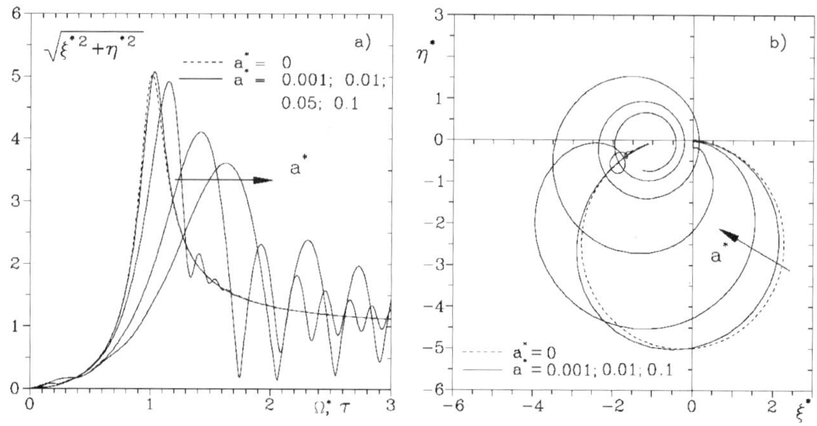

FIGURE 11.2. Motion of a Jeffcott rotor with $c_n^* = 0.1$ and $c_r^* = 0.01$ crossing a critical speed with constant acceleration. Nondimensional plot for some values of the angular acceleration: (a) Time history of the amplitude; (b) trajectory in the rotating plane $\xi^*\eta^*$.

The amplitude of the motion of a Jeffcott rotor with $c_n^* = 0.1$ and $c_r^* = 0.01$ accelerating from standstill to a speed equal to three times the critical speed with constant acceleration is plotted in nondimensional form in Figure 11.2(a). Note that the nondimensional time τ and the nondimensional velocity ω^* coincide, because time has been made nondimensional with reference to time t_1 needed to reach the critical speed. With very low values of the acceleration, the motion follows a pattern very similar to that seen for the steady-state case, as was easily predicted. With increasing values of the acceleration, the peak amplitude is reduced and the self-centered conditions are reached after some oscillations are damped out. The oscillations become stronger with increasing acceleration.

The trajectory of point C in the rotating plane is shown in Figure 11.2(b). From this plot, it is clear that the oscillations are actually the result of a spiral motion of the system taking place about the self-centered position. Note that the threshold of instability is never exceeded and the motion is stable. When the speed becomes higher than the threshold of instability, the spiral stops decaying and starts growing as the unstable behavior develops.

The driving torque needed to perform the acceleration is shown in Figure 11.3.

Actually, the nondimensional torque plotted is only the amount of torque that exceeds the value aJ_p needed to accelerate a perfectly balanced rotor. Also, strong oscillations are visible here in the supercritical field. In Figure 11.3, the curve related to a vanishingly small acceleration computed through Equation (2.176) has also been plotted.

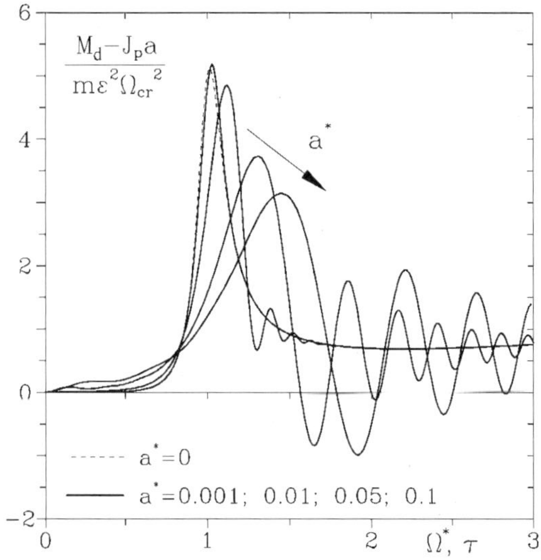

FIGURE 11.3. Nondimensional driving torque needed to perform the acceleration through the critical speed. Same system of Figure 11.2.

11.1.3 Torsionally stiff rotor with imposed torque

A slightly more complex, although more realistic, case is a torsionally stiff rotor that accelerates under the effect of a given torque, or better, of a given law $M_d(t)$. The relevant Equations (11.13), possibly without the terms related to the nonrotating forces, allow us to compute the laws $\Omega(t)$ and $\dot{\Omega}(t)$ and the lateral motions of the rotor caused by the imposed driving torque. Although the only way to solve the resulting nonlinear problem is by numerical integration in the time domain, some general insight on the behavior of the system can be obtained by writing the equations in nondimensional form and studying a number of cases.

The following nondimensional state variables and time can be introduced:

$$\begin{aligned} z_1 &= \frac{\dot{\xi}}{\epsilon\sqrt{k/m}}, \quad z_2 = \frac{\dot{\eta}}{c\sqrt{k/m}}, \quad z_3 = \Omega^* = \frac{\dot{\theta}}{\sqrt{k/m}}, \\ z_4 &= \xi^* = \frac{\xi}{\epsilon}, \quad z_5 = \eta^* = \frac{\eta}{\epsilon}, \quad z_6 = \theta, \quad \tau = t\sqrt{\frac{k}{m}}\,. \end{aligned} \tag{11.16}$$

The behavior of the system is then governed by just four nondimensional parameters:

$$c^* = \frac{c}{2\sqrt{km}}, \quad c_r^* = \frac{c_r}{2\sqrt{km}}, \quad J_p^* = \frac{J_p}{m\epsilon^2}, \quad M_d^* = \frac{M_d}{k\epsilon^2}\,.$$

The nondimensional state equation is then

$$\begin{cases} z_1' = -2c^* z_1 + 2z_2 z_3 - \left(1 - z_3^2\right) z_4 + 2c_n^* z_3 z_5 + z_3^2 + \frac{M_d^*}{J_p^*} z_5 + \frac{2c^*}{J_p^*} z_2 z_5 + \\ \quad + \frac{2c_n^*}{J_p^*} z_3 z_4 z_5 + \frac{1}{J_p^*} z_5^2 \ , \\ z_2' = -2c^* z_2 - 2z_1 z_3 - \left(1 - z_3^2\right) z_5 - 2c_n^* z_3 z_4 + z_3^2 z_5 - \frac{M_d^*}{J_p^*} \left(z_4 + 1\right) + \\ \quad - \frac{2c_n^*}{J_p^*} z_2 \left(z_4 + 1\right) - \frac{2c_n^*}{J_p^*} z_3 z_4 \left(z_4 + 1\right) - \frac{2c_n^*}{J_p^*} z_5 \left(z_4 + 1\right) \ , \\ z_3' = \frac{M_d^*}{J_p^*} + \frac{1}{J_p^*} \left(2c^* z_2 + 2c_n^* z_3 z_4 + z_5\right) \ , \\ z_4' = z_1 \ , \\ z_5' = z_2 \ , \\ z_6' = z_3 \ , \end{cases} \tag{11.17}$$

where prime indicates differentiation with respect to τ. Ratio M_d^*/J_p^* is the nondimensional acceleration that would result by neglecting rotordynamics effects.

The results for an acceleration with constant driving torque for the same system studied in Figures 11.2 and 11.3 ($c_n^* = 0.1$ and $c_r^* = 0.01$) are reported in Figure 11.4. A value $J_p^* = 1000$ has been assumed for the nondimensional moment of inertia, and three values of M_d^*, namely, 100, 5.5, and 4.5, have been considered.

The curves for $M_d^* = 100$ practically coincide with those for $a^* = 0.1$ in the previous analysis. The driving torque is high enough to produce a constant acceleration, and the assumption of constant torque practically coincides with that of constant acceleration.

The minimum value of the torque to allow the system to accelerate beyond the critical speed can be obtained from Equation (2.177)

$$M_{d_{max}} = \frac{k\epsilon^2}{2c_n^*} \ ;$$

i.e., in nondimensional terms,

$$M_{d_{max}}^* = \frac{1}{2c_n^*} \ . \tag{11.18}$$

In the present case, because $c_n^* = 0.1$, the nondimensional driving torque must be equal to at least $M_d^* = 5$, to avoid stalling. Actually, the simulation shows that for $M_d^* = 5.5$, the rotor barely succeeds in accelerating beyond the critical speed, whereas for $M_d^* = 4.5$, the rotor stalls.

11.1.4 Torsionally compliant rotor: small torsional vibrations with imposed acceleration

A further case of some interest is that of a torsionally compliant Jeffcott rotor, driven with a stated law $\theta_0(t)$ ($\Omega(t) = \dot{\theta}_0(t)$). Assume, for instance,

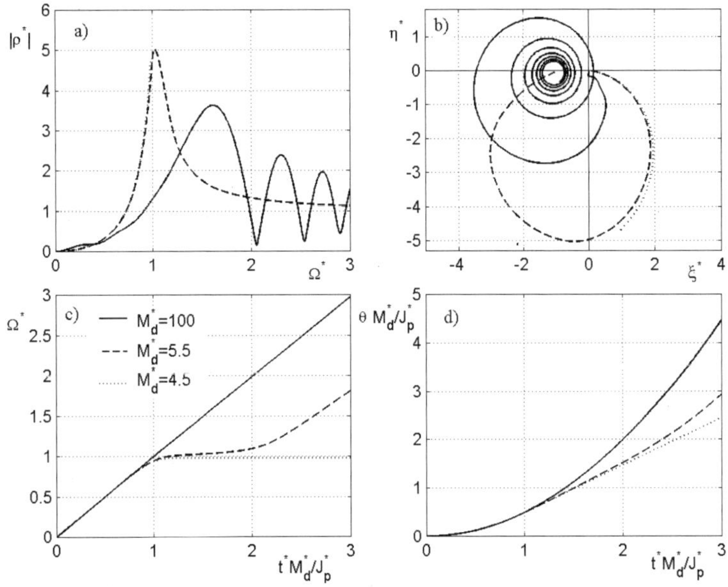

FIGURE 11.4. Constant torque acceleration of the same Jeffcott rotor studied in Figures 11.2 and 11.3. $c_n^* = 0.1$, $c_r^* = 0.01$, $J_p^* = 1000$. The various curves refer to different values of M_d^*, namely, 100, 5.5, and 4.5. (a) Nondimensional amplitude versus nondimenional speed; (b) trajectory in the rotating $\xi^*\eta^*$ plane; (c): speed as a function of time; (d) rotation angle as a function of time.

that the law $\Omega(t)$ is imposed at the upper end of the system of Figure 11.1, and that the torsional stiffness and the damping coefficient of the part of the shaft between point C and the section in which such law is imposed are, respectively, k_t and c_t. The law $\theta(t)$ and its derivatives are linked with the angle of torsion of the shaft ϕ by the relationships

$$\theta = \theta_0 + \phi, \quad \dot{\theta} = \Omega + \dot{\phi}, \quad \ddot{\theta} = \dot{\Omega} + \ddot{\phi} \; .$$

Using rotating coordinates, the equations of motion are Equations (2.169) in which the torsion of the shaft ϕ is used instead of θ as a generalized coordinate and torque M_d has been substituted by $-k_t\phi - c_t\dot{\phi}$. Note also that the rotating frame is obtained from the fixed one by multiplying the coordinate r by $e^{-i\theta_0}$, whereas in the terms that come from $\cos(\theta)$ and $\sin(\theta)$, a term $e^{-i(\theta_0+\phi)} = e^{-i\theta_0}e^{-i\phi}$ is present.

Equations (2.169) become

$$\left\{\begin{array}{l} m\ddot{\rho}+\left[c+2im(\dot{\theta}+\dot{\phi})\right]\dot{\rho}+\left[k-m(\dot{\theta}+\dot{\phi})^2+i(\dot{\theta}+\dot{\phi})c_n+\right. \\ \qquad \left.+im\left(\ddot{\theta}+\ddot{\phi}\right)\right]\rho-m\epsilon\left[(\dot{\theta}+\dot{\phi})^2-i\left(\ddot{\theta}+\ddot{\phi}\right)\right]e^{i\phi}=F_ne^{-i\theta_0}, \\ (J_p+m\epsilon^2)\left(\ddot{\theta}+\ddot{\phi}\right)+m\epsilon\,\mathrm{Im}\left[\ddot{\rho}+2i(\dot{\theta}+\dot{\phi})\dot{\rho}+i\left(\ddot{\theta}+\ddot{\phi}\right)\rho+\right. = \\ \qquad \left.-(\dot{\theta}+\dot{\phi})^2\rho\right]=\epsilon\,\mathrm{Im}\left(F_ne^{-i(\theta_0+\phi)}\right). \end{array}\right. \tag{11.19}$$

Equations (11.19) can be linearized under the assumption that angle ϕ and all its derivatives are small. By neglecting the nonlinear terms and separating the real and the imaginary parts of the equations, it follows that

$$\begin{bmatrix} m & 0 & 0 \\ 0 & m & m\epsilon \\ 0 & m\epsilon & J_{pt} \end{bmatrix}\begin{Bmatrix} \ddot{\xi} \\ \ddot{\eta} \\ \ddot{\phi} \end{Bmatrix}+\begin{bmatrix} c & -2m\Omega & -2m\epsilon\Omega \\ 2m\Omega & c & 0 \\ 2m\epsilon\Omega & 0 & c_t \end{bmatrix}\begin{Bmatrix} \dot{\xi} \\ \dot{\eta} \\ \dot{\phi} \end{Bmatrix}+$$

$$\begin{bmatrix} k-m\Omega^2 & -c_n\Omega-m\dot{\Omega} & -m\epsilon\dot{\Omega} \\ c_n\Omega+m\dot{\Omega} & k-m\Omega^2 & -m\epsilon\Omega^2 \\ m\epsilon\dot{\Omega} & -m\epsilon\Omega^2 & k_t+\epsilon\left[F_x\cos(\theta_0)-F_y\sin(\theta_0)\right] \end{bmatrix}\begin{Bmatrix} \xi \\ \eta \\ \phi \end{Bmatrix}= \tag{11.20}$$

$$=\begin{Bmatrix} m\epsilon\Omega^2+F_x\cos(\theta_0)+F_y\sin(\theta_0) \\ -m\epsilon\dot{\Omega}-F_x\sin(\theta_0)+F_y\cos(\theta_0) \\ -J_{pt}\dot{\Omega}-\epsilon\left[F_x\sin(\theta_0)-F_y\cos(\theta_0)\right] \end{Bmatrix}.$$

From Equations (11.20) it is clear that the first two equations (the lateral behavior of the shaft) are very weakly coupled with the third one, expressing the torsional behavior, because the coupling terms are of the order of ϵ. If the unbalance is considered as small as the displacements and rotations, as it is usually the case, the coupling terms are of the same order as the other terms that are dropped in the linearization process. Furthermore, by stating that Ω is constant ($\dot{\Omega}=0$), Equations (11.20) can be used to study the torsional-flexural coupling at constant speed.

Equations (11.20) can be written in nondimensional form, using the transformations:

$$\xi^*=\frac{\xi}{\epsilon}, \quad \eta^*=\frac{\eta}{\epsilon}, \quad \tau=t\Omega_{cr},$$

where $\Omega_{cr}=\sqrt{k/m}$ is the critical speed of the Jeffcott rotor. The other nondimensional parameters are

$$\begin{array}{rcl} c_n^* & = & \dfrac{c_n}{2\sqrt{km}}, \quad c_r^*=\dfrac{c_r}{2\sqrt{km}}, \quad \Omega^*=\dfrac{\Omega}{\Omega_{cr}}, \quad a^*=\dfrac{\dot{\Omega}}{\Omega_{cr}^2}, \\ c_t^* & = & \dfrac{c_t}{2\sqrt{k_tJ_{pt}}}, \quad \alpha=\sqrt{\dfrac{kJ_{pt}}{mk_t}}, \quad \beta=\dfrac{k_t}{k\epsilon^2}, \quad F^*=\dfrac{F}{k\epsilon}. \end{array}$$

Note that α is the ratio between the torsional and the flexural natural frequency. The nondimensional equation is thus

$$\begin{bmatrix} 1 & 0 & 0 \\ 0 & 1 & 1 \\ 0 & 1 & \beta\alpha^2 \end{bmatrix} \begin{Bmatrix} \xi^{*''} \\ \eta^{*''} \\ \phi'' \end{Bmatrix} + 2 \begin{bmatrix} c^* & -\Omega^* & -\Omega^* \\ \Omega^* & c^* & 0 \\ \Omega^* & 0 & c_t^*\beta\alpha^* \end{bmatrix} \begin{Bmatrix} \xi^{*'} \\ \eta^{*'} \\ \phi' \end{Bmatrix} +$$

$$+ \begin{bmatrix} 1-\Omega^{*2} & -2\Omega^* c_n^* - a^* & -a^* \\ 2\Omega^* c_n^* + a^* & 1-\Omega^{*2} & -\Omega^{*2} \\ a^* & -\Omega^{*2} & \beta + F_x^* \cos(\theta_0) - F_y^* \sin(\theta_0) \end{bmatrix} \begin{Bmatrix} \xi^* \\ \eta^* \\ \phi \end{Bmatrix} = \tag{11.21}$$

$$= \begin{Bmatrix} \Omega^{*2} + F_x^* \cos(\theta_0) + F_y^* \sin(\theta_0) \\ -a^* - F_x^* \sin(\theta_0) + F_y^* \cos(\theta_0) \\ -\beta\alpha^2 a^* - F_x^* \sin(\theta_0) + F_y^* \cos(\theta_0) \end{Bmatrix} .$$

11.2 Nonstationary general Jeffcott rotor

As seen in the previous sections, if the assumption of constant angular velocity is dropped, the rotordynamics equations do not yield closed-form solutions. As a consequence, there is little difficulty in introducing other effects, like nonlinearity or lack of isotropy of both the stator and the rotor, which have the same result of preventing from obtaining closed-form solutions.

The equations of motion can be easily modified to accommodate the terms needed to describe nonlinearity and lack of isotropy. As an example, consider an isotropic Jeffcott rotor with a Duffing-type restoring force and linear damping, performing an acceleration with constant rate $\dot{\Omega} = a$. The equation of motion can be written in nondimensional form, using the same nondimensional variables and constants as seen for the case of Equation (11.15).

The nondimensional equation governing the lateral behavior of the system is then

$$\begin{Bmatrix} \xi^{*''} \\ \eta^{*''} \end{Bmatrix} + \frac{2}{a^*} \begin{bmatrix} c^* & -\tau \\ \tau & c^* \end{bmatrix} \begin{Bmatrix} \xi^{*'} \\ \eta^{*'} \end{Bmatrix} + \frac{1}{a^{*2}} \begin{bmatrix} 1-\tau^2 & -2\tau c_n^* - a^* \\ 2\tau c_n^* + a^* & 1-\tau^2 \end{bmatrix} \times$$

$$\times \begin{Bmatrix} \xi^* \\ \eta^* \end{Bmatrix} + \mu^* \left(\xi^{*2} + \eta^{*2}\right) \begin{Bmatrix} \xi^* \\ \eta^* \end{Bmatrix} = \frac{1}{a^{*2}} \begin{Bmatrix} \tau^2 \\ -a^* \end{Bmatrix} , \tag{11.22}$$

where the nondimensional nonlinearity parameter μ^* is defined as

$$\mu^* = \mu\epsilon^2 .$$

The equation allowing us to compute the driving torque cannot be obtained from the third Equation (11.15) or, which is the same, from Equation (11.12), but the last Equation (11.11) must be used. The latter becomes, in nondimensional form,

$$\frac{M_d - J_{p_{tot}} a}{m\epsilon^2 \Omega_{cr}^2} = a^* \left(\xi^* + a^* \eta^{*\prime\prime} + 2\tau\xi^{*\prime}\right) - \tau^2 \eta^* ,$$

where $J_{p_{tot}} = J_p + m\epsilon^2$.

A typical result of the numerical integration of Equation (11.22) is shown in Figure 11.5, where the results for both an acceleration and a deceleration maneuver are reported. The values of the nondimensional parameters used for the simulation are $a^* = 0.01$, $c^* = 0.11$, $c_n^* = 0.1$, and $\mu^* = 0.03$. In the same figure, the solutions obtained from the usual steady-state approach also are reported (dashed lines). Note that the values of the parameters are such that the system undergoes a downward jump and achieves a self-centered condition. As in the linear case, the presence of the angular acceleration reduces the peak amplitude; however, although in the linear system, the displacement peak occurs at a speed that is higher than that characterizing the maximum steady-state amplitude, the jump here occurs at a lower speed. The curve related to spin down is different from that related to the acceleration phase. This occurs also for linear rotors, when angular acceleration is accounted for, but here this effect is far greater because both acceleration and nonlinearity contribute to it. An oscillation with frequency that increases with increasing speed is present. By computing its period from the numerical results, it is clear that it occurs at a frequency that is almost exactly equal to twice the rotation frequency. This oscillation is present also in the linear case and can create numerical difficulties as it compels us to resort to very short time steps in order to simulate correctly the acceleration phase. This is particularly true in the case of very slow accelerations, when the number of rotations and hence of periods performed during the whole acceleration is very large. It is true that the amplitude of this oscillation proved to be decreasing with decreasing angular acceleration (when the acceleration tends to zero, the same occurs to the amplitude of the oscillation and the response tends to the classic steady-state unbalance response), but the presence of this sort of natural frequency of the system can drive the integration algorithm to instability. Tests performed in different cases with various algorithms (central differences, Newmark, Houbolt, and Wilson theta algorithms) suggest that even unconditionally stable ones are prone to suffer from this problem. This high-frequency oscillation reduces then the advantages of writing the equation of motion in the rotating frame instead of using an inertial one.

The driving torque has been reported in Figure 11.5(b) as a function of the spin speed.

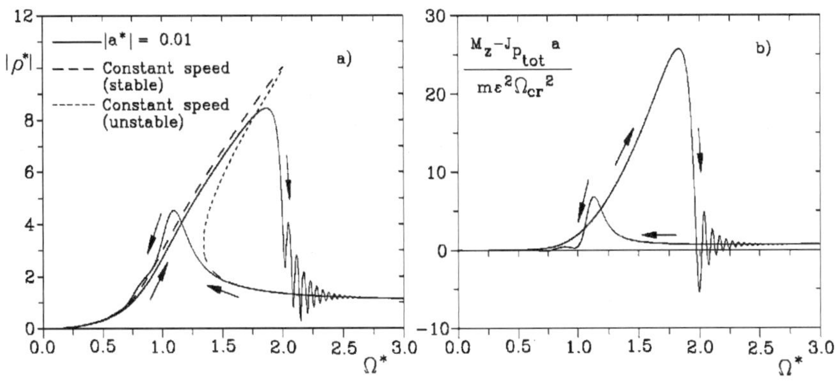

FIGURE 11.5. Rotor response during acceleration and deceleration, compared with the steady-state unbalance response. (a) Amplitude; (b) driving torque. $a^* = 0.01$, $c^* = 0.11$, $c_n^* = 0.1$, and $\mu^* = 0.03$.

Another example is a linear Jeffcott rotor with nonrotating elastic anisotropy undergoing a constant-rate acceleration. Assume that x- and y-axes of the nonrotating reference frame are principal axes of elasticity, and consequently, the terms k_{xy} and k_{yx} of the stiffness matrix vanish. The equation of motion for the flexural behavior can be written in the rotating frame: The fact that some coefficients are time-dependent is not a severe drawback, because the solution must be anyway obtained through numerical integration in time. Written in the same nondimensional form seen for Equation (11.15), it reduces to

$$\left\{ \begin{array}{c} \xi^{*''} \\ \eta^{*''} \end{array} \right\} + \frac{2}{a^*} \left[\begin{array}{cc} c^* & -\tau \\ \tau & c^* \end{array} \right] \left\{ \begin{array}{c} \xi^{*'} \\ \eta^{*'} \end{array} \right\} + \frac{1}{a^{*2}} \left[\begin{array}{cc} 1-\tau^2 & -2\tau c_n^* - a^* \\ 2\tau c_n^* + a^* & 1-\tau^2 \end{array} \right] \times$$
$$\times \left\{ \begin{array}{c} \xi^* \\ \eta^* \end{array} \right\} + \frac{1}{a^{*2}} \frac{\Delta k}{k} \left\{ \begin{array}{c} \xi^* \cos(2\theta) - \eta^* \sin(2\theta) \\ -\xi^* \sin(2\theta) - \eta^* \cos(2\theta) \end{array} \right\} = \frac{1}{a^{*2}} \left\{ \begin{array}{c} \tau^2 \\ -a^* \end{array} \right\}, \tag{11.23}$$

where $\Delta k = k_x - k_y$ is the difference between the stiffness in the xz- and yz-planes and all the nondimensional parameters have been computed with reference to the average stiffness k. The behavior of the system depends again only on four nondimensional parameters, a^*, c^*, c_n^*, and $\Delta k/k$. The equation yielding the driving torque coincides with that for the previous case. If a^* tends to zero, Equation (11.23) has a steady-state solution coinciding with the usual expression yielding the unbalance response of a linear nonisotropic Jeffcott rotor, consisting in an elliptical whirling, with the axes of the ellipses not coinciding with x- and-y axes, owing to the presence of damping.

A typical result of the numerical integration of Equation (11.23) is shown in Figure 11.6, where only an acceleration maneuver is reported. The values

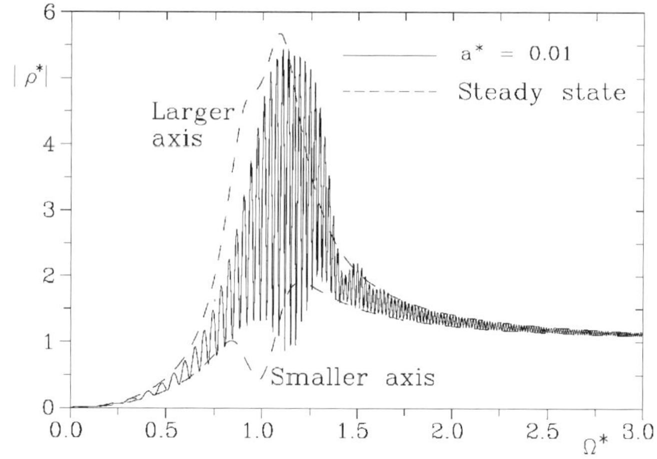

FIGURE 11.6. Acceleration of a Jeffcott rotor running on linear nonsymmetrical supports through its two critical speeds. Amplitude of the motion during the acceleration, compared with the steady-state amplitude. As the steady-state whirling is elliptical, the length of the smaller and larger axes of the orbits are shown. $a^* = 0.01$, $c^* = 0.11$, $c_n^* = 0.1$, and $\Delta k/k = 0.2$.

of the nondimensional parameters used for the simulation are $a^* = 0.01$, $c^* = 0.11$, $c_n^* = 0.1$, and $\Delta k/k = 0.2$. Since the orbits of steady-state solution are elliptical, the larger and smaller axes have been reported (dashed lines). The pattern of the solution related to the accelerating rotor shows very strong oscillations, which in this case are mainly from asymmetry. If the acceleration tends to zero, the pattern would be at any rate oscillatory (in the rotating frame), the oscillations spanning between the smaller and larger axes of the steady-state elliptical solution.

In the high supercritical range, the difference between the steady-state and the accelerating solutions is small, whereas when crossing the critical speed, a larger difference can be found. The frequency of the oscillation proved to be close to twice the rotation frequency, the same as that of the steady-state solution, as dictated by the elliptical whirling.

The trajectories in both the nonrotating plane xy and the rotating plane $\xi\eta$ are shown in Figure 11.7. It is clear that the study of the trajectory in the rotating plane is simpler also in this case, although point P describes complex orbits in both.

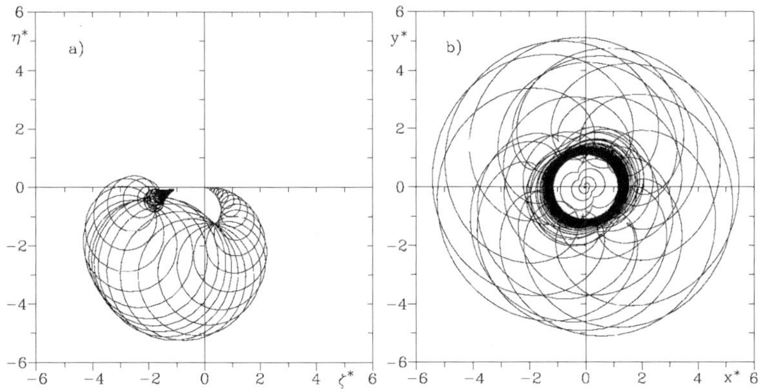

FIGURE 11.7. Trajectories of the same accelerating rotor of Figure 11.6, in the (a) rotating and in the (b) nonrotating planes

11.3 Nonstationary rotor with four degrees of freedom

To take into account the effects of gyroscopic moments on nonstationary rotordynamics, a simple four degrees-of-freedom model, similar to the one studied in Chapter 3, can be considered. However, for greater generality, all six degrees of freedom of the rigid body attached to the shaft will be considered. The expression of the velocity of the center of gravity of the rotor is still Equation (3.9), where, however, $\dot{\theta}$ must be introduced instead of the constant angular velocity Ω.

The translational and rotational kinetic energy are still expressed by Equation (3.10) and Equation (3.13) with some obvious small modifications. Their sum is

$$\begin{aligned}\mathcal{T} = {} & \frac{1}{2}m\left\{\dot{X}^2+\dot{Y}^2+\dot{Z}^2+\epsilon^2\dot{\theta}^2+2\epsilon\dot{\theta}\left[-\dot{X}\sin(\theta+\alpha)+\right.\right.\\ & \left.\left.+\dot{Y}\cos(\theta+\alpha)\right]\right\}+\frac{1}{2}J_t(\dot{\phi}_{X'}^2+\dot{\phi}_y^2+\chi^2\dot{\theta}^2)+ \\ & +\frac{1}{2}J_p(\dot{\theta}^2+2\dot{\theta}\dot{\phi}_{X'}\phi_y)+\dot{\theta}\chi(J_p-J_t)\left[\dot{\phi}_{X'}\cos(\theta)+\dot{\phi}_y\sin(\theta)\right] .\end{aligned} \tag{11.24}$$

The forces and moments acting on the rigid body in point C (Figure 3.2) that will be considered are those caused by the elastic reaction of the shaft. As all six degrees of freedom of the shaft are considered, such forces and moments are F_X, F_Y, F_Z, M_x, M_y, and M_z. The virtual work is

$$\delta\mathcal{L} = \begin{Bmatrix} F_X \\ F_Y \\ F_Z \end{Bmatrix}^T \begin{Bmatrix} \delta X \\ \delta X \\ \delta X \end{Bmatrix} + \\ + \begin{Bmatrix} M_x \\ M_y \\ M_z \end{Bmatrix}^T \left(\mathbf{R}_2 \begin{Bmatrix} \delta\phi_{X'} \\ 0 \\ 0 \end{Bmatrix} + \begin{Bmatrix} 0 \\ \delta\phi_y \\ 0 \end{Bmatrix} + \begin{Bmatrix} 0 \\ 0 \\ \delta\theta \end{Bmatrix} \right) ; \tag{11.25}$$

i.e.

$$\delta\mathcal{L} = F_X\delta X + F_Y\delta Y + F_Z\delta Z + \left[M_x \cos\left(\phi_y\right) + M_z \sin\left(\phi_y\right)\right] \delta\phi_{X'} + \\ + M_y\delta\phi_y + M_z\delta\theta \; . \tag{11.26}$$

The equations of motion can be written using Lagrange equations

$$\frac{d}{dt}\left(\frac{\partial \mathcal{T}}{\partial \dot{q}_i}\right) - \frac{\partial \mathcal{T}}{\partial q_i} = \frac{\partial \delta\mathcal{L}}{\partial \delta q_i} , \tag{11.27}$$

obtaining

$$\begin{cases} m\ddot{X} - m\epsilon\dot{\theta}^2 \left[\ddot{\theta}\sin(\theta+\alpha) + \dot{\theta}^2\cos(\theta+\alpha)\right] = F_X, \\ m\ddot{Y} - m\epsilon\dot{\theta}^2 \left[-\ddot{\theta}\cos(\theta+\alpha) + \dot{\theta}^2\sin(\theta+\alpha)\right] = F_Y, \\ m\ddot{Z} = F_Z, \\ J_t\ddot{\phi}_{X'} + J_p\dot{\theta}\dot{\phi}_y + J_p\dot{\theta}\dot{\phi}_y - \chi(J_t - J_p)\left[\ddot{\theta}\cos(\theta) - \dot{\theta}^2\sin(\theta)\right] = \\ \qquad = M_x + \phi_y M_z, \\ J_t\ddot{\phi}_y - J_p\Omega\dot{\phi}_{X'} - \chi(J_t - J_p)\left[\ddot{\theta}\sin(\theta) + \dot{\theta}^2\cos(\theta)\right] = M_y, \\ \left(J_p + J_t\chi^2 + m\epsilon^2\right)\ddot{\theta} + m\epsilon\left[\ddot{X}\sin(\theta) + \ddot{Y}\cos(\theta)\right] + \\ \qquad -\chi(J_t - J_p)\left[\ddot{\phi}_{X'}\cos(\theta) + \ddot{\phi}_y\sin(\theta)\right] = M_z. \end{cases} \tag{11.28}$$

By computing M_z from the last equation and introducing it into the fourth one, and neglecting nonlinear terms, the latter simplifies as

$$J_t\ddot{\phi}_{X'} + J_p\dot{\theta}\dot{\phi}_y - \chi(J_t - J_p)\left[\ddot{\theta}\cos(\theta) - \dot{\theta}^2\sin(\theta)\right] = M_x \; . \tag{11.29}$$

The third equation is completely uncoupled from the other ones, showing that the axial behavior of the system can be studied separately.

The first two equations are strongly coupled with the fourth and fifth ones, exactly in the same way already seen for the steady-state behavior of the system. By introducing the complex coordinates

$$\begin{cases} r = X + iY\,, \\ \phi = \phi_y - i\phi_{X'}\,, \end{cases}$$

and by linking the forces exerted by the shaft on the rigid body with the displacements through the stiffness matrix, they yield

$$\begin{cases} m\ddot{r} + K_{11}r + K_{12}\phi = m\epsilon\left(\dot{\theta}^2 - i\ddot{\theta}\right)e^{i(\Omega t+\alpha)}\,, \\ J_t\ddot{\phi} - iJ_p\dot{\theta}\dot{\phi} + K_{12}r + K_{22}\phi = \chi\Omega^2(J_t - J_p)\left(\dot{\theta}^2 - i\ddot{\theta}\right)e^{i\Omega t}\,, \end{cases} \tag{11.30}$$

or by introducing the usual matrix notation,

$$\mathbf{q} = \left\{ \begin{array}{c} r \\ \phi \end{array} \right\}, \qquad \mathbf{M} = \left[\begin{array}{cc} m & 0 \\ 0 & J_t \end{array} \right], \qquad \mathbf{G} = \left[\begin{array}{cc} 0 & 0 \\ 0 & J_p \end{array} \right],$$

$$\mathbf{K} = \left[\begin{array}{cc} K_{11} & K_{12} \\ K_{12} & K_{22} \end{array} \right], \qquad \mathbf{f} = \left\{ \begin{array}{c} m\epsilon e^{i\alpha} \\ \chi(J_t - J_p) \end{array} \right\},$$

they can be written in the form

$$\mathbf{M}\ddot{\mathbf{q}} - i\dot{\theta}\mathbf{G}\dot{\mathbf{q}} + \mathbf{K}\mathbf{q} = \left(\dot{\theta}^2 - i\ddot{\theta}\right)\mathbf{f}e^{i\theta}\,. \tag{11.31}$$

Finally, the last equation can be written in the form

$$\left(J_p + J_t\chi^2 + m\epsilon^2\right)\ddot{\theta} = -\operatorname{Im}\left(\overline{\mathbf{f}}^T\ddot{\mathbf{q}}e^{i\theta}\right) + M_z. \tag{11.32}$$

As usual, the presence of damping can be accounted for by introducing the nonrotating and rotating damping matrices into Equation (11.31)

$$\mathbf{M}\ddot{\mathbf{q}} + (\mathbf{C}_n + \mathbf{C}_r - i\dot{\theta}\mathbf{G})\dot{\mathbf{q}} + (\mathbf{K} - i\dot{\theta}\mathbf{C}_r)\mathbf{q} = \left(\dot{\theta}^2 - i\ddot{\theta}\right)\mathbf{f}e^{i\Omega t}\,. \tag{11.33}$$

Often an alternative form of the second Equation (11.30) related to the complex degree of freedom ϕ can be found in the literature. It can be reduced to

$$J_t\ddot{\phi} - iJ_p\left(\dot{\theta}\dot{\phi} + \ddot{\theta}\phi\right) + K_{12}r + K_{22}\phi = \chi\Omega^2(J_t - J_p)\left(\dot{\theta}^2 - i\ddot{\theta}\right)e^{i\Omega t}\,, \tag{11.34}$$

and it is equivalent to the second Equation (11.30) here considered, provided that bending moments are expressed in the XYZ frame instead of xyz. In the case of small displacements, it is customary to refer all forces to the undeflected configuration of the structural member, and the two approaches should lead to equivalent results. Actually, this occurs if the angular acceleration $\ddot{\theta}$ is the second derivative of one of the generalized coordinates and can be considered a small quantity, as in the case of torsional vibrations of a system rotating at constant speed. In this case, term

$J_p\ddot{\theta}\phi$ is one of the terms that are dropped when linearizing the equations. A different situation occurs in the case in which the angular acceleration is imposed and $\ddot{\theta}$ can be theoretically arbitrarily large. Here Equation (11.30) will be used without the modification (11.34) for two reasons: It seems more correct to relate the bending moments to the xyz frame, i.e., to the deflected configuration of the system, when the two approaches lead to different results, and the term $J_p\ddot{\theta}\phi$ is at any rate far smaller than the term $J_p\dot{\theta}\dot{\phi}$, except perhaps if the rotor starts from standstill, and even then only in the first instants of motion. In the case of synchronous whirling, their ratio is equal to $\dot{\Omega}/\Omega^2$. It must be stressed that the second term in the expression $\left(\dot{\theta}^2 - i\ddot{\theta}\right)$ cannot be neglected even if it is far smaller than the first one. Actually, in expression $\left(\dot{\theta}\dot{\phi} + \ddot{\theta}\phi\right)$ there is the arithmetic sum of two terms, one of which is negligible with respect to the other one, whereas in expression $\left(\dot{\theta}^2 - i\ddot{\theta}\right)$ two vectors perpendicular to each other are added, as the imaginary unit suggests. Even if the absolute value of the *out-of-plane* term $\ddot{\theta}$ is very small, its presence can be very important because no other force acts in its direction. Self-centering cannot be explained at all if this term is neglected.

11.4 Generic, torsionally stiff, multi-degrees-of-freedom system

If the elastic behavior of the system is such that axial, torsional, and flexural behavior are uncoupled, the same considerations seen in the previous section for the simple model with six degrees of freedom also hold for more complex models. In particular, within the frame of the linearized theory, the axial degrees of freedom are uncoupled from flexural ones. If the rotor is torsionally stiff, i.e., the torsional rotations of all cross sections are equal, and the acceleration is performed with an imposed law $\theta(t)$ [and then with stated laws $\Omega(t)$ and $\dot{\Omega}(t)$], the flexural behavioral can also be studied independently using equations of the type of Equation (11.33)]. The only equation for the rotational degree of freedom is of the type of Equation (11.32). By introducing the usual complex coordinates, it follows that

$$\left\{\begin{array}{l} \mathbf{M}\ddot{\mathbf{q}} + (\mathbf{C}_n + \mathbf{C}_r - i\Omega\mathbf{G})\dot{\mathbf{q}} + (\mathbf{K} - i\Omega\mathbf{C}_r)\mathbf{q} = (\Omega^2 - i\dot{\Omega})\mathbf{f}e^{i\theta}\,, \\ M_z = J_{p_{tot}}\dot{\Omega} + \Im\left(\overline{\mathbf{f}}^T\ddot{\mathbf{q}}e^{-i\theta}\right)\,, \end{array}\right. \tag{11.35}$$

where $J_{p_{tot}}$ is the total moment of inertia of the rotor about the z-axis, also taking into account the presence of static and couple unbalance. By introducing $\ddot{\theta} = 0$ and the solution for steady-state whirling into the last

Equation (11.35), the driving torque needed to maintain a constant angular velocity is readily obtained. To perform the numerical integration in time of Equation (11.35) it is easier to resort to rotating coordinates, as seen for the case of the Jeffcott rotor. The relevant equations of motion are then

$$\begin{cases} \mathbf{M}\ddot{\mathbf{r}} + \left[\mathbf{C}_n + \mathbf{C}_r - i\Omega(2\mathbf{M} - \mathbf{G})\right]\dot{\mathbf{r}} + \left[\mathbf{K} - \Omega^2(\mathbf{M} - \mathbf{G}) + \right. \\ \qquad \left. + i(\dot{\Omega}\mathbf{M} + \Omega\mathbf{C}_n)\right]\mathbf{r} = (\Omega^2 - i\dot{\Omega})\mathbf{f}\,, \\ M_z = J_{p_{tot}}\dot{\Omega} + \Im\left[\overline{\mathbf{f}}^T\left(\ddot{\mathbf{r}} + i\dot{\Omega}\mathbf{r} + 2i\Omega\dot{\mathbf{r}} - \Omega^2\mathbf{r}\right)\right]. \end{cases} \tag{11.36}$$

Note that the rotating reference frame used for Equation (11.36) rotates at a variable spin speed $\Omega(t)$.

However, as the solution of Equation (11.36) must be at any rate performed by integrating it numerically in the time domain, there is no difficulty in introducing nonlinear terms [a generic vector function $\mathbf{f}(q_i, \dot{q}_i, \theta, t)$] and mean and deviatoric matrices to account for both rotating and nonrotating anisotropy. The resulting equation is

$$\begin{aligned} &+\mathbf{M}_m\ddot{\mathbf{q}} + \left(\mathbf{C}_m - i\Omega\mathbf{G}\right)\dot{\mathbf{q}} + \left(\mathbf{K}_m - i\Omega\mathbf{C}_{r_m}\right)\mathbf{q} + \mathbf{M}_{n_d}\ddot{\overline{\mathbf{q}}} + \\ &+\mathbf{M}_{r_d}e^{2i\theta}\ddot{\overline{\mathbf{q}}} + \mathbf{C}_{n_d}\dot{\overline{\mathbf{q}}} + \left(\mathbf{C}_{r_d} + 2i\Omega\mathbf{M}_{r_d}\right)e^{2i\theta}\dot{\overline{\mathbf{q}}} + \mathbf{K}_{n_d}\overline{\mathbf{q}} + \\ &+\left(\mathbf{K}_{r_d} + i\Omega\mathbf{C}_{r_d}\right)e^{2i\theta}\overline{\mathbf{q}} + \mathbf{f}(q_i, \dot{q}_i, \theta, t) = \left(\Omega^2 - i\dot{\Omega}\right)\mathbf{f}_r e^{i\theta} + \mathbf{f}_n\,. \end{aligned} \tag{11.37}$$

Equation (11.37) has been obtained with the only assumption of uncoupling among flexural, axial, and torsional behavior.

The equation that allows us to describe the rotational degree of freedom of the system, which is just one caused by the assumption of a torsionally rigid rotor, is again the second Equation (11.35), which is not affected by either nonlinearities or deviations from axial symmetry.

In general, it is reasonable to expect that the time history of the accelerating system has a simpler expression in the rotating frame, where a slow variation of generalized coordinates in time should occur, than in the fixed frame, where the relevant quantities vary with a frequency equal to the rotational speed. However, if the system is nonisotropic, the unbalance response is at least polyharmonic and then fast variation can occur in both the rotating and nonrotating frames. By introducing the rotating coordinates into Equation (11.37), it yields

$$\begin{aligned}&\mathbf{M}_m\ddot{\mathbf{r}} + \left[\mathbf{C}_m + i\Omega\left(2\mathbf{M}_m - \mathbf{G}\right)\right]\dot{\mathbf{r}} + \left[\mathbf{K}_m - \Omega^2\left(\mathbf{M}_m - \mathbf{G}\right) + \right.\\ &\left. +i\left(a\mathbf{M}_m + \Omega\mathbf{C}_{r_m}\right)\right]\mathbf{r} + \mathbf{M}_{n_d}e^{-2i\theta}\ddot{\overline{\mathbf{r}}} + \left(\mathbf{C}_{n_d} + 2i\Omega\mathbf{M}_{n_d}\right)e^{-2i\theta}\dot{\overline{\mathbf{r}}} + \\ &+\left[\mathbf{K}_{n_d} - \Omega^2\mathbf{M}_{n_d} - i\left(a\mathbf{M}_{n_d} + \Omega\mathbf{C}_{n_d}\right)\right]e^{-2i\theta}\overline{\mathbf{r}} + \mathbf{M}_{r_d}\ddot{\overline{\mathbf{r}}} + \mathbf{C}_{r_d}\dot{\overline{\mathbf{r}}} + \\ &+\left(\mathbf{K}_{r_d} + \Omega^2\mathbf{M}_{r_d} - i\dot{\Omega}\mathbf{M}_{r_d}\right)\overline{\mathbf{r}} + \mathbf{f}'(r_i, \dot{r}_i, \theta, t)\}e^{-i\theta} = \\ &= \left(\Omega^2 - ia\right)\mathbf{f}_r + \mathbf{f}_n e^{-i\theta}\,.\end{aligned} \tag{11.38}$$

The equation yielding the driving torque is still the second Equation (11.36).

If the angular acceleration is not vanishingly small, Equation (11.38) must be integrated numerically in time, and as a consequence, there is no conceptual difficulty taking into account both nonlinearities and asymmetry. If software allowing us to deal with complex quantities is available, there is no need to split the equation into its real and imaginary parts before integrating: Once laws $\dot{\Omega}(t)$, $\Omega(t)$, and $\theta(t)$ are stated, the numerical integration is straightforward.

Remark 11.1 *Equation (11.38) is a powerful tool to study by numerical integration in time many difficult rotordynamics problems, like constant speed whirling of unsymmetrical rotors or chaotic behavior caused by the simultaneous presence of nonlinearities and asymmetry.*

11.5 Blade loss

Another important case of nonstationary working condition occurs when the unbalance of a rotor changes very quickly, i.e. when the characteristic times of the change of unbalance are comparable with the natural frequencies of the system. An instance in which this occurs is when a small part of the rotor detaches and flies away, leaving the rotor in a strongly unbalanced condition. The release of a blade from a turbine disc (blade loss) is perhaps the most likely example of an occurrence of this type.

An accident in which a small fragment, or a whole blade, detaches from a fast rotating machine has usually severe consequences, which can even include the total destruction of the rotor of the machine, with the projection of fast-moving fragments able to do much damage. If the fragment is small, however, it might be possible that the rotor, the bearings, and the other parts of the system can withstand the added loads for a time sufficient to stop the machine. In this case, the danger may come only from the fragment. A containment structure can be provided, which must fulfill the two tasks of containing the fragment, i.e., of preventing it to fly around, and of nesting it, i.e., stopping it in a position clear from the other parts of the rotor, in such a way that no secondary damage is caused.

The loads on the rotor caused by the loss of a blade or any other part can be easily computed by considering the unbalance as variable in time. In the simple case of a Jeffcott rotor spinning at constant speed Ω, the relevant equation of motion is Equation (2.75), slightly modified to take into account that now ϵ is not constant

$$m\ddot{r}_C + (c_r + c_n)\dot{r}_C + (k - ic_r\Omega)r_C = m\epsilon(t)\Omega^2 e^{i\Omega t} + F_n \ . \tag{11.39}$$

Remark 11.2 *This approach is only a first approximation to the correct solution of the problem, because the mass of the rotor changes in time, owing to the loss of the fragment, which also carries away a part of the energy of the system. To attempt to solve the problem in a more correct way by inserting $\epsilon = \epsilon(t)$ into the expression of the Lagrangian function would not solve the problem. Nevertheless, Equation (11.39) can be used when the mass of the fragment is negligible when compared with the mass of the rotor.*

For an order of magnitude evaluation of the effect of the blade loss, consider a perfectly balanced undamped Jeffcott rotor, which at time $t = 0$ instantly acquires the unbalance $m\epsilon_0$ aligned along ξ-axis. By resorting to real coordinates, the relevant equation of motion is

$$\begin{cases} m\ddot{x}_C(t) + kx_C(t) = m\epsilon_0 u(t)\Omega^2 \cos(\Omega t) \ , \\ m\ddot{y}_C(t) + ky_C(t) = m\epsilon_0 u(t)\Omega^2 \sin(\Omega t) \ , \end{cases} \tag{11.40}$$

where $u(t)$ is the unit step function

$$\begin{cases} u = 0 \quad \text{for } t < 0 \ , \\ u = 1 \quad \text{for } t \geq 0 \ . \end{cases}$$

The solution of Equation (11.40) can be easily obtained by considering that after the time $t = 0$, the unbalance is constant, and hence the solution is the general solution for constant unbalance with the initial conditions $x_C = y_C = 0$ and $\dot{x}_C = \dot{y}_C = 0$ for $t = 0$.

The general solution of the homogeneous equation is

$$\begin{cases} x_C(t) = x_1 \sin(\omega_n t) + x_2 \cos(\omega_n t) \ , \\ y_C(t) = y_1 \sin(\omega_n t) + y_2 \cos(\omega_n t) \ , \end{cases} \tag{11.41}$$

where $\omega_n = \sqrt{k/m}$. The particular solution of the complete equation is Equation (2.28)

$$\begin{cases} x_C(t) = x_0 \cos(\Omega t) \ , \\ y_C(t) = x_0 \sin(\Omega t) \ , \end{cases} \tag{11.42}$$

where x_0 is the steady-state amplitude of the unbalance response

$$x_0 = \epsilon_0 \frac{m\Omega^2}{k - m\Omega^2} = \epsilon_0 \frac{\Omega^2}{\omega_n^2 - \Omega^2} \; . \tag{11.43}$$

The initial conditions on the position yield

$$x_2 + x_0 = 0 \; , \qquad y_2 = 0, \tag{11.44}$$

whereas those on the velocity are

$$\omega_n x_1 = 0 \; , \qquad \omega_n y_1 + \Omega x_0 = 0. \tag{11.45}$$

The solution for the blade loss of an undamped Jeffcott rotor is thus

$$\begin{cases} x_C(t) = x_0 \left[\cos(\Omega t) - \cos(\omega_n t)\right] \; , \\ y_C(t) = x_0 \left[\sin(\Omega t) - \frac{\Omega}{\omega_n} \sin(\omega_n t)\right] \; . \end{cases} \tag{11.46}$$

The response to the blade loss depends on the speed of the rotor or, better, on ratio Ω/ω_n, i.e. Ω/Ω_{cr}. An upper limit for the maximum amplitude in $x-$direction is then twice the steady-state amplitude, whereas an upper limit of that in y-direction is the steady-state amplitude multiplied by $1+ \Omega/\omega_n$. The actual maximum values may be lower than that, depending on whether Ω/ω_n is a rational number. The orbits obtained for four different values of Ω/ω_n are reported in Figure 11.8. The response is made nondimensional by dividing it by the steady-state amplitude. In the subcritical field, the overshoot is about 2 and the maximum amplitude occurs in the x-direction (i.e., in the direction in which the unbalance points at time $t = 0$ or the direction opposite to that in which the blade is cast). In the supercritical field, the maximum amplitude occurs in y-direction and the overshoot is about $1+ \Omega/\omega_n$. Close to the critical speed, the amplitude grows almost linearly in time and the orbit is close to a logarithmic spiral, at least at the beginning. Owing to the very large value of the steady-state amplitude, much time is needed to reach it.

If damping is taken into account, it is advisable to use directly Equation (11.39) and to work using complex coordinates.

Operating in the usual way, the solution can be obtained by adding the general solution of the homogeneous equation to the particular integral of the complete equation

$$r_C(t) = r_1 e^{s_1 t} + r_2 e^{s_2 t} + r_0 e^{i\Omega t} \; , \tag{11.47}$$

where r_1 and r_2 are two complex constants to be computed from the initial conditions, s_1 and s_2 are the two solutions (2.78) of the characteristic Equation (2.77), and r_0 is the complex steady-state amplitude of the unbalance response given by Equation (2.89). The initial conditions yield the following set of two complex equations:

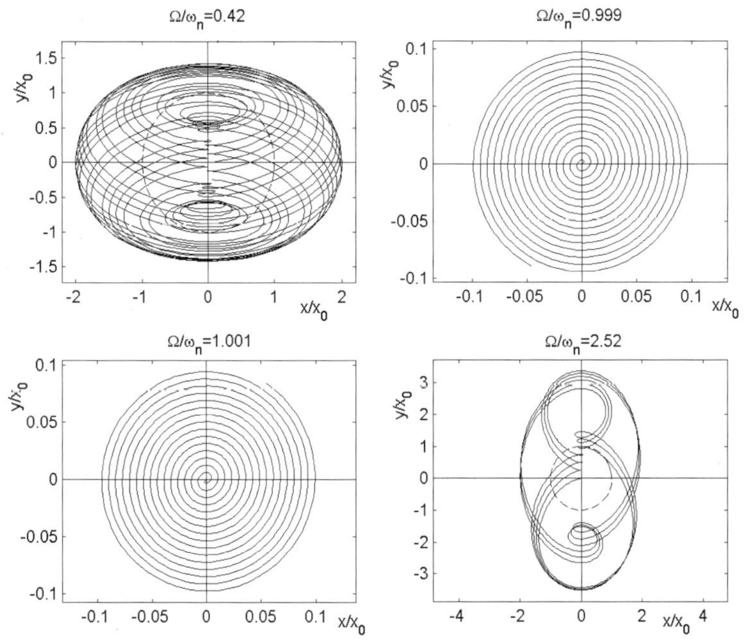

FIGURE 11.8. Orbits following the loss of a blade in an undamped Jeffcott rotor for four different values of Ω/ω_n. A subcritical case ($\Omega/\omega_n = 0.42$), a supercritical case ($\Omega/\omega_n = 2.52$), and two cases close to the critical speed ($\Omega/\omega_n = 0.999$ and $\Omega/\omega_n = 1.001$). The steady-state orbit is reported in two cases with a dashed line.

$$\begin{cases} r_1 + r_2 = -r_0 \ , \\ s_1 r_1 + s_2 r_2 e^t = -i\Omega r_0 \ , \end{cases} \tag{11.48}$$

yielding the constants r_1 and r_2.

The same responses shown in Figure 11.8 but for a damped Jeffcott rotor with $\zeta = 0.04$ and $\zeta_r = 0.01$ are plotted in Figure 11.9. The system reaches the steady-state orbit, but particularly in the case of supercritical working, the overshot may be large.

In case of more complex models, or when also the initial unbalance is accounted for, the equations of motion can be integrated numerically in time introducing a time-varying unbalance.

Example 11.1 *Blade loss on a turbojet.*

Consider the turbojet engine described in Example 8.2. Assume that a blade of the turbine is lost while spinning at 15,000 rpm, and compute the resulting orbit. As the blade has a mass of 0.025 kg and its center of mass is located at a radius of 147 mm, the static unbalance of the turbine wheel after a blade has been lost is $m\epsilon = 3675$ *gmm. It is very large unbalance, corresponding to a balancing grade*

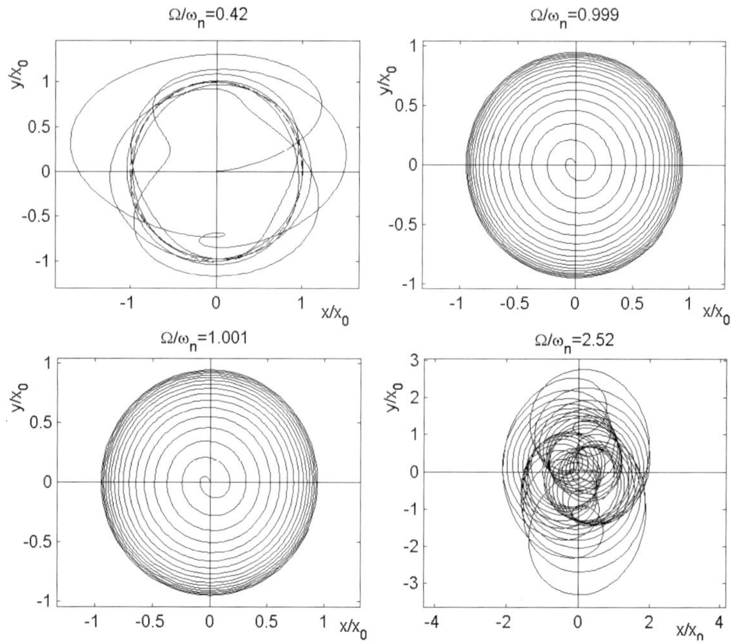

FIGURE 11.9. Same plot as in Figure 11.8 but for a damped Jeffcott rotor with $\zeta = 0.04$ and $\zeta_r = 0.01$.

G1270 at 20,000 rpm. The design unbalance (G2.5) can then be neglected in the study of the blade loss dynamics.

By introducing the relevant value of the unbalance in node 21 (center of mass of the turbine wheel) and integrating numerically in time the equations of motion, the results shown in Figure 11.10 are obtained. The orbit in the fixed frame obtained in the first instants after the loss of the blade (full line) and the orbit after 0.5 s (dashed line) are reported in Figure 11.10(a). The same orbits, but plotted in the rotor-fxed frame, are shown in Figure 11.10(b). Note that after 0.5 s, the steady-state conditions have practically been reached and in the rotating frame the orbit reduces to a point. The time history of the amplitude is reported in Figure 11.10(c).

The whirling amplitude is very large, and it is likely that a failure of the bearings or of the rotor occurs. Even if the rotor can withstand the large unbalance loads, the use of a linearized model for such large amplitudes is questionable.

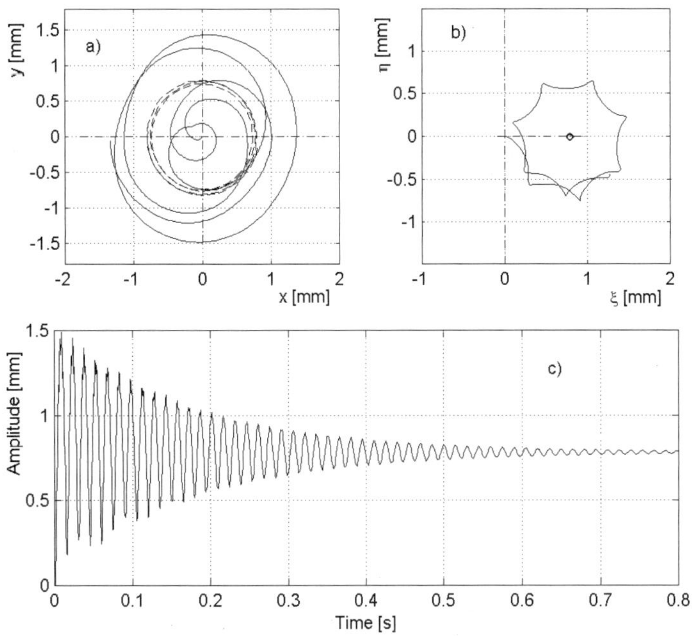

FIGURE 11.10. (a) Orbit of the center of mass of the turbine wheel following the loss of a blade, plotted in the fixed frame. First instants after the loss of the blade (full line) and after 0.5 s (dashed line). (b) Same as (a), but plotted in the rotor-fxed frame. (c) The time history of the amplitide.

12

Dynamic behavior of free rotors

Rotors are defined by the ISO as bodies rotating about a fixed axis, constrained to do so by bearings, acting as cylindrical hinges. Bearings may be more or less compliant, but even in this case, their deformations are considered to be small if compared with the dimensions of the system. The dynamic study of rotors is thus usually performed under the assumptions of small displacements and rotations. These assumptions allow a number of linearizations of both the inertial and the elastic terms in the equations of motion. The assumptions of small displacements and rotations are retained even when studying the behavior of nonlinear rotors, the nonlinearities being usually ascribed to bearings (of the fluid, rolling elements or even magnetic type), dampers, or other causes, as the presence of cracks.

Moreover, the angular velocity is assumed to be constant, or at least a known function of time, because the presence of a driving system that controls the spin speed is postulated.

However, there are rotating systems that are not constrained by bearings, like spinning spacecraft or celestial bodies. They can after all be considered as rotors and are often defined as *free rotors*, as opposed to the more conventional *fixed rotors*, supported in bearings.

Although the absence of the bearings does not change the nature of the problem, the displacements can be far greater in the case of free rotors and this makes the small displacements and, above all, small rotations assumptions more problematic. Moreover, although in the case of fixed rotors the spin speed is assumed to be constant (or, more in general, a known function of time), free rotors are usually studied under the assumption of constant angular momentum. Spinning spacecraft attitude dynamics and rotordynamics are usually seen as two separate branches of dynamics, each one with its own notation, distinctive approach, and limitations, whereas they can be dealt with in a unified approach.

Spacecraft attitude dynamics usually deals with single rigid bodies or with multibody systems, in which the inertial properties of only one of them is considered as relevant for the study of the dynamic behavior of the spacecraft [59, 60]. For the attitude control and guidance, the spacecraft is assumed to be a single rigid body, with its flexible parts not affecting significantly the overall dynamic behavior of the vehicle. However, spacecraft made of several bodies with relevant mass and moments of inertia connected through very compliant structural element require a reconsidering of this assumption. The low stiffness (or better, the low value of the natural frequency) of the structure causes the attitude dynamics to be coupled with the vibration dynamics; such coupling can make the vibration isolation of the payload more difficult and, above all, can cause stability problems.

This issue is usually taken into account using multibody dynamics codes, which are based on the numerical integration in time of a complete nonlinear model of the system. They allow us to simulate the spacecraft attitude and vibration dynamics in detail, but this interaction is still difficult to investigate in a general way.

The experience in the field of conventional fixed rotors can help in clarifying some of these aspects, provided that the consequences of the assumptions of small displacements and, above all, of constant spin speed, which are usually done in rotordynamics, are fully understood. In particular, the constant speed assumption may cause a violation of the conservation of the angular momentum: This issue must be clarified before the usual models seen for fixed rotors can be used for free rotors as well.

12.1 Single rigid-body rotor

12.1.1 General considerations

Consider a free rotor made of a single rotating rigid body free in space. Let $\Omega_{\xi\eta\zeta}$ be the angular velocity expressed in the rotor-fixed frame $G\xi\eta\zeta$, centered in the center of mass, with axes $\xi\eta\zeta$ coinciding with the principal axes of inertia of the rigid body. As no assumption is made on the shape of the body, the moments of inertia about the principal axes of inertia may be all different from each other. Without loss of generality, let $J_\xi < J_\eta < J_\zeta$. The inertia tensor is then

$$\mathbf{J} = \begin{bmatrix} J_\xi & 0 & 0 \\ 0 & J_\eta & 0 \\ 0 & 0 & J_\zeta \end{bmatrix} . \tag{12.1}$$

The angular momentum and the kinetic energy of the rigid body are then

$$\mathbf{H} = \mathbf{J}\,\boldsymbol{\Omega}_{\xi\eta\zeta} = \left\{ \begin{array}{c} J_\xi \Omega_\xi \\ J_\eta \Omega_\eta \\ J_\zeta \Omega_\zeta \end{array} \right\} , \tag{12.2}$$

$$\mathcal{T} = \frac{1}{2}\boldsymbol{\Omega}^T_{\xi\eta\zeta}\,\mathbf{J}\,\boldsymbol{\Omega}_{\xi\eta\zeta} = \frac{1}{2}\left(J_\xi \Omega_\xi^2 + J_\eta \Omega_\eta^2 + J_\zeta \Omega_\zeta^2\right) . \tag{12.3}$$

As no forces and moments act on the body, both the angular momentum and the kinetic energy remain constant.

By writing the kinetic energy in terms of the components of the angular momentum instead of the components of the angular velocity, the condition that it remains constant yields

$$2\mathcal{T} = \left(\frac{H_\xi^2}{J_\xi} + \frac{H_\eta^2}{J_\eta} + \frac{H_\zeta^2}{J_\zeta} \right) = \text{ constant.} \tag{12.4}$$

As the direction of the angular velocity (i.e., the direction of the axis of rotation) and then of the angular momentum is not known in the rotor-fixed frame (in the inertial frame, the direction of the angular momentum is constant, but the rotor-fixed frame moves with respect to the inertial frame during whirling), the condition of constant angular momentum may be written, in the rotor-fixed frame, only by considering its modulus

$$|\mathbf{H}|^2 = H_\xi^2 + H_\eta^2 + H_\zeta^2 = \text{ constant.} \tag{12.5}$$

The last condition states that vector $\mathbf{H}$ is bound to remain on a sphere with radius $|\mathbf{H}|$ in the rotor-fixed frame. The condition on the kinetic energy $\mathcal{T}$ states that vector $\mathbf{H}$ is bound to remain on an ellipsoid in the rotor-fixed frame. Its equation is

$$\frac{H_\xi^2}{a_\xi^2} + \frac{H_\eta^2}{a_\eta^2} + \frac{H_\zeta^2}{a_\zeta^2} = 1 . \tag{12.6}$$

The semi-axes of such an ellipsoid are

$$\left\{ \begin{array}{l} a_\xi = \sqrt{2\mathcal{T} J_\xi} \ , \\ a_\eta = \sqrt{2\mathcal{T} J_\eta} \ , \\ a_\zeta = \sqrt{2\mathcal{T} J_\zeta} \ . \end{array} \right. \tag{12.7}$$

As vector $\mathbf{H}$ must lie on both the sphere and the ellipsoid, it must lie on the intersection between them. The intersection between the sphere and the ellipsoid is then the locus of the points defining the possible trajectories of $\mathbf{H}$ in the rotor-fixed frame. Both the sphere and the ellipsoid, and then the locus of $\mathbf{H}$, depend on the initial condition, which define the values of both $|\mathbf{H}|$ and $\mathcal{T}$.

Rotation about axis	ξ (min. J)	ζ (max. J)	η (interm. J)
Ellipsoid	external to sphere	internal to sphere	intersects sphere
Intersection	2 points	2 points	line
Behavior	stable	stable	indifferent

TABLE 12.1. Behavior of a rigid free rotor without damping.

Consider a rigid body spinning about ξ-axis, i.e., the axis of minimum moment of inertia (like in the case of *long rotors*). Initially, both Ω_η and Ω_ζ vanish; the modulus of the angular momentum is $|\mathbf{H}| = H_{\xi_0}$, and that of the kinetic energy is $2\mathcal{T} = H_{\xi_0}^2/J_\xi$.

The semi-major axes of the ellipsoid are

$$\begin{cases} a_\xi = H_{\xi_0} = |\mathbf{H}| \ , \\ a_\eta = H_{\xi_0}\sqrt{\frac{J_\eta}{J_\xi}} = |\mathbf{H}| \sqrt{\frac{J_\eta}{J_\xi}} \ , \\ a_\zeta = H_{\xi_0}\sqrt{\frac{J_\zeta}{J_\xi}} = |\mathbf{H}| \sqrt{\frac{J_\zeta}{J_\xi}} \ . \end{cases} \tag{12.8}$$

As $J_\xi < J_\eta < J_\zeta$, the ellipsoid is all external to the sphere, being tangent to it on its shortest axis. The intersection is a pair of points, and then the position of $\mathbf{H}$ is fixed also in the rotating frame. The only possible motion is a steady state rotation about ξ-axis, which means that rotation about the axis with minimum moment of inertia is stable.

In a similar way, if the rigid body spins about ζ-axis, i.e., the axis of maximum moment of inertia (like in the case of *disc rotors*), the ellipsoid can be demonstrated to lie all inside the sphere being tangent to it on its longest axis. Again the intersection reduces to a pair of points, and the only possible motion is a steady state rotation about ζ-axis, which means that also rotation about the axis with maximum moment of inertia is stable.

On the contrary, if the rigid body spins about axis η, i.e., the axis of intermediate moment of inertia, the ellipsoid lies partly inside and partly outside the sphere and the intersection is a line. The situation for the case with $J_\xi = 0.5J_\eta$ and $J_\zeta = 2J_\eta$ is reported in Figure 12.1. The angular momentum vector can move freely on the sphere-ellipsoid intersection, or better, all positions, on such an intersection are indifferent equilibrium points. Rotation about the axis with intermediate moment of inertia is then unstable in the sense that any small disturbances will cause the rotor to move in a complex way, with large precessions.

These possibilities are summarized in Table 12.1.

The intersections for different initial conditions are plotted in Figure 12.2(a) for the case of a rotor with $J_\xi = 0.8J_\eta$ and $J_\zeta = 1.2J_\eta$. The plot of the angular velocity is reported in Figure 12.2(b). The intersection of the sphere (ellipsoid in the case of the angular velocity) with the H_ξ-axis represents the stable situation of rotation about the axis with minimum moment of inertia. If the initial conditions are close to a rotation about ξ-

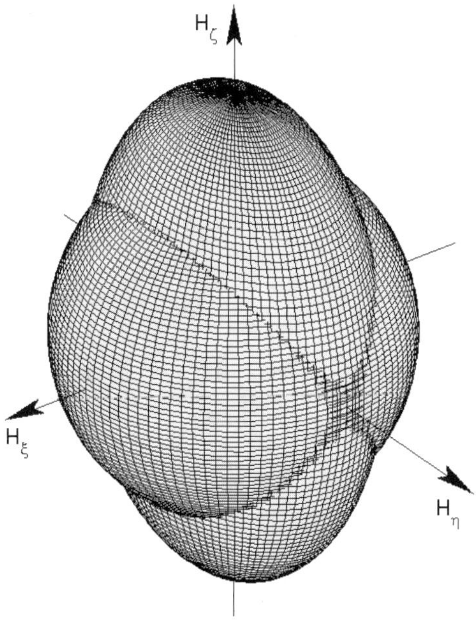

FIGURE 12.1. Intersection between sphere and ellipsoid when the body spins initially about the axis with intermediate moment of inertia. Case with $J_\xi = 0.5J_\eta$ and $J_\zeta = 2J_\eta$.

axis, the locus is a small circle, showing that the rotor whirls with a small amplitude. A similar situation occurs at the intersection with the H_ζ-axis, representing a stable rotation about the axis with maximum moment of inertia.

The intersection with H_η-axis represent the unstable rotation about the axis with intermediate moment of inertia. In this case, the trajectory of **H** in the rotor-fixed frame is large and the rotor may have a complex tumbling motion.

Many rotors have a *gyroscopic* geometry; i.e., two of the principal moments of inertia are equal. As already seen in Chapter 3, two cases are possible: The moment of inertia different from the others may be greater than the other two ($J_\xi = J_\eta < J_\zeta$; disc rotor) or smaller ($J_\xi < J_\eta = J_\zeta$; long rotor). The plots of the angular momentum vector for these two cases are reported in Figures 12.3(a) and (b), respectively. Rotation about the axis with a moment of inertia different from the other two (H_ζ in the first case, H_ξ in the second one) is stable, and rotation about the other two axes is unstable.

Remark 12.1 *The above considerations hold only for the case of a perfectly rigid rotor. Actually, no real-life rotating system is such, and small deformations are always present. These deformations involve always some*

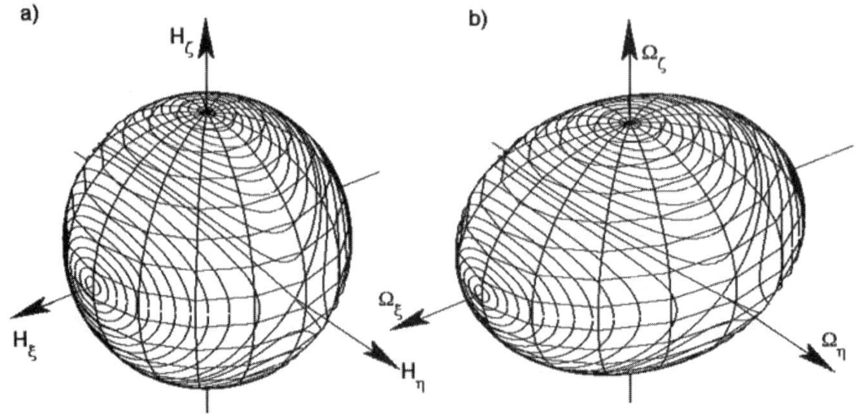

FIGURE 12.2. (a) Loci of the angular momentum for a free rotor with $J_\xi = 0.8 J_\eta$ and $J_\zeta = 1.2 J_\eta$. (b) Loci of the angular velocity for the same rotor.

dissipation of energy and, bacause we have assumed that the rotor does not interact with the outside world (free rotor), this energy can be dissipated only at the expense of the kinetic energy.

As the angular momentum is bound to remain constant, the sphere represented in Figure 12.1 remains constant, whereas the ellipsoid, representing the condition of constant kinetic energy, shrinks in time (the axes expressed by Equation (12.7) decrease with decreasing $\mathcal{T}$).

If the rotor is spinning about the axis with maximum moment of inertia, the ellipsoid is, as already said, all internal to the sphere and is tangent to it along the H_ζ-axis. In this case, the ellipsoid cannot contract any further, because there would be no intersection (i.e., a violation of the condition of constant angular momentum), and the only possible motion is rotation about the axis with maximum moment of inertia, which is stable. No energy dissipation can occur.

If the rotation occurs in a situation close to this one, the rotor whirls with decreasing amplitude while losing energy, until its rotation occurs about ζ-axis (trajectory starting from point A in Figure 12.4). Also, in the case of a quasi-rigid, damped rotor, rotation about ζ-axis is stable. The opposite occurs in the case of a rotation about the axis with the smallest moment of inertia: While becoming smaller, the ellipsoid intersects the sphere more and more far from the intersection with the H_ξ-axis (trajectory starting from point B in Figure 12.4). The rotor whirls with increasing amplitude and finally will end to spin about ζ-axis. In the case of a quasi-rigid, damped rotor, rotation about ξ-axis is unstable.

The various possible situations are summarized in Table 12.2.

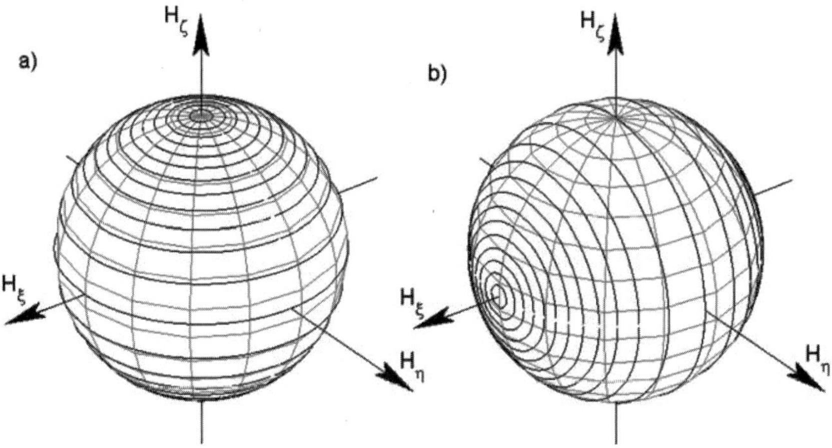

FIGURE 12.3. Plots of the loci of the angular momentum vector for the two cases of (a) $J_\xi = J_\eta < J_\zeta$ (axis ζ is symmetry axis), disc rotor, and (b) $J_\xi < J_\eta = J_\zeta$ (axis ξ is symmetry axis), long rotor.

Rotation about axis	ξ (min. J)	ζ (max. J)	η (interm. J)
Behavior	unstable	stable	unstable

TABLE 12.2. Behavior of a slightly damped quasi-rigid free rotor.

Remark 12.2 *If the presence of internal damping of the rotor is accounted for, the rotation of a free, quasi rigid disc rotor is stable, whereas that of a free long rotor is not.*

12.1.2 Equations of motion

Instead of using the classic approach based on the Euler angles and Euler equations to describe the dynamics of a rigid body, the same generalized coordinates and reference frames seen in Chapter 3 for rotors with four degrees of freedom will be used. In the present case, the assumptions of small displacements and rotations are not adequate, and thus all six degrees of freedom of the rigid body will be accounted for.

The rotor can be sketched as in Figure 3.1(b), where the two springs modeling the bearings are not present. The absence of the bearings allows us to neglect both the static and the couple unbalance, which were defined with reference to geometric center of the shaft C. In the present case, point C cannot be defined and the position of the body is defined by the position of its center of mass, i.e., point P (or G) in Figure 3.2. Also, axes $\xi\eta z$ are directly the principal axes of inertia.

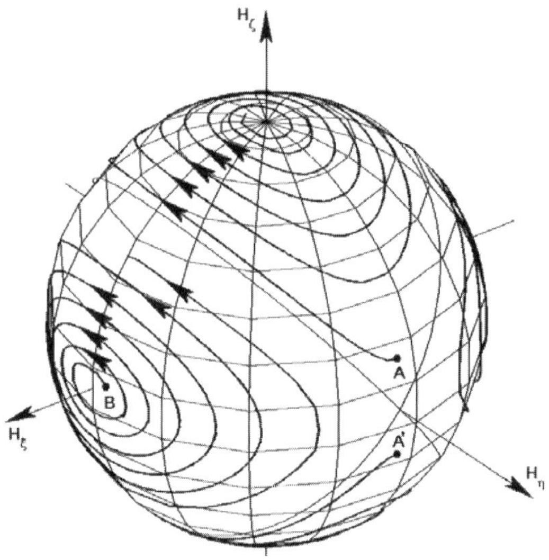

FIGURE 12.4. Loci of the angular momentum for a slightly damped quasi-rigid free rotor with $J_\xi = 0.8J_\eta$ and $J_\zeta = 1.2J_\eta$. Trajectories starting from points A and A' are stable; those starting from B are unstable.

The reference frames $OXYZ$ (inertial frame), OΞHZ (rotating frame), Gxyz (whirling frame), and G$\xi\eta z$ (rotor-fixed frame) are the same as defined in Chapter 3; also the rotation matrices $\mathbf{R}_1$, $\mathbf{R}_2$, and $\mathbf{R}_3$ are defined in the same way. The X-, Y-, and Z-coordinates of point G and angles φ_X, φ_y, and θ are taken as generalized coordinates of the system.

As X, Y, and Z are the coordinates of the center of mass, the translational kinetic energy is simply

$$\mathcal{T}_t = \frac{1}{2}m(\dot{X}^2 + \dot{Y}^2 + \dot{Z}^2) \ . \tag{12.9}$$

For the computation of the rotational kinetic energy, the angular velocity must be expressed in the rotor-fixed reference frame G$\xi\eta z$. Remembering that there is no couple unbalance, Equation (3.11) simplifies as

$$\boldsymbol{\Omega} = \mathbf{R}_3\mathbf{R}_2 \left\{ \begin{array}{c} \dot{\varphi}_X \\ 0 \\ 0 \end{array} \right\} + \mathbf{R}_3 \left\{ \begin{array}{c} 0 \\ \dot{\varphi}_y \\ 0 \end{array} \right\} + \left\{ \begin{array}{c} 0 \\ 0 \\ \dot{\theta} \end{array} \right\} . \tag{12.10}$$

Without resorting to small angles assumptions, Equation (12.10) yields

$$\boldsymbol{\Omega} = \left\{ \begin{array}{c} \dot{\varphi}_X \cos(\theta) \cos\left(\varphi_y\right) + \dot{\varphi}_y \sin(\theta) \\ -\dot{\varphi}_X \sin(\theta) \cos\left(\varphi_y\right) + \dot{\varphi}_y \cos(\theta) \\ \dot{\varphi}_X \sin\left(\varphi_y\right) + \dot{\theta} \end{array} \right\} . \tag{12.11}$$

No assumption of gyroscopic body will be made: The inertia tensor expressed in the same reference frame G$\xi\eta z$ is then

$$\mathbf{J} = \begin{bmatrix} J_\xi & 0 & 0 \\ 0 & J_\eta & 0 \\ 0 & 0 & J_z \end{bmatrix} . \tag{12.12}$$

As the components of $\boldsymbol{\Omega}$ are referred to the principal axes of inertia, the rotational kinetic energy can be easily computed as

$$\begin{gathered}\mathcal{T}_r = \tfrac{1}{2} J_\xi \left[\dot{\varphi}_X \cos(\theta) \cos\left(\varphi_y\right) + \dot{\varphi}_y \sin(\theta)\right]^2 + \\ + \tfrac{1}{2} J_\eta \left[-\dot{\varphi}_X \sin(\theta) \cos\left(\varphi_y\right) + \dot{\varphi}_y \cos(\theta)\right]^2 + \tfrac{1}{2} J_z \left[\dot{\varphi}_X \sin\left(\varphi_y\right) + \dot{\theta}\right]^2 .\end{gathered} \tag{12.13}$$

The equations of motion can be obtained as usual through the Lagrange equation. By performing the relevant derivatives, the six equations of motion are obtained

$$\begin{cases} m\ddot{X} = Q_X , \\ m\ddot{Y} = Q_Y , \\ m\ddot{Z} = Q_Z , \end{cases} \tag{12.14}$$

$$\begin{cases} \left\{\left[J_\xi \cos^2(\theta) + J_\eta \sin^2(\theta)\right] \cos^2(\varphi_y) + J_z \sin^2(\varphi_y)\right\} \ddot{\varphi}_X + \\ \quad + (J_\xi - J_\eta) \sin(\theta) \cos(\theta) \cos(\varphi_y) \ddot{\varphi}_y + J_z \sin(\varphi_y) \ddot{\theta} + \\ \quad -2(J_\xi - J_\eta) \sin(\theta) \cos(\theta) \cos^2(\varphi_y) \dot{\theta} \dot{\varphi}_X + \\ \quad + \left\{(J_\xi - J_\eta)\left[\cos^2(\theta) - \sin^2(\theta)\right] + J_z\right\} \cos(\varphi_y) \dot{\theta} \dot{\varphi}_y + \\ \quad -2\left\{\left[J_\xi \cos^2(\theta) + J_\eta \sin^2(\theta)\right] - J_z\right\} \cos(\varphi_y) \sin(\varphi_y) \dot{\varphi}_X \dot{\varphi}_y + \\ \quad - (J_\xi - J_\eta) \sin(\theta) \cos(\theta) \sin(\varphi_y) \dot{\varphi}_y^2 = Q_{\varphi_X} , \\ \left[J_\xi \sin^2(\theta) + J_\eta \cos^2(\theta)\right] \ddot{\varphi}_y + (J_\xi - J_\eta) \sin(\theta) \cos(\theta) \cos(\varphi_y) \ddot{\varphi}_X + \\ \quad +2(J_\xi - J_\eta) \sin(\theta) \cos(\theta) \dot{\theta} \dot{\varphi}_y + \left\{(J_\xi - J_\eta)\left[\cos^2(\theta) - \sin^2(\theta)\right] + \right. \\ \quad \left. - J_z\right\} \cos(\varphi_y) \dot{\theta} \dot{\varphi}_X + \left[J_\xi \cos^2(\theta) + J_\eta \sin^2(\theta) - J_z\right] \cdot \\ \quad \cdot \cos(\varphi_y) \sin(\varphi_y) \dot{\varphi}_X^2 = Q_{\varphi_y} , \\ J_z \ddot{\theta} + J_z \sin(\varphi_y) \ddot{\varphi}_X - \left\{(J_\xi - J_\eta)\left[\cos^2(\theta) - \sin^2(\theta)\right] - J_z\right\} \cdot \\ \quad \cdot \cos(\varphi_y) \dot{\varphi}_y \dot{\varphi}_X + (J_\xi - J_\eta) \sin(\theta) \cos(\theta) \cos^2(\varphi_y) \dot{\varphi}_X^2 + \\ \quad -(J_\xi - J_\eta) \sin(\theta) \cos(\theta) \dot{\varphi}_y^2 = Q_\theta . \end{cases} \tag{12.15}$$

The three equations governing the translational motion are completely uncoupled from those describing rotations and coincide with the equations of motion of a point mass.

The six generalized forces Q_i vanish in the case of a free rotor on which no external forces are acting. Otherwise, they must be computed once the physical characteristics of the system that controls the motion of the rotor are defined.

In the simplified case of an axially symmetrical (gyroscopic) rotor ($J_\xi = J_\eta = J_t$, $J_z = J_p$), by solving the last Equation (12.15) in $\ddot{\theta}$ and substituting it into the second one, the equations of motion for rotations reduce to

$$\begin{cases} \ddot{\varphi}_X = -\left(\frac{J_p}{J_t} - 2\right)\tan(\varphi_y)\dot{\varphi}_y\dot{\varphi}_X - \frac{J_p}{J_t\cos(\varphi_y)}\dot{\theta}\dot{\varphi}_y + \frac{Q_{\varphi_X} - \sin(\varphi_y)Q_\theta}{J_t\cos^2(\varphi_y)}, \\ \ddot{\varphi}_y = \left(\frac{J_p}{J_t} - 1\right)\sin(\varphi_y)\cos(\varphi_y)\dot{\varphi}_X^2 + \frac{J_p}{J_t}\cos(\varphi_y)\dot{\theta}\dot{\varphi}_X + \frac{Q_{\varphi_y}}{J_t}, \\ \ddot{\theta} = -\sin(\varphi_y)\ddot{\varphi}_X - \cos(\varphi_y)\dot{\varphi}_y\dot{\varphi}_X + \frac{Q_\theta}{J_p}. \end{cases} \tag{12.16}$$

In case of a body rotating about z-axis, with small angular displacements about the two other axes, angles φ_x and φ_y and their derivatives can be regarded as small quantities, whereas no such assumption can be done for θ and $\dot{\theta}$. In this case, Equations (12.16) reduce to

$$\begin{cases} J_t\ddot{\varphi}_X + J_p\dot{\theta}\dot{\varphi}_y = Q_{\varphi_X} - \varphi_y Q_\theta, \\ J_t\ddot{\varphi}_y - J_p\dot{\theta}\dot{\varphi}_X = Q_{\varphi_y}, \\ J_p\ddot{\theta} = Q_\theta. \end{cases} \tag{12.17}$$

By remembering that $Q_{\varphi_X} - \varphi_y Q_\theta = M_X$, $Q_{\varphi_y} = M_y$, and $Q_\theta = M_z$ [a consequence of applying the small angle assumption to Equation (11.26)], Equations (12.17) coincide with Equations (11.30) and (11.32) once that unbalance has been neglected.

In the case of a free rotor, where the generalized forces vanish, the third equation states that the angular velocity is constant

$$\dot{\theta} = \text{constant}. \tag{12.18}$$

Remark 12.3 *This is an important result, because it shows that if angles φ_x and φ_y are assumed to be small, constant angular velocity is a result of constant angular momentum. This allows us to use with confidence for spinning spacecraft dynamics the results obtained for classic rotordynamics, if the amplitude of the precessional motion is small. This assumption is acceptable in most cases, because the attitude control system is designed to keep precession within small limits.*

In the case of constant angular velocity, the rotational dynamics about x- and y- axes (angles φ_X and φ_y) can be studied separately and the first two Equations (12.17) coincide with Equations (11.28), once the elastic reaction of the bearings and unbalance have been neglected.

By introducing the usual complex coordinate

$$\phi = \phi_y - i\phi_{X'} ,$$

and neglecting the external forces, the first two Equations (12.17) reduce to

$$J_t\ddot{\phi} - iJ_p\dot{\theta}\dot{\phi} = 0 \; . \tag{12.19}$$

The free whirling of the free rotor can thus be studied by introducing a solution of the type $\phi = \phi_0 e^{st}$ into Equation (12.19), obtaining the following algebraic linear equation:

$$\phi_0 \left(J_t s^2 + isJ_p\Omega \right) = 0 \, . \tag{12.20}$$

The characteristic equation yielding the whirling frequencies is

$$J_t s^2 + isJ_p\Omega = 0 \, , \tag{12.21}$$

whose solutions are

$$\begin{cases} s_1 = 0 \; , \\ s_2 = i\Omega \frac{J_p}{J_t} \; . \end{cases} \tag{12.22}$$

A free spinning rigid body has then five vanishing natural frequencies, plus a natural frequency that is proportional to the spin speed, the constant of proportionality being ratio J_p/J_t. The Campbell diagram of a free rigid rotor is shown in Figure 12.5 for different values of J_p/J_t. The ω and Ω scales are made nondimensional using an arbitrary value of the spin speed Ω_0.

Remark 12.4 *From Figure 12.5, it is clear that in the case of disc rotors ($J_p > J_t$), the curve lies in the subcritical range, and hence rotating damping cannot destabilize the system. The opposite occurs for long rotors ($J_p < J_t$): This result is the same as that shown in Section 12.1.1, obtained using a different approach.*

Example 12.1 *The Levitron: a case of a quasi-free rotor.*

A good example of the dynamics of a free rigid rotor, very loosely constrained, is the Levitron^{TM}: a small toy built with the aim of demonstrating that gyroscopic moments can, in certain conditions, stabilize passive magnetic levitation. It consists of a permanent magnet in the form of a spinning top, which can float about an equilibrium position located above a heavy base containing a magnetized ceramic square slab. The top is spun by hand on a plastic lifter plate located on the permanent magnet and then slowly raised to levitation height, approximately 40 mm above the surface of the base. At this point, the top leaves the plate and spins in mid-air for over 2 minutes, fully trapped in three dimensions inside the magnetic field $\overrightarrow{B}$ provided by the base, until its spin velocity Ω decreases, owing to air drag, to about 100 rad/s.

Figure 12.6(a) shows a sketch of the Levitron: The small top with mass m, polar and transversal moments of inertia Jp and J_t, is made of a nonmagnetic spindle inserted into a toroidal permanent magnet; the base is a square slab, uniformly magnetized perpendicularly to the plane surface, and it has a large circular hole

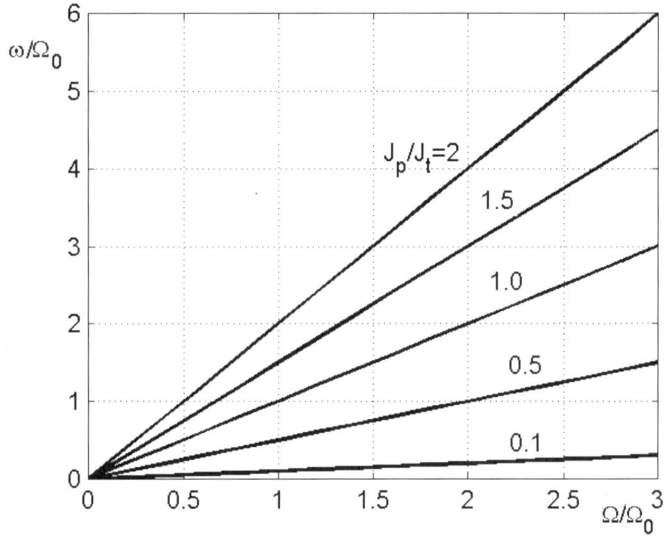

FIGURE 12.5. Campbell diagram of a free rigid rotor for different values of J_p/J_t. The ω and Ω scales are made nondimensional using an arbitrary value of the spin speed Ω_0.

at its center, whose purpose is to provide an approximately field-free region where the top can be spun up before being lifted.

The inertial reference frame $O'X'Y'Z$ *is centered on the center of the circular hole of the magnet [Figure 12.6(b)] with the* Z*-axis being regarded as a symmetry axis for the divergence- and curl-free magnetic field, which at a short distance from this axis, can be considered as completely described by its axial* B_Z *and radial* B_R *components (if the magnet was cylindrical instead of a square slab with a central circular hole, this assumption would have been exact). The angular momentum* $\vec{H}$, *and the angular speed* $\vec{\Omega}$ *are represented in the same figure. The top can be regarded as a magnetic dipole with moment vector* $\vec{\mu}$ *(fixed magnitude* μ*) centered in* G *and directed along its axis of symmetry. The gradients of the magnetic field* $\vec{B}$ *compensate for the gravitational force by generating a repulsive force that acts on* $\vec{\mu}$ *in the presence of* $\vec{H}$ *(whose gyroscopic effect prevents the top from overturning and falling) and must provide the mechanism for the top to levitate in a stable way above the base.*

The magnetic field can be computed by expanding its components about the equilibrium point (which is unstable when the top does not spin), located on the Z*-axis at the height* Z_0 *corresponding to the condition of vertical equilibrium between the gravitational force and the upward magnetic force*

$$mg = -\mu \left. \frac{\partial B_Z}{\partial Z} \right|_{Z=Z_0} . \tag{12.23}$$

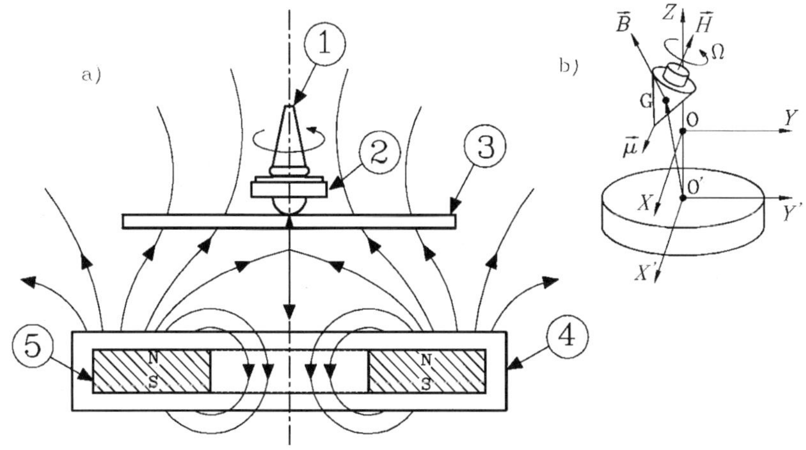

FIGURE 12.6. (a): Sketch of the Levitron: (1) spinning top, (2) toroidal ceramic magnet, (3) lifter plate, (4) fixed base, (5) square ceramic magnet with central hole. (b): Inertial reference frames O'$X'Y'Z$, centered in the center of the magnet, and OXYZ, centered in the static equilibrium position of the top.

The inertial frame $OXYZ$ has its origin O in the equilibrium point ($X' = Y' = 0$, $Z = Z_0$). Using the curl- and divergence-equations $\vec{\nabla} \wedge \vec{B} = \vec{0}$ and $\vec{\nabla} \cdot \vec{B} = 0$, the second-order Taylor's expansion gives [6]

$$
\begin{aligned}
B_X &= -\frac{1}{2} S_0 X - K_0 XZ + ..., \\
B_Y &= -\frac{1}{2} S_0 Y - K_0 YZ + ..., \\
B_Z &= B_0 + S_0 Z + K_0 Z^2 - \frac{1}{2} K_0 \left(X^2 + Y^2\right) + ...,
\end{aligned}
\tag{12.24}
$$

where B_0, S_0, and K_0 are, respectively, the values of the vertical component B_Z, of the axial derivative of B_Z and of the semi-curvature of B_Z at the equilibrium point ($X = Y = Z = 0$)

$$
S_0 = \left. \frac{\partial B_Z}{\partial Z} \right|_{X=Y=Z=0}, \qquad K_0 = \frac{1}{2} \left. \frac{\partial^2 B_Z}{\partial Z^2} \right|_{X=Y=Z=0}. \tag{12.25}
$$

The equations of motion of the system are Equations (12.14) and Equations (12.16), where the generalized forces Q_i can be obtained from the potential energy $\mathcal{U}$, sum of the gravitational term $\mathcal{U}_G = mgZ$, and the magnetic one $\mathcal{U}_M = -\vec{\mu} \cdot \vec{B}$.

The magnetic dipole moment $\vec{\mu}$, expressed in the inertial frame $OXYZ$, must be computed first. Remembering that the components of vector $\vec{\mu}$ with respect to the principal reference frame $G\xi\eta z$ are 0, 0 and $-\mu$, the following expression is obtained:

$$
\vec{\mu} = \left\{ \begin{array}{c} -\mu \sin\left(\varphi_y\right) \\ \mu \sin\left(\varphi_X\right) \cos\left(\varphi_y\right) \\ -\mu \cos\left(\varphi_X\right) \cos\left(\varphi_y\right) \end{array} \right\}. \tag{12.26}
$$

The potential energy is then

$$\mathcal{U} = mgZ - \mu \sin(\varphi_y)\left(\tfrac{S_0}{2} + K_0 Z\right) X + \mu \sin(\varphi_X)\cos(\varphi_y)\left(\tfrac{S_0}{2} + K_0 Z\right) Y +$$
$$+\mu \cos(\varphi_X)\cos(\varphi_y)\left[B_0 + S_0 Z + K_0 Z^2 - \tfrac{K_0}{2}\left(X^2 + Y^2\right)\right]. \tag{12.27}$$

The generalized forces corresponding to the translational and rotational degrees of freedom are

$$\left\{\begin{array}{l}
Q_X = \mu\left[\left(\frac{S_0}{2} + K_0 Z\right)\sin(\varphi_y) + K_0 X \cos(\varphi_X)\cos(\varphi_y)\right],\\
Q_Y = \mu\left[K_0 Y \cos(\varphi_X) - \left(\frac{S_0}{2} + K_0 Z\right)\sin(\varphi_X)\right]\cos(\varphi_y),\\
Q_Z = -\mu\left[K_0\left(Y \sin(\varphi_X)\cos(\varphi_y) - X\sin(\varphi_y)\right) + \right.\\
\qquad \left. + (S_0 + 2K_0 Z)\cos(\varphi_X)\cos(\varphi_y)\right] - mg,\\
Q_{\varphi_X} = \mu \cos(\varphi_y)\left\{-\left(\frac{S_0}{2} + K_0 Z\right) Y \cos(\varphi_X) + \right.\\
\qquad \left. + \sin\varphi_X\left[B_0 + S_0 Z + K_0 Z^2 - \frac{K_0}{2}\left(X^2 + Y^2\right)\right]\right\},\\
Q_{\varphi_y} = \mu\left\{\left(\frac{S_0}{2} + K_0 Z\right)\left[X\cos(\varphi_y) + Y \sin(\varphi_X)\sin(\varphi_y)\right] + \right.\\
\qquad \left. + \cos(\varphi_X)\sin(\varphi_y)\left[B_0 + S_0 Z + K_0 Z^2 - \frac{K_0}{2}\left(X^2 + Y^2\right)\right]\right\},\\
Q_\theta = 0.
\end{array}\right. \tag{13}$$

A set of six nonlinear coupled differential equations is thus obtained; its solution may be attempted only by numerical integration in time.

However, it is possible to linearize the equations about the equilibrium position to study the stability in the small of the system. Such a linearization yields

$$\left\{\begin{array}{l}
\ddot{X} = \frac{\mu K_0}{m} X + \frac{\mu S_0}{2m}\varphi_y,\\
\ddot{Y} = \frac{\mu K_0}{m} Y - \frac{\mu S_0}{2m}\varphi_X,\\
\ddot{Z} = -\frac{2\mu K_0}{m} Z,\\
\ddot{\varphi}_X = -\frac{\mu S_0}{2J_t} Y + \frac{\mu B_0}{J_t}\varphi_X - \frac{J_p\omega}{J_t}\dot{\varphi}_y,\\
\ddot{\varphi}_y = \frac{\mu S_0}{2J_t} X + \frac{\mu B_0}{J_t}\varphi_y - \frac{J_p\omega}{J_t}\dot{\varphi}_X,\\
\ddot{\theta} = 0.
\end{array}\right. \tag{12.28}$$

Owing to linearization, the third Equation (vertical motion) and the last one (spin) uncouple from the equations describing the lateral behavior, which involve *the X, Y, φ_X and φ_y coordinates. Moreover, both the spin velocity $\dot{\theta} = \Omega$ and the angular momentum $\overrightarrow{H}$ are constant.*

By introducing the usual complex coordinates $r = X + iY$ and $\varphi = \varphi_y - i\varphi_X$, the equations of motion for the two (complex) degrees of freedom rotor are obtained

$$\begin{cases} m\ddot{r} - \mu K_0 r - \frac{\mu S_0}{2}\varphi = 0, \\ J_t\ddot{\varphi} - i\Omega J_p\dot{\varphi} - \frac{\mu S_0}{2}r - \mu B_0\varphi = 0. \end{cases} \tag{12.29}$$

The system behaves then as a two degrees-of-freedom rotor with a stiffness matrix

$$\mathbf{K} = -\frac{1}{2}\mu \begin{bmatrix} 2K_0 & S_0 \\ S_0 & 2B_0 \end{bmatrix}. \tag{12.30}$$

The stability in the small can be studied by introducing the time histories $r(t) = r_0 e^{st}$ and $\varphi(t) = \varphi_0 e^{st}$ into the equations of motion. The characteristic equation is

$$det\begin{bmatrix} ms^2 - \mu K_0 & -\frac{\mu S_0}{2} \\ -\frac{\mu S_0}{2} & J_t s^2 - iJ_p\Omega s - \mu B_0 \end{bmatrix} = 0.; \tag{12.31}$$

i.e.,

$$s^4 - i\frac{J_P}{J_t}\Omega s^3 - \mu\left(\frac{B_0}{J_t} + \frac{K_0}{m}\right)s^2 + i\frac{Jp\mu K_0}{J_t m}\Omega s + \mu^2\frac{4K_0B_0 - S_0^2}{4J_t m} = 0. \tag{12.32}$$

Figure 12.7 shows the real and imaginary parts of the values of s obtained from Equations (12.32) as functions of the spin speed, assuming the following data:

$m = 0.02135\ kg$, $Jp = 2.2 \times 10^{-6}\ kgm^2$,
$J_t = 1.32 \times 10^{-6}\ kgm^2$, $\mu = 0.65\ Am^2$,
$B_0 = 0.0136\ T$, $S_0 = -0.322\ T/m$,
$K_0 = 1.12\ T/m^2$.

From the plot of the real part of s, it follows clearly that at low speed and at very high speed the system is unstable, whereas an intermediate speed range in which its behavior is stable exists. In this range, there are four whirling modes, the conical (the fastest modes, s_3 and s_4, both forward) and the cylindrical (the slowest, s_1 and s_2, one backward and one forward) ones. The coupling between the two conical (rotational) modes s_3 and s_4 produces the lower stability limit ω_{min}, whereas the coupling between one of the two cylindrical (translational) modes s_2 and the slow conical one s_3 is responsible for producing the maximum speed for stability $\Omega_{\max}$. The two limits Ω_{min} and $\Omega_{\max}$ are plotted in Figure 12.7 by dashed vertical lines.

The physical reason for the existence of the lower limit Ω_{min} is obvious: Because the effect on stability caused by the gyroscopic moment is proportional to the spin velocity Ω, if the top is spun too slowly, then it becomes unstable against rotation and tips over; hence, the magnetic force acting on it will be oriented downward, i.e., acting in the same direction as the gravitational force, thus pulling quickly the top to the base.

Having previously remarked that the lower stability limit takes its origin from the coupling between the two conical modes, to compute analytically the value of

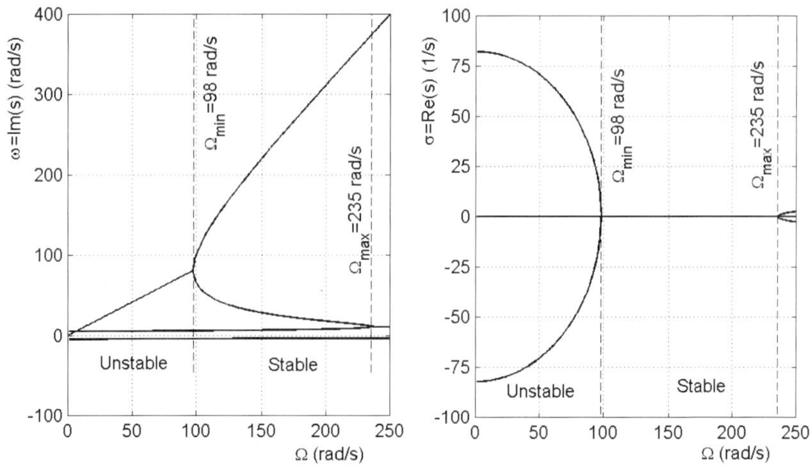

FIGURE 12.7. Real and imaginary parts of the values of s obtained from equations (12.32), assuming the following data: $m = 0.02135$ kg, $J_P = 2.2 \cdot 10^{-6}$ kgm^2, $J_t = 1.32 \cdot 10^{-6}$ kgm^2, $\mu = 0.65$ Am^2, $B_0 = 0.0136$ T, $S_0 = -0.322$ T/m and $K_0 = 1.12$ T/m^2.

Ω_{min}, *it is sufficient to study the stability of the rotational motion (described by the* φ_X *and* φ_y *degrees of freedom), considered as uncoupled with the translational one. Using the complex variable* φ, *the equation of the whirl motion becomes*

$$J_t\ddot{\varphi} - i\Omega J_P\dot{\varphi} - \mu B_0\varphi = 0. \tag{12.33}$$

For the two rotational natural frequencies to be real, the determinant of the characteristic equation must be positive; then the spin speed Ω *must satisfy the condition*

$$\Omega > \Omega_{min} = \sqrt{\frac{4\mu B_0 J_t}{J_p^2}}. \tag{12.34}$$

By using the same numerical values as for Figure 12.7, a value $\Omega_{min} = 98$ *rad/s is obtained.*

If the top is spun too fast, its axis becomes too stiff and cannot respond quickly enough to the changing direction of the magnetic field. Then the magnetic dipole moment, which has the same direction of the axis of the top, can be considered as fixed in space, and according to Earnshaw's theorem, the system becomes unstable for translational motions. The upper stability limit is produced by the coupling between the slow conical mode and the faster of the two cylindrical ones. As the existence of this limit does not depend on the presence of the fast conical mode, it is possible to compute $\Omega_{\max}$ *by using the fast top assumption, i.e., by substituting* $J_t = 0$ *in the equations of motion (12.29).*

Using the variables r and φ, the following third-order set of differential equations is obtained:

$$\begin{cases} m\ddot{r} + \mu K_0 r + \frac{\mu S_0}{2}\varphi = 0, \\ J_p\Omega\dot{\varphi} - i\frac{\mu S_0}{2}r - i\mu B_0\varphi = 0, \end{cases} \tag{12.35}$$

whose characteristic equation is

$$4mJ_p\Omega\lambda^3 - 4m\mu B_0\lambda^2 + 4\mu K_0 J_p\Omega\lambda + m^2g^2 - 4\mu^2 K_0 B_0 = 0. \tag{12.36}$$

The condition for stability is then

$$\Omega < \Omega_{\max} = \frac{\gamma_{\max}}{gJ_p}\sqrt{\frac{\mu^3 B_0^3}{m}}, \tag{12.37}$$

where

$$\gamma_{\max} = \sqrt{\frac{1+\eta}{2}\left[f(\eta) + \sqrt{f(\eta)^2 + 64\eta}\right]}, \tag{12.38}$$

$$f(\eta) = 1 - 18\eta - 27\eta^2 \ , \quad \eta = \frac{S_0^3}{4K_0B_0} - 1.$$

With the same numerical values as for Figure 12.7, $\Omega_{\max} = 235$ rad/s.

As the present model, even in its nonlinear version, is based on an approximate expression of the magnetic field close to the equilibrium position, it cannot simulate the motion of the top under instability conditions, but it just predicts and simulates its behavior in conditions close to the instability limits.

Consider first a numerical simulation at a spin velocity slightly higher than the minimum speed for stability, namely, $\Omega = 99$ rad/s. Assuming that initially the state is close to the equilibrium conditions, the time histories $X(t)$, $Z(t)$, $\varphi_y(t)$, and $\dot{\theta}(t)$ reported in Figure 12.8 are obtained. Also, the results obtained using the fast top assumption, i.e., by neglecting the term in J_t, have been reported, in order to give a comparison between a simplified model often found in the literature and the exact one. The results obtained using the linearized equations of motion are not presented, because they are practically coincident with those obtained using the nonlinear ones.

The results of the numerical simulation performed at a spin velocity roughly halfway between the two stability limits, $\Omega = 150$, rad/s are reported in Figure 12.9.

In the first case ($\Omega = 99$ rad/s), the radial vibrational behavior described by the simplified model is close to that predicted by the exact one [see Figure 12.8(a)], as the translational natural frequencies in the horizontal plane ω_1 and ω_2 computed by the two models are nearly the same. The only difference involves the high-frequency harmonic component, corresponding to the slow whirl mode ω_3, for which the two models give the values of about 40 rad/s and 70 rad/s, respectively. Figure 12.8(b) clearly shows that the fast top model is less accurate than the exact one in simulating the vertical motion of the top; in fact, the first model describes oscillations with constant amplitude (equal to the initial one Z_0) about

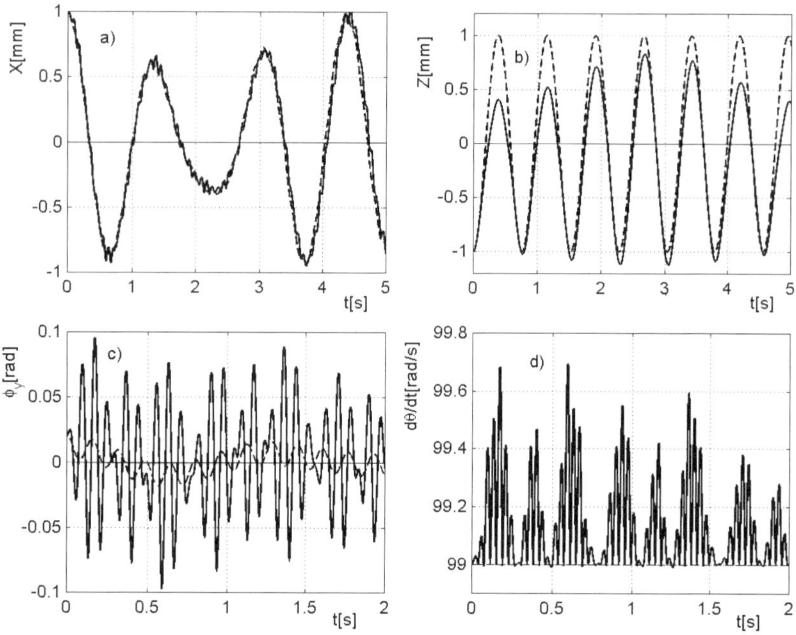

FIGURE 12.8. Results of the numerical simulation for the case $\Omega = 99$ rad/s with initial conditions $X_0 = 1$ mm, $Z_0 = -1$ mm, $v_{z_0} = 10$ mm/s, $\phi_{y_0} = 0.02$ rad. Time histories of (a) coordinate X, (b), coordinate Z, (c) angle ϕ_y, (d) spin velocity $\dot{\theta}$. The nonlinear solution of the exact model (full lines) is compared with that of the fast top model (dashed lines).

the equilibrium position $Z = 0$, and the second one yields a time history characterized by a sinusoidally modulated amplitude, with a maximum value as large as Z_0, but not centered in the position $Z = 0$. More evident differences can be seen in the simulation of the rotational behavior, described by the time history of angle φ_y [Figure 12.8(c)]; such differences can be ascribed to the facts that the simplified model is not able to predict the existence of the lower limit for stability Ω_{min} and that the simulation has been performed at a spin velocity Ω just close to this speed. Although the simulation performed using the fast top model shows a stable behavior of the top, consisting in regular oscillations of its axis with a maximum amplitude as large as the initial one $\varphi_{y0} = 0.02$ rad, the exact solution has a larger amplitude (with a maximum value of about 0.1 rad) and higher frequencies of oscillation (associated with the two whirl modes $\omega_3 = 70$ rad/s and $\omega_4 = 95$ rad/s). Referring to Figure 12.8(d), one notes that although by resorting to the fast top assumption the spin speed is constant, by using the exact model, it oscillates and is, as an average, higher than the value taken at time $t = 0$. The maximum difference between the exact history of $\dot{\theta}$ and the constant value $\Omega = 99$ rad/s predicted by the simplified model is of 0.7%.

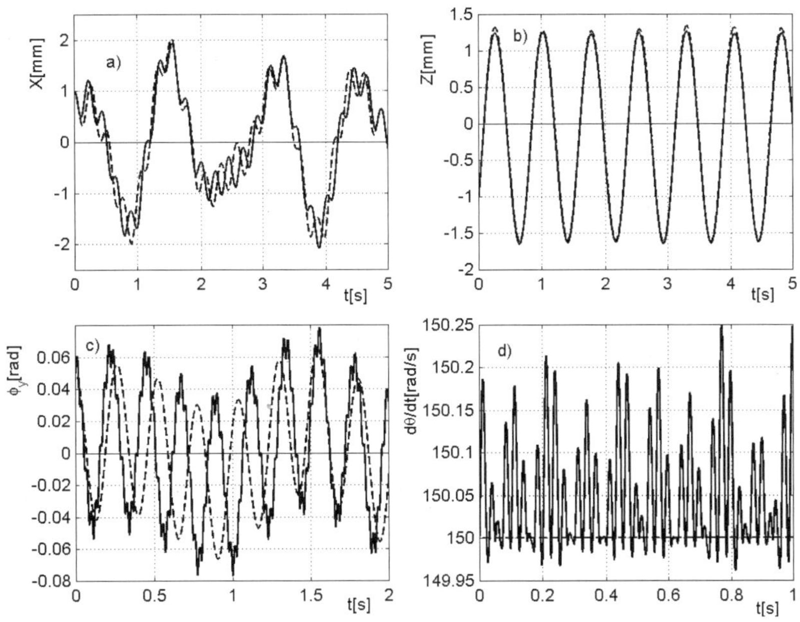

FIGURE 12.9. Same as Figure 12.8, but with $\Omega = 150$ rad/s and initial conditions $X_0 = 1$ mm, $Z_0 = -1$ mm, $v_{z_0} = 10$ mm/s, $\phi_{y_0} = 0.05$ rad, and $\dot{\phi}_{y_0} = 1$ rad/s.

In the case of the spin velocity $\Omega = 150$ rad/s the simplified model gives good results, particularly in predicting the translational behavior (both horizontal and vertical) of the top [see Figures 12.9(a) and 12.9(b)]. More evident differences with respect to the results provided by the exact model can be found in the time history of angle φ_y [Figure 12.9(c)]; the amplitude of the oscillations is nearly the same, whereas the frequency contents are different, owing to the fact that the simplified model does not predict the fast whirl mode and provides the value $\omega_3 = 24$ rad/s for the slow one, and the exact model gives $\omega_3 = 28$ rad/s. The frequency of the oscillations of the spin speed $\dot{\theta}$ is now higher [Figure 12.9(d)]; the maximum difference from the constant value $\Omega = 150$ rad/s is lower than in the first case (0.2%).

12.2 Large amplitude whirling of a linearily constrained rigid rotor

Equations (12.14) and (12.15), or Equations (12.14) and (12.16) for the case of a gyroscopic body, can be used also for the study of large amplitude whirling of a constrained rigid rotor, once the generalized forces caused by the bearings are introduced into the equations of motion.

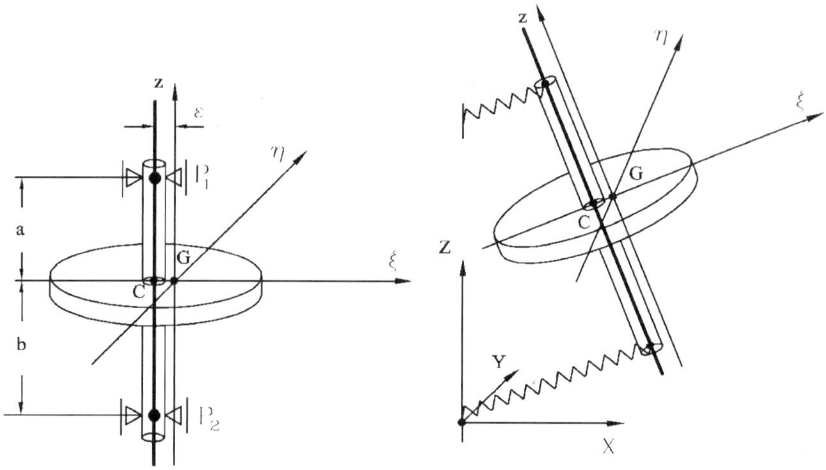

FIGURE 12.10. Elastically constrained rigid rotor. (a) Sketch of the system, and (b) deflected configuration

Consider a rigid body rotating about two cylindrical hinges located on one its principal axes of inertia (z-axis), and assume that the ellipsoid of inertia is round; i.e., its moments of inertia in a plane perpendicular to the rotation axis are equal (Figure 12.10). Assume that the center of mass G is not exactly coincident with point C, located on the axis connecting the two bearings, and that its eccentricity ϵ lies, for simplicity, on its ξ-axis. This does not detract from the generality of the model, because the body is axi-symmetrical. No couple unbalance is assumed. Let J_p and J_t be, respectively, the moments of inertia about the baricentrical principal axis, which coincides with the rotation axis but for the eccentricity ϵ, and any axis in the rotation plane passing through point G.

This model coincides, except for the lack of couple unbalance, with that of Figure 3.1(b). The only differences are that here axes $\xi\eta z$ are centered in the center of mass G instead of point C and that no assumption of small displacements and rotations is made. Assume that the elastic and damping behavior of the bearings is linear and isotropic, with stiffness k and damping c. This implies that the axial stiffness is equal to the radial one, but this does not constitute a limitation of the present model. If the usual assumptions of small displacements and rotations are made, the lateral, axial, and torsional behaviors can be shown to be uncoupled and the former can be studied by resorting to four degrees of freedom, which can be coupled two by two in just two complex coordinates. The model coincides, apart from the couple unbalance here neglected, with the four degrees-of-freedom model described in Chapter 3.

The coordinates of points P_{ir} (with $i = 1, 2$) in which the bearings are located on the rotor (expressed in the rotor-fixed frame) are (Figure 12.10) as follows:

$$\overline{(P_{1r} - G)} = \begin{bmatrix} -\epsilon & 0 & a \end{bmatrix}^T, \qquad \overline{(P_{2r} - G)} = \begin{bmatrix} -\epsilon & 0 & -b \end{bmatrix}^T. \tag{12.39}$$

The couple unbalance could be introduced into the model by simply assuming different vales of the ξ- and η-coordinates of points P_{1r} and P_{2r}. This would complicate formally the equations, but no conceptual difficulty would result.

The coordinates of the first (upper) point in the fixed reference frame are

$$\overline{(P_{1r} - O)} = \mathbf{R}_1^T \mathbf{R}_2^T \mathbf{R}_3^T \begin{Bmatrix} -\epsilon \\ 0 \\ a \end{Bmatrix} + \begin{Bmatrix} X \\ Y \\ Z \end{Bmatrix}. \tag{12.40}$$

The coordinates of the second point are expressed by the same equation in which $-b$ is introduced instead of a.

The coordinates of the corresponding points of the stator P_{is} (with $i = 1, 2$) are, in the inertial frame,

$$\overline{(P_{1s} - O)} = \begin{bmatrix} 0 & 0 & a \end{bmatrix}^T, \qquad \overline{(P_{2s} - O)} = \begin{bmatrix} 0 & 0 & -b \end{bmatrix}^T.$$

The potential energy stored in the ith spring of stiffness k is

$$\mathcal{U}_i = \frac{1}{2}k\left[\overline{(P_{ir} - O)} - \overline{(P_{is} - O)}\right]^T \left[\overline{(P_{ir} - O)} - \overline{(P_{is} - O)}\right]. \tag{12.41}$$

The complete nonlinear expression of the potential energy can be obtained by introducing the expressions of the rotation matrices into Equation (12.40) and the latter into Equation (12.41). This yields

$$\begin{aligned}
\mathcal{U}_1 = {} & \frac{1}{2}k\Big(X^2 + Y^2 + Z^2 + \epsilon^2 + 2a^2 + 2X\left[-\epsilon\cos(\theta)\cos(\varphi_y) + \right. && (12.42)\\
& \left. + a\sin(\varphi_y)\right] + 2Y\left[-\epsilon\cos(\theta)\sin(\varphi_y)\sin(\varphi_x) - \epsilon\sin(\theta)\cos(\varphi_x) + \right. \\
& \left. - a\cos(\varphi_y)\sin(\varphi_x)\right] + 2Z\left[\epsilon\cos(\theta)\sin(\varphi_y)\cos(\varphi_x) + \right. \\
& \left. - \epsilon\sin(\theta)\sin(\varphi_x) + + a\cos(\varphi_y)\cos(\varphi_x) - a\right] + \\
& -2\epsilon a\left[\cos(\theta)\sin(\varphi_y)\cos(\varphi_x) - \sin(\theta)\sin(\varphi_x)\right] + && (12.43)\\
& -2a^2\cos(\varphi_y)\cos(\varphi_x)\Big). && (12.44)
\end{aligned}$$

The generalized forces are then

$$\left\{\begin{aligned}
Q_{X_1} &= -\frac{\partial \mathcal{U}}{\partial X_1} = -k\left[X - \epsilon\cos(\theta)\cos(\varphi_y) + a\sin(\varphi_y)\right],\\
Q_{Y_1} &= -\frac{\partial \mathcal{U}}{\partial Y_1} = -k\left[Y - \epsilon\cos(\theta)\sin(\varphi_y)\sin(\varphi_x) - \epsilon\sin(\theta)\cos(\varphi_x) + \right.\\
&\qquad\qquad \left. - a\cos(\varphi_y)\sin(\varphi_x)\right],\\
Q_{Z_1} &= -\frac{\partial \mathcal{U}}{\partial Z_1} = -k\left[Z + \epsilon\cos(\theta)\sin(\varphi_y)\cos(\varphi_x) - \epsilon\sin(\theta)\sin(\varphi_x) + \right.\\
&\qquad\qquad \left. + a\cos(\varphi_y)\cos(\varphi_x) - a\right],
\end{aligned}\right. \tag{12.45}$$

$$\left\{\begin{aligned}
Q_{\varphi_{X_1}} &= -\frac{\partial \mathcal{U}}{\partial \phi_{X_1}} = -k\Big\{Y[-\epsilon\cos(\theta)\sin(\varphi_y)\cos(\varphi_x) + \\
&+\epsilon\sin(\theta)\sin(\varphi_x) - a\cos(\varphi_y)\cos(\varphi_x)] + Z[-\epsilon\cos(\theta)\sin(\varphi_y)\sin(\varphi_x) + \\
&-\epsilon\sin(\theta)\cos(\varphi_x) - a\cos(\varphi_y)\sin(\varphi_x)] + \epsilon a[\cos(\theta)\sin(\varphi_y)\sin(\varphi_x) + \\
&+\sin(\theta)\cos(\varphi_x)] + a^2\cos(\varphi_y)\sin(\varphi_x)\Big\},\\
Q_{\varphi_{y_1}} &= -\frac{\partial \mathcal{U}}{\partial \phi_{y_1}} = -k\Big\{X\left[\epsilon\cos(\theta)\sin(\varphi_y) + a\cos(\varphi_y)\right] + \\
&+Y\left[-\epsilon\cos(\theta)\cos(\varphi_y)\sin(\varphi_x) + a\sin(\varphi_y)\sin(\varphi_x)\right] + \\
&+Z\left[\epsilon\cos(\theta)\cos(\varphi_y)\cos(\varphi_x) - a\sin(\varphi_y)\cos(\varphi_x)\right] + \\
&-\epsilon a\cos(\theta)\cos(\varphi_y)\cos(\varphi_x) + a^2\sin(\varphi_y)\cos(\varphi_x)\Big\},
\end{aligned}\right.$$

$$\left\{\begin{aligned}
Q_{\theta_1} &= -\frac{\partial \mathcal{U}}{\partial \theta_1} = -k\Big\{X\epsilon\sin(\theta)\cos(\varphi_y) + Y_1[\epsilon\sin(\theta)\sin(\varphi_y)\sin(\varphi_x) + \\
&-\epsilon\cos(\theta)\cos(\varphi_x)] + Z[-\epsilon\sin(\theta)\sin(\varphi_y)\cos(\varphi_x) + \\
&-\epsilon\cos(\theta)\sin(\varphi_x)] + \epsilon a[\sin(\theta)\sin(\varphi_y)\cos(\varphi_x) + \cos(\theta)\sin(\varphi_x)]\Big\}.
\end{aligned}\right. \tag{12.46}$$

By substituting $-b$ for a, the generalized forces caused by the second spring are readily obtained.

To account for damping, the Rayleigh dissipation function can be written. Consider a linear viscous damper located between points P_{ir} (on the rotor) and P_{is} (on the stator). Assuming that the element in which the energy dissipation occurs is stationary, the generalized damping forces caused by the ith damper can be computed from the Rayleigh dissipation function

$$\mathcal{F}_i = \frac{1}{2}c\left(\overline{\mathrm{P}_{ir} - \mathrm{O}}\right)^{\dot{}\,\mathrm{T}}\left(\overline{\mathrm{P}_{ir} - \mathrm{O}}\right)^{\dot{}}, \tag{12.47}$$

where $\dot{\left(\overline{\mathrm{P}_{ir} - \mathrm{O}}\right)}$ is equal to the relative velocity of points P_{ir} and P_{is} (in the fixed reference frame) because the latter is stationary. By differentiating Equation (12.40), the velocity of the upper point P_{1r} is

$$\dot{\left(\overline{\mathrm{P}_{1r} - \mathrm{O}}\right)} = \left(\dot{\varphi}_x \boldsymbol{\gamma}^{\mathrm{T}}\mathbf{R}_2^{\mathrm{T}}\mathbf{R}_3^{\mathrm{T}} + \dot{\varphi}_y \mathbf{R}_1^{\mathrm{T}}\boldsymbol{\beta}^{\mathrm{T}}\mathbf{R}_3^{\mathrm{T}} + \dot{\theta}\mathbf{R}_1^{\mathrm{T}}\mathbf{R}_2^{\mathrm{T}}\boldsymbol{\alpha}^{\mathrm{T}}\right)\left\{\begin{array}{c} -\epsilon \\ 0 \\ a \end{array}\right\} + \\ + \left\{\begin{array}{c} \dot{X} \\ \dot{Y} \\ \dot{Z} \end{array}\right\}; \tag{12.48}$$

i.e.,

$$\dot{\left(\overline{\mathrm{P}_{1r} - \mathrm{O}}\right)} = \dot{\varphi}_x \mathbf{V} + \dot{\varphi}_y \mathbf{W} + \dot{\theta}\mathbf{P} + \left\{\begin{array}{c} \dot{X} \\ \dot{Y} \\ \dot{Z} \end{array}\right\}, \tag{12.49}$$

where

$$\mathbf{V} = \boldsymbol{\gamma}^{\mathrm{T}}\mathbf{R}_2^{\mathrm{T}}\mathbf{R}_3^{\mathrm{T}}\left\{\begin{array}{c} -\epsilon \\ 0 \\ a \end{array}\right\}, \qquad \mathbf{W} = \mathbf{R}_1^{\mathrm{T}}\boldsymbol{\beta}^{\mathrm{T}}\mathbf{R}_3^{\mathrm{T}}\left\{\begin{array}{c} -\epsilon \\ 0 \\ a \end{array}\right\},$$

$$\mathbf{P} = \mathbf{R}_1^{\mathrm{T}}\mathbf{R}_2^{\mathrm{T}}\boldsymbol{\alpha}^{\mathrm{T}}\left\{\begin{array}{c} -\epsilon \\ 0 \\ a \end{array}\right\},$$

$$\boldsymbol{\alpha} = \begin{bmatrix} -\sin(\theta) & \cos(\theta) & 0 \\ -\cos(\theta) & -\sin(\theta) & 0 \\ 0 & 0 & 0 \end{bmatrix},$$

$$\boldsymbol{\beta} = \begin{bmatrix} -\sin(\varphi_y) & 0 & -\cos(\varphi_y) \\ 0 & 0 & 0 \\ \cos(\varphi_y) & 0 & -\sin(\varphi_y) \end{bmatrix},$$

$$\boldsymbol{\gamma} = \begin{bmatrix} 0 & 0 & 0 \\ 0 & -\sin(\varphi_x) & \cos(\varphi_x) \\ 0 & -\cos(\varphi_x) & -\sin(\varphi_x) \end{bmatrix},$$

By introducing the column of the generalized coordinates,

$$\mathbf{q} = \left[\begin{array}{cccccc} X & Y & Z & \varphi_x & \varphi_x & \theta \end{array}\right]^T, \tag{12.50}$$

the usual expression of the Rayleigh dissipation function for the damper at point P_1 is obtained:

$$\mathcal{F}_i = \frac{1}{2}\dot{\mathbf{q}}^T \mathbf{C}_1 \dot{\mathbf{q}} \,, \tag{12.51}$$

where the time-dependent matrix $\mathbf{C}$ is

$$\mathbf{C}_1 = c \begin{bmatrix} 1 & 0 & 0 & \mathbf{V}^T\mathbf{I}_1 & \mathbf{W}^T\mathbf{I}_1 & \mathbf{P}^T\mathbf{I}_1 \\ & 1 & 0 & \mathbf{V}^T\mathbf{I}_2 & \mathbf{W}^T\mathbf{I}_2 & \mathbf{P}^T\mathbf{I}_2 \\ & & 1 & \mathbf{V}^T\mathbf{I}_3 & \mathbf{W}^T\mathbf{I}_3 & \mathbf{P}^T\mathbf{I}_3 \\ & & & \mathbf{V}^T\mathbf{V} & \mathbf{W}^T\mathbf{V} & \mathbf{P}^T\mathbf{V} \\ & \text{Symm.} & & & \mathbf{W}^T\mathbf{W} & \mathbf{P}^T\mathbf{W} \\ & & & & & \mathbf{P}^T\mathbf{P} \end{bmatrix} ,$$

and $\mathbf{I}_1$, $\mathbf{I}_2$, and $\mathbf{I}_3$ are the first, second, and third columns of the identity matrix (i.e., the unit vectors directed along X-, Y-, and Z-axes), respectively.

The generalized forces caused by the presence of this damper, to be inserted in the equations of motion, are

$$\mathbf{Q}_1 = -\mathbf{C}_1\dot{\mathbf{q}}. \tag{12.52}$$

To obtain the forces caused by the damper located between points P_{2r} and P_{2s}, $-b$ must be substituted for a in all equations.

Equations (12.14) and (12.16), together with the expressions of the generalized forces Q_i, are thus a set of six nonlinear differential equations allowing us to simulate the behavior of the system during whirling with large amplitude. It is easy to verify that these equations can be linearized, under the assumption of small displacements X, Y, Z and rotations φ_X and φ_y, whereas rotation θ and the spin speed $\dot{\theta}$ are arbitrary large, and small unbalance ϵ, obtaining

$$\left\{ \begin{array}{l} m\ddot{X} + C_{11}[\dot{X} + \epsilon\dot{\theta}\sin(\theta)] + C_{12}\dot{\varphi}_y + K_{11}[X - \epsilon\cos(\theta)] + K_{12}\varphi_y = 0, \\ m\ddot{Y} + C_{11}[\dot{Y} - \epsilon\dot{\theta}\cos(\theta)] - C_{12}\dot{\varphi_X} + K_{11}[Y - \epsilon\sin(\theta)] - K_{12}\varphi_X = 0, \\ m\ddot{Z} + C_{11}\dot{Z} + K_{11}Z = 0 \,, \\ J_t\ddot{\varphi}_X + J_p\dot{\theta}\dot{\varphi}_y - C_{21}[\dot{Y} - \epsilon\dot{\theta}\cos(\theta)] + C_{22}\dot{\varphi_X} + -K_{21}[Y - \epsilon\sin(\theta)] + \\ \qquad\qquad + K_{22}\varphi_X = 0, \\ J_t\ddot{\varphi}_y - J_p\dot{\theta}\dot{\varphi}_X + C_{21}[\dot{X} + \epsilon\dot{\theta}\sin(\theta)] + C_{22}\dot{\varphi}_y + K_{21}[X - \epsilon\cos(\theta)] + \\ \qquad\qquad + K_{22}\varphi_y = 0, \\ \ddot{\theta} = 0, \end{array} \right. \tag{12.53}$$

where

$$\mathbf{K} = k \begin{bmatrix} 2 & a-b \\ a-b & a^2+b^2 \end{bmatrix}, \qquad \mathbf{C} = c \begin{bmatrix} 2 & a-b \\ a-b & a^2+b^2 \end{bmatrix}.$$

Apart from the differences caused by referring the equation of motion to point G instead of point C, Equations (12.53) are those commonly used in linearized rotordynamics. It is easy to verify that the uncoupling among axial, torsional, and lateral behavior holds and that the last equation states that the angular velocity $\dot{\theta}$ is constant. In the linearized system, there is no difference between the assumption of constant angular momentum and that of constant spin speed.

As the system is damped and unbalanced, there is energy dissipation in the bearings (even if they are assumed to be frictionless) caused by the whirling motion. If this effect is accounted for as shown in Chapter 11, the last equation of motion becomes

$$\ddot{\theta} + \Omega^2 \operatorname{Im}\left\{\overline{\mathbf{F}}^{\mathrm{T}}\left[\Omega^4(\mathbf{M}-\mathbf{G}) + i\Omega\mathbf{C} + \mathbf{K}\right]^{-1}\mathbf{F}\right\} = 0, \tag{12.54}$$

where the symbols are referred to the complex notation for steady state whirling. The term expressing the rotordynamic drag in the bearings is then quadratic in the eccentricity (included in vector $\mathbf{F}$), which is a small parameter, and hence in a linearized model, it should be dropped. However, it will be retained in the following numerical simulations to show the importance of such phenomenon in practical applications.

The number of independent parameters involved in the equations of motion is large, and hence, it is impossible to draw general conclusions: Only a few numerical simulations will be reported here.

Consider a rigid rotor with the following characteristics: mass $m = 10$ kg, moments of inertia $J_t = 0.1$ kgm^2, $J_p = 0.15$ kgm^2, distances between the bearings and the center of mass $a = 100$ mm and $b = 200$ mm, and stiffness and damping of the bearings $k = 5 \times 10^6$ N/m and $c = 100$ Ns/m, respectively. The linearized analysis yields a single critical speed at 957 rad/s (9139 rpm) and two natural frequencies at standstill equal to 921.2 rad/s (146.6 Hz) and 1628 rad/s (259.1 Hz).

Three values of the eccentricity will be considered, namely, $\epsilon = 1$ μm, 0.1 mm and 1 mm. Assuming a maximum speed of 20,000 rpm, they correspond to balancing grades $G = 2$, $G = 200$ and $G = 2,000$: The first one is a fairly accurate balancing, as common in gas turbines; the second is a rough balancing, as for crankshafts of car engines; and the third is so rough that it is not included in ISO 1940 standards. Actually the last value, corresponding to an unbalance of 10,000 g mm, is too high for any practical application; it has been chosen as a sort of limiting case.

The simulations were performed using a standard fourth order Runge-Kutta algorithm with adaptive timestep at speeds close to the critical speed, one slightly lower (942.5 rad/s = 9,000 rpm) and one slightly higher (963.4 rad/s = 9,200 rpm). Some typical results are reported in Table 12.2.

In the case of small unbalance, the results from the nonlinear model are practically coincident with those obtained from the linearized theory. The

ϵ [μm]	Ω [rpm]	Orbit radius [μm]		$\Delta\Omega/\Omega$ (in 1 s)		z_{max} [μm]	
		linear	nonlin.	linear	nonlin.	lin.	nonlin.
1	9,000	28.9	28.9	0.94×10^{-6}	1.12×10^{-6}	0	1.14×10^{-4}
	9,200	42.6	42.6	2.11×10^{-6}	2.35×10^{-6}	0	3.07×10^{-4}
100	9,000	2,880	2,880	6.8×10^{-3}	7.7×10^{-3}	0	1.14
	9,200	4,260	4,260	21.1×10^{-3}	23.5×10^{-3}	0	3.08
1000	9,000	28,800	28,800	0.074	0.074	0	114
	9,200	45,500	44,200	0.094	0.108	0	308

TABLE 12.3. Some results of the simulation of the behavior of a fixed rotor with different values of the eccentricity and at different speeds close to the critical speed. The radius of the orbit, the speed reduction $\Delta\Omega$ in 1 s, and the amplitude of the axial displacement z_{max} are reported.

slowing of the rotor is almost negligible, and close to that predicted by the linearized model and the amplitude of the axial vibration is very small: The flexural-torsional and flexural-axial coupling is then completely negligible.

Also in the case of the intermediate unbalance (100 μm), the results obtained are still close to those obtained from the linearized model, as also shown by the increase of the orbit radius that is proportional to the increase of eccentricity. The amplitude of the axial vibration is still negligible, and the slowing down, although no more negligible, is still small and close to that predicted by Equation (12.54): The flexural-torsional and flexural-axial coupling is still negligible.

With the largest value of the unbalance, the rotor slows at a high rate because much energy is dissipated by the supports. When the initial spin speed is higher than the critical speed, the rotor slows quickly and enters the subcritical regime with a sudden drop of speed (see Figure 12.11, in which an initial speed of 1000 rad/s has been assumed). The radius of the orbit oscillates in time during the critical speed crossing as described in Chapter 10. However, even in this case, the time history of the lateral displacement is close to that computed using the linearized model, and the amplitude of the axial displacement is small if compared with that of the lateral one (about 0.26%), showing a very weak flexural-axial coupling. The time history of the speed shows that a certain flexural-torsional coupling is present, as the system undergoes some torsional oscillations that are not predicted by the linearized model [Equation (12.54)]; however, the approximation of the linearized model is generally not bad even for what the decrease of speed is concerned (note that in a strictly linearized model even the drop of speed caused by damping vanishes). The time histories of the orbit amplitude computed using the nonlinear and the linearized models cannot be compared directly, because they are much influenced by the decrease of the spin speed, but the latter model yields still reasonable results, even if the amplitude is definitely very large.

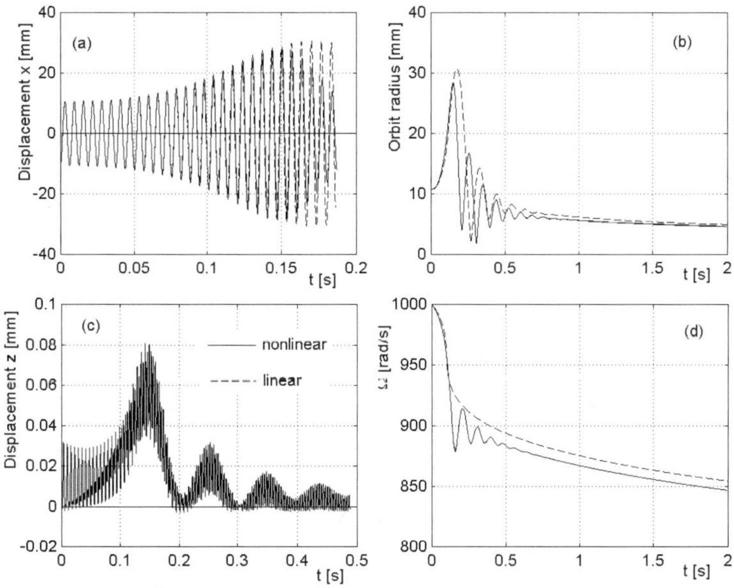

FIGURE 12.11. Single rigid-body fixed rotor; results of the numerical simulation with an eccentricity of 1 mm and starting speed 9200 rpm. Time history of (a) lateral displacement in x-direction, (b) radius of the orbit, (c) axial displacement, and (d) spin speed.

Remark 12.5 *The example shows that the usual linearized model is still applicable even in the case of large unbalances and large whirling orbits. Note that the simulations have been performed in the worst conditions, i.e., with an unrealistically high unbalance and at speeds very close to the critical speed (in Figure 12.11, the critical speed is actually crossed).*

12.3 Twin rigid-bodies free rotor

The aim of this section is to show that the concepts of critical speeds and of self-centering apply as well to multibody free rotors, caused by the presence of elastic and damped connections between the carrier and the inner masses of the system. Also, the role of damping (nonrotating, rotating, synchronous) on the stability of multibody free rotors can be fully understood only if it is studied with reference to a frame fixed to the element in which the energy dissipation occurs. However, the lack of a stator (a set of nonrotating parts) makes a deformable free rotor prone to instability, because all damping is of the rotating type.

Consider a free rotor, made of two rigid axi-symmetrical bodies connected with each other through linear elastic springs and viscous dampers. This

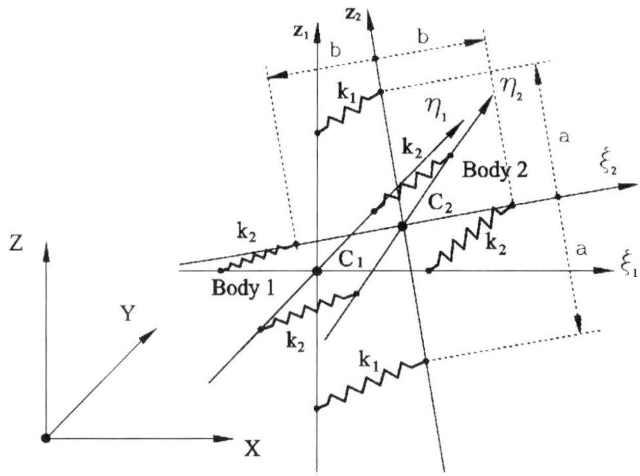

FIGURE 12.12. Twin rigid-bodies free rotor. Reference frames and position of the springs. The dampers (not represented) are located in parallel to the springs.

model is the simplest, allowing us to study a spinning spacecraft: In this case, it is possible to identify a part, the *carrier*, which can be dealt with as a rigid body, to which all other parts, themselves modeled as rigid bodies, are connected. Assume that the centers of mass of the two bodies are coincident in the undeflected position and that they are linked together, for instance, by six linear springs and viscous dampers arranged in the way shown in Figure 12.12. The values of the stiffness and damping coefficient of the links located on the x- and y-axes are equal (k_2 and c_2), in such a way that the system possesses axial symmetry. Also, the inertial properties of the two rigid bodies are axially symmetrical. Reference frames $\xi_1\eta_1z_1$ and $\xi_2\eta_2z_2$ are fixed to the first and second body, respectively, whereas frame XYZ is an inertial frame.

The two bodies are assumed to be perfectly balanced, i.e., the geometrical centers C_i of the connection system coincide with the mass centers G_i and the principal axes of inertia coincide with the principal axes of elasticity.

12.3.1 Linearized approach

By resorting to the usual assumptions of linearized, steady state rotordynamics, the lateral behavior uncouples from the axial and torsional dynamics and the equation of motion for the former is

$$\mathbf{M}\ddot{\mathbf{q}} - i\Omega\mathbf{G}\dot{\mathbf{q}} + \mathbf{C}\dot{\mathbf{q}} + (\mathbf{K} - i\Omega\mathbf{C}_r)\,\mathbf{q} = \mathbf{F}\ , \tag{12.55}$$

where the vector of the complex coordinates

$$\mathbf{q} = \{r_1\ r_2\ \phi_1\ \phi_2\}^T$$

contains the complex displacements and rotations (defined in the usual way) of both bodies, $\mathbf{C} = \mathbf{C}_n + \mathbf{C}_r$ is the total damping matrix and the mass and gyroscopic matrices are

$$\mathbf{M} = \begin{bmatrix} m_1 & 0 & 0 & 0 \\ 0 & m_2 & 0 & 0 \\ 0 & 0 & J_{t1} & 0 \\ 0 & 0 & 0 & J_{t2} \end{bmatrix} ,$$

$$\mathbf{G} = \begin{bmatrix} 0 & 0 & 0 & 0 \\ 0 & 0 & 0 & 0 \\ 0 & 0 & J_{p1} & 0 \\ 0 & 0 & 0 & J_{p2} \end{bmatrix} .$$

The stiffness and rotating damping matrices are usually full, symmetrical matrices, whereas the nonrotating damping matrix vanishes, because there is no nonrotating element in the system. Assuming that the positions of the centers of mass of the two bodies are coincident in the undeflected position and that the spring and damper systems are symmetrical with respect to the $\xi\eta$ plane, the stiffness matrix reduces to

$$\mathbf{K} = \begin{bmatrix} k & -k & 0 & 0 \\ -k & k & 0 & 0 \\ 0 & 0 & \kappa & -\kappa \\ 0 & 0 & -\kappa & \kappa \end{bmatrix} .$$

The rotating damping matrix has the same structure, with the damping coefficient for translations c and for rotations Γ instead of k and κ. With reference to the spring and damper arrangement shown in Figure 12.12, the values of the overall stiffness and damping are

$$\begin{cases} k = 2k_1 + 4k_2 \ , \\ \kappa = 2(k_1 a^2 + k_2 b^2) \ , \end{cases} \tag{12.56}$$

$$\begin{cases} c = 2c_1 + 4c_2 \ , \\ \Gamma = 2(c_1 a^2 + c_2 b^2) \ . \end{cases} \tag{12.57}$$

The set of four equations uncouples into two sets, one for cylindrical (translational) whirling

$$\begin{cases} m_1 \ddot{r}_1 + c\,(\dot{r}_1 - \dot{r}_2) + (k - i\Omega c)\,(r_1 - r_2) = F_1 \ , \\ m_2 \ddot{r}_2 + c\,(\dot{r}_2 - \dot{r}_1) + (k - i\Omega c)\,(r_2 - r_1) = F_2 \ , \end{cases} \tag{12.58}$$

and one for conical (rotational) whirling

$$\begin{cases} J_{t1}\ddot{\phi}_1 - i\Omega J_{p1}\dot{\phi}_1 + \Gamma\left(\dot{\phi}_1 - \dot{\phi}_2\right) + (\chi - i\Omega\Gamma)\,(\phi_1 - \phi_2) = M_1 \ , \\ J_{t2}\ddot{\phi}_2 - i\Omega J_{p2}\dot{\phi}_2 + \Gamma\left(\dot{\phi}_2 - \dot{\phi}_1\right) + (\chi - i\Omega\Gamma)\,(\phi_2 - \phi_1) = M_2 \ . \end{cases} \tag{12.59}$$

By using as complex coordinates the displacement of the center of mass of the system r_G and the relative displacement Δr

$$\begin{cases} r_G = \dfrac{m_1 r_1 + m_2 r_2}{m_1 + m_2} , \\ \Delta r = r_2 - r_1 , \end{cases} \tag{12.60}$$

the equations for translational motions reduce to

$$\begin{cases} m_{tot} \ddot{r}_G = F_{tot} , \\ m_{eq} \Delta \ddot{r} + c \Delta \dot{r} + (k - i\Omega c) \Delta r = F_{eq} . \end{cases} \tag{12.61}$$

The two equations uncouple from each other. The first one describes the motion of a point mass $m_{tot} = m_1 + m_2$ located in the center of mass of the system under the action of the force $F_{tot} = F_1 + F_2$. The second one is that of a Jeffcott rotor with mass m_{eq} moving under the action of force F_{eq}, where

$$m_{eq} = \frac{m_1 m_2}{m_1 + m_2} , \qquad F_{eq} = \frac{m_1 F_2 - m_2 F_1}{m_1 + m_2} . \tag{12.62}$$

The model described by Equation (12.61) has then two rigid-body modes with natural frequency equal to zero and two modes with nonvanishing natural frequency, one forward and one backward. The former may be stable or unstable, and the second one is always stable. A plot of the nondimensional natural frequency ω^* and nondimensional decay rate σ^* defined as

$$\omega^* = \frac{\omega}{\sqrt{k/m_{eq}}} = \mathrm{Im}\left(\frac{s}{\sqrt{k/m_{eq}}}\right)$$

and

$$\sigma^* = \frac{\sigma}{\sqrt{k/m_{eq}}} = \mathrm{Re}\left(\frac{s}{\sqrt{k/m_{eq}}}\right)$$

as functions of the nondimensional spin speed $\Omega^* = \Omega/\sqrt{k/m_{eq}}$ is reported in Figure 12.13. The various curves refer to values of the rotating damping ratio $\zeta = c/\sqrt{4km_{eq}}$ spanning from 0 to 1. Note that the forward mode is unstable in the whole supercritical range because the only damping present is rotating damping.

A similar coordinate transformation can be introduced also for the rotational degrees of freedom

$$\begin{cases} \phi_G = \dfrac{J_{t1}\phi_1 + J_{t2}\phi_2}{J_{t1} + J_{t2}} , \\ \Delta\phi = \phi_2 - \phi_1 . \end{cases} \tag{12.63}$$

The equations for rotations, however, do not uncouple in general. They reduce to

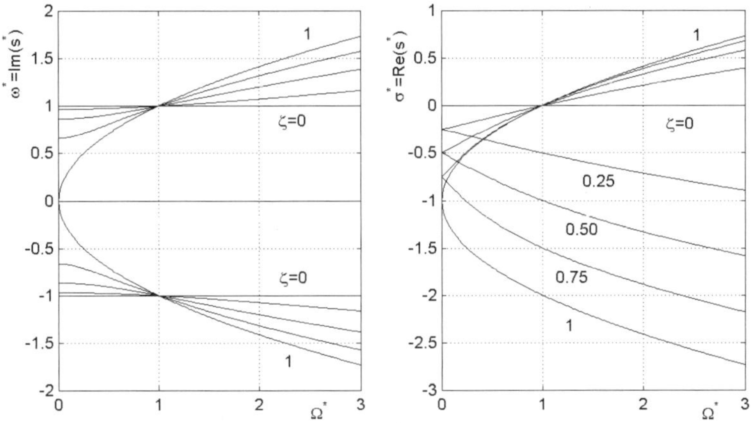

FIGURE 12.13. Plot of the nondimensional natural frequency ω^* and nondimensional decay rate σ^* as functions of the nondimensional spin speed $\Omega^* = \Omega/\sqrt{k/m_{eq}}$. The various curves refer to values of the rotating damping ratio $\zeta = c/\sqrt{4km_{eq}}$ spanning from 0 to 1.

$$\begin{cases} (J_{t1}+J_{t2})\ddot{\phi}_G - i\Omega\,(J_{p1}+J_{p2})\,\dot{\phi}_G - i\Omega J_{t_{eq}}\,(\delta_2-\delta_1)\,\Delta\dot{\phi} = M_1+M_2\;, \\ J_{t_{eq}}\Delta\ddot{\phi} - i\Omega J_{t_{eq}}\,(\delta_2-\delta_1)\,\dot{\phi}_G - i\Omega R\Delta\dot{\phi} + \Gamma\Delta\dot{\phi} + (\chi - i\Omega\Gamma)\,\Delta\phi = M_{eq}\;, \end{cases} \tag{12.64}$$

where

$$J_{t_{eq}} = \frac{J_{t_1}J_{t_2}}{J_{t_1}+J_{t_2}}\;,\qquad M_{eq} = \frac{J_{t_1}M_2 - J_{t_2}M_1}{J_{t_1}+J_{t_2}}\;,$$

$$\delta_1 = \frac{J_{p1}}{J_{t_1}}\;,\qquad \delta_2 = \frac{J_{p2}}{J_{t_2}}\;,\qquad R = \frac{J_{t_1}J_{t_2}\,(J_{p1}+J_{p2})}{(J_{t_1}+J_{t_2})^2}\;.$$

If $\delta_1 = \delta_2$, the two equations uncouple from each other. The first one describes the *attitude* motions of a rigid body with transversal and polar moments of inertia equal to $J_{t_1} + J_{t_2}$ and $J_{p1} + J_{p2}$ under the action of the moment $M_1 + M_2$. The model described by the first equation has then two rigid body modes, one with natural frequency equal to zero, and one with natural frequency proportional to the spin speed (Figure 12.5). The latter may be stable or unstable depending on the value of δ, i.e., depending whether the system as a whole is a disc or a long rotor.

The second Equation (12.64) describes the conical whirling of an uncoupled rotor with four degrees of freedom [see Chapter 3, Figure 3.3(b)]. The data of the equivalent rotor are transversal moment of inertia: $J_{t_{eq}}$; polar moment of inertia $\delta_1 J_{t_{eq}}$; stiffness χ; and rotating damping coefficient Γ. This system has two nonvanishing natural frequencies related to deforma-

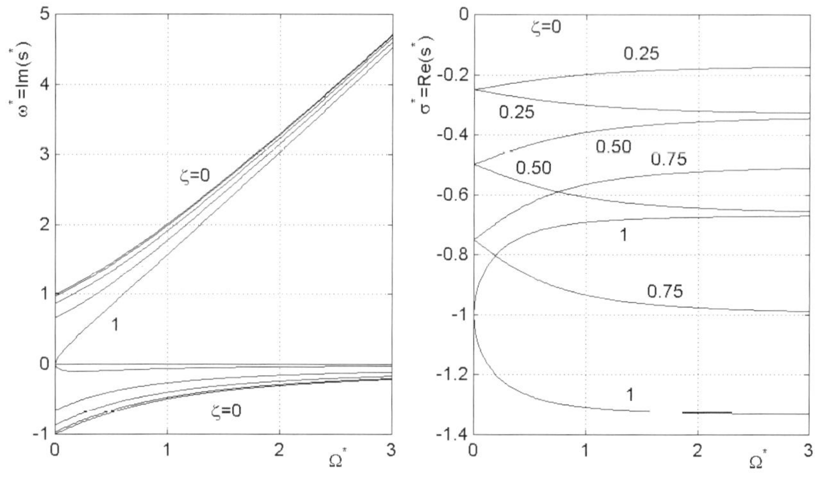

FIGURE 12.14. Disc rotor with $\delta_1 = \delta_2 = 1.5$. Plot of the nondimensional natural frequency ω^* and nondimensional decay rate σ^* as functions of the nondimensional spin speed $\Omega^* = \Omega/\sqrt{\chi/J_{t_{eq}}}$. The various curves refer to values of the rotating damping ratio $\zeta = \Gamma/\sqrt{4\chi J_{t_{eq}}}$ spanning from 0 to 1.

tion modes, one forward (stable or unstable) and one backward (always stable). The results for the undamped system are shown in Figure 3.4.

Two plots of the nondimensional natural frequency ω^* and nondimensional decay rate σ^*, now defined as

$$\omega^* = \frac{\omega}{\sqrt{\chi/J_{t_{eq}}}} = \mathrm{Im}\left(\frac{s}{\sqrt{\chi/J_{t_{eq}}}}\right)$$

and

$$\sigma^* = \frac{\sigma}{\sqrt{\chi/J_{t_{eq}}}} = \mathrm{Re}\left(\frac{s}{\sqrt{\chi/J_{t_{eq}}}}\right)$$

as functions of the nondimensional spin speed $\Omega^* = \Omega/\sqrt{\chi/J_{t_{eq}}}$ are reported in Figures 12.14 (disc rotor with $\delta_1 = \delta_2 = 1.5$) and 12.15 (long rotor with $\delta_1 = \delta_2 = 0.5$). The various curves refer to values of the rotating damping ratio $\zeta = \Gamma/\sqrt{4\chi J_{t_{eq}}}$ spanning from 0 to 1. Note that in the case of the long rotor the forward mode is unstable in the whole supercritical range, whereas in the case of the disc rotor, stability is always assured (there is no critical speed related to conical whirling, so the system never works in the supercritical range).

Owing to the large number of independent parameters, the only way to compare the results obtained through the linearized approach with those that will be obtained using the full nonlinear equations is to perform a large number of numerical simulations. Only four cases are reported here,

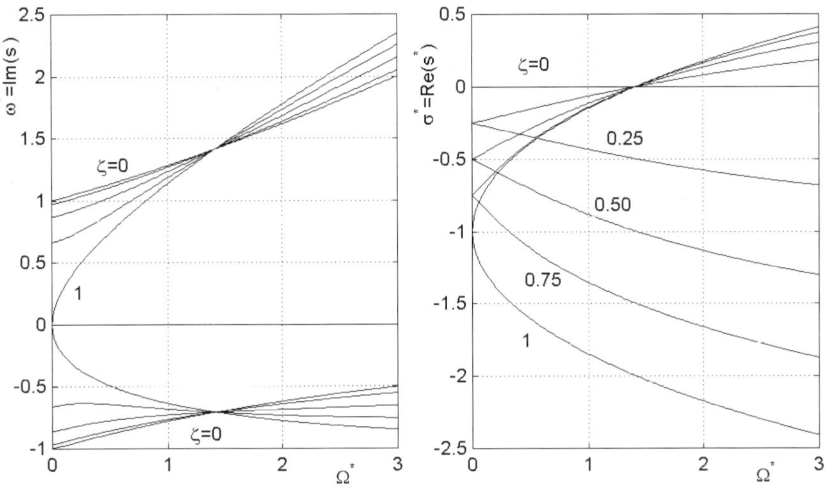

FIGURE 12.15. Same as Figure 12.14 but for a long rotor with $\delta_1 = \delta_2 = 0.5$.

Case n.	1	2	3	4
$m_1 = m_2$ [kg]	20	20	20	20
J_{p1}/J_{p2} [kgm^2]	0.015 / 0.014	0.015 / 0.014	0.010 / 0.010	0.010 / 0.010
J_{t1}/J_{t2} [kgm^2]	0.010 / 0.010	0.010 / 0.010	0.015 / 0.014	0.015 / 0.014
$a = b$ [m]	0.25	0.25	0.25	0.25
$k_1 = k_2$ [N/m]	4×10^6	2,000	4×10^6	4,000
c_1/c_2 [Ns/m]	45.1 / 68	1.01 /1.52	79.5 / 56.6	2.51 / 1.79
k [N/m]	24×10^6	12,000	24×10^6	24,000
κ[Nm/rad]	1×10^6	500	1×10^6	1,000
c [Ns/m]	362.2	8.1	385.6	13.6
Γ [Nms/rad]	14.14	0.316	17.02	0.538

TABLE 12.4. Data for the four cases considered.

to obtain some indications that can be shown to yield a good qualitative understanding of the effect of such assumptions. In all cases, the values of the damping coefficients have been chosen in such a way that the rotational flexural and torsional vibrations are characterized by a damping ratio $\zeta = 0.1$ (quality factor $Q \approx 5$). The relevant data are listed in Table 12.4. Cases number 1 and 2 refer to free rotors with polar moment of inertia larger than the transversal moment of inertia (disc rotors). As a consequence, they have no critical speeds related to rotational modes. On the contrary, cases 3 and 4 deal with rotors with a polar moment of inertia smaller than the transversal one (long rotors). In this case, two critical speeds are present, one linked to the rotational and one to the translational mode.

The main results of the linearized analysis are reported in Tables 12.5 and 12.6, separately for translational and rotational modes, which are de-

	Case n.	1	2	3	4
Undamped	ω_n [rad/s]	1550	34.64	1550	48.99
Damped, $\Omega = 0$	ω [rad/s]	1549	34.64	1549	48.99
	σ [1/s]	-18.1	-0.41	-19.13	-0.68
Damped, $\Omega = 1000$	ω_B [rad/s]	-1549	-36.38	-1549	-50.78
	σ_B [1/s]	-29.8	-11.5	-31.7	-14.1
	ω_F [rad/s]	1549	36.38	1549	50.78
	σ_F [1/s]	-6.4	**10.7**	-6.8	**12.7**

TABLE 12.5. Natural frequencies of the undamped system ω_n and complex eigenfrequencies at standstill (Ω= 0) and a spin speed of 1000 rad/s (backward and forward whirling) of the four free rotors described in Table 12.4; cylindrical whirling (bold figures designate instability).

	Case n.	1	2	3	4
Undamped	ω_n [rad/s]	14140	316	11751	372
	ω_{nt} [rad/s]	11751	263	14140	447
	Ω_{cr} [rad/s]	–	–	21213	671
Damped, $\Omega = 0$	ω [rad/s]	14071	315	11692	369.8
	σ [1/s]	-1414	-31.6	-1175	-85
Damped, $\Omega = 1000$	ω_B [rad/s]	-13360	-65.5	-11352	-163
	σ_B [1/s]	-1442i	-42.6	-1241	-85
	ω_{F1} [rad/s]	1450	65.5	690	687
	σ_{F1} [1/s]	-1.15×10^{-8}	-3.9	$\mathbf{1.49 \times 10^{-9}}$	**0.44**
	ω_{F2} [rad/s]	14815	1541	12043	857
	σ_{F2} [1/s]	-1386	-16.6	-1109	**10.3**

TABLE 12.6. Results for the linearized analysis of the conical whirling of the four free rotors described in Table 12.4. Flexural and torsional undamped natural frequencies, critical speed, and complex eigenfrequencies at standstill ($\Omega = 0$) and at a spin speed of 1000 rad/s (bold figures designate instability).

coupled. The rotors have been considered both at standstill and spinning at 1000 rad/s.

The flexural (translational) and axial natural frequencies of the undamped system ω_n, which coincide with each other and with the critical speed Ω_{cr} are shown in Table 12.5 together with the frequency ω and the decay rate σ of the damped whirl motion (real and imaginary parts of s) at standstill (Ω= 0) and at a spin speed of 1000 rad/s. At Ω= 1000 rad/s, both the forward and backward cylindrical modes are considered

The results reported in the table show clearly that the forward cylindrical whirl mode is unstable in cases 2 and 4: Owing to the low value of the stiffness, at 1000 rad/s, the system is in the supercritical range and hence rotating damping makes it unstable.

The results for the linearized analysis of the conical whirling modes of the four free rotors described in Table 12.4 are reported in Table 12.6.

As $\delta_1 \neq \delta_2$ the attitude modes and the deformation (differential) modes are coupled (the two equations (12.64) do not uncouple). The flexural and torsional natural frequencies ω_n and ω_{n_t} and the critical speed Ω_{cr} of the undamped system are reported together with the complex eigenfrequencies at standstill (Ω= 0) and at a spin speed of 1000 rad/s for backward and forward whirling. Note that cases 1 and 2 refer to disc rotors, and there is no critical speed related to conical whirling. At standstill, there is only one natural frequency different from zero. Although the system spins, there is only one backward mode with nonvanishing whirl speed, whereas both forward modes have a whirl speed different from zero. Instability develops at 1000 rad/s in cases 3 and 4.

As a conclusion of the linearized analysis, it follows that:

- Case 1: very stiff disc rotor. At a spin speed of 1000 rad/s, the system operates in the subcritical range and all modes are stable.
- Case 2: soft disc rotor. At 1000 rad/s, the system operates in the supercritical range and the forward cylindrical mode is unstable.
- Case 3: very stiff long rotor. At 1000 rad/s, the system operates in the subcritical range and all deformation modes are stable. However, owing to the fact that $J_t > J_p$, the forward conical attitude mode is slightly unstable. The value of the decay rate, although positive, is very low and the time needed for the instability to develop may be very long.
- Case 4: soft long rotor. At 1000 rad/s, the system operates in the supercritical range with respect to both the cylindrical mode and the conical mode; both deformation modes are then unstable. Also, the conical attitude mode is unstable and the positive value of the decay rate is fairly high.

12.3.2 Nonlinear approach

As the two rigid bodies constituting the system of Figure 12.12 are axially symmetrical about z-axis, their dynamic behavior can be studied using the nonlinear Equations (12.14) and (12.16), written for each one of the bodies. The expression of the potential energy and of the Rayleigh dissipation function caused by the springs and dampers and of the generalized forces are so long that it is not worthwhile to report here their complete expression. The reader can refer to [4].

A set of 12 coupled, nonlinear differential equations is thus obtained. Owing to their complexity, only numerical integration is possible. However, a simplification that does not detract much from the generality of the model can be introduced. Although the displacements and rotations of each individual body with respect to the inertial frame are generally large owing to the free-rotor nature of the system, the differential displacements and rotations of one with respect to the other are small. A model based on this assumption will be referred to as a semilinearized model.

This approach allows the generalized forces caused by the springs and the dampers to be computed in closed form, obtaining [4]

$$\left\{\begin{array}{l}
Q_{X_1} = k(X_2 - X_1) + c\Big\{\dot{X}_2 - \dot{X}_1 + \Big[(Y_2 - Y_1)\sin(\varphi_{X_1}) - (Z_2 - Z_1)\cos(\varphi_{X_1} \\
\qquad \dot{\varphi}_{y_1} + \cos(\varphi_{y_1})\Big[(Y_2 - Y_1)\cos(\varphi_{X_1}) + (Z_2 - Z_1)\sin(\varphi_{X_1})\Big]\dot{\theta}_1\Big\} , \\
Q_{Y_1} = k(Y_2 - Y_1) + c\Big\{\dot{Y}_2 - \dot{Y}_1 + (Z_2 - Z_1)\dot{\varphi}_{X_1} - (X_2 - X_1)\sin(\varphi_{X_1}) \\
\qquad \dot{\varphi}_{y_1} + \Big[-(X_2 - X_1)\cos(\varphi_{X_1})\cos(\varphi_{y_1}) + (Z_2 - Z_1)\sin(\varphi_{y_1})\Big]\dot{\theta}_1\Big\} , \\
Q_{Z_1} = k(Z_2 - Z_1) + c\Big\{\dot{Z}_2 - \dot{Z}_1 - (Y_2 - Y_1)\dot{\varphi}_{X_1} + (X_2 - X_1)\cos(\varphi_{X_1})\dot{\varphi}_{y_1} + \\
\qquad -\Big[(X_2 - X_1)\sin(\varphi_{X_1})\cos(\varphi_{y_1}) + (Z_2 - Z_1)\sin(\varphi_{y_1})\Big]\dot{\theta}_1\Big\} , \\
Q_{\varphi_{X_1}} = 2\Big\{ka^2\cos^2(\varphi_{y_1}) + k_t b^2\Big[1 + \sin^2(\varphi_{y_1})\Big]\Big\}(\varphi_{x_2} - \varphi_{x_1}) + \\
\qquad +4k_t b^2\sin(\varphi_{y_1})(\theta_2 - \theta_1) + 2\Big\{ca^2\cos^2(\varphi_{y_1}) + k_t b^2\Big[1 + \sin^2(\varphi_{y_1})\Big]\Big\} \\
\qquad (\dot{\varphi}_{x_2} - \dot{\varphi}_{x_1}) + 2(ca^2 - 2c_t b^2)\sin(\varphi_{y_1})\cos(\varphi_{y_1})(\varphi_{x_1} - \varphi_{x_2})\dot{\varphi}_{y_1} + \\
\qquad +4c_t b^2\sin(\varphi_{y_1})(\dot{\theta}_2 - \dot{\theta}_1) + \Gamma\cos(\varphi_{y_1})(\varphi_{y_2} - \varphi_{y_1})\dot{\theta}_1 , \\
Q_{\varphi_{y_1}} = \chi(\varphi_{y_2} - \varphi_{y_1}) + 2(ca^2 + c_t b^2)(\dot{\varphi}_{y_2} - \dot{\varphi}_{y_1}) + \Gamma\cos(\varphi_{y_1})(\varphi_{x_1} - \varphi_{x_2})\dot{\theta}_1 , \\
Q_{\theta_1} = 4k_t b^2(\theta_2 - \theta_1) + 4k_t b^2\sin(\varphi_{y_1})(\varphi_{x_2} - \varphi_{x_1}) + 4c_t b^2\Big[(\dot{\theta}_2 - \dot{\theta}_1) + \\
\qquad +\sin(\varphi_{y_1})(\dot{\varphi}_{x_2} - \dot{\varphi}_{x_1}) - \cos(\varphi_{y_1})(\varphi_{x_2} - \varphi_{x_1})\dot{\varphi}_{y_1}\Big] ,
\end{array}\right. \tag{12.65}$$

$$\left\{\begin{array}{l}
Q_{X_2} = -Q_{X_1} , \\
Q_{Y_2} = -Q_{Y_1} , \\
Q_{Z_2} = -Q_{Z_1} ,
\end{array}\right. \qquad
\left\{\begin{array}{l}
Q_{\varphi_{X_2}} = -Q_{\varphi_{X_1}} , \\
Q_{\varphi_{y_2}} = -Q_{\varphi_{y_1}} , \\
Q_{\theta_2} = -Q_{\theta_1} .
\end{array}\right. \tag{12.66}$$

Note that a further linearization, obtained by assuming that also the absolute displacements and rotations are small, yields the usual equations of rotordynamics. Also, in this case, the two assumptions of constant spin speed and constant angular momentum coincide in the linearized model. The destabilizing effect of rotating damping is clearly present in the equations. The four cases seen in the previous section will be again dealt with.

Case 1

Case 1 refers to a disc rotor with a stiff connection between the two bodies, i.e., with flexural natural frequencies at standstill far higher than the maximum spin speed. The linearized results at 1000 rad/s show that the rigid-body conical precessional motion, occurring at 1450 rad/s (i.e., at the frequency $\Omega J_p/J_t$) is completely uncoupled from the mode involving relative motions of the two parts of the system. The former is very little damped, with a decay rate of about 10^{-8} 1/s, and the second one is very much damped (decay rate of 1386 1/s) and its frequency is not much affected by the spin speed up to above 1000 rad/s. Also, the translational mode is always damped in the whole speed range considered.

The results of a numerical simulation performed using the nonlinear, the semilinearized, and the fully linearized models are reported in Figure 12.16. All initial generalized displacements and velocities are equal to zero, except for $\varphi_{y_1} = 0.12$ rad, $\varphi_{y_2} = 0.08$ rad, $\dot{\varphi}_{X_1} = \dot{\varphi}_{X_2} = -145$ rad/s, and $\dot{\theta}_1 = \dot{\theta}_2 = 1000$ rad/s. Note that the displacements remain vanishingly small as the coupling terms of the damping matrix depend linearly on the displacements, and hence, the initial conditions assumed make them to remain equal to zero for the whole integration time. The result is a whirling motion with a peak-to-peak amplitude of about 0.2 rad (more than 10°) in which the two bodies move together, after a short transient in which the oscillations of one relative to the other are quickly damped out. The frequency of the motion resulting from the nonlinear model is 1436 rad/s, whereas that of the semilinearized model is 1450 rad/s. The latter value coincides with that obtained from the frequency domain computation for the linearized model. They differ from each other by less than 1% even if the amplitude of the motion is large. Note the quick damping of differential motions of the two bodies, whereas the overall whirling of the system is almost undamped (this could be easily expected owing to the values of the decay rate of the linearized model, see Table 12.6); the oscillations in the angular velocity $d\theta/dt$ show clearly the lateral-torsional coupling, not predicted by the linearized model; such coupling is, however, not strong even if the amplitude is large and does not affect the lateral behavior. Figure 12.16(d) shows that the energy dissipation caused by the damping of the differential oscillations is large but the results of the fully nonlinear solution almost coincide with those of the semilinearized solution.

The results shown in Figure 12.16 support the claim that the linearized solution is still accurate in predicting the lateral behavior even at amplitudes as large as 10° (peak to peak) and that the torsional-lateral coupling is not too strong.

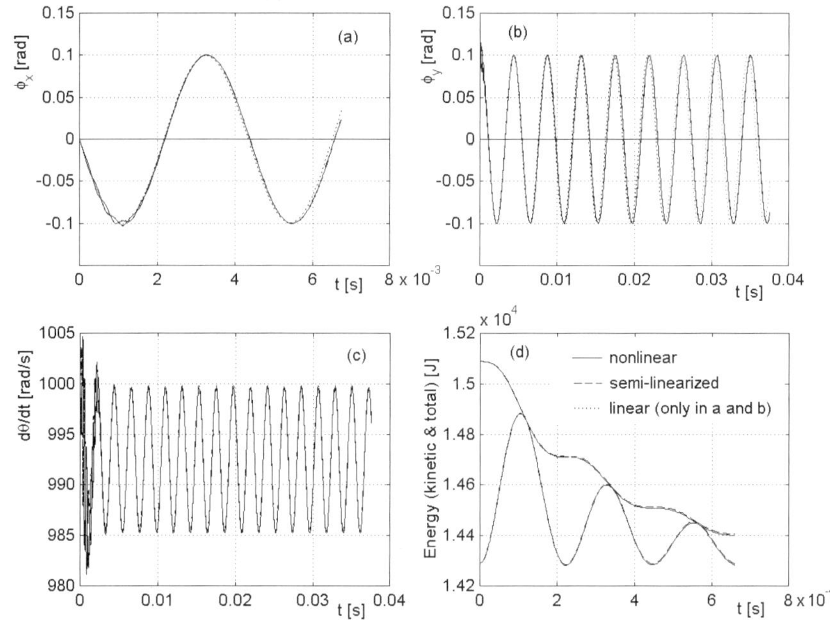

FIGURE 12.16. Twin rigid-bodies rotor; results of the numerical simulation for case 1. Time history of (a) angles φ_{X_i}, (b) angles φ_{y_i}, (c) spin speed of the two rotors $d\theta_i/dt$, and (d) total and kinetic energy. The plots are related to different time intervals.

Case 2

Case 2 deals with a rotor having the same inertial properties of the previous one, but with a very soft connection. The cylindrical (translational) mode is a supercritical one (i.e., the corresponding whirl speed is smaller than the spin speed or, which is the same, the spin speed is higher than the critical speed related to the translational mode), and the linearized analysis shows that it is unstable. Owing to uncoupling, the cylindrical modes are not excited, and the related instability does not show up. The conical (rotational) modes are subcritical because there is no critical speed linked to rotational degrees of freedom.

The linearized results at 1000 rad/s show that the rigid-body precessional motion occurs at 1425 rad/s (i.e., at a frequency slightly lower than $\Omega J_p/Jt$) and is strongly coupled with the forward mode involving relative motions of the two parts of the system, which occurs at 1541 rad/s. As a result, both modes are very strongly damped, with decay rates of the order of 4 and 17 1/s. The presence of the damper between the two bodies cause not only the differential motions to die out quickly, but also the overall whirling to decay. The results of a numerical simulation performed using the nonlinear, the

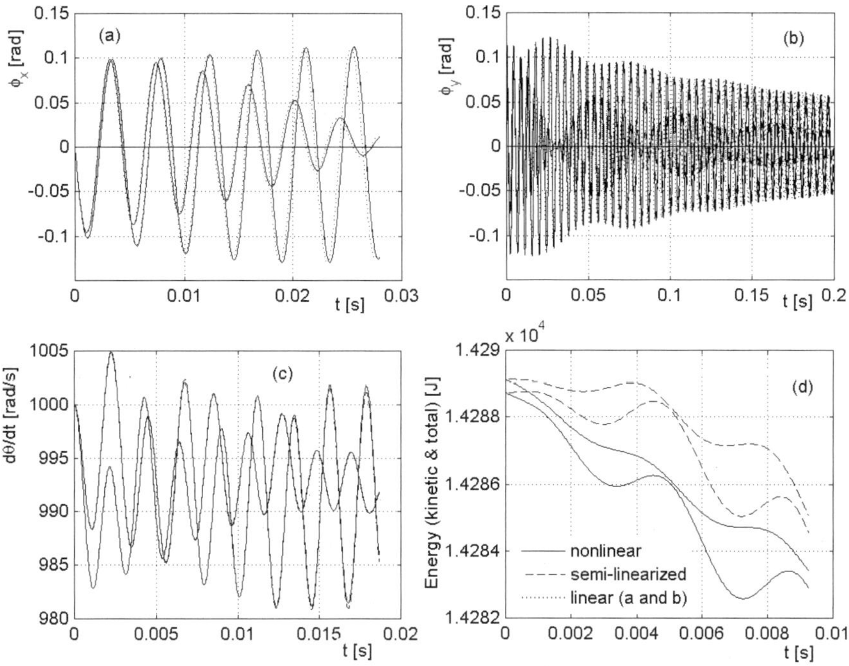

FIGURE 12.17. Same as Figure 12.16, but for case 2.

semilinearized, and the fully linearized models are reported in Figure 12.17. The initial conditions are the same as in the previous example, except for $\dot{\varphi}_{X_1} = \dot{\varphi}_{X_2} = -142.5$ rad/s. The result is a decaying whirling motion with an initial peak-to-peak amplitude of about 0.2 rad (more than 10°). Note that the instability of the translational mode would, in an actual case, drive to instability also the rotational one; this effect is not accounted for in the simulation owing to the initial displacements and velocities that have been assumed to be exactly equal to zero.

The time history of angles φ_{X_i} and φ_{y_i} show that a sort of beat is present, owing to the fact that two natural frequencies are close. The overall motion is damped. Figure 12.17(c), related to the spin speed of the rotor $d\theta_i/dt$, shows that a lateral-torsional coupling, not predicted by the linearized model, is again present but is not very strong even if the amplitude is large and does not affect the lateral behavior. The plot of the time history of the total and kinetic energy shows that the fully nonlinear solution gives results that are different from the semilinearized solution; the difference between the two is, however, magnified by the very expanded scale and the relative error is still small. Except for this last result, the fully nonlinear solution is completely superimposed to the semilinearized solution and the

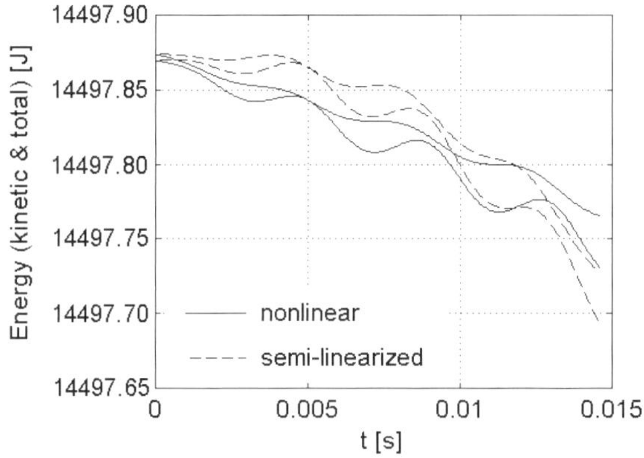

FIGURE 12.18. Same as Figure 12.17(d), but for an amplitude of the motion divided by 10.

linearized model yields similar results, but overestimates the frequency by about 1%.

To investigate on the discrepancies between the time history of the total and kinetic energy computed using the nonlinear and semilinearized solutions the simulation has been repeated with initial conditions on the displacements and velocities (except the spin speed) divided by 10 (Figure 12.18). The peak-to peak amplitude is now of about 1°, and the results of the two model are very close to each other.

Case 3

The rotor of case 3 has a very stiff connection between the bodies and operates at the nominal speed (1000 rad/s) in the subcritical regime. The linearized analysis shows that it is very weakly unstable in the small: The imaginary part of the whirl speed of the first forward mode is negative, but its absolute value is so small (1.49×10^{-9} 1/s) that a very long time is expected to be needed to develop an actual unstable behavior (doubling the amplitude takes more than 14 years). This type of behavior is typical of all free rotors in which $J_p < J_t$.

The rigid-body precessional motion, occurring at 690 rad/s (i.e., at the frequency $\Omega J_p/Jt$) is completely uncoupled from the mode involving relative motions of the two parts of the system. The former is slightly unstable, as said above, and the second one is very much damped and its frequency is not much affected by the spin speed. The results of a numerical simulation performed using the nonlinear, the semilinearized, and the fully linearized models are reported in Figure 12.19. The initial conditions are the same as for the previous cases, except for $\dot{\varphi}_{X_1} = \dot{\varphi}_{X_2} = -69$ rad/s. The result is

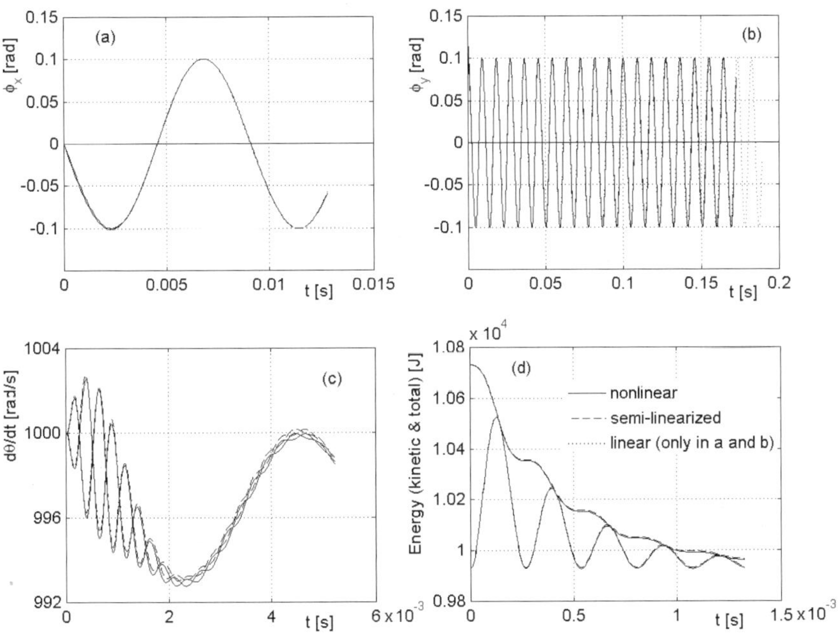

FIGURE 12.19. Same as Figure 12.16, but for case 3.

a whirling motion with a peak-to-peak amplitude of about 0.2 rad (more than 10°) with the two bodies moving together, after a short transient in which the oscillations of one relative to the other are quickly damped out.

Figure 12.19(b) shows that the buildup of the amplitude is so slow that the time history is practically coincident with that of an undamped system. From Figure 12.19(d) it is clear that the time history of the total and kinetic energy computed using the fully nonlinear solution is almost superimposed to the semilinearized solution as in case 1.

Case 4

Owing to the very soft springs, at 1000 rad/s, the rotor of case 4 operates in supercritical conditions for both modes. The linearized analysis shows that the rotor of case 4 is very unstable in all forward modes, particularly in the translational and second rotational ones. However, also the first rotational forward mode is unstable, possibly owing to the closeness of the two rotational modes.

The results of a numerical simulation performed using the nonlinear, the semilinearized and the fully linearized models are reported in Figure 12.20. The initial conditions are the same as for case 3. The result is a whirling motion with an initial peak-to-peak amplitude of about 0.2 rad (more than 10°); the amplitude quickly grows to very large values. Note

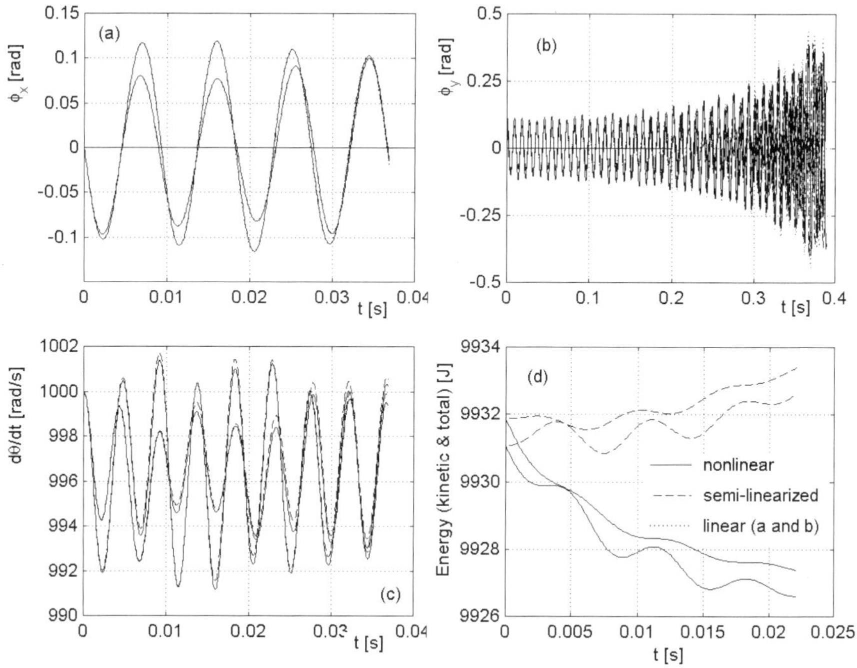

FIGURE 12.20. Same as Figure 12.16, but for case 4.

that in practice, the instability of the translational modes would add to that of the rotational modes owing to the coupling that would start as soon as the amplitude of the former starts to be non-negligible.

The time history of the total and kinetic energy shows that in this case, as in case 2, the nonlinear solution gives results that are not in good accordance with the semilinearized solution, but the same considerations on the smallness of the relative error still hold.

To show the effects of the nonlinearities caused by large amplitudes, the simulation for case 2 has been repeated with initial conditions with larger values of φ_{y_i} and $\dot{\varphi}_{X_i}$: $\varphi_{y_1} = 0.48$ rad, $\varphi_{y_2} = 0.32$ rad, and $\dot{\varphi}_{X_1} = \dot{\varphi}_{X_2} = -570$ rad/s (Figure 12.21).

Owing to the very large amplitude (about 60° peak to peak), the linearized solution gives results that are quantitatively incorrect, while being qualitatively correct. The semilinearized solution, is on the contrary reliable in predicting the behavior of the system, except for the kinetic energy, as it was already noted in Figure 12.17.

Remark 12.6 *The assumption of constant angular velocity can be obtained from that of constant angular momentum, typical of free rotor dynamics, when small amplitude motion is considered. The stability consid-*

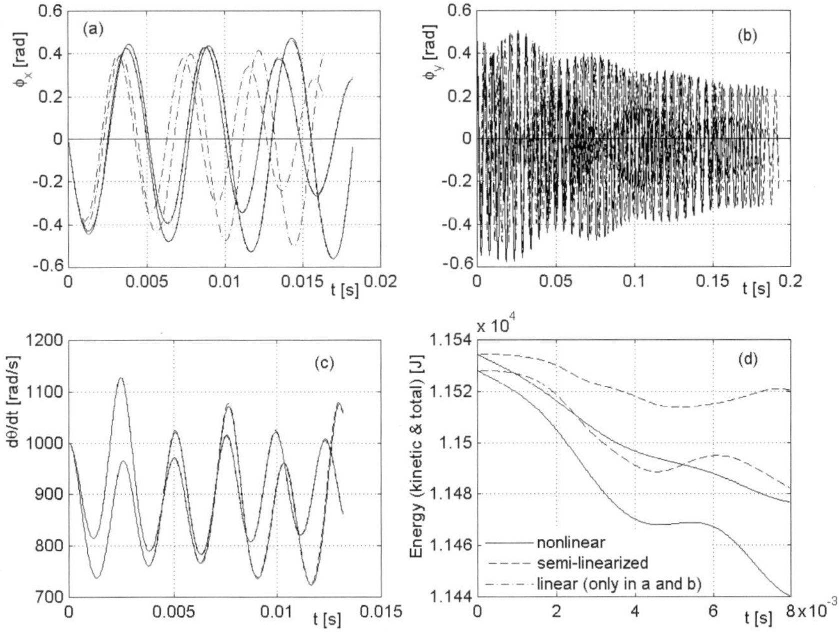

FIGURE 12.21. Same as Figure 12.17 (case 2), but with different initial conditions. The very large amplitude makes the linearized solution incorrect, whereas the semilinearized one retains its applicability.

erations drawn from classic rotordynamic analysis hold for the study of the stability in the small of free rotors and spacecraft.

Remark 12.7 *The axial and torsional motions are coupled with the lateral behavior of the rotor. However, this coupling becomes vanishingly small with reducing amplitude of the latter (for the torsional motion, this statement can be seen as a consequence of the previous point). The axial-flexural and torsional-flexural decoupling is then applicable to the motion in the small of both fixed and free rotors.*

Remark 12.8 *The linearized analysis holds with good precision even for motions occurring with angular amplitudes up to several degrees. This allows us to use the classic linearized rotordynamic approach to the study of all fixed rotors (which is fairly obvious) but also in general to free rotors, at least if they are provided of an attitude control system (which can be introduced into the model) preventing large amplitude motion from occurring.*

	m [kg]	J_p [kg m^2]	J_t [kg m^2]
Upper mass	10	0.1	0.08
Central body	100	28	16
Lower mass	5	0.06	0.04

TABLE 12.7. Inertial data for the small satellite.

12.4 Multibody free rotors

In the case of multibody rotors the nonlinear equations become so complicated that it is useless to write explicitly a nonlinear or semilinearized model. Actually, this occurs even in the case of a twin-bodies rotor when the centers of mass do not coincide in the undeflected position, as it was assumed in the previous sections.

As a consequence, the study of the motion in the large of the system can be performed only by numerical integration in time of a model built through the usual multibody approach and no general solution can be achieved.

However, if the assumption of small displacements and rotations (for displacements, it is enough that the relative displacements are small, whereas all rotations must be small) holds, it is possible to write an explicit linear differential equation in the generalized coordinates and to obtain an approximate frequency-domain solution.

A consequence of the linearization is the uncoupling of the axial, torsional, and flexural behavior, which holds if the geometry of the system complies with the conditions seen for this uncoupling when dealing with the beam theory. The linearized equation of motion for an axi-symmetrical system, written with reference to the complex coordinates, is still Equation (12.55), where matrix $\mathbf{K}$ is singular, owing to the fact that the rotor is free in space, and matrix $\mathbf{C}$ coincides with $\mathbf{C}_r$ (because $\mathbf{C}_n$ vanishes) and is singular.

The fact that the centers of mass of the various bodies do not coincide makes it impossible to uncouple translational modes from rotational ones, and attitude modes can be separated from deformation modes only through the methods of modal analysis, and with the relevant limitations related to gyroscopic and damped system. This feature will be shown clearly in the following example.

Example 12.2 *Example: small spinning satellite.*

Consider a small spinning satellite made by a central axi-symmetrical body carrying at the poles two long slender booms, whose longitudinal axis lies along the spin axis. Two concentrated masses are located at the ends of booms. A sketch is inset in Figure 12.22. The inertial data of the system are summarized in Table 12.7.

The two booms are 5 m long, and the cross-sectional area and the transversal area moment of inertia are $A = 400$ mm^2 and $I = 2 \times 10^{-6}$ m^4. The material *of the beams have a Young modulus* $E = 7.2 \times 10^{10}$ *N/m²*, *an apparent density* $\rho = 4000$ *kg/m³*, *and an average loss factor* η *=0.01. Note that a system of this kind is unrealistic, and its purpose is only to show some properties of multibody free rotors.*

By modeling the system using three mass and two beam elements, the following matrices are readily obtained:

$$\mathbf{M} = \begin{bmatrix} 11.486 & 1.047 & 0.514 & -0.620 & 0 & 0 \\ & 1.033 & 0.620 & -0.715 & 0 & 0 \\ & & 102.971 & 0 & 0.514 & -0.620 \\ & & & 17.906 & 0.620 & -0.715 \\ & symm. & & & 6.486 & -1.047 \\ & & & & & 0.993 \end{bmatrix},$$

$$\mathbf{G} = \begin{bmatrix} 0.00095 & 0.00039 & -0.00095 & 0.00039 & 0 & 0 \\ & 0.10266 & -0.00039 & -0.00067 & 0 & 0 \\ & & 0.00191 & 0 & -0.00095 & 0.00039 \\ & & & 28.00532 & -0.00039 & -0.00067 \\ & symm. & & & 0.00095 & -0.00039 \\ & & & & & 0.06266 \end{bmatrix},$$

$$\mathbf{K} = \begin{bmatrix} 0.3445 & 0.8612 & -0.3445 & 0.8612 & 0 & 0 \\ & 2.8731 & -0.8612 & 1.4331 & 0 & 0 \\ & & 0.68900 & 0 & -0.3445 & 0.8612 \\ & & & 5.7462 & -0.8612 & 1.4331 \\ & symm. & & & 0.3445 & -0.8612 \\ & & & & & 2.8731 \end{bmatrix} \times 10^4 .$$

The imaginary part of the complex stiffness matrix (caused by the presence of structural damping) is $\mathbf{K}" = 0.01\mathbf{K}$.

The rigid-body inertial property of the system are $m = 123$ *kg*, $J_p = 28.2$ *kg m²*, *and* $J_t = 452$ *kg m² and the position of the center of mass is at 4,797 mm from the upper end. The satellite behaves as a long rotor; hence, a rigid-body instability can be expected. There are five rigid body modes with a natural frequency equal to zero (three lateral, one axial and one torsional) plus a rigid-body mode with natural frequency* $\omega = \Omega Jp/J_t = 0.0624\Omega$.

The damped eigenfrequencies at standstill are reported in Table 12.8.

The axial and torsional dynamics will be dealt no further.

The Campbell diagram is reported in Figure 12.22. The system has a critical speed $\Omega_{cr} = 11.45$ *rad/s (109 rpm), at which the first deformation mode becomes unstable, with a decay rate* $\text{Re}(s) = 0.057$ *1/s. The following critical speed, at which another mode becomes unstable, is high, namely at 162 rad/s.*

The whirl frequencies and decay rate of the first four forward and backward modes at $\Omega = 10$ *rad/s are reported in Table 12.9.*

The speed is lower than the first critical speed, and then all deformation modes are stable. As expected, the second rigid-body mode is slightly unstable, and its frequency is very close to that computed using a rigid-body model (0.6206 against 0.624).

Flexural	
Frequency	Decay rate
0 (double)	0
11.41 rad/s (1.82 Hz)	−0.057 1/s
48.75 rad/s (7.76 Hz)	−0.244 1/s
175.17 rad/s (27.88 Hz)	−0.876 1/s
190.62 rad/s (30.34 Hz)	−0.953 1/s

Axial		Torsional	
Frequency	Decay rate	Frequency	Decay rate
0	0	0	0
529.55 rad/s (84.28 Hz)	−2.65	318.88 rad/s (50.75 Hz)	−1.59
703.01 rad/s (111.89 Hz)	−3.51	403.07 rad/s (64.15 Hz)	−2.02

TABLE 12.8. Damped eigenfrequencies at standstill.

	Forward		Backward	
Mode	ω [rad/s]	Decay rate [1/s]	ω [rad/s]	Decay rate [1/s]
1	0	0	0	0
2	0.6206	2.295×10^{-5}	0	0
3	11.44	−0.0570	−11.38	−0.0570
4	55.92	−0.2412	−42.62	−0.2411

TABLE 12.9. Whirl frequencies and decay rate of the first four forward and backward modes at $\Omega = 10$ rad/s.

Instead of using consistent mass and gyroscopic matrices, it was possible to lump the mass of the beam at its ends, a procedure yielding diagonal matrices

$$\mathbf{M} = diag\left(\begin{bmatrix} 12 & 0.08 & 104 & 16 & 7 & 0.04 \end{bmatrix}\right) ,$$

$$\mathbf{G} = diag\left(\begin{bmatrix} 0 & 0.1 & 0 & 28 & 0 & 0.06 \end{bmatrix}\right) .$$

The rigid-body properties are slightly different, because a larger value of the transversal moment of inertia is found ($J_t = 486$ kg m^2), the mass is the same, and the polar moment of inertia and the position of the center of mass are only marginally affected. The whirl frequencies of the first four forward and backward modes at $\Omega = 10$ rad/s modify as shown in Table 12.10

To uncouple rigid-body and deformation modes, a coordinate transformation similar to the one seen in Equations (12.60) and (12.63) can be performed:

$$\begin{cases} r_G = \dfrac{m_1 r_1 + m_2 r_2 + m_3 r_3}{m_T} , \\ \Delta r_1 = r_1 - r_2 , \\ \Delta r_3 = r_3 - r_2 , \end{cases} \tag{12.67}$$

$$\begin{cases} \phi_G = \frac{J_{t1}\phi_1 + J_{t2}\phi_2 + J_{t3}\phi_3}{J_{ttT}} , \\ \Delta\phi_1 = \phi_1 - \phi_2 , \\ \Delta\phi_3 = \phi_3 - \phi_2 , \end{cases} \tag{12.68}$$

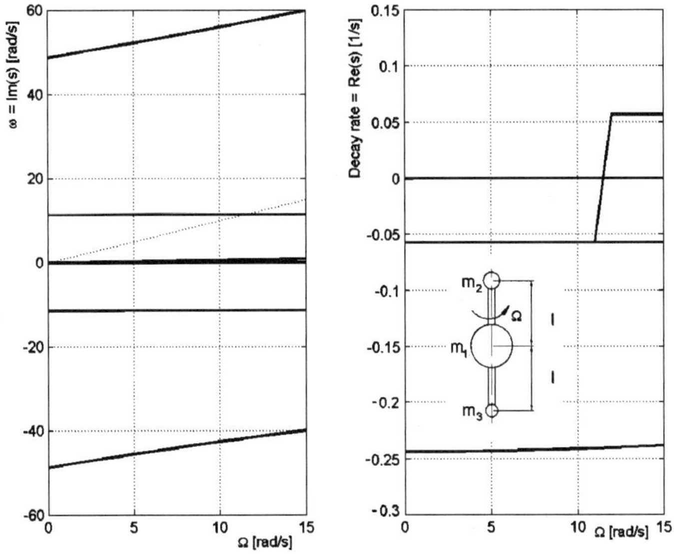

FIGURE 12.22. Cambell diagram and decay rate plot for the satellite of Example 12.1.

	Forward		Backward	
Mode	ω [rad/s]	Decay rate [1/s]	ω [rad/s]	Decay rate [1/s]
1	0	0	0	0
2	0.5773	2.085×10^{-5}	0	0
3	10.70	−0.0534	−10.65	−0.0534
4	62.03	−0.2603	−45.14	−0.2603

TABLE 12.10. Whirl frequencies and decay rate of the first four forward and backward modes at $\Omega = 10$ rad/s for the lumped parameters model.

where

$$m_T = m_1 + m_2 + m_3 \ , \qquad J_{ttT} = J_{t1} + J_{t2} + J_{t3} \ . \tag{12.69}$$

The coordinate transformation matrix **T** *is defined by the relationship*

$$\left\{ \begin{array}{c} r_G \\ \phi_G \\ \Delta r_1 \\ \Delta r_2 \\ \Delta \phi_1 \\ \Delta \phi_3 \end{array} \right\} = \left[\begin{array}{cccccc} \frac{m_1}{m_T} & 0 & \frac{m_2}{m_T} & 0 & \frac{m_3}{m_T} & 0 \\ 0 & \frac{J_{t1}}{J_{ttT}} & 0 & \frac{J_{t21}}{J_{ttT}} & 0 & \frac{J_{t3}}{J_{ttT}} \\ 1 & 0 & -1 & 0 & 0 & 0 \\ 0 & 0 & -1 & 0 & 1 & 0 \\ 0 & 1 & 0 & -1 & 0 & 0 \\ 0 & 0 & 0 & -1 & 0 & 1 \end{array} \right] \left\{ \begin{array}{c} r_1 \\ \phi_1 \\ r_2 \\ \phi_2 \\ r_3 \\ \phi_3 \end{array} \right\} .$$

By applying this transformation to the matrices related to the lumped parameters system, it follows that

$$\mathbf{M} = \begin{bmatrix} 123 & 0 & 0 & 0 & 0 & 0 \\ & 16.12 & 0 & 0 & 0 & 0 \\ & & 10.83 & -0.683 & 0 & 0 \\ & & & 6.601 & 0 & 0 \\ & symm. & & & 0.0796 & -1.985 \times 10^{-4} \\ & & & & & 0.0399 \end{bmatrix},$$

$$\mathbf{G} = \begin{bmatrix} 0 & 0 & 0 & 0 & 0 & 0 \\ & 28.16 & 0 & 0 & -0.0398 & -0.0099 \\ & & 0 & 0 & 0 & 0 \\ & & & 0 & 0 & 0 \\ & symm. & & & 0.0997 & -0.00020 \\ & & & & & 0.0599 \end{bmatrix},$$

$$\mathbf{K} = \begin{bmatrix} 0 & 0 & 0 & 0 & 0 & 0 \\ & 17.225 & 1.722 & 1.722 & 4.221 & 4.263 \\ & & 0.344 & 0 & 0.853 & -0.004 \\ & & & 0.344 & 0.009 & -0.857 \\ & symm. & & & 2.831 & -0.032 \\ & & & & & 2.852 \end{bmatrix} \times 10^4 .$$

The imaginary part of the stiffness matrix (hysteretic damping matrix) is proportional to the stiffness matrix, because the loss factor has been assumed to be constant in the structure. Any coordinate transformation yields a damping matrix proportional to the stiffness matrix, so the former will be dealt with no further.

By inspecting the matrices, it is clear that:

- *The element 11 of all matrices is uncoupled from all the others: The rigid-body translational motion is fully uncoupled and is that of a point mass free in space.*
- *The structure of the mass matrix is such that also the rotational motion is uncoupled, and that the differential translational modes are uncoupled from differential rotational modes.*
- *The structure of the gyroscopic matrix shows that rotational differential modes are coupled with the rigid-body rotational mode.*
- *The last two points are the same as seen in Section 12.2.1 for a twin-bodies system with coinciding centers of mass.*
- *The stiffness matrix couples all equations, except for the first one, related to the rigid-body translational motion.*
- *The matrix transformation shown leads then to a strong elastic coupling. This could be expected, because rotation ϕ_G, with all Δr_i equal to zero does not describe a rigid-body motion of the system, owing to the fact that the centers of mass of the various bodies are not coincident.*

A transformation in which ϕ_G is really a rigid-body motion is that in which Equation (12.67) is substituted by

$$\begin{cases} r_G = \dfrac{m_1 r_1 + m_2 r_2 + m_3 r_3}{m_1 + m_2 + m_3} \ , \\ \Delta r_1 = r_1 - r_2 + a\phi_G \ , \\ \Delta r_3 = r_3 - r_2 - b\phi_G \ , \end{cases} \tag{12.70}$$

where a and b are the distances from the centers of mass of the first and the third body to that of the second.

The coordinate transformation is then

$$\begin{Bmatrix} r_G \\ \phi_G \\ \Delta r_1 \\ \Delta r_2 \\ \Delta \phi_1 \\ \Delta \phi_3 \end{Bmatrix} = \begin{bmatrix} \frac{m_1}{m_T} & 0 & \frac{m_2}{m_T} & 0 & \frac{m_3}{m_T} & 0 \\ 0 & \frac{J_{t1}}{J_{ttT}} & 0 & \frac{J_{t21}}{J_{ttT}} & 0 & \frac{J_{t3}}{J_{ttT}} \\ 1 & a & -1 & 0 & 0 & 0 \\ 0 & -b & -1 & 0 & 1 & 0 \\ 0 & 1 & 0 & -1 & 0 & 0 \\ 0 & 0 & 0 & -1 & 0 & 1 \end{bmatrix} \begin{Bmatrix} r_1 \\ \phi_1 \\ r_2 \\ \phi_2 \\ r_3 \\ \phi_3 \end{Bmatrix} .$$

The mass matrix becomes

$$\mathbf{M} = \begin{bmatrix} 123 & 0 & 0 & 0 & 0 & 0 \\ & 486.04 & -57.56 & 36.42 & 467.59 & -1.166 \\ & & 10.83 & -0.683 & -57.28 & 0.143 \\ & & & 6.601 & 36.24 & -0.0904 \\ & symm. & & & 465.35 & -1.161 \\ & & & & & 0.0428 \end{bmatrix} ,$$

the gyroscopic matrix is the same as that for the previous computation, and the stiffness matrix becomes

$$\mathbf{K} = \begin{bmatrix} 0 & 0 & 0 & 0 & 0 & 0 \\ & 0 & 0 & 0 & 0 & 0 \\ & & 0.344 & 0 & 0.861 & 0 \\ & & & 0.344 & 1.722 & -0.861 \\ & symm. & & & 11.486 & -4.306 \\ & & & & & 2.873 \end{bmatrix} \times 10^4 .$$

Now the rotational motion is elastically uncoupled, but the mass matrix couples all motions except rigid-body translation.

Consider again the consistent parameter system. The alternative of resorting to modal analysis does not allow a much better uncoupling. The modal transformation can be started as usual, by computing the eigenvectors of the undamped nongyroscopic system and then normalizing them in such a way that all elements of the modal mass matrix on the main diagonal are equal to 1. In the present case, however, this does not guarantee that the modal mass matrix is an identity matrix ($\overline{\mathbf{M}}$− $\mathbf{I}$), because there are two modes (the first two, if they are ordered in ascending order) with equal eigenfrequency (in the present case with vanishing eigenfrequency, because they are rigid-body modes). When this occurs, all linear *combinations of the eigenvectors are eigenvectors, and then the two eigenvectors that are found are not necessarily the two that are m-orthogonal.*

As a consequence, the element 12 of the modal mass matrix is not in general null ($\overline{\mathbf{M}}_{12} \neq 0$).

Let the two rigid-body eigenvectors be $\mathbf{q}_1$ *and* $\mathbf{q}_2$, *and the sought m-orthogonal eigenvectors be*

$$\begin{cases} \mathbf{q}_3 = a\mathbf{q}_1 + b\mathbf{q}_2 \ , \\ \mathbf{q}_4 = c\mathbf{q}_1 + d\mathbf{q}_2 \ . \end{cases}$$

The conditions leading to unit modal masses and to $\overline{\mathbf{M}}_{12} = 0$ *are*

$$\begin{cases} \mathbf{q}_3^T\mathbf{M}\mathbf{q}_3 = (a\mathbf{q}_1 + b\mathbf{q}_2)^T \mathbf{M}\,(a\mathbf{q}_1 + b\mathbf{q}_2)^T = 1 \ , \\ \mathbf{q}_4^T\mathbf{M}\mathbf{q}_4 = (c\mathbf{q}_1 + d\mathbf{q}_2)^T \mathbf{M}\,(c\mathbf{q}_1 + d\mathbf{q}_2)^T = 1 \ , \\ \mathbf{q}_3^T\mathbf{M}\mathbf{q}_4 = (a\mathbf{q}_1 + b\mathbf{q}_2)^T \mathbf{M}\,(c\mathbf{q}_1 + d\mathbf{q}_2)^T = 0 \ . \end{cases}$$

As the unknown coefficients are 4, a further condition can be added. In this case k-orthogonality is at any rate verified, because the first two modes are rigid-body modes, and the first two lines and columns of the modal stiffness matrix are at any rate equal to zero. As a fourth condition, the first modal gyroscopic term will be assumed to vanish

$$\mathbf{q}_3^T\mathbf{G}\mathbf{q}_3 = (a\mathbf{q}_1 + b\mathbf{q}_2)^T \mathbf{G}\,(a\mathbf{q}_1 + b\mathbf{q}_2)^T = \overline{\mathbf{G}}_{11} = 0 \ ,$$

which means that the first mode is a nongyroscopic one; i.e., the first modal system is a free Jeffcott rotor.

By remembering that $\mathbf{q}_1^T\mathbf{M}\mathbf{q}_1 = \mathbf{q}_2^T\mathbf{M}\mathbf{q}_2 = 1$, *the four conditions become*

$$\begin{cases} a^2 + b^2 + 2ab\overline{\mathbf{M}}_{12} = 1 \ , \\ a^2\overline{\mathbf{G}}_{11} + b^2\overline{\mathbf{G}}_{22} + 2ab\overline{\mathbf{G}}_{12} = 0 \ , \\ c^2 + d^2 + 2cd\overline{\mathbf{M}}_{12} = 1 \ , \\ ac + bd + (ad + bc)\,\overline{\mathbf{M}}_{12} = 0 \ . \end{cases}$$

Performing the relevant computations, the (complex conjugate) values of the four constants are $a = 0.274 + 0.394\ i$, $b = 0.274 - 0.3945i$, $c = 0.2585 - 0.4177i$, *and* $d = 0.2585 + 0.4177\ i$.

The matrix of the normalized eigenvectors is then

$$\mathbf{\Phi} = \begin{bmatrix} -0.0902 & -0.2254 & -0.1726 & 0.0546 & 0.0577 & 0.0297 \\ 0 & 0.0470 & 0.0696 & -0.1454 & -0.8983 & -0.5141 \\ -0.0902 & 0.0096 & 0.0376 & 0.0005 & 0.0118 & -0.0033 \\ 0 & 0.0470 & -0.0140 & 0.2178 & -0.0270 & -0.0930 \\ -0.0902 & 0.2445 & -0.3012 & -0.1035 & 0.0676 & -0.1090 \\ 0 & 0.0470 & -0.0936 & -0.1566 & 0.5403 & -0.9610 \end{bmatrix} .$$

	Mode	$\overline{M}_i$	$\overline{K}_i$	$\overline{G}_i$	$\overline{C}_i$	$\overline{C}_{r_i}$
	1	1	0	0.0000	0	0
	2	1	0	0.0621	-0.0000	0.0000
$\Omega \to 0$	3	1	130	0.0067	0.1141	0.0004
	4	1	2377	1.3319	0.4875	0.3246
	5	1	30686	0.1214	1.7517	0.1064
	6	1	36335	0.3270	1.9061	0.3116
	1	1	0	0.0000	0	0
	2	1	0	0.0621	-0.0000	-0.0000
$\Omega = 10$	3	1	130	0.0067	0.1141	0.0004
	4	1	2383	1.3306	0.4824	0.3216
	5	1	30685	0.1214	1.7517	0.1064
	6	1	36374	0.3285	1.9050	0.3099

TABLE 12.11. Modal parameters at standstill ($\Omega = 0$) and at a spin speed $\Omega = 10$ rad/s for the spining satellite.

The modal mass matrix is an identity matrix, and the gyroscopic and stiffness matrices are

$$\overline{\mathbf{G}} = \begin{bmatrix} 0 & 0 & 0 & 0 & 0 & 0 \\ & 0.0623 & -0.0184 & 0.2854 & -0.0382 & -0.1275 \\ & & 0.0067 & -0.0856 & 0.0011 & 0.0384 \\ & & & 1.3319 & -0.1563 & -0.5499 \\ & symm. & & & 0.1214 & 0.0851 \\ & & & & & 0.3270 \end{bmatrix},$$

$$\overline{\mathbf{K}} = diag\left(\begin{bmatrix} 0 & 0 & 130 & 2,377 & 30,684 & 36,333 \end{bmatrix}\right) .$$

The first two modes are now rigid-body modes, and the first one is, as expected, a purely cylindrical mode, and the second is a conical mode with the apex in the center of mass. In this way, the first mode is a translational mode, and the first row and column of the modal gyroscopic matrix vanishes. Otherwise, the gyroscopic matrix is full, showing a complete gyroscopic coupling of all other modes, as easily expected in a gyroscopic system.

To uncouple the equations of motion, it is possible to resort to the methods seen in Section 3.7.

By applying the state-space, complex coordinates approach, the values of the complex frequencies can be computed as functions of speed, and from them, through Equation (3.87), the frequency-dependent modal parameters (modal mass, gyroscopic term, stiffness, rotating and nonrotating damping) can be obtained. The system can thus be decoupled in a number of uncoupled, speed-dependent gyroscopic systems with two degrees of freedom.

By operating in this way, the values reported in Table 12.11 are obtained.

13

Dynamics of rotating beams and blades

The rotors studied in the previous sections were modeled as beamlike structures, spinning about the axis of the beam. In many machines, however, there are elongated elements, which can often be modeled as beams, with their axis lying in radial direction, as for instance turbine or propeller blades. In the simplest models, the dynamics of these elements is neglected, and the blades are assumed to be rigid bodies, which contribute to the inertia of the system but not to its compliance. The dynamics of the blades is then studied separately.

When the aim of the study is to compute the critical speeds, the approximations linked with neglecting blade (and disc) compliance are usually acceptable, because the slope of the branches of the Campbell diagram linked with blades (and discs) vibration is higher than that of the $\omega = \Omega$ line and they do not give way to additional critical speeds.

There are, however, cases where the dynamics of the blades cannot be neglected and influences the dynamics of the whole machine. Moreover, the dynamic behavior of the blades is influenced by the rotational speed and the rotor-blades interaction may give way to dangerous phenomena such as instability.

The present chapter is devoted to give a qualitative understanding of the main phenomena related to the blade dynamics, whereas a more quantitative approach will be seen in the chapter dealing with the tridimensional modeling of rotors.

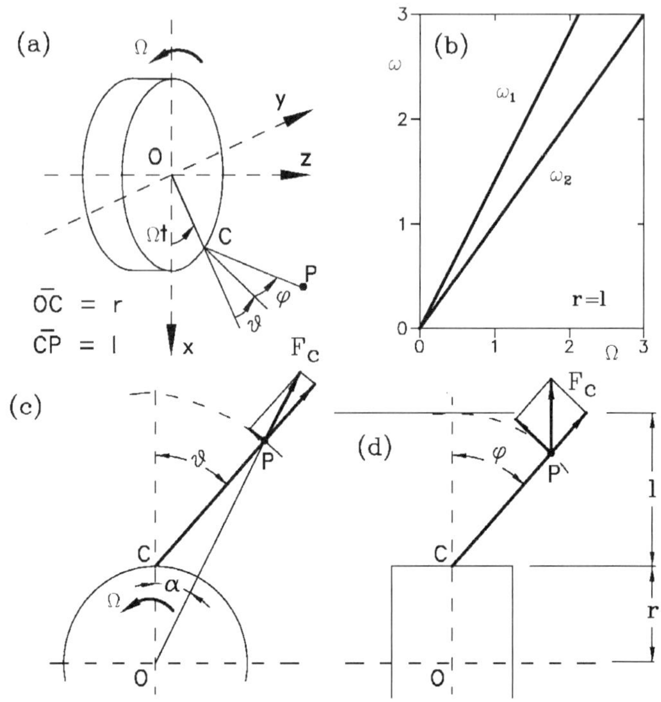

FIGURE 13.1. Rotating pendulum. (a) Sketch of the system and generalized coordinates, (b) Campbell diagram, (c) situation in the xy-plane, and (d) situation in a plane containing the z-axis.

13.1 Rotating pendulum

The simplest model for an unidimensional object attached to a rotor in a radial direction is a rotating pendulum. Although having little in common with a blade (e.g., a turbine or compressor blade), it can shed some light on the gyroscopic and centrifugal stiffening effects. Moreover, rotating pendulums have been used as torsional vibration dampers and constitute a good first approximation model for the blades of helicopter rotors; their study has then a practical interest.

A rotating pendulum can be described as a pendulum attached to the outer radius of a rotating disc, whose angular velocity will be considered as constant [Figure 13.1(a)].

As the angular velocity Ω of the disc is imposed, the system has only two degrees of freedom, and angles θ (between the projection of line PC on the plane of the disc and radius OC) and ϕ (between line PC and the mentioned plane) can be assumed to be the generalized coordinates. All other force fields except the centrifugal field (e.g., the gravitational field) are neglected. The position of point P is

$$(\overline{\text{P-O}}) = \left\{ \begin{array}{c} r\cos(\Omega t) + l\cos(\phi)\cos(\Omega t+\theta) \\ r\sin(\Omega t) + l\cos(\phi)\sin(\Omega t+\theta) \\ l\sin(\phi) \end{array} \right\} . \tag{13.1}$$

By differentiating the expressions of the coordinates with respect to time, the velocity of point P is readily obtained

$$\vec{V}_{\text{P}} = \left\{ \begin{array}{c} -\Omega r\sin(\Omega t) - \dot{\phi} l\sin(\phi)\cos(\Omega t+\theta) - l(\Omega+\dot{\theta})\cos(\phi)\sin(\Omega t+\theta) \\ \Omega r\cos(\Omega t) - \dot{\phi} l\sin(\phi)\sin(\Omega t+\theta) + l(\Omega+\dot{\theta})\cos(\phi)\cos(\Omega t+\theta) \\ \dot{\phi} l\cos(\phi) \end{array} \right\} . \tag{13.2}$$

The kinetic energy of the mass located in point P is then

$$\begin{aligned} \mathcal{T} = \frac{1}{2}m|\vec{V}_{\text{P}}|^2 = \frac{1}{2}m\big[&\Omega^2 r^2 + \dot{\phi}^2 l^2 + l^2(\Omega+\dot{\theta})^2\cos^2(\phi) + \\ &-2\Omega r l\dot{\phi}\sin(\phi)\sin(\theta) + 2\Omega r l(\Omega+\dot{\theta})\cos(\phi)\cos(\theta)\big] . \end{aligned} \tag{13.3}$$

The equations of motion can be obtained by resorting to Lagrange equations. They are clearly nonlinear, because of the presence of trigonometric functions of the generalized coordinates. The derivatives appearing in the first equation, the one related with the motion in the plane of rotation, are

$$\begin{aligned} \frac{\partial \mathcal{T}}{\partial \dot{\theta}} &= m\left[l^2(\Omega+\dot{\theta})\cos^2(\phi) + \Omega r l\cos(\phi)\cos(\theta)\right] , \\ \frac{d}{dt}\left(\frac{\partial \mathcal{T}}{\partial \dot{\theta}}\right) &= m\left[l^2\ddot{\theta}\cos^2(\phi) - 2l^2(\Omega+\dot{\theta})\dot{\phi}\cos(\phi)\sin(\phi) + \right. \\ &\left. -\Omega r l\dot{\phi}\sin\phi)\cos(\theta) - \Omega r l\dot{\theta}\cos(\phi)\sin(\theta)\right] , \\ \frac{\partial \mathcal{T}}{\partial \theta} &= m\left[-\Omega r l\dot{\phi}\sin(\phi)\cos(\phi) - \Omega r l(\Omega+\dot{\theta})\cos(\phi)\sin(\theta)\right] . \end{aligned} \tag{13.4}$$

Operating in the same way also for the second equation, the following nonlinear equations of motion are obtained:

$$\left\{ \begin{array}{l} l\ddot{\theta}\cos^2(\phi) - 2l(\Omega+\dot{\theta})\dot{\phi}\cos(\phi)\sin(\phi) + \Omega^2 r\cos(\phi)\sin(\theta) = 0 , \\ l\ddot{\phi} + l(\Omega+\dot{\theta})^2\cos(\phi)\sin(\phi) + \Omega^2 r\sin(\phi)\cos(\theta) = 0 . \end{array} \right. \tag{13.5}$$

The equations of motion can be linearized for the study of the small oscillations of the pendulum about the static equilibrium position

$$\left\{ \begin{array}{l} l\ddot{\theta} + \Omega^2 r\theta = 0 , \\ l\ddot{\phi} + \Omega^2(r+l)\phi = 0 . \end{array} \right. \tag{13.6}$$

As usual, the expression of the kinetic energy can be subdivided into three terms:

$$\mathcal{T}_0 = \frac{1}{2} m\Omega^2 \left[r^2 + l^2 \cos^2(\phi) + 2rl \cos(\phi)\cos(\theta) \right] \tag{13.7}$$

is independent from the generalized velocities. Its role is similar to that of the potential energy and is usually referred to as *dynamic potential*. Apart from a constant term, whose derivatives are nil, it yields the so-called geometric stiffness terms in the linearized equation of motion. As usual in rotating systems, they constitute a centrifugal stiffening and are proportional to the square of the spin speed.

$$\mathcal{T}_1 = m\Omega l \left[l\dot{\theta}\cos^2(\phi) - r\dot{\phi}\sin(\phi)\sin(\theta) + r\dot{\theta}\cos(\phi)\cos(\theta) \right] \tag{13.8}$$

is linear in the generalized velocities.

$$\mathcal{T}_2 = \frac{1}{2} m l^2 \left[\dot{\phi}^2 + \dot{\theta}^2 \cos^2(\phi) \right] \tag{13.9}$$

is quadratic in the generalized velocities and yields the inertial terms of the equations of motion, which are also present in natural systems.

The linearized equations can also be obtained directly from an expression of the kinetic energy truncated after quadratic terms. By introducing the series for the sine and cosine or the generalized coordinates in which terms of order greater than two are discarded, the kinetic energy can be written as

$$\mathcal{T} = \frac{1}{2} m \left[\Omega^2 (r+l)^2 + \dot{\phi}^2 l^2 + \dot{\theta}^2 l^2 - \Omega^2 l (r+l)\phi^2 - \Omega^2 r l \theta^2 + 2\Omega l (r+l)\dot{\theta} \right] . \tag{13.10}$$

The linearized expressions for three terms $\mathcal{T}_0$, $\mathcal{T}_1$, and $\mathcal{T}_2$ are

$$\mathcal{T}_0 = \frac{1}{2} m\Omega^2 \left[(r+l)^2 - l\left(r + \frac{l}{2} \right)\phi^2 - rl\theta^2 \right] , \tag{13.11}$$

$$\mathcal{T}_1 = m\Omega l (r+l)\dot{\theta} . \tag{13.12}$$

Note that $\mathcal{T}_1$ is independent of the displacements, and its derivatives in the equations of motion are nil (the derivative with respect to $\dot{\theta}$ is constant, and hence, it yields no term once it has been differentiated with respect to time): There is no gyroscopic term in the linearized equations of the rotating pendulum.

$$\mathcal{T}_2 = \frac{1}{2} m l^2 \left[\dot{\phi}^2 + \dot{\theta}^2 \right] . \tag{13.13}$$

Remark 13.1 *As the linearized equations of motion are decoupled, the motion in the rotation plane xy (in-plane motion) is uncoupled, within the validity of the linearization of the equations of motion, from the motion* *in axial direction z (out-of-plane motion). The former is the motion of a pendulum whose length is l within a constant force field whose acceleration is $r\Omega^2$, and the latter is the motion of the same pendulum within a constant force field whose acceleration is $(r+l)\Omega^2$.*

The natural frequencies of the motions outside and within the rotation plane are, respectively,

$$\omega_1 = \Omega\sqrt{1+\frac{r}{l}} = \Omega\sqrt{1+\delta}\,, \qquad \omega_2 = \Omega\sqrt{\frac{r}{l}} = \Omega\sqrt{\delta}\,, \tag{13.14}$$

where

$$\delta = \frac{r}{l} \qquad (0 < \delta < \infty)\ , \tag{13.15}$$

where the limiting cases are $\delta = 0$, infinitely long pendulum or infinitely small disc, and $\delta \to \infty$, vanishingly short pendulum.

As the natural frequencies are proportional to the spin speed, it is possible to introduce ratio β between the frequency and the speed:

$$\beta = \frac{\omega}{\Omega}. \tag{13.16}$$

The frequency of the out-of-plane oscillations ω_1 is always larger than the spin speed ($\beta_1 = \sqrt{1+\delta} > 1$). In the case of in-plane oscillations, two cases can be defined:

- $0 < \delta < 1$, i.e., the pendulum is longer than the radius at which it is attached. In this case, here referred to as *long pendulum*, the frequency is lower than the spin speed ($\beta_2 < 1$).
- $\delta > 1$, i.e., the pendulum is shorter than the radius at which it is attached. In this case, here referred to as *short pendulum*, the frequency is higher than the spin speed ($\beta_2 > 1$).

Instead of ratio δ, it is possible to use ratio

$$\gamma = \frac{r}{r+l} = \frac{\delta}{\delta+1} \tag{13.17}$$

between the radius r of the disc and the outer radius $r_o = r + l$ of the disc-pendulum assembly The values of β_1 and β_2 are

$$\beta_1 = \frac{\omega_1}{\Omega} = \sqrt{\frac{1}{1-\gamma}}\,, \qquad \beta_2 = \frac{\omega_2}{\Omega} = \sqrt{\frac{\gamma}{1-\gamma}}\,. \tag{13.18}$$

Remark 13.2 *The natural frequencies are referred to the rotating frame, i.e., are those seen by an observer that rotates at the speed Ω. This applies everywhere in the present chapter, unless otherwise specified.*

The Campbell diagram, shown in Figure 13.1(b) for the case in which $r = l$, is then made of two straight lines. A simple explanation of the different behavior of the system in a plane containing the axis of rotation and a plane perpendicular to it (xy-plane) is shown in Figures 13.1(c) and (d). In the former, the restoring force acting on the pendulum in a direction perpendicular to line PC is

$$F_c \sin(\phi) \approx m\omega^2(r+l)\phi\,.$$

In the xy-plane, the restoring force is

$$F_c \sin(\theta-\alpha) \approx m\omega^2(r+l)(\theta-\alpha)\,.$$

As $\alpha \approx \theta l/(r+l)$, the restoring force is $\approx m\Omega^2 r\theta$.

Equations (13.18) are plotted in Figure 13.2. From the figure, it is clear that the frequency of the in-plane oscillations vanishes when $\gamma \to 0$; i.e., the pendulum is hinged on the spin axis, and the frequency of the out-of-plane oscillations tends to be equal to the spin speed in the same conditions. This means that point P moves on a circle inclined with respect to the spin axis. When the pendulum becomes very short ($\gamma \to 1$, $\beta \to \infty$), the frequencies of the in-plane and out-of-plane oscillations tend to coincide and become much higher than the spin speed.

Remark 13.3 *Although in the case of the spherical pendulum, the oscillations can take the form of any combination of harmonic motions in two perpendicular planes, and hence can follow a straight line, an ellipse (obviously projected on the spherical surface to which the bob of the pendulum is constrained) or a circle, the rotating pendulum can oscillate only in the two mentioned planes. This occurs because when two natural frequencies are equal, any linear combination of the corresponding mode shapes is itself a mode shape (case of the spherical pendulum); the rotating pendulum does not have a pair of identical eigenfrequencies, and hence, the two modes (in-plane and out-of-plane oscillations) cannot combine if not as the mixing of two oscillations at different frequencies, giving way to a sort of Lissajous figures.*

13.2 Rotating pendulum constrained to oscillate in a plane

If the pendulum is constrained to oscillate in a plane that neither is perpendicular to the spin axis nor contains it, angles θ and ϕ are not independent. If the plane of oscillation makes and angle ψ with the axis of rotation (Figure 13.3), they can be expressed as

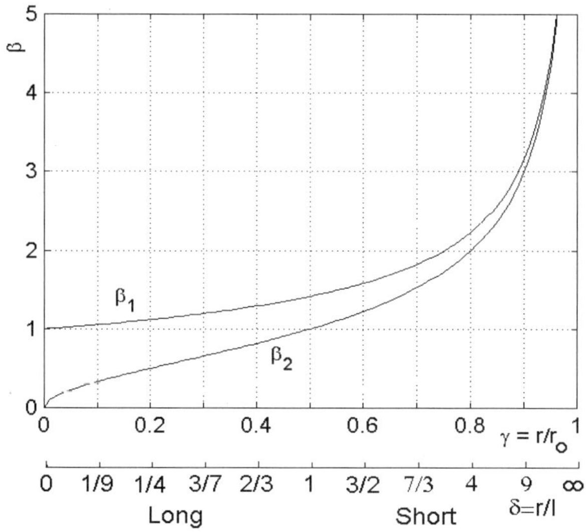

FIGURE 13.2. Ratios β_1 and β_2 between the natural frequencies of the pendulum and the spin speed as functions of ratio $\gamma = r/r_o$ between the radius r of the disc and the outer radius $r_o = r + l$ of the disc-pendulum assembly and as function of ratio $\delta = r/l$.

$$\begin{cases} \theta = \theta_1 \sin(\psi) \ , \\ \phi = \theta_1 \cos(\psi) \ , \end{cases} \tag{13.19}$$

where θ_1 is the angle of oscillation of the pendulum in its plane.

The equation of motion can be obtained simply by substituting the Expressions (13.19) into Equation (13.3). To avoid long computations, only the linearized approach is here followed, and Equation (13.10) is used to express the kinetic energy. The substitution yields

$$\mathcal{T} = \frac{1}{2}m\left[\Omega^2(r+l)^2 + \dot{\theta}_1 l^2 - \Omega^2 l\left[r + l\cos^2(\psi)\right]\theta_1{}^2 + 2\Omega l(r+l)\dot{\theta}_1\sin(\psi)\right] . \tag{13.20}$$

The linearized equation of motion is then

$$l\ddot{\theta}_1 + \Omega^2\left[r + l\cos^2(\psi)\right]\theta_1 = 0 , \tag{13.21}$$

yielding a natural frequency

$$\omega = \Omega\sqrt{\frac{r}{l} + \cos^2(\psi)} = \Omega\sqrt{\delta + \cos^2(\psi)} \ . \tag{13.22}$$

The limiting cases $\psi = 0$ and $\psi = 90°$ yield, respectively, the out-of-plane and the in-plane natural frequency.

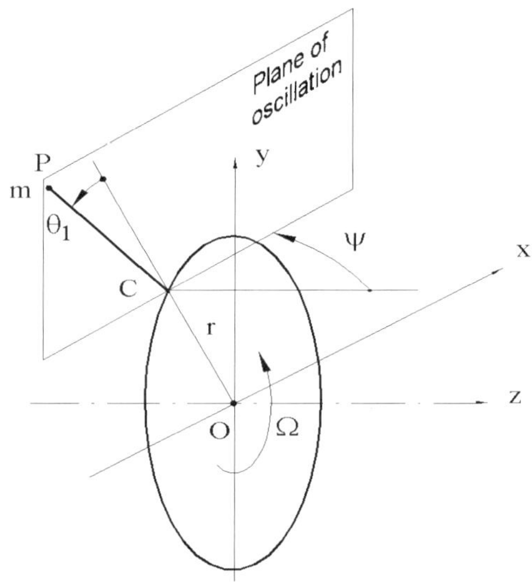

FIGURE 13.3. Rotating pendulum constrained to oscillate in a plane making angle ψ with the axis of rotation.

13.3 Spring-loaded rotating pendulum

In the rotating pendulums studied in the previous two sections, the restoring effect is only caused by the centrifugal field and the natural frequency vanishes at zero speed. Consider now the case of a rotating pendulum in which there are two springs supplying a restoring force toward the radial direction. Let k_θ and k_ϕ be the stiffness of the springs acting in-plane and out of the rotation plane. The potential energy caused by the springs is

$$\mathcal{U} = \frac{1}{2}k_\phi\phi^2 + \frac{1}{2}k_\theta\theta^2 . \tag{13.23}$$

By introducing the derivatives of the potential energy into the linearized equations of motion (13.6), the latter reduce to

$$\begin{cases} ml^2\ddot{\theta} + \left(m\Omega^2 rl + k_\theta\right)\theta = 0 , \\ ml^2\ddot{\phi} + \left[m\Omega^2 l(r+l) + k_\phi\right]\phi = 0 . \end{cases} \tag{13.24}$$

The out-of-plane and in-plane natural frequencies are then

$$\omega_1 = \sqrt{\frac{k_\phi}{ml^2} + \Omega^2\frac{r+l}{l}}, \qquad \omega_2 = \sqrt{\frac{k_\theta}{ml^2} + \Omega^2\frac{r}{l}} . \tag{13.25}$$

By introducing the natural frequencies at standstill

$$\omega_\phi = \sqrt{\frac{k_\phi}{ml^2}} , \qquad \omega_\theta = \sqrt{\frac{k_\theta}{ml^2}} , \tag{13.26}$$

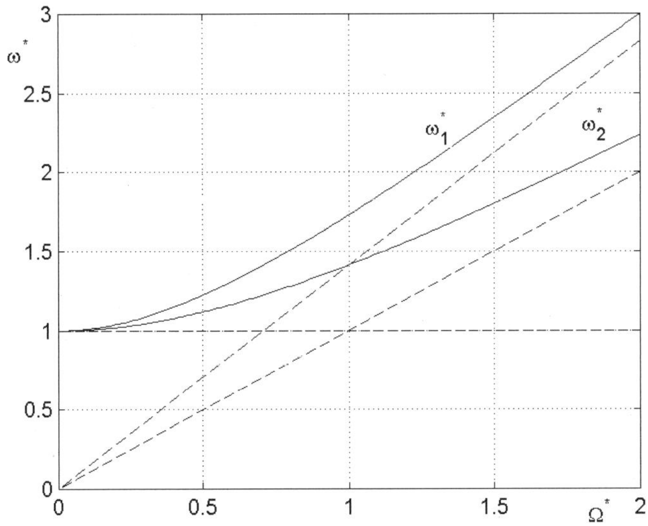

FIGURE 13.4. Nondimensional natural frequencies of a spring-loaded rotating pendulum as functions of the nondimensional speed. The dashed lines refer to the case in Figure 13.1(b) and to the constant natural frequency of the nonrotating pendulum.

and remembering the definitions of ratios β_1 and β_2 between the natural frequencies of the (springless) pendulum and the spin speed expressed by Equation (13.18), the expressions of the natural frequencies reduce to

$$\omega_1 = \sqrt{\omega_\phi^2 + \Omega^2 \beta_1^2}, \qquad \omega_2 = \sqrt{\omega_\theta^2 + \Omega^2 \beta_2^2} \, . \tag{13.27}$$

Equations (13.27) are plotted in nondimensional form in Figure 13.4. Both the natural frequency ω^* and the spin speed Ω^* have been made nondimensional by dividing them by ω_ϕ for out-of-plane natural frequencies and by ω_θ for in-plane oscillations.

13.4 Rotating string

Consider a string, i.e., a one-dimensional body whose bending stiffness is vanishingly small, attached at the outer radius of a rotating disc [Figure 13.5(a)]. The generic point P of the string, located at radius r, moves to P'; its displacement in the rotating frame $\xi\eta z$ is $\mathbf{u} = [u, v, w]^T$. The radius of the disc is r_i, the outer radius is r_o, and the length of the string is $l = (r_o - r_i)$.

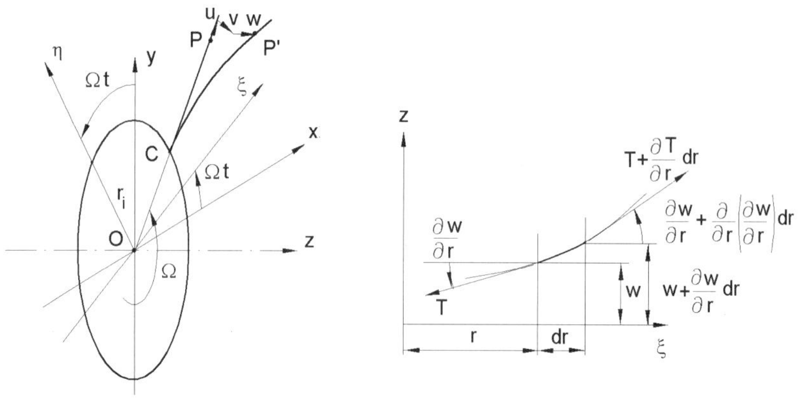

FIGURE 13.5. Rotating string. (a) Sketch and reference frame. P is the position of a generic point of the string in its undeflected position; it moves to P' during deformation. (b) Force acting on the length dr of the string; situation in ξz-plane.

The coordinates of point P' in the inertial frame are

$$(\overline{\text{P'-O}}) = \left\{ \begin{array}{c} (r+u)\cos(\Omega t) - v\sin(\Omega t) \\ (r+u)\sin(\Omega t) + v\cos(\Omega t) \\ w \end{array} \right\} . \tag{13.28}$$

By differentiating the expressions of the coordinates with respect to time and assuming that the spin speed Ω is constant, the velocity of point P is readily obtained

$$\vec{V}_{\text{P'}} = \left\{ \begin{array}{c} -\Omega\,(r+u)\sin(\Omega t) + \dot{u}\cos(\Omega t) - \Omega v\cos(\Omega t) - \dot{v}\sin(\Omega t) \\ \Omega\,(r+u)\cos(\Omega t) + \dot{u}\sin(\Omega t) - \Omega v\sin(\Omega t) + \dot{v}\cos(\Omega t) \\ \dot{w} \end{array} \right\} . \tag{13.29}$$

The kinetic energy of the length dr of the string located at point P' [Figure 13.5(b)] is

$$d\mathcal{T} = \tfrac{1}{2}dm|\vec{V}_{\text{P'}}|^2 = \tfrac{1}{2}\rho A dr\left[\Omega^2\,(r+u)^2 + \Omega^2 v^2 + \dot{u}^2 + \dot{v}^2 + \dot{w}^2 + \right. \\ \left. +2\Omega\,(r+u)\,\dot{v} - 2\Omega\dot{u}v\right] . \tag{13.30}$$

The equations of motion of the length dr of the string can be obtained through the Lagrange equations. By performing the relevant derivatives, it follows that

$$\left\{ \begin{array}{l} \rho A dr\left[\ddot{u} - \Omega^2\,(r+u) - 2\Omega\dot{v}\right] = Q_\xi \ , \\ \rho A dr\left(\ddot{v} - \Omega^2 v + 2\Omega\dot{u}\right) = Q_\eta \ , \\ \rho A dr\ddot{w} = Q_z \ . \end{array} \right. \tag{13.31}$$

The generalized forces acting on the length dr of the string are caused by the tensile force T in the string. With reference to Figure 13.5(b), in which the situation in the ξz-plane is shown (but an identical sketch could be plotted for the $\xi\eta$-plane), it follows that

$$\left\{\begin{array}{l} Q_\xi = \frac{\partial T}{\partial r} dr \; , \\ Q_\eta = \left(\frac{\partial T}{\partial r}\frac{\partial v}{\partial r} + T\frac{\partial^2 v}{\partial r^2} \right) dr = \frac{\partial}{\partial r}\left(T\frac{\partial v}{\partial r} \right) dr, \\ Q_z = \left(\frac{\partial T}{\partial r}\frac{\partial w}{\partial r} + T\frac{\partial^2 w}{\partial r^2} \right) dr = \frac{\partial}{\partial r}\left(T\frac{\partial w}{\partial r} \right) dr \; . \end{array}\right. \tag{13.32}$$

Assume that the string is infinitely stiff in radial direction and that the radial displacement and its derivatives are negligible. Under these conditions, the first Equation (13.31) reduces to

$$-\rho\Omega^2 Ar dr = \frac{\partial T}{\partial r} dr \, , \tag{13.33}$$

which can be readily integrated, yielding the tensile force in the string

$$T = C - \frac{1}{2}\rho\Omega^2 Ar^2 \; . \tag{13.34}$$

The integration constant C can be computed by stating that the tensile force must vanish at the end of the string ($r = r_e$), obtaining

$$T = \frac{1}{2}\rho\Omega^2 A \left(r_o^2 - r^2 \right) \; . \tag{13.35}$$

The two equations describing the motion in $\xi\eta$ and ξz planes are then

$$\left\{\begin{array}{l} \ddot{v} - \Omega^2 v + \Omega^2 r\frac{\partial v}{\partial r} - \frac{1}{2}\Omega^2 \left(r_o^2 - r^2 \right) \frac{\partial^2 v}{\partial r^2} = 0 \; , \\ \ddot{w} + \Omega^2 r\frac{\partial w}{\partial r} - \frac{1}{2}\Omega^2 \left(r_o^2 - r^2 \right) \frac{\partial^2 w}{\partial r^2} = 0 \; . \end{array}\right. \tag{13.36}$$

Remark 13.4 *The equations describing the motion in the plane of rotation ($\xi\eta$-plane) and in a plane containing the axis of rotation (ξz-plane) are uncoupled but different from each other. Like in the case of the rotating pendulum, the behavior in the two planes is different.*

By introducing the nondimensional variable

$$\chi = \frac{r}{r_o} \, ,$$

and by assuming that the motion is harmonic in time

$$\left\{\begin{array}{l} v = u_\eta(r)\cos(\omega_2 t) \; , \\ w = u_z(r)\cos(\omega_1 t) \; , \end{array}\right. \tag{13.37}$$

Equation (13.36) becomes

$$\left\{\begin{array}{l} \left(\omega_2^2 + \Omega^2 \right) u_\eta - \Omega^2 \chi \frac{du_\eta}{d\chi} + \frac{1}{2}\Omega^2 \left(1 - \chi^2 \right) \frac{d^2 u_\eta}{d\chi^2} = 0 \; , \\ \omega_1^2 u_z - \Omega^2 \chi \frac{du_z}{d\chi} + \frac{1}{2}\Omega^2 \left(1 - \chi^2 \right) \frac{d^2 u_z}{d\chi^2} = 0 \; . \end{array}\right. \tag{13.38}$$

The equations describing the shapes taken in the out-of-plane and in-plane vibration of the rotating string are then

$$\begin{cases} \left(1-\chi^2\right)\frac{d^2u_z}{d\chi^2}-2\chi\frac{du_z}{d\chi}+2\beta_1^2u_z=0\ , \\ \left(1-\chi^2\right)\frac{d^2u_\eta}{d\chi^2}-2\chi\frac{du_\eta}{d\chi}+2\left(\beta_2^2+1\right)u_\eta=0\ , \end{cases} \tag{13.39}$$

where

$$\beta_1=\frac{\omega_1}{\Omega}\ ,\qquad \beta_2=\frac{\omega_2}{\Omega} \tag{13.40}$$

are the ratios between the frequencies of the out-of-plane and in-plane oscillations and the spin speed.

The solution of the first Equation (13.39) can be assumed to be

$$u_z=\sum_0^\infty C_i\chi^i\ . \tag{13.41}$$

By introducing the Solution (13.41) into Equation (13.39) and balancing the powers of χ, it follows that

$$C_{i+2}=C_i\frac{i(i+1)-2\beta_1^2}{(i+1)(i+2)}\ . \tag{13.42}$$

Solution (13.41) contains only two arbitrary coefficients, C_0, which appears in all even terms of the series, and C_1, which appears in all odd terms. As the problem is a homogeneous one, the amplitude of the inflected shape is not determined, and one coefficient, say, C_1, can be left undetermined. The ratio C_0/C_1 can be determined using one of the boundary conditions, e.g., the condition at the inner radius r_i. As the string is attached to the disc, at the inner radius, u_z must vanish. This condition yields

$$\frac{C_0}{C_1}=-\frac{\gamma+\frac{C_3}{C_1}\gamma^3+\frac{C_5}{C_1}\gamma^5+\frac{C_7}{C_1}\gamma^7+\cdots}{1+\frac{C_2}{C_0}\gamma^2+\frac{C_4}{C_0}\gamma^4+\frac{C_6}{C_0}\gamma^6+\cdots}\ , \tag{13.43}$$

where $\gamma=r_i/r_e$.

The second boundary condition must be stated at the outer edge, which is assumed to be free. From Figure 13.5(b) it follows that at a free edge, the slope du_z/dr must be equal to the ratio between the acceleration caused by vibration and the centrifugal acceleration

$$\left(\frac{du_z}{d\chi}\right)_{\chi=1}=\beta_1^2\left(u_z\right)_{\chi=1}\ ; \tag{13.44}$$

i.e.,

$$\frac{C_0}{C_1}\left(2\frac{C_2}{C_0}+4\frac{C_4}{C_0}+6\frac{C_6}{C_0}+\cdots\right)+1+3\frac{C_3}{C_1}+5\frac{C_5}{C_1}+7\frac{C_7}{C_1}+\cdots+$$

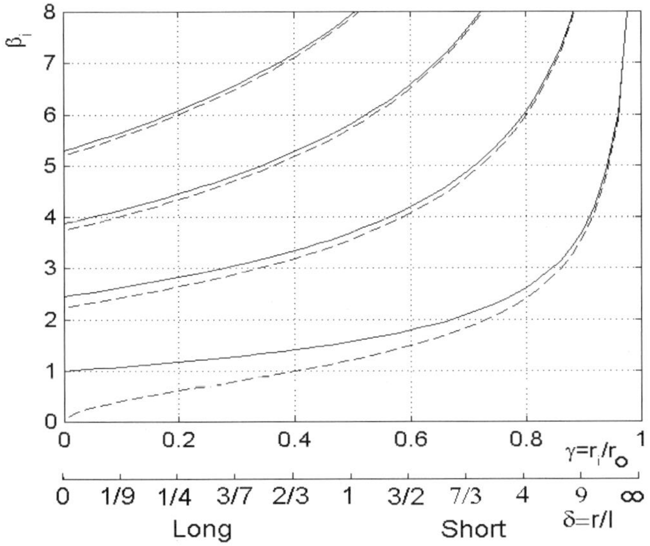

FIGURE 13.6. Values of β as functions of the radii ratio γ for out-of-plane oscillations (full lines) and in-plane oscillation (dashes lines) and of ratio δ.

$$-\beta_1^2\left[\frac{C_0}{C_1}\left(1+\frac{C_2}{C_0}+\frac{C_4}{C_0}+\frac{C_6}{C_0}+\cdots\right)+1+\frac{C_3}{C_1}+\frac{C_5}{C_1}+\frac{C_7}{C_1}+\cdots\right]=0\,. \tag{13.45}$$

As the various ratios between coefficients that appear in Equation (13.45) are functions of β_1, it can be regarded as an equation in β_1 yielding the natural frequencies of the system.

An exact solution can be found for the case $\gamma = 0$, i.e., of a string reaching the axis of rotation. Constant C_0 vanishes, and only a limited number of terms of the series can be shown to be present. The values of β_1 obtained from Equation (13.45) are

$$\beta_{1_i} = \sqrt{i(2i-1)} \qquad \text{with } i = 1, 2, 3, 4, \ldots\,. \tag{13.46}$$

No exact solution can be found for other values of γ, but the series converges quickly and numerical solutions can be obtained by truncating it after a number of terms. The results of a number of numerical computations are reported in Figure 13.6, full lines. Also, in this case, as in that of the rotating pendulum, the first natural frequency is equal to the spin speed when $\gamma = 0$ ($\beta_1 = 1$) and then increases with increasing inner radius.

A similar computation can be performed for the case of in-plane oscillations. The two Equations (13.39) are identical, if $1 + \beta_2^2$ is substituted for β_1^2. Also the boundary conditions are identical; it follows that if β_1^2 is an eigenvalue for the equation of the out-of-plane oscillations, $\beta_2^2 = \beta_1^2 - 1$ is an eigenvalue for the equation of the in-plane oscillations. For each out-of-plane natural frequency, there is an in-plane natural frequency that can be obtained as

$$\omega_2 = \sqrt{\omega_1^2 - \Omega^2}\ . \tag{13.47}$$

The nondimensional natural frequencies for the in-plane oscillation are reported as functions of γ in Figure 13.6, dashed lines. The first natural frequency vanishes for $\gamma = 0$, and the in-plane frequencies get closer and closer to the out-of-plane frequencies with increasing γ.

Also, in this case, it is possible to distinguish between *long* strings, with $\beta < 1$, and *short* strings, with $\beta > 1$.

13.4.1 Rotating string constrained to oscillate in a plane

If the string is constrained to oscillate in a plane that is neither perpendicular to the spin axis nor contains it, displacements v and w are not independent. If the plane of oscillation makes and angle ψ with the axis of rotation, the displacement in ξ- and z-directions are linked with the displacement v^* by the relationships:

$$\begin{cases} v = v^* \sin(\psi)\ , \\ w = v^* \cos(\psi)\ . \end{cases} \tag{13.48}$$

The expression of the kinetic energy of a length dr of the string is then

$$dT = \tfrac{1}{2}dm|\vec{V}_\mathrm{P}|^2 = \tfrac{1}{2}\rho A dr\left[\Omega^2(r+u)^2 + \Omega^2 v^{*2}\sin^2(\psi) + \dot{u}^2 + \dot{v}^{*2} - 4\Omega(r+u)\dot{v}^*\sin(\psi)\right]. \tag{13.49}$$

Operating in the same way as seen for rotating pendulums, the equation describing the motion in a plane inclined of angle ψ with the ξz-plane is

$$\ddot{v}^* - \Omega^2 v^* \sin^2(\psi) + \Omega^2 r\frac{\partial v^*}{\partial r} - \frac{1}{2}\Omega^2\left(r_e^2 - r^2\right)\frac{\partial^2 v^*}{\partial r^2} = 0\,, \tag{13.50}$$

or, in nondimensional terms

$$\left(1-\chi^2\right)\frac{d^2u^*}{d\chi^2} - 2\chi\frac{du^*}{dr} + 2\left[\beta^2 + \sin^2(\psi)\right]u^* = 0\,. \tag{13.51}$$

The natural frequency of the string is thus

$$\omega = \sqrt{\omega_1^2 - \Omega^2\sin^2(\psi)}\ , \tag{13.52}$$

where $\omega_1 = \Omega\beta_1$ and $\beta_1$is the ratio between the out-of-plane natural frequency and the spin speed [Equation (13.40)].

A constrained string can be considered as an idealized model of a very thin blade, with a negligible bending stiffness in the direction perpendicular to its plane and a very high stiffness in its plane (Figure 13.7). The case

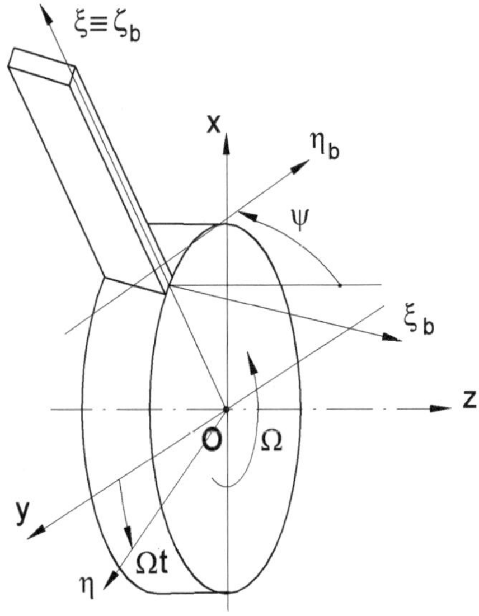

FIGURE 13.7. Rotating beam: sketch and reference frames

with $\psi = 0$ is then the case of a infinitely thin blade lying in the plane of the disc (like the blade of a helicopter rotor with zero pitch), whereas $\psi = 90°$ is the case of a blade parallel to the spin axis. The two cases coincide with the out-of-plane and the in-plane oscillations of a string seen in the previous section. The natural frequencies in all intermediate cases are included between the limits so identified (i.e., the curves $\beta(\gamma)$ are included between the dashed and the full lines in Figure 13.6).

13.4.2 Rotating beam

Consider a beam attached radially to a rotating disc. Assume that all assumptions yielding a complete uncoupling among axial, bending, and torsional behavior are satisfied and that the cross section has two symmetry axes (Figure 13.7). The principal axis of elasticity of the cross section corresponding to the largest moment of inertia, axis η_b, makes an angle ψ with the plane of rotation and the beam is not twisted.

The limiting case of the previous section corresponds to the case in which the moments of inertia of the cross section about the principal axes ξ_b and η_b are $I_{\xi b} \to 0$ and $I_{\eta b} \to \infty$.

Assume that the beam is slender enough to approximate its behavior using the Euler-Bernoulli approach, i.e., neglecting the shear deformation and the kinetic energy caused by rotation of the cross section about their principal axes of inertia. The equation of motion for bending in $\eta_b\zeta_b$-plane (but the same holds also for $\xi_b\zeta_b$-plane) can be obtained by adding the restoring force caused by the bending moment in the beam to the equation of motion of the string (Equation 13.50). Using the Euler-Bernoulli approach, such a term can be expressed as

$$\frac{\partial^2}{\partial r^2}\left(EI_{\xi b}\frac{\partial^2 v_b}{\partial r^2}\right) .$$

Remembering that Equation (13.50) has been divided by ρA and releasing the assumption of prismatic homogeneous beam, the equation of motion of the length $d\zeta$ of the beam becomes

$$\ddot{v}_b - \Omega^2 v_b \sin^2(\psi) - \frac{\partial}{\partial r}\left(T\frac{\partial v_b}{\partial r}\right) + \frac{1}{\rho A}\frac{\partial^2}{\partial r^2}\left(EI_{\xi b}\frac{\partial^2 v_b}{\partial r^2}\right) = 0 . \qquad (13.53)$$

If the beam is homogeneous and prismatic, the equation motion simplifies as

$$\ddot{v}_b - \Omega^2 v_b \sin^2(\psi) + \Omega^2 r\frac{\partial v_b}{\partial r} - \frac{1}{2}\Omega^2\left(r_o^2 - r^2\right)\frac{\partial^2 v_b}{\partial r^2} + \frac{EI_{\xi b}}{\rho A}\frac{\partial^4 v_b}{\partial r^4} = 0 . \qquad (13.54)$$

By introducing the nondimensional variable $\chi = r/r_o$ and assuming that the motion is harmonic in time

$$v_b = u(r)\cos(\omega t) , \qquad (13.55)$$

Equation (13.54) becomes

$$\left[\frac{\omega^2}{\Omega^2} + \sin^2(\psi)\right]u - \chi\frac{du}{d\chi} + \frac{1}{2}\left(1 - \chi^2\right)\frac{d^2u}{d\chi^2} - \frac{EI_{\xi b}}{\Omega^2 \rho A r_o^4}\frac{d^4u}{d\chi^4} = 0 . \qquad (13.56)$$

The boundary conditions depend on how the beam is constrained to the disc and whether it is constrained at the outer edge. In the case of Figure 13.7, the outer edge ($\chi = 1$) is free; i.e., the bending moment and the shear force are vanishingly small. This yields

$$\left(\frac{\partial^2 u}{\partial \chi^2}\right)_{\chi=1} = 0 , \quad \left(\frac{\partial^3 u}{\partial \chi^3}\right)_{\chi=1} = 0. \qquad (13.57)$$

The beam is clamped to the disc at its inner edge ($\chi = \gamma = r_i/r_o$), so that its displacement and rotation vanish

$$(u)_{\chi=\gamma} = 0 , \quad \left(\frac{\partial u}{\partial \chi}\right)_{\chi=\gamma} = 0. \qquad (13.58)$$

By introducing the nondimensional speed and frequency

$$\Omega^* = \Omega\frac{\rho A r_o^4}{EI_{\xi b}} , \quad \omega^* = \omega\frac{\rho A r_o^4}{EI_{\xi b}} , \qquad (13.59)$$

the solution of the eigenproblem associated with Equation (13.41), i.e., the value of the natural frequency ω^*, can be shown to depend, for each value of ratio $\gamma = r_i/r_o$, angle ψ, and each set of boundary conditions, only on a single parameter, namely, the spin speed Ω^*. The Campbell diagram $\omega^*(\Omega^*)$ is then unique for a given geometrical configuration of the beam.

The solution of Equation (13.56) can be attempted by using a series of the type of Equation (13.41). Relationships of the type of Equation (13.42) can be found also in the present case, but here the number of coefficients C_i to be found using the boundary conditions is four (instead of two) and the complexity of this approach is overwhelming.

For $\Omega^* = 0$, the natural frequencies of the nonrotating beam are obtained. Their value is given by Equation (5.29)

$$\omega_{i_n} = \frac{\beta_i^2}{(r_o - r_i)^2}\sqrt{\frac{EI_{\xi b}}{\rho A}}\,, \tag{13.60}$$

i.e.,

$$\omega^*_{i_n} = \frac{\beta_i^2}{(1-\gamma)^2}\,, \tag{13.61}$$

where the values of constants β_i depend on the boundary conditions (Table 5.2).

If the beam is clamped at one end (inner edge) and free at the other one (outer edge), as it often occurs for rotating blades, the mode shapes are [17]

$$u(r) = \sin(\beta_i\chi_1) - \sinh(\beta_i\chi_1) - N\left[\cos(\beta_i\chi_1) - \cosh(\beta_i\chi_1)\right]\,, \tag{13.62}$$

where

$$N = \frac{\sin(\beta_i) + \sinh(\beta_i)}{\cos(\beta_i) + \cosh(\beta_i)} \tag{13.63}$$

and $\chi_1 = (\chi - \gamma)/(1-\gamma)$ is a variable that takes the value 0 at the inner end of the beam and 1 at the outer end. From the mentioned tables the values of β_i for the first four natural frequencies are 1.875, 4.694, 7.855, and 10.996.

To obtain the structure of the equation yielding ω, taking rotation into account without actually solving a very complex eigenproblem, Equation (13.56) can be multiplied by $u(r)$ and integrated along the beam

$$\int_{r_i}^{r_e}\left\{\left[\frac{\omega^2}{\Omega^2} + \sin^2(\psi)\right]u^2 - \chi u\frac{du}{d\chi} + \frac{1}{2}\left(1-\chi^2\right)u\frac{d^2u}{d\chi^2} + \right. \\ \left. -\frac{EI_{\xi b}}{\Omega^2\rho A r_o^4}u\frac{d^4u}{d\chi^4}\right\}dr = 0\,. \tag{13.64}$$

By introducing variable χ instead of r and performing some computation, it follows that

$$\left[\frac{\omega^2}{\Omega^2} + \sin^2(\psi)\right] \int_\gamma^1 u^2 d\chi + \frac{1}{2} \int_\gamma^1 u \frac{d}{d\chi}\left[\left(1-\chi^2\right)\frac{d^2u}{d\chi^2}\right] d\chi + \\ - \frac{EI_{\xi b}}{\Omega^2 \rho A r_o^4} \int_\gamma^1 u \frac{d^4u}{d\chi^4} d\chi = 0 . \tag{13.65}$$

The last integral can be shown to be

$$\int_\gamma^1 u \frac{d^4u}{d\chi^4} d\chi = \left| u \frac{d^3u}{d\chi^3} \right|_\gamma^1 - \int_\gamma^1 \frac{du}{d\chi}\frac{d^3u}{d\chi^3} d\chi =$$

$$= \left| u \frac{d^3u}{d\chi^3} \right|_\gamma^1 - \left| \frac{du}{d\chi}\frac{d^2u}{d\chi^2} \right|_\gamma^1 + \int_\gamma^1 \left(\frac{d^2u}{d\chi^2}\right)^2 d\chi = \int_\gamma^1 \left(\frac{d^2u}{d\chi^2}\right)^2 d\chi . \tag{13.66}$$

As both $\left| u \frac{d^3u}{d\chi^3} \right|_\gamma^1$ and $\left| \frac{du}{d\chi}\frac{d^2u}{d\chi^2} \right|_\gamma^1$ vanish because of the boundary conditions, it follows that

$$\int_\gamma^1 u \frac{d^4u}{d\chi^4} d\chi = \int_\gamma^1 \left(\frac{d^2u}{d\chi^2}\right)^2 d\chi . \tag{13.67}$$

In the same way, the second integral can be shown to be

$$\int_\gamma^1 u \frac{d}{d\chi}\left[\left(1-\chi^2\right)\frac{d^2u}{d\chi^2}\right] d\chi = \left| \left(1-\chi^2\right)\frac{du}{d\chi} \right|_\gamma^1 - \int_\gamma^1 \left(1-\chi^2\right)\left(\frac{du}{d\chi}\right)^2 d\chi =$$

$$= - \int_\gamma^1 \left(1-\chi^2\right)\left(\frac{du}{d\chi}\right)^2 d\chi \tag{13.68}$$

because also $\left| \left(1-\chi^2\right)\frac{du}{d\chi} \right|_\gamma^1$ vanishes because of the boundary conditions.

Equation (13.65) then simplifies as

$$\left[\frac{\omega^2}{\Omega^2} + \sin^2(\psi)\right] \int_\gamma^1 u^2 d\chi - \frac{1}{2} \int_\gamma^1 \left(1-\chi^2\right)\left(\frac{du}{d\chi}\right)^2 d\chi + \\ - \frac{EI_{\xi b}}{\Omega^2 \rho A r_o^4} \int_\gamma^1 \left(\frac{d^2u}{d\chi^2}\right)^2 d\chi = 0 , \tag{13.69}$$

which yields the natural frequency of the beam

$$\omega^2 = \frac{EI_{\xi b}}{\rho A r_o^4} \frac{\int_\gamma^1 \left(\frac{d^2u}{d\chi^2}\right)^2 d\chi}{\int_\gamma^1 u^2 d\chi} + \Omega^2 \left[\frac{1}{2} \frac{\int_\gamma^1 \left(1-\chi^2\right)\left(\frac{du}{d\chi}\right)^2 d\chi}{\int_\gamma^1 u^2 d\chi} - \sin^2(\psi) \right] = 0 . \tag{13.70}$$

Equation (13.65) yields the natural frequency of the rotating beam, once the mode shape $u(r)$ is known, which implies solving the eigenproblem mentioned above. It can be used, however, to obtain approximated solutions.

As a first point, if the angular velocity vanishes, Equation (13.65) coincides with Equation (13.60). This can be demonstrated by introducing the mode shape of the nonrotating beam [Equation (13.62)] and integrating numerically. This can be a way to compute the values of β_i for a clamped-free beam reported in Table 5.2.

The expression for the natural frequency then reduces to

$$\omega_i = \sqrt{\omega_{i_n}^2 + \Omega^2 \left(c_i - \sin^2(\psi)\right)}, \tag{13.71}$$

where

$$c = \frac{1}{2} \frac{\int_\gamma^1 \left(1 - \chi^2\right) \left(\frac{du}{d\chi}\right)^2 d\chi}{\int_\gamma^1 u^2 d\chi}. \tag{13.72}$$

A lower bound for c is easily computed by following the suggestion given by Southwell for rotating plates [6], and computing an approximated value of the natural frequency as

$$\omega \approx \sqrt{\omega_n^2 + \omega_s^2}, \tag{13.73}$$

where ω_n and ω_s are the natural frequencies of the nonrotating beam and of the rotating strings.[1] Operating in this way, the value of c is the square of β reported in Figure 13.6, full line.

An upper bound can be obtained by introducing into Equation (13.72) the inflected shape computed for a nonrotating beam and integrating numerically. In the case of a beam starting from the axis of rotation ($\gamma = 0$), the values of the lower and upper bounds of c for the first four modes are

mode #	1	2	3	4
c (lower bound)	1	6	15	28
c (upper bound)	1.1933	6.4782	17.8596	36.0554

In spite of the analytical complexity, a numerical solution, e.g., obtained using the finite element method, is straightforward.

The nondimensional Campbell diagram for the first natural frequency of the out-of-plane vibrations ($\psi = 0$) of a rotating beam with $\gamma = r_i/r_e = 0.5$, the inner edge clamped and the outer one free, is reported in Figure 13.8.

The lower and upper bounds for c are in this case 2.5265 and 2.7642, respectively. The corresponding values of the natural frequency obtained using Equation (13.71) are reported, together with the value obtained using the FEM, in Figure 13.8.

[1] A similar equation holds exactly in the case of a spring-loaded rotating pendulum.

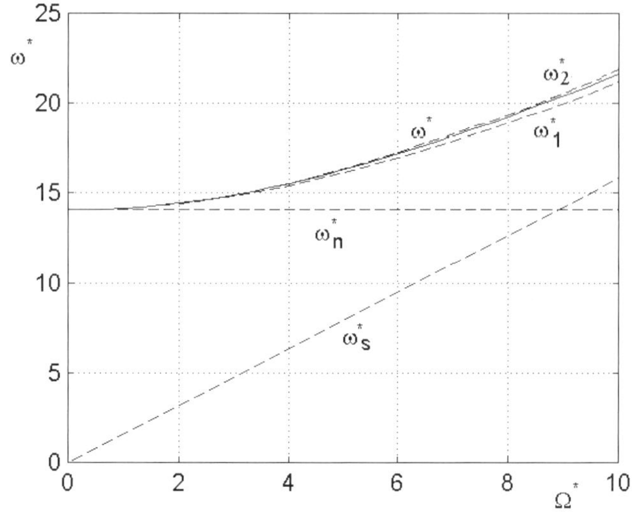

FIGURE 13.8. Nondimensional Campbell diagram for the first out-of-plane mode of a clamped-free beam with $\gamma = r_i/r_o = 0.5$ (full line). Also the value ω_n^* for a nonrotating beam and ω_s^* for the rotating strings are reported. ω_1^* and ω_2^* are the lower and upper bounds given by Equation (13.71).

In-plane blade vibration is seen, at a rotor level, as a torsional vibration, whereas out-of-plane blade vibration is essentially an axial vibration of the rotor. If the blades have an angle ψ different from 0° or 90°, their presence couple axial and torsional modes. Moreover, in most cases, blades are twisted; i.e., angle ψ is not constant.

13.5 Dynamics of a row of rotating pendulums

13.5.1 Pendulums on a rigid support

In the previous sections, the dynamic behavior of the blades has been studied at the level of a single unit. In real machinery, usually the blades are integrated in rows and their global behavior interacts with the dynamics of the disc (or discs) on which they are mounted and of the shaft. The dynamic behavior of all parts must then be studied at a system level.

The simplest model for a row of blades is a system made of a number n of identical pendulums evenly spaced at the outer radius of a rotating disc. Assume that they can oscillate only in axial direction. The system has then a number n of degrees of freedom, and angles ϕ_i [Figure 13.1(a)] can be used as generalized coordinates. However, a coordinate transformation is possible:

$$\phi_i = u_0 + u_1 \cos(\psi_i) + u_2 \sin(\psi_i) + u_3 \cos(2\psi_i) + u_4 \sin(2\psi_i) + \dots \,, \tag{13.74}$$

where angle

$$\psi_i = 2\pi \frac{i-1}{n} \tag{13.75}$$

is the angle at which the ith pendulum is attached. The coordinate transformation can thus be written in the form

$$\boldsymbol{\phi} = \left\{ \begin{array}{c} \phi_1 \\ \phi_2 \\ \phi_3 \\ \dots \end{array} \right\} = \left[\begin{array}{cccc} 1 & \cos(\psi_1) & \sin(\psi_1) & \dots \\ 1 & \cos(\psi_2) & \sin(\psi_2) & \dots \\ 1 & \cos(\psi_3) & \sin(\psi_3) & \dots \\ \dots & \dots & \dots & \dots \end{array} \right] \left\{ \begin{array}{c} u_0 \\ u_1 \\ u_2 \\ \dots \end{array} \right\} = \mathbf{Tu} \,. \tag{13.76}$$

The transformation matrix $\mathbf{T}$ must be square, so that the number of harmonics that must be considered depends on the number of the blades. Consider for instance the case of four pendulums. Angles ψ_i are 0, π, 2π, and $3\pi/2$, and matrix $\mathbf{T}$ is

$$\mathbf{T} = \left[\begin{array}{cccc} 1 & 1 & 0 & 1 \\ 1 & 0 & 1 & -1 \\ 1 & -1 & 0 & 1 \\ 1 & 0 & -1 & -1 \end{array} \right] .$$

By applying the coordinate transformation (13.76) to the linearized equation of motion for the in-plane oscillations of the four pendulums [second Equation (13.6)], it follows that

$$l\mathbf{T}^T\mathbf{T}\ddot{\mathbf{u}} = \Omega^2 (r + l) \mathbf{T}^T\mathbf{T}\mathbf{u} \,.$$

The result is a set of four identical equations of the type

$$l\ddot{u}_i = \Omega^2 (r + l) u_i \,.$$

As expected, the natural frequencies are all equal and are equal to the frequencies of the pendulums, but the meaning of the equations is different: Now they describe the motion of the row of pendulums as a single entity. The first equation, in u_0, describes the synchronous motion of all pendulums [Figure 13.9(a)]. This mode is coupled with the axial vibrations of the disc.

The second and the third modes [Figures 13.9(b and (c)] regard two of the pendulums that move with 180° phasing with each other, whereas the other two do not move. These modes are coupled with rotations ϕ_y and ϕ_x of the disc about y- and x-axes, respectively, and hence are coupled with the flexural behavior of the shaft. Even in this case it is possible to define a complex coordinate, in the same way already seen for the rotational degrees of freedom of the shaft:

$$\phi_{row} = u_1 - iu_2 \,.$$

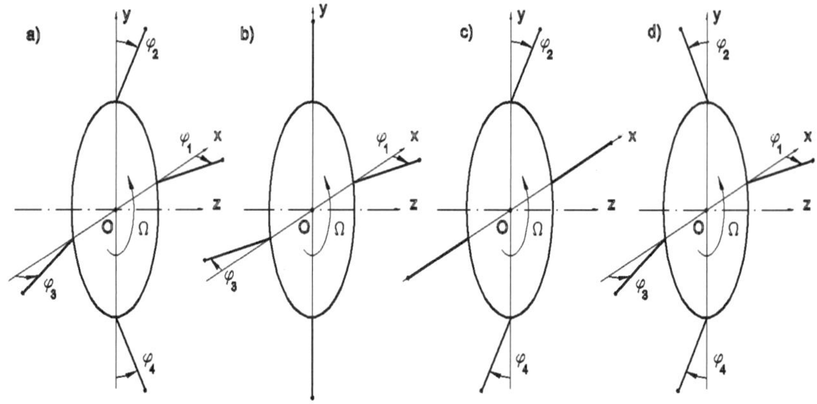

FIGURE 13.9. Modes of the row of four pendulums oscillating in a direction perpendicular to the rotation plane of the disc.

The fourth mode [Figure 13.9(d)] is a mode in which the pendulums move at 180° phasing with each other. No inertia reaction is exerted on the disc (if it is considered as a rigid body) or on the shaft, and hence it is uncoupled with the dynamics of the latter. More modes would have been found if more pendulums were present, but none of them would have been coupled with the overall dynamics of the rotor (always assuming that the disc to which they are attached is rigid).

Consider now four identical pendulums that can perform in-plane oscillations. Operating in the same way, the row of pendulums can be considered as a single entity, and four modes are readily identified [Figure 13.10(a)]. The first mode, with all pendulums oscillating in phase, is coupled with the torsional oscillations of the disc and of the shaft.

The second and the third modes [Figures 13.10(b) and (c)] regard two of the pendulums that move with 180° phasing with each other, whereas the other two do not move. These modes cause a shift of the center of mass in y- and x-directions and are coupled with displacements Y_C and X_C of the disc along y- and x-axes, respectively, and hence are again coupled with the flexural behavior of the shaft. It is possible to define a complex coordinate, in the same way already seen for the displacement of the shaft:

$$r_{row} = -u_2 + iu_1 \ .$$

The fourth mode [Figure 13.10(d)] is a mode in which the pendulums move at 180° phasing with each other. No inertia reaction is exerted on the disc (if it is considered as a rigid body) or on the shaft, and hence it is uncoupled with the dynamics of the latter. Also in this case other modes would have been found if more pendulums were present, but none of them would have been coupled with the overall dynamics of the rotor.

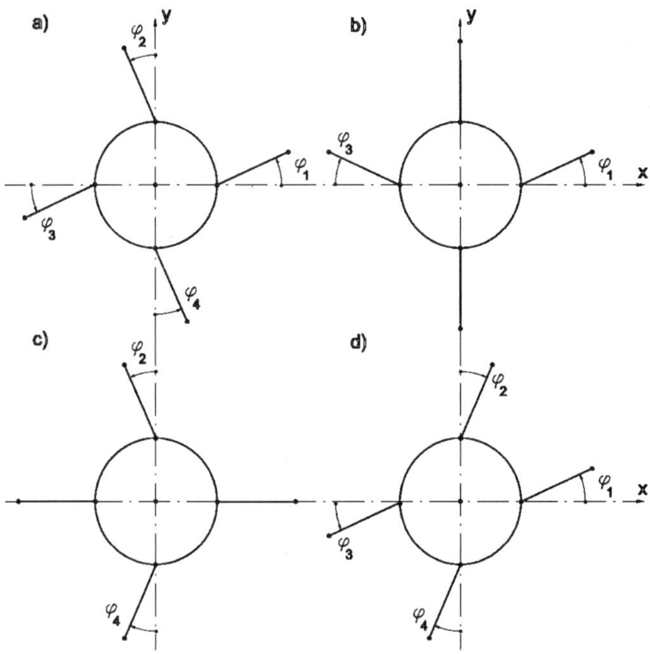

FIGURE 13.10. Modes of the row of four pendulums oscillating in the rotation plane of the disc.

In the cases studied above, the pendulums could oscillate either in the rotation plane or in a plane containing the spin axis. If the oscillation plane is inclined with respect to the spin axis (by angle ψ, Figure 13.7), the zero-order mode couples with both the axial and the torsional vibrations of the rotor. The first-order modes are coupled also in this case with the flexural behavior of the rotor, and they do not introduce further coupling. The higher modes are uncoupled from the dynamics of the shaft.

Remark 13.5 *The presence of blades that are set at an angle with the axial direction or that are twisted cause a torsional-axial coupling of the system. It, however, does not cause any coupling between the flexural dynamics and the resulting axial-torsional dynamics.*

What has been said above applies to the case in which the pendulums are identical and equispaced. If the first condition is not satisfied, each pendulum has its own natural frequency and the transformation (13.76) does not yield any more a set of uncoupled equations. Consider, for example, the case of four equispaced pendulums of different lengths oscillating in axial direction. Assume that the natural frequency of the ith pendulum is

$$\omega_i^2 = \frac{(r+l)}{l} = \overline{\omega}^2 (1 + \epsilon_i) \ ,$$

where $\overline{\omega}$ is the root mean square (rms) value of the various natural frequencies. From this definition, it follows that $\sum_i^4 \epsilon_i = 0$.

By applying the transformation (13.76) to the equation of motion, it follows that

$$\text{diag}\begin{bmatrix} 4 & 2 & 2 & 4 \end{bmatrix}\ddot{\mathbf{u}} =$$
$$= \overline{\omega}^2 \begin{bmatrix} 4 & \epsilon_1 - \epsilon_3 & \epsilon_2 - \epsilon_4 & \epsilon_1 + \epsilon_3 - \epsilon_2 - \epsilon_4 \\ & 2 + \epsilon_1 + \epsilon_3 & 0 & \epsilon_1 - \epsilon_3 \\ & & 2 + \epsilon_2 + \epsilon_4 & \epsilon_4 - \epsilon_2 \\ & \text{symm} & & 4 \end{bmatrix} \mathbf{u} \ . \tag{13.77}$$

The equations of motion are all coupled, and as a result, the dynamics of the pendulums couples with the axial and the flexural behavior of the rotor (and with the torsional behavior, if the in-plane oscillations were accounted for). Coordinates u_i are then not modal coordinates, and it is no more possible to speak of zero-order, first-order, etc. modes. The modes, i.e., the eigenvectors that can be computed from Equation (13.77), correspond to the vibrations of the single pendulums.

This case is of practical interest, because in most applications, the turbine blades are *detuned*, i.e., are not exactly equal with each other. These small differences in cross section, i.e., in mass and elasticity, are caused by machining tolerances or even purposely introduced to avoid very strong resonances of the row of blades as a single entity. The natural frequencies of the blades are spread in a range, and the coupling causes the vibration of the blades to be damped by the same damping mechanisms that are aimed to damp the vibration of the rotor as a whole. This coupling, however, is not so strong to prevent from studying the axial, torsional, and flexural vibrations except when dealing with the damping of the vibration of the blades.

13.5.2 In-plane oscillations of pendulums on elastic supports

Consider now the same row of rotating pendulums seen in the previous section, but now assume that the rigid disc is soft-mounted on an elastic support; assume also that the support is isotropic in the rotation plane and that its radial stiffness is k in any direction. Consider only the in-plane vibrations of the system, and assume that the rotational speed Ω is constant.

The disc at which the pendulums are attached has a mass m_d, and the coordinates of its center are x and y [Figure 13.11(a)]. Neglecting the kinetic energy caused by rotation, because it is constant and does not yield any term of the equations of motion, its kinetic energy is

$$\mathcal{T}_d = \frac{1}{2} m_d |\vec{V}_C|^2 = \frac{1}{2} m_d \left(\dot{x}^2 + \dot{y}^2\right) \ . \tag{13.78}$$

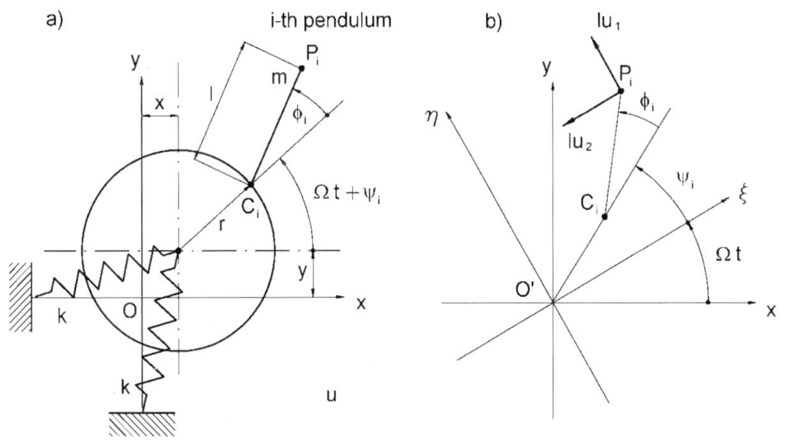

FIGURE 13.11. (a) Sketch of the position of the ith pendulum of the row. (b) Geometrical interpretation of coordinates u_1 and u_2.

The potential energy of the spring is

$$\mathcal{U}_d = \frac{1}{2}k\left(x^2 + y^2\right) . \tag{13.79}$$

The position in xy-plane of point P_i in which the bob of the ith pendulum is located is

$$(\overline{\mathrm{P}_i\text{-O}}) = \left\{ \begin{array}{c} x + r\cos(\theta_i) + l\cos(\theta_i + \phi_i) \\ y + r\sin(\theta_i) + l\sin(\theta_i + \phi_i) \end{array} \right\} , \tag{13.80}$$

where angle θ is

$$\theta_i = \Omega t + \psi_i \tag{13.81}$$

and ψ_i is the angle defining the position of the ith pendulum in the row. As the latter is a constant, $\dot{\theta}_i = \Omega$.

The velocity of point P_i can be obtained by differentiating the expressions of the coordinates with respect to time

$$\vec{V}_{\mathrm{P}_i} = \left\{ \begin{array}{c} \dot{x} - \Omega r\sin(\theta_i) - l(\Omega + \dot{\phi}_i)\sin(\theta_i + \phi_i) \\ \dot{y} + \Omega r\cos(\theta_i) + l(\Omega + \dot{\phi}_i)\cos(\theta_i + \phi_i) \end{array} \right\} . \tag{13.82}$$

The kinetic energy of the mass located in point P_i is then

$$\begin{gathered} \mathcal{T}_{\mathrm{P}_i} = \tfrac{1}{2}m|\vec{V}_{\mathrm{P}_i}|^2 = \tfrac{1}{2}m\left\{\dot{x}^2 + \dot{y}^2 + \Omega^2\left[r^2 + l^2 + 2rl\cos(\phi_i)\right] + \dot{\phi}_i^2 l^2 + \right. \\ +2\Omega\dot{\phi}_i l\left[r\cos(\phi_i) + l\right] - 2\dot{x}\left[\Omega r\sin(\theta_i) + l\left(\Omega + \dot{\phi}_i\right)\sin(\theta_i + \phi_i)\right] + \\ \left. +2\dot{y}\left[\Omega r\cos(\theta_i) + l\left(\Omega + \dot{\phi}_i\right)\cos(\theta_i + \phi_i)\right]\right\} . \end{gathered} \tag{13.83}$$

The total kinetic energy is then

$$\mathcal{T} = \mathcal{T}_d + \sum_{i=1}^{n} \mathcal{T}_{\mathrm{P}_i} . \tag{13.84}$$

Note that, as usual, the expression of the kinetic energy contains a term that is independent from the spin speed ($\mathcal{T}_2$), a second term linear in Ω ($\mathcal{T}_1$), and finally a term containing the square of the speed ($\mathcal{T}_0$).

By performing the relevant derivatives of the kinetic and potential energy, and linearizing the results, the following $n+2$ linear equations of motion can be obtained:

- First Equation (translation along x-axis)

$$\begin{array}{c} m_T\ddot{x} + m\sum_{i=1}^{n}\Big[-\ddot{\phi}_i l\sin(\theta_i) - \Omega^2\,(r+l)\cos(\theta_i) + \\ +\Omega^2 l\phi_i \sin(\theta_i) - 2\Omega l\dot{\phi}_i\cos(\theta_i)\Big] + kx = 0\ , \end{array} \tag{13.85}$$

where

$$m_T = m_d + nm \tag{13.86}$$

is the total mass of the system. By remembering some trigonometric identities and noting that

$$\sum_{i=1}^{n} \sin(\psi_i) = \sum_{i=1}^{n} \cos(\psi_i) = 0\ , \tag{13.87}$$

it is possible to write the first equation in the form

$$\begin{array}{c} m_T\ddot{x} + ml\sum_{i=1}^{n}\left[\left(-\ddot{\phi}_i + \Omega^2\phi_i\right)\cos(\psi_i) + 2\Omega\dot{\phi}_i\sin(\psi_i)\right]\sin(\Omega t) + \\ +ml\sum_{i=1}^{n}\left[\left(-\ddot{\phi}_i + \Omega^2\phi_i\right)\sin(\psi_i) - 2\Omega\dot{\phi}_i\cos(\psi\theta_i)\right]\cos(\Omega t) + kx = 0\ . \end{array} \tag{13.88}$$

It is possible to simplify substantially the first equation of motion by applying the coordinate transformation (13.76). By inverting the transformation matrix, it is possible to write

$$\left\{\begin{array}{c} u_0 \\ u_1 \\ u_2 \\ \dots \end{array}\right\} = \frac{1}{n}\left[\begin{array}{ccccc} 1 & 1 & 1 & \dots \\ h\cos(\psi_1) & h\cos(\psi_2) & h\cos(\psi_3) & \dots \\ h\sin(\psi_1) & h\sin(\psi_2) & h\sin(\psi_3) & \dots \\ h\cos(2\psi_1) & h\cos(2\psi_2) & h\cos(2\psi_3) & \dots \\ \dots & \dots & \dots & \dots \end{array}\right]\left\{\begin{array}{c} \phi_1 \\ \phi_2 \\ \phi_3 \\ \dots \end{array}\right\} , \tag{13.89}$$

where $h = 2$ everywhere, except for the last row in which $h = 2$ if n is odd and $n = 1$ if n is even.

By inspecting Equation (13.89), it is clear that the sums included into Equation (13.88) are coordinates u_1 and u_2 and their time derivatives. Equation (13.88) reduces to

$$\begin{array}{c} m_T\ddot{x} + m\frac{nl}{2}\left(-\ddot{u}_1 + \Omega^2 u_1 + 2\Omega\dot{u}_2\right)\sin(\Omega t) + \\ +m\frac{nl}{2}\left(-\ddot{u}_2 + \Omega^2 u_2 - 2\Omega\dot{u}_1\right)\cos(\Omega t) + kx = 0\ . \end{array} \tag{13.90}$$

Note that in this equation, only three generalized coordinates (namely, x, u_1, and u_2) are present. Moreover, the equation is linear, but the coefficients are periodic in time through angle Ωt.

- Second Equation (translation along y-axis)

$$\begin{array}{c} m_T\ddot{y} + m\sum_{i=1}^{n}\left[\ddot{\phi}_i l\cos(\theta_i) - \Omega^2\left(r+l\right)\sin(\theta_i) + \right. \\ \left. -\Omega^2 l\phi_i\cos(\theta_i) - 2\Omega l\dot{\phi}_i\sin(\theta_i)\right] + ky = 0\ . \end{array} \tag{13.91}$$

By operating in the same way seen for the first equation and using the same coordinate transformation, Equation (13.91) reduces to

$$\begin{array}{c} m_T\ddot{y} - m\frac{nl}{2}\left(-\ddot{u}_1 + \Omega^2 u_1 + 2\Omega\dot{u}_2\right)\cos(\Omega t) + \\ +m\frac{nl}{2}\left(-\ddot{u}_2 + \Omega^2 u_2 - 2\Omega\dot{u}_1\right)\sin(\Omega t) + ky = 0\ . \end{array} \tag{13.92}$$

- Last n equations (rotations ϕ_i)

$$-\ddot{x}\sin(\theta_i) + \ddot{y}\cos(\theta_i) + \ddot{\phi}_i l + \Omega^2 r\phi_i = 0 \qquad \text{for } i = 1,\ ...,\ n\ ; \tag{13.93}$$

i.e.,

$$\begin{array}{c} \left[-\ddot{x}\sin(\psi_i) + \ddot{y}\cos(\psi_i)\right]\cos(\Omega t) + \left[-\ddot{x}\cos(\psi_i) - \ddot{y}\sin(\psi_i)\right]\sin(\Omega t) + \\ +\ddot{\phi}_i l + \Omega^2 r\phi_i = 0\ . \end{array} \tag{13.94}$$

If the generic ith equation of this group is written in the form $p_i = 0$, they can be substituted by their linear combinations

$$\mathbf{T}^{-1}\mathbf{p} = 0\ , \tag{13.95}$$

where $\mathbf{T}$ is the coordinate transformation matrix defined in Equation (13.76).

The first of these equation is simply

$$\ddot{u}_0 l + \Omega^2 r u_0 = 0\ . \tag{13.96}$$

It decouples from the other equations and states that the motion with all pendulums oscillating in phase following a pattern of the type shown in Figure 13.10(a) is not affected by the lateral motion of the disc. It is, however, coupled with the torsional torsional behavior of the disc, which has not been considered here. This motion will not be dealt with any further.

The following two equations yield:

$$\begin{aligned} -\ddot{x}\sin(\Omega t)+\ddot{y}\cos(\Omega t)+\ddot{u}_1 l+\Omega^2 r u_1=0\ , \\ -\ddot{x}\cos(\Omega t)-\ddot{y}\sin(\Omega t)+\ddot{u}_2 l+\Omega^2 r u_2=0\ . \end{aligned} \tag{13.97}$$

They are coupled with the first two equations, showing that the motion with all pendulums following a pattern of the type shown in Figure 13.10(b) and (c) is coupled with the motion of the disc.

All of the following equations do not contain either x or y. They are all of the type

$$\ddot{u}_i l+\Omega^2 r u_i=0 \qquad \text{for } i=3,\ ...,\ n-1\ . \tag{13.98}$$

They show that all motions following a pattern of the type shown in Figure 13.10(d), i.e., with the center of mass of the system of the pendulums stationary in the center of the disc, are uncoupled with the motion of the latter. These equations will not be dealt with any further.

The disc-pendulums interaction can thus be studied using a set of four linear equations (with periodic coefficients) in the coordinates x, y, u_1, and u_2

$$\left\{\begin{array}{l} m_T\ddot{x}+m\frac{nl}{2}\left(-\ddot{u}_1+\Omega^2 u_1+2\Omega\dot{u}_2\right)\sin(\Omega t)+ \\ \qquad +m\frac{nl}{2}\left(-\ddot{u}_2+\Omega^2 u_2-2\Omega\dot{u}_1\right)\cos(\Omega t)+kx=0\ , \\ m_T\ddot{y}-m\frac{nl}{2}\left(-\ddot{u}_1+\Omega^2 u_1+2\Omega\dot{u}_2\right)\cos(\Omega t)+ \\ \qquad +m\frac{nl}{2}\left(-\ddot{u}_2+\Omega^2 u_2-2\Omega\dot{u}_1\right)\sin(\Omega t)+ky=0\ , \\ -\ddot{x}\sin(\Omega t)+\ddot{y}\cos(\Omega t)+\ddot{u}_1 l+\Omega^2 r u_1=0\ , \\ -\ddot{x}\cos(\Omega t)-\ddot{y}\sin(\Omega t)+\ddot{u}_2 l+\Omega^2 r u_2=0\ . \end{array}\right. \tag{13.99}$$

The physical meaning of u_1 and u_2 is straightforward. Each one of the terms that are added to yield u_1 and u_2 in Equation (13.89) are the projection of the displacements in the direction of the rotating axes ξ and η [Figure 13.11(b)] of the displacement of point P_i (divided by $nl/2$), when the center of the support is locked in its rest position. After adding all of the terms, what is obtained is the displacement (divided by $l/2$) of the center of mass of the pendulums when they move following the pattern of Figure 13.10(b) and (c).

Equations (13.99) can be transformed into a set of constant coefficients equations by resorting to a further change of coordinates:

$$\left\{\begin{array}{l} x_1=-lu_2\cos(\Omega t)-lu_1\sin(\Omega t)\ , \\ y_1=lu_1\cos(\Omega t)-lu_2\sin(\Omega t)\ , \end{array}\right. \tag{13.100}$$

which amounts to writing the displacement of the center of mass of the pendulums in an inertial frame instead of writing it with reference to the rotating frame.

By performing the relevant differentiations with respect to time (e.g.,

$$\ddot{x}_1 = l\left(-\ddot{u}_2 + \Omega^2 u_2 - 2\Omega \dot{u}_1\right)\cos(\Omega t) + l\left(-\ddot{u}_1 + \Omega^2 u_1 + 2\Omega \dot{u}_2\right)\sin(\Omega t) \tag{13.101}$$

and so on), the equations of motions transform into

$$\begin{cases} m_T\ddot{x} + \frac{nm}{2}\ddot{x}_1 + kx = 0\ , \\ m_T\ddot{y} + \frac{nm}{2}\ddot{y}_1 + ky = 0\ , \\ \ddot{x} + \ddot{x}_1 + 2\Omega\dot{y}_1 - \Omega^2\left(1-\delta\right)x_1 = 0\ , \\ \ddot{y} + \ddot{y}_1 - 2\Omega\dot{x}_1 - \Omega^2\left(1-\delta\right)y_1 = 0\ . \end{cases} \tag{13.102}$$

The set of equations can be further simplified by resorting to complex coordinates

$$\begin{cases} r_d = x + iy\ , \\ r_p = x_1 + iy_1\ . \end{cases} \tag{13.103}$$

By adding the first equation to the second multiplied by i and the third one to the fourth multiplied by i, it follows that

$$\begin{cases} m_T\ddot{r}_d + \frac{nm}{2}\ddot{r}_p + kr_d = 0\ , \\ \ddot{r}_d + \ddot{r}_p - 2i\Omega\dot{r}_p - \Omega^2\left(1-\delta\right)r_p = 0\ . \end{cases} \tag{13.104}$$

By introducing the nondimensional parameters

$$\alpha = \frac{nm}{2m_T} \quad (0 < \alpha < \frac{1}{2})\ , \tag{13.105}$$

the equation of motion can be written in the form

$$\begin{aligned} &\begin{bmatrix} 1 & \alpha \\ \alpha & \alpha \end{bmatrix}\begin{Bmatrix} \ddot{r}_d \\ \ddot{r}_p \end{Bmatrix} - 2i\Omega\alpha\begin{bmatrix} 0 & 0 \\ 0 & 1 \end{bmatrix}\begin{Bmatrix} \dot{r}_d \\ \dot{r}_p \end{Bmatrix} + \\ &+\begin{bmatrix} \omega_1 & 0 \\ 0 & \Omega^2\alpha\left(\delta-1\right) \end{bmatrix}\begin{Bmatrix} r_d \\ r_p \end{Bmatrix} = \mathbf{0}\ , \end{aligned} \tag{13.106}$$

where $\omega_1 = \sqrt{k/m_T}$ is the natural frequency of mass m_T on a spring with stiffness k.

By introducing the nondimensional speed and Laplace variable

$$\Omega^* = \frac{\Omega}{\omega_1}\ , \qquad s^* = \frac{s}{\omega_1}\ , \tag{13.107}$$

a nondimensional response depending only on α and δ can be obtained from the characteristic equation

$$\det\begin{bmatrix} s^{*2}+1 & s^{*2}\alpha \\ s^{*2}\alpha & s^{*2}\alpha - 2i\Omega^*\alpha s^* + \Omega^{*2}\alpha\left(\delta-1\right) \end{bmatrix} = 0\ . \tag{13.108}$$

By inspecting the stiffness matrix of the system, it is clear that the condition $\delta = 1$, i.e., $r = l$, is peculiar, because it corresponds to a singular stiffness matrix. Two cases can be defined:

- Short pendulums, $l < r$ (i.e., $\delta > 1$), where the stiffness matrix is positive defined,
- Long pendulums, $l > r$ (i.e., $\delta < 1$), where the stiffness matrix is negative defined.

Before attempting to solve Equation (13.108), an approximate solution can be obtained by neglecting the terms outside the main diagonal. A first equation

$$s^{*2} + 1 = 0 \tag{13.109}$$

yields

$$s^* = \pm i \ , \tag{13.110}$$

which corresponds to the natural frequencies of the system oscillating on its supports with the pendulum locked.

The second equation

$$s^{*2} - 2i\Omega^* s^* + \Omega^{*2}(\delta - 1) = 0 \tag{13.111}$$

yields the solutions

$$s^* = i\Omega^* \left(1 \pm \sqrt{\delta}\right) \tag{13.112}$$

corresponding to the natural frequencies of the pendulums oscillating on the locked support.

This is fully consistent with what has been obtained earlier on rotating pendulums: The natural frequency for in-plane motion of a single rotating pendulum is [Equation (13.14)]

$$\omega_{p_{rf}} = \Omega\sqrt{\frac{r}{l}} = \Omega\sqrt{\delta}\,. \tag{13.113}$$

This value is computed with reference to the rotating frame. In the rotating frame, the global motion of the pendulum row can be seen as a forward moving wave, propagating at an angular velocity $\omega_{p_{rf}}$ plus a backward moving wave, propagating at an angular velocity $-\omega_{p_{rf}}$. The forward moving wave is seen in the fixed frame as a wave propagating in forward direction with a speed $\Omega + \omega_{p_{rf}} = \Omega\left(1 + \sqrt{\delta}\right)$, which coincides with the solution with (+) in Equation (13.112).

The backward wave is seen as in the fixed frame as a wave propagating with a speed $\Omega - \omega_{p_{rf}} = \Omega\left(1 - \sqrt{\delta}\right)$, coinciding with the solution with (−) in Equation (13.112). This wave propagates in forward direction if $\delta < 1$, i.e., in the case of long pendulums, and in backward direction if $\delta > 1$, i.e., in the case of short pendulums.

The nondimensional Campbell diagram and the decay rate plot for the case of long pendulums with $\alpha = 0.1$ and $\delta = 0.2$ are shown in Figure 13.12(a) and (b). A value $\alpha = 0.1$ corresponds to light pendulums, having a

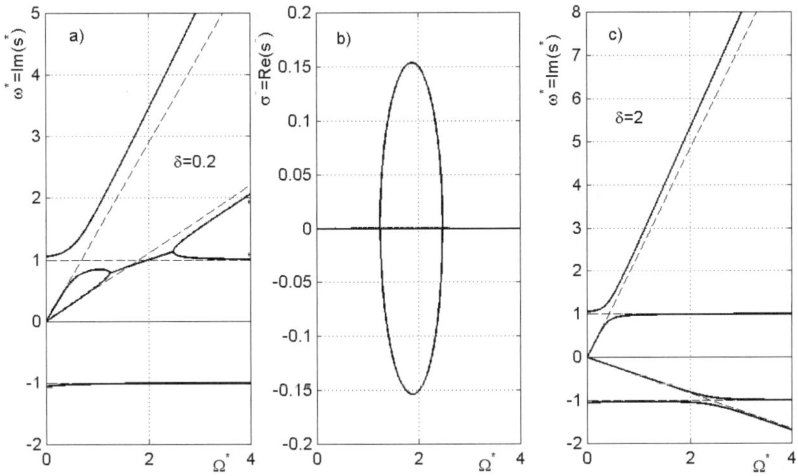

FIGURE 13.12. (a) and (b): Campbell diagram and the decay rate plot for the case of long pendulums with $\alpha = 0.1$ and $\delta = 0.2$. Note the field of instability. (c) Campbell diagram and the decay rate plot for the case of short pendulums with $\alpha = 0.1$ and $\delta = 2$. In this case no field of instability exists.

total mass nm equal to one-quarter of the mass of the disc ($nm = 0.25m_d$). A field of instability, with a threshold at $\Omega^* = 1.24$ is clearly present. Note that, as expected, the threshold of instability is located in the supercritical field.

A similar nondimensional Campbell diagram, but for the case of short pendulums with $\alpha = 0.1$ and $\delta = 2$, is reported in Figure 13.12(c). Here no instability range is present.

Remark 13.6 *Numerical experimentation has shown that the field of instability is present wherever the traveling wave, which moves backwards in the rotating frame, moves forward in the fixed frame, i.e., in the case of long pendulums ($\delta < 1$). The value of α affects the width of the instability range, but not its presence. No instability was found for the case of short pendulums ($\delta > 1$).*

13.5.3 Spring-loaded pendulums on elastic supports

Consider now the case in which the pendulums studied in the previous section are loaded by a spring supplying a restoring force toward the radial position. The terms of the equations of motion and the coordinate transformations still hold, but some terms linked with the stiffness of the springs must be added.

The potential energy of the spring is

$$\mathcal{U}_p = \frac{1}{2} k_p \sum_{i=1}^{n} \phi_i^2 \,, \tag{13.114}$$

where k_p is the torsional stiffness of each spring.

The first two equations are not affected by the presence of the springs, whereas all other ones [Equations (13.93) or (13.94)] are easily modified by adding a term coming from the derivatives of the potential energy with respect to ϕ_i

$$-\ddot{x}\sin(\theta_i) + \ddot{y}\cos(\theta_i) + \ddot{\phi}_i l + \left(\Omega^2 r + \frac{k_p}{ml}\right)\phi_i = 0 \qquad \text{for } i = 1, \; ..., \; n \,. \tag{13.115}$$

By performing the various changes of coordinates, the final equations to study the interaction between the motion of the pendulums and that of the suspension system [Equations (13.104)] become

$$\begin{cases} m_T \ddot{r}_d + \frac{nm}{2} \ddot{r}_p + k r_d = 0 \,, \\ \ddot{r}_d + \ddot{r}_p - 2i\Omega\dot{r}_p + \left[\Omega^2\left(\frac{r}{l} - 1\right) + \frac{k_p}{ml^2}\right] r_p = 0 \,. \end{cases} \tag{13.116}$$

Note that the term k_p/ml^2 is just the square of the natural frequency ω_p of the pendulums at standstill (neglecting the gravitational acceleration as consistent with the whole study). A further nondimensional parameter can thus be added to the two seen above:

$$\omega_p^* = \frac{\omega_p}{\omega_1} \,, \tag{13.117}$$

namely, the ratio between the natural frequency of the pendulums at standstill and that of the whole system (with locked pendulums) on its supports.

The characteristic equation yielding the nondimensional natural frequencies then becomes

$$\det \begin{bmatrix} s^{*2} + 1 & s^{*2}\alpha \\ s^{*2}\alpha & s^{*2}\alpha - 2i\Omega^* \alpha s^* + \alpha\left[\Omega^{*2}(\delta - 1) + \alpha\omega_p^{*2}\right] \end{bmatrix} = 0 \,. \tag{13.118}$$

Again, the uncoupled dynamics of the system can be studied by neglecting the terms outside the main diagonal. The first equation, dealing with the system oscillating on its supports with the pendulum locked, is the same as that of the previous case.

The second equation becomes

$$s^{*2} - 2i\Omega^* s^* + \Omega^{*2}(\delta - 1) + \omega_p^{*2} = 0 \tag{13.119}$$

and yields the solutions

$$s^* = i\Omega^* \left(1 \pm \sqrt{\delta + \left(\frac{\omega_p^*}{\Omega^*}\right)^2}\right) \tag{13.120}$$

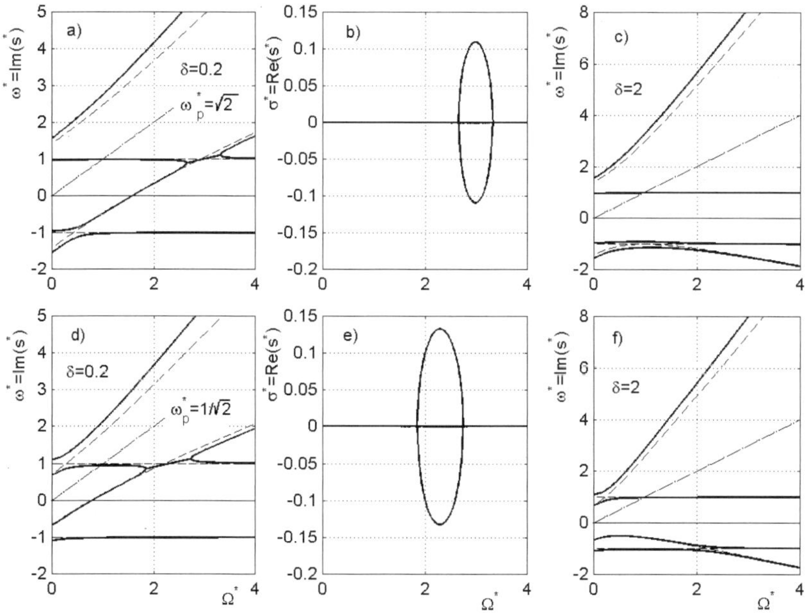

FIGURE 13.13. (a) and (b) Campbell diagram and decay rate plot for the case of long pendulums with $\alpha = 0.1$ and $\delta = 0.2$. Note the field of instability. The same holds for (d) and (e), in which a different value of ω_p^* was used. (c) and (f) Campbell diagram for the case of short pendulums with $\alpha = 0.1$ and $\delta = 2$. In these cases, no field of instability exists.

corresponding to the natural frequencies of the pendulums oscillating on the locked support. This result is again consistent with what has been obtained earlier on rotating pendulums on a fixed support.

For vanishing small speeds, two values of the frequency are found: $\omega^* = \pm\omega_p^*$, i.e., two waves traveling in opposite directions (at standstill, the words forward and backward lose meaning). With increasing speed, the wave traveling in forward direction in the rotating frame travels always in forward direction. The backward wave is seen in the fixed frame as a wave traveling backward in the case of short pendulums ($\delta > 1$). In the case of long pendulums ($\delta < 1$), the backward wave is seen in the fixed frame as a backward wave at low speed and as a forward wave at high speed.

The nondimensional Campbell diagram and the decay rate plot for the case of long pendulums with $\alpha = 0.1$ and $\delta = 0.2$ are shown in Figure 13.13(a) and (b) for a system with $\omega_p^* = \sqrt{2}$ and (d) and (e) for a system with $\omega_p^* = 1/\sqrt{2}$. A value of $\omega_p^* > 1$ means that the frequency of the pendulums is higher than that of the system on its supports.

In both cases, a field of instability is present and is located in the supercritical field.

The cases reported in Figure 13.13(c) and (f) are related to short pendulums with $\alpha = 0.1$ and $\delta = 2$ with $\omega_p^* = \sqrt{2}$ and $\omega_p^* = 1/\sqrt{2}$, respectively.

Remark 13.7 *The presence of the restoring force does not change the conclusion that a field of instability is present in the case of long pendulums ($\delta < 1$) and not in the case of the short pendulums ($\delta > 1$).*

13.5.4 Damped pendulums on elastic supports

Consider now the same system studied in the previous section, but with damping added. Two viscous dampers with damping coefficient c_d act in parallel with the springs of stiffness k constraining the disc, whereas a number n of dampers with damping coefficient l^2c_p act in parallel with the springs with stiffness k_p, constraining the motion of the pendulums (the coefficient has been multiplied by l^2 so that dimensionally all damping coefficients are equivalent). A further number n of viscous dampers with damping coefficient c_i are located between the bobs of two subsequent pendulums. The latter have been introduced to simulate the damping acting between the blades.

To obtain the equations of motion, a Rayleigh dissipation function

$$\mathcal{F} = \frac{1}{2}c_d\left(\dot{x}^2 + \dot{y}^2\right) + \frac{1}{2}l^2c_p\sum_{i=1}^{n}\dot{\phi}_i^2 + \frac{1}{2}l^2c_i\sum_{i=1}^{n}\left(\dot{\phi}_i + \dot{\phi}_{i+1}\right)^2 , \qquad (13.121)$$

in which $\dot{\phi}_{n+1} = \dot{\phi}_1$, must be introduced into the Lagrange equations.

The first two Equations (13.90) and (13.92) are modified by simply adding a term $c_d\dot{x}$ and $c_d\dot{y}$, respectively, to the left-hand side.

The other n Equations (13.115) become

$$\begin{aligned}&-\ddot{x}\sin(\theta_i) + \ddot{y}\cos(\theta_i) + \ddot{\phi}_i l + \left(\Omega^2 r + \frac{k_p}{ml}\right)\phi_i + \\ &+\frac{l}{m}\left(c_p + 2c_i\right)\dot{\phi}_i - \frac{lc_i}{m}\dot{\phi}_{i-1} - \frac{lc_i}{m}\dot{\phi}_{i+1} = 0 \qquad \text{for } i = 1,\ ...,\ n\ ,\end{aligned} \qquad (13.122)$$

where $\dot{\phi}_0 = \dot{\phi}_n$.

These n equations can be transformed using Equation (13.95). The first transformed equation is of the same type of Equation (13.96), with a few additional terms

$$\ddot{u}_0 l + \frac{l}{m}c_p\dot{u}_0 + \left(\Omega^2 r + \frac{k_p}{ml}\right)u_0 = 0\ , \qquad (13.123)$$

Not only does it decouple from the other equations, like in the previous cases, but it does not contain any term in c_i: The dampers located between the pendulums do not affect the motion of the row when it oscillates with all pendulums in phase.

The following two equations yield

$$\begin{aligned}&-\ddot{x}\sin(\Omega t)+\ddot{y}\cos(\Omega t)+\ddot{u}_1 l+\tfrac{l}{m}\left(c_p+2c_i\right)\dot{u}_1+\left(\Omega^2 r+\tfrac{k_p}{ml}\right)u_1+\\&\quad-\tfrac{l}{m}\textstyle\sum_{i=1}^{n}\dot{\phi}_i\cos\left(\psi_{i+1}\right)-\tfrac{l}{m}\sum_{i=1}^{n}\dot{\phi}_i\cos\left(\psi_{i-1}\right)=0\;,\\&-\ddot{x}\cos(\Omega t)-\ddot{y}\sin(\Omega t)+\ddot{u}_2 l+\tfrac{l}{m}\left(c_p+2c_i\right)\dot{u}_2+\left(\Omega^2 r+\tfrac{k_p}{ml}\right)u_2+\\&\quad-\tfrac{l}{m}\textstyle\sum_{i=1}^{n}\dot{\phi}_i\sin\left(\psi_{i+1}\right)-\tfrac{l}{m}\sum_{i=1}^{n}\dot{\phi}_i\sin\left(\psi_{i-1}\right)=0\;.\end{aligned}\tag{13.124}$$

By remembering that $\psi_{i+1} = \psi_i + \Delta\psi$ and $\psi_{i-1} = \psi_i - \Delta\psi$, where $\Delta\psi = 2\pi/n$ is the angle between two subsequent pendulums, Equation (13.124) reduces to

$$\left\{\begin{array}{l}\ddot{x}+\ddot{x}_1+2\Omega\dot{y}_1+c\dot{x}_1+\left[\Omega^2\left(\frac{r}{l}-1\right)+\frac{k_p}{ml^2}\right]x_1+\Omega c y_1=0\;,\\ \ddot{y}+\ddot{y}_1-2\Omega\dot{x}_1+c\dot{y}_1+\left[\Omega^2\left(\frac{r}{l}-1\right)+\frac{k_p}{ml^2}\right]y_1-\Omega c x_1=0\;,\end{array}\right.\tag{13.125}$$

where

$$c=\frac{1}{m}\left\{c_p+2c_i\left[1-\cos\left(\Delta\psi\right)\right]\right\}=\frac{1}{m}\left\{c_p+4c_i\sin^2\left(\frac{\Delta\psi}{2}\right)\right\}\;.$$

The set of equations can be further simplified by resorting to complex coordinates r_d and r_p. By adding the first equation to the second multiplied by i and the third one to the fourth multiplied by i, it follows that

$$\left\{\begin{array}{l}m_T\ddot{r}_d+\frac{nm}{2}\ddot{r}_p+c_d\dot{r}_d+kr_d=0\;,\\ \ddot{r}_d+\ddot{r}_p+\left(c-2i\Omega\right)\dot{r}_p+\left[\Omega^2\left(\frac{r}{l}-1\right)+\frac{k_p}{ml^2}-i\Omega c\right]r_p=0\;.\end{array}\right.\tag{13.126}$$

Apart from parameters α, δ, and ω_p^*, there are two further parameters linked with damping

$$\zeta_d=\frac{c_d}{2\sqrt{km_T}}\quad\text{and}\quad\zeta_p=\frac{m_T}{ml}\frac{c_p+4c_i\sin^2\left(\frac{\Delta\psi}{2}\right)}{2\sqrt{km_T}}\;.\tag{13.127}$$

The characteristic equation yielding the nondimensional natural frequencies then becomes

$$\det\left[\begin{array}{cc}s^{*2}+2s^*\zeta_d+1 & s^{*2}\alpha\\ s^{*2}\alpha & \alpha f(s^*)\end{array}\right]=0\;,\tag{13.128}$$

where

$$f(s^*)=s^{*2}+2s^*\zeta_p+\Omega^{*2}\left(\delta-1\right)+\omega_p^{*2}-2i\Omega^*\left(s^*+\zeta_p\right)\;.$$

The uncoupled dynamics is easily studied. The motion of the system with the pendulums locked is equivalent to that of a Jeffcott rotor with no rotating damping, and assuming that the system is underdamped ($\zeta_d < 1$), the solutions for s^* are

$$s^* = -\zeta_d \pm i\sqrt{1-\zeta_d^2} \,. \tag{13.129}$$

The real part of s^* is always negative, and the system is stable.

The motion of the pendulums on the locked support are governed by the equation

$$s^{*2} + 2s^*\zeta_p + \Omega^{*2}(\delta - 1) + \omega_p^{*2} - 2i\Omega^*\left(s^* + \zeta_p\right) = 0 \,, \tag{13.130}$$

yielding

$$s^* = -\zeta_p + i\Omega^* \pm \sqrt{\zeta_p^2 - \delta\Omega^{*2} - \omega_p^{*2}} \,. \tag{13.131}$$

Here the cases are two: If the damping is low, i.e., if $\zeta_p^2 < \delta\Omega^{*2} - \omega_p^{*2}$, the solutions for s are

$$s^* = -\zeta_p + i\left(\Omega^* \pm \sqrt{\delta\Omega^{*2} + \omega_p^{*2} - \zeta_p^2}\right) \,. \tag{13.132}$$

On the contrary, if the damping is high, i.e., if $\zeta_p^2 > \delta\Omega^{*2} - \omega_p^{*2}$, the solutions for s are

$$s^* = -\zeta_p \pm \sqrt{\zeta_p^2 - \delta\Omega^{*2} - \omega_p^{*2}} + i\Omega^* \,. \tag{13.133}$$

In all cases, the real part of s^* is always negative, and the system is stable.

Remark 13.8 *This is a very remarkable result, because one would expect that the damping applied to the pendulums, and above all that located between them, would have a destabilizing effect, being rotating damping. However, it must be remembered that this applies to the uncoupled system, and that the instability caused by the coupling of the disc and pendulum dynamics is not included into the uncoupled model.*

To study the interaction, the coupled system must be studied. Some numerical results are shown in Figure 13.14: The nondimensional Campbell diagram and the decay rate plot reported deal with the case of long pendulums with $\alpha = 0.1$, $\delta = 0.2$, and $\omega_p^* = \sqrt{2}$, the same system studied in Figure 13.13(a) and (b), but with damping added. Cases with different combinations of nonrotating ζ_d and rotating ζ_p damping are shown. The effect of damping on the Campbell diagram is not large, because all values of ζ used are smaller than unity, whereas it is important on the decay rate plot.

From the figure, it is clear that with a low value of both nonrotating (disc) and rotating (pendulums) damping, the field of instability, although reduced in comparison with the undamped case, is still present. An increase

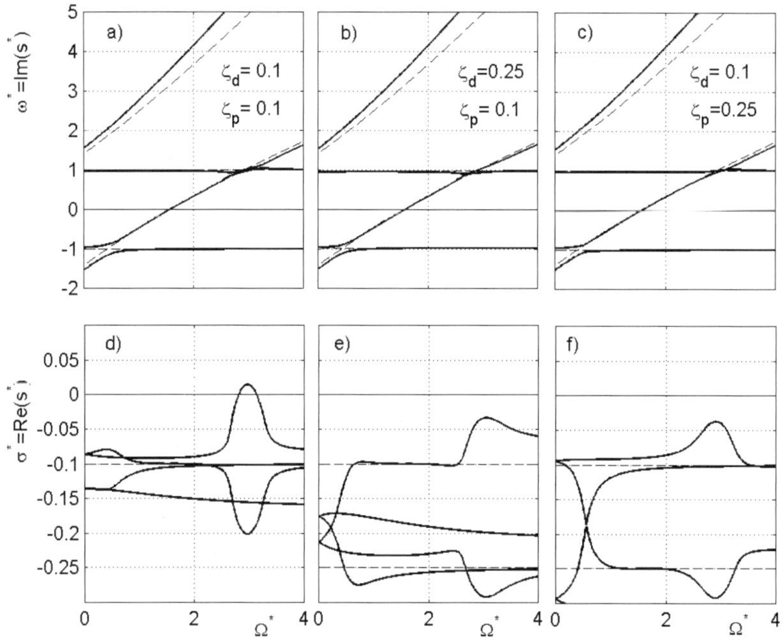

FIGURE 13.14. Campbell diagram (a, b, c) and decay rate plot (d, e, f) for damped long pendulums with $\alpha = 0.1$, $\delta = 0.2$. and $\omega_p^* = \sqrt{2}$. Different combinations of nonrotating ζ_d and rotating ζ_p damping.

of any form of damping is effective in stabilizing the system, although a large damping is needed to cancel completely the instability range.

The remaining equations, from the fourth one to the nth are of the type

$$\left\{\begin{array}{l} \ddot{u}_3 l + \frac{l}{m}\left[c_p + 4c_i \sin^2\left(\Delta\psi\right)\right]\dot{u}_3 + \left(\Omega^2 r + \frac{k_p}{ml}\right)u_3 = 0\ , \\ \ddot{u}_4 l + \frac{l}{m}\left[c_p + 4c_i \sin^2\left(\Delta\psi\right)\right]\dot{u}_4 + \left(\Omega^2 r + \frac{k_p}{ml}\right)u_4 = 0\ , \\ \ddot{u}_5 l + \frac{l}{m}\left[c_p + 4c_i \sin^2\left(\frac{3}{2}\Delta\psi\right)\right]\dot{u}_5 + \left(\Omega^2 r + \frac{k_p}{ml}\right)u_5 = 0\ . \\ \dots\dots\dots\dots\dots\dots\dots\dots\dots \end{array}\right. \tag{13.134}$$

These equations are uncoupled and describe the motion of the pendulums with different phasing. No instability can be found from these equations. Note that the equations are all equal, except for the role played by damping c_i, which increases with increasing order of the equations. If the pendulums are many, the damping coefficient appearing in the last equation, that dealing with the variable u_{n-1}, is $c_p + 4c_i$.

Remark 13.9 *The study of the in-plane vibrations of a row of rotating pendulums apply to cases in which there are at least three pendulums: If they are only two, the ellipsoid of inertia of the system thought as a rigid* *body is no more axially symmetrical, and a further cause of instability is present, namely, rotor anisotropy. This study has been used in the past to deal with the instability of propellers and helicopter rotors.*

13.5.5 Out-of-plane oscillations of pendulums on elastic supports

To study the out-of-plane oscillations of the row of rotating pendulums seen in the previous sections, assume that the rigid disc on which the pendulums are located is rigidly constrained against translational motions along x-, y- and z-axes but can rotate about x- and y-axes being constrained in its rotations by identical linear torsional springs. Assume also that the disc is axially symmetrical and its moments of inertia are J_t about x- and y-axes and J_p about z-axis [Figure 13.11(a)]. The dynamics of the disc alone is that of the uncoupled gyroscopic system studied in Section 3.2. No couple unbalance is considered.

Assuming as generalized coordinates for the disc the angles ϕ_X and ϕ_y defined in Chapter 3, the kinetic energy caused by rotations of the disc is expressed by Equation (3.13)

$$\mathcal{T}_d = \frac{1}{2}\left[J_t\left(\dot{\phi}_X^2 + \dot{\phi}_y^2\right) + J_p\left(\Omega^2 + 2\Omega\dot{\phi}_X\phi_y\right)\right] . \tag{13.135}$$

The potential energy of the springs with torsional stiffness k_d, is

$$\mathcal{U}_d = \frac{1}{2}k_d\left(\phi_X^2 + \phi_y^2\right) . \tag{13.136}$$

The position in xy-plane of point P_i where the bob of the ith pendulum is located is

$$\left(\overline{P_i - O}\right) = \mathbf{R}_1^T\mathbf{R}_2^T\mathbf{R}_3^T\left\{\begin{array}{c} r + l\cos(\phi_i) \\ 0 \\ l\sin(\phi_i)\end{array}\right\} , \tag{13.137}$$

where now angles ϕ_i are the out-of-plane oscillation angles of the pendulums, matrices $\mathbf{R}_1$, $\mathbf{R}_2$, and $\mathbf{R}_3$ are defined by Equations from (3.1) to (3.3), and angle θ_i included in matrix $\mathbf{R}_3$ is

$$\theta_i = \Omega t + \psi_i . \tag{13.138}$$

By performing the relevant matrix multiplications, it follows that

$$\left(\overline{P_i - O}\right) = \left\{\begin{array}{c} [r + l\cos(\phi_i)]\cos(\theta_i)\cos(\phi_y) + l\sin(\phi_i)\sin(\phi_y) \\ [r + l\cos(\phi_i)]\sin(\theta_i)\cos(\phi_X) + [r + l\cos(\phi_i)]\cos(\theta_i)\cdot \\ \cdot\sin(\phi_y)\sin(\phi_X) - l\sin(\phi_i)\cos(\phi_y)\sin(\phi_X) \\ [r + l\cos(\phi_i)]\sin(\theta_i)\sin(\phi_X) - [r + l\cos(\phi_i)]\cos(\theta_i)\cdot \\ \cdot\sin(\phi_y)\cos(\phi_X) + l\sin(\phi_i)\cos(\phi_y)\cos(\phi_X)\end{array}\right\} . \tag{13.139}$$

By differentiating the position of point P_i with respect to time and computing the square of the speed, the kinetic energy of the ith pendulum can be computed

$$\begin{aligned}\mathcal{T}_p = \tfrac{1}{2}m\Big[&\Omega^2(r+l)^2+\dot{\phi}_X^2(r+l)^2\sin^2(\theta_i)+\dot{\phi}_y^2(r+l)^2\cos^2(\theta_i)\\ &-\phi^2\Omega^2 l(r+l)+2\dot{\phi}_X\phi_y\Omega(r+l)^2-2\dot{\phi}_X\phi\Omega l(r+l)\cos(\theta_i)+\\ &+\dot{\phi}^2 l^2-2\dot{\phi}_y\phi\Omega l(r+l)\sin(\theta_i)+2\dot{\phi}\dot{\phi}_X l(r+l)\sin(\theta_i)+\\ &-2\dot{\phi}_X\dot{\phi}_y(r+l)^2\cos(\theta_i)\sin(\theta_i)-2\dot{\phi}\dot{\phi}_y l(r+l)\cos(\theta_i)\Big]\ .\end{aligned} \tag{13.140}$$

As seen for the in-plane oscillations of the pendulums, if they are provided with a restoring spring with stiffness k_p, a damper with damping coefficient c_i between the pendulums and the disc, and a further damper with damping c_i between each other, a potential energy and a Rayleigh dissipation function can be defined

$$\mathcal{U}_p = \frac{1}{2}k_p\sum_{i=1}^{n}\phi_i^2\ , \tag{13.141}$$

$$\mathcal{F} = \frac{1}{2}c_d\left(\dot{\phi}_y^{\,2}+\dot{\phi}_X^{\,2}\right)+\frac{1}{2}c_p\sum_{i=1}^{n}\dot{\phi}_i^2+\frac{1}{2}c_i\sum_{i=1}^{n}\left(\dot{\phi}_i+\dot{\phi}_{i+1}\right)^2\ . \tag{13.142}$$

Nonrotating dampers with damping coefficients c_d have also been included into the system.

By adding the kinetic and potential energies of the disc and of the pendulums, performing the relevant derivatives and remembering that because the pendulums are equispaced,

$$\textstyle\sum_{i=1}^{n}\cos^2(\theta_i)=\sum_{i=1}^{n}\cos^2(\theta_i)=\frac{n}{2}\ ,$$

$$\textstyle\sum_{i=1}^{n}\sin(\theta_i)\cos(\theta_i)=\sum_{i=1}^{n}\sin(\theta_i)=\sum_{i=1}^{n}\cos(\theta_i)=0\ ,$$

the first two equations, those related with ϕ_X and ϕ_y, are

$$\left\{\begin{aligned}&\ddot{\phi}_y\left[J_t+\tfrac{nm}{2}(r+l)^2\right]-\Omega\dot{\phi}_X\left[J_p+nm(r+l)^2\right]+c_d\dot{\phi}_y+k_d\phi_y+\\ &\qquad -ml(r+l)\textstyle\sum_{i=1}^{n}\left[\ddot{\phi}_i+\Omega^2\phi_i\right]\cos(\theta_i)=0\ ,\\ &\ddot{\phi}_X\left[J_t+\tfrac{nm}{2}(r+l)^2\right]+\Omega\dot{\phi}_y\left[J_p+nm(r+l)^2\right]+c_d\dot{\phi}_X+k_d\phi_X+\\ &\qquad +ml(r+l)\textstyle\sum_{i=1}^{n}\left[\ddot{\phi}_i+\Omega^2\phi_i\right]\sin(\theta_i)=0\ ;\end{aligned}\right. \tag{13.143}$$

i.e.,

$$\begin{cases} \ddot{\phi}_y J_{t_T} - \Omega\,\dot{\phi}_X J_{p_T} + c_d\,\dot{\phi}_y + k_d\,\phi_y - ml\,(r+l)\cos(\Omega t)\sum_{i=1}^{n}\Big[\ddot{\phi}_i + \\ \qquad +\Omega^2\phi_i\Big]\cos(\psi_i) + ml\,(r+l)\sin(\Omega t)\sum_{i=1}^{n}\Big[\ddot{\phi}_i + \Omega^2\phi_i\Big]\sin(\psi_i) = 0\ , \\ \ddot{\phi}_X J_{t_T} + \Omega\,\dot{\phi}_y J_{p_T} + c_d\,\dot{\phi}_X + k_d\,\phi_X + ml\,(r+l)\sin(\Omega t)\sum_{i=1}^{n}\Big[\ddot{\phi}_i + \\ \qquad +\Omega^2\phi_i\Big]\cos(\psi_i) + ml\,(r+l)\cos(\Omega t)\sum_{i=1}^{n}\Big[\ddot{\phi}_i + \Omega^2\phi_i\Big]\sin(\psi_i) = 0\ , \end{cases} \tag{13.144}$$

where

$$J_{t_T} = J_t + \frac{nm}{2}(r+l)^2\ , \qquad J_{p_T} = J_p + nm\,(r+l)^2\ . \tag{13.145}$$

By remembering the coordinate transformation expressed by Equation (13.89), it follows that

$$\begin{cases} \ddot{\phi}_y J_{t_T} - \Omega\,\dot{\phi}_X J_{p_T} + c_d\,\dot{\phi}_y + k_d\,\phi_y - \frac{mnl}{2}(r+l)\left(\ddot{u}_1 + \Omega^2 u_1\right)\cos(\Omega t) + \\ \qquad + \left(\ddot{u}_2 + \Omega^2 u_2\right)\sin(\Omega t) = 0\ , \\ \ddot{\phi}_X J_{t_T} + \Omega\,\dot{\phi}_y J_{p_T} + c_d\,\dot{\phi}_X + k_d\,\phi_X + \frac{mnl}{2}(r+l)\left(\ddot{u}_1 + \Omega^2 u_1\right)\sin(\Omega t) + \\ \qquad + \frac{mnl}{2}(r+l)\left(\ddot{u}_2 + \Omega^2 u_2\right)\cos(\Omega t) = 0\ . \end{cases} \tag{13.146}$$

The following n equations, those related with degrees of freedom ϕ_i, are

$$\begin{aligned} &\ddot{\phi} - \ddot{\phi}_y\cos(\theta_i)\left(\tfrac{r}{l}+1\right) + \ddot{\phi}_X\left(\tfrac{r}{l}+1\right)\sin(\theta_i) + 2\,\Omega\,\dot{\phi}_y\left(\tfrac{r}{l}+1\right)\sin(\theta_i) + \\ &\quad +2\,\Omega\,\dot{\phi}_X\left(\tfrac{r}{l}+1\right)\cos(\theta_i) + \phi_i\left[\Omega^2\left(\tfrac{r}{l}+1\right) + \tfrac{k_p}{ml^2}\right] + \\ &+\tfrac{1}{ml^2}(c_p + 2c_i)\,\dot{\phi}_i - \tfrac{c_i}{ml^2}\dot{\phi}_{i-1} - \tfrac{c_i}{ml^2}\dot{\phi}_{i+1} = 0 \qquad \text{for } i = 1,\ ...,\ n\ ; \end{aligned} \tag{13.147}$$

i.e.,

$$\begin{aligned} &(\delta+1)\left[-\left(\ddot{\phi}_y - 2\,\Omega\,\dot{\phi}_X\right)\cos(\psi_i) + \left(\ddot{\phi}_X + 2\,\Omega\,\dot{\phi}_y\right)\sin(\psi_i)\right]\cos(\Omega t) + \\ &+(\delta+1)\left[\left(\ddot{\phi}_y - 2\,\Omega\,\dot{\phi}_X\right)\sin(\psi_i) + \left(\ddot{\phi}_X + 2\,\Omega\,\dot{\phi}_y\right)\cos(\psi_i)\right]\sin(\Omega t) + \\ &+\ddot{\phi} + \phi_i\left[\Omega^2(\delta+1) + \tfrac{k_p}{ml^2}\right] + \tfrac{1}{ml^2}(c_p + 2c_i)\,\dot{\phi}_i - \tfrac{c_i}{ml^2}\dot{\phi}_{i-1} - \tfrac{c_i}{ml^2}\dot{\phi}_{i+1} = 0\ , \end{aligned} \tag{13.148}$$

where, as usual, $\delta = r/l$.

These n equations can be combined linearly as seen for the case of in-plane oscillations (13.95).

The first one of the combined equations decouples from all the others and describes the motion with all pendulums oscillating in phase

$$\ddot{u}_0 + \frac{c_p}{ml^2}\dot{u}_0 + u_0\left[\Omega^2(\delta+1) + \frac{k_p}{ml^2}\right] = 0\ . \tag{13.149}$$

Actually the complete uncoupling is because the disc was assumed to be fixed in axial direction: In a general case, the equation for u_0 is coupled with the axial dynamics of the system.

The following two equations are coupled with the first two equations (those dealing with the motion of the disc):

$$
\begin{array}{c}
(\delta+1)\left[-\left(\ddot{\phi}_y - 2\,\Omega\,\dot{\phi}_X\right)\right]\cos(\Omega t) + (\delta+1)\left[\left(\ddot{\phi}_X + 2\,\Omega\,\dot{\phi}_y\right)\right]\sin(\Omega t) + \\
+\ddot{u}_1 + u_1\left[\Omega^2\,(\delta+1) + \frac{k_p}{ml^2}\right] + \frac{c}{ml^2}\dot{u}_1 = 0\;, \\
(\delta+1)\left[\left(\ddot{\phi}_X + 2\,\Omega\,\dot{\phi}_y\right)\right]\cos(\Omega t) + (\delta+1)\left[\left(\ddot{\phi}_y - 2\,\Omega\,\dot{\phi}_X\right)\right]\sin(\Omega t) + \\
+\ddot{u}_2 + u_2\left[\Omega^2\,(\delta+1) + \frac{k_p}{ml^2}\right] + \frac{c}{ml^2}\dot{u}_2 = 0\;,
\end{array}
\tag{13.150}
$$

where

$$
c = c_p + 4c_i\sin^2\left(\frac{\Delta\psi}{2}\right)\;.
$$

All of the following equations do not contain either ϕ_X or ϕ_y. They are all of the type

$$
\ddot{u}_i + u_i\left[\Omega^2\,(\delta+1) + \frac{k_p}{ml^2}\right] + \frac{c^*}{ml^2}\dot{u}_i = 0 \qquad \text{for } i = 3,\ ...,\ n-1\;, \tag{13.151}
$$

where c^* is a damping coefficient of the type of those appearing in Equation (13.134), showing that the damping located between the pendulums is more important in the equations with a high value of i. As they are uncoupled with the other ones, these equations will not be dealt with any further.

The disc-pendulums interaction can thus be studied using a set of four linear equations (with periodic coefficients) in the coordinates ϕ_X, ϕ_y, u_1, and u_2. They can be transformed into a set of constant coefficients equations by resorting to a further change of coordinates:

$$
\left\{
\begin{array}{l}
\phi_{y1} = -u_1\cos(\Omega t) + u_2\sin(\Omega t)\;, \\
\phi_{X1} = u_1\sin(\Omega t) + u_2\cos(\Omega t)\;,
\end{array}
\right.
\tag{13.152}
$$

obtaining

$$
\left\{
\begin{array}{l}
\ddot{\phi}_y J_{t_T} - \Omega\,\dot{\phi}_X J_{p_T} + \ddot{\phi}_{y1} J_p - 2\Omega\,\dot{\phi}_{X1} J_p + c_d\,\dot{\phi}_y + k_d\,\phi_y = 0\;, \\
\ddot{\phi}_X J_{t_T} + \Omega\,\dot{\phi}_y J_{p_T} + \ddot{\phi}_{X1} J_p + 2\Omega\,\dot{\phi}_{y1} J_p + c_d\,\dot{\phi}_X + k_d\,\phi_X = 0\;, \\
(\delta+1)\left(\ddot{\phi}_y - 2\,\Omega\,\dot{\phi}_X\right) + \ddot{\phi}_{y1} - 2\,\Omega\,\dot{\phi}_{X1} + \frac{c}{ml^2}\left(\dot{\phi}_{y1} - \Omega\,\phi_{X1}\right) + \\
\quad + \left[\Omega^2\delta + \frac{k_p}{ml^2}\right]\phi_{y1} = 0\;, \\
(\delta+1)\left(\ddot{\phi}_X + 2\,\Omega\,\dot{\phi}_y\right) + \ddot{\phi}_{X1} + 2\,\Omega\,\dot{\phi}_{y1} + \frac{c}{ml^2}\left(\dot{\phi}_{X1} + \Omega\,\phi_{y1}\right) + \\
\quad + \left[\Omega^2\delta + \frac{k_p}{ml^2}\right]\phi_{X1} = 0\;.
\end{array}
\right.
\tag{13.153}
$$

where $J_p = nml\,(r+l)/2$.

This set of equations can be further simplified by resorting to complex coordinates

$$\begin{cases} \phi_d = \phi_y - i\,\phi_X \, , \\ \phi_p = \phi_{y1} - i\,\phi_{X1} \, . \end{cases} \tag{13.154}$$

By adding the first equation to the second multiplied by $-i$ and the third one to the fourth multiplied by $-i$, it follows that

$$\begin{cases} \ddot{\phi}_d J_{t_T} - i\Omega\,\dot{\phi}_d J_{p_T} + \ddot{\phi}_p J_p - 2i\Omega\,\dot{\phi}_p J_p + c_d\,\dot{\phi}_d + k_d\,\phi_d = 0 \, , \\ (\delta+1)\left(\ddot{\phi}_d - 2i\Omega\,\dot{\phi}_d\right) + \ddot{\phi}_p - 2i\Omega\,\dot{\phi}_p + \\ \qquad + \frac{c}{ml^2}\left(\dot{\phi}_p - i\,\Omega\,\phi_p\right) + \left[\Omega^2\delta + \frac{k_p}{ml^2}\right]\phi_p = 0 \, . \end{cases} \tag{13.155}$$

By introducing, together with the already mentioned parameter δ, the nondimensional parameters

$$\alpha = \frac{J_p}{J_{t_T}} = \frac{nml\,(r+l)}{2J_t + nm\,(r+l)^2} \, , \quad \beta = \frac{J_{p_T}}{J_{t_T}} \, , \tag{13.156}$$

$$\zeta_d = \frac{c_d}{2\sqrt{k_d J_{t_T}}} \, , \quad \zeta_p = \frac{c\sqrt{J_{t_T}}}{2ml^2\sqrt{k_d}} \, ,$$

and the natural frequencies $\omega_1 = \sqrt{k_d/J_{t_T}}$ and $\omega_p = \sqrt{k_p/ml^2}$, the equation of motion can be written in the form

$$\begin{bmatrix} 1 & \alpha \\ \alpha & \frac{\alpha}{1+\delta} \end{bmatrix} \begin{Bmatrix} \ddot{\phi}_d \\ \ddot{\phi}_p \end{Bmatrix} + \left(2\omega_1 \begin{bmatrix} \zeta_d & 0 \\ 0 & \frac{\alpha}{1+\delta}\zeta_p \end{bmatrix} + \right.$$

$$\left. -i\Omega \begin{bmatrix} \beta & 2\alpha \\ 2\alpha & \frac{2\alpha}{1+\delta} \end{bmatrix}\right) \begin{Bmatrix} \dot{\phi}_p \\ \dot{\phi}_d \end{Bmatrix} + \left(\omega_1^2 \begin{bmatrix} 1 & 0 \\ 0 & \frac{\alpha}{\omega_1^2(1+\delta)}\left(\delta\Omega^2 + \omega_p^2\right) \end{bmatrix} + \right.$$

$$\left. + -i\Omega\omega_1 \begin{bmatrix} 0 & 0 \\ 0 & \frac{2\alpha}{1+\delta}\zeta_p \end{bmatrix}\right) \begin{Bmatrix} \phi_d \\ \phi_p \end{Bmatrix} = \mathbf{0} \, . \tag{13.157}$$

By introducing the nondimensional speed, Laplace variable, and frequency

$$\Omega^* = \frac{\Omega}{\omega_1} \, , \qquad s^* = \frac{s}{\omega_1} \, , \qquad \omega_p^* = \frac{\omega_p}{\omega_1} \, , \tag{13.158}$$

a nondimensional response can be obtained from the characteristic equation

$$\det \begin{bmatrix} s^{*2} + s^*\left(2\zeta_d - i\Omega^*\beta\right) + 1 & s^{*2}\alpha - 2i\Omega^* s^*\alpha \\ s^{*2}\alpha - 2i\Omega^* s^*\alpha & \frac{\alpha}{1+\delta} f(s^*) \end{bmatrix} = 0 \, , \tag{13.159}$$

where

$$f(s^*) = s^{*2} + 2s^*\left(\zeta_p - i\Omega^*\right) + \Omega^{*2}\delta + \omega_p^{*2} - 2i\Omega^*\zeta_p \, .$$

In this case, there is no value of δ , causing the stiffness matrix to be singular. In this condition, there is no need in distinguishing between the cases of short and long pendulums.

As usual, the uncoupled problem can be solved before attempting to solve Equation (13.159). The first equation is

$$s^{*2} + s^* \left(2\zeta_d - i\Omega^* \beta\right) + 1 = 0 \; . \tag{13.160}$$

It coincides with the equation of motion of an uncoupled system with four degrees of freedom in which only nonrotating damping is present. Its solutions are

$$s^* = -\zeta_d + i\Omega^* \frac{\beta}{2} \pm \sqrt{\left(\zeta_d - i\Omega^* \frac{\beta}{2}\right)^2 - 1} \; , \tag{13.161}$$

and no instability is obtained.

The second equation

$$s^{*2} + 2s^* \left(\zeta_p - i\Omega^*\right) + \Omega^{*2}\delta + \omega_p^{*2} - 2i\Omega^*\zeta_p = 0 \tag{13.162}$$

yields the solutions

$$s^* = -\zeta_p + i\Omega^* \pm \sqrt{\zeta_p^2 - \Omega^{*2}\left(1 + \delta\right) - \omega_p^{*2}} \; . \tag{13.163}$$

If damping and elastic restoring forces are neglected, Equation (13.163) reduces to

$$s^* = i\Omega^* \left(1 \pm \sqrt{1 + \delta}\right) \; , \tag{13.164}$$

which is consistent with what has been obtained earlier on rotating pendulums: the natural frequency for out-of-plane motion, referred to the rotating frame, of a single rotating pendulum is [Equation (13.14)]

$$\omega_{p_{rf}} = \Omega \sqrt{\frac{r + l}{l}} = \Omega \sqrt{1 + \delta} \, . \tag{13.165}$$

As in the previous case, the global motion of the pendulum row can be seen in the rotating frame as a forward-moving wave, propagating at an angular velocity $\omega_{p_{rf}}$, plus a backward-moving wave, propagating at an angular velocity $-\omega_{p_{rf}}$. The forward-moving wave is seen in the fixed frame as a wave propagating in forward direction with a speed $\Omega + \omega_{p_{rf}} = \Omega\left(1 + \sqrt{\delta}\right)$, which coincides with the solution with (+) in Equation (13.163). What is different from the in-plane oscillations of the pendulums, in the present case, the backward wave is always seen in the fixed frame as a wave propagating in backward direction. Its speed is $\Omega - \omega_{p_{rf}} = \Omega\left(1 - \sqrt{\delta + 1}\right)$, coinciding with the solution with (−) in Equation (13.163).

The nondimensional Campbell diagram a for the case of long pendulums with $\alpha = 0.1$ and $\delta = 0.2$ are shown in Figure 13.15 for two different values

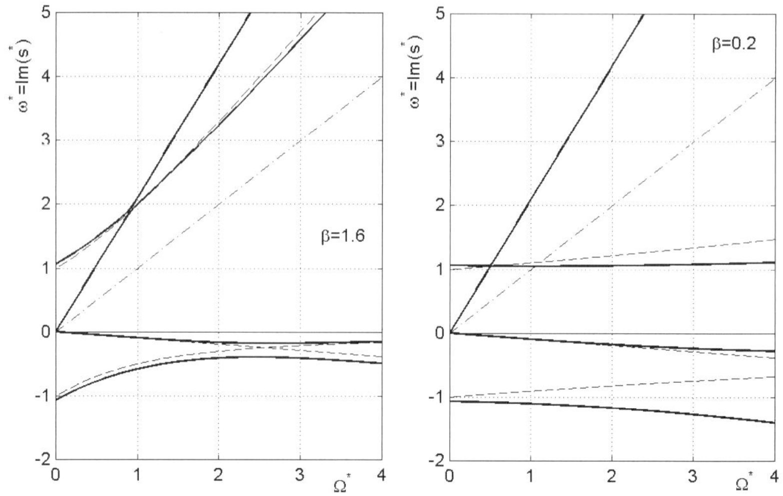

FIGURE 13.15. Campbell diagrams for the case of the out-of-plane oscillations of a row of undampd pendulums with $\alpha = 0.1$, $\delta = 0.2$, and $\beta = 1.6$ and $\beta = 0.2$. Note that no field of instability is present.

of β, namely, 1.6 (disc rotor) and 0.2 (long rotor; such a low value has been used, even though it has little sense for a rotor with many pendulums, as a limiting case). The nondimensional parameters are the same (except for β) as those used in Figure 13.12(a).

The decay rate plots are not shown because the real part of s is identically zero: No instability range is present in this case.

Remark 13.10 *Although in the case of in-plane oscillations of the row of pendulums an instability range was found in the case of long pendulums owing to the coupling of the motion of the pendulums and of the supporting disc, no such phenomenon was found for the case of out-of-plane oscillations.*

Some numerical results for the complete system, with restoring force and damping, are shown in Figure 13.16: The nondimensional Campbell diagram and the decay rate plot reported deal with the same case in Figure 13.15, with $\alpha = 0.1$, $\delta = 0.2$, and two values of β, 1.6 and 0.2. The restoring spring has a stiffness leading to $\omega_p^* = \sqrt{2}$, and the damping parameters are $\zeta_d = 0.05$ and $\zeta_p = 0.1$.

From the figure, it is clear that although rotating damping is much higher than nonrotating damping, no field of instability is present.

Remark 13.11 *Also in the case of out-of-plane oscillations, rotating damping of the pendulums does not trigger instability. In this case, instability*

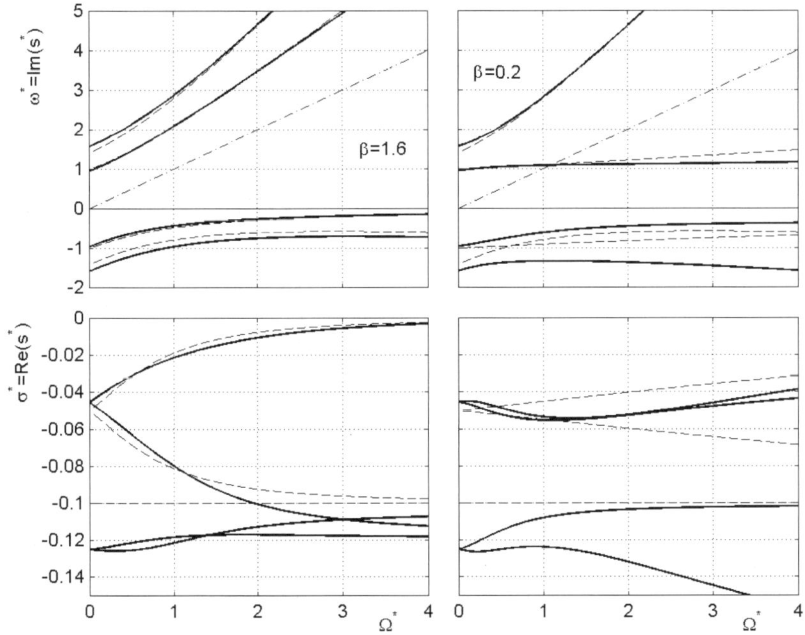

FIGURE 13.16. Campbell diagram and the decay rate plot for the out-of-plane oscillations of a row of damped pendulums with $\alpha = 0.1$, $\delta = 0.2$ and two values of β, 1.6 and 0.2. The other data are $\omega_p^* = \sqrt{2}$, $\zeta_d = 0.05$ and $\zeta_p = 0.1$.

cannot be caused by the interaction between the motion of the pendulums and of the supports, so that the system is always stable.

13.6 Interaction between the dynamics of the blades and the dynamics of the shaft

As the study of rotating pendulums has shown, the dynamics of an array of blades that may surround a rotor can interact with the dynamics of the rotor as a whole, to the point that instability ranges may occur. In particular, the study of the behavior of pendulums suggests that in-plane vibration is more dangerous, from this viewpoint, than out-of-plane vibration.

Unfortunately, there is no simple model to substantiate these statements. The present section will be devoted to the construction and the solution of a few highly idealized numerical models, attempting to extract some general rules. These models will be based on the finite element method, using the formulation that will be described in some details in Section 15.2. In a way that is similar to what has been done for the study of pendulums arrays, the dynamic behavior of the whole bladed array will be considered:

for each mode of the single blades, a number n (where n is the number of blades in the array) of harmonics in the angle will be considered. The zeroth-order harmonics [like the one shown in Figures 13.9(a) or 13.10(a)] couples with the torsional and axial dynamics of the shaft, whereas the first-order harmonics [Figure 13.9(b) and (c) or Figure 13.10(b) and (c)] are coupled with the flexural behavior. Higher order harmonics do not affect the behavior of the rotor as a whole, and they will not be considered here.

As the number of parameters is high, no attempt has been done to extract nondimensional parameters. Consider a system made by a rigid disc, suspended on a spring system, with a number of prismatic blades. The data are summarized in the following table:

disc	inertial properties m, J_t, J_p	8 kg, 2.6 kg m^2, 4 kg m^2
	outer radius r_o	100 mm
	suspension stiffness k_{rad}, k_{bend}	10^7 N/m, 10^7 Nm/rad
blades	number n	10
	square cross section	10 mm × 10 mm
	length l	500 mm
material	E	2.1× 10^{11} N/m^3
	ρ	7810 kg/m^3

As the length of the blades is greater than the radius of the disc, they can be defined as *long blades*, to use a term similar to that used for rotating pendulums. In the case of pendulums, this leads to the existence of a field of instability.

The model is made by a rigid mass element, a spring element, and ten row-of-blades elements. The uncoupled dynamics is studied in Figure 13.17. If the blades are rigid bodies, the behavior is that of an uncoupled four degrees-of-freedom (real) system with the following inertial properties: $m =$ 11.92 kg, $J_t = 2.88$ kg m^2, and $J_p = 4.57$ kg m^2. The results are plotted as dashed lines in the Campbell diagram of Figure 13.17; owing to uncoupling, the translational modes are those of a Jeffcott rotor.

The full lines in the same figures are related to the dynamics of the flexible blades on a rigid support. As the cross section of the blades is square, at standstill the in-plane natural frequencies coincide with the out-of-plane ones. As soon as the spin speed is nonvanishing, however, they differentiate and the curves related to the various modes split into two lines. The out-of-plane natural frequencies (particularly the first one) increase more sharply with the speed than the in-plane ones. Like in the case of long pendulums, the lowest backward mode in the rotating frame becomes a forward mode in the fixed frame. The simplified blade dynamics, computed using the natural frequency at standstill and then through Equation (13.73), is reported in the same figure with dot-and-dash curves. Note that the results obtained through Equation (13.73) are remarkably close to those obtained through the FEM.

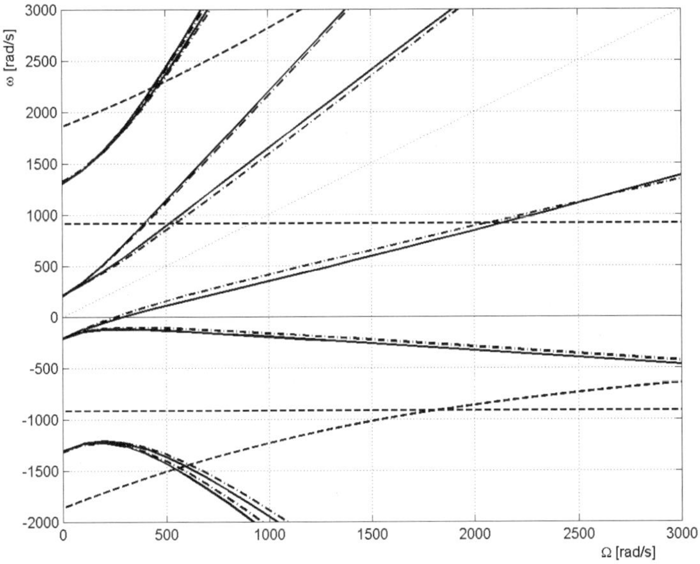

FIGURE 13.17. Uncoupled dynamics of a bladed disc on elastic supports. Full lines: dynamics of the blades on a stiff support; dot and dash curves: same as before but computed through an approximated formula [Equation (13.73)]; dashed lines: dynamics of the rigid system on elastic supports.

The coupled dynamics is shown in Figure 13.18. The curves related to the disc and blade dynamics interact with each other, and the first backward mode (actually it is backward only in the rotating frame, in the inertial frame, it is a forward mode) of the bladed array interacts with the forward translational mode of the rigid system on elastic supports giving way to a field of instability. The latter is very clearly evidenced by the decay rate plot.

Consider now the case of blades with rectangular cross section, so that there are no coincident natural frequencies at standstill. The cross section of the blades in the model was modified, keeping the cross-sectional area constant: A rectangle 2.5 mm $\times$ 40 mm is assumed. A different value of the material density, namely, $\rho = 4,000$ kg/m^3, has been assumed.

The dynamic behavior is then much influenced by the angle the blade makes with the axial direction: If $\psi = 90°$ (see Figure 13.7 for the definition of ψ), the blade is aligned along the axial direction (lowest natural frequency is in-plane), whereas if $\psi = 0$, the blade is perpendicular to the axial direction.

The Campbell diagram and the decay rate plot for the two limiting cases, $\psi = 90°$ and $\psi = 0$, are reported in Figure 13.19. The instability range shifts toward higher speeds when ψ decreases from 90° [Figure 13.19(a) and (b)] to 0 [Figure 13.19(c) and (d)].

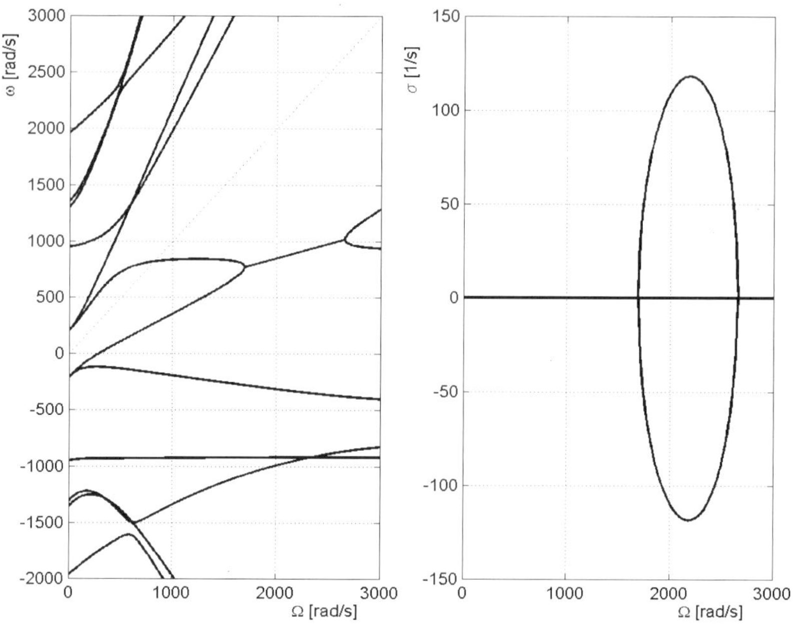

FIGURE 13.18. Coupled dynamics of a bladed disc on elastic supports: Campbell diagram and decay rate plot.

The speeds at which the field of instability starts and ends are reported in Figure 13.20 as a function of angle ψ.

From the plots of Figure 13.19, it is clear that in the case in which the cross section of the blades is aligned with the axial direction, the frequency of the in-plane modes are lower than those of the out-of-plane ones and the relevant branches in the Campbell diagram do not cross. The opposite occurs when $\psi = 90°$.

To investigate the effect of damping in blades, the same model was modified by assigning a loss factor to the various elements. All rotating elements were given a very high damping ($\eta = 0.1$), whereas the only nonrotating element, the spring supporting the system, was given a very low damping ($\eta = 0.005$). The results for the case $\psi = 90°$ are reported in Figure 13.21.

As expected, the effect of damping on the Campbell diagram is not large, and it influences deeply the decay rate plot. In a way that is similar to what has been seen for the rotating pendulum, the damping associated with the blades does not have a destabilizing effect, although being a form of rotating damping. The instability range is only slightly reduced by the presence of damping, but all blade vibrations have a decay rate that is negative.

Note that to show the danger of instability, a speed range well in excess to practical limits has been considered in this example. The stresses at the root of the blades are reported in Figure 13.22: It is easy to see that they

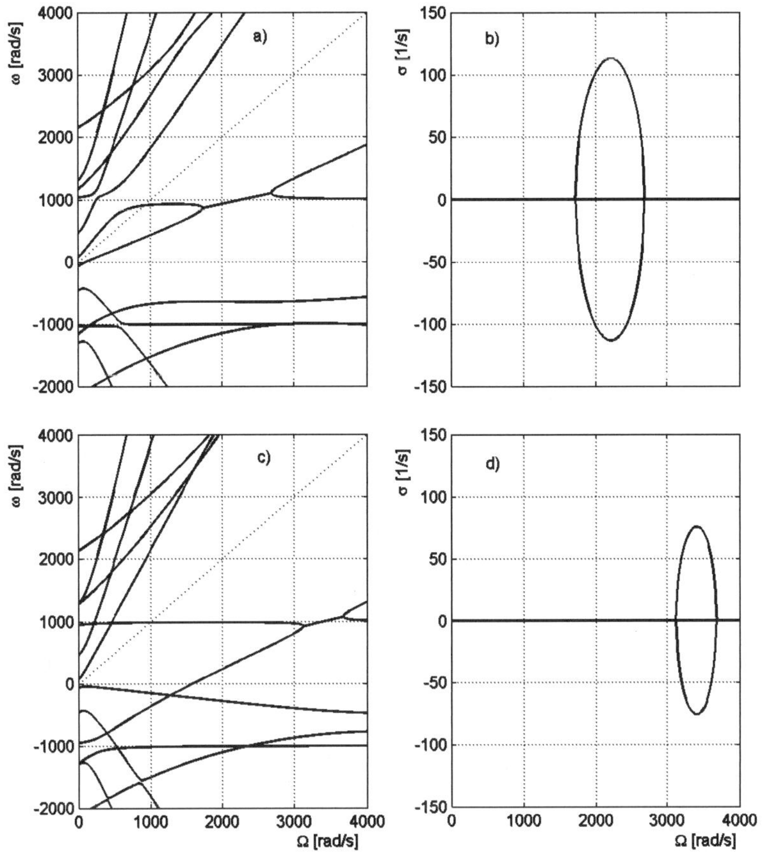

FIGURE 13.19. Bladed disc with blades with rectangular cross section. Campbbell diagram and decay rate plot for the two limiting cases, $\psi = 90°$ (a) and (b) and $\psi = 0$ (c) and (d).

are way too large even if the material density has been lowered. In an actual case, this bladed disc would be limited to speeds well lower than 1000 rad/s and no instability would have been present.

Remark 13.12 *The damping associated with the blades, although rotating, does not trigger instability even for modes in the supercritical range.*

Remark 13.13 *The instability threshold caused by the blades is very often in excess of the maximum speed, which can be reached for strength consideration and then is more of academic than of practical interest. However, with the tendency toward high speeds made possible by stronger and lighter materials, stability issues cannot be dismissed lightly.*

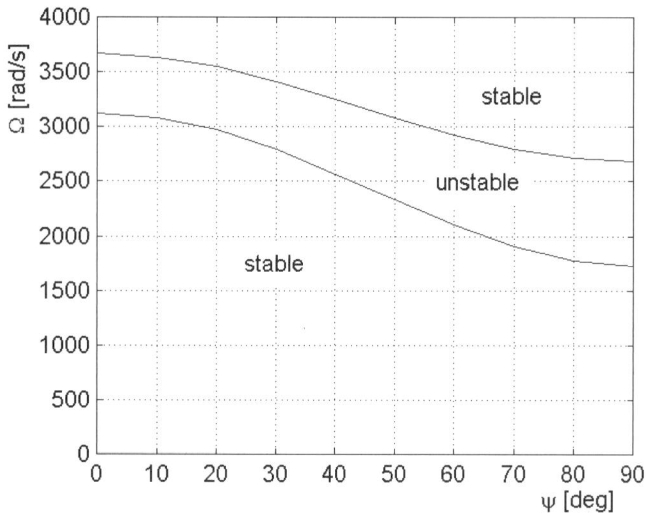

FIGURE 13.20. Bladed disc with blades with rectangular cross section. Speeds at which the instability range starts and ends as functions of angle ψ.

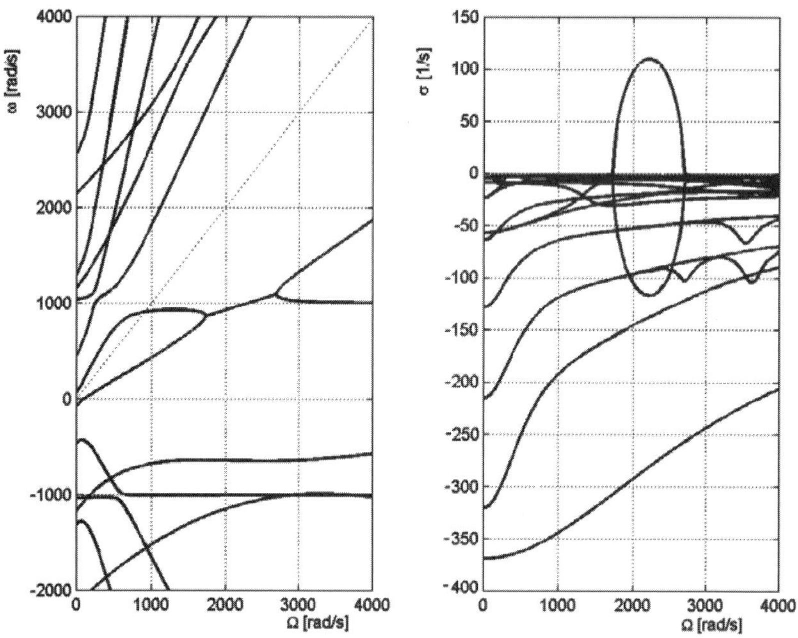

FIGURE 13.21. Campbbell diagram and decay rate plot for the bladed disc studied in Figure 13.19 (a) and (b) ($\psi = 90°$), but with hysteretic damping added. Rotating damping with loss factor $\eta = 0.1$; nonrotating damping with $\eta = 0.005$.

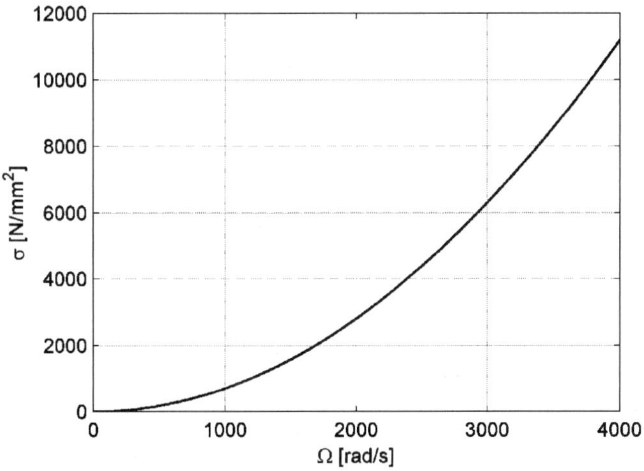

FIGURE 13.22. Centrifugal stressing at the root of the blades as a function of the speed.

14

Dynamics of rotating discs and rings

Many rotors, particularly many turbine rotors, are provided with thin, large diameter parts, sometimes carrying one or more row of blades. If the ratio between the outer diameter and the thickness is large, they are usually referred to as discs.

In elementary rotordynamics, the discs are considered as rigid bodies attached to the otherwise beam-like rotor. There are, however, cases in which the dynamics of the discs becomes important, and the rigid-body assumption is too rough for a detailed dynamic analysis. The simplest model for a thin rotating disc is the rotating membrane.

14.1 Rotating membranes

A membrane is a very thin two-dimensional object, so thin that its bending stiffness is vanishingly small and relies to in-plane stretching to maintain its shape. In a rotating membrane, the in-plane stretching is supplied by the centrifugal field. The dynamics of rotating membranes was studied by Southwell and Lamb in 1921 [6].

Consider a circular membrane with constant thickness rotating about its axis (Figure 14.1). The dynamic equilibrium equation for displacements in axial direction of the element of area located in the position (r, θ) is

$$\rho\frac{\partial^2 w}{\partial t^2} = \frac{1}{r}\frac{\partial}{\partial r}\left(r\sigma_r\frac{\partial w}{\partial r}\right) + \frac{\sigma_c}{r^2}\frac{\partial^2 w}{\partial \theta^2}\,, \tag{14.1}$$

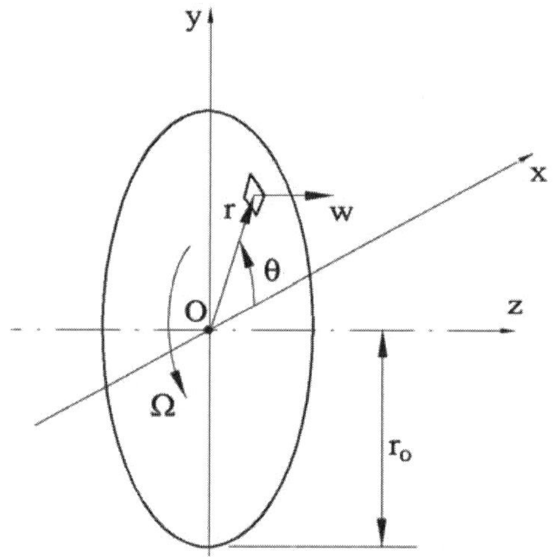

FIGURE 14.1. Rotating membrane. Sketch and reference frames.

where the radial and circumferential stresses σ_r and σ_c are functions of the radius only.

By introducing the nondimensional radial coordinate

$$\chi = r_i/r_o \ ,$$

Equation (A.53) yields

$$r_o^2\rho\frac{\partial^2 w}{\partial t^2} = \sigma_r\frac{\partial^2 w}{\partial \chi^2} + \left(\frac{\partial \sigma_r}{\partial \chi} + \frac{1}{\chi}\sigma_r\right)\frac{\partial w}{\partial \chi} + \frac{\sigma_c}{\chi^2}\frac{\partial^2 w}{\partial \theta^2} \ . \tag{14.2}$$

The solution of Equation (14.2) is of the type

$$w = u(\chi)\cos(m\theta)\cos(\omega t) \ , \tag{14.3}$$

where m is an integer.

By introducing the Solution (14.3) into Equation (14.2), the latter yields

$$\sigma_r\frac{d^2 u}{d\chi^2} + \left(\frac{d\sigma_r}{d\chi} + \frac{1}{\chi}\sigma_r\right)\frac{du}{d\chi} - \left(m^2\frac{\sigma_c}{\chi^2} - \rho\omega^2 r_o^2\right)u = 0 \ . \tag{14.4}$$

The circumferential and radial stresses can be caused by different causes, like centrifugal stressing, temperature gradients in the membrane, and many other ones. If the stresses are only caused by rotation, their expression is [61]

$$\left\{\begin{array}{l} \sigma_r = \dfrac{\rho\Omega^2 r_o^2(3+\nu)}{8}\left(1+\gamma^2+\frac{\gamma^2}{\chi^2}-\chi^2\right) \ , \\ \sigma_c = \dfrac{\rho\Omega^2 r_o^2(3+\nu)}{8}\left(1+\gamma^2+\frac{\gamma^2}{\chi^2}-\frac{1+3\nu}{3+\nu}\chi^2\right) \ , \end{array}\right. \tag{14.5}$$

where ν is the Poisson's ratio of the material and

$$\gamma = r_i/r_o .$$

The stress distribution expressed by Equation (14.5) has been obtained for a membrane with a central hole, which does not exchange forces in radial direction with the shaft [61].

By introducing Equation (14.5) into Equation (14.4), it follows

$$\left(1+\gamma^2+\frac{\gamma^2}{\chi^2}-\chi^2\right)\frac{d^2u}{d\chi^2}+\left(\frac{1+\gamma^2}{\chi^2}-\frac{\gamma^2}{\chi^3}-3\chi\right)\frac{du}{d\chi}+ \qquad (14.6)$$

$$-\left[m^2\left(\frac{1+\gamma^2}{\chi^2}+\frac{\gamma^2}{\chi^4}-\frac{1+3\nu}{3+\nu}\right)-\frac{8}{3+\nu}\frac{\omega^2}{\Omega^2}\right]u=0 .$$

The natural frequency of the membrane is then proportional to the spin speed, i.e.,

$$\omega = \beta\Omega , \qquad (14.7)$$

where β is a constant.

If there is no central hole, Equations (14.5) and (14.6) reduce to

$$\begin{cases} \sigma_r = \dfrac{\rho\Omega^2 r_o^2(3+\nu)}{8}\left(1-\chi^2\right) \\ \sigma_c = \dfrac{\rho\Omega^2 r_o^2(3+\nu)}{8}\left(1-\frac{1+3\nu}{3+\nu}\chi^2\right) \end{cases} \qquad (14.8)$$

$$\left(1-\chi^2\right)\left(\frac{d^2u}{d\chi^2}+\frac{1}{\chi}\frac{du}{d\chi}\right)-2\chi\frac{du}{d\chi}+\left[-m^2\left(\frac{1}{\chi^2}-\frac{1+3\nu}{3+\nu}\right)+\frac{8}{3+\nu}\frac{\omega^2}{\Omega^2}\right]u=0 . \qquad (14.9)$$

The solution of Equation (14.9) can be assumed to be

$$u=\sum_0^\infty C_i\chi^i . \qquad (14.10)$$

Operating in the same way as seen for the rotating string, it follows that

$$C_{i+2}=C_i\frac{i(i+2)-m^2\frac{1+3\nu}{3+\nu}-\beta^2\frac{8}{3+\nu}}{(i+2)^2-m^2} . \qquad (14.11)$$

Southwell demonstrated that the solution of Equation (14.9) is a hypergeometric series and solved the equation for the natural frequencies, obtaining [6]

$$\beta_i=\sqrt{\frac{3+\nu}{8}(m+2n)(m+2n+2)-m^2\frac{1+3\nu}{8}} , \qquad (14.12)$$

where m is the number of nodal diameters and n is the number of nodal circles. The Solution (14.12) has been obtained by assuming that only a

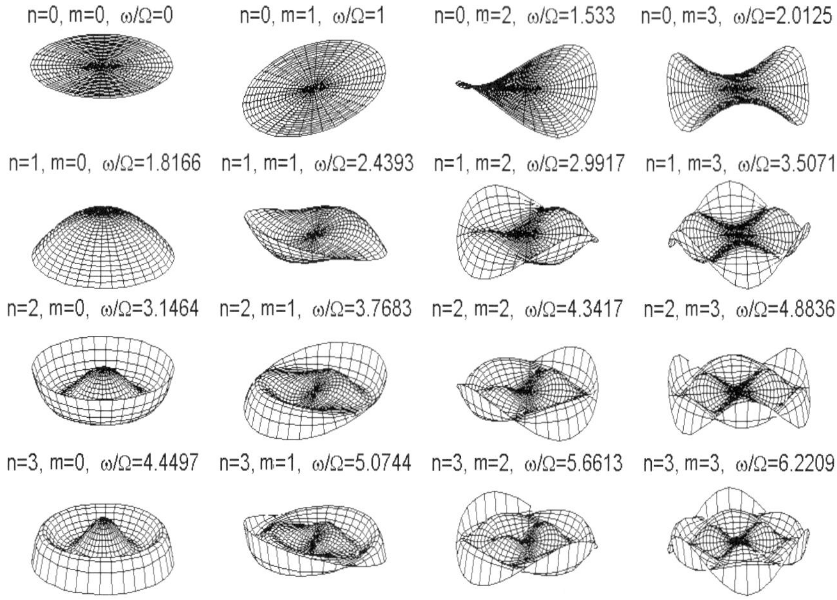

FIGURE 14.2. Rotating membrane: values of $\beta = \omega/\Omega$ and mode shapes for a number of values of n and m.

limited number of terms of the series (14.10) do not vanish. The result is that only even terms are present if m is even and odd terms if m is odd; the first term is that in χ^m, and the last one is that in χ^{m+2n}. The values of $\beta = \omega/\Omega$ for a number of values of n and m and the corresponding mode shapes are reported in Figure 14.2.

The solution obtained by Southwell refers to a free rotating unpierced membrane. The mode with $m = 0$ and $n = 0$ is a rigid-body mode in which the membrane translates along z-axis and its natural frequency vanishes. In all modes with $m = 0$, the center of the membrane undergoes a displacement. This detracts from the applicability of this solution, but it must be noted that it is impossible to consider a membrane constrained in its center: a force applied to the center of a membrane causes a singularity to be present. Note also that in all modes with $m > 0$, the center has zero displacement: These modes apply to a constrained membrane as well.

The mode with $m = 1$ and $n = 0$ is a rigid-body mode, but its natural frequency does not vanish owing to rotation.

The vibration of the membrane so obtained must be regarded as a vibration occurring in a rotating reference frame; i.e., the nodal lines are fixed to the latter and rotate, with respect to the inertial frame, with the spin speed Ω. However, the vibration of the membrane can be seen as the superimposition of two waves that travel in forward and backward direction

n	$m=0$	$m=1$	$m=2$	$m=3$	$m=4$
0	0	2.000	3.533	5.013	6.470
1	1.817	3.439	4.992	6.507	8.000
2	3.146	4.768	6.342	7.884	9.404
3	4.450	6.074	7.661	9.221	10.760
4	5.745	7.372	8.968	10.540	12.093
5	7.036	8.665	10.267	11.849	13.413
6	8.325	9.955	11.563	13.151	14.724

TABLE 14.1. Values of ratios $\beta = \omega_{ff}/\Omega$ for forward modes with a different number of diametral (m) and radial (n) nodes.

with speed ω/m. This can be shown by noting that Solution (14.3) can be written as

$$w = u(\chi)\cos(m\theta)\cos(\omega t) = \frac{1}{2}u(\chi)\left\{\cos\left[m\left(\theta + \frac{\omega}{m}t\right)\right] + \cos\left[m\left(\theta - \frac{\omega}{m}t\right)\right]\right\} . \tag{14.13}$$

The wave velocity $\pm\omega/m$ is computed in the rotating frame; the value computed in the fixed frame is then

$$\frac{\omega_{ff}}{m} = \Omega \pm \frac{\omega_{rf}}{m} . \tag{14.14}$$

The wave traveling in forward direction in the rotating frame gives way to forward whirling of the membrane also in the fixed frame, with a frequency (seen in the fixed frame) equal to

$$\omega_{Fff} = m\Omega + \omega_{rf} . \tag{14.15}$$

The frequency (in the fixed frame) related to the wave traveling backwards is

$$\omega_{Bff} = m\Omega - \omega_{rf} \tag{14.16}$$

and yields a forward whirling if $\omega_{rf} < m\Omega$ and to backward whirling otherwise. The values of ratios $\beta = \omega_{ff}/\Omega$ for forward and backward whirling corresponding to the modes shown in Figure 14.2 are reported in Tables 14.1 and 14.2, respectively.

The speed at which the frequency in the fixed frame vanishes is often referred to as *critical speed*, (is a sort of secondary critical speed) and is

$$\Omega_{cr} = -\frac{\omega_{rf}}{m} . \tag{14.17}$$

To check whether the Southwell solution can be used for an actual membrane, a numerical computation was performed using a rotordynamics FEM code. The membrane was modeled using ten annular disc elements, with very low thickness/outer radius ratio (10^{-7}). The results are reported in

n	$m=0$	$m=1$	$m=2$	$m=3$	$m=4$
0	0	0	0.467	0.988	1.530
1	1.817	-1.439	-0.992	-0.507	0.000
2	3.146	-2.768	-2.342	-1.884	-1.404
3	4.450	-4.074	-3.661	-3.221	-2.760
4	5.745	-5.372	-4.968	-4.540	-4.093
5	7.036	-6.665	-6.267	-5.849	-5.413
6	8.325	-7.955	-7.563	-7.151	-6.724

TABLE 14.2. Values of ratios $\beta = \omega_{ff}/\Omega$ for backward modes with a different number of diametral (m) and radial (n) nodes.

n	Free			Constrained		
	$m=0$	$m=1F$	$m=1B$	$m=0$	$m=1F$	$m=1B$
0	0	2	0	0.445	2	0
1	1.817	3.439	-1.439	1.998	3.439	-1.439
2	3.146	4.769	-2.769	3.345	4.769	-2.769
3	4.450	6.076	-4.076	4.666	6.076	-4.076
4	5.745	7.375	-5.375	5.978	7.375	-5.375
5	7.040	8.677	-6.677	7.289	8.677	-6.677

TABLE 14.3. Values of ratios $\beta = \omega/\Omega$ for backward and forward modes with a different number of diametral (m) and radial (n) nodes. Finite element solution for a free and a constrained membrane.

Table 14.3. By comparing Table 14.3 with Tables 14.1 and 14.2, it is clear that the natural frequencies for the modes with $m = 0$ of the Southwell solution coincide with those obtained from the FEM model in which the membrane has been assumed to be free, whereas the effect of a constraint near the center somewhat increases the natural frequencies. However, except for the case of the case with $n = 0$, the difference is not very great. The computations for $m = 1$ shows that the presence of the constraint at the center does not affect the value of the frequency and the Southwell solution holds in general.

14.2 Rotating circular plate

Consider a circular plate with uniform thickness, rotating about its symmetry axis. By accepting the Kirkchoff assumption and neglecting shear deformations (which holds for fairly thin plates), the equation of motion of the area element located in the position (r, θ) in a direction perpendicular to its plane (Figure 14.1) is

$$\rho\frac{\partial^2 w}{\partial t^2} = \frac{1}{r}\frac{\partial}{\partial r}\left(r\sigma_r\frac{\partial w}{\partial r}\right) + \frac{\sigma_c}{r^2}\frac{\partial^2 w}{\partial \theta^2} - \frac{Eh^2}{3\left(1-\nu^2\right)}\nabla^4 w\;, \tag{14.18}$$

n	m= 0	m= 1	m= 2	m= 3	m= 4
0	1.876	0.896	2.63	6.12	10.80
1	10.46	10.26	17.63	26.46	36.55
2	30.60	29.93	41.95	55.65	71.40
3	60.30	59.50	77.00	96.05	116.15
4	99.99	99.10	121.35	145.35	170.20

TABLE 14.4. Values of β_f for a nonrotating circular plate clamped at its center and free at the outer edge, with different number of diametral (m) and radial (n) nodes (the value for $n = 0$, $m = 1$ was not found in the literature and was computed numerically).

where the biharmonic operator, in polar coordinates, is

$$\nabla^4 = \left(\frac{\partial^2}{\partial r^2} + \frac{1}{r}\frac{\partial}{\partial r} + \frac{1}{r^2}\frac{\partial^2}{\partial \theta^2}\right)\left(\frac{\partial^2}{\partial r^2} + \frac{1}{r}\frac{\partial}{\partial r} + \frac{1}{r^2}\frac{\partial^2}{\partial \theta^2}\right) \tag{14.19}$$

and h is the thickness of the plate.

The first terms at the right-hand side are those linked to the stiffening effect caused by stressing (centrifugal or caused by other causes) that were present also in the case of the membrane. The following term is linked with the bending stiffness of the plate.

The solution of Equation (14.18) can in general be written in the form

$$w = u(r, \theta)\cos(\omega t) \ , \tag{14.20}$$

obtaining

$$\frac{1}{r}\frac{\partial}{\partial r}\left(r\sigma_r\frac{\partial u}{\partial r}\right) + \frac{\sigma_c}{r^2}\frac{\partial^2 u}{\partial \theta^2} - \frac{Eh^2}{3(1-\nu^2)}\nabla^4 u + \rho\omega^2 u = 0 \ . \tag{14.21}$$

The natural frequency of a nonrotating circular plate can be written in the form

$$\omega = \beta_f\sqrt{\frac{Eh^2}{3\rho r_o^4(1-\nu^2)}} \ , \tag{14.22}$$

where coefficients β_f for the various modes depend on the boundary conditions and on the ratio between the inner and outer radius. For instance, the values of β for a plate clamped at the outer edge and free at the center for various values of m and n are reported in Table 14.4 [62].

The frequencies so computed are referred to a frame that rotates together with the disc. The natural frequencies in a fixed reference frame are then computed using Equation (14.14): When the result is negative, a backward whirl mode is found.

If only centrifugal stressing is considered, Equation (14.21) can be written in nondimensional form

$$\frac{1}{\chi}\frac{\partial}{\partial \chi}\left(\chi\sigma_r^*\frac{\partial u}{\partial \chi}\right) + \frac{\sigma_c^*}{\chi^2}\frac{\partial^2 u}{\partial \theta^2} - \frac{1}{\Omega^{*2}}\nabla^4 u + \left(\frac{\omega^*}{\Omega^*}\right)^2 u = 0 \ , \tag{14.23}$$

where the biharmonic operator refers to variables χ and θ and the nondimensional quantities are

$$\sigma_r^* = \frac{\sigma_r}{\rho\Omega^2 r_o^2}, \quad \sigma_c^* = \frac{\sigma_c}{\rho\Omega^2 r_o^2}, \tag{14.24}$$

$$\Omega^* = \Omega\sqrt{\frac{3\rho r_o^4 (1-\nu^2)}{Eh^2}}, \quad \omega^* = \omega\sqrt{\frac{3\rho r_o^4 (1-\nu^2)}{Eh^2}}. \tag{14.25}$$

For a given geometrical configuration (i.e., ratio $\gamma = r_i/r_o$ and boundary conditions), the nondimensional frequency ω^* depends on a single parameter, namely, the nondimensional speed Ω^*. The nondimensional Campbell diagram for a disc clamped at its center is shown Figure 14.3. The first two modes with $m = 0$ are shown in Figure 14.3(a), whereas the first two backward and forward modes with $m = 1$ are shown in Figure 14.3(b). Together with the FEM solution, an approximation obtained through Equation (13.73) is also reported (dashed lines). The latter equation can be written in the present case as

$$\omega \approx \sqrt{\omega_n^2 + \omega_m^2}, \tag{14.26}$$

where ω_n is the natural frequencies of the nonrotating disc and ω_m is the natural frequency of the membrane rotating at the spin speed Ω.

The natural frequency at speed Ω can be rewritten with reference to the fixed frame as

$$\omega_{ff} \simeq m\Omega \pm \sqrt{\omega_n^2 + \omega_m^2} = m\Omega \pm \sqrt{\omega_n^2 + \Omega^2\beta_m^2}, \tag{14.27}$$

where β_m is the ratio between the natural frequency of the rotating membrane and the spin speed.

Equation (14.27) yields a very good approximation for $m = 1$; the approximation for the modes with $m = 0$ is far worse.

Within the validity of Equation (14.27), the speed at which the backward traveling wave is stationary in the fixed frame is

$$\Omega_{cr} \simeq \frac{\omega_n}{\sqrt{m^2 - \beta_m^2}} \tag{14.28}$$

and clearly exists only for the modes with $\beta_m < m$.

From Figure 14.2, it is clear that, at least in the case of a membrane without a central hole, no critical speed [in the sense defined by Equation (14.28)] exists for modes with $m = 0$ or $m = 1$. For $m = 2$, the first mode has a critical speed because $\beta_m = 1.533 < m$.

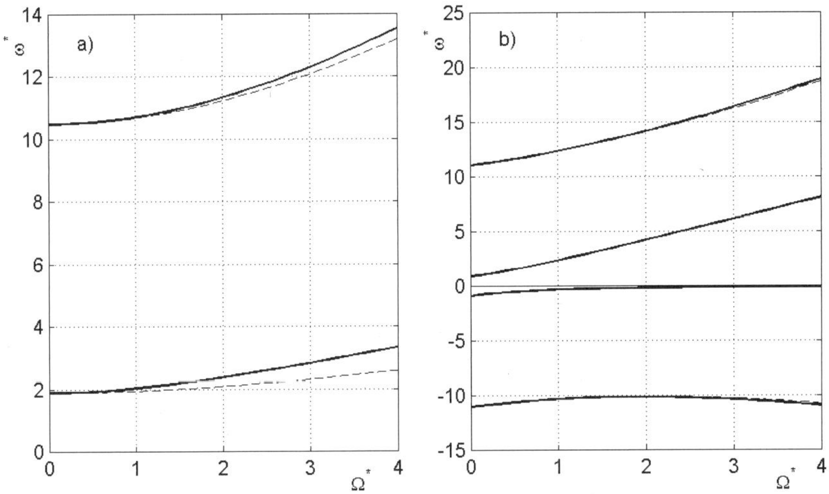

FIGURE 14.3. Nondimensional Campbell diagram for a disc clamped at its center. (a) First two modes with $m = 0$; (b) first two backward and forward modes with $m = 1$. The dashed lines have been obtained using Equation (14.14).

14.3 Disc-shaft interaction (modes with $m = 0$ or $m = 1$)

The vibrations of discs studied in the previous section are flexural, out-of-plane vibrations. Clearly a disc can vibrate also in its plane, but at least in the case of thin discs, the in-plane natural frequencies are far higher than those related to bending. If the disc is very thick, this may be no more the case, but the disc is very stiff and its dynamics can usually be reduced to that of a rigid body.

The flexural modes with $m = 0$ involve only axial motion of the disc and couple with the axial vibration of the rotor: They do not affect lateral behavior. Modes with $m = 1$ on the contrary can be likened to rotations of the disc about x- and y-axes and hence couple with the bending behavior of the shaft. All other modes with $m > 1$ can be considered as local modes, with little influence on the global behavior of the system, and will be neglected in the study of the disc-shaft interaction.

A similar consideration holds also for in-plane modes, if they are considered at all. They too can be expressed in the form of Equation (14.13) and subdivided following the value of m. The in-plane modes with $m = 0$ couple with the torsional vibration of the rotor and have no effect on its lateral behavior. Modes with $m = 1$ can be likened to displacements of the disc in x- and y-directions and couple with the bending behavior of the shaft. Again, all modes with $m > 1$ can be considered as local modes, with little

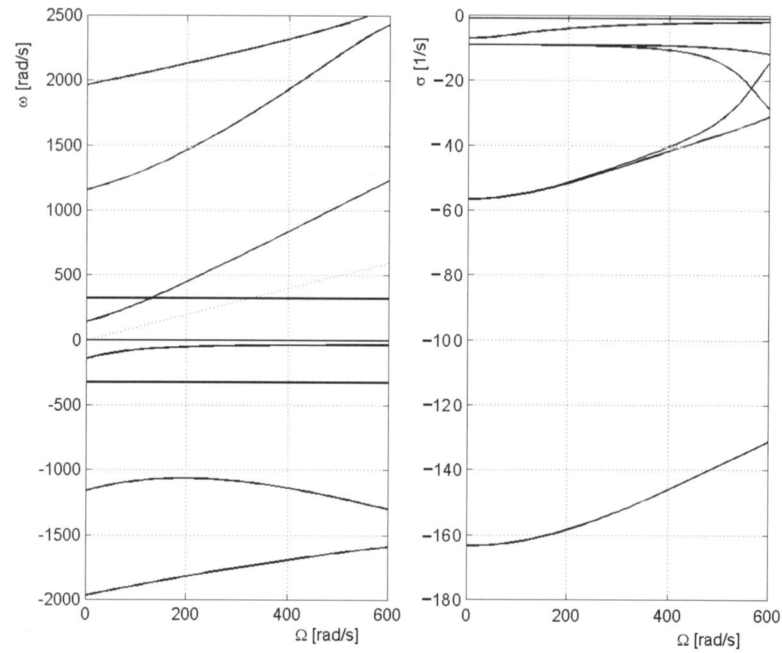

FIGURE 14.4. Campbell diagram and decay rate plot of a thin disc on elastic supports; same system studied in Section 15.6, but with a disc with thickness $t = 10$ mm instead of the blades. The system has a high rotating damping (loss factor $\eta = 0.1$ for the disc) and low nonrotating damping ($\eta = 0.005$ for the springs).

influence on the global behavior of the system. They too will be neglected in the study of the disc-shaft interaction.

Owing to the analytical complexity, it is impossible to solve the interaction problem in a general way. A very simple case will be studied in some detail in this section. Consider the bladed disc studied in Section 15.6, but substitute a thin disc (thickness $t = 10$ mm) for the blades, keeping all other parameters at the same value. The Campbell diagram is reported in Figure 14.4, together with a decay rate plot computed assuming a loss factor $\eta = 0.1$ for the disc and $\eta = 0.005$ for the springs supporting the system.

The Campbell diagram shows that the only in-plane mode, a rigid-body translation of the whole system on the suspension spring, does not interact with the bending modes, and its natural frequency, independent from the speed, coincides with the value computed using the Jeffcott rotor model. The frequencies of the bending modes increase quickly with the speed, and they do not give way to any critical speed.

The decay rate plot shows that the system is stable, in spite of the high rotating damping and almost vanishing nonrotating damping. This, however, can be expected: The damping of the disc has little destabilizing effect because the modes in which the disc deformations are involved are subcritical (are above the $\omega = \Omega$ line).

The example studied shows that the disc deformation has a large effect on the Campbell diagram as a whole, but practically no effect whatsoever on the critical speed, even if the disc is a very thin and flexible one. This is, however, mostly because of the uncoupling between translational and rotational degrees of freedom (i.e., between cylindrical and conical modes), with the first, not affected by the disc flexibility, giving way to critical speeds and the latter, highly affected by the behavior of the disc, which yield no critical speed.

Remark 14.1 *As the forward modes involving mostly the discs and the blades have a frequency that increases fast with the speed, if they remain above the $\omega = \Omega$ line, they are essentially subcritical, even if the rotor as a whole works in the supercritical range. A consequence is that rotating damping associated with the discs and the blades have little destabilizing effect. The backward modes (or better, the modes that cause a wave moving backward in the rotating frame) may in the fixed frame be seen as a backward or a forward mode. In the former case, there is usually no problem, whereas in the second one, they may lie in the supercritical range and may trigger instability. As seen above, this is usually not the case for disc modes with $m = 0$ or $m = 1$.*

14.4 Uncoupled modes (modes with $m \geq 2$)

Some modes with $m \geq 2$ may cause a critical speed [in the sense defined by Equation (14.28)] to be present. After the critical speed the wave traveling backward in the rotating frame is seen as a low-frequency forward mode in the fixed frame. Such mode is supercritical in the sense that the whirl speed is lower than the spin speed. The nondimensional Campbell diagram for the modes with $m = 2$, $n = 0$ and $m = 3$, $n = 0$ for a disc without a central hole, computed using the approximated Formula (14.27), is reported in Figure 14.5.

As expected, after the critical speed, a forward whirling with low frequency occurs.

At the critical speed (always in the sense outlined above), a sort of resonance with a static force occurs. As the mode is an out-of-plane bending, this force must be applied in a direction perpendicular to the disc rotation plane and its modal force for the relevant mode must be nonvanishing. An axi-symmetrical force distribution will not excite this mode, but a concentrated force on any given point (or short arc) of the periphery of the disc

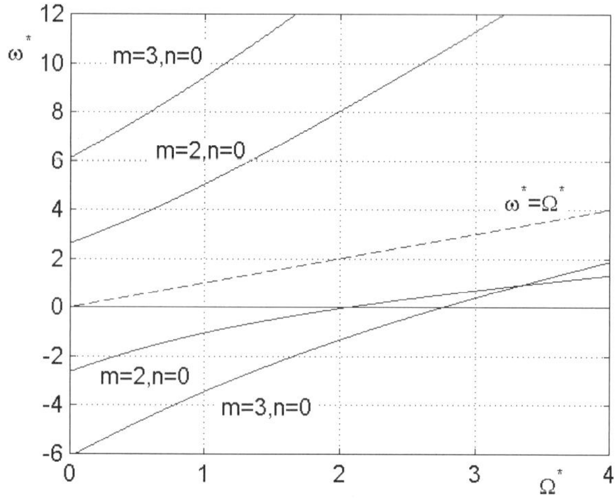

FIGURE 14.5. Same as Figure 14.4,. but for modes with $m = 2$, $n = 0$ and $m = 3$, $n = 0$.

will. This is the well-known case of the lateral force exerted on the blade of a circular saw, or the pressure of the reading head on a computer floppy disc or hard disc. This resonance occurs at high speed, particularly for high values of m.

Note that the value of the critical speed given by Equation (14.27) is only an approximation, which can be demonstrated to be a lower bound to the correct value.

In the case in which not only a lateral force, but also a nonrotating mass is applied on the periphery of the disc, as in the case of a computer disc drive in which the reading head can follow the disc in its lateral vibration, two fields of instability can be identified. The lowest one starts immediately after the critical speed [63]. If damping is also considered, the fields of instability occurring above the critical speed merge and all speeds above ω_{cr} give way to unstable vibration. This occurs not only if rotating damping is considered, but also if nonrotating damping is supplied, for instance, by the system supporting the load.

Remark 14.2 *The speed defined by Equation (14.27) is, in the case of a disc with a lateral loading, a true threshold of instability. The term critical speed is then in a sense a misnomer for it.*

14.5 Vibration of rotating circular rings

A circular ring can vibrate in different ways, and extensional, torsional, in-plane flexural, and out-of-plane flexural modes are usually identified. The

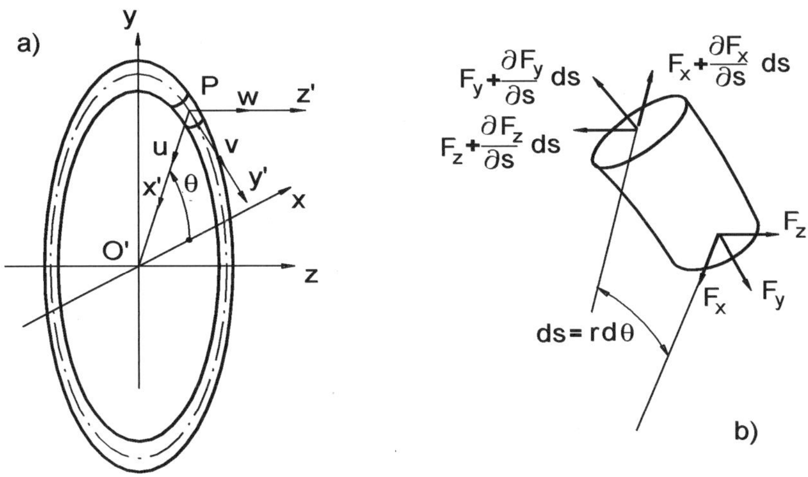

FIGURE 14.6. Circular ring. (a) Sketch and reference frames; (b) forces acting on a length ds of the ring.

natural frequencies of nonrotating circular rings are well described in the literature, e.g., in [42, 64]. The rotation of the ring about its axis, however, affects its natural frequencies. To take this effect into account, consider a ring with mean radius r and axial and radial thickness small when compared with the radius. A generic point P of the ring has a displacement whose components in radial, circumferential, and axial directions are u, v, and w with reference to the rotating frame (Figure 14.6). Note that to be consistent with [64], the radial displacement u is positive inward.

The equilibrium equation of the length $ds = rd\theta$ of the ring can be expressed in the form

$$\begin{cases} \dfrac{\partial F_x}{\partial \theta} - rF_z\kappa_y + rF_y\kappa_z + rf_{mx} = 0 \ , \\[2ex] \dfrac{\partial F_y}{\partial \theta} - rF_x\kappa_z + rF_z\kappa_x + rf_{my} = 0 \ , \\[2ex] \dfrac{\partial F_z}{\partial \theta} - rF_y\kappa_x + rF_x\kappa_y + rf_{mz} = 0 \ , \end{cases} \tag{14.29}$$

$$\begin{cases} \dfrac{\partial M_x}{\partial \theta} - rM_z\kappa_y + rM_y\kappa_z - rF_z + rm_{mx} = 0 \ , \\[2ex] \dfrac{\partial M_y}{\partial \theta} - rM_x\kappa_z + rM_z\kappa_x + rm_{my} = 0 \ , \\[2ex] \dfrac{\partial M_z}{\partial \theta} - rM_y\kappa_x + rM_x\kappa_y + rF_x + rm_{mz} = 0 \ , \end{cases} \tag{14.30}$$

where f_{mi} and m_{mi} are the mass forces and moments per unit length acting on the length ds of the ring, κ_x and κ_z are the out-of-plane and in-plane curvature of the deformed ring, and κ_y is the twist of the ring.

Equation (14.29) and (14.30) hold in general for a curved and twisted beam. In the present case, an initial curvature is present only in the xy-plane ($\kappa_{x0} = 0$ and $\kappa_{z0} = 1/r$) and there is no initial twist ($\kappa_{y0} = 0$). Under these assumptions, the total curvatures and twist are

$$\begin{cases} \kappa_x = \frac{1}{r^2}\left(r\phi_y - \frac{\partial^2 w}{\partial \theta^2}\right) , \\ \kappa_z = \frac{1}{r^2}\left(r + \frac{\partial v}{\partial \theta} + \frac{\partial^2 u}{\partial \theta^2}\right) , \\ \kappa_y = \frac{1}{r^2}\left(r\frac{\partial \phi_y}{\partial \theta} + \frac{\partial w}{\partial \theta}\right) , \end{cases} \tag{14.31}$$

where ϕ_y is the rotation of the cross section of the ring.

The bending and torsional moments are linked with the curvatures and twist of the beam by the usual formulas

$$\begin{cases} M_x = EI_x\left(\kappa_x - \kappa_{x0}\right) = \frac{EI_x}{r^2}\left(r\phi_y - \frac{\partial^2 w}{\partial \theta^2}\right) , \\ M_z = EI_z\left(\kappa_z - \kappa_{z0}\right) = \frac{EI_z}{r^2}\left(\frac{\partial v}{\partial \theta} + \frac{\partial^2 u}{\partial \theta^2}\right) , \\ M_y = GI'_p\left(\kappa_y - \kappa_{y0}\right) = \frac{GI'_p}{r^2}\left(r\frac{\partial \phi_y}{\partial \theta} + \frac{\partial w}{\partial \theta}\right) . \end{cases} \tag{14.32}$$

Remembering that the radial displacement u is positive toward the center and assuming that the ring rotates at a constant speed Ω, the position of P in the fixed reference frame at time t is

$$(\overline{\text{P-O}}) = \begin{Bmatrix} (r-u)\cos(\Omega t) + v\sin(\Omega t) \\ (r-u)\sin(\Omega t) - v\cos(\Omega t) \\ w \end{Bmatrix} . \tag{14.33}$$

By differentiating the expressions of the coordinates with respect to time and assuming that the spin speed Ω is constant, the velocity of point P is readily obtained

$$\vec{V}_{\text{P}} = \begin{Bmatrix} -\Omega(r-u)\sin(\Omega t) - \dot{u}\cos(\Omega t) + \Omega v\cos(\Omega t) + \dot{v}\sin(\Omega t) \\ \Omega(r-u)\cos(\Omega t) - \dot{u}\sin(\Omega t) + \Omega v\sin(\Omega t) - \dot{v}\cos(\Omega t) \\ \dot{w} \end{Bmatrix} . \tag{14.34}$$

The kinetic energy of the length $rd\theta$ of the ring with cross-section area A located at point P is

$$dT = \frac{1}{2}dm|\vec{V}_{\text{P}}|^2 = \frac{1}{2}\rho A r d\theta\left[\Omega^2(r-u)^2 + \Omega^2 v^2 + \dot{u}^2 + \dot{v}^2 + \right. \\ \left. + \dot{w}^2 - 2\Omega\dot{u}v - 2\Omega(r-u)\dot{v}\right] . \tag{14.35}$$

The rotational kinetic energy of the length $ds = rd\theta$ of the ring about its center of mass is here neglected, in a way that is consistent with the

Euler-Bernoulli approach to the dynamics of beams. This is consistent also with having neglected the shear deformation of the ring. As a consequence, the mass moments per unit length m_{mi} acting on the length $rd\theta$ of the ring vanish.

The mass forces per unit length f_{mi_0} expressed in the $x'y'z'$ reference frame can be easily computed through Lagrange equations. By performing the relevant derivatives of the kinetic energy, it follows that

$$\left\{\begin{array}{l} f_{mx_0}=-\dfrac{1}{rd\theta}\left[\dfrac{d}{dt}\left(\dfrac{\partial dT}{\partial \dot{u}}\right)-\dfrac{\partial dT}{\partial u}\right]=-\rho A\left[\ddot{u}-2\Omega\dot{v}+\Omega^2\left(r-u\right)\right] \ , \\ f_{my_0}=-\dfrac{1}{rd\theta}\left[\dfrac{d}{dt}\left(\dfrac{\partial dT}{\partial \dot{v}}\right)-\dfrac{\partial dT}{\partial v}\right]=-\rho A\left[\ddot{v}+2\Omega\dot{u}-\Omega^2 v\right] \ , \\ f_{mz_0}=-\dfrac{1}{rd\theta}\left[\dfrac{d}{dt}\left(\dfrac{\partial dT}{\partial \dot{w}}\right)-\dfrac{\partial dT}{\partial w}\right]=-\rho A\ddot{w} \, . \end{array}\right. \tag{14.36}$$

The forces f_{mi_0} appearing in Equations (14.29) are not expressed in the $x'y'z'$ frame, but in a frame whose axes are the local principal torsion-flexure axes (to use the definition in [64]). Using the derivations in [64], they can be computed by multiplying forces f_{mi_0} by a suitable rotation matrix

$$\left\{\begin{array}{c} f_{mx} \\ f_{my} \\ f_{mz} \end{array}\right\}=\left[\begin{array}{ccc} 1 & -\frac{1}{r}\left(v+\frac{\partial u}{\partial \theta}\right) & \phi_y \\ \frac{1}{r}\left(v+\frac{\partial u}{\partial \theta}\right) & 1 & \frac{\partial v}{\partial \theta} \\ -\phi_y & -\frac{\partial v}{\partial \theta} & 1 \end{array}\right]\left\{\begin{array}{c} f_{mx_0} \\ f_{my_0} \\ f_{mz_0} \end{array}\right\} . \tag{14.37}$$

By neglecting all small terms, it follows that

$$\left\{\begin{array}{l} f_{mx}=-\rho A\left[\ddot{u}-2\Omega\dot{v}+\Omega^2\left(r-u\right)\right] \ , \\ f_{my}=-\rho A\left(\ddot{v}+2\Omega\dot{u}+\Omega^2\frac{\partial u}{\partial \theta}\right) \ , \\ f_{mz}=-\rho A\left(\ddot{w}-\phi_y r\Omega^2\right) \, . \end{array}\right. \tag{14.38}$$

By inspecting the various terms of the equations of motion, once the mass forces and the expressions of the curvatures and twist have been inserted into Equations (14.29) and (14.30), it is clear that all of them contain at least one small quantity, except two terms of the first Equation (14.29), which yield the equation

$$F_{y0}=\rho A\Omega^2 r^2 \, . \tag{14.39}$$

Equation (14.39) yields the value of the circumferential force in the ring caused by the centrifugal field and coincides with the usual value of the stress in a rotating ring. In the following developments, the circumferential force to be introduced into Equations (14.29) will be written in the form

$$F_y+\rho A\Omega^2 r^2 \ ,$$

where F_y is only the dynamic component of the force and is considered as a small quantity.

If all quantities containing more than one small quantity are neglected, the set of six equations (14.29) and (14.30) splits in two separate sets of three equations, describing separately the out-of-plane and the in-plane behavior of the ring.

14.5.1 Out-of plane flexural vibrations

The out-of-plane behavior of the ring can be studied using the third Equation (14.29) and the first and second Equations (14.30). By introducing the expressions of the curvature and twist and the inertia forces into the equations of motion, it follows that

$$\begin{cases} \dfrac{\partial F_z}{\partial \theta} - \rho A \Omega^2 r \left(r\phi_y - \dfrac{\partial^2 w}{\partial \theta^2} \right) = \rho A r \left(\ddot{w} - \phi_y r \Omega^2 \right) , \\ \dfrac{\partial M_x}{\partial \theta} + M_y - rF_z = 0 , \\ \dfrac{\partial M_y}{\partial \theta} - M_x = 0 . \end{cases} \tag{14.40}$$

By introducing the Expressions (14.32) for the bending and torsional moments into Equation (14.40) and eliminating one equation between the first two, it follows that

$$\begin{cases} -\dfrac{\partial^4 w}{\partial \theta^4} + \chi \dfrac{\partial^2 w}{\partial \theta^2} + r(1+\chi)\dfrac{\partial^2 \phi_y}{\partial \theta^2} + \dfrac{\rho A r^4}{EI_x} \left[\Omega^2 \dfrac{\partial^2 w}{\partial \theta^2} - \ddot{w} \right] = 0 , \\ \chi \left(r\dfrac{\partial^2 \phi_y}{\partial \theta^2} + \dfrac{\partial^2 w}{\partial \theta^2} \right) + \left(\dfrac{\partial^2 w}{\partial \theta^2} - r\phi_y \right) = 0 , \end{cases} \tag{14.41}$$

where

$$\chi = \frac{GI'_p}{EI_x} . \tag{14.42}$$

If the cross section is circular or annular,

$$\chi = \frac{1}{1+\nu} , \tag{14.43}$$

where ν is the Poisson's ratio of the material.

The out-of-plane bending and the torsional deformation of the ring are thus coupled with each other. The solutions $w(\theta, t)$ and $\phi_y(\theta, t)$ can be expressed as

$$\begin{cases} w = \cos(\omega t) \sum_{s=1}^{\infty} [a_s \cos(s\theta) + b_s \sin(s\theta)] \;, \\ \\ \phi_y = \cos(\omega t) \sum_{s=1}^{\infty} [c_s \cos(s\theta) + d_s \sin(s\theta)] \;, \end{cases} \tag{14.44}$$

i.e., by an harmonic motion in time and by a trigonometric polynomial (without the constant term) in angle θ.

By introducing the expressions of w and ϕ_y into Equation (14.41) and balancing the various harmonics of the solution, it follows, for the generic mth harmonic, that

$$\begin{bmatrix} -m^4 - \chi m^2 + \frac{\rho A r^4}{EI_x}\left(-m^2\Omega^2 + \omega^2\right) & -rm^2(1+\chi) \\ -m^2(1+\chi) & -r\left(\chi m^2 + 1\right) \end{bmatrix} \begin{Bmatrix} a_m \\ c_m \end{Bmatrix} = \mathbf{0} \,. \tag{14.45}$$

Identical expression can be obtained for the component in sine of the solution, i.e., for the amplitudes b_m and d_m. By equating to zero the determinant of the coefficients of Equation (14.45), the characteristic equation allowing us to compute the natural frequency ω is obtained

$$\omega = \sqrt{\frac{EI_x}{\rho A r^4}\chi \frac{m^2(1-m^2)^2}{(1+\chi m^2)} + \Omega^2 m^2} \,. \tag{14.46}$$

At standstill ($\Omega = 0$) the natural frequency of the ring is

$$\omega_{\Omega=0} = \sqrt{\frac{EI_x}{\rho A r^4}\chi \frac{m^2(1-m^2)^2}{(1+\chi m^2)}} \,. \tag{14.47}$$

In the case of a circular or annular cross section, the usual expression of the natural frequency is obtained

$$\omega_{\Omega=0} = \sqrt{\frac{EI_x}{\rho A r^4} \frac{m^2(1-m^2)^2}{(1+m^2+\nu)}} \,. \tag{14.48}$$

In the case of a rotating string, i.e., of a ring with negligible bending stiffness, the expression of the natural frequency reduces to

$$\omega = \Omega m \,. \tag{14.49}$$

By introducing the nondimensional spin speed and natural frequency

$$\Omega^* = \frac{\Omega}{\omega_{\Omega=0}} \,, \qquad \omega^* = \frac{\omega}{\omega_{\Omega=0}} \,, \tag{14.50}$$

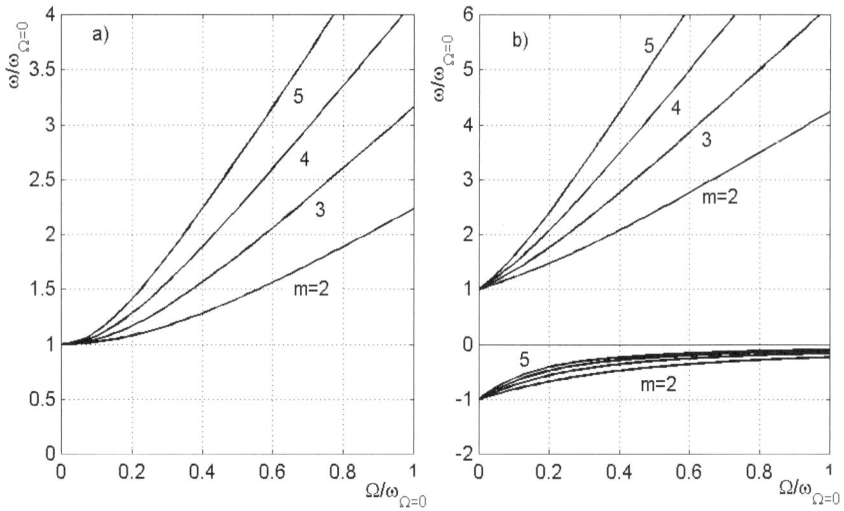

FIGURE 14.7. Nondimensional natural frequency as a function of the nondimensional spin speed for the out-of-plane vibrations of a rotating ring; mode shapes with $m = 2$, 3, 4 and 5. (a) Frequencies in the rotating frame; (b) forward and backward frequencies in the fixed frame.

the frequency-speed relationship reduces to

$$\omega^* = \sqrt{1 + \Omega^{*2} m^2} \; . \tag{14.51}$$

The above-mentioned natural frequencies are expressed with reference to the rotating frame. Each value of the frequency produces two natural frequencies, one related to a forward traveling wave and one to a backward one, in the fixed frame. By using Equations (14.15) and (14.16), the forward frequency is

$$\omega_{Fff} = m\Omega + \omega_{rf} = m\Omega + \sqrt{\frac{EI_x}{\rho A r^4} \chi \frac{m^2(1-m^2)^2}{(1+\chi m^2)} + \Omega^2 m^2} \; , \tag{14.52}$$

whereas the backward one is

$$\omega_{Bff} = m\Omega - \omega_{rf} = m\Omega - \sqrt{\frac{EI_x}{\rho A r^4} \chi \frac{m^2(1-m^2)^2}{(1+\chi m^2)} + \Omega^2 m^2} \; . \tag{14.53}$$

The nondimensional natural frequency (in the rotating frame) is plotted as a function of the nondimensional spin speed in Figure 14.7(a). The same plot, but plotted with reference to the fixed frame, is reported in Figure 14.7(b).

14.5.2 In-plane flexural vibrations

The in-plane behavior of the ring can be studied using the first and second Equations (14.29) and the third Equation (14.30). Operating in the same way seen for the out-of-plane behavior, the equations of motion become

$$\begin{cases} \dfrac{\partial F_x}{\partial \theta} + F_y + \rho A \Omega^2 r \left(\dfrac{\partial v}{\partial \theta} + \dfrac{\partial^2 u}{\partial \theta^2} \right) = \rho A r \left[\ddot{u} - 2\Omega \dot{v} - u\Omega^2 \right] \ , \\ \dfrac{\partial F_y}{\partial \theta} - F_x = \rho A r \left[\ddot{v} + 2\Omega \dot{u} + \Omega^2 \dfrac{\partial u}{\partial \theta} \right] \ , \\ \dfrac{\partial M_z}{\partial \theta} + r F_x = 0 \ . \end{cases} \tag{14.54}$$

By introducing the Expression (14.32) for the bending moment M_z into the third Equation (14.54), the latter becomes

$$\frac{EI_z}{r^2} \left(\frac{\partial^2 v}{\partial \theta^2} + \frac{\partial^3 u}{\partial \theta^3} \right) + r F_x = 0 \ . \tag{14.55}$$

By eliminating F_x and F_y, a single equation of motion is then obtained

$$\begin{gathered} \frac{\partial^2 v}{\partial \theta^2} + \frac{\partial^4 v}{\partial \theta^4} + \frac{\partial^3 u}{\partial \theta^3} + \frac{\partial^5 u}{\partial \theta^5} + \frac{\rho A r^4}{EI_z} \left[\frac{d^2}{dt^2} \left(-v + \frac{\partial u}{\partial \theta} \right) + \right. \\ \left. -2\Omega \frac{d}{dt} \left(u + \frac{\partial v}{\partial \theta} \right) - \Omega^2 \left(\frac{\partial^2 v}{\partial \theta^2} + 2\frac{\partial u}{\partial \theta} + \frac{\partial^3 u}{\partial \theta^3} \right) \right] = 0 \ . \end{gathered} \tag{14.56}$$

In the study of the flexural vibrations of the ring, it can be considered as inextensible, i.e., the circumferential strain can be assumed to vanish. In such a case, the radial and circumferential displacements are linked by the equation

$$u = \frac{\partial v}{\partial \theta} \ , \tag{14.57}$$

and Equation (14.56) reduces to

$$\begin{gathered} \frac{\partial^6 v}{\partial \theta^6} + 2\frac{\partial^4 v}{\partial \theta^4} + \frac{\partial^2 v}{\partial \theta^2} + \frac{\rho A r^4}{EI_z} \left[\frac{d^2}{dt^2} \left(-v + \frac{\partial^2 v}{\partial \theta^2} \right) + \right. \\ \left. -4\Omega \frac{d}{dt} \left(\frac{\partial v}{\partial \theta} \right) - \Omega^2 \left(3\frac{\partial^2 v}{\partial \theta^2} + \frac{\partial^4 v}{\partial \theta^4} \right) \right] = 0 \ . \end{gathered} \tag{14.58}$$

Actually, the in-plane flexural modes can be subdivided in modes in which the axial deformations are negligible (inextensional modes, similar to the bending vibrations of beams), and other ones in which the axial deformations are predominant (extensional modes, similar to the axial vibrations of beams). The natural frequencies of the latter are one or two orders of magnitude higher than the frequencies of the former (at equal number of circumferential nodes) and are usually of small practical importance. The natural frequencies of the former, computed by keeping into account the extensional deformation, differ by far less than 1% from those obtained by the inextensional approach [i.e., by using Equation (14.57)] [64]. In the following sections, only the inextensional approach will be followed.

The circumferential displacement v can be expressed by a trigonometric polynomial (without the constant term) in angle θ and by a harmonic motion in time

$$\begin{aligned} v = &\cos(\omega t)\sum_{m=1}^{\infty}\left[a_m\cos(m\theta)+b_m\sin(m\theta)\right]+\\ &+\sin(\omega t)\sum_{m=1}^{\infty}\left[c_m\cos(m\theta)+d_m\sin(m\theta)\right]\ . \end{aligned} \tag{14.59}$$

By introducing the expression for v into Equation (14.56) and balancing the various harmonics of the solution, it follows, for the generic mth harmonic, that

$$\begin{bmatrix} f(m)+\omega^2\left(m^2+1\right) & -4m\Omega\omega \\ -4m\Omega\omega & f(m)+\omega^2\left(m^2+1\right) \end{bmatrix}\begin{Bmatrix} a_m \\ d_m \end{Bmatrix} = \mathbf{0}\ , \tag{14.60}$$

$$\begin{bmatrix} f(m)+\omega^2\left(m^2+1\right) & 4m\Omega\omega \\ 4m\Omega\omega & f(m)+\omega^2\left(m^2+1\right) \end{bmatrix}\begin{Bmatrix} c_m \\ b_m \end{Bmatrix} = \mathbf{0}\ , \tag{14.61}$$

where

$$f(m) = -\frac{EI_z}{\rho A r^4}m^2\left(m^2-1\right)^2 - \Omega^2 m^2\left(m^2-3\right)\ . \tag{14.62}$$

The characteristic equations allowing us to compute the natural frequency ω are identical:

$$\begin{aligned} &\left[\omega^2\left(m^2+1\right)+4m\Omega\omega-\tfrac{EI_z}{\rho A r^4}m^2\left(m^2-1\right)^2-\Omega^2m^2\left(m^2-3\right)\right]\cdot\\ &\cdot\left[\omega^2\left(m^2+1\right)-4m\Omega\omega-\tfrac{EI_z}{\rho A r^4}m^2\left(m^2-1\right)^2-\Omega^2m^2\left(m^2-3\right)\right]=0. \end{aligned} \tag{14.63}$$

The four solutions are

$$\omega = \pm\left[\frac{2m\Omega}{1+m^2}\pm\sqrt{\frac{EI_z}{\rho A r^4}\frac{m^2(1-m^2)^2}{(1+m^2)}+\Omega^2\frac{m^2(1-m^2)^2}{\left(1+m^2\right)^2}}\right]\ . \tag{14.64}$$

From the study of the eigenvectors [65, 66], it is possible to observe that the solutions are actually just two: The one with the highest absolute value (the one with the + sign within brackets) is a backward traveling wave, whereas that with the lowest absolute value (the one with the – sign within brackets) is a forward traveling wave. The corresponding values in the fixed frame are

$$\omega_{Fff} = m\Omega\frac{3+m^2}{1+m^2}-\sqrt{\frac{EI_z}{\rho A r^4}\frac{m^2(1-m^2)^2}{(1+m^2)}+\Omega^2\frac{m^2(1-m^2)^2}{\left(1+m^2\right)^2}}\ , \tag{14.65}$$

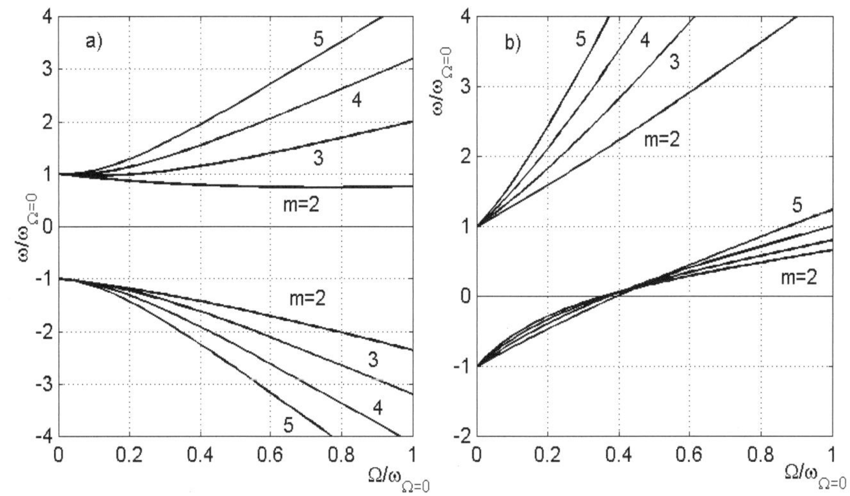

FIGURE 14.8. Nondimensional natural frequency as a function of the nondimensional spin speed for the in-plane vibrations of a rotating ring; mode shapes with $m = 2$, 3, 4 and 5. (a) Frequencies of the forward and backward traveling wave in the rotating frame; (b) forward and backward whirling frequencies in the fixed frame.

$$\omega_{Bff} = m\Omega\frac{m^2-1}{1+m^2} + \sqrt{\frac{EI_z}{\rho A r^4}\frac{m^2(1-m^2)^2}{(1+m^2)} + \Omega^2\frac{m^2(1-m^2)^2}{(1+m^2)^2}}\ . \qquad (14.66)$$

At standstill ($\Omega = 0$), the usual expression of the natural frequency is obtained

$$\omega_{\Omega=0} = \sqrt{\frac{EI_z}{\rho A r^4}\frac{m^2(1-m^2)^2}{(1+m^2)}}\ . \qquad (14.67)$$

Also, in the present case, it is possible to express the frequency and the spin speed in nondimensional form, by dividing them by the natural frequency of the nonrotating ring. With reference to the rotating frame, it follows that

$$\omega^* = \pm\left[\frac{2m\Omega^*}{1+m^2} \pm \sqrt{1 + \Omega^{*2}\frac{m^2(1-m^2)^2}{(1+m^2)^2}}\right]\ . \qquad (14.68)$$

The nondimensional natural frequencies for the forward and backward traveling waves (in the rotating frame) are plotted as functions of the nondimensional spin speed in Figure 14.8(a). The same plots, but with reference to the fixed frame, are reported in Figure 14.8(b).

Remark 14.3 *Also, in the case of rotating rings, in-plane vibration may give way to a wave that travels in backward direction in the rotating frame* *but in the forward one in the fixed frame. This feature is potentially destabilizing.*

In the case of a rotating string loop, i.e., of a ring with negligible bending stiffness, the solutions of the characteristic equation are, in the rotating frame,

$$\begin{aligned} \omega &= \Omega m \, , \\ \omega &= \Omega m \tfrac{3-m^2}{m^2+1} \, . \end{aligned} \tag{14.69}$$

Note that the results reported above are not the only ones reported in the literature. Different formulae can be obtained by accounting for some small terms which appear in the equation of motion. Actually the differences among them are so small that it is a common opinion that *from the physical point of view, one cannot judge which theoretical model is better* [65], [67], in spite of a number of experimental investigations on the subject.

14.6 Vibration of thin-walled, rotating cylinders

Often the shafts of rotating machines are made of thin-walled tubes of fairly large diameter (if compared with their thickness). In this case, apart from the flexural, torsional, and axial vibrations, which can be dealt with (more or less accurately depending on the slenderness of the shaft) using the beam theory, there may be also vibrations of the cross section of the rotating cylindrical shell, very similar to the in-plane vibrations of rings.

If the tube is very long (i.e., the length is large if compared with its diameter), it is possible to model it as an infinitely long cylindrical tube, and each cross section can be assimilated to a vibrating ring. By resorting to the inxtensional theory, the frequencies of the forward and backward traveling waves are still given by Equation (14.68), the only difference being in the value of the natural frequency at standstill ($\Omega = 0$), which in the present case must account for the added stiffness because Poisson's effect tends to constrain (i.e., to increase the natural frequencies of) any flexural deformation of the shell. The natural frequency of the tube at $\Omega = 0$ is

$$\omega_{\Omega=0} = \sqrt{\frac{Eh^2}{12\rho r^4 \left(1-\nu^2\right)} \frac{m^2(1-m^2)^2}{(1+m^2)}} \, , \tag{14.70}$$

where ν is the Poisson's ratio.

If the length of the cylinder is finite, the boundary conditions at the ends must be introduced, and the natural frequencies depend not only on the number of nodes in circumferential direction, i.e., on the order of the mode m based on the Fourier series in which the displacement is developed in circumferential direction, but also on the number of nodes in axial direction, i.e., on the axial order of the node n.

Although solutions exist for the simplest cases, like a constant thickness cylinder with simply supported ends [68, 67], their complexity suggests to resort to numerical solutions, particularly in the case of complex shapes, like those occurring in real-life machinery.

14.7 Instability of rotating cylinders partially filled with liquid

There are instances in which some fluid is trapped inside a rotor, either by chance, like in the case of a steam turbine rotor, or by design, as in centrifuges. Owing to the centrifugal field, the liquid is "plastered" against the inside walls of the rotor, and it can move more or less freely; the combined dynamics of the rotor and the fluid can lead to instability of the system.

The simplest model, which can apply to simple centrifuges, is that of a Jeffcott rotor made of a rigid hollow cylinder suspended on linear springs, in which a layer of fluid of thickness h hangs against the bore of the cylinder. The thickness is kept constant by the centrifugal field (if the inner radius is constant), but waves can form on the free surface (inner surface) of the fluid.

To study the uncoupled dynamics, consider first the cylinder running on stiff bearings. The motion of the fluid can be seen as a wave propagating in forward direction and one propagating in backward direction. The frequency of these waves is [11]

$$\omega = \Omega \frac{1 \pm \sqrt{1+\delta}}{\delta} , \tag{14.71}$$

where

$$\delta = \frac{1+\gamma^2}{1-\gamma^2}$$

and $\gamma = r_i/r_o$ is the ratio between the inner and outer radii of the fluid layer. If the liquid layer is very thin, i.e., if γ is close to unity, it follows that

$$\delta \simeq \frac{r_o}{h} - 1 \quad , \quad \omega \simeq \Omega \frac{h \pm \sqrt{r_o h}}{r_o - h} . \tag{14.72}$$

In the fixed frame, they are two waves propagating with angular velocity

$$\omega = \Omega + \omega = \Omega \frac{1 + \delta \pm \sqrt{1+\delta}}{\delta} . \tag{14.73}$$

The wave velocity is always positive; i.e., both waves are seen in the fixed frame as propagating in forward direction. The solution with $(-)$ is then a wave that moves backward in the rotating frame but forward in the fixed one. As seen in many other cases, this is prone to produce instability.

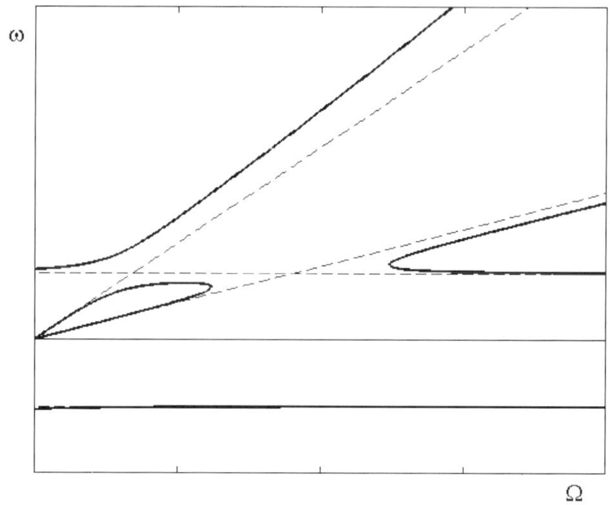

FIGURE 14.9. Qualitative Campbell diagram for a cylindrical Jeffcott rotor partially filled with liquid. Dashed lines: uncoupled dynamics. Full lines: coupled dynamics.

By freezing the fluid in a cylindrical layer and freeing the suspension springs (whose isotropic stiffness is k), and modeling the system as a Jeffcott rotor, the natural frequency of the whole system on the springs is obviously

$$\omega = \sqrt{\frac{k}{m_t}} , \tag{14.74}$$

where m_t is the total mass of the rotor, fluid plus structural parts.

If the coupled dynamics is considered, a plot of the type of the qualitative one shown in Figure 14.9 is then obtained.

A field of instability, located about the intersection of the lowest branch of the dynamics of the fluid layer with the upper branch of the Jeffcott rotor line is clearly visible.

15
Three-dimensional modeling of rotors

As already stated, in many cases rotors are modeled as one-dimensional bodies, made basically of beam-like shafts with rigid bodies attached to them. This approach is typical of rotordynamics, both analytical and numerical, and several rotordynamic codes, either based on the transfer matrices approach or on the finite element method, follow this trend.

The one-dimensional mathematical models usually yield results that are accurate enough for most practical purposes while simple enough to allow relatively straightforward computations to be performed. However, rotors are intrinsically more complicated than assemblies of slender shafts and rigid bodies, and there are cases in which true three-dimensional modelling is required. There are cases in which the shafts are very stub or have a thin-walled tubular cross section, and neither the Euler-Bernoulli nor the Timoshenko beam theory is adequate to model their behavior in detail or thin bladed discs do not behave as rigid bodies. The effects linked with these deviation from the classic shafts-rigid-bodies models are usually felt more strongly on the high-frequency modes, and the very reason for which they are usually neglected is that the first critical speed and the lowest vibration frequency (or, better, the low-frequency part of the Campbell diagram) are little affected by them.

There are, however, cases in which the search for a better agreement between simulation and experimental results; the presence of high-frequency excitation that compels us to take into account also the high-frequency response, the presence of very compliant parts like thin discs or long, slender blades, or simply an unconventional rotor geometry makes the classic approach inadequate. Moreover, in the previous two chapters it has been shown that the presence of flexible blades or discs may cause a rotating system to become unstable.

Another reason for resorting to true 3D modeling is a practical one: Although a beam-model must be created more or less "by hands", a tridimensional model can be obtained automatically from the CAD drawings of the machine. Usually automatic modeling results in a large number of degrees of freedom, but this can be at least partially dealt with by using reduction techniques, so that the construction of a 3D model can be today less expensive (in particular for what the men-hours are concerned) than that of a beam-like model. No doubt that this is the trend in the numerical analysis community.

There is no difficulty in modeling a rotor using a standard 3D FEM code to study its dynamic behavior, but such codes usually do not take into account that a rotor is a rotating structure and the very presence of a spin speed induces several effects that are not present in general structural dynamics. They can be summarized as follows:

- Gyroscopic (and Coriolis),
- Centrifugal stiffening, and
- Rotating damping effects.

The former, which in elementary rotordynamics is dealt with in the context of rigid-body dynamics, affects also the behavior of compliant rotating bodies, as seen dealing with the dynamics of blades and discs. The second, together with the stiffening caused by other load conditions like thermal stressing in gas turbine or brake discs, cannot be accounted for in simple beam-like models.

Many FEM codes can deal with centrifugal stiffening and similar effects because they can compute static stressing and stresses caused by rotation and then use the geometric matrices approach to evaluate their effects on the natural frequencies. For gyroscopic and rotating damping effects, things are more complex. Many times it has been proposed to play some tricks to account for these features within the frame of standard finite element codes, but the only correct way for dealing with them is to develop element formulations in which they are included since the beginning. So correct 3-D approach rotordynamics is just a matter of developing adequate formulations and implementing them in suitable FEM codes.

15.1 Symmetry of the rotor

The rotors studied in classic rotordynamics are mostly axi-symmetrical. Even in this case, although it is possible to assume that axial symmetry holds for the geometrical configuration, the displacement field has no particular symmetry.

When both the rotor and the stator are symmetrical with respect to the rotation axis, the study can be performed indifferently using a fixed frame (inertial frame) or a rotating frame. The former is usually selected because it simplifies the analytical formulation.

In Chapter 6, it has been stated that when the stator has no axial symmetry, the use of the inertial frame is mandatory (or, better, that if the equations are written in a rotating frame, they transform into periodic coefficient differential equations). In a similar way, when the rotor is not axially symmetrical, although axial symmetry holds for the stator, the study can be performed in the rotating frame using a set of differential equations whose coefficients are constant in time. It is only when both stator and rotor (or, in case of a multirotor machine, more than one element) have no axial symmetry that an actual periodic coefficient differential equation must be dealt with, because no frame in which a constant coefficients equation can be written exists.

When either the stator or the rotor is axially symmetrical, it is a common practice in finite element modeling to use elements that are axially symmetrical for what their geometry and properties are concerned, but whose displacement field is expressed along the rotational angle θ by trigonometric polynomial. This approach allows us to deal with problems of higher dimensionality without the need of a computationally complex fully multidimensional approach.

As an example, a thin disc whose geometry is axially symmetrical can be dealt with by neglecting its thickness, locating a certain number of nodes along the radius and expressing the displacement field using any type of shape function along the radius and by a trigonometric polynomial along the polar angle. This approach, which will be dealt with in detail in Section 15.2, is often referred to as a 1 and one half dimensional approach. Its advantages are mainly caused by the consideration that the coefficients of the trigonometric polynomials that are taken as generalized coordinates are much fewer than the generalized coordinates linked with the displacements of nodes distributed along the polar angle as it would occur if a full two dimensional approach was followed.

If the body cannot be modeled as a thin disc, it is possible to exploit its symmetry by using annular (axy-symmetrical) elements and then approximating the dependence of the displacements on the polar angle by trigonometric polynomials. Also, in this case, the degrees of freedom of such a 2 and one half dimensional model are much fewer than those of a full three dimensional model.

If the stator is not axially symmetrical, the equations of motion can be written with reference to the inertial frame, whereas in the case of a nonisotropic rotor, the use of a rotor fixed frame is mandatory. Up to this

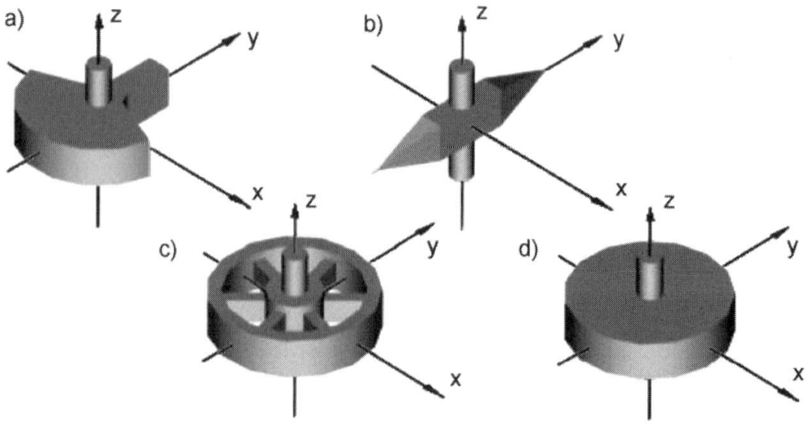

FIGURE 15.1. Possible symmetry classes for rotors. (a) No symmetry about the rotational axis; (b) two orthogonal symmetry planes (symmetry of order 2); (c) cyclic symmetry; (d) axial symmetry (symmetry of order ∞). (a) is equivalenmt to (b) and (d) to (c).

point, there is no difference in the general approach whether a beam-like, a 2 and one half dimensional, or a full 3D model is used.

However, there is a symmetry class that is in a way intermediate between an axial symmetry and no symmetry at all (note that in the case of rotors, a symmetry with respect to one or two planes is equivalent to a complete lack of symmetry): Cyclic symmetry [Figure 15.1(c)].

Many rotors, including bladed turbine rotors, display a cyclic symmetry, i.e., can be considered as made of a number of identical sectors. Apart from the obvious issue that modeling of such a rotor by exploiting this peculiar symmetry makes the construction of the model far simpler, there are some added considerations of great importance.

Firstly, in the case of a rigid body, any cyclic symmetry with order equal or greater than three is equivalent to a symmetry of order infinity, i.e., to axial symmetry (the ellipsoid of inertia becomes an ellipsoid of rotation). A rigid three-bladed propeller is then dynamically equivalent to a disc. In the same way, a planar linear elastic system with cyclic symmetry with order equal to, or greater than, three is equivalent to an isotropic elastic system (the ellipse of elasticity degenerates into a circle). A triangular beam, whose cross section is an equilateral triangle, behaves elastically as a circular beam.

In the case of flexible body with cyclic symmetry of order n, a similar property holds. This can be proven by considering that any possible eigenvector of the whole structure can be described in terms of the vectors describing the deformations of the n identical substructures

$$\mathbf{u} = \begin{bmatrix} \mathbf{u}_1 & \mathbf{u}_2 & \mathbf{u}_3 & \dots & \mathbf{u}_n \end{bmatrix}^T . \tag{15.1}$$

D. L.Thomas [69] showed that the eigenvectors fall into three classes:

1. Eigenvectors in which all substructures displace in an identical way ($\mathbf{u}_1 = \mathbf{u}_2 = \dots = \mathbf{u}_n$).
2. Eigenvectors in which all substructures displace in an identical way, but with a phasing of 180° from each other ($\mathbf{u}_1 = -\mathbf{u}_2 = \mathbf{u}_3 = \dots$).
3. Eigenvectors of other types.

The eigenvectors of the first type, for example, zero-order axial, torsional or radial extensional modes of disc-type structures, do not exhibit in general degeneracy and result in distinct eigenfrequencies.

Eigenvectors of the second type can exist only if n is even (otherwise the last substructure would have the same deformation of the first one, violating the continuity at the interface between them). They have the same properties as those of the first type.

The eigenvectors of the third type exhibit a two-fold degeneracy, and the eigenfrequencies come in pairs with identical values. This characteristic is identical to what is typical for axi-symmetrical structures. An eigenvector of this type can be rotated of any angle about the symmetry axis (i.e., the spin axis), and the result is still an eigenvector with the same eigenvalue.

This does not mean that the stiffness, mass, gyroscopic, etc. matrices are not affected by an arbitrary rotation about the spin axis, but that the eigenvalues and the eigenvectors are such. The dynamics of a rotor with cyclic symmetry can be performed by using the nonrotating frame and, as a consequence, that a rotor of this type running on an anisotropic stator can be studied without resorting to an equation with periodic coefficients. In this sense the rotor of Figure 15.1(c) is equivalent to that shown in Figure 15.1(d), whereas those shown in (a) and (b) are equivalent.

All this holds if the cyclic symmetry is at least of order 3: If the rotor is made of two identical substructures rotated by 180° with respect to each other the situation is that of Figure 15.1(b) and the eigenvectors do not exhibit any particular degeneracy. As a result, the equation of motion must be written in the rotating frame, and if also the stator is not isotropic, no constant coefficient equation can be obtained.

Clearly, what has been said for a rotor with cyclic symmetry can be repeated also for the case of a stator with cyclic symmetry of order 3 or higher.

Example 15.1 *Rim-with-spokes flywheel. Compute the out-of-plane natural frequencies at standstill of a rim-with spokes flywheel on nonisotropic supports using*

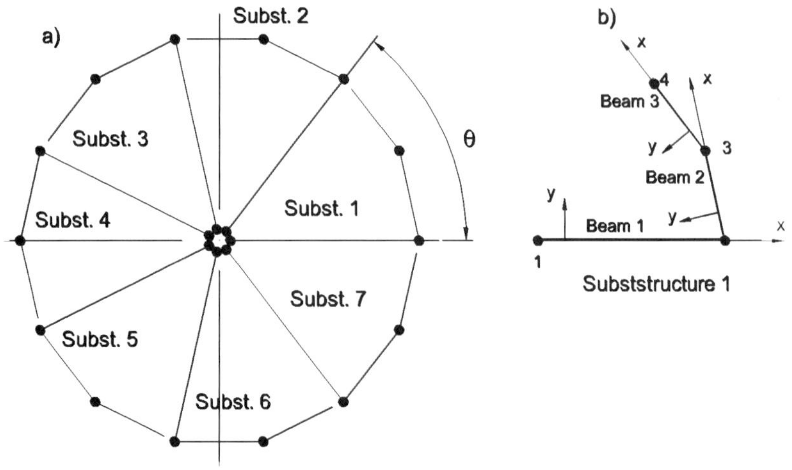

FIGURE 15.2. Rim-with-spokes flywheel modelled using a number of beam elements. (a) Structure with symmetry of order 7; (b) one of the substructures.

a very simple FEM model based on straight beam elements to show that the behavior of a cyclic symmetry structure is the same as that of an axi-symmetric structure.

A very simple model for a rim flywheel with seven spokes is shown in Figure 15.2. The rim is modeled with 14 straight Euler Bernoulli beam elements, whereas the spokes consist of other seven beam elements of the same type. The degrees of freedom for each node involved in the out-of-plane behavior are the axial displacement u_z*, the bending rotation* ϕ_y*, and the torsional rotation* ϕ_x *(*x*-axis lies along the axis of the beam, whereas* z*-axis of each beam is parallel to the rotation axis).*

The structure can be subdivided into n *(seven in the figure) identical sectors, each one made by three beam elements. Each sector contains four nodes, node 1 being in common with all other sectors (and with the shaft), node 2 with the preceding sector, and node 4 with the following one. The total number of degrees of freedom for each sector is 12, and the whole structure contains* $6n+3$ *degrees of freedom (45 in the case of the figure).*

Assume that the rim has a circular cross section with a diameter of 50 mm and a mean diameter of 1 m, the spokes have a circular cross section with a diameter of 30 mm and that the whole structure is made of steel ($E = 2.1 \times 10^{11}$ *N/m*^{2}*,* $\rho = 7810$ *kg/m*^{3}*,* $\nu = 0.3$*). The stiffness of the support is* $K_z = 10^8$ *N/m in axial direction,* $K_x = 10^{11}$ *Nm/rad for rotations about* x*-axis, and* $K_y = 10^6$ *Nm/rad for rotations about* y*-axis.*

Angle θ *and the length of the beams modeling the rim are, for* $n = 7$*,*

$$\theta = \frac{2\pi}{n} = 0.8976 \ rad \ , \qquad l = 2R\sin\left(\frac{\theta}{4}\right) = 222.5 \ mm \ .$$

By remembering that in case of circular cross sections $I'_{\text{p}} = I_x = 2I_y$, *ordering the generalized coordinates of each node as* u_z, ϕ_x, *and* ϕ_y, *the stiffness matrix of each beam of the rim is*

$$\mathbf{K}_{rim} = \frac{EI_y}{l^3}\begin{bmatrix} 12 & 0 & -6l & -12 & 0 & -6l \\ & \frac{l^2}{1+\nu} & 0 & 0 & -\frac{l^2}{1+\nu} & 0 \\ & & 4l^2 & 6l & 0 & 2l^2 \\ & & & 12 & 0 & 6l \\ & & & & \frac{l^2}{1+\nu} & 0 \\ symm. & & & & & 4l^2 \end{bmatrix}.$$

Taking into account both the translational inertia and the rotational inertia of the cross section, the mass matrix of each beam of the rim is

$$\mathbf{M}_{rim} = \frac{\rho Al}{420}\begin{bmatrix} 156 & 0 & -22l & 54 & 0 & 13l \\ & 140\frac{I_x}{A} & 0 & 0 & 70\frac{I_x}{A} & 0 \\ & & 4l^2 & -13l & 0 & -3l^2 \\ & & & 156 & 0 & 22l \\ & & & & 140\frac{I_x}{A} & 0 \\ symm. & & & & & 4l^2 \end{bmatrix} +$$

$$+\frac{\rho I_y}{30l}\begin{bmatrix} 36 & 0 & -3l & -36 & 0 & -3l \\ & 0 & 0 & 0 & 0 & 0 \\ & & 4l^2 & 3l & 0 & -l^2 \\ & & & 36 & 0 & 3l \\ & & & & 0 & 0 \\ symm. & & & & & 4l^2 \end{bmatrix}.$$

Similar expressions hold for the spokes, once the radius is substituted for the length l *and the relevant cross-section characteristics have been introduced.*

The three beams constituting a substructure are then rotated by the relevant angles (0 for the spoke, $\frac{1}{4}(\theta + 2\pi)$ *and* $\frac{1}{4}(3\theta + 2\pi)$ *for the rim elements) and assembled in the usual way. The stiffness and mass matrices of each substructure have 12 rows and 12 columns.*

To assemble the various substructures, it is useful to use a map, i.e., a table showing the correspondence between the generalized coordinates of the various substructures and those of the whole system. Such a map is reported in Table 15.1.

The first substructure is already oriented in the right way, each of the following ones is rotated by an angle θ *with respect to the previous one. The matrices of the structure are thus obtained by rotating and assembling those of the substructures. The stiffness of the constraints is then added to the elements 11, 22, and 33 of the stiffness matrix.*

A first computation is then performed without adding the stiffness of the constraints, to compute the free-free natural frequencies. The first three modes are rigid-body modes with natural frequency equal to zero. The eigenfrequencies from the fourth to the twelfth are reported in Table 15.2.

Substr.	Generalized coordinate #											
1	1	2	3	4	5	6	7	8	9	10	11	12
2	1	2	3	10	11	12	13	14	15	16	17	18
3	1	2	3	16	17	18	19	20	21	22	23	24
4	1	2	3	22	23	24	25	26	27	28	29	30
5	1	2	3	28	29	30	31	32	33	34	35	36
6	1	2	3	34	35	36	37	38	39	40	41	42
7	1	2	3	40	41	42	43	44	45	4	5	6

TABLE 15.1. Map for assembling the substructures.

Mode #	4	5	6	7	8	9	10	11	12
ω (rad/s)	1390	1390	1604	3774	3774	5531	5531	7205	7205
Type	c	c	a	c	c	c	c	c	c

TABLE 15.2. Values of the first 12 free-free natural frequencies.

Type "a" designates eigenvectors in which all substructures displace in an identical way (the one reported is an axial "umbrella" mode) and "c" the eigenvectors with a twofold degeneracy, identical to those typical for axi-symmetrical structures. As the symmetry is of odd order (7), there are no modes in which all substructures displace in an identical way, but with a phasing of 180° from each other.

The natural frequencies of the system on the elastic supports were then computed (Table 15.3). To verify that the system behaves as an axi-symmetrical structure, the system has been modeled six times, with the flywheel rotated by 0, $\theta/5$, $2\theta/5$, $3\theta/5$, etc. with respect to the axes of elasticity of the supports. As the rotational stiffness about y-axis is much lower than that about x-axis, the first two natural frequencies differ greatly from each other. The third mode is an axial umbrella mode. Then some modes with two-fold degeneracy are maintained: They are the modes in which no rotation about x- and y-axis occurs, and the two different stiffnesses can play no role. All results are reported with eight figures, to check numerical approximations.

15.2 Simplified FEM elements for thin bladed-discs modeling

If the rotor contains a thin axi-symmetrical bladed disc, it is possible to extend the simple one-dimensional models in a straightforward way ($1\frac{1}{2}$ dimensional models). The shafts can be modeled by using beams and the discs can be assimilated to annular plates. As they are geometrically axially symmetrical, only the cross section is meshed as in the case of axi-symmetrical plates, but the displacement field needs to allow considering non-axi-symmetrical deformations. This can be easily done by assuming

	Angle [deg]					
Mode	0	10.29	20.57	30.86	41.14	51.43
1	175.050	175.050	175.050	175.050	175.050	175.050
2	403.793	403.793	403.793	403.793	403.793	403.793
3	414.474	414.474	414.474	414.474	414.474	414.474
4	1390.328	1390.328	1390.328	1390.328	1390.328	1390.328
5	1390.328	1390.328	1390.328	1390.328	1390.328	1390.328
6	2347.196	2347.196	2347.196	2347.196	2347.196	2347.196
4	3773.669	3773.669	3773.669	3773.669	3773.669	3773.669
5	3773.669	3773.669	3773.669	3773.669	3773.669	3773.669
6	5971.796	5971.796	5971.796	5971.796	5971.796	5971.796
7	7204.965	7204.965	7204.965	7204.965	7204.965	7204.965
8	7204.965	7204.965	7204.965	7204.965	7204.965	7204.965
9	8097.518	8097.518	8097.518	8097.518	8097.518	8097.518

TABLE 15.3. Values of the first nine free-free natural frequencies of the constrained rotor, computed in different positions.

that the axial, radial, and circumferential displacements of all nodes of the reference cross section are expressed by a Fourier series in angle θ (Figure 14.1).

In a similar way, the blades are modeled as rows of radial beams, in a way similar to the rows of pendulums described in Chapter 13.

The beam elements modeling the shaft can be Timoshenko or Euler-Bernoulli beam elements, with two end nodes. Each node has six degrees of freedom: three displacements in x-, y- and z- (axial) direction, and three rotations about the same axes.

Each disc is modeled by a number of annular plates, located at a certain axial coordinate (identified by the position of the midplane of the disc) and extending radially. As the disc is reduced to its midplane, the displacement field is a function of just two coordinates, the radius r and the polar angle θ. As already stated, the disc is assumed to be axially symmetrical and the displacement field can be approximated along angle θ by a Fourier series.

The disc is located in a given point of the shaft [point C in Figure (15.3b)]: If it was a rigid disc, like those modeled by a mass element in Chapter 4, its midplane would lie in the plane containing the cross section of the beam element. Such plane is identified by axes x and y in Figure 15.3. The inflected configuration of the beam element is shown in Figure 15.3(a). The disc is attached to one end of the beam, which in the undeflected position is located in O and in C in the deformed configuration [7]. The displacement of point C has then components X, Y, and Z in the inertial frame.

15.2.1 Kinematics

The deflected configuration of the disc is sketched in Figure 15.3(b) (in the figure an "umbrella" deformation, i.e., a deformation that is independent

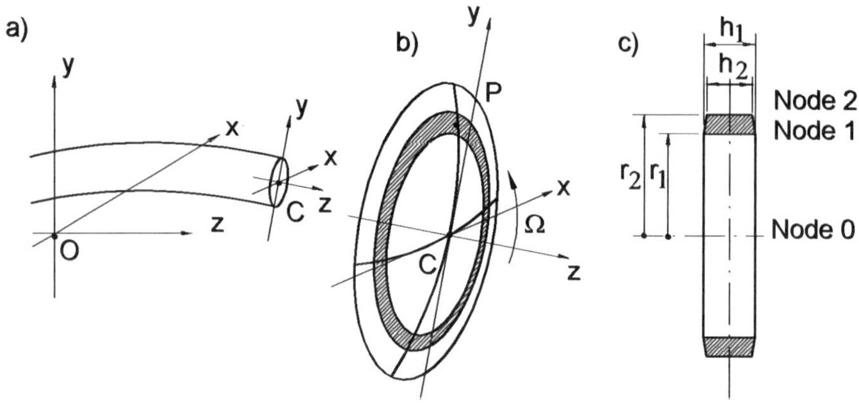

FIGURE 15.3. Simplified disc element. (a) Inflected configuration of the beam element. The disc is attached to one end of the beam (point C). (b) Sketch of the disc. Its deflected configuration (in the figure an "umbrella" deformation, i.e., a deformation that is independent from the polar angle θ is represented) is referred to the plane xy identified by the deflected position of the beam cross section in C. The annular element is the dashed part of the disc, between radius r_1 and radius r_2. (c) Geometrical definitions and nodes.

from the polar angle θ is represented). The axial displacements w are referred to the xy-plane. The annular element is the dashed part of the disc, between radius r_1 and radius r_2. Point P is a generic point of the element belonging to the midplane of the disc.

The element has three nodes: a central node (referred to as node 0) located in point C, center of the cross section of the shaft at the shaft-disc interface, node 1 and node 2 at the inner and outer radii [Figure 15.3(c)]. Note that all of the elements in which a disc is subdivided have in common the same central node (node 0), which guarantees that the rigid-body behavior of all elements is consistent.

The $Cxyz$-frame coincides with the whirling frame defined in Chapter 3. It has been obtained by rotating the inertial frame $OXYZ$ about the X-axis by an angle ϕ_X and then about the y-axis by an angle ϕ_y. The rotation matrices, defined in Chapter 3, are here reported

$$\mathbf{R}_1 = \begin{bmatrix} 1 & 0 & 0 \\ 0 & \cos(\phi_X) & \sin(\phi_X) \\ 0 & -\sin(\phi_X) & \cos(\phi_X) \end{bmatrix}, \tag{15.2}$$

$$\mathbf{R}_2 = \begin{bmatrix} \cos(\phi_y) & 0 & -\sin(\phi_y) \\ 0 & 1 & 0 \\ \sin(\phi_y) & 0 & \cos(\phi_y) \end{bmatrix}. \tag{15.3}$$

To take into account also a torsional rotation by an angle ϕ_z of the end of the beam where the disc is attached, the following rotation matrix is also introduced:

$$\mathbf{R}_3 = \begin{bmatrix} \cos(\phi_z) & \sin(\phi_z) & 0 \\ -\sin(\phi_z) & \cos(\phi_z) & 0 \\ 0 & 0 & 1 \end{bmatrix}. \tag{15.4}$$

Note that frame xyz follows the rotor during its whirling motion but does not rotate. As the disc rotates with respect to the whirling frame, the angle line PC makes with x-axis is $\theta + \Omega t$, where angle theta defines the angular position of P on the disc at time $t = 0$. A fourth rotation matrix, defining the position of P in the $Cxyz$-frame can then be defined

$$\mathbf{R}_4 = \begin{bmatrix} \cos(\theta+\Omega t) & \sin(\theta+\Omega t) & 0 \\ -\sin(\theta+\Omega t) & \cos(\theta+\Omega t) & 0 \\ 0 & 0 & 1 \end{bmatrix}. \tag{15.5}$$

The position of point P can thus be expressed as

$$(\overline{\mathrm{P}-\mathrm{O}}) = \begin{Bmatrix} X \\ Y \\ Z \end{Bmatrix} + \mathbf{R}_1^T \mathbf{R}_2^T \mathbf{R}_3^T \mathbf{R}_4^T \begin{Bmatrix} r+u \\ v \\ w \end{Bmatrix}, \tag{15.6}$$

where u, v, and w are the displacements in radial, circumferential, and axial direction of point P, with respect to the position of the same point, as belonging to a rigid disc.

The small-displacement assumption allows us to simplify the trigonometric functions of angles ϕ_X, ϕ_y, and ϕ_z by neglecting all terms beyond the second one in their series expansion $[\sin(\phi) \approx \phi,\ \cos(\phi) \approx (1 - \phi^2/2)]$. Equation (15.6) then reduces to

$$(\overline{\mathrm{P}-\mathrm{O}}) = \begin{Bmatrix} X + \left[\left(1 - \frac{\phi_y^2}{2}\right) r + u\right] \cos(\phi_z + \theta + \Omega t) + \\ \qquad -v \sin(\phi_z + \theta + \Omega t) + \phi_y w \\ Y + \left[\left(1 - \frac{\phi_y^2}{2}\right) r + u\right] \sin(\theta + \Omega t) + v \cos(\phi_z + \theta + \Omega t) + \\ \qquad -\phi_X w + r\phi_X \phi_y \cos(\phi_z + \theta + \Omega t) \\ Z + \left[(r+u)\,\phi_X + v\phi_y\right] \sin(\phi_z + \theta + \Omega t) + \\ \qquad - \left[(r+u)\,\phi_y - v\phi_X\right] \cos(\phi_z + \theta + \Omega t) + w \end{Bmatrix}. \tag{15.7}$$

The six generalized coordinates X, Y, Z, ϕ_X, ϕ_y, and ϕ_z describe the rigid-body motion of the disc and are the same for all the disc elements that share the same central node, whereas $u(r,\ \theta,\ z,\ t)$, $v(r,\ \theta,\ z,\ t)$, and $w(r,\ \theta,\ z,\ t)$ describe the deformation of the given disc element. Note that the same description can be used also for row of blades elements [8].

15.2.2 Shape functions

As the disc element is based on the plate theory, i.e., the disc is assumed to be a two-dimensional object, displacements u, v, and w of the midplane of the disc are just functions of r, θ, and t.

A nondimensional radial coordinate

$$\chi = \frac{r - r_1}{\Delta r} , \tag{15.8}$$

where $\Delta r = r_2 - r_1$ is the radial width of the element, is introduced. The values of χ span from 0 at the inner edge of the element ($r = r_1$) to 1 at the outer edge ($r = r_2$).

As already stated, the displacements u, v, and w are expressed by trigonometric polynomials in angle θ

$$\begin{cases} u\left(r,\theta,t\right) = u_0\left(r,t\right) - \frac{1}{2}r\phi_0^2\left(r,t\right) + \\ \qquad\qquad + \sum_{i=1}^{n}\left[u_{ic}\left(r,t\right)\cos(i\theta) + u_{is}\left(r,t\right)\sin(i\theta)\right] , \\ v\left(r,\theta,t\right) = r\phi_0\left(r,t\right) + \sum_{i=1}^{n}\left[v_{ic}\left(r,t\right)\cos(i\theta) - v_{is}\left(r,t\right)\sin(i\theta)\right] , \\ w\left(r,\theta,t\right) = w_0\left(r,t\right) + \sum_{i=1}^{n}\left[w_{ic}\left(r,t\right)\cos(i\theta) + w_{is}\left(r,t\right)\sin(i\theta)\right] . \end{cases} \tag{15.9}$$

The zero-order harmonic includes a radial extensional displacement u_0, a torsional rotation ϕ_0, and an axial (out-of-plane bending) displacement w_0, all functions of the radius and time. As the radial displacement is linked only with very high frequency modes, uncoupled with the modes of the shaft, the first term is usually neglected. This assumption is similar to that seen for the case of the dynamics of rings.

Note that the displacements u, v, and w are displacements of point P from the position the same point would take if the disc was a rigid body. The rotation ϕ_0 is then an angular displacement at point P, caused by the torsional deformation of the disc, but has nothing to do with the torsional rotation of the shaft. The displacements caused by the latter are included in the terms in ϕ_z in Equation (15.7).

The first order harmonics ($i = 1$) in Equation (15.7) corresponds to a rigid displacement of a circumference made by all points with a given value of r in xy-plane (a combination of displacements u and v) and a rigid rotation of the circumference about X- and y-axes (displacement w) as shown in Figure 15.4. The components u_{ic} and v_{ic} have the same effect: A displacement of the circle in x-direction, whereas u_{is} and v_{is} are displacements of the circle in y-direction. By convention, the displacement is ascribed half to the radial u and half to the circumferential v displacement, by assuming $u_{ic} = v_{ic}$ and $u_{is} = v_{is}$. The axial displacement w_{ic} corresponds to a rotation about $-y$-axis, whereas w_{is} to a rotation about X-axis.

Figure 15.5 shows the displacements corresponding to the second-order harmonics [$i = 2$ in Equation (15.7)]. The components u_{ic} and u_{is} are two-lobed flexural deformations, and v_{ic} and v_{is} are purely circumferential extensional displacements. The axial displacements w_{ic} and w_{is} are flexural

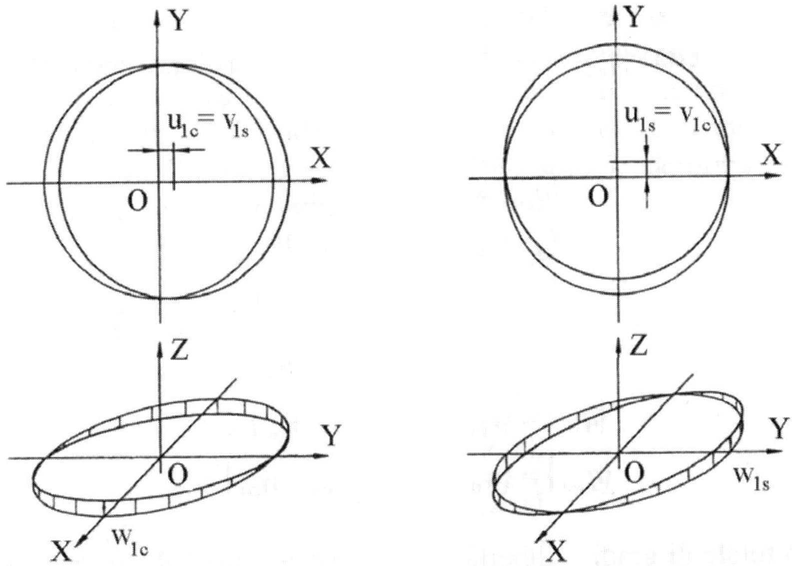

FIGURE 15.4. First harmonic component of the displacement of a circle.

out-of-plane displacements. Note that all second-order harmonics terms yield a displacement pattern that does not displace the center of mass or change the direction of the rotation axis.

The same holds for all harmonics of higher order. As the various harmonics are uncoupled with each other, only the zero- and first-order harmonics need to be considered in the study of the dynamic behavior of the rotor as a whole.

Coefficients u_0, u_{ic}, u_{is}, etc. are functions of the radial coordinate r, or better χ. As usual with the FEM, they are approximated by using shape functions.

The expressions for the coefficients of the zero-order harmonics (as seen above, u_0 is here neglected) are

$$\phi_0(r,t) = \mathbf{N}_1\mathbf{q}_{\phi_0}(t) \ , \qquad w_0(r,t) = \mathbf{N}_3\mathbf{q}_{w_0}(t) \ , \tag{15.10}$$

where the shape functions are

$$\begin{array}{l} \mathbf{N}_1 = \left[\begin{array}{cc} 1-\chi & \chi \end{array}\right] , \\ \mathbf{N}_3 = \left[\begin{array}{cccc} 2\chi^3-3\chi^2+1 & \chi^2(2\chi-3) & -\chi^3+2\chi^2-\chi & \chi\left(1-\chi^2\right) \end{array}\right] . \end{array} \tag{15.11}$$

The use of cubic shape functions (instead of linear ones) for out-of-plane displacement compels us to resort to four generalized coordinates per element in vector $\mathbf{q}_{w_0}$

$$\mathbf{q}_{\phi_0} = \left[\begin{array}{cc} \phi_{z1} & \phi_{z2} \end{array}\right]^T , \qquad \mathbf{q}_{w_0} = \left[\begin{array}{cccc} \phi_{01} & \beta_{01} & \phi_{02} & \beta_{02} \end{array}\right]^T . \tag{15.12}$$

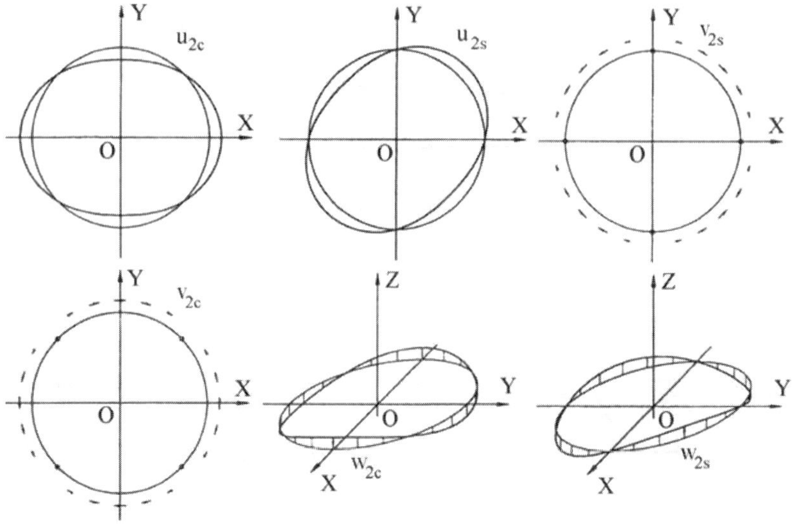

FIGURE 15.5. Second harmonic component of the displacement of a circle.

where $\phi_{0j} = w_{0j}/r_j$ and $\beta_{0j} = (\partial w_0/\partial r)_{r=r_j}$ for $j = 1, 2$.

The expressions for the coefficients of the generic ith order harmonic are

$$\begin{array}{ll} u_{ic}(r,t) = \mathbf{N}_1 \mathbf{q}_{iuc}(t) \ , & u_{is}(r,t) = \mathbf{N}_1 \mathbf{q}_{ius}(t) \ , \\ v_{ic}(r,t) = \mathbf{N}_1 \mathbf{q}_{ivc}(t) \ , & v_{is}(r,t) = \mathbf{N}_1 \mathbf{q}_{ivs}(t) \ , \\ w_{ic}(r,t) = \mathbf{N}_3 \mathbf{q}_{iwc}(t) \ , & w_{ic}(r,t) = \mathbf{N}_3 \mathbf{q}_{iws}(t) \ , \end{array} \tag{15.13}$$

where the shape functions are the same as above and the generalized coordinates are

$$\begin{array}{ll} \mathbf{q}_{iuc} = \left[\begin{array}{cc} u_{xi1} & u_{xi1} \end{array} \right]^T , & \mathbf{q}_{ivs} = \left[\begin{array}{cc} v_{yi1} & v_{yi1} \end{array} \right]^T , \\ \mathbf{q}_{ius} = \left[\begin{array}{cc} u_{yi1} & u_{yi1} \end{array} \right]^T , & \mathbf{q}_{iwc} = \left[\begin{array}{cccc} \phi_{yi1} & \beta_{yi1} & \phi_{yi2} & \beta_{yi2} \end{array} \right]^T , \\ \mathbf{q}_{ivc} = -\left[\begin{array}{cc} v_{xi1} & v_{xi1} \end{array} \right]^T , & \mathbf{q}_{iws} = -\left[\begin{array}{cccc} \phi_{xi1} & \beta_{xi1} & \phi_{xi2} & \beta_{xi2} \end{array} \right]^T , \end{array} \tag{15.14}$$

and

$$\begin{array}{ll} \phi_{xij} = w_{yi}/r_j \ , & \phi_{yij} = -w_{xi}/r_j \ , \\ \beta_{xij} = (\partial w_y/\partial r)_{r=rj} \ , & \beta_{0i} = -\left(\partial w_x/\partial r\right)_{r=r_j} \ , \end{array}$$

for $j = 1, 2$. If the Fourier series is limited to two harmonics, the zeroth- and the first-order ones, each element has 28 degrees of freedom: the 6 degrees of freedom of node 0, plus the 11 degrees of freedom of nodes 1 and 2. Of the latter, 3 are related to the zeroth-order harmonic and 8 to the first-order one.

The two elements of $\mathbf{q}_{\phi_0}$, together with the torsional degree of freedom of node 0, constitute the generalized coordinates for torsional behavior. The four elements of $\mathbf{q}_{w_0}$, together with the axial degree of freedom of node 0, constitute the generalized coordinates for axial behavior. The remaining 20 degrees of freedom related to flexural behavior can be grouped obtaining 10 complex degrees of freedom: $\mathbf{q}_{iuc} + i\mathbf{q}_{ius}$ and $\mathbf{q}_{ivc} + i\mathbf{q}_{ivs}$, together with the complex coordinate $X + iY$ of node 0, are the five complex coordinates expressing the in-plane deformation of the element and $\mathbf{q}_{iws}(t) + i\mathbf{q}_{iwc}$, together with the complex coordinate $\phi_y - i\phi_X$ of node 0, are the five complex coordinates expressing the out-of-plane deformation of the element.

As a general approach, this holds also for defining the kinematics of the array of blades, with an important difference: although in the case of the disc element it is justified to use different types of shape functions for the in-plane displacement v and the out-of-plane displacement w, in the case of the blades, they are both caused by blade bending and it is advisable to use cubic, beam-like, shape functions for both. In this way, one additional degree of freedom must be used at each one of nodes 1 and 2 for the zeroth-order harmonics and two for the first-order one. This leads to a total of 34 degrees of freedom per element, 5 for torsional and axial behavior and 24 (or 12 complex) for flexural behavior.

15.2.3 Kinetic and potential energy

The kinetic energy of the annular disc element with axial thickness $h(r)$ is

$$\mathcal{T} = \frac{1}{2} \int_{r_1}^{r_2} \int_0^{2\pi} \rho h (\dot{\overline{\mathrm{P} - \mathrm{O}}})^T (\dot{\overline{\mathrm{P} - \mathrm{O}}}) r \, dr \, d\theta \, . \tag{15.15}$$

In the same way, the kinetic energy of the part of a row of n blades, with cross-section area A, extending from radius r_1 to radius r_2 is

$$\mathcal{T} = \frac{1}{2} \sum_{j=1}^{n} \int_{r_1}^{r_2} \rho A (\dot{\overline{\mathrm{P} - \mathrm{O}}})^T (\dot{\overline{\mathrm{P} - \mathrm{O}}}) \, dr \, . \tag{15.16}$$

The potential energy of the disc annular element is the sum of two terms. The first one is directly caused by the deformation of the plate

$$\mathcal{U}_e = \frac{1}{2} \int_{r_1}^{r_2} \int_0^{2\pi} \rho h \boldsymbol{\epsilon}^T \mathbf{D} \boldsymbol{\epsilon} r \, dr \, d\theta \, , \tag{15.17}$$

where the strain vector $\boldsymbol{\epsilon}$ is

$$\boldsymbol{\epsilon} = \left\{ \begin{array}{c} \dfrac{\partial u}{\partial r} \\ \dfrac{u}{r} + \dfrac{1}{r}\dfrac{\partial v}{\partial \theta} \\ \dfrac{1}{r}\dfrac{\partial v}{\partial \theta} + \dfrac{\partial v}{\partial r} - \dfrac{v}{r} \\ \dfrac{\partial^2 w}{\partial r^2} \\ -\dfrac{1}{r}\left(\dfrac{\partial w}{\partial r} + \dfrac{1}{r}\dfrac{\partial^2 w}{\partial \theta^2}\right) \\ 2\left(\dfrac{1}{r^2}\dfrac{\partial w}{\partial \theta} - \dfrac{1}{r}\dfrac{\partial^2 w}{\partial r \partial \theta}\right) \end{array} \right\} \tag{15.18}$$

and matrix $\mathbf{D}$ is

$$\mathbf{D} = \begin{bmatrix} \dfrac{Eh}{1-\nu^2}\mathbf{D}_1 & \mathbf{0} \\ \mathbf{0} & \dfrac{Eh^3}{12\left(1-\nu^2\right)}\mathbf{D}_1 \end{bmatrix} , \tag{15.19}$$

where

$$\mathbf{D}_1 = \begin{bmatrix} 1 & \nu & 0 \\ \nu & 1 & 0 \\ 0 & 0 & \frac{1}{2}\left(1-\nu\right) \end{bmatrix} . \tag{15.20}$$

The second term, usually referred to as *geometric effect*, is caused by the stress field present in the undeflected element (σ_r, σ_c) caused by centrifugal stressing but also by other causes. Note that only σ_r and σ_c are considered owing to the plane stress and axial symmetry assumptions; they can be computed using any closed-form (if possible) or numerical model currently used in static stress analysis. The expression of the geometric effect is

$$\begin{aligned} \mathcal{U}_g = \frac{1}{2}\int_{r_1}^{r_2}\int_0^{2\pi} & \left[\sigma_r\left(\frac{\partial w}{\partial r}\right)^2 + \sigma_c\left(\frac{1}{r}\frac{\partial w}{\partial \theta}\right)^2 + \right. \\ & \left. +\sigma_r\left(\frac{\partial v}{\partial r}\right)^2 + \sigma_c\left(\frac{1}{r}\frac{\partial u}{\partial \theta}\right)^2\right] r\,dr\,d\theta \, . \end{aligned} \tag{15.21}$$

In case of centrifugal stressing, both σ_r and σ_c are proportional to Ω^2, whereas they are independent of Ω if other load conditions are considered. The corresponding expressions for the element used to model a row of blades can be obtained using expressions similar to those used for beam elements, with the important difference that in general, the cross section of the blade is set at an angle with respect to the axial direction and the displacement components v and w do not occur along the principal axes of elasticity of the blades.

Remark 15.1 *When the blade is set at an angle with respect to the axial or circumferential direction, the torsional and the axial behavior of the rotor are coupled with each other. However, the flexural behavior remains uncoupled because its uncoupling is due to the fact that the various harmonics of the Fourier decomposition of the displacements act independently from each other. This feature does not depend on the orientation of the blades.*

15.2.4 Element matrices

The equation of motion of the element can be obtained as usual through Lagrange equations. The equation describing the torsional-axial behavior is

$$\mathbf{M}_{ta}\ddot{\mathbf{q}}_{ta} + \left(\mathbf{K}_{ta} + \Omega^2\mathbf{K}_{\Omega_{ta}}\right)\mathbf{q}_{ta}=\mathbf{0} \ , \tag{15.22}$$

where the vector of the torsional-axial generalized coordinates $\mathbf{q}_{ta}$ has eight elements in the case of disc elements and ten for blade arrays elements and matrix $\mathbf{K}_{ta}$ contains both the elastic terms and the geometric terms independent from Ω (e.g., those caused by thermal stressing). In the case of disc elements, or for blades in which the cross section is either perpendicular or parallel to the axial direction, the torsional-axial equation further uncouples into two separate equations for torsional and axial dynamics.

The equation describing the flexural behavior can be written in terms of complex coordinates

$$\mathbf{M}_f\ddot{\mathbf{q}}_f - i\Omega\mathbf{G}_f\dot{\mathbf{q}}_f + \left(\mathbf{K}_f + \Omega^2\mathbf{K}_{g_f}\right)\mathbf{q}_f=\mathbf{0} \ . \tag{15.23}$$

Complex vector $\mathbf{q}_f$ has eight elements in the case of disc elements and ten for blade elements.

The expressions of the various matrices in Equations (15.22) and (15.23), or better of the functions that, once integrated numerically along the radius, yield the matrices, are too complicated to be reported here. Complete derivations for the disc and the blade elements are reported in [7] and [8], respectively.

Hysteretic damping matrices can be obtained in the usual way, remembering that only the elastic stiffness matrix must be multiplied by the loss factor, whereas geometric matrices do not enter the computation of damping matrices. Obviously, if the element rotates, its damping must be considered as rotating damping and assembled accordingly.

Example 15.2 *Thin disc on a compliant spindle. Consider a very thin disc (e.g., the blade of a circular saw) mounted on a compliant spindle. Let the data of the system be as follows:*

- *Disc: outer radius $r_0 = 300$ mm, thickness $h = 1.5$ mm.*
- *Spindle: shaft diameter $\phi = 30$ mm, length $l = 600$ mm, disc position at 200 mm from the end.*

- *Material:* $E = 2.1 \times 10^{11}$ N/m^2, $\rho = 7810$ kg/m^3, $\nu = 0.3$.
- *Bearings: stiff bearings at the ends of the shaft.*

The system is modeled using the $1\frac{1}{2}$ dimensional approach through 6 Timoshenko beam elements and 13 thin disc elements of the type described above.

As a first thing, consider the system as made of a rigid disc on a compliant spindle. The inertial properties of the disc assumed as a rigid body are $m = 3.312$ *kg,* $J_p = 0.1491$ *kg m*^{2}, *and* $J_t = 0.0745$ *kg m*^{2}. *The Campbell diagram for the compliant spindle with a rigid disc is shown in Figure 15.6(a), full lines. The critical speed is* $\Omega_{cr} = 694.3$ *rad/s* $= 6629$ *rpm.*

The dynamics of the thin disc, modeled as a membrane is easily studied through Equations (14.7)

$$\omega = \beta\Omega,$$

where coefficients β are reported in Table 14.3. The results are reported in Figure 15.6(a), dashed lines. No critical speed is caused by the flexibility of the membrane.

The coupled dynamics is studied in Figure 15.6(b). The critical speed is now sightly lower, as expected: $\Omega_{cr} = 664.4$ *rad/s* $= 6434$ *rpm. Note that the first forward modes, of both the disc and the spindle, are fairly uncoupled, whereas the other modes are more strongly affected by coupling. At any rate, the flexibility of the disc does not give way to critical speeds, which are dominated by the spindle behavior.*

15.3 General finite element discretization

The solutions seen in the previous section can be only used to model a thin bladed disc. If the disc is thick and cannot be assimilated to a plate or if the rotor is an actual three-dimensional structure, it must be modeled following the practice of general solid modeling, adding Coriolis and gyroscopic effects, centrifugal stiffening, and a distinction between rotating and nonrotating damping that are essential in rotordynamics but are usually not accounted for in general-purpose FEM codes.

The kinematic description of the deformation of a rotating body is difficult, in particular for what the definition of rotation is concerned. When the rotor is modeled using beam-like elements, the cross sections were considered as rigid bodies and there was no problem in defining their rotation and their angular velocity. If the body is a general flexible body, the deformation parameter and then the angular velocity change from point to point: There is no way of defining exactly what the spin speed of the rotor is. Nevertheless, an approximate approach is possible: The displacements caused by deformations are assumed to be infinitesimal and a spin axis is defined, at least for each point of the rotor. As a consequence of this assumption, the Lagrangian and the Eulerian approaches differ from each

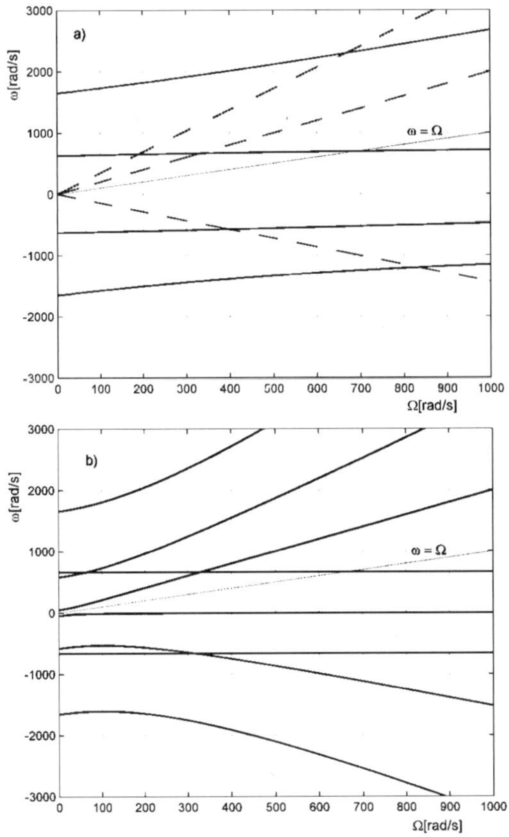

FIGURE 15.6. Campbell diagram for a flexible disc on a compliant spindle. (a) Uncoupled dynamics. Full lines: rigid dics on compliant spindle; dashed lines: flexible membrane on a rigid spindle. (b) Coupled dynamics.

other by small quantities and are considered as coincident [70]. Many difficulties, however, originate in the definitions of the exact kinematics, and different choices in the details of the model lead to different results, some of which are unacceptable, because they neglect gyroscopic effects or other points that are known to be important in rotordynamics.

The approach here reported has the advantage of being not too complicated while leading to results that are in good accordance with the few closed-form solutions (when available) and with the experimental results.

15.3.1 *Kinematics of the deformation of a rotating body*

Consider a flexible body rotating about an axis whose direction is fixed in space, except for small displacements and rotations, and assume that

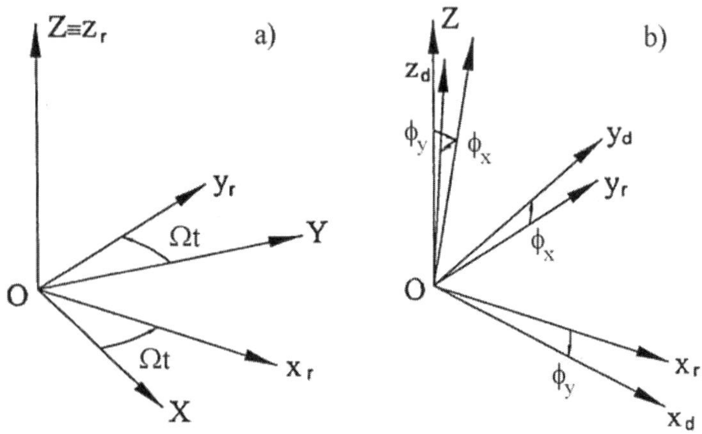

FIGURE 15.7. Reference frames; definition of rotations. (a) Inertial and rotating frames; (b) rotating and dynamic frames.

its angular velocity Ω is constant. Let $OXYZ$ be an inertial frame, with Z-axis coinciding with the spin axis of the rotor in its undeflected position. Define a rotating reference frame $\mathrm{O}x_r y_r z_r$ with its z_r-axis coinciding with the Z-axis of the inertial frame and with axes x_r and y_r rotating in the XY plane at speed Ω [Figure 15.7(a)].

The rotation matrix allowing us to express a vector in the rotating frame from the components of the same vector in the inertial frame is

$$\mathbf{R}_1 = \begin{bmatrix} \cos(\Omega t) & \sin(\Omega t) & 0 \\ -\sin(\Omega t) & \cos(\Omega t) & 0 \\ 0 & 0 & 1 \end{bmatrix} . \tag{15.24}$$

Define then a further reference frame, $\mathrm{O}x_d y_d z_d$, by rotating the $\mathrm{O}x_r y_r z_r$-frame about its y_r- and x_r-axes by angles ϕ_y and ϕ_x [Figure 15.7(b)]. The latter frame, which is often referred to as the *dynamic frame* [70], has the same origin as the other frames and has its axes oriented in such a way that it follows the deformation of the body. Apart from the position of its origin, it is similar to the rotor-fixed frame defined in Chapter 3, with the difference that in that case, there was only one rotor-fixed frame, because the rotor was assumed to be rigid, whereas here there is a dynamic frame for each point of the flexible body.

The rotation matrices allowing us to express a vector in the dynamic frame from the components of the same vector in the rotating frame are

$$\mathbf{R}_2 = \begin{bmatrix} \cos(\phi_y) & 0 & -\sin(\phi_y) \\ 0 & 1 & 0 \\ \sin(\phi_y) & 0 & \cos(\phi_y) \end{bmatrix} , \tag{15.25}$$

$$\mathbf{R}_3 = \begin{bmatrix} 1 & 0 & 0 \\ 0 & \cos(\phi_x) & \sin(\phi_x) \\ 0 & -\sin(\phi_x) & \cos(\phi_x) \end{bmatrix} .$$

A vector $\mathbf{r}_d$ in the dynamic frame is thus obtained from the its expression in the inertial frame $\mathbf{r}_i$ as

$$\mathbf{r}_d = \mathbf{R}_3\mathbf{R}_2\mathbf{R}_1\mathbf{r}_i = \mathbf{R}\mathbf{r}_i \; . \tag{15.26}$$

Note that this way of defining the reference frames follows closely what was seen in Chapter 3 for a rigid rotor, with the difference that here there is an infinity of dynamic frames, and in that case there was only one. Furthermore, here the frames are all centered in O, and there the rotating frame was centered in point C, fixed to the rotor. The order of the rotations is also different, because in the present case, the first rotation occurs about Z-axis, and then rotations about y- and x-axes (obviously in their rotated positions) are performed, whereas in Chapter 3, the order was exactly the opposite.

The position of a generic point P of the rotor can be considered as the sum of a vector $\mathbf{x}$ defining its position in the undeformed body and a vector $\mathbf{u}$ defining its displacement caused by deformation. If the system was rigid and the only motion was rotation about the spin axis, the position of any point of the rotor would be a constant in the rotating frame. If deformation is considered, the coordinates of point P are, in the dynamic frame,

$$(\overline{\mathrm{P}-\mathrm{O}})_d = \mathbf{x}_d + \mathbf{u}_d = \left\{ \begin{array}{c} X + u(t) \\ Y + v(t) \\ Z + w(t) \end{array} \right\}_d . \tag{15.27}$$

In the dynamic frame, the components X, Y. and Z of $\mathbf{x}$ are constant, and the components of the displacement u, v, and w, are functions of time.

The position of point P in the inertial frame is then

$$(\overline{\mathrm{P'}-\mathrm{O}})_i = \mathbf{R}^T(\mathbf{x}_d + \mathbf{u}_d) = \mathbf{R}^T \left\{ \begin{array}{c} X + u(t) \\ Y + v(t) \\ Z + w(t) \end{array} \right\}_d . \tag{15.28}$$

By differentiating the expressions of the coordinates with respect to time, the velocity of point P in the inertial frame is obtained

$$\vec{V}_{\mathrm{P}_i} = \mathbf{R}^T(\dot{\mathbf{x}}_d + \dot{\mathbf{u}}_d) + \dot{\mathbf{R}}^T(\mathbf{x}_d + \mathbf{u}_d) \; . \tag{15.29}$$

Provided that it is correctly computed, i.e., the differentiation is performed in an inertial frame, the velocity of point P can be expressed by its components in any frame (for the computation of the kinetic energy is the modulus of the velocity that matters, not how it is decomposed) and the vectors included in its expression may be expressed in any reference frame. The choice of these frames is arbitrary, and it may be guided by how simply the results are obtained. It is expedient to express the velocity vector in the rotating frame and to use the same frame to express vectors $\mathbf{x}$ and $\mathbf{u}$.

As

$$(\mathbf{x}_d + \mathbf{u}_d) = \mathbf{R}_3\mathbf{R}_2(\mathbf{x}_r + \mathbf{u}_r) , \tag{15.30}$$

Equation (15.29) can be written as

$$\vec{V}_{\mathrm{P}_i} = \mathbf{R}^T \frac{d}{dt}[\mathbf{R}_3\mathbf{R}_2(\mathbf{x}_r + \mathbf{u}_r)] + \dot{\mathbf{R}}^T\mathbf{R}_3\mathbf{R}_2(\mathbf{x}_r + \mathbf{u}_r) . \tag{15.31}$$

The velocity of point P can then be expressed in the rotating frame

$$\vec{V}_{\mathrm{P}_r} = \mathbf{R}_1\vec{V}_{\mathrm{P}_i} = \mathbf{R}_1\mathbf{R}^T \frac{d}{dt}[\mathbf{R}_3\mathbf{R}_2(\mathbf{x}_r + \mathbf{u}_r)] + \mathbf{R}_1\dot{\mathbf{R}}^T\mathbf{R}_3\mathbf{R}_2(\mathbf{x}_r + \mathbf{u}_r) ; \tag{15.32}$$

i.e.,

$$\begin{aligned}\vec{V}_{\mathrm{P}_r} &= \mathbf{R}_2^T\mathbf{R}_3^T \tfrac{d}{dt}[\mathbf{R}_3\mathbf{R}_2(\mathbf{x}_r + \mathbf{u}_r)] + \mathbf{R}_1\dot{\mathbf{R}}^T\mathbf{R}_3\mathbf{R}_2(\mathbf{x}_r + \mathbf{u}_r) = \\ &= \dot{\mathbf{u}}_r + \mathbf{R}_1\dot{\mathbf{R}}_1^T(\mathbf{x}_r + \mathbf{u}_r) .\end{aligned} \tag{15.33}$$

By remembering that

$$\mathbf{R}_1\dot{\mathbf{R}}_1^T = \Omega\mathbf{B} , \tag{15.34}$$

where

$$\mathbf{B} = \begin{bmatrix} 0 & -1 & 0 \\ 1 & 0 & 0 \\ 0 & 0 & 0 \end{bmatrix} ,$$

the expression of the velocity of point P is

$$\vec{V}_{\mathrm{P}_r} = \dot{\mathbf{u}}_r + \Omega\mathbf{B}(\mathbf{x}_r + \mathbf{u}_r) ; \tag{15.35}$$

i.e.,

$$\vec{V}_{\mathrm{P}_r} = \left\{ \begin{array}{c} \dot{u}_r - \Omega(y + v_r) \\ \dot{v}_r + \Omega(x + u_r) \\ \dot{w}_r \end{array} \right\} . \tag{15.36}$$

Note that the axial position of point P does not appear in the expression of the velocity.

15.3.2 *Kinetic energy*

The kinetic energy can thus be expressed as

$$\mathcal{T} = \frac{1}{2}\int_V |\vec{V}_{\mathrm{P}}|^2 dm = \tag{15.37}$$

$$= \frac{1}{2}\int_V \left[\rho\dot{\mathbf{u}}_r^T\dot{\mathbf{u}}_r dV + 2\Omega\rho\dot{\mathbf{u}}_r^T\mathbf{B}(\mathbf{x}_r + \mathbf{u}_r) + \Omega^2\rho(\mathbf{x}_r + \mathbf{u}_r)^T\mathbf{B}^T\mathbf{B}(\mathbf{x}_r + \mathbf{u}_r)\right] dV ,$$

where ρ is the density of the material and V is the volume of the body. Note that the volume V over which the integration must be performed does not change in time, because the integration is performed with reference to the rotating frame, in which the body has just small vibrations but not large displacements and rotations.

The kinetic energy can thus be considered as the sum of three terms: A translational term,

$$\mathcal{T}_0 = \frac{1}{2}\int_V \rho\dot{\mathbf{u}}_r^T\dot{\mathbf{u}}_r dV \ , \tag{15.38}$$

which does not depend on the spin speed and yields the usual mass matrix; a gyroscopic term,

$$\mathcal{T}_1 = \Omega\int_V \rho\dot{\mathbf{u}}_r^T\mathbf{B}\left(\mathbf{x}_r + \mathbf{u}_r\right) dV \ , \tag{15.39}$$

linear in Ω, linked with the Coriolis and gyroscopic effects, yielding the gyroscopic matrix; and a centrifugal term

$$\mathcal{T}_2 = \frac{1}{2}\Omega^2\int_V \rho\left(\mathbf{x}_r + \mathbf{u}_r\right)^T\mathbf{A}\left(\mathbf{x}_r + \mathbf{u}_r\right) dV \ , \tag{15.40}$$

where

$$\mathbf{A} = \mathbf{B}^T\mathbf{B} = \begin{bmatrix} 1 & 0 & 0 \\ 0 & 1 & 0 \\ 0 & 0 & 0 \end{bmatrix} \ , \tag{15.41}$$

caused by the noninertial terms from rotation, other than the one considered above and depending on the square of the spin speed.

The expressions of the three terms of the kinetic energy can be written explicitly as

$$\begin{aligned} \mathcal{T}_0 &= \frac{1}{2}\int_V \rho\left(\dot{u}_r^2 + \dot{v}_r^2 + \dot{w}_r^2\right) dV \ , \\ \mathcal{T}_1 &= \Omega\int_V \rho\left[\dot{v}_r\left(x + u_r\right) - \dot{u}_r\left(y + v_r\right)\right] dV \ , \\ \mathcal{T}_2 &= \frac{1}{2}\Omega^2\int_V \rho\left[\left(x + u_r\right)^2 + \left(y + v_r\right)^2\right] dV \ . \end{aligned} \tag{15.42}$$

If the displacement is expressed in cylindrical coordinates and u, v, and w are the displacements in radial, circumferential, and axial directions, the three contributions to the kinetic energy are

$$\mathcal{T}_0 = \frac{1}{2}\int_V \rho\left(\dot{u}^2+\dot{v}^2+\dot{w}^2\right)dV\ ,$$

$$\mathcal{T}_1 = \Omega\int_V \rho\left[\dot{v}\left(r+u\right)-\dot{u}v\right]dV\ , \tag{15.43}$$

$$\mathcal{T}_2 = \frac{1}{2}\Omega^2\int_V \rho\left[\left(r+u\right)^2+v^2\right]dV\ ,$$

where r is the radius of point P.

The finite element discretization allows us to write the displacements of the points belonging to any element as

$$\mathbf{u}_r = \mathbf{N}\mathbf{q}\ , \tag{15.44}$$

where $\mathbf{N}$ is the matrix of the shape functions and $\mathbf{q}$ is the vector of the generalized coordinates of the element.

The first term of the kinetic energy $\mathcal{T}_0$ can thus be written in the form

$$\mathcal{T}_0 = \frac{1}{2}\int_V \rho\dot{\mathbf{q}}^T\mathbf{N}^T\mathbf{N}\dot{\mathbf{q}}dV = \frac{1}{2}\dot{\mathbf{q}}^T\mathbf{M}\dot{\mathbf{q}}\ , \tag{15.45}$$

where the mass matrix is.

$$\mathbf{M} = \int_V \rho\mathbf{N}^T\mathbf{N}dV\ . \tag{15.46}$$

The second term of the kinetic energy $\mathcal{T}_1$ is

$$\mathcal{T}_1 = \Omega\int_V \rho\dot{\mathbf{q}}^T\mathbf{N}^T\mathbf{B}\left(\mathbf{x}_r+\mathbf{N}\mathbf{q}\right)dV = \Omega\dot{\mathbf{q}}^T\mathbf{G}\mathbf{q}+\Omega\dot{\mathbf{q}}^T\mathbf{f}_1 \tag{15.47}$$

and contains the gyroscopic matrix

$$\mathbf{G} = \int_V \rho\mathbf{N}^T\mathbf{B}\mathbf{N}dV \tag{15.48}$$

plus a vector

$$\mathbf{f}_1 = \int_V \rho\mathbf{N}^T\mathbf{B}\mathbf{x_r}dV\ . \tag{15.49}$$

The term in $\mathbf{f}_1$, once differentiated with respect to $\dot{\mathbf{q}}$, yields constant terms and vanishes in the following differentiation with respect to time included in the Lagrange equations, yielding no terms in the equations of motion.

The third term

$$\mathcal{T}_2 = \frac{1}{2}\Omega^2\int_V \rho\left(\mathbf{x}_r+\mathbf{N}\mathbf{q}\right)^T\mathbf{A}\left(\mathbf{x}_r+\mathbf{N}\mathbf{q}\right)dV = \frac{1}{2}\Omega^2E+\frac{1}{2}\Omega^2\mathbf{q}^T\mathbf{f}_2+\frac{1}{2}\Omega^2\mathbf{q}^T\mathbf{M}_{ni}\mathbf{q} \tag{15.50}$$

where

$$E=\int_V \rho \mathbf{x_r^T A x_r} dV\ ,\quad \mathbf{f}_2=2\int_V \rho \mathbf{N^T A x_r} dV\ ,\quad \mathbf{M}_{ni}=\int_V \rho \mathbf{N^T A N} dV\ , \tag{15.51}$$

is made by three parts. The first one is a constant, the kinetic energy caused by rotation of the rigid body, and yields no terms in the equation of motion. The second one once differentiated with respect to $\mathbf{q}$ yields terms not containing the generalized coordinates. The last term yields a noninertial term that enters the equations of motion.

The final expression of the kinetic energy is thus

$$\mathcal{T}=\frac{1}{2}\dot{\mathbf{q}}^T\mathbf{M}\dot{\mathbf{q}}+\frac{1}{2}\Omega\dot{\mathbf{q}}^T\mathbf{G}\mathbf{q}+\frac{1}{2}\Omega\dot{\mathbf{q}}^T\mathbf{f}_1+\frac{1}{2}\Omega^2 E+\frac{1}{2}\Omega^2\mathbf{q}^T\mathbf{f}_2+\frac{1}{2}\Omega^2\mathbf{q}^T\mathbf{M}_{ni}\mathbf{q}\ . \tag{15.52}$$

15.3.3 Potential energy

As the displacements u_r, v_r, and w_r were defined with reference to a rotating frame, the potential energy coincides with that of a nonrotating body. To take into account that the body can be loaded by static stresses while vibrating, the potential energy of the elementary volume dV can be written in the form

$$d\mathcal{U}=\int \boldsymbol{\sigma}^T d\boldsymbol{\epsilon}=\int \left[\boldsymbol{\sigma}_0^T+\left(\boldsymbol{\epsilon}^T-\boldsymbol{\epsilon}_0^T\right)\mathbf{E}\right] d\boldsymbol{\epsilon}\ , \tag{15.53}$$

where $\boldsymbol{\sigma}_0$ and $\boldsymbol{\epsilon}_0$ are the pre-stress and per-strain fields, respectively. As they are constant during deformation, while $\boldsymbol{\epsilon}$ grows linearly, by performing the integration, it follows that

$$d\mathcal{U}=\boldsymbol{\sigma}_0^T\boldsymbol{\epsilon}+\frac{1}{2}\boldsymbol{\epsilon}^T\mathbf{E}\boldsymbol{\epsilon}-\boldsymbol{\epsilon}_0^T\mathbf{E}\boldsymbol{\epsilon}\ . \tag{15.54}$$

Although the term in $\boldsymbol{\epsilon}^T\mathbf{E}\boldsymbol{\epsilon}$ is quadratic, the other two are linear in the strains $\boldsymbol{\epsilon}$. In their computation, the second-order (nonlinear) terms of the strains $\boldsymbol{\epsilon}_{nl}$ must be accounted for together with the usual terms $\boldsymbol{\epsilon}_l$

$$\boldsymbol{\epsilon}=\left\{\begin{array}{c}\epsilon_x\\ \epsilon_y\\ \epsilon_z\\ \gamma_{xy}\\ \gamma_{xz}\\ \gamma_{yz}\end{array}\right\}=\left\{\begin{array}{c}\frac{\partial u}{\partial x}\\ \frac{\partial v}{\partial y}\\ \frac{\partial w}{\partial z}\\ \frac{\partial u}{\partial y}+\frac{\partial v}{\partial x}\\ \frac{\partial u}{\partial z}+\frac{\partial w}{\partial x}\\ \frac{\partial v}{\partial z}+\frac{\partial w}{\partial y}\end{array}\right\}+\left\{\begin{array}{c}\frac{1}{2}\left[\left(\frac{\partial u}{\partial x}\right)^2+\left(\frac{\partial v}{\partial x}\right)^2+\left(\frac{\partial w}{\partial x}\right)^2\right]\\ \frac{1}{2}\left[\left(\frac{\partial u}{\partial y}\right)^2+\left(\frac{\partial v}{\partial y}\right)^2+\left(\frac{\partial w}{\partial y}\right)^2\right]\\ \frac{1}{2}\left[\left(\frac{\partial u}{\partial z}\right)^2+\left(\frac{\partial v}{\partial z}\right)^2+\left(\frac{\partial w}{\partial z}\right)^2\right]\\ \frac{\partial u}{\partial x}\frac{\partial u}{\partial y}+\frac{\partial v}{\partial x}\frac{\partial v}{\partial y}+\frac{\partial w}{\partial x}\frac{\partial w}{\partial y}\\ \frac{\partial u}{\partial x}\frac{\partial u}{\partial z}+\frac{\partial v}{\partial x}\frac{\partial v}{\partial z}+\frac{\partial w}{\partial x}\frac{\partial w}{\partial z}\\ \frac{\partial u}{\partial y}\frac{\partial u}{\partial z}+\frac{\partial v}{\partial y}\frac{\partial v}{\partial z}+\frac{\partial w}{\partial y}\frac{\partial w}{\partial z}\end{array}\right\}. \tag{15.55}$$

The final expression of the potential energy is then

$$\mathcal{U} = \frac{1}{2}\int_V \boldsymbol{\epsilon}_l^T \mathbf{E}\boldsymbol{\epsilon}_l dV + \int_V \left(\boldsymbol{\sigma}_0^T \boldsymbol{\epsilon}_l + \boldsymbol{\sigma}_0^T \boldsymbol{\epsilon}_{nl} - \boldsymbol{\epsilon}_0^T \mathbf{E}\boldsymbol{\epsilon}_l - \boldsymbol{\epsilon}_0^T \mathbf{E}\boldsymbol{\epsilon}_{nl}\right) dV \ . \quad (15.56)$$

The first contribution is the elastic energy: It contains the strain tensor $\boldsymbol{\epsilon}$ in which the strains caused by the dynamic displacement are listed, and the elasticity tensor **E**, expressing the stress-strain relationship. The second term, the so-called geometrical potential energy, is caused by the nonlinear components of the strain field $\boldsymbol{\epsilon}_{nl}$ and the pre-stresses $\boldsymbol{\sigma}_0$ or pre-strain $\boldsymbol{\epsilon}_0$. Note that in the case of rotors, $\boldsymbol{\sigma}_0$ has usually a component linked with centrifugal stressing, which is proportional to the square of the spin speed (centrifugal stiffening) and a component that has no direct link with the spin speed and is caused by other causes, like, for example, thermal stressing. In most cases, no pre-strain $\boldsymbol{\epsilon}_0$ is present.

The strains of the points belonging to any element can be expressed as

$$\boldsymbol{\epsilon} = \mathbf{B}\mathbf{q} \ , \quad (15.57)$$

where **B** is a matrix containing suitable space derivatives of the shape functions and $\boldsymbol{\epsilon}$ coincides with $\boldsymbol{\epsilon}_l$, the vector containing the strains as in the usual linear approximation.

The potential energy is then

$$\mathcal{U} = \frac{1}{2}\mathbf{q}^T\mathbf{K}\mathbf{q} + \frac{\mathbf{1}}{\mathbf{2}}\mathbf{f}_3^T\mathbf{q} + \frac{1}{2}\mathbf{q}^T\mathbf{K}_g\mathbf{q} \ , \quad (15.58)$$

where the first term is the usual one caused by $\boldsymbol{\epsilon}_l$, and the following ones are linked with prestress. The one containing $\mathbf{f}_3$ is caused by pre-stresses $\boldsymbol{\sigma}_0$ or pre-strain $\boldsymbol{\epsilon}_0$ and by $\boldsymbol{\epsilon}_l$, whereas the last one, which is quadratic in $\mathbf{q}$, is caused by pre-stresses $\boldsymbol{\sigma}_0$ or pre-strain $\boldsymbol{\epsilon}_0$ and by $\boldsymbol{\epsilon}_{nl}$.

15.3.4 Equations of motion of the element

The derivatives that are included into the Lagrange equations are

$$\frac{\partial(\mathcal{T}-\mathcal{U})}{\partial\dot{\mathbf{q}}} = \mathbf{M}\dot{\mathbf{q}} + \frac{1}{2}\Omega\mathbf{G}\mathbf{q} + \frac{1}{2}\Omega\mathbf{f}_1 \ , \quad (15.59)$$

$$\frac{d}{dt}\left(\frac{\partial(\mathcal{T}-\mathcal{U})}{\partial\dot{\mathbf{q}}}\right) = \mathbf{M}\ddot{\mathbf{q}} + \frac{1}{2}\Omega\mathbf{G}\dot{\mathbf{q}} \ , \quad (15.60)$$

$$\frac{\partial(\mathcal{T}-\mathcal{U})}{\partial\mathbf{q}} = \frac{1}{2}\Omega\mathbf{G}^T\dot{\mathbf{q}} + \frac{1}{2}\Omega^2\mathbf{f}_2\Omega^2 + \Omega^2\mathbf{M}_{ni}\mathbf{q} - \mathbf{K}\mathbf{q} - \frac{1}{2}\mathbf{f}_3^T - \mathbf{K}_g\mathbf{q} \ . \quad (15.61)$$

By remembering that **G** is skew-symmetric ($\mathbf{G}^T = -\mathbf{G}$), the equations of motion are then

$$\mathbf{M}\ddot{\mathbf{q}} + \Omega\mathbf{G}\dot{\mathbf{q}} + \left(\mathbf{K} + \mathbf{K}_g - \Omega^2\mathbf{M}_{ni}\right)\mathbf{q} = \frac{1}{2}\Omega^2\mathbf{f}_2 - \frac{1}{2}\mathbf{f}_3^T + \mathbf{f}(t) \ , \quad (15.62)$$

where $\mathbf{f}(t)$ is a generic vector of time-dependent nodal forces applied to the element.

As the problem is linear, the constant terms can be neglected in the dynamic study, and the equation of motion reduces to

$$\mathbf{M}\ddot{\mathbf{q}}+\Omega\mathbf{G}\dot{\mathbf{q}}+\left(\mathbf{K}+\mathbf{K}_g-\Omega^2\mathbf{M}_{ni}\right)\mathbf{q}=\mathbf{f}(t)\ . \tag{15.63}$$

The geometric matrix can be usually subdivided in two parts, a first one linked with centrifugal stiffening, which is proportional to the square of the spin speed, and can be written as $\Omega^2\mathbf{K}_\Omega$, and a second one that is caused by non-speed-dependent effects, like thermoelastic stresses, which is constant. If only the latter is referred to as $\mathbf{K}_g$, the equation of motion becomes

$$\mathbf{M}\ddot{\mathbf{q}}+\Omega\mathbf{G}\dot{\mathbf{q}}+\left[\mathbf{K}+\mathbf{K}_g+\Omega^2\left(\mathbf{K}_\Omega-\mathbf{M}_{ni}\right)\right]\mathbf{q}=\mathbf{f}(t)\ . \tag{15.64}$$

15.4 Equation of motion in the inertial frame

15.4.1 Velocity

When the rotor has axial symmetry, the equation of motion is best expressed with reference to the inertial frame.

To write the equations of motion in the inertial frame, it is expedient to perform the rotations in a different order, the same as seen in Chapter 3: First, the inertial frame is rotated about x-axis (matrix $\mathbf{R}_3$), then it is rotated about y-axis (matrix $\mathbf{R}_2$), and finally about z-axis (matrix $\mathbf{R}_1$). Let the frame resulting after the first two rotations be referred to as the whirling frame and addressed to using subscript w. By remembering that angles ϕ_x and ϕ_y are small, the position of the point P in the whirling frame is then

$$\overline{(\text{P-O})}_w=\mathbf{R}_2\mathbf{R}_3\left\{\begin{array}{c}X_i+u_i\\Y_i+v_i\\Z_i+w_i\end{array}\right\}\approx\left\{\begin{array}{c}X_i+u_i\\Y_i+v_i\\Z_i+w_i+\phi_yX_i-\phi_xY_i\end{array}\right\}. \tag{15.65}$$

The position of the same point in the dynamic frame is then

$$\overline{(\text{P-O})}_d\approx\mathbf{R}_1\left\{\begin{array}{c}X_i+u_i\\Y_i+v_i\\Z_i+w_i+\phi_yX_i-\phi_xY_i\end{array}\right\}. \tag{15.66}$$

By premultiplying the expression for $\overline{(\text{P-O})}_d$ by the transpose of matrix $\mathbf{R}=\mathbf{R}_3\mathbf{R}_2\mathbf{R}_1$, the displacement expressed in the inertial frame is obtained. Its derivative with respect to time is the absolute velocity of point P expressed in the inertial frame

$$V_{\text{P}_i}\approx\left(\mathbf{R}^T\dot{\mathbf{R}}_1+\dot{\mathbf{R}}^T\mathbf{R}_1\right)\left\{\begin{array}{c}X_i+u_i\\Y_i+v_i\\Z_i+w_i+\phi_yX_i-\phi_xY_i\end{array}\right\}+ \tag{15.67}$$

$$+\mathbf{R}^T\mathbf{R}_1\left\{\begin{array}{c}\dot{u}_i\\ \dot{v}_i\\ \dot{w}_i+\dot{\phi}_yX_i-\dot{\phi}_xY_i\end{array}\right\}.$$

As for the computation of the kinetic energy the velocity of point P can be expressed in any reference frame, Equation (15.67) can be premultiplied by matrix $\mathbf{R}_1^T\mathbf{R}$, obtaining

$$V_{\mathrm{P}_i}\approx\mathbf{R}_1^T\left(\dot{\mathbf{R}}_1+\mathbf{R}\dot{\mathbf{R}}^T\mathbf{R}_1\right)\left\{\begin{array}{c}X_i+u_i\\ Y_i+v_i\\ Z_i+w_i+\phi_yX_i-\phi_xY_i\end{array}\right\}+ \tag{15.68}$$

$$+\left\{\begin{array}{c}\dot{u}_i\\ \dot{v}_i\\ \dot{w}_i+\dot{\phi}_yX_i-\dot{\phi}_xY_i\end{array}\right\};$$

i.e., after performing all of the required simplifications linked with the small displacements and small rotations assumptions,

$$V_{\mathrm{P}}\approx\left\{\begin{array}{c}\dot{u}_i+Y_i\left(\dot{\phi}_y\phi_x-\Omega\right)+\dot{\phi}_y\left(Z_i+w_i+\phi_yX_i-\phi_xY_i\right)\\ \dot{v}_i-X_i\left(\dot{\phi}_y\phi_x-\Omega\right)-\dot{\phi}_x\left(Z_i+w_i+\phi_yX_i-\phi_xY_i\right)\\ \dot{w}_i+\dot{\phi}_yX_i-\dot{\phi}_xY_i\end{array}\right\}. \tag{15.69}$$

15.4.2 Kinetic energy

Also, in this case the kinetic energy can thus be considered as the sum of three terms

$$\mathcal{T}=\frac{1}{2}\int_V|\vec{V}_{\mathrm{P}}|^2dm=\mathcal{T}_0+\mathcal{T}_1+\mathcal{T}_2\ .$$

By introducing the expression for the velocity of point P into the expression of the kinetic energy, the first and the last terms are easily computed

$$\begin{array}{l}\mathcal{T}_0=\frac{1}{2}\int_V\rho\left(\dot{u}_i^2+\dot{v}_i^2+\dot{w}_i^2\right)dV\ ,\\ \mathcal{T}_2=\frac{1}{2}\Omega^2\int_V\rho\left(X_i^2+Y_i^2\right)dV\ .\end{array} \tag{15.70}$$

The second term, linked with Coriolis and gyroscopic effects, is much more complicated

$$\mathcal{T}_1=-\Omega\int_V\rho\left[w_i\left(\dot{\phi}_xX_i+\dot{\phi}_yY_i\right)+X_i^2\left(\dot{\phi}_x\phi_y-\dot{\phi}_y\phi_x\right)+\right. \tag{15.71}$$

$$\left.+X_iY_i\left(\dot{\phi}_y\phi_y-\dot{\phi}_x\phi_x\right)\right]dV\ .$$

Angles ϕ_x and ϕ_y are different from point to point, and they can be computed from the displacement field

$$\begin{aligned}\phi_x &= \frac{1}{2}\left(\frac{\partial u_i}{\partial z} - \frac{\partial w_i}{\partial x}\right) , \\ \phi_y &= \frac{1}{2}\left(\frac{\partial w_i}{\partial y} - \frac{\partial v_i}{\partial z}\right) .\end{aligned} \tag{15.72}$$

As this approach is useful in the case of rotors displaying axial symmetry, the integration included in the expression of the kinetic energy can be performed in the inertial frame instead of resorting to the rotating frame (which is here not expressly defined) as it was the case for the previous approach.

By introducing the shape functions to discretize the system following the finite element approach, the expression of the kinetic energy reduces to

$$\mathcal{T} = \frac{1}{2}\dot{\mathbf{q}}^T\mathbf{M}\dot{\mathbf{q}}+\frac{1}{2}\Omega\dot{\mathbf{q}}^T\mathbf{G}\mathbf{q}+\frac{1}{2}\Omega^2 E . \tag{15.73}$$

Note that in the present case, there is no term in Ω^2 containing the generalized coordinates. This could be expected, because an inertial reference frame was used. Note also that while performing the relevant integrals, most terms in the expression for $\mathcal{T}_1$ [Equation (15.71)] vanish.

15.4.3 Equations of motion of the element

The expression for the potential energy is the same as that seen in the previous section, and hence the equation of motion for the dynamic analysis reduces to

$$\mathbf{M}\ddot{\mathbf{q}}+\Omega\mathbf{G}\dot{\mathbf{q}} + (\mathbf{K} + \mathbf{K}_g)\,\mathbf{q} = \mathbf{f}(t) . \tag{15.74}$$

Also, in this case, the geometric matrix can be subdivided into two parts, a first one proportional to the square of the spin speed and a second one that is constant:

$$\mathbf{M}\ddot{\mathbf{q}}+\Omega\mathbf{G}\dot{\mathbf{q}} + \left(\mathbf{K} + \mathbf{K}_g + \Omega^2\mathbf{K}_\Omega\right)\mathbf{q} = \mathbf{f}(t) . \tag{15.75}$$

15.5 Axi-symmetrical annular elements

15.5.1 Shape functions

If the geometrical configuration of the rotor is axially symmetrical, it can be modeled using annular axi-symmetric elements in which the displacements are expressed as functions of the polar angle, as well as of the axial and radial coordinates. This approach has been defined above as a two and one half dimensional approach.

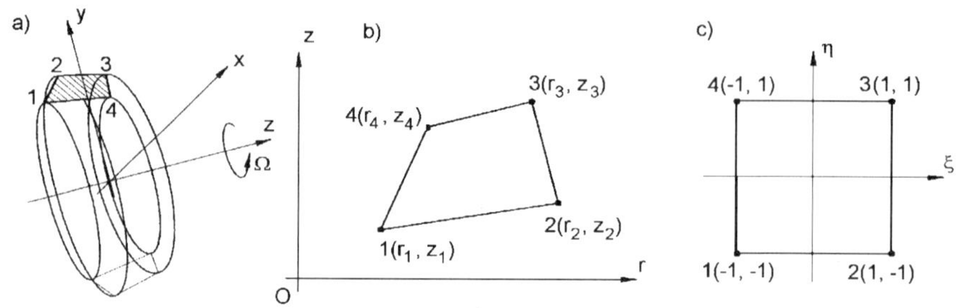

FIGURE 15.8. Four-nodes isoparametric axi-symmetrical element. (a) Sketch of the element; (b) cross section in rZ-plane; (c) element in the auxiliary frame $\xi\eta$ in which the cross section is square.

In the present section, only the case of a simple four-nodes isoparametric element [Figure 15.8(a)] will be described, but simpler (three nodes) or more complex elements having 8, 9, 12, ... nodes can be defined. The cross section of the element [Figure 15.8(b)] is a general quadrilateral in any plane through the axis of rotation; the four nodes have been referred to as node 1, 2, 3, and 4. To make it easier to define the position of a point in the cross section of the element, an auxiliary reference frame $\xi\eta$ has been defined [Figure 15.8(c)], in which the element is transformed into a square, with sides with lengths equal to 2 units.

As the body is axially symmetrical, it is expedient to use cylindrical coordinates $r\theta Z$ for both the definition of the geometry of the element and the displacement field. In the present section, the displacements u, v, and w will then be defined in radial, circumferential, and axial directions.

The relationship allowing us to obtain the coordinates of a point in the physical plane rZ from that in the auxiliary frame $\xi\eta$ is

$$\begin{cases} r = \frac{(1-\xi)(1-\eta)}{4} r_1 + \frac{(1+\xi)(1-\eta)}{4} r_2 + \frac{(1+\xi)(1+\eta)}{4} r_3 + \frac{(1-\xi)(1+\eta)}{4} r_4 \, , \\ Z = \frac{(1-\xi)(1-\eta)}{4} Z_1 + \frac{(1+\xi)(1-\eta)}{4} Z_2 + \frac{(1+\xi)(1+\eta)}{4} Z_3 + \frac{(1-\xi)(1+\eta)}{4} Z_4 \, ; \end{cases} \tag{15.76}$$

i.e., in matrix notation,

$$r = \mathbf{N} \left[\begin{array}{cccc} r_1 & r_2 & r_3 & r_4 \end{array} \right]^T , \qquad Z = \mathbf{N} \left[\begin{array}{cccc} Z_1 & Z_2 & Z_3 & Z_4 \end{array} \right]^T , \tag{15.77}$$

where the shape functions matrix is

$$\mathbf{N}\left(\xi,\eta\right) = \frac{1}{4} \left[\begin{array}{cccc} (1-\xi)(1-\eta) & (1+\xi)(1-\eta) & (1+\xi)(1+\eta) & (1-\xi)(1+\eta) \end{array} \right] . \tag{15.78}$$

As already stated, even if the shape of the rotor is axially symmetrical, the displacement field is not such, and as a consequence, the problem does not reduce to a two-dimensional problem having as generalized coordinates just the radial and axial coordinate: The displacement field depends on r, Z, and θ coordinates and on time. A very convenient way to approximate the displacement field is to use a polynomial expansion along the radial and axial coordinates and a trigonometric expansion along angle θ

$$\left\{\begin{array}{l} u(r,Z,\theta,t)=u_0(r,Z,t)+\sum_{i=1}^{n}\left[u_{ic}(r,Z,t)\cos(i\theta)+u_{is}(r,Z,t)\sin(i\theta)\right] , \\ v(r,Z,\theta,t)=-v_0(r,Z,t)+\sum_{i=1}^{n}\left[-v_{ic}(r,Z,t)\cos(i\theta)+v_{is}(r,Z,t)\sin(i\theta)\right] , \\ w(r,Z,\theta,t)=w_0(r,Z,t)+\sum_{i=1}^{n}\left[w_{ic}(r,Z,t)\cos(i\theta)+w_{is}(r,Z,t)\sin(i\theta)\right] , \end{array}\right. \tag{15.79}$$

where functions u_{ic}, u_{is}, etc. are polynomial expansions in r and Z (or, better, of ξ and η), whose coefficients are the generalized coordinates of the problem and hence are functions of time:

$$\begin{array}{ll} u_{ic}=\mathbf{N}(\xi,\eta)\mathbf{q}_{u_{ic}}(t) , & u_{is}=\mathbf{N}(\xi,\eta)\mathbf{q}_{u_{is}}(t) , \\ v_{ic}=\mathbf{N}(\xi,\eta)\mathbf{q}_{v_{ic}}(t) , & v_{is}=\mathbf{N}(\xi,\eta)\mathbf{q}_{v_{is}}(t) , \\ w_{ic}=\mathbf{N}(\xi,\eta)\mathbf{q}_{w_{ic}}(t) , & w_{is}=\mathbf{N}(\xi,\eta)\mathbf{q}_{w_{is}}(t) . \end{array} \tag{15.80}$$

As the same shape functions are used for expressing both the distorsion of the element (i.e., the relationship between xy- and $\xi\eta$-coordinates) and the displacement field, the element is an isoparametric one.

The minus sign in the second equation (circumferential displacements) has been introduced to have equal formulations for the matrices for sine and cosine components of the displacements.

Each one of the vectors $\mathbf{q}_{k_{ij}}(t)$ has four terms, related to the four nodes of the element: The element has then 24 degrees of freedom for each one of the harmonics of the trigonometric polynomial, except for the zero-order harmonics, which has only the cosine component and hence just 12 degrees of freedom. If a number of harmonics up to the nth is considered, the element has a total of $12 + 24n$ degrees of freedom.

The situation is similar to that seen in Section 15.3 for the simplified elements for the study of bladed discs. The zero-order harmonics models a radial displacement $u_0(r, Z, t)$, a circumferential displacement $v_0(r, Z, t)$, and an axial displacement $w_0(r, Z, t)$ that are constant along angle θ. The corresponding generalized coordinates $\mathbf{q}_{u_{0c}}(t)$, $\mathbf{q}_{v_{0c}}(t)$ and $\mathbf{q}_{w_{0c}}(t)$ (those with subscript s do not exist) are radial, circumferential, and axial displacements at the four nodes that are independent from θ. The first-order harmonics [$i = 1$ in Equation (15.79)] corresponds to rigid displacements of a circumference made by all points with a given value of r and Z in the XY-plane (a combination of displacements u and v) and in a rigid rotation of the circumference about X- and Y-axes (displacement w), as shown in Figure 15.4. The components u_{ic} and v_{ic} have the same effect: a displacement of the circle in X-direction, whereas u_{is} and v_{is} are displacements of the circle in Y-direction. By convention, the displacement is ascribed half to the radial u and half to the circumferential v displacement, by assuming $u_{ic} = v_{ic}$ and $u_{is} = v_{is}$. The axial displacement w_{ic} corresponds to a rotation about $-Y$-axis, and w_{is} to a rotation about X-axis.

The higher order harmonics correspond to modes in which the cross section is deformed; they are likely to be little excited by unbalance and not to have strong effects on the overall behavior of the rotor. Even if these higher order modes are accounted for, the number of harmonics needs generally not to be high and the number of degrees of freedom is much smaller than what is common in full tridimensional modeling.

15.5.2 Kinetic and potential energy

The kinetic energy of the element can be expressed with reference to the fixed frame. The expressions are Equations (15.70) and (15.71) transformed into polar coordinates, with the difference that the integration must be performed in the auxiliary frame and the elementary volume is

$$dV_{r,Z} = \det(\mathbf{J})\, dV_{\xi\eta} \ , \qquad \text{where } \mathbf{J} = \begin{bmatrix} \dfrac{\partial r}{\partial \xi} & \dfrac{\partial r}{\partial \eta} \\ \dfrac{\partial Z}{\partial \xi} & \dfrac{\partial Z}{\partial \eta} \end{bmatrix} \tag{15.81}$$

is the Jacobian matrix of the shape functions.

By introducing Equation (15.79) into (15.70) and (15.71) and integrating, the final expression for the kinetic energy is obtained. The integration along angle θ can be performed in closed form, because only integrals of $\sin(i\theta)$, of $\cos(i\theta)$, and of their products and powers are present. Moreover, because

$$\begin{aligned} &\int_0^{2\pi} \sin(i\theta)\sin(j\theta) = \begin{cases} 0 & \text{for } i \neq j \text{ or } i = j = 0 \ , \\ \pi & \text{for } i = j \neq 0 \ , \end{cases} \\ &\int_0^{2\pi} \cos(i\theta)\cos(j\theta) = \begin{cases} 0 & \text{for } i \neq j \ , \\ \pi & \text{for } i = j \neq 0 \ , \\ 2\pi & \text{for } i = j = 0 \ , \end{cases} \\ &\int_0^{2\pi} \sin(i\theta)\cos(j\theta) = \ 0 \qquad \forall i,j \ , \end{aligned} \tag{15.82}$$

each harmonics of the Fourier series for the displacement yields an independent expression of the kinetic energy. The integration along ξ- and η-coordinates, on the contrary, needs to be performed numerically, but this is a general rule with isoparametric elements, and it adds very little to the complexity of the analysis.

The kinetic energy caused by the ith harmonic of the displacement is thus

$$\mathcal{T}^i = \frac{1}{2}\dot{\mathbf{q}}^{iT}\mathbf{M}^i\dot{\mathbf{q}}^i + \Omega\dot{\mathbf{q}}^{iT}\mathbf{G}^i\mathbf{q}^i \ , \tag{15.83}$$

where $\mathbf{M}^i$, $\mathbf{G}^i$, and $\mathbf{M}^i_{ni}$ are the mass and the gyroscopic matrices. No matrix $\mathbf{M}_{ni}$ is present, because the equation is written in an inertial frame. They are all referred to the ith harmonics and are all symmetric, except for the gyroscopic matrix, which is skew-symmetric. Their size is 12 × 12 for the zeroth-order harmonics and 24 × 24 for the other ones.

The expression of the elastic potential energy is the usual one

$$\mathcal{U}_e = \frac{1}{2}\int_0^{2\pi}\int_{-1}^{1}\int_{-1}^{1} \mathbf{q}^T\mathbf{B}^T\mathbf{E}\mathbf{B}\mathbf{q}\cdot\mathbf{N}\mathbf{r}\det(\mathbf{J})\,d\xi\,d\eta\,d\theta\ ,$$

where **B** and **E** are, respectively, the matrix expressing the relationship between the strains and the displacements, containing the derivatives of the shape functions, and the matrix expressing the stress-strain relationship of the material. Their derivation is not shown here, because they are dealt with in many texts on the finite element method, when dealing with axi-symmetrical elements with harmonic displacements (see, for example, [36]). Also in this case, the various harmonics uncouple, and it is possible to state the elastic potential energy harmonics by harmonics in the form

$$\mathcal{U}_e^i = \frac{1}{2}\mathbf{q}^{iT}\mathbf{K}_e^i\mathbf{q}^i\ . \tag{15.84}$$

Matrix $\mathbf{K}_e^0$ is a 12 × 12 symmetrical matrix, whereas all other matrices $\mathbf{K}_e^i$ are 24 × 24 matrices, all different from each other.

The expression of the geometric potential energy caused by pre-stress $\boldsymbol{\sigma}_0$ for the element in cylindrical coordinates is

$$\mathcal{U}_g = \frac{1}{2}\int_0^{2\pi}\int_{-1}^{1}\int_{-1}^{1} 2\boldsymbol{\epsilon}_{NL}^T\boldsymbol{\sigma}_0\cdot\mathbf{N}\mathbf{r}\det(\mathbf{J})\,d\xi\,d\eta\,d\theta\ ,$$

where the vector of the static stresses can be expressed as

$$\boldsymbol{\sigma}_0 = \Omega^2\boldsymbol{\sigma}_{0\Omega} + \boldsymbol{\sigma}_{01}\ .$$

It is the sum of a term caused by centrifugal forces ($\boldsymbol{\sigma}_{0\Omega}$ is the vector of the centrifugal stresses at unit speed) and a term that is independent from speed (typically, is caused by thermal stresses). The static stress distribution must be computed, e.g., by using the FEM, in advance, and the results are then introduced into the dynamic analysis.

The various harmonics uncouple also in this case, so that the geometric potential energy can be expressed for each harmonics as

$$\mathcal{U}_g^i = \frac{1}{2}\mathbf{q}^{iT}\left(\Omega^2\mathbf{K}_{g\Omega}^i + \mathbf{K}_{g1}^i\right)\mathbf{q}^i\ . \tag{15.85}$$

The matrices $\mathbf{K}_{g\Omega}^0$ and $\mathbf{K}_{g1}^0$, related to the zero-order harmonics, are 12 × 12 symmetrical matrices, and all other matrices are 24 × 24 matrices, all different from each other [71]. The same uncoupling between the sine and cosine components of the deformation seen for the elastic potential energy holds also for the geometric potential energy.

15.5.3 Equations of motion

As the various harmonic components are uncoupled in all matrices, the equations of motion of the element can be obtained separately for each harmonics through Lagrange equations. The equation for the ith harmonic $(i = 0, ..., n)$ is

$$\mathbf{M}^i\ddot{\mathbf{q}}^i + \Omega\mathbf{G}^i\dot{\mathbf{q}}^i + \left(\mathbf{K}^i + \mathbf{K}^i_{g1} + \Omega^2\mathbf{K}^i_{g\Omega}\right)\mathbf{q}^i = \mathbf{0} \,. \qquad (15.86)$$

Note that the equation of motion of the element is of the same type of the equations seen for a general multi-degrees-of-freedom undamped rotor in Chapter 4. Moreover, also in the present cases, all matrices are symmetrical, except for $\mathbf{G}^i$,which is skew-symmetrical.

Also in this case, it is possible to resort to complex coordinates. By stating

$$\mathbf{q}^i = \left\{ \begin{array}{c} \mathbf{q}_{u_{ic}} + i\mathbf{q}_{u_{is}} \\ \mathbf{q}_{v_{is}} + i\mathbf{q}_{v_{ic}} \\ \mathbf{q}_{w_{ic}} + i\mathbf{q}_{w_{is}} \end{array} \right\} , \qquad (15.87)$$

Equation (15.86) becomes

$$\mathbf{M}^i\ddot{\mathbf{q}}^i - i\Omega\mathbf{G}^i\dot{\mathbf{q}}^i + \left(\mathbf{K}^i + \mathbf{K}^i_{g1} + \Omega^2\mathbf{K}^i_{g\Omega}\right)\mathbf{q}^i = \mathbf{0} \,, \qquad (15.88)$$

where all matrices have 12 rows and columns and are symmetrical. As usual, the skew-symmetrical terms in the equation written in real coordinates yield imaginary terms when complex coordinates are used. The number of degrees of freedom of the element is now 12 for each harmonics.

The assembling procedure and the introduction of the constraints proceeds in the same way as for the standard FEM formulation, with the only difference that each harmonics is assembled separately. When using real coordinates, the total number of degrees of freedom of the model is thus $(3 + 6n)\, n_n$, where n is the number of harmonics and n_n is the number of nodes of the model. This is a large number of equations, but the uncoupling between the various harmonics allows us to study separately a problem with $3n_n$ degrees of freedom (zero-order harmonics) plus n problems with $6n_n$ degrees of freedom (all other harmonics). When using complex coordinates, the size of all problems is $3n_n$.

15.6 Axi-symmetrical shell element

If the rotor is made of a thin-walled tube, the use of the axi-symmetrical elements seen in the previous section is questionable. To approximate satisfactorily all deformations in which the wall is inflected, a number of elements should be used in the thickness. As the wall is thin, the elements must be very thin, leading to much distorted elements unless a very fine

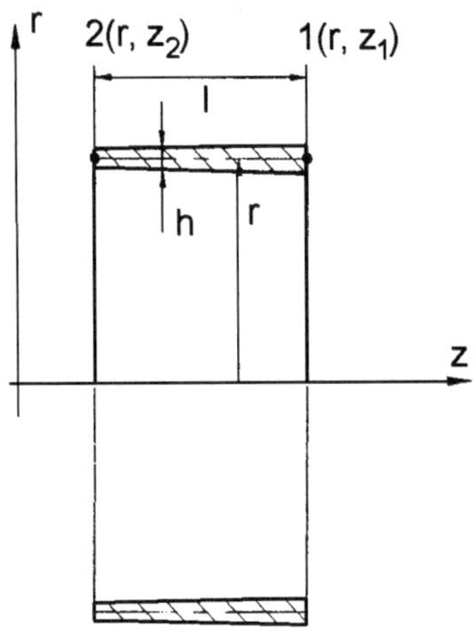

FIGURE 15.9. Schematic cross section of the shell element.

mesh is used. In these cases, it is advisable to use a shell element, so that a single element can be used in the thickness.

A sketch of a two-nodes axi-symmetrical shell element is shown in Figure 15.9. The geometrical parameters (radius r, thickness h) are assumed to be linear functions of the axial coordinate Z, or better of the nondimensional axial coordinate $\zeta = Z/L$

$$r = \mathbf{N}_1\mathbf{r} , \qquad h = \mathbf{N}_1\mathbf{h} , \tag{15.89}$$

where the shape functions $\mathbf{N}_1$, $\mathbf{r}$, and $\mathbf{h}$ are

$$\mathbf{N}_1 = \begin{bmatrix} 1-\zeta & \zeta \end{bmatrix} , \qquad \mathbf{r} = \begin{bmatrix} r_1 & r_2 \end{bmatrix} , \qquad \mathbf{h} = \begin{bmatrix} h_1 & h_2 \end{bmatrix} . \tag{15.90}$$

The displacement field can still be expressed by Equation (15.79), but in this case, the coefficients of the various harmonic components are functions of Z and t only

$$\begin{aligned} u(Z,\theta,t) &= u_0(Z,t) + \textstyle\sum_{i=1}^{n} \left[u_{ic}(Z,t)\cos(i\theta) + u_{is}(Z,t)\sin(i\theta)\right] , \\ v(Z,\theta,t) &= v_0(Z,t) + \textstyle\sum_{i=1}^{n} \left[v_{ic}(Z,t)\cos(i\theta) + v_{is}(Z,t)\sin(i\theta)\right] , \\ w(Z,\theta,t) &= w_0(Z,t) + \textstyle\sum_{i=1}^{n} \left[w_{ic}(Z,t)\cos(i\theta) + w_{is}(Z,t)\sin(i\theta)\right] . \end{aligned} \tag{15.91}$$

Functions u_{ic}, u_{is}, etc. can be expressed by polynomial expansions in Z (or, better, in ζ). The radial displacement is assumed to be a cubic function of ζ, whereas the axial and circumferential displacements are assumed to be linear

$$\begin{aligned} u_{ic} &= \mathbf{N}_3(\zeta)\mathbf{q}_{u_{ic}}(t) , & u_{is} &= \mathbf{N}_3(\zeta)\mathbf{q}_{u_{is}}(t) , \\ v_{ic} &= \mathbf{N}_1(\zeta)\mathbf{q}_{v_{ic}}(t) , & v_{is} &= \mathbf{N}_1(\zeta)\mathbf{q}_{v_{is}}(t) , \\ w_{ic} &= \mathbf{N}_1(\zeta)\mathbf{q}_{w_{ic}}(t) , & w_{is} &= \mathbf{N}_1(\zeta)\mathbf{q}_{w_{is}}(t) . \end{aligned} \tag{15.92}$$

where the shape functions $\mathbf{N}_1$ are those defined above and

$$\mathbf{N}_3 = \left[\begin{array}{cccc} 1 - 3\zeta^2 + 2\zeta^3 & L\zeta\left(1 - 2\zeta + \zeta^2\right) & \zeta\left(3\zeta - 2\zeta^2\right) & L\zeta\left(-\zeta + \zeta^2\right) \end{array} \right] . \tag{15.93}$$

Operating in this way, vectors $\mathbf{q}_{v_{ic}}(t)$, $\mathbf{q}_{w_{ic}}(t)$, $\mathbf{q}_{v_{is}}(t)$, and $\mathbf{q}_{w_{is}}(t)$ contain the cosine and sine displacements in v- and w-directions at the two nodes (they have two elements), whereas $\mathbf{q}_{u_{ic}}(t)$ and $\mathbf{q}_{u_{is}}(t)$ contain both the displacements and the rotations at the two nodes, and then have four components. The total number of degrees of the element is then $8 + 16n$, where n is the number of harmonics considered (the zero-order harmonics has only the cosine components, and hence only 8 degrees of freedom, whereas all other harmonics have 16). Clearly this element allows a far smaller model to be built than the previous element, because each element has a smaller number of degrees of freedom, but, above all, because there are less elements in the model.

The derivation of the matrices of the element follows the same lines seen for the solid element seen in the previous section and are not repeated here (see [71]). Also the final equations of motion are the same, i.e., Equation (15.86) if real coordinates are used, or Equation (15.88) if the complex-coordinates approach is followed.

15.6.1 Brick elements

A different approach, which does not require that the rotor is axially symmetrical, is that of subdividing the rotor in elements in the usual way as in standard solid modeling. A variety of brick elements can be used, with a different number of sides (tetrahedron, hexahedron, hexagonal prism, etc.) and with a different number of nodes (rectangular brick with 8, 20, 27, ...nodes). The mesh can be generated using the standard FEM procedures, but general-purpose commercial codes are usually not able to deal with the specific problems encountered in rotordynamics.

As this type of solid modeling is best suited to deal with rotors that do not show axial symmetry, the problem needs to be studied with reference to the rotating frame.

The expression of the kinetic energy is Equation (15.43) and contains the term in $\Omega^2\mathbf{q}^T\mathbf{M}_{ni}\mathbf{q}$, as usual when the rotating frame approach is used.

The advantage of the present approach is that of allowing us to model all details of the rotor, particularly in the case of bladed rotor in which the blades can be modeled with the required accuracy. This, however, has a very heavy cost in terms of size of the model and hence in computer time. Although in the harmonic annular element the number of degrees of freedom required depends largely on the number of harmonics that are considered, when brick elements are used the number of degrees of freedom depends on the number of element that are present along the angular coordinate θ. Although a small number of harmonics is usually sufficient, the number of elements required can be quite large, and it is very large if an attempt is made to model fine details or if the presence of the blades is accounted for. Moreover, the latter approach yields a single problem with a very large number of degrees of freedom, whereas the harmonic axi-symmetrical approach allows the problem to be split in a number of independent smaller problems, one for each harmonic considered.

An advantage of the true three-dimensional modeling is the possibility of dealing with irregularities like blade mistuning: Some parameters (e.g., the thickness of the blades) may vary slightly, perhaps in a random way, from sector to sector.

An alternative is that of using the cyclic nature of the geometry of most rotors. Only a sector of the geometry is modeled, and the relevant matrices are computed. However, although in static analysis it is possible to model just one sector of the structure, at least if the loads are symmetric or antisymmetric, in dynamics, the whole structure must be studied. The presence of a cyclic symmetry allows anyway a substantial reduction of the complexity of the problem. Condensation of the model, using either Guyan reduction or the component mode synthesis approach, may allow a further reduction of the size of the problem.

As already stated, a rotor with cyclic symmetry behaves like an axi-symmetrical rotor (in the sense that the structure of its eigenvalues and eigenvectors is the same as those of an isotropic rotor), so that it is possible to study the dynamics of a rotor of this kind supported by an asymmetric stator obtaining closed-form solutions in the inertial frame. This is clearly not possible in the case of a general non-axi-symmetrical rotor, as stated in Chapter 9.

When the dynamics of a rotor is studied in the rotating frame, the solution obtained can be interpreted in terms of traveling waves. As to plot the Campbell diagram the whirl frequencies in an inertial reference frame must be obtained, a transformation of the type of Equation (14.14), yielding the forward and backward whirling frequencies, must be used. The interpretation in terms of traveling waves of the solution in the rotating frame allows us to simplify the problem when the structure is subdivided into n equal sectors because the sectors deform in the same way, with only a phase lag from one to another [72].

Example 15.3 *Rotating beam.*

Compute the Campbell diagram of a simple rotating cantilever beam on stiff bearings using a number of brick elements, and compare the results with the upper and lower bound closed-form solutions shown in Section 13.4.2.

The beam, whose dimensions are $10 \times 5 \times 200$ *mm and rotates about an axis passing for the center of one of the end cross-sections (Figure 15.10), is made of aluminium* ($E = 7.31 \times 10^{10} N/m^2$; $\rho = 2770 kg/m^3$; $\nu = 0.33$)*. A single row of 20 brick elements with 20 nodes built following the theory shown in this chapter were used [9]. This very simple mesh proved to be sufficient for obtaining accurate*

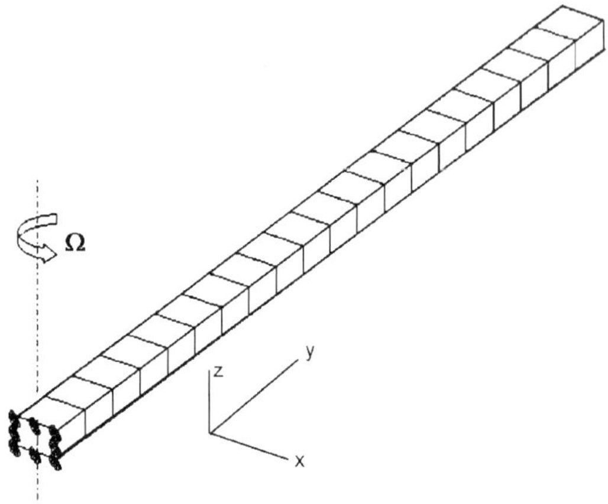

FIGURE 15.10. Mesh for the rotating beam.

results for the flexural natural frequencies, both in-plane and out-of-plane, whereas it is too rough for the torsional modes. As the bearings are stiff, the end of the beam corresponding to the axis of rotation is considered as clamped in the rotating frame.

The flexural natural frequencies are plotted in Figure 15.11. Their increase with increasing spin speed is clearly visible. The torsional and axial natural frequencies are not plotted because they lie outside the range considered, and they are not affected by the speed. The results obtained through the present FEM formulation are enclosed within the narrow band defined by the theoretical upper and lower bounds. Note that the maximum speed reported on the plot is much greater than the maximum allowable speed for a beam of this kind (at 5,000 rad/s, the stress at the root of the beam would be 3116 N/mm^2). This has been done to test the numerical solution against the closed- form ones even at very high (although unrealistic) speeds.

Example 15.4 *Rotating ring*

Consider a rotating ring of 0.5 m mean diameter with rectangular cross section (20 $\times$ 50 mm), and compare the results obtained using a three-dimensional FEM model with the closed-form solution shown in Section 14.5. The mesh shown in Figure 15.12 is made by a single row of 40 elements with 20 nodes of the same type used in the previous example. No attempt to exploit the axial symmetry of the structure was made. The Campbell diagram showing the forward and backward whirl frequencies in the speed range from 0 to 1000 rad/s is reported in Figure 15.13. eight flexural frequencies plus the only nonzero rigid-body frequency, all computed in the rotating frame, are shown. The agreement between the FEM and the closed-form solutions is very good, except for the eighth in-plane mode, where the discretization errors start to be felt and a finer mesh would be required.

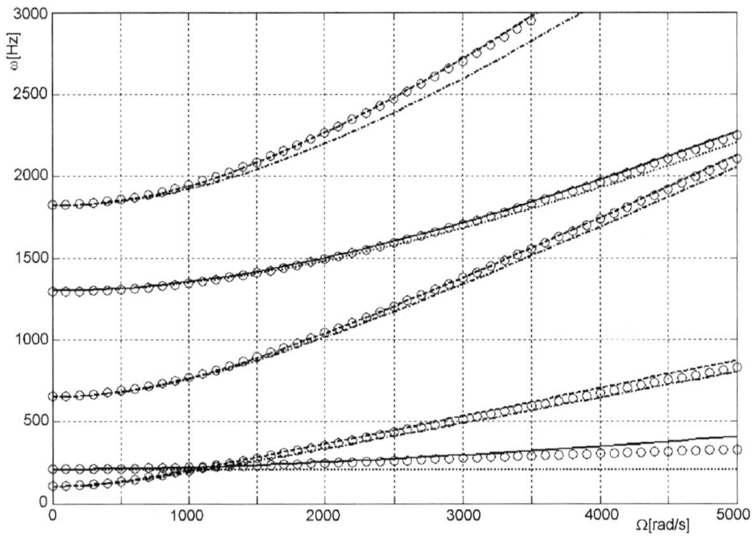

FIGURE 15.11. Campbell diagram for the first five flexural natural frequencies of a rotating beam (dots: FEM results; full lines: theoretical in-plane upper bound; dotted lines: theoretical in-plane lower bound; dashed lines: theoretical out-of-plane upper bound; dash and dots lines: theoretical out-of-plane lower bound).

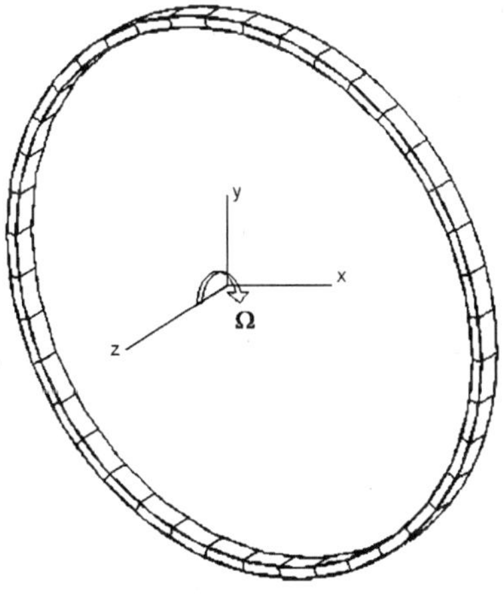

FIGURE 15.12. Mesh for the rotating ring.

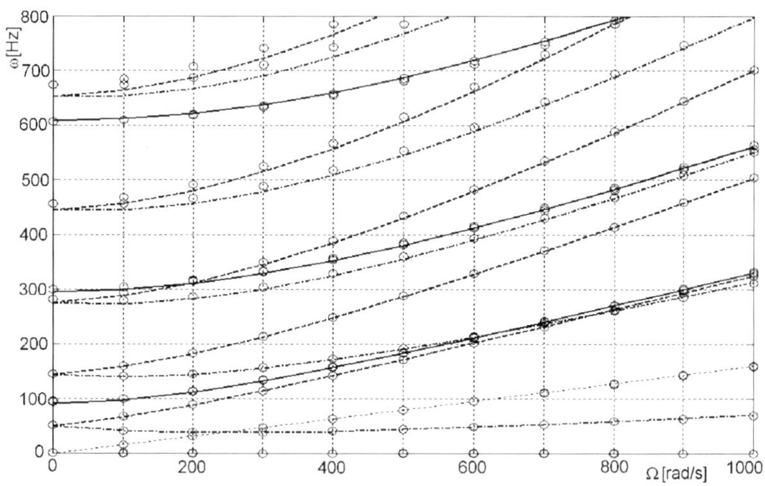

FIGURE 15.13. Cambell diagram for a rotating ring (circles: FEM results; full lines: theoretical out-of-plane; dashed lines: theoretical forward in-plane; dash and dot lines: theoretical backward in-plane; dotted lines: nonzero rigid-body frequency).

16

Dynamics of controlled rotors

Rotors can be provided with sensors and actuators that, through a suitable control system, can perform different tasks, from relieving the loads from the bearings to performing an active control of the dynamic behavior of the system. The control system, with its sensors and actuators, can act on only a few of the degrees of freedom of the rotor, for example to supply an additional damping, or it can fully control its position in space, as it occurs in a fully active magnetic suspension.

Magnetic bearings are perhaps the most common example of active control applied to the dynamics of rotors, but there are also many other applications, like controlled hydrodynamic supports, electrostatic bearings used in micromachines, or controlled pneumostatic bearings. In the following sections, only a brief introduction to the dynamics of controlled rotors will be presented, with particular reference to magnetic bearings, but the relevant equations can be applied to other devices.

A simplified linearized study of the behavior of rotors on magnetic bearings was presented in Section 8.5. There each magnetic bearing was considered as a linear spring-damper system, neglecting the dynamics of the sensors, actuators, amplifiers, and controller and neglecting the fact that in most applications, the sensors and the actuators are not colocated; i.e., the force exerted by the latter is applied in a point that does not coincide with the point where the sensor reads the displacement. These assumptions will be dropped in the present section.

16.1 Open-loop equations of motion

16.1.1 Real coordinates

The general equations of motion of a rotating system that includes a number of actuators can be written in the form

$$\mathbf{M}\ddot{\mathbf{x}} + (\mathbf{C}_n + \mathbf{C}_r + \Omega\mathbf{G})\dot{\mathbf{x}} + (\mathbf{K} + \mathbf{K}_\Omega\Omega^2 + \Omega\mathbf{C}_r)\mathbf{x} = \mathbf{f}_c + \mathbf{f}_e \ , \tag{16.1}$$

where $\mathbf{f}_c$ and $\mathbf{f}_n$ are forcing vector functions of time due, respectively, to the actuators and to other causes. The latter includes also unbalance forces, which are harmonic in time with an amplitude proportional to Ω^2 and a frequency equal to Ω. Such an equation may contain the lateral behavior as well as the axial behavior in the case of a beam-like model, or may be much more complex in the case of $2\frac{1}{2}$ D or 3D modeling. At any rate, at least the rotor must possess axial (or cyclic) symmetry because the equation has been written with reference to the fixed frame and has constant coefficients.

If the rotor is supported only by the actuators, like in the case of a rotor on active magnetic bearings, the stiffness matrix $\mathbf{K}$ is singular with four (six, if also the axial and the torsional behavior is included) vanishing eigenvalues because the four (six) rigid-body motions of the rotor are unconstrained. Also the rotating damping matrix $\mathbf{C}_r$ is singular. If it was not for the control forces, the rotor would behave like a free rotor. If, on the contrary, the rotor is supported in another way and the actuators are used to control its dynamic behavior, such matrices are not singular.

Equation (16.1) can be written with reference to the state space as

$$\dot{\mathbf{z}} = \mathbf{A}\mathbf{z} + \mathbf{B}_c\mathbf{u}_c(t) + \mathbf{B}_e\mathbf{u}_e(t) \ , \tag{16.2}$$

where

- Vector $\mathbf{z}$ contains the n complex state variables $\dot{\mathbf{x}}$ and $\mathbf{x}$.
- Vector $\mathbf{u}_c(t)$ contains the r control input functions. Vector $\mathbf{u}_e$ contains all external inputs, related to nonrotating and rotating forces, which are usually (at least the rotating ones) functions of time.
- $\mathbf{A}$ is the dynamic matrix of the system,

$$\mathbf{A} = \begin{bmatrix} -\mathbf{M}^{-1}(\mathbf{C}_n + \mathbf{C}_r + \Omega\mathbf{G}) & -\mathbf{M}^{-1}(\mathbf{K}+\Omega\mathbf{C}_r) \\ \mathbf{I} & \mathbf{0} \end{bmatrix} . \tag{16.3}$$

- Matrices $\mathbf{B}_c$ and $\mathbf{B}_e$ are the input gain matrices, respectively, for the control and external inputs.

If the rotating system is provided with sensors, an output equation

$$\mathbf{y} = \mathbf{C}\mathbf{z} + \mathbf{D}\mathbf{u} \tag{16.4}$$

must be added to the state equation. In the output equation

- Vector $\mathbf{z}$ contains the n complex state variables $\dot{\mathbf{x}}$ and $\mathbf{x}$.
- Vector $\mathbf{y}(t)$ contains the s outputs of the system.
- $\mathbf{C}$ is the output gain matrix, linking the outputs to the state vector.
- $\mathbf{D}$ is the matrix linking directly the inputs and the outputs of the system. It has been introduced for completeness, but it is usually nil and, as a consequence, is seldom considered.

16.1.2 Complex coordinates

The lateral equation of motion of an axi-symmetrical rotor-stator system can be written also in terms of complex coordinates even if it includes a number of actuators. By separating the forcing function caused by unbalance $\mathbf{f}_r$ from the other forcing functions $\mathbf{f}_e$, it can be written in the form

$$\mathbf{M}\ddot{\mathbf{q}} + (\mathbf{C}_n + \mathbf{C}_r - i\Omega\mathbf{G})\dot{\mathbf{q}} + (\mathbf{K} - i\Omega\mathbf{C}_r)\mathbf{q} = \tag{16.5}$$
$$= \mathbf{f}_c + \mathbf{f}_e + \Omega^2\mathbf{f}_r e^{i\Omega t} \,,$$

where the forces exerted by the actuators on the rotor and the stator are listed in vector $\mathbf{f}_c$. The real part of vector $\mathbf{f}_c$ refers to the forces caused by the actuators in the xz-plane, whereas its imaginary part is linked with the forces exerted in the yz-plane.

Equation (16.5) can be written with reference to the state space as

$$\dot{\mathbf{z}} = \mathbf{A}\mathbf{z} + \mathbf{B}_c\mathbf{u}_c(t) + \mathbf{B}_n\mathbf{u}_n(t) + \mathbf{B}_r\mathbf{u}_r(t)e^{i\Omega t} \,, \tag{16.6}$$

where

- Vector $\mathbf{z}$ contains the n complex state variables $\dot{\mathbf{q}}$ and $\mathbf{q}$.
- Vector $\mathbf{u}_c(t)$ contains the r control input functions. It can be written in complex form, and the real and imaginary parts of its components are related to the forces in the two coordinate planes xz and yz. Vectors $\mathbf{u}_n(t)$ and $\mathbf{u}_r(t)$ are input vectors related to nonrotating and rotating forces, which, in the most general case, can be functions of time. They can be real vectors, in some cases simply scalar quantities, but it is possible to formulate the equations in such a way that they are complex.

- $\mathbf{A}$ is the complex dynamic matrix of the system,

$$\mathbf{A} = \begin{bmatrix} -\mathbf{M}^{-1}(\mathbf{C}_n + \mathbf{C}_r - i\Omega\mathbf{G}) & -\mathbf{M}^{-1}(\mathbf{K} - i\Omega\mathbf{C}_r) \\ \mathbf{I} & \mathbf{0} \end{bmatrix} . \tag{16.7}$$

- Matrices $\mathbf{B}_i$ are the input gain matrices, respectively, for the control, nonrotating, and rotating inputs.

Also in this case, if the rotating system is provided with sensors, the output equation

$$\mathbf{y} = \mathbf{C}\mathbf{z} + \mathbf{D}\mathbf{u} \tag{16.8}$$

can be associated with the state equation. In this case, $\mathbf{y}$ is the complex output vector, and its real and imaginary parts are the sensor outputs related to the two coordinate planes xz and yz.

16.2 Closed-loop equations of motion

16.2.1 Ideal proportional control

An ideal proportional feedback control system supplies control inputs that are proportional to the difference between the actual outputs of the system $\mathbf{y}$ and the reference values $\mathbf{r}$, which can be also functions of time:

$$\mathbf{u}_c = -\mathbf{K}_y\mathbf{y}(t) + \mathbf{K}_r\mathbf{r}(t) . \tag{16.9}$$

By introducing the feedback and the output equations into the state Equation (it is immaterial whether reference is made to the real or complex coordinates approach, apart from the fact that in the second case, the system matrices can be complex), the following closed-loop equation is obtained:

$$\dot{\mathbf{z}} = \left(\mathbf{A} - \mathbf{B_c}\mathbf{K}_y\mathbf{C}\right)\mathbf{z} + \mathbf{B}_c\mathbf{K}_r\mathbf{r}(t) + \mathbf{B}_e\mathbf{u}_e(t) . \tag{16.10}$$

Assigning given laws $\mathbf{r}(\mathbf{t})$, it is possible to introduce a feedforward control, for instance, to compensate for known disturbances, like unbalance.

In case of an MIMO (multiple input-multiple output control system) matrices $\mathbf{K}_r$ and $\mathbf{K}_y$ have as many rows as the number of control inputs and as many columns as the number of references and outputs, respectively. There are cases in which the control system is decentralized (i.e., each control input is determined by a single output independently), and the reference signals are supplied for each output of the system separately. In this case, matrices $\mathbf{K}_r$ and $\mathbf{K}_y$ are square and diagonal.

Remark 16.1 *The proportional ideal controller defined above may have also a derivative action; i.e., it may introduce also a damping into the system. As also the generalized velocities are included among the states of the system, they can be made to affect the outputs and the proportional control operates actually as a PD (proportional-derivative) control.*

16.2.2 Ideal PID control

The proportional action (in the sense that the control inputs of the system are proportional to the displacements $\mathbf{x}$, or $\mathbf{q}$) supplies a stiffness to the system. The derivative action, which can be caused by a true derivative control in which the control inputs are proportional to the derivatives of the outputs or simply by using a proportional control in which the structure of matrices $\mathbf{C}$ and $\mathbf{K}_y$ are proportional to the velocities included in the state vector, supplies a damping action. If the rotor is completely suspended by the control forces, both actions are needed.

However, a PD control allows the position of the rotor to be displaced from the required position by static forces. To achieve a centered position in presence of static forces, a control action proportional to the integral of the position error

$$\mathbf{e} = \mathbf{y}(t) - \mathbf{r}(t) \tag{16.11}$$

is required. A PID (proportional-integrative-derivative) control is so obtained, where the control inputs u_c depend on the position error e through the relationship (written for a SISO system)

$$u_c = -k_c \left(e + T_d \frac{de}{dt} + \frac{1}{T_i} \int_0^t e(u)du \right) , \tag{16.12}$$

where k_c is the overall gain, T_d is the prediction or derivative time, and T_i is the reset time.

By resorting to the Laplace transforms of the control input and of the error, the transfer function of an ideal PID control is

$$\frac{U(s)}{E(s)} = -k_c \left(1 + T_d s + \frac{1}{sT_i} \right) . \tag{16.13}$$

In terms of state space approach, integral control can be introduced by augmenting the state vector through a number of additional states $\mathbf{z}_I$ defined as

$$\dot{\mathbf{z}}_I = \mathbf{S}\mathbf{e} = \mathbf{S}\left(\mathbf{y} - \mathbf{r}_I\right) , \tag{16.14}$$

where the selection matrix $\mathbf{S}$ has been introduced to allow us to use the integrals of a reduced number of outputs (if needed) for the control and the references $\mathbf{r}_I$ may be different from the corresponding references $\mathbf{r}$. The additional states $\mathbf{z}_I$ defined in Equation (16.14) are the integrals of the errors.

The control inputs are thus

$$u_c = -\mathbf{K}_y\mathbf{y} + \mathbf{K}_r\mathbf{r} - \mathbf{K}_{Iy}\mathbf{z}_I \ , \tag{16.15}$$

where the derivative action is accounted for through the velocities included into the states and the various gain matrices include all of the constants of the control system.

The state equation of the closed loop system is thus

$$\begin{cases} \dot{\mathbf{z}} = \mathbf{A}\mathbf{z} + \mathbf{B}_c(-\mathbf{K}_y\mathbf{C}\mathbf{z} + \mathbf{K}_r\mathbf{r} - \mathbf{K}_{Iy}\mathbf{z}_I) + \mathbf{B}_e\mathbf{u}_e \ , \\ \dot{\mathbf{z}}_I = \mathbf{S}\mathbf{C}\mathbf{z} - \mathbf{S}\mathbf{r}_I \ ; \end{cases} \tag{16.16}$$

i.e.,

$$\begin{Bmatrix} \dot{\mathbf{z}} \\ \dot{\mathbf{z}}_I \end{Bmatrix} = \begin{bmatrix} \mathbf{A}\mathbf{z} - \mathbf{B}_c\mathbf{K}_y\mathbf{C} & -\mathbf{B}_c\mathbf{K}_{Iy} \\ \mathbf{S}\mathbf{C} & \mathbf{0} \end{bmatrix} \begin{Bmatrix} \mathbf{z} \\ \mathbf{z}_I \end{Bmatrix} + \tag{16.17}$$

$$+ \begin{bmatrix} \mathbf{B}_c\mathbf{K}_r & \mathbf{0} \\ \mathbf{0} & -\mathbf{S} \end{bmatrix} \begin{Bmatrix} \mathbf{r} \\ \mathbf{r}_I \end{Bmatrix} + \begin{bmatrix} \mathbf{B}_e \\ \mathbf{0} \end{bmatrix} \mathbf{u}_e \ .$$

Example 16.1 *Jeffcott rotor made of a rigid body suspended on a magnetic bearing with ideal PID controller.*

Assume the following data: mass $m = 3$ *kg; actuator constant* $K = 1.5246 \times 10^{-6}$ *Nm²/A²; nominal air gap* $c = 0.75$ *mm; bias current* $i_0 = 3$ *A; overall gain of the controller* $k_c = 1.25$ *V/m; reset time* $T_i = 6.14$ *s; derivative time* $T_d = 0.9035 \times 10^{-3}$ *s; gain of the sensor 7500 V/m; gain of the power amplifier 1.027 A/V.*

Using real coordinates, the equation of motion of the rotor is

$$\begin{bmatrix} \mathbf{m} & 0 \\ \mathbf{0} & \mathbf{m} \end{bmatrix} \begin{Bmatrix} \ddot{x} \\ \ddot{y} \end{Bmatrix} = \begin{Bmatrix} \mathbf{f}_{cx} \\ \mathbf{f}_{cy} \end{Bmatrix} + m\epsilon\Omega^2 \begin{Bmatrix} \cos(\Omega t) \\ \sin(\Omega t) \end{Bmatrix} + \begin{Bmatrix} \mathbf{f}_{nx} \\ \mathbf{f}_{ny} \end{Bmatrix} \ .$$

The static forces $\mathbf{f}_{nx}$ *and* $\mathbf{f}_{ny}$ *are equal to the weight of the rotor in vertical direction and 0 in horizontal direction.*

The open-loop state equation is

$$\begin{Bmatrix} \dot{v}_x \\ \dot{v}_y \\ \dot{x} \\ \dot{y} \end{Bmatrix} = \begin{bmatrix} 0 & 0 & 0 & 0 \\ 0 & 0 & 0 & 0 \\ 1 & 0 & 0 & 0 \\ 0 & 1 & 0 & 0 \end{bmatrix} \begin{Bmatrix} v_x \\ v_y \\ x \\ y \end{Bmatrix} + \begin{bmatrix} \frac{1}{m} & 0 \\ 0 & \frac{1}{m} \\ 0 & 0 \\ 0 & 0 \end{bmatrix} \begin{Bmatrix} \mathbf{f}_{cx} \\ \mathbf{f}_{cy} \end{Bmatrix} +$$

$$+\epsilon\Omega^2 \begin{bmatrix} 1 & 0 \\ 0 & 1 \\ 0 & 0 \\ 0 & 0 \end{bmatrix} \begin{Bmatrix} \cos(\Omega t) \\ \sin(\Omega t) \end{Bmatrix} + \begin{bmatrix} \frac{1}{m} & 0 \\ 0 & \frac{1}{m} \\ 0 & 0 \\ 0 & 0 \end{bmatrix} \begin{Bmatrix} \mathbf{f}_{nx} \\ \mathbf{f}_{ny} \end{Bmatrix} \ .$$

The output is directly the state of the system, consisting of its positions and velocities, and hence the output gain matrix $\mathbf{C}$ *is the identity matrix and the direct link matrix is equal to zero. The forces caused by the actuators of the bearing can be expressed by the linearized Equation (8.42)*

$$\left\{\begin{array}{c}\mathbf{f}_{cx}\\ \mathbf{f}_{cy}\end{array}\right\}=\left\{\begin{array}{c}F_{0x}+K_{ix}i_x+K_{ux}x\\ F_{0y}+K_{iy}i_y+K_{uy}y\end{array}\right\},$$

where the linearized characteristics of the bearings are linked to the bias and compensation currents through the relationships (8.44)

$$F_0=4K\frac{i_0 i_0'}{c^2}\,,\qquad K_i=4K\frac{i_0}{c^2}\,,\qquad K_u=4K\frac{i_0^2+i_0'^2}{c^3}.$$

The static force correspond to the weight in vertical direction (x-axis) and 0 in horizontal direction (y axis). The corresponding compensation currents are

$$i_{0x}'=F_0\frac{c^2}{4Ki_0}=0.9048\ A\,,\qquad i_{0y}'=0.$$

The current and displacement bearing stiffness are (see Example 8.5, where, however, the bearing has been assumed to be isotropic)

$$K_{ix}=K_{iy}=32.525\ \ N/A,\qquad K_{ux}=1.419\times10^5\ N/m,\qquad K_y=1.301\times10^5\ N/m.$$

The bearing is almost isotropic.

Assume that the sensors are ideal and read the actual positions and velocities x, y, $\dot{x}$, and $\dot{y}$ through a gain k_s. Moreover, the reference position is with the rotor at the center of the bearing, i.e., $r=0$.

Assume that the controller is isotropic. The x channel supplies a current

$$i_x=-k_ck_a\left(x+T_dv_x+\frac{1}{T_i}\int_0^t x(u)du\right)\,,$$

where k_a is the gain of the power amplifier. The equation describing the behavior of the y channel is identical, with the obvious change of the symbols.

By augmenting the state vector introducing the additional states z_x and z_y

$$\left\{\begin{array}{c}\dot{z}_x\\ \dot{z}_y\end{array}\right\}=\left[\begin{array}{cccc}0&0&1&0\\0&0&0&1\end{array}\right]\left\{\begin{array}{c}v_x\\ v_y\\ x\\ y\end{array}\right\},$$

the closed-loop state equation becomes

$$\left\{\begin{array}{c}\dot{v}_x\\ \dot{v}_y\\ \dot{x}\\ \dot{y}\\ \dot{z}_x\\ \dot{z}_y\end{array}\right\}=\left[\begin{array}{cccccc}-\frac{K_{Tx}T_d}{m}&0&\frac{K_{ux}-K_{Tx}}{m}&0&-\frac{K_{Tx}}{mT_i}&0\\ 0&-\frac{K_{Ty}T_d}{m}&0&\frac{K_{uy}-K_{Ty}}{m}&0&-\frac{K_{Ty}}{mT_i}\\ 1&0&0&0&0&0\\ 0&1&0&0&0&0\\ 0&0&1&0&0&0\\ 0&0&0&1&0&0\end{array}\right]\times$$

$$\times \begin{Bmatrix} v_x \\ v_y \\ x \\ y \\ z_x \\ z_y \end{Bmatrix} + \epsilon\Omega^2 \begin{bmatrix} 1 & 0 \\ 0 & 1 \\ 0 & 0 \\ 0 & 0 \\ 0 & 0 \\ 0 & 0 \end{bmatrix} \begin{Bmatrix} \cos(\Omega t) \\ \sin(\Omega t) \end{Bmatrix},$$

where the total gain of the control loop is, for the x- and y-directions

$$K_{Tx} = K_{ix}k_c k_a k_s \;, \qquad K_{Ty} = K_{iy}k_c k_a k_s \;.$$

Note that the static force does not appear in the equation of motion.

The total gain of the control loop is

$$K_{Tx} = K_{Ty} = 3.1315 \times 10^5 \;.$$

The closed-loop dynamic matrix of the system is

$$\begin{bmatrix} -94 & 0 & -57,073 & 0 & -17,001 & 0 \\ 0 & -94 & 0 & -61,018 & 0 & -17,001 \\ 1 & 0 & 0 & 0 & 0 & 0 \\ 0 & 1 & 0 & 0 & 0 & 0 \\ 0 & 0 & 1 & 0 & 0 & 0 \\ 0 & 0 & 0 & 1 & 0 & 0 \end{bmatrix}.$$

The eigenvalues of the dynamic matrix are $-47.01 \pm 234.17\,i$; $-47.02 \pm 242.45i$ (poles of the PD section); -0.30; and -0.28, (poles of the integrative section). The two complex pairs correspond to the two damped eigenfrequencies in the vertical and horizontal plane, whereas the two real ones are nonoscillatory modes. The rotor behaves as a Jeffcott rotor on slightly nonisotropic supports.

The unbalance response is easily computed: As the excitation is harmonic with frequency Ω, also the response is harmonic with the same frequency. The augmented state can thus be expressed as the following function of time:

$$\mathbf{z} = \mathbf{z}_c \cos(\Omega t) + \mathbf{z}_s \sin(\Omega t)\,.$$

As

$$\dot{\mathbf{z}} = -\Omega\mathbf{z}_c \sin(\Omega t) + \Omega\mathbf{z}_s \cos(\Omega t)\,,$$

by introducing the solution into the state equation, it follows that

$$-\Omega\mathbf{z}_c \sin(\Omega t) + \Omega\mathbf{z}_s \cos(\Omega t) = \mathbf{A}\mathbf{z}_c \cos(\Omega t) +$$

$$+\mathbf{A}\mathbf{z}_s \sin(\Omega t) + \epsilon\Omega^2\mathbf{f}_c \cos(\Omega t) + \epsilon\Omega^2\mathbf{f}_s \sin(\Omega t)\,,$$

where

$$\mathbf{f}_c = \begin{bmatrix} 1 & 0 & 0 & 0 & 0 & 0 \end{bmatrix}^T, \qquad \mathbf{f}_c = \begin{bmatrix} 0 & 1 & 0 & 0 & 0 & 0 \end{bmatrix}^T.$$

By balancing the components in sine and cosine of the equation, it follows that

$$\begin{cases} \Omega\mathbf{z}_s - \mathbf{A}\mathbf{z}_c = \epsilon\Omega^2\mathbf{f}_c \;, \\ -\Omega\mathbf{z}_c - \mathbf{A}\mathbf{z}_s = \epsilon\Omega^2\mathbf{f}_s \;; \end{cases}$$

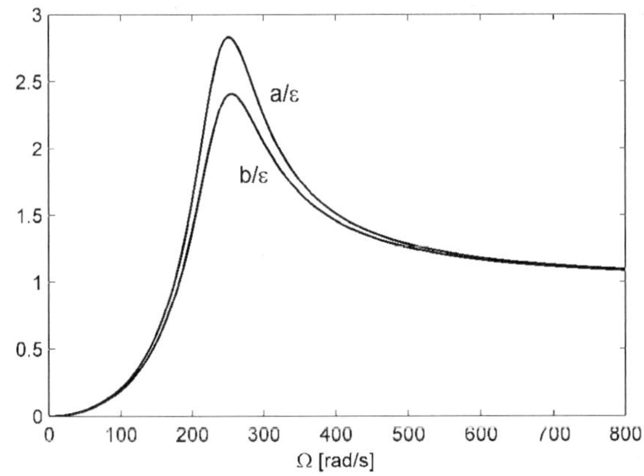

FIGURE 16.1. Nondimensional unbalance response of the Jeffcott rotor on magnetic bearings. Larger axis a and smaller axis b of the elliptical response, divided by the eccentricity as functions of the speed.

i.e.,

$$\begin{bmatrix} -\mathbf{A} & \Omega\mathbf{I} \\ -\Omega\mathbf{I} & -\mathbf{A} \end{bmatrix} \begin{Bmatrix} \mathbf{z}_c \\ \mathbf{z}_s \end{Bmatrix} = \epsilon\Omega^2 \begin{Bmatrix} \mathbf{f}_c \\ \mathbf{f}_s \end{Bmatrix} .$$

The unbalance response so obtained is reported in nondimensional form in Figure 16.1. The orbits are slightly elliptical, but the small anisotropy and the high damping make backward whirling impossible.

16.2.3 Dynamics of the control system

Usually the control system is not an ideal one and the various components of the control loop (sensors, controller, power amplifiers, actuators) have their own dynamics. Moreover, a compensator may be present. To take this into account, the differential equations governing the dynamics of all elements must be associated with the rotordynamics Equation (16.1).

The dynamics of the control system can be expressed in terms of its transfer function transforming the input into the control system (which coincides with the output $\mathbf{y}$ of the controlled system or *plant*), into the output of the control system, which coincides with the control input $\mathbf{u}_c$ of the plant. Assuming that the transfer function is expressed as a ratio of two polynomials in s

$$\frac{U(s)}{Y(s)} = \frac{b_1 s^n + b_2 s^{n-1} + ... + b_{n+1}}{s^m + a_1 s^{m-1} + ... + a_m}, \tag{16.18}$$

where $m \geq n$, it is possible to write a state-space representation of the controller using what is usually called the *control canonical form*

$$\begin{aligned}\dot{\mathbf{z}}_c &= \mathbf{A}_c\mathbf{z}_c + \mathbf{B}_y\mathbf{y} \ ,\\ \mathbf{u}_c &= \mathbf{C}_c\mathbf{z}_c + \mathbf{D}_c\mathbf{y} \ ,\end{aligned} \tag{16.19}$$

where $\mathbf{z}_c$ is the vector containing the states of the controller, and if $m = n + 1$,

$$\mathbf{A}_c = \begin{bmatrix} -a_1 & -a_2 & \dots & -a_m \\ 1 & 0 & \dots & 0 \\ 0 & 1 & \dots & 0 \\ \dots & \dots & \dots & \dots \end{bmatrix} , \tag{16.20}$$

$$\mathbf{B}_y = \begin{bmatrix} 1 \\ 0 \\ 0 \\ \dots \end{bmatrix} , \qquad \mathbf{C}_c = \begin{bmatrix} b_1 \\ b_2 \\ \dots \\ b_{n+1} \end{bmatrix}^T . \tag{16.21}$$

$\mathbf{A}_c$, $\mathbf{B}_y$, and $\mathbf{C}_c$ are the dynamic, input gain, and output gain matrices of the control system.

The closed-loop state equation of the system is then

$$\left\{ \begin{array}{c} \dot{\mathbf{z}} \\ \dot{\mathbf{z}}_c \end{array} \right\} = \begin{bmatrix} \mathbf{A} & -\mathbf{B}_c\mathbf{C}_c \\ \mathbf{B}_y\mathbf{C} & \mathbf{A}_c \end{bmatrix} \left\{ \begin{array}{c} \mathbf{z} \\ \mathbf{z}_c \end{array} \right\} + \begin{bmatrix} \mathbf{B}_e \\ \mathbf{0} \end{bmatrix} \mathbf{u}_e \, . \tag{16.22}$$

For instance, the transfer function of an actual PID controller is not Equation (16.13), but it can be expressed as

$$\frac{U(s)}{E(s)} = k_c \left(1 + \frac{T_d s}{1 + s\frac{T_d}{N}} + \frac{1}{sT_i} \right) , \tag{16.23}$$

in which also the ratio N between zero and pole of the PD section has been introduced.

Assuming that the reference is 0, the error coincides with the displacement measured by the sensor. The filter on the sensor output can be assumed to have the following transfer function:

$$\frac{E(s)}{Y(s)} = \frac{1}{1 + s\tau} , \tag{16.24}$$

where τ is the time constant of the sensor.

The total transfer function from the displacement to the force is the product of the two transfer functions written above, i.e., written in terms of the ratio between two polynomials,

$$\frac{U(s)}{Y(s)} = k_c \frac{s^2 \frac{1+N}{\tau} + s\frac{1}{\tau}\left(\frac{N}{T_d} + \frac{1}{T_i}\right) + \frac{N}{\tau T_d T_i}}{s^3 + s^2 \left(\frac{N}{T_d} + \frac{1}{\tau}\right) + s\frac{N}{\tau T_d}} , \tag{16.25}$$

and the state-space realization of the control system can then be expressed by the quadruple

$$\mathbf{A}_c = \begin{bmatrix} -\left(\frac{N}{T_d}+\frac{1}{\tau}\right) & -\frac{N}{\tau T_d} & 0 \\ 1 & 0 & 0 \\ 0 & 1 & 0 \end{bmatrix}, \qquad \mathbf{B}_y = \begin{bmatrix} 1 \\ 0 \\ 0 \end{bmatrix}, \tag{16.26}$$

$$\mathbf{C}_c = k_c \begin{bmatrix} \frac{1+N}{\tau} & \frac{1}{\tau}\left(\frac{N}{T_d}+\frac{1}{T_i}\right) & \frac{N}{\tau T_d T_i} \end{bmatrix}, \qquad \mathbf{D}_c = 0\,.$$

The equations of motion listed above implicitly rely on the linearity of the whole system (plant and controller). If the control loop contains elements that are intrinsically nonlinear or that cannot be modeled accurately enough using linearized models, the only possible approach is to simulate the behavior of the closed-loop system in the time domain. The most common causes of nonlinearity in controlled rotors are the nonlinear behavior of the actuators, which becomes more pronounced if a low bias current is used, and the saturation of the power amplifiers.

Example 16.2 *Jeffcott rotor supported by a magnetic bearing with real PID controllers.*

Assume the same data seen in Example 16.1, with the following characteristics added: ratio between zero and pole of the PD section $N = 18.07$*, time constant of the sensor electronics* $\tau = 10^{-6}\,s$

The forces exerted by the actuators are made of three components

$$\left\{ \begin{array}{c} \mathbf{f}_{cx} \\ \mathbf{f}_{cy} \end{array} \right\} = \left\{ \begin{array}{c} F_{0x} + K_{ix} i_x + K_{ux} x \\ F_{0y} + K_{iy} i_x + K_{uy} y \end{array} \right\}.$$

The static component of the force is assumed to balance the static loads, equal to the weight in vertical direction and to zero in horizontal direction, and hence it cancels with external nonrotating forces f_{nx} *and* f_{ny}*. The third component is proportional to the displacements and then can enter directly the dynamic matrix. By keeping the two control systems acting in* x*- and* y*-directions separate, the open-loop state-space equation of the system is then*

$$\left\{ \begin{array}{c} \dot{v}_x \\ \dot{v}_y \\ \dot{x} \\ \dot{y} \end{array} \right\} = \begin{bmatrix} 0 & 0 & \frac{K_{ux}}{m} & 0 \\ 0 & 0 & 0 & \frac{K_{uy}}{m} \\ 1 & 0 & 0 & 0 \\ 0 & 1 & 0 & 0 \end{bmatrix} \left\{ \begin{array}{c} v_x \\ v_y \\ x \\ y \end{array} \right\} + \begin{bmatrix} \frac{K_{ix}}{m} \\ 0 \\ 0 \\ 0 \end{bmatrix} i_x +$$

$$\begin{bmatrix} 0 \\ \frac{K_{ix}}{m} \\ 0 \\ 0 \end{bmatrix} i_y + \epsilon\Omega^2 \begin{bmatrix} 1 & 0 \\ 0 & 1 \\ 0 & 0 \\ 0 & 0 \end{bmatrix} \left\{ \begin{array}{c} \cos(\Omega t) \\ \sin(\Omega t) \end{array} \right\}.$$

This equation defines matrices $\mathbf{A}$*,* $\mathbf{B}_{cx}$*,* $\mathbf{B}_{cy}$*, and* $\mathbf{B}_e$*, and vectors* $\mathbf{z}$*,* $\mathbf{u}_{cx}$*,* $\mathbf{u}_{cy}$ *(containing the control currents), and* $\mathbf{u}_e$*. As the sensors in the actual application* *read only the displacements of the rotor and not its velocities, the output equation is*

$$\begin{cases} x = \begin{bmatrix} 0 & 0 & 1 & 0 \end{bmatrix} \mathbf{z} \,, \\ y = \begin{bmatrix} 0 & 0 & 0 & 1 \end{bmatrix} \mathbf{z} \,. \end{cases}$$

This equation defines matrices $\mathbf{C}_x$ and $\mathbf{C}_y$ and vector $\mathbf{y}$.

The control system is made by two distinct PID controllers operating in x- and y-direction. Their dynamics can be summarized as

$$\begin{Bmatrix} \dot{\mathbf{z}}_{cx} \\ \dot{\mathbf{z}}_{cy} \end{Bmatrix} = \begin{bmatrix} \mathbf{A}_c & 0 \\ 0 & \mathbf{A}_c \end{bmatrix} \begin{Bmatrix} \mathbf{z}_{cx} \\ \mathbf{z}_{cy} \end{Bmatrix} + \begin{bmatrix} \mathbf{B}_y & 0 \\ 0 & \mathbf{B}_y \end{bmatrix} \begin{Bmatrix} x \\ y \end{Bmatrix} ,$$

where matrices $\mathbf{A}_c$ and $\mathbf{B}_y$ are given by Equation (16.26). Note that the two controllers are assumed to be identical.

The outputs of the controllers are the control currents, so that the output equation is

$$\begin{Bmatrix} i_x \\ i_x \end{Bmatrix} + \begin{bmatrix} \mathbf{C}_c & 0 \\ 0 & \mathbf{C}_c \end{bmatrix} \begin{Bmatrix} \mathbf{z}_{cx} \\ \mathbf{z}_{cy} \end{Bmatrix} ,$$

where matrix $\mathbf{C}_c$ is again given by Equation (16.26), with gain k_c substituted by the total gain of the control loop (controller, amplifier, sensor)

$$K_T = k_c k_a k_s \,.$$

The closed-loop state-space equation is then

$$\begin{Bmatrix} \dot{\mathbf{z}} \\ \dot{\mathbf{z}}_{cx} \\ \dot{\mathbf{z}}_{cy} \end{Bmatrix} = \begin{bmatrix} \mathbf{A} & \mathbf{B}_{cx}\mathbf{C}_c & \mathbf{B}_{cx}\mathbf{C}_c \\ \mathbf{B}_y\mathbf{C}_x & \mathbf{A}_c & \mathbf{0} \\ \mathbf{B}_y\mathbf{C}_y & \mathbf{0} & \mathbf{A}_c \end{bmatrix} \begin{Bmatrix} \mathbf{z} \\ \mathbf{z}_{cx} \\ \mathbf{z}_{cy} \end{Bmatrix} + \epsilon\Omega^2 \begin{bmatrix} \mathbf{B}_e \\ \mathbf{0} \\ \mathbf{0} \end{bmatrix} \begin{Bmatrix} \cos(\Omega t) \\ \sin(\Omega t) \end{Bmatrix} .$$

The order of the system is thus 10.

The eigenvalues of the dynamic matrix are $-47.17 \pm 234.73\, i$, $-47.19 \pm 243.03i$, -0.30, -0.28, 1.99×10^4.(double value), and 10^6 (double value). The first six eigenvalues are almost exactly the same obtained in Example 16.1 for the ideal PID controller, whereas the following four are high and do not influence the dynamics of the system at the frequencies of interest.

Also the results for the unbalance response are practically identical and are not reported here.

16.3 Rigid rotor on magnetic linearized bearings

A rigid rotor on a four-active-axis magnetic suspension is shown in Figure 8.21; the fifth degree of freedom, axial translation, may be constrained by either a further active suspension or a passive magnetic bearing but is at any rate considered decoupled from the four lateral degrees of freedom. In Section 8.5, the bearings were modeled as linear springs and dampers, which

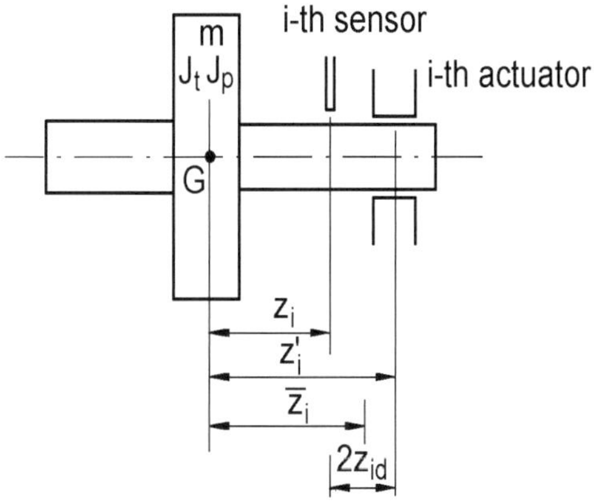

FIGURE 16.2. Rotor with four degrees of freedom on magnetic bearings. Sketch of the position of the ith sensor-actuator pair.

implies that each actuator has a colocated sensor and that each control loop is an ideal SISO (Single Input, Single Output) PD (Proportional Derivative) control. Now the control loop is still modeled as an ideal SISO PD system, but no colocation assumption is made.

16.3.1 Equations of motion

The system is modeled as a rotor with four real (two complex) degrees of freedom (see Chapter 3), and the displacements x and y of the center of mass and the rotations ϕ_y and ϕ_x are taken as generalized coordinates. Its lateral behavior can be modeled using Equation (16.5) written in terms of complex coordinates, in which no stiffness and damping matrix is present owing to the rigid-body assumption

$$\mathbf{M}\ddot{\mathbf{q}} - i\Omega\mathbf{G}\dot{\mathbf{q}} = \mathbf{f}_c + \mathbf{f}_n + \Omega^2\mathbf{f}_r, \tag{16.27}$$

where

$$\mathbf{M} = \begin{bmatrix} m & 0 \\ 0 & J_t \end{bmatrix}, \ \mathbf{G} = \begin{bmatrix} 0 & 0 \\ 0 & J_p \end{bmatrix}, \ \mathbf{q} = \left\{ \begin{array}{c} x + iy \\ \phi_y - i\phi_x \end{array} \right\}$$

and $\mathbf{F}_c$, $\mathbf{F}_n$, and $\mathbf{F}_r$ are the control, nonrotating, and the rotating (unbalance) forces.

For sake of generality, assume that the rotor is supported by n (usually 2) active magnetic bearings (AMB). The geometry of the generic ith bearing is sketched in Figure 16.2. The axial coordinates of the centre of the sensor and actuator arc, respectively, z_i and z'_i.

With reference to Equations (8.41) and (8.42), assuming that the controller is an ideal decentralized PD controller, the law expressing the force

$\mathbf{F}_i$ (in complex notation) exerted by the ith actuator as a function of the displacements $x_i + iy_i$ and $x_i' + iy_i'$ at the ith sensor and actuator locations and the velocity at the ith sensor is

$$\mathbf{F}_i = -k_i(x_i + iy_i) - k_i c_i(\dot{x}_i + i\dot{y}_i) + k_{u_i}(x_i' + iy_i') \, , \tag{16.28}$$

where and $k_i = K_i K_{cp} K_a K_s$ is the proportional gain of the ith control loop, $c_i = K_{cd}/K_{cp}$ is the ratio between the derivative and the proportional gains, and $k_{u_i} = K_u$ is the open-loop destabilizing stiffness of the ith bearing.

The control force vector $\mathbf{f}_c$ caused by a number n of actuators is then

$$\begin{array}{c} \mathbf{f}_c = -\begin{bmatrix} \sum_{i=1}^{n} c_i k_i & \sum_{i=1}^{n} c_i k_i z_i' \\ \sum_{i=1}^{n} c_i k_i z_i & \sum_{i=1}^{n} c_i k_i z_i z_i' \end{bmatrix} \dot{\mathbf{q}} + \\ - \begin{bmatrix} \sum_{i=1}^{n} (k_i - k_{u_i}) & \sum_{i=1}^{n} (k_i z_i' - k_{u_i} z_i') \\ \sum_{i=1}^{n} (k_i z_i - k_{u_i} z_i') & \sum_{i=1}^{n} \left(k_i z_i z_i' - k_{u_i} z_i'^2\right) \end{bmatrix} \mathbf{q} \, . \end{array} \tag{16.29}$$

Owing to noncolocation ($z_i \neq z_i'$), the matrices in Equation (16.29) are non symmetrical and can be nonpositive defined. The presence of the negative terms caused by K_{u_i} is usually not causing problems in colocated systems, owing to their smallness, but in the present case, they can contribute to make the stiffness matrix nonpositive defined.

The equation of motion (16.27) can be rewritten in real coordinates, obtaining

$$\mathbf{M}\ddot{\mathbf{x}} + (\mathbf{C} + \Omega\mathbf{G})\dot{\mathbf{x}} + \mathbf{K}\mathbf{x} = \mathbf{f}_n + \Omega^2 \mathbf{f}_r \, , \tag{16.30}$$

where

$$\mathbf{M} = \begin{bmatrix} \mathbf{M}^* & 0 \\ 0 & \mathbf{M}^* \end{bmatrix}, \qquad \mathbf{G} = \begin{bmatrix} 0 & \mathbf{G}^* \\ -\mathbf{G}^* & 0 \end{bmatrix},$$

$$\mathbf{C} = \begin{bmatrix} \mathbf{C}^* & 0 \\ 0 & \mathbf{C}^* \end{bmatrix}, \qquad \mathbf{K} = \begin{bmatrix} \mathbf{K}^* & 0 \\ 0 & \mathbf{K}^* \end{bmatrix},$$

$$\mathbf{x} = \begin{bmatrix} x & \phi_y & y & -\phi_x \end{bmatrix}^T,$$

and $\mathbf{M}^*$ and $\mathbf{G}^*$ are the two matrices in Equation (16.27) whereas $\mathbf{C}^*$ and $\mathbf{K}^*$ are those in Equation (16.29).

Matrix $\mathbf{M}$ is symmetric and positive defined, $\mathbf{G}$ is skew symmetric, and $\mathbf{C}$ and $\mathbf{K}$ are generally nonsymmetric and can be nonpositive defined. Note that if the sensors and actuators were colocated, the last two matrices would have been symmetric and positive defined and the rotordynamic behavior of the system would have been described by the usual "rotor with four degrees of freedom" model described in Chapter 3 with the stiffness and damping of the supports expressed by Equation (8.41).

Matrices ($\mathbf{C}+\Omega\mathbf{G}$) and $\mathbf{K}$ can be easily decomposed into their symmetric and skew-symmetric parts by introducing the average distance of the ith sensor-actuator pair $\overline{z}_i = \frac{1}{2}(z_i + z_i')$ and the noncolocation $z_{i_d} = \frac{1}{2}(z_i - z_i')$ of the same pair.

The stiffness matrix $\mathbf{K}$ can thus be split into a symmetrical stiffness matrix plus a circulatory matrix

$$\mathbf{K} = \begin{bmatrix} k_1 & k_2 & 0 & 0 \\ k_2 & k_3 & 0 & 0 \\ 0 & 0 & k_1 & k_2 \\ 0 & 0 & k_2 & k_3 \end{bmatrix} + \begin{bmatrix} 0 & -k4 & 0 & 0 \\ k4 & 0 & 0 & 0 \\ 0 & 0 & 0 & -k4 \\ 0 & 0 & k4 & 0 \end{bmatrix}, \tag{16.31}$$

where

$$k1 = \sum_{i=1}^{n}(k_i - k_{u_i}), \qquad k2 = \sum_{i=1}^{n}(k_i\overline{z}_i - k_{u_i}z_i'),$$

$$k3 = \sum_{i=1}^{n}\left[k_i\left(\overline{z}_i^2 - z_{i_d}^2\right) - k_{u_i}z_i'^2\right], \qquad k_4 = \sum_{i=1}^{n}k_i z_{i_d}.$$

In the same way, matrix $\mathbf{C}+\Omega\mathbf{G}$ can be split into a symmetrical damping matrix plus a gyroscopic matrix

$$\mathbf{C}+\Omega\mathbf{G} = \begin{bmatrix} c_1 & c_2 & 0 & 0 \\ c_2 & c_3 & 0 & 0 \\ 0 & 0 & c_1 & c_2 \\ 0 & 0 & c_2 & c_3 \end{bmatrix} + \begin{bmatrix} 0 & -c_4 & 0 & 0 \\ c_4 & 0 & 0 & +\Omega J_p \\ 0 & 0 & 0 & -c_4 \\ 0 & -\Omega J_p & c_4 & 0 \end{bmatrix}, \tag{16.32}$$

where

$$c_1 = \sum_{i=1}^{n}c_i k_i, \qquad c_2 = \sum_{i=1}^{n}c_i k_i \overline{z}_i,$$

$$c_3 = \sum_{i=1}^{n}c_i k_i(\overline{z}_i^2 - z_{i_d}^2), \qquad c_4 = \sum_{i=1}^{n}c_i k_i z_{i_d}.$$

The circulatory matrix contains only distances z_{i_d}, whereas the gyroscopic matrix depends on the noncolocation as well as on the product ΩJ_p expressing the gyroscopic effect.

To study the stability of the system, consider the homogeneous equation associated with Equation (16.35). Assuming a solution of the type $\mathbf{q} = \mathbf{q}_0 e^{st}$, where vector $\mathbf{q}$ contains the complex coordinates $x + iy$ and $\phi_y - i\phi_x$, and solving the related eigenproblem, the following nondimensional characteristic equation allowing us to compute the whirling frequencies is obtained:

$$s^{*4} - i\left[\Omega^*\delta + 2i\left(\zeta + \eta\alpha\right)\right]s^{*3} + \left[1 + \alpha + 2\zeta\eta\alpha - 2\theta\kappa(\beta^2 - \gamma^2) + \right.$$

$$-2i\Omega^*\zeta\delta]\,s^{*2} - \imath\left[\Omega^*\delta + 2i\alpha(\zeta+\eta) - 2i(\beta^2-\gamma^2)(\theta+\kappa)\right]s^* + \qquad (16.33)$$
$$+\alpha - \beta^2 + \gamma^2 = 0\ ,$$

where the nondimensional complex whirl frequency $s^* = s/\omega_1$ and the nondimensional spin speed $\Omega^* = \Omega/\omega_1$ have been defined with reference to the natural frequency

$$\omega_1 = \sqrt{\frac{K_{11}^*}{m}}$$

of a Jeffcott rotor with the same mass and a stiffness equal to K_{11}^*.

Equation (16.33) depends only on eight nondimensional parameters, namely,

- *Elastic* parameters:

$$\alpha = \alpha' + \alpha''\ ,$$

$$\alpha' = \frac{m}{J_t}\frac{\sum_{i=1}^{n}(k_i\overline{z}_i^2 - k_{u_i}z_i'^2)}{\sum_{i=1}^{n}(k_i - k_{u_i})}\ ,\quad \alpha'' = -\frac{m}{J_t}\frac{\sum_{i=1}^{n}k_i z_{i_d}^2}{\sum_{i=1}^{n}(k_i - k_{u_i})}\ ,$$

$$\beta = \frac{\sum_{i=1}^{n}(k_i\overline{z}_i - k_{u_i}z_i')}{\sum_{i=1}^{n}(k_i - k_{u_i})}\sqrt{\frac{m}{J_t}}\ ,\quad \gamma = \frac{\sum_{i=1}^{n}k_i z_{i_d}}{\sum_{i=1}^{n}(k_i - k_{u_i})}\sqrt{\frac{m}{J_t}}\ .$$

- *Inertial* parameter:

$$\delta = \frac{J_p}{J_t}\ .$$

- *Damping* parameters:

$$\zeta = \frac{c_{11}^*}{2k_{11}^*}\omega_1,\qquad \eta = \frac{c_{22}^*}{2k_{22}^*}\omega_1,$$

$$\theta = \frac{c_{12}^*}{2k_{12}^*}\omega_1,\qquad \kappa = \frac{c_{21}^*}{2k_{21}^*}\omega_1.$$

Note that:

- α is made of two parts, namely, α' and α''. The first one does not depend on noncolocation but only on the average positions $\overline{z}_i$ and is always positive; α'' vanishes for colocated systems and is always negative.

- β does not depend on noncolocation like α', can be either positive or negative and vanishes for symmetrical systems (see below), but its sign has no effect on the behavior of the system, because only its square is included in the equations.

- γ can be positive or negative and vanishes for either colocated or symmetrical systems. Its sign too has no effect on the behavior of the system.

- δ is the usual parameter for gyroscopic effects; its value can span from 0 (long rotors) to 2 (disc rotors); however, a smaller variability range is expected in actual applications.

- ζ coincides with the damping ratio of the above-mentioned Jeffcott rotor. If all bearings have the same derivative gain $c_i k_i$ and the contributions caused by the terms k_{u_i} are small enough to be neglected, $\eta = \theta = \kappa = \zeta$ and the number of relevant nondimensional parameters reduces to five.

As the equation has complex coefficients, the solutions are complex but not conjugate. Although little can be said in general on the stability of the system, Equation (16.33) allows us to assess numerically the stability in any given case.

In the case of the undamped system, Equation (16.33) reduces to

$$s^{*4} - i\Omega^* \delta s^{*3} + (1 + \alpha) s^{*2} - i\Omega\delta s^* + \alpha - \beta^2 + \gamma^2 = 0, \qquad (16.34)$$

which depends on just four nondimensional parameters. Solutions with positive real part (unstable solutions) may exist.

16.3.2 Symmetrical system

Consider a rotor on two equal bearings and the center of mass at midspan. Assume that also the sensors are symmetrically located. The equations of motion for the translational and rotational degrees of freedom uncouple (only four nondimensional parameters are different from zero, namely, α, δ, ζ, and η) and the characteristic Equation (16.33) splits into two independent equations:

$$\begin{cases} s^{*2} + 2\zeta s^* + 1 = 0 \, , \\ s^{*2} + (2\eta\alpha - i\Omega^* \delta) \, s^* + \alpha = 0 \, . \end{cases} \qquad (16.35)$$

Cylindrical whirling is governed by the same equation of the well-known equation of motion of the Jeffcott rotor. The equation describing the conical whirling differs from the usual equation dealing with colocated systems because the product $z_1 z_1'$ can be negative in the case in which the actuator on one side is connected with the sensor on the other one. In this case, α is negative.

If α is positive, the behavior of the system is equal to that of a colocated system with the actuator in the position $z_1^* = \sqrt{z_1 z_1'}$. The case with negative α has very little practical interest, because the system is unstable at standstill, behaving like a spring, mass, damper system with negative stiffness and damping coefficient. However, the gyroscopic moment can stabilize the undamped system.

The solution of the second Equation (16.35) is

$$s^* = \frac{-2\eta\alpha + i\Omega^*\delta \pm \sqrt{(i\Omega^*\delta - 2\eta\alpha)^2 - 4\alpha}}{2}\,, \tag{16.36}$$

which holds for both α positive or negative.

If α is negative, it follows that:

$$\omega^* = \mathrm{Im}\,(s^*) = \frac{\Omega^*\delta}{2} \pm \sqrt{\frac{\sqrt{a^2+b^2}-a}{2}}, \tag{16.37}$$

$$\sigma^* = \mathrm{Re}\,(s^*) = \eta|\alpha| \mp \sqrt{\frac{\sqrt{a^2+b^2}+a}{2}}, \tag{16.38}$$

where

$$a = 4\delta^2\alpha^2 - \Omega^{*2}\delta^2 + 4|\alpha|\,, \qquad b = 4\Omega^*\delta\eta|\alpha|\,.$$

In the case of the undamped system, stability occurs if

$$\Omega^* > \frac{2\sqrt{|\alpha|}}{\delta}. \tag{16.39}$$

However, the presence of damping makes the system unstable at all speeds because the real part of one of two values of s' is always positive for any value of the spin speed Ω^*.

16.3.3 Nonsymmetrical system

If the center of mass of the rotor is not at midspan or if the symmetry assumed in the previous section is violated, the two equations of motion do not uncouple and the modes do not reduce to conical and cylindrical ones. Nevertheless they are often still referred to as conical or cylindrical, but only in a general way, because the first one does not have its vertex in the center of mass and in the latter, the rotor axis does not remain exactly parallel to itself.

The condition for stability of the undamped system at standstill is

$$(1-\alpha)^2 + 4\beta^2 - 4\gamma^2 > 0\,, \tag{16.40}$$

which is obviously verified for $\beta > \gamma$, although being less restrictive than that.

The equations become complicated enough to prevent from performing a closed-form general study of the stability, even in the undamped case.

Some typical plots and conclusions drawn from numerical experimentation on undamped systems will be reported here. The nondimensional Campbell diagram and the decay rate plot of a system with $\alpha' = 0.5$,

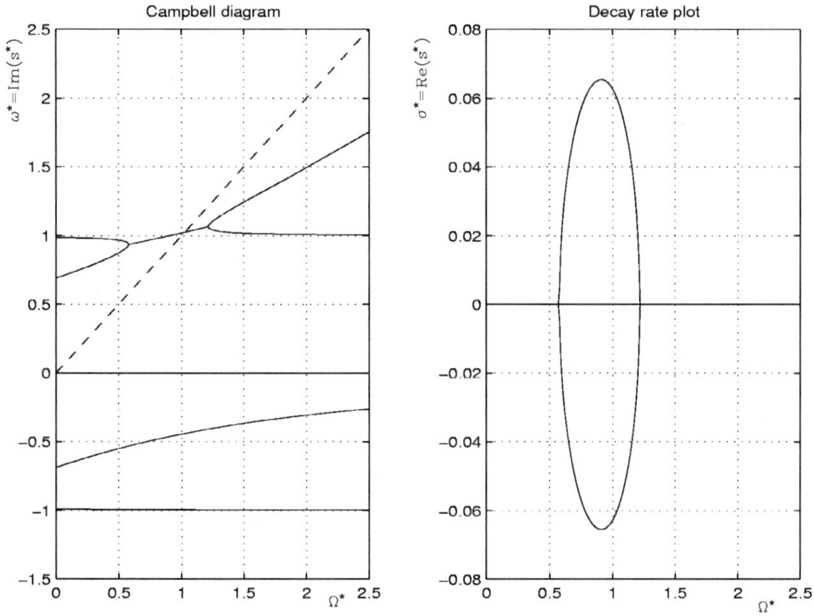

FIGURE 16.3. Nondimensional campbell diagram and decay rate plot of a system with $\alpha' = 0.5$, $\alpha'' = 0.05$, $\beta = 0.1$, $\gamma = 0.15$, and $\delta = 0.6$.

$\alpha'' = 0.05$, $\beta = 0.1$, $\gamma = 0.15$, and $\delta = 0.6$ are reported in Figure 16.3. The system is stable for $\Omega = 0$, because

$$(1 - \alpha)^2 + 4\beta^2 - 4\gamma^2 = 0.2525 > 0 \; .$$

The curves related to cylindrical and conical whirling cross in the first quadrant, and where they meet, a field of instability starts. The unstable conditions persist up to a certain speed, which is beyond the crossing of the Campbell diagram with the line $\omega^* = \Omega^*$.

The plot is repeated in Figure 16.4, with the same values of the parameters, but with $\delta = 1.5$ instead of $\delta = 0.6$, i.e., with a disc rotor instead of a long rotor.

The results are similar to the ones previously seen, with the difference that the curve related to the conical mode in forward whirling (whose asymptote is the straight line with equation $\omega^* = \Omega^* J_p/J_t$) has a greater slope. As a result, the instability range moves toward lower speeds and lies all in the subcritical range (on the left of the line $\omega^* = \Omega^*$).

The plot of Figurc 16.5 dcals with the same values of the parameters as in Figure 2 (long rotor), but for the value of α', which is now greater than 1 (2 instead of 0.5). The curves related to cylindrical and conical whirling

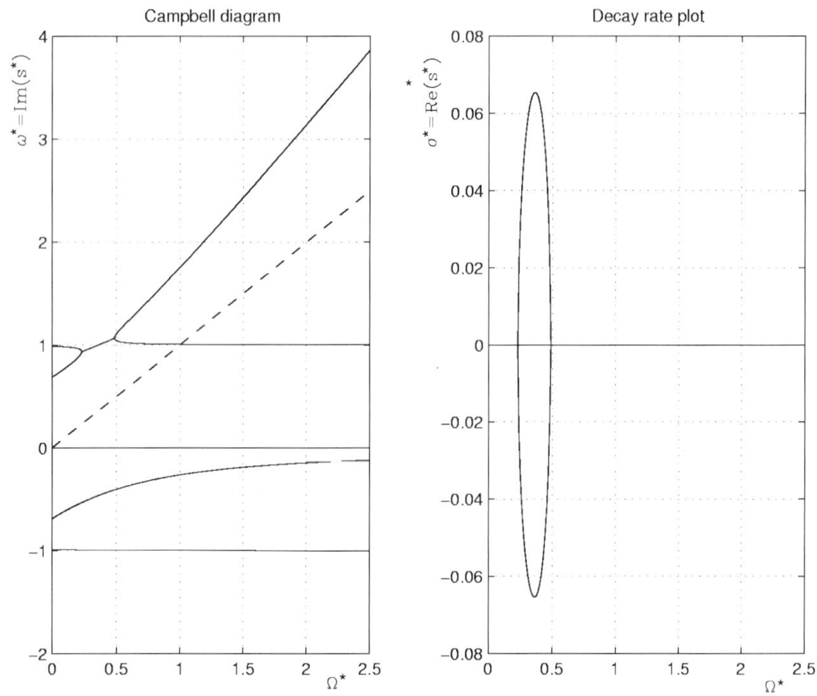

FIGURE 16.4. Nondimensional campbell diagram and decay rate plot of a system with $\alpha' = 0.5$, $\alpha'' = 0.05$, $\beta = 0.1$, $\gamma = 0.15$, and $\delta = 1.5$.

now cross in the fourth quadrant, and consequently, the field of instability occurs in backward whirling conditions.

The plot of Figure 16.6 refers to the same case of Figure 16.5, but for a disc rotor ($\delta = 1.5$ instead of $\delta = 0.6$). As $\alpha' > 1$, the instability range lies in the backward whirl zone of the plot, but it is displaced toward lower values of the speed.

The roots loci of the four cases studied in figures from 2 to 5 are reported in Figure 16.7 from (a) to (d).

Remark 16.2 *In all of the cases studied above, an instability range was present. Further numerical investigation showed that this is because $\gamma > \beta$. If, on the contrary, $\gamma < \beta$, no instability range was encountered, at least unless $\alpha'' > \alpha'$. The conclusions drawn from the numerical experiments run on undamped systems are reported in the following table:*

	Crossing in	$\gamma < \beta$	$\gamma > \beta$
$\alpha' < 1$	*first quadrant*	*no instability*	*unstable forward modes*
$\alpha' > 1$	*fourth quadrant*	*no instability*	*unstable backward modes*

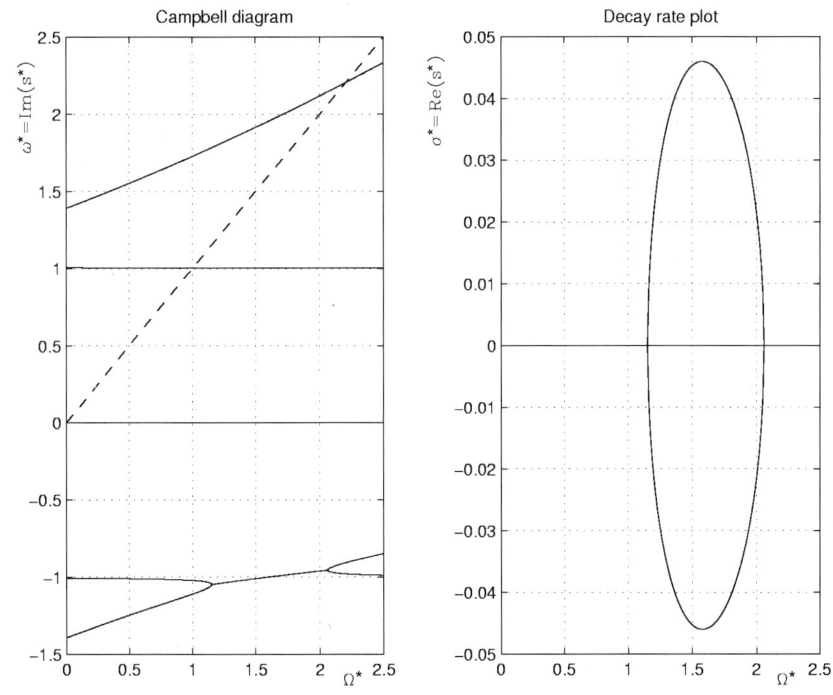

FIGURE 16.5. Nondimensional campbell diagram and decay rate plot of a system with $\alpha' = 2$, $\alpha'' = 0.05$, $\beta = 0.1$, $\gamma = 0.15$, and $\delta = 0.6$.

A further case, with the same parameters of that studied in Figure 16.3, but with $\alpha' = 1.1$, is shown in Figure 16.8. Note that now

$$(1 - \alpha)^2 + 4\beta^2 - 4\gamma^2 = -0.0475 < 0 :$$

The system is unstable even at standstill, for both forward and backward modes, to be stabilized at high speed by the gyroscopic effect.

The effect of damping is that of reducing the width of the instability range, and if the system is damped enough, no instability is encountered.

16.3.4 Geometric re-colocation

Consider a rigid rotor running on two magnetic bearings. Owing to the rigid-body assumption, the noncolocation effect can be compensated for by using a centralized control system; i.e., it is possible to design a centralized control system that causes the actuators to produce forces that are proportional to the displacements (or the velocities, for the derivative branch of the control loop), at the locations of the actuators instead of that of the sensors.

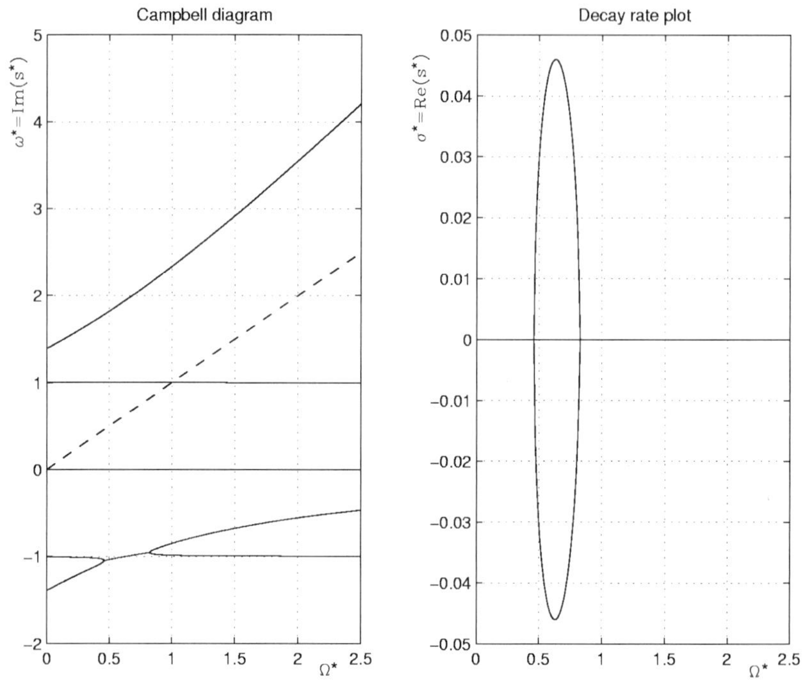

FIGURE 16.6. Nondimensional campbell diagram and decay rate plot of a system with $\alpha' = 2$, $\alpha'' = 0.05$, $\beta = 0.1$, $\gamma = 0.15$, and $\delta = 1.5$.

The complex displacements at the sensor and actuator locations can be expressed as functions of the displacement and rotation at the center of gravity as

$$\left\{ \begin{array}{c} x_1 + iy_1 \\ x_2 + iy_2 \end{array} \right\} = \left[\begin{array}{cc} 1 & z_1 \\ 1 & z_2 \end{array} \right] \left\{ \begin{array}{c} x + iy \\ \phi_y - i\phi_x \end{array} \right\} \ddot{y} = \tag{16.41}$$

$$= \mathbf{T} \left\{ \begin{array}{c} x + iy \\ \phi_y - i\phi_x \end{array} \right\} ,$$

$$\left\{ \begin{array}{c} x_1' + iy_1' \\ x_2' + iy_2' \end{array} \right\} = \left[\begin{array}{cc} 1 & z_1' \\ 1 & z_2' \end{array} \right] \left\{ \begin{array}{c} x + iy \\ \phi_y - i\phi_x \end{array} \right\} = \tag{16.42}$$

$$= \mathbf{T}' \left\{ \begin{array}{c} x + iy \\ \phi_y - i\phi_x \end{array} \right\} .$$

The proportional part of the forces exerted by the actuators are proportional to the displacements at the actuator location if

$$\left\{ \begin{array}{c} F_{x_1} + iF_{y_1} \\ F_{x_2} + iF_{y_2} \end{array} \right\} = \left[\begin{array}{cc} k_1 & 0 \\ 0 & k_2 \end{array} \right] \left\{ \begin{array}{c} x_1' + iy_1' \\ x_2' + iy_2' \end{array} \right\} = \tag{16.43}$$

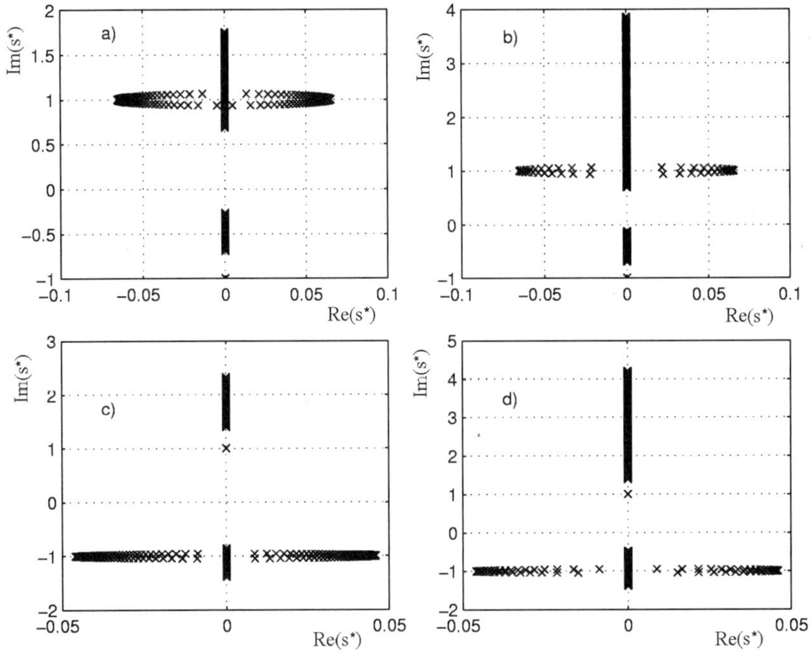

FIGURE 16.7. Roots loci for the cases in Figures 16.3 to 16.6.

$$= \begin{bmatrix} k_1 & 0 \\ 0 & k_2 \end{bmatrix} \mathbf{T}'\mathbf{T}^{-1} \begin{Bmatrix} x_1 + iy_1 \\ x_2 + iy_2 \end{Bmatrix} .$$

The matrix of the gains of the control system required perform the re-colocation is thus

$$\mathbf{K} = \begin{bmatrix} k_1 & 0 \\ 0 & k_2 \end{bmatrix} \mathbf{T}'\mathbf{T}^{-1} = \tag{16.44}$$

$$= \frac{1}{z_1 - z_2} \begin{bmatrix} k_1(z_1' - z_2) & k_1(z_1 - z_1') \\ k_2(z_2' - z_2) & k_2(z_1 - z_2') \end{bmatrix} .$$

The matrix of the derivative gains can be obtained in the same way by substituting $c_i k_i$ for k_i.

Example 16.3 *Turbomolecular pump on magnetic bearings.*

Consider the rotor of a turbomolecular pump with the following inertial data: $m = 9.270$ *kg;* $J_t = 0.0800$ *kg* m^2*;* $J_p = 0.0337$ *kg* m^2*. The center of mass of the rotor is at 134.5 mm from one end of the shaft, whereas the actuators and sensors are at 128.7 mm, 250.8 mm (actuators), 90.1 mm, and 219.5 mm (sensors) respectively. The gains of the sensor-actuator loop of the bearings are* $k_1 = 2.2 \times 10^6$ *N/m and* $k_2 = 0.6 \times 10^6$ *N/m,* $k_{u_1} = 32,000$ *N/m, and* $k_{u_2} = 36,000$ *N/m.*

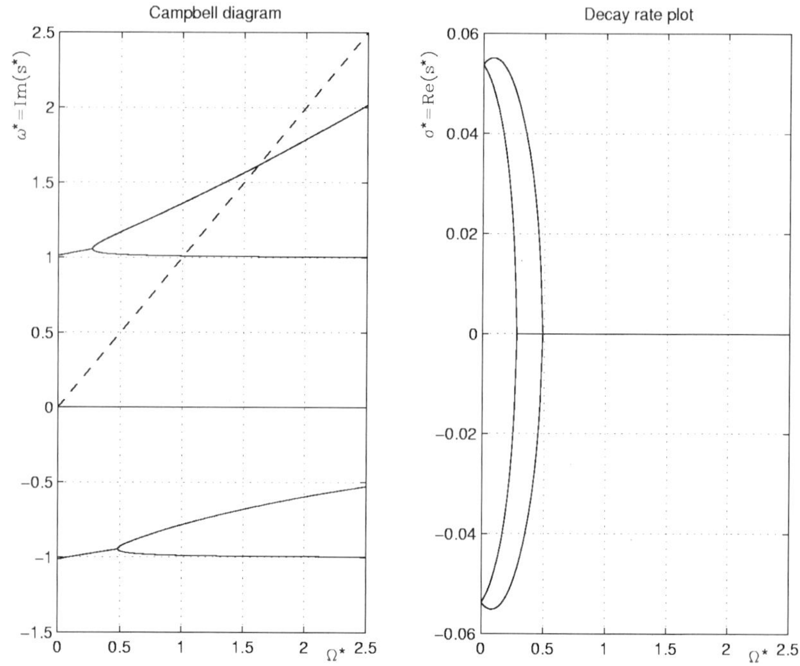

FIGURE 16.8. Nondimensional Campbell diagram and decay rate plot of a system with $\alpha' = 1.1$, $\alpha'' = 0.05$, $\beta = 0.1$, $\gamma = 0.15$, and $\delta = 0.6$. As $(1-\alpha)^2 + 4\beta^2 - 4\gamma^2 = -0.0475 < 0$, the system is unstable for $\Omega^* = 0$.

The nondimensional parameters of the undamped system are: $\alpha = 0.2620$ $(\alpha' = 0.3030$, $\alpha'' = 0.0410)$, $\beta = 0.0199$, $\gamma = 0.2043$ *and* $\delta = 0.4212$. *The value of* ω_1 *is* $\omega_1 = 542.8$ *rad/s.*

The system is stable for $\Omega = 0$, *because* $(1-\alpha)^2 + 4\beta^2 - 4\gamma^2 = 0.378 > 0$. *The Campbell diagram of the undamped system is shown in Figure 16.9.*

As expected, the branches of the Campbell diagram meet in the first quadrant $(\alpha' < 1)$, *a field of instability exists* $(\gamma > \beta)$ *and it is located mainly in the supercritical field* $(\delta < 1)$.

The computation of the Campbell diagram was repeated with different non-colocation schemes and values of the damping to obtain stability maps with the aim of assessing stability boundaries. The results are reported in Figure 16.10 where the spin speeds at which the rotor becomes unstable and then stable again are plotted as functions of the distance d between the sensors and the actuators.

The various curves have been obtained for different values of the damping ratio ζ. *Note that the sensor-actuator distance has been assumed to be the same for the two bearings (which is not the case in the actual system) and the controllers have been assumed to supply the same derivative action (equal* c_i*). Strictly speaking, the values of* ζ, η, θ, *and* κ *are not exactly equal.*

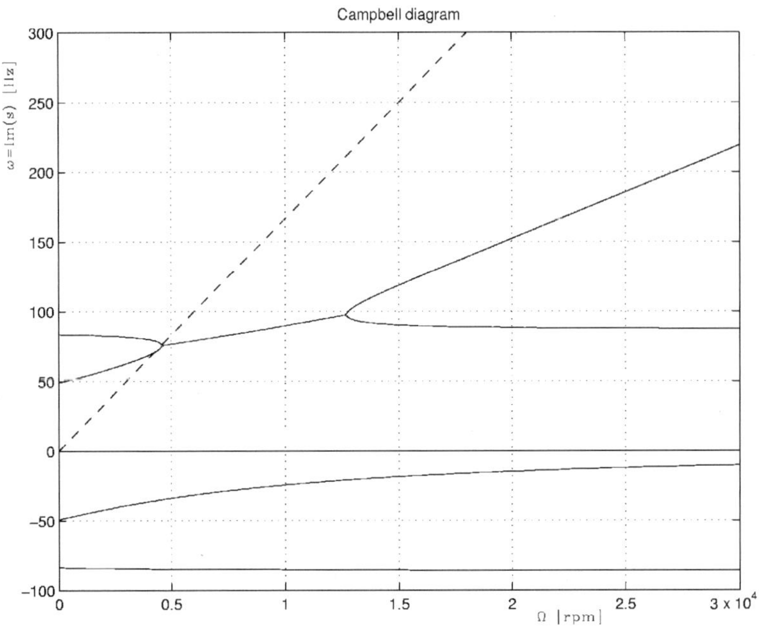

FIGURE 16.9. Campbell diagram of the system studied in the example (undamped system).

If the sensor-actuator distance is smaller than 20 mm, no instability occurs even if the system is undamped, whereas larger sensor-actuator distances lead to increasingly large instability ranges.

By adding damping, the maximum value of d for which the system is stable increases, and if the unstable range is at any rate found, the threshold of instability increases with damping. The value of the upper limit of the instability range has a more complex behavior: The presence of damping causes it to increase, but then it decreases with further increases of damping.

As the average sensor-actuator distance is of 35 mm, a damping ratio in excess of 0.175 is required to guarantee stability. A larger value of damping, i.e., $c_1 = c_2 = 1 \times 10^{-3}$ *is assumed, to account for the fact that the larger bearing has a larger non-colocation (due to a greater bulk of the actuator). It leads to* $\zeta = 0.278$, $\eta = 0.286$, $\theta = 0.262$, *and* $\kappa = 0.279$. *The result of the analysis of the damped system is reported in Figure 16.11: the Campbell diagram and the decay rate plot are reported together with the roots locus.*

The matrix of the gains of a centralized control system able to recolocate the system is

$$\begin{bmatrix} 2.896 & -0.696 \\ 0.154 & 0.446 \end{bmatrix} \times 10^6 \ N/m \ .$$

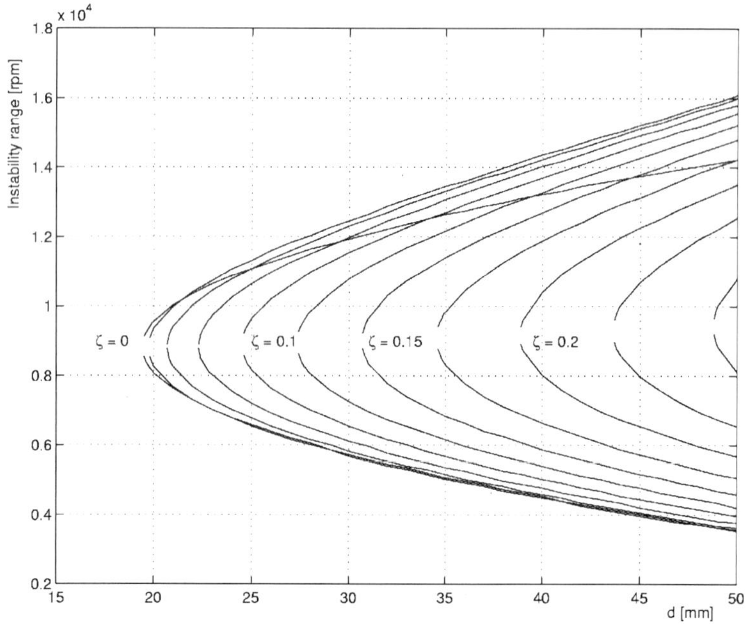

FIGURE 16.10. Lower and upper limits of the instability range as functions of the sensor-actuator distance, for various values of the damping ratio ζ.

The Campbell diagram of the undamped system is reported in Figure 16.12: Its overall pattern is that of a conventional rotor on soft bearings, and no non-colocation effect is present.

16.4 Modal control of rotors

If the model of the rotor contains a large number of degrees of freedom, the dynamic model of the closed-loop controlled system can be very complex. In this case, as well as when a small model is required to allow real-time simulation of the system, it is possible to resort to a reduced-order model by means of any one of the reduction technique seen for rotor modeling.

Guyan reduction, for instance, allows us to obtain fairly small models without losing too much precision, in particular when eliminating the rotational degrees of freedom in beam-like models.

Modal reduction is particularly suitable for the control of nonrotating structures, and it is interesting to assess whether modal control can be applied also to rotors. The point is that rotors are non-natural, and often circulatory, systems, with a stiffness variable with the speed if centrifugal stiffening is present. The standard approach of using the eigenvectors of the

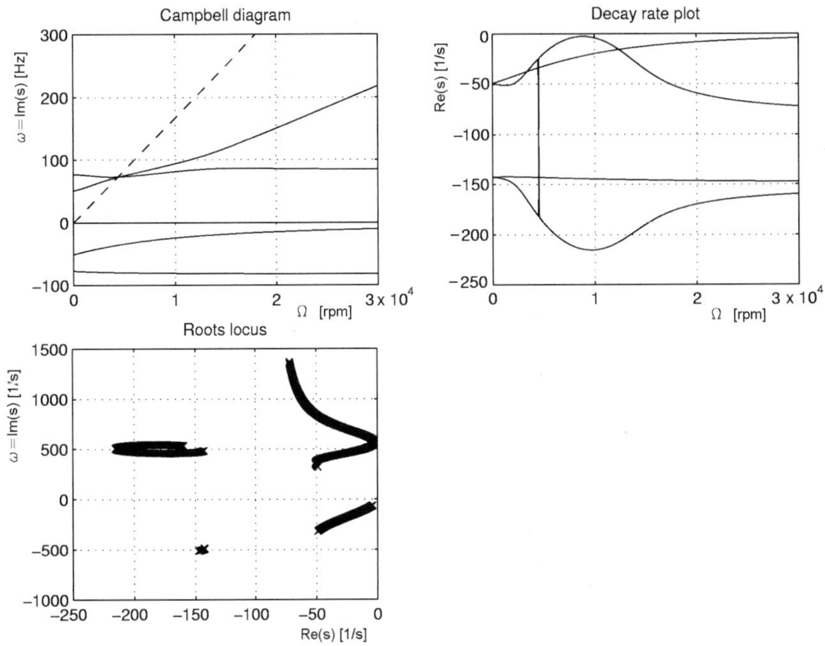

FIGURE 16.11. Campbell diagram, decay rate plot and roots locus for the damped system ($c_1 = c_2 = 1 \times 10^{-3}$, i.e., $\zeta = 0.275$).

undamped system for uncoupling the equations of motion is in this case questionable and can lead to a very large spillover.

There are then two different possible approaches:

- To use the eigenvectors of the undamped system at standstill.
- To use any one of the approaches yielding speed-dependent modal parameters.

The second way is clearly much more complex, and it is not suitable for designing the control system of complex rotors, but it is much more accurate. It will not be dealt with any deeper here.

The first approach is simple and straightforward but can be questionable in the case of rotors in which gyroscopic or circulatory effects or centrifugal stiffening are large. In this case, a better precision can be obtained by using a larger number of modes, but this reduces the advantages of the modal approach.

Let $\mathbf{\Phi}^*$ be the reduced matrix of the eigenvectors, normalized in such a way that the modal masses have a unit value. It is a $n \times m$ matrix, where n is the number of degrees of freedom and m is the number of modes considered.

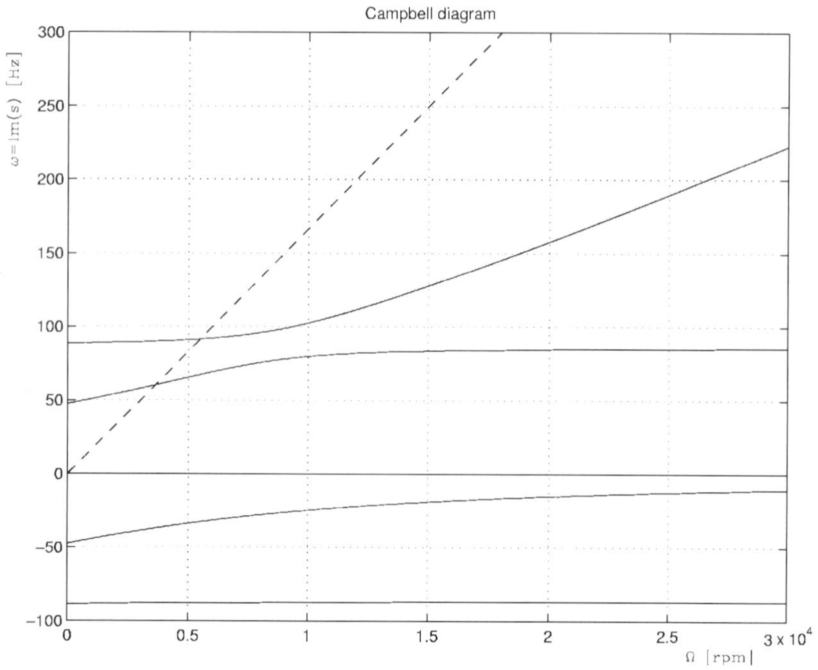

FIGURE 16.12. Campbell diagram of the same system of Figure 16.9, but with a centralized controller that re-colocates sensors and actuators (undamped system).

The modal equations of motion of a rotating system that includes a number of actuators can be written in the form

$$\ddot{\boldsymbol{\eta}} + (\overline{\mathbf{C}}_n + \overline{\mathbf{C}}_r + \Omega\overline{\mathbf{G}})\dot{\boldsymbol{\eta}} + (\overline{\mathbf{K}}+\overline{\mathbf{K}}_\Omega\Omega^2 + \Omega\overline{\mathbf{C}}_r)\boldsymbol{\eta} = \overline{\mathbf{f}}_c + \overline{\mathbf{f}}_e \;, \tag{16.45}$$

where η_i are the m modal coordinates and the modal matrices and vectors are obtained in the usual way. The modal stiffness matrix is diagonal, whereas the others are not.

The modal state space and output equations are then

$$\begin{aligned} \dot{\overline{\mathbf{z}}} &= \overline{\mathbf{A}}\overline{\mathbf{z}} + \overline{\mathbf{B}}_c\mathbf{u}_c + \overline{\mathbf{B}}_e\mathbf{u}_e \;, \\ \mathbf{y} &= \overline{\mathbf{C}}\mathbf{z} + \mathbf{D}\mathbf{u} \;, \end{aligned} \tag{16.46}$$

where

- Vector $\overline{\mathbf{z}}$ contains the n complex state variables $\dot{\boldsymbol{\eta}}$ and $\boldsymbol{\eta}$.
- Vectors $\mathbf{u}_c$ and $\mathbf{u}_e$ contains control and external inputs, in the standard form.

- $\overline{\mathbf{A}}$ is the modal dynamic matrix of the system

$$\overline{\mathbf{A}} = \begin{bmatrix} -\overline{\mathbf{C}}_n - \overline{\mathbf{C}}_r - \Omega\overline{\mathbf{G}} & -\overline{\mathbf{K}} - \Omega\overline{\mathbf{C}}_r \\ \mathbf{I} & \mathbf{0} \end{bmatrix} . \tag{16.47}$$

- Matrices $\overline{\mathbf{B}}_c$ and $\overline{\mathbf{B}}_e$ are the input gain matrices, respectively, for the control and external inputs. They can be obtained from those related to the nonmodal state space equation as

$$\overline{\mathbf{B}}_c = \begin{bmatrix} \mathbf{\Phi}^{*T}\mathbf{T}_c \\ \mathbf{0} \end{bmatrix} , \qquad \overline{\mathbf{B}}_e = \begin{bmatrix} \mathbf{\Phi}^{*T}\mathbf{T}_c \\ \mathbf{0} \end{bmatrix} , \tag{16.48}$$

  where matrices $\mathbf{T}_c$ and $\mathbf{T}_e$ are selection matrices, with n rows and as many columns as there are inputs, showing the degrees of freedom on which the inputs act.

- Vector $\mathbf{y}(t)$ contains the outputs of the system.
- $\overline{\mathbf{C}}$ is the modal output gain matrix,

$$\overline{\mathbf{C}} = \mathbf{C} \begin{bmatrix} \mathbf{\Phi}^* \\ \mathbf{\Phi}^* \end{bmatrix} . \tag{16.49}$$

The control and external inputs and the outputs are the same as for the case of nonmodal control, and hence the dynamic equations of the control system and the operations required for closing the loop are identical to what has already been seen.

Example 16.4 *Consider the spindle running on five active axes magnetic bearings shown in Figure 16.13. The shaft is supported by two radial magnetic bearings (rad. 1 and rad. 2 in the figure), constrained axially by an axial magnetic bearing (ax.) and operated by an induction motor (mot.). An FEM model of the rotor is shown in Figure 16.14(a). The shaft is modeled using 38 Timoshenko beam elements, which connect 39 nodes to each other. In the figure, the laminations of the bearings and of the motor are shown with dotted lines, because they contribute to the mass but not to the stiffness.*

The disc of the axial bearing has been modeled using three thin disc elements (or better a beam-disc transition element and two disc elements) that are connected to three further nodes. The total number of degrees of freedom involved in the lateral behavior is 90, reduced then to 42 through Guyan reduction. A larger reduction could have been used; the size of the model is, however, fairly small, and no further reduction was contemplated. The inertial properties computed by the FEM model are $m = 5.608$ *kg,* $J_t = 0.138$ *kg m*^{2}*, and* $J_p = 0.0019$ *kg m*^{2}.

A free body analysis was first performed. The first two critical speeds, corresponding to rigid-body modes, have a zero value, whereas the two following ones are $\Omega_{cr3} = 2,109$ *rad/s (*$20,143$ *rpm) and* $\Omega_{cr4} = 4,651$ *rad/s (*$44,412$ *rpm).*

FIGURE 16.13. Spindle on a five-active-axes magnetic suspension.

The corresponding mode shapes are reported in Figure 16.14(b). The Campbell diagram is practically flat, and hence, it is not reported.

The radial magnetic bearings are identical. The data are: Actuator constant $K = 1.5246 \times 10^{-6}$ *Nm^2/A^2, nominal air gap* $c = 0.75$ *mm, and bias current* $i_0 = 3$ *A. No allowance is taken for the effect of the radial load, which causes the bearings to behave in an anisotropic way.*

The control loop is based on four independent identical PID controllers with the following data: overall gain $k_c = 1.25$ *V/m, reset time* $T_i = 6.14$ *s, derivative time* $T_d = 0.9035 \times 10^{-3}$ *s, ratio between zero and pole of the PD section* $N = 18.07$*, time constant of the sensor electronics* $\tau = 10^{-6}$*s, gain of the sensor 7500 V/m, time constant of the sensor* 15.92×10^{-6} *s, gain of the power amplifier 1.027 A/V, and time constant of the power amplifier* 50×10^{-6} *s.*

The first four critical speeds of the rotor on active magnetic bearings are $\Omega_{cr1} = 249.1$ *rad/s (2,380 rpm),* $\Omega_{cr2} = 415.6$ *rad/s (3,966 rpm),* $\Omega_{cr3} = 2,157$ *rad/s (20,601 rpm), and* $\Omega_{cr4} = 4,679$ *rad/s (44,686 rpm). Note that the first two critical speeds are very low and are determined by the characteristics of the bearings, whereas the other two are essentially caused by the rotor. The suspension behaves like a very soft mechanical suspension.*

A Campbell diagram computation was then performed by using a modal approach. As the whole system is axially symmetrical, the complex coordinates approach was used. By using four modes, the order of the open-loop system is eight and relevant matrices system are

$$\mathbf{A} = \begin{bmatrix} 0.0116i\Omega & 0.00626i\Omega & 0.00229i\Omega & 0.00503i\Omega & 77080 & 24980 & 12240 & -9523 \\ 0.00344i\Omega & 0.00185i\Omega & 0.000677i\Omega & 0.00149i\Omega & 13310 & 56060 & 20460 & 20720 \\ 0.00522i\Omega & 0.00281i\Omega & 0.0441i\Omega & 0.00635i\Omega & 26230 & 90720 & -42200 & 32300 \\ 0.0103i\Omega & 0.00553i\Omega & 0.00568i\Omega & 0.0853i\Omega & -21540 & 84150 & 28880 & -1.977 \times 10^7 \\ 1 & 0 & 0 & 0 & 0 & 0 & 0 & 0 \\ 0 & 1 & 0 & 0 & 0 & 0 & 0 & 0 \\ 0 & 0 & 1 & 0 & 0 & 0 & 0 & 0 \\ 0 & 0 & 0 & 1 & 0 & 0 & 0 & 0 \end{bmatrix},$$

$$\mathbf{B} = \begin{bmatrix} -19.645 & 4.188 & 5.359 & 18.840 & 0 & 0 & 0 & 0 \\ 13.505 & 13.987 & 23.114 & 16.748 & 0 & 0 & 0 & 0 \end{bmatrix}^T,$$

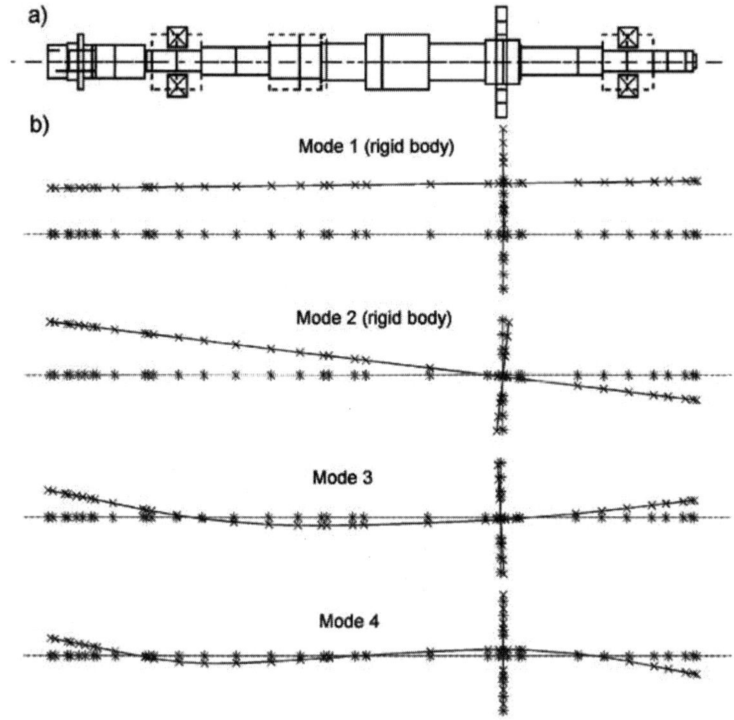

FIGURE 16.14. Spindle on five-active-axes magnetic suspension. (a) FEM model of the rotor; (b) first four mode shapes corresponding to the lowest critical speeds (the rigid-body modes correspond to null critical speeds).

$$\mathbf{C} = \begin{bmatrix} 0 & 0 & 0 & 0 & -0.8376 & 0.2217 & 0.5199 & -0.2388 \\ 0 & 0 & 0 & 0 & 0.5810 & 0.9853 & 0.5875 & 0.8803 \end{bmatrix} .$$

The Campbell diagram and the decay rate plot computed using four modes in each plane are reported in Figure 16.15. Also in this case, the gyroscopic effect is not very strong and the Campbell diagram is almost flat. The suspension is stable at all speeds.

To check the approximations linked with the modal computation, the results obtained at standstill and at a speed of 1500 rad/s using four and ten modes are reported in Table 16.1.

The results show that all frequencies and decay rates are not much affected by the speed and are computed with good approximation using just four modes. The first four forward and backward modes are shown at the top of the table: They include two rigid-body modes, mostly from the bearings, and two deformation modes. The situation is much similar to that shown in Figure 16.14(b).

Then two real nonoscillatory modes linked mostly with the integrative action of the controller follow. The other eight modes reported, mostly caused by the control

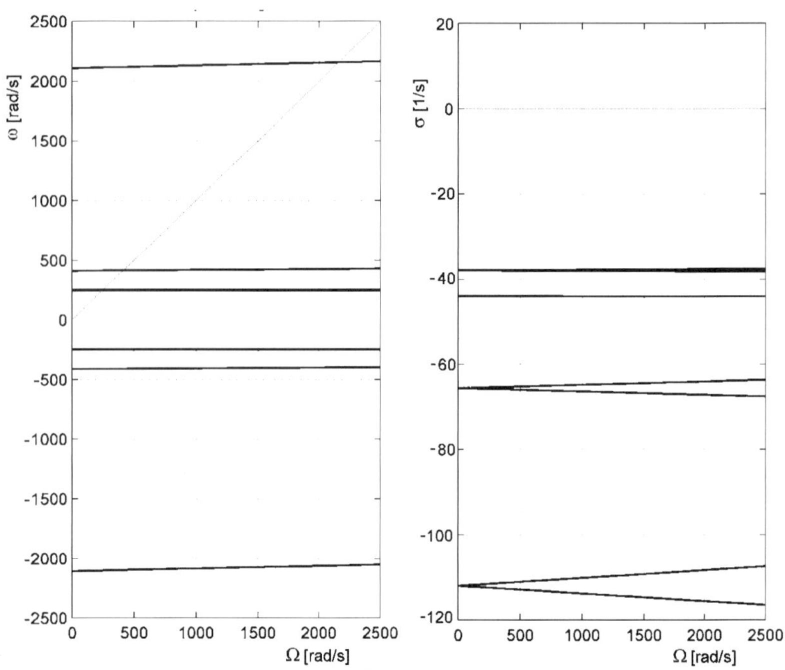

FIGURE 16.15. Campbell diagram and decay rate plot of the rotor on active magnetic bearings.

system are so much damped to be of little practical interest. At standstill, they are nonoscillatory, whereas at 1500 rad/s, they have a nonvanishing frequency, but the accuracy of such a small imaginary part (if compared with the real part) is doubtful.

$\Omega = 0$		$\Omega - 1500$ rad/s	
4 modes	10 modes	4 modes	10 modes
-38.0 -4477.9 i	-38.2 -4477.8 i	-37.8 -4413.9 i	-38.3 -4413.9 i
-65.0 -2108.5 i	-65.7 -2108.3 i	-63.8 -2074.9 i	-64.6 -2074.7 i
-112.0 -412.6 i	-112.1 -412.6 i	-109.3 -402.4 i	-109.3 -402.5 i
-43.9 -248.8 i	-44.11 -249.1 i	-43.9 -248.8 i	-44.1 -249.1 i
-43.9 +248.8 i	-44.11 +249.1 i	-43.9 +248.9 i	-44.1 +249.2 i
-112.0 +412.6 i	-112.1 +412.6 i	-114.8 +423.0 i	-114.8 +423.1 i
-65.0 +2108.5 i	-65.7 +2108.3 i	-66.2 +2142.6 i	-66.9 +2142.3 i
-38.0 +4477.9 i	-38.2 +4477.8 i	-38.2 +4542.9 i	-38.0 +4542.1 i
-0.242	-0.242	-0.242	-0.242
-0.279	-0.279	-0.279	-0.279
-15913	-17602	-15913 -8.449 i	-17602 +10.212 i
-17706	-17878	-17705 +1.042 i	-17878 +3.760 i
-22230	-21779	-22230 -0.841 i	-21779 -17.936 i
-23920	-22033	-23920 +5.318 i	-22033 -3.325 i
-62548	-62940	-62548 -0.286 i	-62760 +0.273 i
-62730	-62761	-62730 +0.025 i	-62940 +2.062 i
-100015	-99969	-100015 -0.003 i	-99969 -0.302 i
-100048	-100009	-100048 +0.032 i	-100009 -0.041 i

TABLE 16.1. Eigenfrequencies of the spindle on magnetic bearings at standstill and at 1500 rad/s. Results obtained through a modal computation with four and ten modes.

Appendix A

Vectors, matrices, and equations of motion

The study of the equations of motion for a general space discretized system can be performed with reference to either the *frequency domain* or the *time domain*, yielding the natural frequencies, the mode shapes, and time free and forced responses. When the system is linear and the sets of equations are written in terms of vectors and matrices, the tools of linear algebra can be used to cope with the problems of motion (for example, see [19] or [18] for a general introduction to lumped parameters system, and [73] or [74] for the related theory of matrices, including numerical aspects).

In particular, for what the lateral dynamics of rotating systems is concerned, a further analytical tool is introduced and used throughout the present book, namely, the *complex coordinates* approach, in which complex numbers are used to represent generalized displacement vectors. This approach proved to be very expedient for modeling both single- and multi-degrees-of-freedom rotors, particularly when the whole system is axially symmetrical and the forcing functions acting in the rotation plane in the direction of the coordinate axes are in *quadrature*, i.e., with a phase delay of $\pi/2$ with respect to each other, like unbalance forces.

A.1 Equation of motion

Lumped parameter systems may be related to mechanical, electrical, or electromechanical systems (see, for example, [75] or [76]). Very often, when the motion is confined to small variations, the general equation of motion for lumped parameters structures can be linearized and may be conveniently expressed in the following matrix form:

$$[M]_{n\times n}\left\{\ddot{q}(t)\right\}_n+[C+G]_{n\times n}\left\{\dot{q}(t)\right\}_n+[K+H]_{n\times n}\left\{q(t)\right\}_n=[S_f]_{n\times m}\left\{f(t)\right\}_m \;, \tag{A.1}$$

where n is the number of the degrees of freedom listed in the generalized displacement vector $\mathbf{q}(t)$, $\mathbf{M}$ is the real symmetric *mass matrix*, $\mathbf{C}$ is the real symmetric *damping matrix*, $\mathbf{G}$ is the real skew-symmetric *gyroscopic matrix*, $\mathbf{K}$ is the real symmetric *stiffness matrix*, $\mathbf{H}$ is the real skew-symmetric *circulatory matrix*, and $\mathbf{S}_f$ is an influence matrix of m external generalized forces $\mathbf{f}(t)$ acting on the system. For its relevance, Equation (A.1) is often simply referred to as the *equation of motion*.

Remark A.1 *Actually it is possible to write the set of linear differential Equations (A.1) in a way in which no matrix is either symmetric or skew symmetric (it is enough to multiply one of the equations by a constant different from 1). A better definition would be to say that* $\mathbf{M}$, $\mathbf{C}$, *and* $\mathbf{K}$ *can be reduced to symmetrical matrices by the same linear transformation that reduces* $\mathbf{G}$ *and* $\mathbf{H}$ *into skew-symmetric matrices.*

Remark A.2 *The same form of Equation (A.1) may result from mathematical modeling of physical systems whose equations of motion are obtained by means of space discretization techniques, such as the well-known finite elements method.*

Remark A.3 *The role played by the signs of the coefficients in scalar polynomial for the assessment of stability is here played by the definition or semidefinition of coefficient matrices that, in turn, is strictly related to the definition of energy-based Lyapunov functions (see [19], Chapter 4).*

A.1.1 Associated eigenproblem

As any nonsymmetric matrix $\mathbf{A}$ can be decomposed as the sum of a symmetric $\mathbf{A}_{\text{sym}}=\mathbf{A}_{\text{sym}}^T$ and a skew-symmetric $\mathbf{A}_{\text{skew}}=-\mathbf{A}_{\text{skew}}^T$ part

$$\mathbf{A}=\mathbf{A}_{\text{sym}}+\mathbf{A}_{\text{skew}} \;, \tag{A.2}$$

where

$$\mathbf{A}_{\text{sym}} = \frac{(\mathbf{A}+\mathbf{A}^T)}{2} \;, \tag{A.3}$$

$$\mathbf{A}_{\text{skew}} = \frac{(\mathbf{A}-\mathbf{A}^T)}{2} \;, \tag{A.4}$$

it follows that the matrices of the coefficients of the equation of motion (A.1) may be written as

$$\mathbf{A}_2 = \mathbf{A}_{2,\mathrm{sym}} = \mathbf{M} \ , \tag{A.5}$$

$$\mathbf{A}_1 = \mathbf{A}_{1,\mathrm{sym}} + \mathbf{A}_{1,\mathrm{skew}} = \mathbf{C} + \mathbf{G} \ , \tag{A.6}$$

$$\mathbf{A}_0 = \mathbf{A}_{0,\mathrm{sym}} + \mathbf{A}_{0,\mathrm{skew}} = \mathbf{K} + \mathbf{H} \ , \tag{A.7}$$

where symmetric and skew-symmetric matrices have a specific physical meaning that is almost self-explanatory in the case of rotating systems.

The general time domain solution of the ordinary differential Equation (A.1) is the sum of the general solution of the homogeneous equation

$$\mathbf{A}_2\ddot{\mathbf{q}}(t) + \mathbf{A}_1\dot{\mathbf{q}}(t) + \mathbf{A}_0\mathbf{q}(t) = 0 \tag{A.8}$$

plus a particular integral of the complete Equation (A.1).

Assuming a solution of Equation (A.8) of the kind

$$\mathbf{q}(t) = \mathbf{q}_0 \, e^{st} \ , \tag{A.9}$$

where $\{\mathbf{q}_0\}_n \in C^n$ and $s \in C$ are a constant complex vector and complex scalar, respectively, the following algebraic equation is obtained:

$$\left(\mathbf{A}_2 s^2 + \mathbf{A}_1 s + \mathbf{A}_0\right)\mathbf{q}_0 = \mathbf{D}_2(s)\mathbf{q}_0 = 0 \ , \tag{A.10}$$

where $\mathbf{D}_2(s)$ is a special case of the algebraic problem known as *lambda matrix*[1] of order two (see [73] for the general theory of lambda matrices and in particular Chapter 7 of [73] for its application to vibrating systems). As $[\mathbf{D}_2(s)]_{n\times n}$ is square and $\mathbf{A}_2 = \mathbf{M}$ is nonsingular, the lambda matrix associated with the equation of motion is said to be *regular*.

In addition, if for each s_i satisfying the *generalized eigenproblem*

$$\left(\mathbf{A}_2 s_i^2 + \mathbf{A}_1 s_i + \mathbf{A}_0\right)\mathbf{q}_i = \mathbf{D}_2(s_i)\mathbf{q}_i = 0 \ , \tag{A.11}$$

the rank of $\mathbf{D}_2(s_i)$ is equal to $n - \alpha_i$, where α_i is the multiplicity of the latent eigenvalue s_i, the lambda matrix is also said to be *simple*.

In the case of a regular and simple lambda matrix of order two, the latent characteristic equation

$$|\mathbf{D}_2(s_i)| = 0 \tag{A.12}$$

yields $2n$ latent eigenvalues $(s_i)_{i=1}^{2n}$ and $2n$ latent (right) eigenvectors $(\{\mathbf{q}_i\}_n)_{i=1}^{2n}$.

[1] The term *lambda matrix* comes from the habit of using symbol λ for expressing the solution $\mathbf{q}(t) = \mathbf{q}_0 \, e^{\lambda t}$. However, here symbol s is used instead of λ, following a more modern tradition.

Conservative nongyroscopic (natural) systems

If $\mathbf{C} = \mathbf{G} = \mathbf{H} = \mathbf{0}$, i.e., in the case of the so-called MK systems, the related eigenproblem

$$\left(\mathbf{M}s_i^2 + \mathbf{K}\right)\mathbf{q}_i = \mathbf{0} \tag{A.13}$$

reduces to the standard form of the eigenproblem

$$\mathbf{A}\mathbf{q}_i = \mu_i \mathbf{q}_i \tag{A.14}$$

(so-called regular pencil in the lambda matrix denomination) when the following positions are taken:

$$\mathbf{A} = \mathbf{M}^{-1}\mathbf{K}\,, \tag{A.15}$$

$$\mu_i = -s_i^2\,. \tag{A.16}$$

Remark A.4 *The positive definiteness (semipositiveness) of matrices* $\mathbf{M}$ *and* $\mathbf{K}$ *implies that the eigenvalues* μ_i *are* n *real and positive (zero) scalars, and hence,* s_i *are* $2n$ *purely imaginary scalars that come in complex conjugate pairs*

$$(s_i, \overline{s}_i) = \pm i\sqrt{\mu_i}\,. \tag{A.17}$$

It should be noted that also the n *eigenvectors of dimension* n $\left(\{\mathbf{q}_i\}_n\right)_{i=1}^n$ *are real vectors.*

Remark A.5 *As matrix* $\mathbf{A} = \mathbf{M}^{-1}\mathbf{K}$ *is nonsymmetric*[2], *the most efficient numerical algorithms for the solution of the associated eigenproblem that are based on real symmetric matrices cannot be used.*

Anyhow, as both $\mathbf{M}$ and $\mathbf{K}$ are real and symmetric and furthermore $\mathbf{M}$ is positive definite, it is possible to recast the eigenproblem of Equation (A.13) in terms of a single real symmetric matrix by means of the so-called Cholesky decomposition:

$$\mathbf{M} = \mathbf{L}\mathbf{L}^T\,, \tag{A.18}$$

where $\mathbf{L}$ is a lower triangular nonsingular real matrix. Then, introducing the linear transformation

$$\mathbf{q}_i = \mathbf{L}^{-T}\mathbf{p}_i\,, \tag{A.19}$$

where $\mathbf{L}^{-T} = \left(\mathbf{L}^T\right)^{-1} = (\mathbf{L}^{-1})^T$, Equation (A.14) can be reduced to

$$\mathbf{A}^*\mathbf{p}_i = \mu_i \mathbf{p}_i\,, \tag{A.20}$$

where

$$\mathbf{A}^* = \mathbf{L}^{-1}\mathbf{K}\mathbf{L}^{-T} = \mathbf{A}^{*T} \tag{A.21}$$

is a real symmetric matrix.

---

[2] Often symbol $\mathbf{D}$ is used for the dynamic matrix $\mathbf{M}^{-1}\mathbf{K}$ written with reference to the configuration space, whereas $\mathbf{A}$ is commonly used for the dynamic matrix in the state space.

A.1.2 Free response

The time solution of the homogeneous dynamic equation, based on a regular and simple lambda matrix, may be expressed in terms of latent eigenvalues and eigenvectors

$$\mathbf{q}(t) = \sum_{i=1}^{2n} c_i \mathbf{q}_i e^{s_i t} , \tag{A.22}$$

where the $2n$ complex constants c_i are to be found from the initial conditions (n generalized displacements and n generalized velocities).

As all of the matrix coefficients of the dynamic equation are real, the solution is real and the $2n$ complex quantities are n complex conjugate pairs:

$$(s_i = \sigma_i + j\omega_i)_{i=1}^{n} \quad \text{and} \quad (\overline{s}_i = \sigma_i - j\omega_i)_{i=1}^{n} \ , \text{with} \ \sigma_i \in R \ , \ \omega_i \in R \ ,$$

$$\left(c_i = c_i^R + jc_i^I\right)_{i=1}^{n} \quad \text{and} \quad \left(\overline{c}_i = c_i^R - jc_i^I\right)_{i=1}^{n} \ , \text{with} \ c_i^R \in R \ , \ c_i^I \in R \ ,$$

$$\left(\mathbf{q}_i = \mathbf{q}_i^R + j\mathbf{q}_i^I\right)_{i=1}^{n} \quad \text{and} \quad \left(\overline{\mathbf{q}}_i = \mathbf{q}_i^R - j\mathbf{q}_i^I\right)_{i=1}^{n} \ ,$$

with $\left\{\mathbf{q}_i^R\right\}_n \in R^n$, $\left\{\mathbf{q}_i^I\right\}_n \in R^n$.

It follows that the expression of the time solution (A.22) may be written as a purely real expression

$$\mathbf{q}(t) = \sum_{i=1}^{n} \left(c_i \mathbf{q}_i e^{s_i t} + \overline{c}_i \overline{\mathbf{q}}_i e^{\overline{s}_i t}\right) =$$

$$= 2\sum_{i=1}^{n} e^{\sigma_i t} \left[\left(c_i^R \mathbf{q}_i^R - c_i^I \mathbf{q}_i^I\right)\cos(\omega_i t) + \left(c_i^R \mathbf{q}_i^I + c_i^I \mathbf{q}_i^R\right)\sin(\omega_i t)\right] \ , \tag{A.23}$$

where $\left(c_i = c_i^R + jc_i^I\right)_{i=1}^{n}$ and $\left(\overline{c}_i = c_i^R - jc_i^I\right)_{i=1}^{n}$ with $c_i^R \in R$, $c_i^I \in R$ are $2n$ coefficients to be obtained from the $2n$ equations corresponding to the $2n$ initial conditions $(\{\mathbf{q}_i(0)\}_n)_{i=1}^{n}$ and $(\{\dot{\mathbf{q}}_i(0)\}_n)_{i=1}^{n}$.

Underdamped systems

In general, any vector and thus the time solution vector $\{\mathbf{q}(t)\}_n$ may be represented as a linear combination of n independent vectors $\{\mathbf{p}_i\}_n$

$$\mathbf{q}(t) = \sum_{i=1}^{n} \mathbf{p}_i \alpha_i(t) , \tag{A.24}$$

with $\alpha_i(t)$ the coefficients of the linear combination.

Many structures are characterized by small and usually uncertain damping. If this is the case, Equation (A.24) may be written as

$$\begin{aligned} \mathbf{q}(t) &= \sum_{i=1}^{n} \mathbf{q}_i \left(c_i e^{s_i t} + \overline{c}_i e^{\overline{s}_i t}\right) && \text{(A.25)} \\ &= \sum_{i=1}^{n} \mathbf{q}_i e^{\sigma_i t} \left[a_i \cos(\omega_i t) + b_i \sin(\omega_i t)\right] && \text{(A.26)} \\ &= \sum_{i=1}^{n} \mathbf{q}_i e^{\sigma_i t} \alpha_i (\cos(\omega_i t + \phi_i) \; , && \text{(A.27)} \end{aligned}$$

where $\mathbf{q}_i$ are the (real) eigenvectors of the MK eigenproblem (A.13), $\sigma_i + j\omega_i = s_i$ are the complex (and conjugate) eigenvalues of the damped system, and $(a_i)_{i=1}^n$ and $(b_i)_{i=1}^n$ are $2n$ coefficients to be obtained from the $2n$ equations corresponding to the $2n$ initial conditions $(\{\mathbf{q}_i(0)\}_n)_{i=1}^n$ and $(\{\dot{\mathbf{q}}_i(0)\}_n)_{i=1}^n$.

Remark A.6 *If all the eigenvalues are either complex or imaginary, the system is said to be underdamped. When at least one of the eigenvalues is real, the system is overdamped.*

Remark A.7 *The time Solution (A.24) may also be used in general when a modal base made of eigenvectors is not readily available. It is sufficient to adopt any base made of n linear-independent vectors, e.g., the principal one of orthogonal unit vectors, of the same dimension of vector $\{\mathbf{q}(t)\}_n$ and then substitute and solve for the initial conditions (n positions and n velocities) to find the $2n$ coefficient of the time Solutions (A.25), (A.26), or (A.27).*

Overdamped systems

In the case of overdamped systems, a number $2m$ of the latent eigenvalues may be real numbers and Equation (A.22) may be partitioned accordingly:

$$\mathbf{q}(t) = \sum_{i=1}^{2m} c_i \mathbf{q}_i e^{-t/\tau_i} + \sum_{i=2s+1}^{2n} c_i \mathbf{q}_i e^{s_i t} \; , \tag{A.28}$$

where $\tau_i = -1/s_i$ for all the real-valued s_i (the corresponding $c_i \mathbf{q}_i$ are also real valued).

A.1.3 Forced response

The steady-state time solution of the dynamic Equation (A.1) for a sinusoidal forcing function

$$\mathbf{f}(t) = \mathbf{f}_0 \cos(\omega_f t + \phi_f) \tag{A.29}$$

is a sinusoidal function $\mathbf{q}(t)$ with the same frequency ω_f of the forcing function

$$\mathbf{q}(t) = \mathbf{q}\cos(\omega_f t + \boldsymbol{\phi}) \ , \tag{A.30}$$

where the $2n$ elements of $\{\mathbf{q}\}_n$ and $\{\boldsymbol{\phi}\}_n$ can be computed substituting Solution (A.30) for two different times t_1 and t_2 into Equation (A.1).

A.1.4 State-space representation

As it is well known from elementary calculus, a set of n equations of order p can always be recast into a set of $n \times p$ of first-order equations. If the derivative term is left in monic form on the left-hand side of each first order equation, a *normalized* set of equation is obtained known as state-space representation.

With regard to the equation of motion (A.1), the *state vector* $\mathbf{x}(t)$ is a $2n$-dimensional vector usually defined in terms of generalized displacements $\mathbf{q}(t)$ and velocities $\dot{\mathbf{q}}(t)$

$$\mathbf{z}(t) = \left\{ \begin{array}{c} \{\dot{\mathbf{q}}(t)\}_n \\ \{\mathbf{q}(t)\}_n \end{array} \right\}_{2n} , \tag{A.31}$$

and the *input vector* $\mathbf{u}(t)$ is a m-dimensional vector collecting all of the external forces acting on the system. With these positions, a new set of $2n$ equations of motion is obtained,

$$\dot{\mathbf{z}}(t) = \left\{ \begin{array}{c} \ddot{\mathbf{q}}(t) \\ \dot{\mathbf{q}}(t) \end{array} \right\} = \mathbf{A}\,\mathbf{x}(t) + \mathbf{B}\,\mathbf{u}(t) \ , \tag{A.32}$$

where

$$\mathbf{A} = \left[\begin{array}{cc} \left[-\mathbf{M}^{-1}(\mathbf{C}+\mathbf{G})\right]_{n\times n} & \left[-\mathbf{M}^{-1}(\mathbf{K}+\mathbf{H})\right]_{n\times n} \\ [\mathbf{I}]_{n\times n} & [\mathbf{0}]_{n\times n} \end{array} \right]_{2n\times 2n} \tag{A.33}$$

and

$$\mathbf{B} = \left[\begin{array}{c} [\mathbf{S}_f]_{n\times m} \\ [\mathbf{0}]_{n\times m} \end{array} \right]_{2n\times m} \tag{A.34}$$

are the so-called *dynamic* or *space matrix* and *input matrix*, respectively.

The solution of the associated homogeneous state equation

$$\dot{\mathbf{z}}(t) = \mathbf{A}\,\mathbf{z}(t) \tag{A.35}$$

is obtained assuming a solution of the kind

$$\mathbf{z}(t) = \mathbf{z}_0\, e^{st} \ , \tag{A.36}$$

where $\{\mathbf{z}_0\}_{2n} \in C^{2n}$ and $s \in C$ are a constant complex vector and a scalar, whose substitution leads to a standard eigenproblem:

$$\mathbf{A}\,\mathbf{z}_i = s_i \mathbf{z}_i \ .$$

Remark A.8 *As compared with the so-called* configuration space *representation of Equation (A.1), the* state-space *representation is particularly useful when* $\mathbf{C}+\mathbf{G}\neq\mathbf{0}$ *and* $\mathbf{H}\neq\mathbf{0}$, *i.e., when the dynamic equation cannot be solved in* s^2 *[biquadratic approach of Equation (A.13)]. An alternative but seldom used approach is to tackle the numerical problem in terms of lambda matrix (see [73], Chapter 5).*

The state-space representation is usually completed by a so-called *measure* or *output* equation consisting of a collection of p linear combinations of states, referred to as *outputs* of the system:

$$\mathbf{y}(t)=\mathbf{C}\,\mathbf{z}(t)=[\mathbf{S}_q]_{p\times n}\{\mathbf{q}(t)\}_n+[\mathbf{S}_{\dot q}]_{p\times n}\{\dot{\mathbf{q}}(t)\}_n=$$

$$=\left[\begin{array}{cc}\mathbf{S}_{\dot q} & \mathbf{S}_q\end{array}\right]_{p\times 2n}\left\{\begin{array}{c}\dot{\mathbf{q}}(t)\\ \mathbf{q}(t)\end{array}\right\}_{2n}, \tag{A.37}$$

where $\mathbf{C}$ is the so-called output matrix (not to be confused with the damping matrix for which the same symbol is normally used, but whose number of columns is half).

Conservative nongyroscopic systems (state-space approach)

In the case of $\mathbf{C}=\mathbf{G}=\mathbf{H}=\mathbf{0}$, the state matrix reduces to

$$\mathbf{A}=\left[\begin{array}{cc}[\mathbf{0}]_{n\times n} & \left[-\mathbf{M}^{-1}\mathbf{K}\right]_{n\times n}\\ [\mathbf{I}]_{n\times n} & [\mathbf{0}]_{n\times n}\end{array}\right]_{2n\times 2n} \tag{A.38}$$

and the associated eigenproblem can be solved by substituting the solution of Equation (A.9) used in the configuration space into the state-space vector of Equation (A.31), obtaining

$$\mathbf{z}(t)=\left\{\begin{array}{c}s\{\mathbf{q}\}_n\\ \{\mathbf{q}\}_n\end{array}\right\}_{2n}e^{st}\,. \tag{A.39}$$

By introducing Equation (A.39) into Equation (A.35), with the state matrix as in Equation (A.38), and neglecting the identity expressed by the last n equation, the eigenproblem reduces to

$$s^2\mathbf{q}=-\,\mathbf{M}^{-1}\mathbf{K}\,\mathbf{q}\,, \tag{A.40}$$

i.e., the same eigenproblem already stated in Equation (A.14).

A.1.5 Frequency response

The frequency domain approach is an alternative method of solving linear differential equations using the Laplace transform. As the equation of motion (A.1) is linear, it is possible to obtain the time response for a generic

Laplace transformable forcing function, i.e., *any function* $\mathbf{f}(t)$ for which a Laplace transform $\mathcal{L}\{\mathbf{f}(t)\} = \hat{\mathbf{f}}(s)$ can be computed. Once the Laplace transform of the equation of motion (A.1) is obtained

$$\mathcal{L}\{\mathbf{M}\ddot{\mathbf{q}}(t) + [\mathbf{C}+\mathbf{G}]\,\dot{\mathbf{q}}(t) + [\mathbf{K}+\mathbf{H}]\,\mathbf{q}(t)\} =$$

$$= \mathbf{M}\left(s^2\hat{\mathbf{q}}(s) - s\dot{\mathbf{q}}(0) - \mathbf{q}(0)\right) + [\mathbf{C}+\mathbf{G}]\,(s\hat{\mathbf{q}}(s) - \mathbf{q}(0)) + [\mathbf{K}+\mathbf{H}]\,\hat{\mathbf{q}}(s) =$$

$$= \left(\mathbf{M}s^2 + [\mathbf{C}+\mathbf{G}]\,s + [\mathbf{K}+\mathbf{H}]\right)\hat{\mathbf{q}}(s) - \mathbf{M}s\dot{\mathbf{q}}(0) - [\mathbf{M}+\mathbf{C}+\mathbf{G}]\,\mathbf{q}(0) = \hat{\mathbf{f}}(s) \tag{A.41}$$

and with the following position:

$$\mathbf{D}_2(s) = \mathbf{M}s^2 + [\mathbf{C}+\mathbf{G}]\,s + [\mathbf{K}+\mathbf{H}] \ , \tag{A.42}$$

it is possible to write

$$\mathbf{D}_2(s)\hat{\mathbf{q}}(s) = \mathbf{M}s\dot{\mathbf{q}}(0) + [\mathbf{M}+\mathbf{L}+\mathbf{G}]\,\mathbf{q}(0) + \hat{\mathbf{f}}(s) \ . \tag{A.43}$$

The differential problem of the equation of motion (A.1) is therefore transformed into an algebraic problem and hence

$$\hat{\mathbf{q}}(s) = \mathbf{D}_2^{-1}(s)\,(\mathbf{M}s\dot{\mathbf{q}}(0) + [\mathbf{M}+\mathbf{C}+\mathbf{G}]\,\mathbf{q}(0)) + \mathbf{D}_2^{-1}(s)\hat{\mathbf{f}}(s) \tag{A.44}$$

from which the general time solution is obtained by taking the appropriate inverse Laplace transform $\mathbf{q}(t) = \mathcal{L}^{-1}\{\hat{\mathbf{q}}(s)\}$.

Remark A.9 *As already stated, the general time Solution (A.44) is the superposition of the general free response and of the forced response, i.e., transient and steady-state response, respectively.*

Remark A.10 *Taking the Laplace transform of the so-called output Equation (A.37)*

$$\mathcal{L}\{\mathbf{y}(t)\} = \hat{\mathbf{y}}(s) = \mathbf{S}_q\hat{\mathbf{q}}(s) + \mathbf{S}_{\dot{q}}s\hat{\mathbf{q}}(s) \tag{A.45}$$

and zero initial conditions, the so-called transfer function $\mathbf{H}(s)$ *between the forcing inputs and measured outputs of the dynamic system is obtained*

$$\hat{\mathbf{y}}(s) = \mathbf{H}(s)\hat{\mathbf{f}}(s) = [\mathbf{S}_q + \mathbf{S}_{\dot{q}}s]\,\mathbf{D}_2^{-1}(s)\hat{\mathbf{f}}(s) \ . \tag{A.46}$$

A.2 Rotating systems

Consider an axi-symmetrical rotor modeled using the one-dimensional, i.e., beam-like, approach (possibly extended to the $1\frac{1}{2}$ D approach). When it is at standstill, it may be modeled as a beam whose displacements in the xy-plane, say along x-axis and y-axis, are uncoupled: Gyroscopic and circulatory terms are then responsible for their coupling at nonzero spin speed.

The displacement at each node i along the spin axis may be expressed in vector form

$$\vec{r}_i(t) = x_i(t)\vec{i} + y_i(t)\vec{j} \tag{A.47}$$

and its equivalent matrix form

$$\mathbf{r}_i(t) = \left\{ \begin{array}{c} x_i(t) \\ y_i(t) \end{array} \right\} . \tag{A.48}$$

An alternative is to express the displacement in the section plane as a complex number

$$r(t) = x_i(t) + iy_i(t) , \tag{A.49}$$

where $i = \sqrt{-1}$. This kind of notation may be conveniently used in rotordynamics to obtain a rather compact notation in studying the basic behavior *when both the rotor and the stator are isotropic with respect to rotation axis.*

Remark A.11 *As seen in Chapter 6, the complex-coordinates approach can be used also for nonisotropic rotors, but some of its advantages are lost.*

A.2.1 Real coordinates

If the rotor generalized displacements are expressed in the Form (A.48), i.e., the displacements and rotations in xz- and yz-planes are directly taken as generalized coordinates

$$\{\mathbf{q}(t)\}_{2n} = \left\{ \begin{array}{c} \{\mathbf{q}_x(t)\}_n \\ \{\mathbf{q}_y(t)\}_n \end{array} \right\}_{2n} , \tag{A.50}$$

the general equation of motion is

$$\begin{bmatrix} [\mathbf{M}']_{n\times n} & \mathbf{0} \\ \mathbf{0} & [\mathbf{M}']_{n\times n} \end{bmatrix}_{2n\times 2n} \{\ddot{\mathbf{q}}(t)\}_{2n} + \left[\begin{bmatrix} [\mathbf{C}']_{n\times n} & \mathbf{0} \\ \mathbf{0} & [\mathbf{C}']_{n\times n} \end{bmatrix}_{2n\times 2n} + \right.$$
$$\left. + \begin{bmatrix} \mathbf{0} & [-\mathbf{G}']_{n\times n} \\ [\mathbf{G}']_{n\times n} & \mathbf{0} \end{bmatrix}_{2n\times 2n} \right] \{\dot{\mathbf{q}}(t)\}_{2n} + \left[\begin{bmatrix} [\mathbf{K}']_{n\times n} & \mathbf{0} \\ \mathbf{0} & [\mathbf{K}']_{n\times n} \end{bmatrix}_{2n\times 2n} + \right.$$
$$\left. + \begin{bmatrix} \mathbf{0} & [-\mathbf{H}']_{n\times n} \\ [\mathbf{H}']_{n\times n} & \mathbf{0} \end{bmatrix}_{2n\times 2n} \right] \{\mathbf{q}(t)\}_{2n} = \{\mathbf{f}(t)\}_{2n} , \tag{A.51}$$

where n is the number of generalized coordinates in each inflection plane and the dependence of the gyroscopic and circulatory terms from the spin speed is implicit. Equation (A.51) holds in the case of an axially symmetrical rotor. In case the nonrotating parts of the system are nonsymmetrical, the structure of the equation is still the same, but the matrices related to the two planes are not identical, and the coupling matrices are not zero. If the rotating parts of the system are nonsymmetrical, things are more complicated as shown in Chapters 6 and 9.

Remark A.12 *If the system is axially symmetrical, all eigenvalues have multiplicity 2, since the behaviour in the two planes is identical. This property holds also for systems with cyclic symmetry of order 3 or greater.*

A.2.2 Complex coordinates

The use of complex coordinates of the Form (A.49) yields an equation of motion whose size is half that obtained using real coordinates. In the case of an isotropic system, the equation in complex coordinates is

$$[\mathbf{M}']_{n\times n}\{\ddot{\mathbf{q}}(t)\}_n + \left[[\mathbf{C}']_{n\times n} - i\,[\mathbf{G}']_{n\times n}\right]\{\dot{\mathbf{q}}(t)\}_{2n} + \\ + \left[[\mathbf{K}']_{n\times n} - i\,[\mathbf{H}']_{n\times n}\right]\{\mathbf{q}(t)\}_n = \{\mathbf{f}(t)\}_n \ . \tag{A.52}$$

Remark A.13 *The matrices that are skew-symmetric in real coordinates are symmetric but imaginary when complex coordinates are used.*

Remark A.14 *In complex coordinates, a rotation of* 90° *is equivalent to multiplication by the imaginary unit i. In general, rotation of angle α is obtained by multiplication by the exponential $e^{i\alpha}$: e.g., $\mathbf{q}(t) = \mathbf{q}_0 e^{i\omega t}$ is a vector rotating in time at the angular velocity ω, with initial position $\mathbf{q}_0$.*

The number of eigenvalues s is halved with respect to the real coordinate approach, because they do not come in pairs of conjugate values. Moreover, the sign of their imaginary part yields the information on the whirling direction.

If the system is not isotropic, also the conjugates of the complex coordinates appear into the equation of motion (Chapters 6 and 9), and a doubling of the equations of motion is needed. The simplifications with respect to the real coordinates approach are thus lost.

A.3 Circulatory and noncirculatory coupling

The equations of motion of an isotropic undamped and nongyroscopic rotor are two identical uncoupled equations related to the behavior in xz- and yz-planes. If the characteristics in the two planes are not identical and the stiffness matrix does not provide any coupling between the inflection planes, the two sets of equations are not identical but still uncoupled. In the case of the Jeffcott rotor, the relevant homogeneous equation of motion are

$$\begin{cases} m\ddot{x} + k_1 x = 0 \ , \\ m\ddot{y} + k_2 y = 0 \ , \end{cases} \tag{A.53}$$

or, by dividing the equations by m,

$$\begin{cases} \ddot{x} + \omega_1^2 x = 0 \ , \\ \ddot{y} + \omega_2^2 y = 0 \ , \end{cases} \tag{A.54}$$

By introducing the mean and deviatoric stiffness and using matrix notation, it follows that

$$\left\{ \begin{array}{c} \ddot{x} \\ \ddot{y} \end{array} \right\} + \omega_0^2 \left[\begin{array}{cc} 1-\alpha & 0 \\ 0 & 1+\alpha \end{array} \right] \left\{ \begin{array}{c} x \\ y \end{array} \right\} = \mathbf{0} \ , \tag{A.55}$$

where the mean and deviatoric stiffness are

$$\omega_0^2 = \frac{\omega_1^2 + \omega_2^2}{2} \ , \qquad \omega_d^2 = \frac{\omega_2^2 - \omega_1^2}{2} \ , \tag{A.56}$$

and parameter α is defined as

$$\alpha = \frac{\omega_d^2}{\omega_0^2} \ . \tag{A.57}$$

and is included in the interval $[-1, \ 1]$. When α takes the values -1 or 1, one of the two stiffnesses vanish and the system becomes statically undetermined.

By introducing a solution of the type

$$x = x_0 e^{st} \ , \qquad y = y_0 e^{st} \ , \tag{A.58}$$

it is possible to check that the natural frequencies of the system are ω_1 and ω_2, or in nondimensional terms,

$$\frac{\omega_{1,2}}{\omega_0} = \sqrt{1 \pm \alpha} \ . \tag{A.59}$$

The values of the nondimensional natural frequencies are plotted as functions of parameter α in Figure A.1(a) or (b), curves labeled as $\epsilon = 0$.

As seen in Section 6.1, Equation (A.54) can be used to describe the free motion of a perfectly balanced Jeffcott rotor on anisotropic supports and can be written using complex coordinates in the form

$$\ddot{r} + \omega_0^2 r - \omega_d^2 \bar{r} = 0 \ , \tag{A.60}$$

where $r = x_1 + ix_2$.

Now introduce a coupling term into the equation of motion (A.55)

$$\left\{ \begin{array}{c} \ddot{x} \\ \ddot{y} \end{array} \right\} + \omega_0^2 \left[\begin{array}{cc} 1-\alpha & \epsilon \\ \epsilon & 1+\alpha \end{array} \right] \left\{ \begin{array}{c} x \\ y \end{array} \right\} = \mathbf{0} \ . \tag{A.61}$$

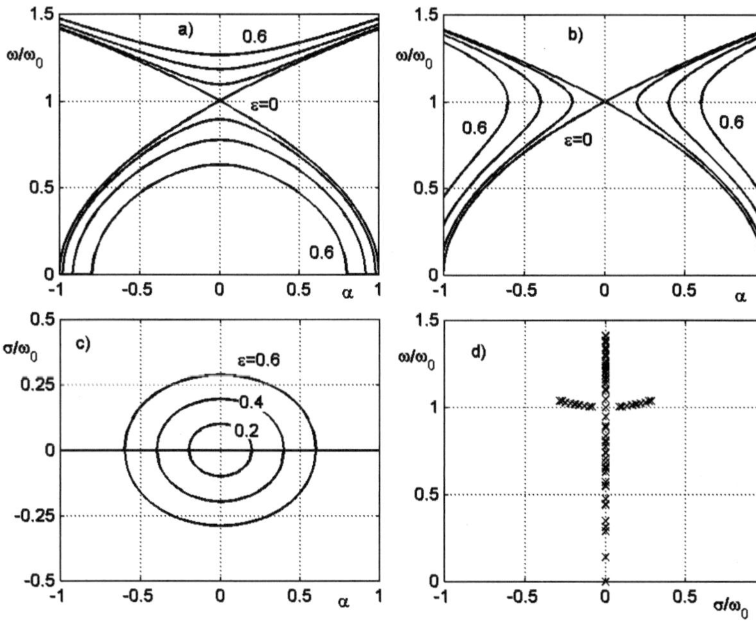

FIGURE A.1. (a) Natural frequencies of a system with two degrees of freedom with noncirculatory coupling as functions of parameter α for different values of ϵ ($\epsilon = 0$, 0.2, 0.4, 0.6). (b) Same as (a), but for a circulatory system. (c) and (d) Decay rate as a function of α and roots locus for a circulatory system.

Such coupling will be defined as noncirculatory, or conservative [11], coupling, because the terms added outside the main diagonal of the stiffness matrix are equal. In the case of the Jeffcott rotor, such a coupling can be caused by elastic (and hence conservative) terms like those introduced by rotating the reference frame in such a way that x- and y-axes are no more principal axes of elasticity. The natural frequencies are plotted in Figure A.1(a) as functions of α for different values of ϵ. The two curves of the plot (that for $\omega > \omega_0$ and that for $\omega < \omega_0$) get less close to each other with respect to the uncoupled system; hence, this type of coupling is also said to be repulsive.

The value of s obtained from the eigenproblem is imaginary for all values of α; as a consequence, the behavior of the system is that of an undamped system.

The stability is linked to the fact that the stiffness matrix is symmetrical, and there is no circulatory matrix.

A second type of coupling, usually referred to as circulatory, or nonconservative, coupling is

$$\left\{ \begin{array}{c} \ddot{x} \\ \ddot{y} \end{array} \right\} + \omega_0^2 \left[\begin{array}{cc} 1-\alpha & \epsilon \\ -\epsilon & 1+\alpha \end{array} \right] \left\{ \begin{array}{c} x \\ y \end{array} \right\} = \mathbf{0} \, . \tag{A.62}$$

The out-of-diagonal terms are here equal in absolute value, but opposite in sign, so that a circulatory matrix is present. This is, for instance, the case of the effect of rotating damping. The natural frequencies are plotted in Figure A.1(b) as functions of α for the same values of ϵ used in the previous computations. The two curves now get closer to each other with respect to the uncoupled system, and starting at $\alpha = -1$, for a certain value of α included in the interval $(-1,\ 0)$, they meet. There is a range centered in $\alpha = 0$ in which the solutions of the eigenproblem in s are complex. Beyond this value, the two distinct curves detach from each other. As the upper curve gets closer and closer to the lower one until they meet, this type of coupling is also said to be attractive.

In the range in which the values of s are complex, one of the two solutions has a positive real part: An unstable solution exists, as shown by the decay rate plot of Figure A.1(c) and by the roots locus of Figure A.1(d).

Instability is linked to the fact that coupling gives way to a nonsymmetric part of the stiffness matrix; hence, there is a nonvanishing circulatory matrix. Note that if circulatory coupling is caused by rotating damping, the same has also a stabilizing effect because of its presence in the damping matrix, here neglected.

Circulatory coupling is not infrequent in rotordynamics and may be from causes different from rotating damping: Anytime it is present, the presence of an instability range must be checked.

Another form of coupling is that caused by the gyroscopic terms, but they are conservative and do not give way to instability.

Appendix B
An outline on rotor balancing

Rotor balancing is a highly specialized technology, which has been dealt with in several books and handbooks, sometimes published by firms that build balancing machines [77, 78]. This appendix will only summarize some of the basic concepts, without any attempt in dealing with this subject in any depth. The interested reader can find all of the required details in specialized texts, in particular, those mentioned above.

All rotors, particularly those intended to operate at high rotational speed, must be balanced before they start their service life. Sometimes balancing procedures must be repeated during the life of a machine. The designer must then take into account this requirement to provide the possibility of removing or adding masses in proper locations from the early design stages. Balancing must be regarded as one of the construction stages, to be performed after assembling the whole rotor or before, on its component, if they must be balanced separately (which usually does not avoid a further balancing process on the assembly), and balancing tolerances must be stated in a way that is not conceptually different from what is done for other types of tolerances, dimensional or geometrical. The balance conditions of a rotor can change in time, and periodic rebalancing may be needed. In some cases, this phenomenon can be severe and is usually referred to as *wandering unbalance*. It can be caused by thermal deformations of the rotor, material inhomogeneity, cracks, loose tolerances in built-up rotors, and the like. Some rotors must be balanced several times during the first runs at subsequent higher speeds, in order to reach good balancing conditions at operating speed and running temperature. Poor balance conditions can be encountered during startup, until steady-state conditions are reached.

Rotor balancing has been the object of standardization, and designers must refer to the standards in stating balancing tolerance at the design stage. Standards are stated for the various types of machines, but it is the duty of the designer to verify that the stresses and deformations caused by the maximum residual unbalance prescribed are not beyond allowable limits. He must also be sure that the prescribed balancing tolerances are strict enough to prevent the rotor from being a source of unwanted vibration and noise for the surrounding environment. As with all tolerances, it must be remembered that it is impossible to reach a perfect balancing and that it is not necessary and, generally, not advisable (at least from the economical point of view) to impose too strict balancing requirements.

From the point of view of balancing, rotors are usually divided into two categories: rigid and deformable rotors. This subdivision, which is accepted by ISO standards, is in a certain sense arbitrary, because no rigid body exists in the real word. A rotor can belong to either class, depending on the speed at which it is supposed to operate and, in particular, a speed at which any rigid rotor ceases to behave as such always exists. As already stated, the balancing of rigid and deformable rotors will be only briefly summarized in the following sections.

B.1 Rigid rotors

Following the ISO 1925 standard, a rotor can be considered rigid if it can be balanced by adding or removing mass in two arbitrarily chosen planes perpendicular to the rotation axis and if its balance conditions are practically independent from speed up to the maximum allowable speed. If this condition is satisfied, the inertia of the stator is neglected and its elastic properties, written with reference to the center of gravity of the rotor, are summarized in the stiffness matrix **K**, the rotor can be assimilated to a rigid body and modeled as a rotor with four (real) degrees of freedom. The balance conditions of the machine can then be summarized by two parameters: the eccentricity ϵ (or the static unbalance $m\epsilon$) and angle χ between the principal axis of inertia of the rotor and the rotation axis [or the couple unbalance $(J_p - J_t)\chi$].

Standards use the peripheral velocity of the center of mass of the rotor as a measure for the static unbalance

$$V_G = \Omega_{max}\epsilon_{max} \tag{B.1}$$

and define a quality grade for balancing, usually referred to as G, as the maximum allowable peripheral velocity of the center of mass, expressed in millimeters/second. A rotor that has to be balanced in the class $G = 2.5$ at 10,000 rpm, for example, must have an eccentricity smaller than $\epsilon_{max} =$

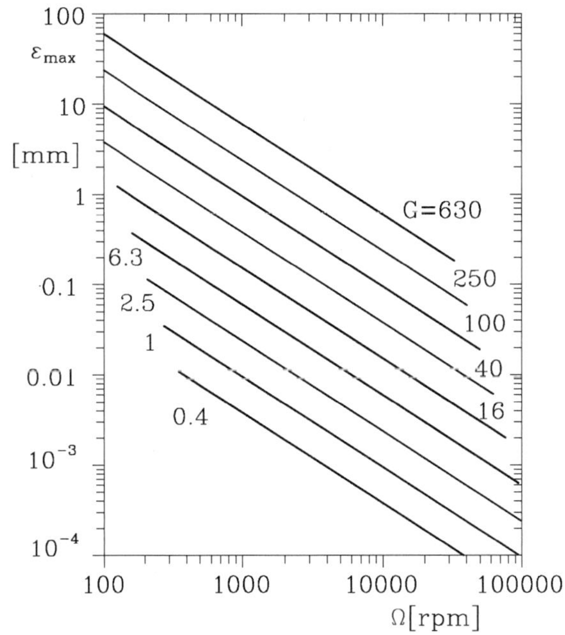

FIGURE B.1. Maximum residual eccentricity as a function of the maximum operating speed for quality grades between $G = 0.4$ and $G = 630$ mm/s.

$V_G/\Omega_{max} = 2.38$ μm, which results in a peripheral velocity of the mass center of 2.5 mm/s.

The maximum values of the residual eccentricity are plotted as functions of the maximum operating speed for different values of the quality grade in Figure B.1. The quality grades suggested by ISO standard 1940 for different types of rotors are reported in Table B.1.

The mentioned standards consider only eccentricity and then static unbalance. In the case of a rigid rotor running on two bearings, a couple unbalance caused by two static unbalances, each equal to half the maximum allowable static unbalance placed at the bearing locations and phased at 180°, is considered a limit. If d is the distance between the bearings, it follows that

$$|\chi(J_p - J_t)|_{max} = m\epsilon_{max}\frac{d}{2} \,. \tag{B.2}$$

Rigid rotors are normally balanced using balancing machines. They can be either of the high-stiffness or the low-stiffness type. Rigid (high-stiffness) balancing machines are machines in which the rotor can be spun on two very stiff supports provided with force transducers. The unbalance condition of the rotor is obtained from the measurement of the forces it exerts on its

Grade	Examples
G 4000	Crankshaft drives of rigidly mounted slow marine diesel engines with an uneven number of cylinders.
G 1600	Crankshaft drives of rigidly mounted large two-cycle engines.
G 630	Crankshaft drives of rigidly mounted large four-cycle engines, crankshaft drives of elastically mounted marine diesel engines.
G 250	Crankshaft drives of rigidly mounted fast four-cylinder diesel engines.
G 100	Crankshaft drives of fast diesel engines with six or more cylinders, complete engines (gasoline or diesel) for cars, trucks, and locomotives.
G 40	Car wheels, wheel rims, wheel sets, drive shafts, crankshaft drives of elastically mounted fast four-cycle engines (gasoline or diesel) with six or more cylinders, crankshaft drives for engines of cars, trucks, and locomotives.
G 16	Drive shafts (propeller shafts, cardan shafts) with special requirements, parts of crushing machinery, parts of agricultural machinery, individual components of engines (gasoline or diesel) for cars, trucks, and locomotives, crankshaft drives of engines with six or more cylinders under special requirements.
G 6.3	Parts of process plant machinery, marine main turbine gears (merchant service), centrifuge drums, fans, assembled aircraft gas turbine rotors, flywheels, pump impellers, machine tools and general machinery parts, normal electrical armatures, individual components of engines under special requirements.
G 2.5	Gas and steam turbines, including marine main turbines (merchant service), rigid turbogenerator rotors, rotors, turbocompressors, machine-tool drives, medium and large electrical armatures with special requirements, small electrical armatures, turbine-driven pumps.
G 1	Tape recorder and phonograph (gramophone) drives, grinding machine drives, small electrical armatures with special requirements.
G 0.4	Spindles, discs and armatures of precision grinders, gyroscopes.

TABLE B.1. Quality grades G suggested for different types of rotors (ISO 1940).

supports. Low-stiffness machines are similar to the former, but the supports are more or less free to move and the transducers measure a quantity linked with their motion, usually the acceleration but sometimes the displacement or velocity. The unbalance of the rotor is obtained from the displacement measurements.

Two planes perpendicular to the rotation axis are chosen, as far from each other as possible, on which masses can be either added or removed. The maximum values of the residual unbalance on the two balancing planes are computed from the allowable residual unbalance corresponding to the required quality grade. The actual unbalance state of the rotor is then determined, spinning the rotor on a balancing machine. Modern balancing machines are supplied with data-acquisition systems that perform all needed computations and directly supply the values of the unbalance on the two correction planes and all information on the masses to be added or removed in them (amount of mass, radius, and angular position). Once the correction has been performed, a further measurement aimed at checking whether the required tolerance has been achieved usually follows.

To satisfy the needs of high-volume mass production, automatic balancing machines that perform all operations without human intervention are used. They are a part of the assembly line, and provide to measure the unbalance, to make the required corrections usually removing material by drilling and then to check the final balancing grade obtained.

The need to add or remove mass in the two planes at the same angular position is a clear symptom of a static unbalance. If the corrections must be phased 180° from each other, the unbalance is purely a couple unbalance. Generally, the phasing is neither at 0° nor at 180°, corresponding to a general state of dynamic unbalance, defined as the sum of static plus couple unbalance.

In some cases, the various parts that constitute a rotor must be balanced separately. In this case, the balancing tolerances of the various parts must be stated, remembering that unbalance is a vector quantity and that in the assembly process, they usually add to each other in a random way. The absolute value of the unbalance is, in the most unfavorable case, equal to the sum of the absolute values of the unbalances of the parts. Dimensional tolerances and their effects on the relative positions of the various parts must also be considered. If possible, the rotor must be balanced after assembly, possibly on its bearings in such a way that the tolerances of the bearings and their seats are also accounted for. In some cases (rotors whose size exceeds the possibilities of available balancing machines, high-precision machinery), the rotor is balanced directly on-site. The machine is instrumented, and the synchronous component of the vibration of the machine is monitored at different speeds. The amplitude and phase of the synchronous component give all of the information needed to identify and correct the unbalance.

Remark B.1 *Disc rotors often need only to be balanced with respect to static unbalance. The conical mode has no critical speed and tends to remain self-centered in a wide speed range. "Vertical" balancing machines that deal only with static unbalance are then often used for rigid rotors with* $J_p > J_t$.

B.2 Flexible rotors

Balancing flexible rotors is much more difficult than balancing rigid rotors, and strictly speaking, it cannot be performed on balancing machines. The stiffness of the supports of the machine has, in fact, great influence on the deflected shape of the rotor, and balancing should be performed directly on the whole machine. This process is generally referred to as *field balancing*. It is, however, often still possible to resort to a balancing machine, provided that its stiffness is not too different from that of the stator of the actual machine and that proper allowance is taken for the unavoidable differences.

Strictly speaking, a flexible rotor is balanced only if each of its cross sections is statically and dynamically balanced. In practice, it is not necessary that this condition is met to achieve the required aim, which is that of maintaining the effects of unbalance (vibrations, stresses, noise, etc.) within tolerable limits in the whole working range of the machine.

The flexible nature of the rotor and the difficulties that can be encountered in the balancing process are directly linked with the ratio between the maximum operating speed and the first flexural critical speed caused by a bending mode of the rotor. If the rotor has to operate at speeds far lower than the first critical speed (below about half of it), it can be assumed to be rigid. When the operating speed is close to the first critical speed, the possibility that the rotor inflects, assuming a shape not far from the first mode shape, must be taken into account. Correspondingly, near the nth critical speed, the inflected shape is not dissimilar from the nth mode shape. If the rotor has to operate above the nth critical speed, it must be balanced with reference to the first $n + 1$ mode shapes to be balanced in the whole working range .

From a practical point of view, ISO standards subdivide rotors into five classes. Rigid rotors, as described in the preceding section, belong to the first class. Rotors of the second class are defined as semirigid, i.e., rotors that cannot be considered rigid but can be balanced in a low-speed balancing machine. This class is subdivided into eight subclasses, from 2a to 2h, as shown in Figure B.2. The third class contains the true flexible rotors, which cannot be balanced using a low-speed balancing machine, but require balancing at high speed. From this class, the rotors belonging to classes 4 and 5 are excluded. Rotors of the fourth class are said to be special flexible rotors, which are defined as rotors that would be rigid or semirigid but to which one or more flexible components have been added. The fifth class contains flexible rotors, which would belong to class 3 or 4, but for which balancing at a single speed (generally at the operating speed) is required.

Two procedures are usually considered for field balancing of flexible rotors: modal balancing and the influence coefficients method.

B.2.1 Modal balancing

As described in detail in Sections 3.6 and 4.9, the unbalance response of a rotor can be computed through a modal approach, and in case the rotor is undamped, the modes uncouple as in the case of vibrating systems. In the case of a damped system, the modes can still be uncoupled in an approximate way, provided that damping is small enough. The problems caused by unbalance are mainly linked with the strong response of the modes that "resonate" at the critical speeds lying within the working range, and as a consequence, only the modes whose response shows a resonance

$$\begin{aligned} &\begin{bmatrix} \mathbf{M}' & \mathbf{0} \\ \mathbf{0} & \mathbf{M}' \end{bmatrix} \ddot{\mathbf{q}}(t) + \left(\begin{bmatrix} \mathbf{C}' & \mathbf{0} \\ \mathbf{0} & \mathbf{C}' \end{bmatrix} + \begin{bmatrix} \mathbf{0} & -\mathbf{G}' \\ \mathbf{G}' & \mathbf{0} \end{bmatrix} \right) \dot{\mathbf{q}}(t) + \\ &+ \left(\begin{bmatrix} \mathbf{K}' & \mathbf{0} \\ \mathbf{0} & \mathbf{K}' \end{bmatrix} + \begin{bmatrix} \mathbf{0} & -\mathbf{H}' \\ \mathbf{H}' & \mathbf{0} \end{bmatrix} \right) = \mathbf{f}(t)\ . \end{aligned} \tag{1.7}$$

$$\mathbf{r}_1 = \left\{ \begin{array}{c} 1 \\ 0 \end{array} \right\} , \qquad \mathbf{r}_2 = \left\{ \begin{array}{c} 0 \\ 1 \end{array} \right\} , \tag{2.24}$$

$$\left\{ \begin{array}{l} x_{\mathrm{C}}(t) = x_{\mathrm{C}_0} \cos(\Omega t) \ , \\ y_{\mathrm{C}}(t) = y_{\mathrm{C}_0} \sin(\Omega t) = y_{\mathrm{C}_0} \cos(\Omega t + \pi/2) \ . \end{array} \right. \tag{2.26}$$

$$\mathbf{q} = \begin{Bmatrix} r \\ \phi \end{Bmatrix}, \quad \mathbf{M} = \begin{bmatrix} m & 0 \\ 0 & J_t \end{bmatrix}, \quad \mathbf{G} = \begin{bmatrix} 0 & 0 \\ 0 & J_p \end{bmatrix},$$

$$\mathbf{K} = \begin{bmatrix} K_{11} & K_{12} \\ K_{12} & K_{22} \end{bmatrix}, \quad \mathbf{f} = \begin{Bmatrix} m\epsilon e^{i\alpha} \\ \chi(J_t - J_p) \end{Bmatrix}. \tag{3.22}$$

$$\mathbf{B} = \frac{1}{3EIl} \begin{bmatrix} a^2 (l-a)^2 & a(l-a)(l-2a) \\ a(l-a)(l-2a) & l^2 - 3al + 3a^2 \end{bmatrix} . \tag{3.59}$$

$$\mathbf{\Phi} = \left[\begin{array}{cc} \mathbf{q}_1 & \mathbf{q}_2 \end{array} \right] , \tag{3.72}$$

$$\begin{aligned} \omega_{n1} &= \sqrt{\frac{K_{22} + \beta K_{11} + D}{2\beta m}}, \\ \omega_{n2} &= \sqrt{\frac{K_{22} + \beta K_{11} - D}{2\beta m}}, \end{aligned} \tag{3.77}$$

$$\mathbf{T}_{ni}=\begin{bmatrix} 1 & 0 & 0 & 0 \\ 0 & 1 & 0 & 0 \\ s^2 m_i + k_i + s c_i & 0 & 1 & 0 \\ 0 & s^2 J_{t_i} + is\Omega J_{p_i} + \kappa_i + s\Gamma_i & 0 & 1 \end{bmatrix} . \tag{4.25}$$

$$\Delta = \det \begin{bmatrix} T_{12} & T_{13} \\ T_{42} & T_{43} \end{bmatrix} \tag{4.26}$$

$$\mathbf{q} = \begin{bmatrix} r_1 & \phi_1 & r_2 & \phi_2 & \dots & r_n & \phi_n \end{bmatrix}^T , \tag{4.31}$$

$$\mathbf{M} = \text{diag} \begin{bmatrix} m_1 & J_{t_1} & m_2 & J_{t_2} & \dots & m_n & J_{t_n} \end{bmatrix},$$

$$\mathbf{G} = \text{diag} \begin{bmatrix} 0 & J_{p_1} & 0 & J_{p_2} & \dots & 0 & J_{p_n} \end{bmatrix}, \tag{4.32}$$

$$\left\{ \begin{array}{c} u_x \\ \phi_y \end{array} \right\} = \left[\begin{array}{cccc} N_{11} & N_{12} & N_{13} & N_{14} \\ N_{21} & N_{22} & N_{23} & N_{24} \end{array} \right] \left\{ \begin{array}{c} u_{x_1} \\ \phi_{y_1} \\ u_{x_2} \\ \phi_{y_2} \end{array} \right\},$$
$$\left\{ \begin{array}{c} u_y \\ -\phi_x \end{array} \right\} = \left[\begin{array}{cccc} N_{11} & N_{12} & N_{13} & N_{14} \\ N_{21} & N_{22} & N_{23} & N_{24} \end{array} \right] \left\{ \begin{array}{c} u_{y_1} \\ -\phi_{x_1} \\ u_{y_2} \\ -\phi_{x_2} \end{array} \right\}. \tag{4.47}$$

$$\mathbf{a}_2 = \frac{\rho I_y}{12(1-\Phi)} \begin{bmatrix} -6 & -6 \\ l(1+4\Phi) & l(-1+2\Phi) \\ 6 & 6 \\ l(-1+2\Phi) & l(1+4\Phi) \end{bmatrix}.$$

$$\begin{bmatrix} \mathbf{K}'+\Omega\mathbf{C}'_r - \Omega_x^2(\mathbf{M}' - \mathbf{K}'_\Omega) & \Omega(\mathbf{C}' + \Omega\mathbf{G}') \\ -\Omega(\mathbf{C}' + \Omega\mathbf{G}') & \mathbf{K}' + \Omega\mathbf{C}'_r - \Omega_x^2(\mathbf{M}' - \mathbf{K}'_\Omega) \end{bmatrix} \begin{Bmatrix} \mathbf{x}_{0c} \\ \mathbf{x}_{0s} \end{Bmatrix} =$$
$$= \Omega^2 \begin{Bmatrix} \mathbf{f}_{rc} \\ \mathbf{f}_{rs} \end{Bmatrix}\ . \tag{4.92}$$

$$\mathbf{q} = \left[\begin{array}{cccccc} r_A & \phi_A & r_G & \phi_G & r_B & \phi_B \end{array}\right]^T . \tag{4.99}$$

$$\mathbf{M}_{cond} = diag \left\{ \begin{matrix} 0 & m & J_t & 0 \end{matrix} \right\} , \quad \mathbf{G}_{cond} = diag \left\{ \begin{matrix} 0 & 0 & J_p & 0 \end{matrix} \right\} \tag{4.108}$$

$$\mathbf{q}_1' = \begin{bmatrix} r_G & \phi_G \end{bmatrix}^T , \quad \mathbf{q}_2' = \begin{bmatrix} r_A & r_B \end{bmatrix}^T . \tag{4.109}$$

$$\mathbf{M}_{cond} = diag \left\{ \begin{matrix} m & J_t \end{matrix} \right\} , \quad \mathbf{G}_{cond} = diag \left\{ \begin{matrix} 0 & J_p \end{matrix} \right\} . \tag{4.113}$$

$$\begin{bmatrix} \sin(al) + \sinh(al) & \cos(al) + \cosh(al) \\ \cos(al) + \cosh(al) & -\sin(al) + \sinh(al) \end{bmatrix} \begin{Bmatrix} C_1 \\ C_2 \end{Bmatrix} = \begin{Bmatrix} 0 \\ 0 \end{Bmatrix} \ . \tag{5.26}$$

$$\begin{Bmatrix} F_x \\ F_y \end{Bmatrix} = - \begin{bmatrix} k_x & k_{xy} \\ k_{yx} & k_y \end{bmatrix} \begin{Bmatrix} x \\ y \end{Bmatrix} \tag{6.5}$$

$$s'^{*^4} + 4s'^{*3}(\xi_r + \xi_n) + 2s'^{*^2}\left[(1+\Omega^{*^2}) + 2(\xi_r + \xi_n)^2 - 2\beta^{*2}\xi_r^2\right] + \\ + 4s'^{*}\left[(\xi_r + \xi_n)(1-\Omega^{*^2}) - \beta^*\alpha^*\xi_r + 2\Omega^{*^2}\xi_n\right] + \\ + (1-\Omega^{*^2})^2 - \alpha^{*2} + 4\Omega^{*^2}\xi_n^2 = 0\,. \tag{6.60}$$

$$\begin{array}{l} \mathcal{T} = \frac{1}{2} J_i \dot{\phi}_i^2 = \frac{1}{2} J_i^* \dot{\phi}_i^{*2} , \\ \mathcal{U} = \frac{1}{2} k_i \left(\phi_{i+1}^2 - \phi_i^2 \right) = \frac{1}{2} k_i^* \left(\phi_{i+1}^{*2} - \phi_i^{*2} \right) , \end{array} \tag{7.23}$$

$$\mathbf{f} = \mathbf{f}_{st} - \left[\begin{array}{cc} \frac{\partial F_x}{\partial \dot{x}} & \frac{\partial F_x}{\partial \dot{y}} \\ \frac{\partial F_y}{\partial \dot{x}} & \frac{\partial F_y}{\partial \dot{y}} \end{array} \right]_{x_{st}, y_{st}} \left\{ \begin{array}{c} \dot{x} \\ \dot{y} \end{array} \right\} - \left[\begin{array}{cc} \frac{\partial F_x}{\partial x} & \frac{\partial F_x}{\partial y} \\ \frac{\partial F_y}{\partial x} & \frac{\partial F_y}{\partial y} \end{array} \right]_{x_{st}, y_{st}} \left\{ \begin{array}{c} \Delta x \\ \Delta y \end{array} \right\} . \tag{8.36}$$

$$\mathbf{C} = -\left[\left(\frac{\partial F_i}{\partial \dot{x}_j} \right) \right]_{x_{st}, y_{st}} = \frac{6\mu R_j^3 l}{n^2 c^3} \left[\begin{array}{cc} I_7 & I_8 \\ I_9 & I_{10} \end{array} \right], \tag{8.37}$$

$$\mathbf{M} = diag \left\{ \; 0.001 \quad 20 \quad 0.2 \quad 0.001 \; \right\} , \quad \mathbf{G} = diag \left\{ \; 0 \quad 0 \quad 0.25 \quad 0 \; \right\} .$$

$$\begin{bmatrix} m & 0 \\ 0 & J_t \end{bmatrix} \left\{ \begin{array}{c} \ddot{r} \\ \ddot{\phi} \end{array} \right\} - i\Omega \begin{bmatrix} 0 & 0 \\ 0 & J_p \end{bmatrix} \left\{ \begin{array}{c} \dot{r} \\ \dot{\phi} \end{array} \right\} = \mathbf{f}_c + m\epsilon\Omega^2 \left\{ \begin{array}{c} 1 \\ 0 \end{array} \right\} e^{i\Omega t} + \mathbf{f}_n \ ,$$

$$\left\{ \begin{array}{c} f_{c1} \\ f_{c2} \end{array} \right\} = -\begin{bmatrix} k_{d1} & 0 \\ 0 & k_{d2} \end{bmatrix} \left\{ \begin{array}{c} \dot{r}_1 \\ \dot{r}_2 \end{array} \right\} - \begin{bmatrix} k_{p1} & 0 \\ 0 & k_{p2} \end{bmatrix} \left\{ \begin{array}{c} r_1 \\ r_2 \end{array} \right\} ,$$

$$\left\{ \begin{array}{c} r_1 \\ r_2 \end{array} \right\} = \begin{bmatrix} 1 & -a \\ 1 & b \end{bmatrix} \left\{ \begin{array}{c} r \\ \phi \end{array} \right\} = \mathbf{T} \left\{ \begin{array}{c} r \\ \phi \end{array} \right\} .$$

$$\mathbf{f}_c = -\mathbf{T}^T \begin{bmatrix} k_{d1} & 0 \\ 0 & k_{d2} \end{bmatrix} \mathbf{T} \left\{ \begin{array}{c} \dot{r} \\ \dot{\phi} \end{array} \right\} - \mathbf{T}^T \begin{bmatrix} k_{p1} & 0 \\ 0 & k_{p2} \end{bmatrix} \mathbf{T} \left\{ \begin{array}{c} r_1 \\ r_2 \end{array} \right\} .$$

$$\begin{array}{ll} k_n = \frac{1}{2}\left(k_x + k_y\right) , & k_{d_n} = \frac{1}{2}\left(k_x - k_y\right) , \\ k_r = \frac{1}{2}\left(k_\xi + k_\eta\right) , & k_{d_r} = \frac{1}{2}\left(k_\xi - k_\eta\right) , \end{array} \tag{9.2}$$

$$\begin{array}{ll} c = \cos\left(2\Omega t\right) , & R = \beta\left(1+\alpha_n\right) + 1 + \alpha_r c , \\ s = \sin\left(2\Omega t\right) , & S = \beta\left(1-\alpha_n\right) + 1 - \alpha_r c . \end{array}$$

$$x_1 = \frac{\left[1 - \alpha_r^2 + \beta\left(1-\alpha_n\right) + \beta\alpha_r\left(1-\alpha_n\right)\cos\left(2\Omega t\right)\right] x_2 + \beta\alpha_r\left(1-\alpha_n\right)\sin\left(2\Omega t\right) y_2}{\left(1+\beta\right)^2 - \beta^2\alpha_n^2 - \alpha_r^2 - 2\beta\alpha_r\alpha_n\cos\left(2\Omega t\right)} ,$$

$$y_1 = \frac{\beta\alpha_r\left(1+\alpha_n\right)\sin\left(2\Omega t\right) x_2 + \left[1 - \alpha_r^2 + \beta\left(1+\alpha_n\right) - \beta\alpha_r\left(1+\alpha_n\right)\cos\left(2\Omega t\right)\right] y_2}{\left(1+\beta\right)^2 - \beta^2\alpha_n^2 - \alpha_r^2 - 2\beta\alpha_r\alpha_n\cos\left(2\Omega t\right)} . \tag{9.5}$$

$$\left[1 + U\cos\left(2\Omega t\right)\right] \begin{bmatrix} m & 0 \\ 0 & m \end{bmatrix} \begin{Bmatrix} \ddot{x}_2 \\ \ddot{y}_2 \end{Bmatrix} +$$

$$+k_r \begin{bmatrix} R+S+T\cos\left(2\Omega t\right) & T\sin\left(2\Omega t\right) \\ T\sin\left(2\Omega t\right) & R-S-T\cos\left(2\Omega t\right) \end{bmatrix} \begin{Bmatrix} x_2 \\ y_2 \end{Bmatrix} = \tag{9.6}$$

$$= \left[1 + U\cos\left(2\Omega t\right)\right] \left[\begin{Bmatrix} f_{n_x} \\ f_{n_y} \end{Bmatrix} + m\epsilon\Omega^2 \begin{Bmatrix} \cos\left(\Omega t\right) \\ \sin\left(\Omega t\right) \end{Bmatrix} \right] ,$$

$$\left[\begin{array}{cc}\mathbf{M}^{-1}\mathbf{G} & \mathbf{M}^{-1}\mathbf{K}\\ \mathbf{I} & \mathbf{0}\end{array}\right].$$

$$\begin{Bmatrix} F_x \\ F_y \end{Bmatrix}_{rn} = -\begin{Bmatrix} f_x(x,\dot{x})x \\ f_y(y,\dot{y})y \end{Bmatrix} , \quad \begin{Bmatrix} F_x \\ F_y \end{Bmatrix}_{dn} = -\begin{Bmatrix} g_x(x,\dot{x})\dot{x} \\ g_y(y,\dot{y})\dot{y} \end{Bmatrix} , \tag{10.33}$$

$$\begin{bmatrix} \pm 1 - \Omega^{*2} + \mu_\xi^*\xi^{*2} & 2c_n^*\Omega^* \\ 2c_n^*\Omega^* & \pm\alpha - \Omega^{*2} + \alpha\mu_\eta^*\eta^{*2} \end{bmatrix} \left\{\begin{array}{c} \xi_0^* \\ \eta_0^* \end{array}\right\} = \left\{\begin{array}{c} \Omega^{*2}\cos(\beta) \\ \Omega^{*2}\sin(\beta) \end{array}\right\}, \tag{10.40}$$

$$\left\{\begin{array}{c} f_x^*\\ f_y^*\end{array}\right\}=-\frac{\sqrt{\left(x^*-2y^{*'}\right)^2+\left(y^*+2x^{*'}\right)^2}}{1-x^{*2}-y^{*2}}\left\{\begin{array}{c}3x^*V-\sin\left(\alpha\right)G-2\cos\left(\alpha\right)S\\ 3y^*V+\cos\left(\alpha\right)G-2\sin\left(\alpha\right)S\end{array}\right\}\ ,$$

$$\begin{array}{l} V\left(x^*,\ y^*,\ \alpha\right)=\dfrac{2+G\left[y^*\cos\left(\alpha\right)-x^*\sin\left(\alpha\right)\right]}{1-x^{*2}-y^{*2}}\ ,\\ G\left(x^*,\ y^*,\ \alpha\right)=\dfrac{2}{\sqrt{1-x^{*2}-y^{*2}}}\left\{\dfrac{\pi}{2}+artg\left[\dfrac{y^*\cos\left(\alpha\right)-x^*\sin\left(\alpha\right)}{\sqrt{1-x^{*2}-y^{*2}}}\right]\right\}\ ,\\ S\left(x^*,\ y^*,\ \alpha\right)=\dfrac{x^*\cos\left(\alpha\right)+y^*\sin\left(\alpha\right)}{1-\left[x^*\cos\left(\alpha\right)+y^*\sin\left(\alpha\right)\right]^2}\ .\end{array}$$

$$\begin{bmatrix} m & 0 \\ 0 & m \end{bmatrix}\begin{Bmatrix} \ddot{\xi} \\ \ddot{\eta} \end{Bmatrix} + \begin{bmatrix} c & -2m\dot{\theta} \\ 2m\dot{\theta} & c \end{bmatrix}\begin{Bmatrix} \dot{\xi} \\ \dot{\eta} \end{Bmatrix} + \begin{bmatrix} k - m\dot{\theta}^2 & -\dot{\theta}c_n - m\ddot{\theta} \\ \dot{\theta}c_n + m\ddot{\theta} & k - m\dot{\theta}^2 \end{bmatrix}\cdot$$

$$\cdot\begin{Bmatrix} \xi \\ \eta \end{Bmatrix} - m\epsilon\begin{Bmatrix} \dot{\theta}^2 \\ -\ddot{\theta} \end{Bmatrix} = \begin{Bmatrix} F_x\cos(\theta) + F_y\sin(\theta) \\ -F_x\sin(\theta) + F_y\cos(\theta) \end{Bmatrix}, \tag{11.11}$$

$$\begin{aligned} c_n^* &= \frac{c_n}{2\sqrt{km}}\,,\quad c_r^*=\frac{c_r}{2\sqrt{km}}\,,\quad \Omega^*=\frac{\Omega}{\Omega_{cr}}\,,\quad a^*=\frac{\dot{\Omega}}{\Omega_{cr}^2}\,,\\ c_t^* &= \frac{c_t}{2\sqrt{k_tJ_{pt}}}\,,\quad \alpha=\sqrt{\frac{kJ_{pt}}{mk_t}}\,,\quad \beta=\frac{k_t}{k\epsilon^2}\,,\quad F^*=\frac{F}{k\epsilon}\,. \end{aligned}$$

$$\delta\mathcal{L} = \left\{ \begin{array}{c} F_X \\ F_Y \\ F_Z \end{array} \right\}^T \left\{ \begin{array}{c} \delta X \\ \delta X \\ \delta X \end{array} \right\} +$$

$$+ \left\{ \begin{array}{c} M_x \\ M_y \\ M_z \end{array} \right\}^T \left(\mathbf{R}_2 \left\{ \begin{array}{c} \delta\phi_{X'} \\ 0 \\ 0 \end{array} \right\} + \left\{ \begin{array}{c} 0 \\ \delta\phi_y \\ 0 \end{array} \right\} + \left\{ \begin{array}{c} 0 \\ 0 \\ \delta\theta \end{array} \right\} \right) ; \tag{11.25}$$

$$\mathbf{q} = \begin{Bmatrix} r \\ \phi \end{Bmatrix}, \quad \mathbf{M} = \begin{bmatrix} m & 0 \\ 0 & J_t \end{bmatrix}, \quad \mathbf{G} = \begin{bmatrix} 0 & 0 \\ 0 & J_p \end{bmatrix},$$

$$\mathbf{K} = \begin{bmatrix} K_{11} & K_{12} \\ K_{12} & K_{22} \end{bmatrix}, \quad \mathbf{f} = \begin{Bmatrix} m\epsilon e^{i\alpha} \\ \chi(J_t - J_p) \end{Bmatrix},$$

$$\begin{aligned}(\dot{\overline{P_{1r} - O}}) = \left(\dot{\varphi}_x \boldsymbol{\gamma}^T \mathbf{R}_2^T \mathbf{R}_3^T + \dot{\varphi}_y \mathbf{R}_1^T \boldsymbol{\beta}^T \mathbf{R}_3^T + \dot{\theta} \mathbf{R}_1^T \mathbf{R}_2^T \boldsymbol{\alpha}^T\right) \left\{ \begin{array}{c} -\epsilon \\ 0 \\ a \end{array} \right\} + \\ + \left\{ \begin{array}{c} \dot{X} \\ \dot{Y} \\ \dot{Z} \end{array} \right\} ;\end{aligned} \tag{12.48}$$

$$\mathbf{q} = \begin{bmatrix} X & Y & Z & \varphi_x & \varphi_x & \theta \end{bmatrix}^T, \tag{12.50}$$

$$\mathbf{M} = diag\left(\left[\ 12 \quad 0.08 \quad 104 \quad 16 \quad 7 \quad 0.04\ \right]\right) ,$$

$$\mathbf{G} = diag\left(\left[\ 0 \quad 0.1 \quad 0 \quad 28 \quad 0 \quad 0.06\ \right]\right) .$$

$$\phi = \begin{Bmatrix} \phi_1 \\ \phi_2 \\ \phi_3 \\ \ldots \end{Bmatrix} = \begin{bmatrix} 1 & \cos(\psi_1) & \sin(\psi_1) & \ldots \\ 1 & \cos(\psi_2) & \sin(\psi_2) & \ldots \\ 1 & \cos(\psi_3) & \sin(\psi_3) & \ldots \\ \ldots & \ldots & \ldots & \ldots \end{bmatrix} \begin{Bmatrix} u_0 \\ u_1 \\ u_2 \\ \ldots \end{Bmatrix} = \mathbf{T}\mathbf{u} \,. \tag{13.76}$$

$$\mathbf{T} = \begin{bmatrix} 1 & 1 & 0 & 1 \\ 1 & 0 & 1 & -1 \\ 1 & -1 & 0 & 1 \\ 1 & 0 & -1 & -1 \end{bmatrix} .$$

$$\begin{aligned} & m_T\ddot{y} + m\sum_{i=1}^{n}\Big[\ddot{\phi}_i l\cos(\theta_i) - \Omega^2\left(r+l\right)\sin(\theta_i) + \\ & -\Omega^2 l\phi_i\cos(\theta_i) - 2\Omega l\dot{\phi}_i\sin(\theta_i)\Big] + ky = 0 \ . \end{aligned} \tag{13.91}$$

$$\begin{aligned} & [-\ddot{x}\sin(\psi_i) + \ddot{y}\cos(\psi_i)]\cos(\Omega t) + [-\ddot{x}\cos(\psi_i) - \ddot{y}\sin(\psi_i)]\sin(\Omega t) + \\ & + \ddot{\phi}_i l + \Omega^2 r\phi_i = 0 \ . \end{aligned} \tag{13.94}$$

$$\left\{ \begin{aligned} x_1 &= -lu_2\cos(\Omega t) - lu_1\sin(\Omega t) \ , \\ y_1 &= lu_1\cos(\Omega t) - lu_2\sin(\Omega t) \ , \end{aligned} \right. \qquad (13.100)$$

$$\begin{aligned} d\mathcal{T} = \tfrac{1}{2}dm|\vec{V}_\text{P}|^2 = \tfrac{1}{2}\rho A r d\theta\big[\Omega^2\left(r-u\right)^2 + \Omega^2 v^2 + \dot{u}^2 + \dot{v}^2 + \\ + \dot{w}^2 - 2\Omega\dot{u}v - 2\Omega\left(r-u\right)\dot{v}\big] . \end{aligned} \tag{14.35}$$

$$\begin{aligned} \mathbf{N}_1 &= \begin{bmatrix} 1-\chi & \chi \end{bmatrix} , \\ \mathbf{N}_3 &= \begin{bmatrix} 2\chi^3 - 3\chi^2 + 1 & \chi^2(2\chi - 3) & -\chi^3 + 2\chi^2 - \chi & \chi(1-\chi^2) \end{bmatrix} . \end{aligned} \tag{15.11}$$

$$\mathbf{q}_{\phi_0} = \begin{bmatrix} \phi_{z1} & \phi_{z2} \end{bmatrix}^T , \qquad \mathbf{q}_{w_0} = \begin{bmatrix} \phi_{01} & \beta_{01} & \phi_{02} & \beta_{02} \end{bmatrix}^T . \tag{15.12}$$

$$\overline{(\mathrm{P}-\mathrm{O})}_d = \mathbf{x}_d + \mathbf{u}_d = \begin{Bmatrix} X + u(t) \\ Y + v(t) \\ Z + w(t) \end{Bmatrix}_d . \tag{15.27}$$

$$\overline{(\mathrm{P'}-\mathrm{O})}_i = \mathbf{R}^T(\mathbf{x}_d + \mathbf{u}_d) = \mathbf{R}^T \begin{Bmatrix} X + u(t) \\ Y + v(t) \\ Z + w(t) \end{Bmatrix}_d . \tag{15.28}$$

$$\left\{ \begin{array}{c} \dot{\mathbf{z}}_{cx} \\ \dot{\mathbf{z}}_{cy} \end{array} \right\} = \begin{bmatrix} \mathbf{A}_c & 0 \\ 0 & \mathbf{A}_c \end{bmatrix} \left\{ \begin{array}{c} \mathbf{z}_{cx} \\ \mathbf{z}_{cy} \end{array} \right\} + \begin{bmatrix} \mathbf{B}_y & 0 \\ 0 & \mathbf{B}_y \end{bmatrix} \left\{ \begin{array}{c} x \\ y \end{array} \right\} ,$$

$$\left\{ \begin{array}{c} i_x \\ i_x \end{array} \right\} + \begin{bmatrix} \mathbf{C}_c & 0 \\ 0 & \mathbf{C}_c \end{bmatrix} \left\{ \begin{array}{c} \mathbf{z}_{cx} \\ \mathbf{z}_{cy} \end{array} \right\} ,$$

$$\left\{ \begin{array}{l} s^{*2} + 2\zeta s^* + 1 = 0 \ , \\ s^{*2} + (2\eta\alpha - i\Omega^*\delta)\, s^* + \alpha = 0 \ . \end{array} \right. \tag{16.35}$$

$$= \begin{bmatrix} k_1 & 0 \\ 0 & k_2 \end{bmatrix} \mathbf{T}'\mathbf{T}^{-1} \left\{ \begin{array}{c} x_1 + iy_1 \\ x_2 + iy_2 \end{array} \right\} .$$

$$\mathbf{y}(t) = \mathbf{C}\,\mathbf{z}(t) = [\mathbf{S}_q]_{p\times n}\{\mathbf{q}(t)\}_n + [\mathbf{S}_{\dot{q}}]_{p\times n}\{\dot{\mathbf{q}}(t)\}_n = \\ = \begin{bmatrix} \mathbf{S}_{\dot{q}} & \mathbf{S}_q \end{bmatrix}_{p\times 2n} \begin{Bmatrix} \dot{\mathbf{q}}(t) \\ \mathbf{q}(t) \end{Bmatrix}_{2n}, \tag{A.37}$$

$$\mathbf{A} = \begin{bmatrix} [\mathbf{0}]_{n\times n} & \left[-\mathbf{M}^{-1}\mathbf{K}\right]_{n\times n} \\ [\mathbf{I}]_{n\times n} & [\mathbf{0}]_{n\times n} \end{bmatrix}_{2n\times 2n} \tag{A.38}$$

$$\mathbf{z}(t) = \begin{Bmatrix} s\{\mathbf{q}\}_n \\ \{\mathbf{q}\}_n \end{Bmatrix}_{2n} e^{st}. \tag{A.39}$$

Class	Description	Example
2a	A rotor with a single transverse plane of unbalance, e.g., a single mass on a shaft with negligible unbalance.	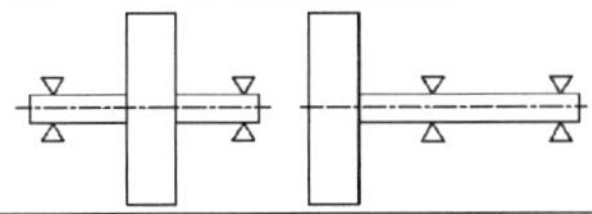
2b	A rotor with two transverse planes of unbalance, e.g., two masses on a shaft with negligible unbalance.	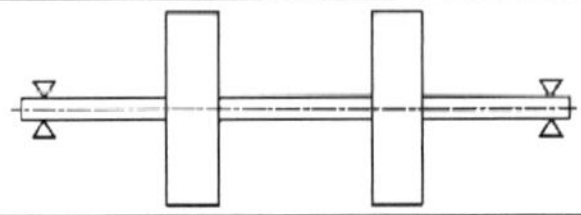
2c	A rotor with more than two transverse planes of unbalance and flexible shaft.	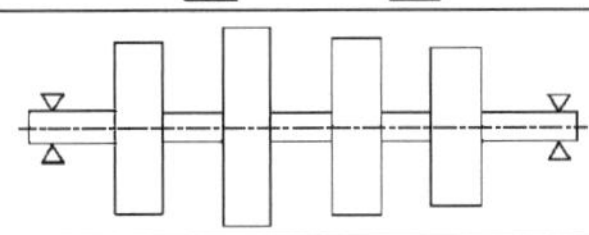
2d	A rotor with uniformly distributed unbalance.	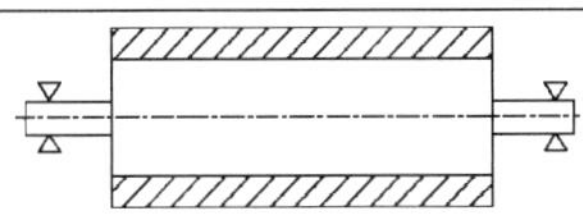
2e	A rotor with more than two transverse planes of unbalance and flexible shaft.	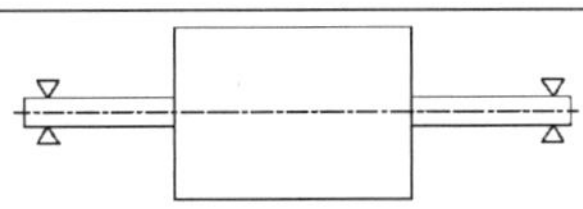
2f	A symmetrical rotor with two end correction planes, whose maximum speed does not significantly approach the second critical speed, whose service speed range does not contain first critical speed and with component balanced separately.	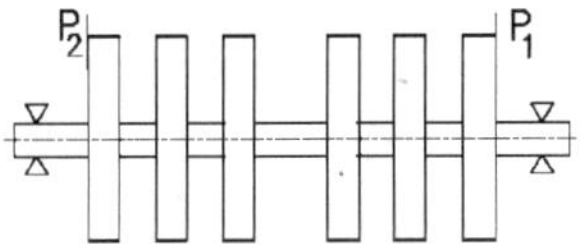
2g	As 2f, but with a third intermediate correction plane.	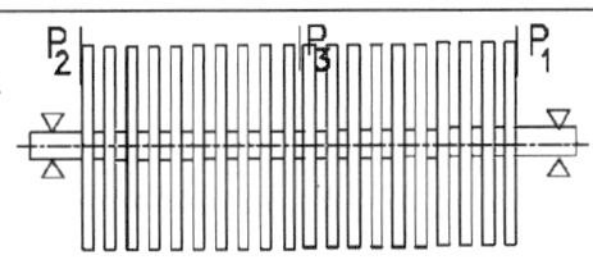
2h	As 2f, but without a symmetry plane at midspan.	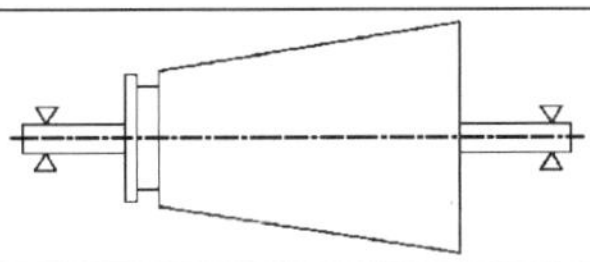

FIGURE B.2. Rotors of class 2 (semirigid rotors).

peak (i.e., the modes that give way to a real critical speed) need to be considered.

Modal balancing is based on balancing the rotor at the various critical speeds lying within the working range, starting from the lowest. The correction masses must be located in the planes in which the relevant mode shape has large displacements. Generally speaking, the number of the correction planes must be equal to the order of the critical speed. It is possible to demonstrate that it is possible to correct the unbalance at the general ith critical speed without disturbing the balance conditions of the previous $i-1$ modes, which have already been balanced.

Consider a rotor modeled as a discretized, multi-degree-of-freedom system. The unbalance response can be computed using Equation (4.90) or by recombining the modal responses. To balance the system at the first critical speed means to add to the unbalance distribution $\mathbf{f}$ another unbalance distribution $\mathbf{f}_{b1}$, which causes the modal force from unbalance related to the first mode to vanish

$$\overline{f}_1 = \mathbf{q}_1^T(\mathbf{f} + \mathbf{f}_{b1}) = 0\,. \tag{B.3}$$

If balancing is performed by adding the unbalance $m_1\epsilon_1$ at the jth generalized coordinate (vector $\mathbf{f}_{b1}$ has all elements equal to zero except the jth, whose value is $m_1\epsilon_1\Omega^2$), the modal unbalance that has been added is

$$\mathbf{q}_1^T\mathbf{f}_{b1} = m_1\epsilon_1 q_{j1}\Omega^2. \tag{B.4}$$

From Equation (B.3) the value of the unbalance to be added to balance the first mode shape is immediately obtained

$$m_1\epsilon_1 = -\mathbf{q}_1^T\mathbf{f}\,\frac{1}{q_{j1}\Omega^2}\,. \tag{B.5}$$

To balance the second mode, the two unbalances $m_{2r}\epsilon_{2r}$ and $m_{2s}\epsilon_{2s}$ are added corresponding to the rth and sth generalized coordinates. The second mode is balanced if

$$(m_{2r}\epsilon_{2r}q_{r2} + m_{2s}\epsilon_{2s}q_{s2})\Omega^2 = -\mathbf{q}_2^T\left(\mathbf{f} + \mathbf{f}_{b1}\right)\,. \tag{B.6}$$

The second mode can be balanced without disturbing the balancing, already achieved, of the first mode if the modal force corresponding to the first mode remains equal to zero after the addition of the new balancing masses at the generalized coordinates r and s

$$m_{2r}\epsilon_{2r}q_{r1} + m_{2s}\epsilon_{2s}q_{s1} = 0\,. \tag{B.7}$$

Equations (B.6) and (B.7) allow computation of the two correction unbalances $m_{2r}\epsilon_{2r}$ and $m_{2s}\epsilon_{2s}$.

The correction of the third mode can be performed in the same way: The three correction unbalances can be computed using an equation of the type of Equation (B.6), stating that the modal force corresponding to the third mode must vanish, and two equations of the type of Equation (B.7) stating that the balancing of the third mode does not affect the balancing conditions at the previously balanced modes. In a similar way, all other modes can be balanced. What has been shown is actually a demonstration that modal balancing is possible, not the procedure to practically implement the balancing process. The knowledge of the unbalance distribution **f**, which is generally unknown, and of the mode shapes is not required. It is, however, clear that the computation of the mode shapes allows an easier way to identify the planes in which the correcting action is most effective, because the balancing masses must be located at the loops of the mode shapes, or not too near to the nodes, where they would be ineffective.

Before starting modal balancing, the rotor can be balanced as a rigid body on a balancing machine. There is no agreement on the advisability of this practice, but, generally speaking, rigid-body balancing can be omitted if the starting unbalance is not severe enough to prevent operating near the first critical speed, as needed to start the modal balancing procedure.

It can happen that a rotor, which has been balanced at the first n critical speeds, causes strong vibrations at the maximum operating speed, which is between the nth and $(n+1)$th critical speeds, without reaching the latter to complete the balancing procedure. The strong vibrations are caused, in this case, by the unbalanced higher modes and, in particular, the $(n+1)$th mode. In this case, the last corrections are implemented, taking into account the mode shape that is predicted at the $(n+1)$th critical speed, even if it is not materially possible to reach it.

B.2.2 Influence coefficients method

The influence coefficients method is based on the observation that the vibrations detected in a number of measuring points can be considered the effect of concentrated unbalances in a number of arbitrarily located planes. This statement is based on the assumption that the behavior of the system is linear. Let the number of measuring points be n, the number of speeds at which the balancing process is performed be m, and the number of planes in which the correction masses are to be located be q. Because of the linearity of the system, the $m \cdot n$ responses r_i obtained in the n measuring points at the m test speeds are linked to the q unknown unbalances $m_i\epsilon_i$ in the q correction planes by the general linear relationship

$$\{r\}_{(m \cdot n)} = [A]_{(m \cdot n) \times q}\{m\epsilon\}_q . \tag{B.8}$$

Note that both responses r_i, which can be displacements, as implicitly assumed in Equation (B.8), but also accelerations or velocities, and the unbalances $m_i\epsilon_i$ are vector quantities and coincide with the complex quantities r and $m\epsilon$ defined in Chapter 2. If the influence coefficients in matrix

$\mathbf{A}$, whose size is $(m \cdot n) \times q$, were known, Equation (B.8) could be used directly to compute the unknown unbalances from the vibration measurements, provided that matrix $\mathbf{A}$ is square; i.e., the number of correction planes q is equal to the product $m \cdot n$ of the number of test speeds m by the number of measuring points n, and it is nonsingular.

The coefficients of influence are easily determined. A known unbalance $m_j\epsilon_j$ is introduced in a generic correction plane (the jth), in a known angular position. The tests are repeated, and a set of new $m \cdot n$ responses r_i^* are measured. They are linked to vector $\{m\epsilon\}^*$ through Equation (B.8): $\mathbf{r}^* = \mathbf{A}\{m\epsilon\}^*$, where vector $\{m\epsilon\}^*$ has all elements coinciding with those of the vector in Equation (B.8) except the jth element, which has been incremented by the known unbalance $m_j\epsilon_j$.

By subtracting Equation (B.8) written for unbalances $\{m\epsilon\}$ and $\{m\epsilon\}^*$, it follows that

$$\mathbf{r}^* - \mathbf{r} = \mathbf{A} \begin{bmatrix} 0 & 0 & \dots & m_j\epsilon_j & \dots & 0 \end{bmatrix}^T . \tag{B.9}$$

The jth column of matrix $\mathbf{A}$ is then simply obtained by dividing the difference between vectors $\mathbf{r}^*$ and $\mathbf{r}$ by $m_j\epsilon_j$. The procedure can be repeated by removing the previously added unbalance from the jth plane and by adding a new known unbalance in another plane. The matrix of the coefficients of influence is thus obtained, column after column, and the unbalance distribution $\{m\epsilon\}$ is obtained:

$$\{m\epsilon\} = \mathbf{A}^{-1}\mathbf{r} . \tag{B.10}$$

The rotor is then balanced by introducing suitable unbalances equal and opposite to the ones computed in the q correction planes. The balanced conditions of the rotor are thus achieved, at least at the speeds at which the tests have been performed.

The procedure described here is the extension to flexible rotors of the usual procedure for the calibration of the balancing machine for rigid rotors. In the latter case, only one test speed is chosen ($m = 1$), because the balance conditions are independent of the speed. By using two measuring points ($n = 2$), the number of the required correction planes reduces to 2 ($q = 2$).

The quality of the balancing of a flexible rotor can only be stated from the amplitude of the synchronous component of the vibrations measured on the running machine. Table B.2, from an ISO recommendation, can supply detailed indications.

	Range of effective pedestal synchronous vibration velocity (mm/s), r.m.s.													Correction factor		
	.28	.45	.71	1.12	1.8	2.8	4.5	7.1	11.2	18	28	45	71	C_1	C_2	C_3
I	A			B	C		D									
	Small electric motors (up to 15 kW)													.63		
	Superchargers													.63	2	
	Gyroscopes													.63	2	
II		A		B		C			D							
	Paper making machines													.63		
	Med. size elec. motors and gener. (17-75 kW), norm. foundations													.63	4	
	Electric motors and gener. up to 3000 kW, special foundations													.63	4	20
	Pumps and compressors													.63	8	15
	Small turbines													.63	4	8
III			A		B		C		D							
	Large electric motors													.63	5	
	Turbines and generators, rigid and heavy foundations													.63	5	20
IV			A			B			C	D						
	Large electric motors, turbines and generators, light foundations													.63	3	10
	Small jet engines													.63		
V				A				B	C		D					
	Jet engines larger than cathegory IV													.63	2	10

Balance quality
A = Precision quality
B = Commercially acceptable
C = In need of attention at next overhaul
D = In need of imediate attention

Correction factors
C_1 = Measurement in high speed balancing machine at service speed where bearing conditions are different from service conditions
C_2 = Shaft vibrations measured in or adjacent to the bearings
C_3 = Shaft vibrations measured at location of maximum shaft lateral deflection.

TABLE B.2. Balance quality criteria for flexible rotors.

Example B.1 *Turbomolecular pump.*

Compute the unbalance response at operating speed (30,000 rpm) of the turbomolecular pump whose schematic cross section is reported in Figure B.3(a), together with the few basic dimensions that are required. The balancing quality grade required is $G = 2.5$.

The rotor of the pump can be modeled as a rigid body, with its center of mass at the connection with the shaft, attached to a massless shaft. The inertial characteristics of the rigid body are $m = 6$ kg, $J_p = 0.035$ kg m^2, and $J_t = 0.055$ kg m^2. The material of the shaft has a Young's modulus $E = 2.1 \times 10^{11}$ N/m^2, and the shaft diameter is constant and equal to 24 mm (moment of inertia is $I_y = 1.628 \times 10^{-8}$ m^4).

The mass of the shaft has been neglected, and the model with two complex degrees of freedom shown in Figure B.3(b) can be used.

The stiffness matrix can be computed using the FEM. However, because of the simple geometry, it is possible to immediately obtain the compliance matrix, which can then be inverted, yielding the stiffness matrix. By applying a unit force and a unit moment in point P and computing the displacement and rotation in the same point (for simple geometries like the current one, the relevant formulas are reported in many handbooks), the elements of the compliance matrix $\boldsymbol{\beta}_{ij}$ are obtained:

$$\beta_{11} = l_2^2(l_1 + l_2)/3EI\ , \qquad \beta_{12} = l_2(2l_1 + 3l_2)/6EI,$$

$$\beta_{22} = (l_1 + 3l_2)/3EI\ .$$

The stiffness matrix $\mathbf{K}$ *is obtained by inverting the compliance matrix* $\mathbf{B}$

$$\mathbf{K} = \frac{6EI}{l_2^3(4l_1 + 3l_2)} \begin{bmatrix} 2(l_1 + 3l_2) & -l_2(2l_1 + 3l_2) \\ -l_2(2l_1 + 3l_2) & 2l_2^2(l_1 + l_2) \end{bmatrix} =$$

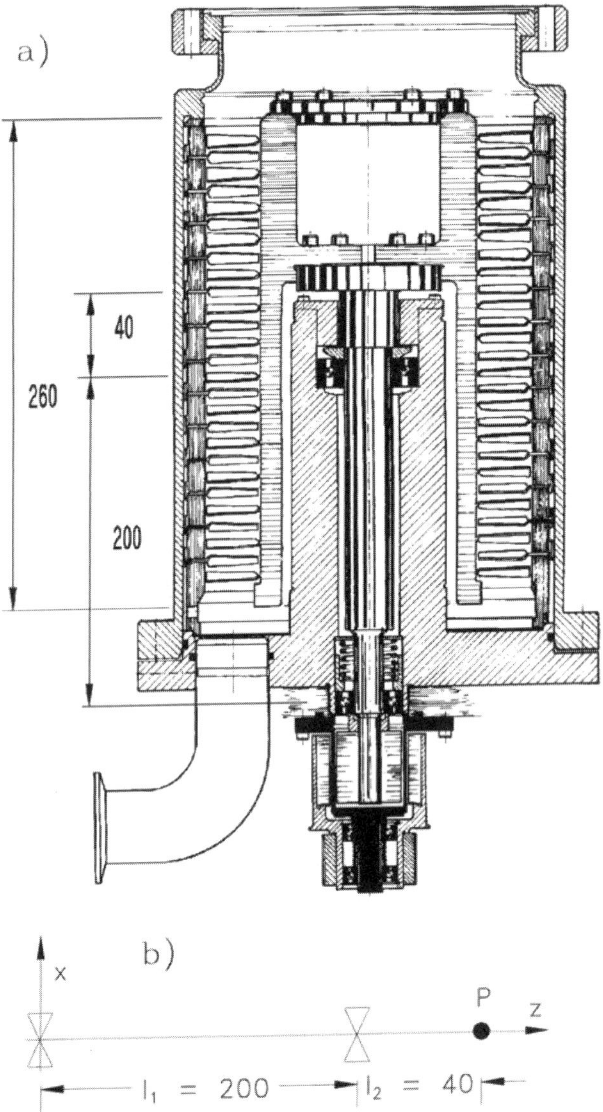

FIGURE B.3. Turbomolecular pump; (a) schematic drawing; (b) model of the rotor (*Vacuum Technology; Its Foundation, Formulae and Tables*, Leybold-Heraeus, Köln, Germany). The drawing refers to a machine no longer in production.

$$= \begin{bmatrix} 2.2297 & -0.0724 \\ -0.0724 & 0.00268 \end{bmatrix} \times 10^8 .$$

The critical speeds are $\Omega_{cr_I} = 10,490$ *rpm, and* $\Omega_{cr_{II}} = 67,080$ *rpm. The second value is high, and it is likely that other critical speeds are located between the computed values. A more realistic model, which also takes into account the mass of the shaft and the compliance of the bearings, must, however, be used to obtain them.*

The peripheral velocity of the center of mass corresponding to a quality grade G = 2.5 is 2.5 mm/s. The maximum value of the eccentricity is then $\epsilon_{max} = V_G/\Omega_{max} = 0.00714$ *mm. Because the distance between the two balancing planes is 260 mm, the maximum value of the couple unbalance is* $|\chi(J_p - J_t)|_{max} = m\epsilon_{max}d/2 = 5.57 \times 10^{-6}$ *kgm*^{2}*.*

The response to a static and a couple unbalance at 30,000 rpm can be computed from the equations

$$\begin{bmatrix} 5.715 & -5.591 \\ -5.591 & 129 \end{bmatrix} \times 10^7 \mathbf{q} = \left\{ \begin{array}{c} 422.9 \\ 0 \end{array} \right\},$$

$$\begin{bmatrix} 5.715 & -5.591 \\ -5.591 & 129 \end{bmatrix} \times 10^7 \mathbf{q} = \left\{ \begin{array}{c} 0 \\ 54.99 \end{array} \right\},$$

which yield the following values of the displacements and rotations:

$$\mathbf{q} = \left\{ \begin{array}{c} -4.0990 \\ -0.3088 \end{array} \right\} \times 10^{-6} \qquad \textit{(static unbalance)},$$

$$\mathbf{q} = \left\{ \begin{array}{c} -4.001 \\ 4.089 \end{array} \right\} \times 10^{-8} \qquad \textit{(couple unbalance)}.$$

From the results obtained, it is clear that at 30,000 rpm, the rotor is almost completely self-centered, where static unbalance is concerned, and couple unbalance causes very small deformations of the shaft.

Appendix C
Rotordynamics videos

In the CD included in this book, the reader can find two videos, *Gyroscopic Effect and Damping in Rotordynamics* (14 min) and *Dynamic Behaviour of Rotors on Anisotropic Supports* (20 min).

The first describes some experiments on simple demonstrators, and their aim is showing the students some physical evidence of gyroscopic effects and of the different roles played by rotating and nonrotating damping in rotordynamics. The demonstrators are based on rotors with very low stiffness in such a way that whirling motions can be seen without the need of instruments. The amplitude of the whirling motion is large enough to enable us to distinguish between forward and backward whirling with ease.

The author is indebted to the late Ing. Edoardo Rava, of Elettrorava, for having suggested a simplified form of the demonstrator, which was later developed with the help of some students.

The aim of the second video is to show the effect of stator anisotropy in causing backward whirling. The demonstrator is based on a five-active-axes magnetic suspension, which has the advantage of allowing us to tailor the stiffness and damping in the different direction, so that experiments with different anisotropy and damping can be performed. The demonstrator is now more similar to an actual machine than the previous ones, and the low amplitude of the whirling motion compels us to use instrumentation to observe the relevant phenomena.

The funding for producing the videos was supplied by the Politecnico di Torino, in its effort of promoting a higher quality of teaching, and they are currently shown as a part of dynamic structural analysis courses. Most of the work was done by Cristiana Delprete and Eugenio Brusa, with assistance from some students.

Appendix D
DYNROT LIGHT rotordynamics code

The DYNROT Finite Element code was developed by the author beginning in 1976. The present version is based on the MATLAB interactive software package and consequently can be used on any hardware on which MatLab has been installed. Version 4.0 or higher of MatLab is required.

The original code was written by the author together with Antonio Gugliotta, both of the Department of Mechanics of Politecnico di Torino, using HPL and then HP-BASIC language, for desktop HP 9800 computers. Subsequent versions written in Fortran and C languages were developed in the 1980s, and finally the present version using the MatLab package was evolved. Its superiority lays mainly in the ability of MatLab to deal with complex arithmetics, its graphic tools, and the possibility of easily obtaining output ASCII files.

At present, the DYNROT project is coordinated by the author. Its aim is to continue the development of the code to widen its capabilities and to make it an even more powerful tool for the dynamic analysis of rotating machinery. A number of persons have worked on it and are still working on it, mostly undergraduate and postgraduate students. Among them Giacomo Brussino, Philip Miller, Domenico Bassani, Cristiana Delprete, Stefano Carabelli, Andrea Tonoli, Nicola Amati, and Mario Silvagni must be mentioned.

A simplified version DYNROT LIGHT is here reported; it contains only the basic elements and is based on the axi-symmetrical, 1D approach seen in Chapter 4 (whereas the full-blown DYNROT has capabilities also for studying nonsymmetric rotors, bladed disks dynamics, with the possibility of time domain study of nonlinear and nonstationary rotors). A basic advantage of a MatLab code is its openness, so that the reader can understand how it works and can modify all points.

Although DYNROT can deal with the lateral, torsional, and axial dynamics of the rotor, the "light" version is designed for the lateral dynamics only. As a consequence, only a small number of the solution routines described in the handbook of version DYNROT 8.3, included in the CD, are available.

As described in detail in the handbook of the DYNROT LIGHT version, also included in the CD, the elements available are as follows:

1) beam
2) tapered beam
3) spring
4) damper
5) mass

Several input files, dealing with some examples presented in the book, are included: Readers can use them as guidelines to build their own input files.

Appendix E

Books on rotordynamics

So many papers and books have been published on rotordynamics that any attempt to supply a complete or even representative bibliography is bound to fail. Loewi and Piarulli, for instance, in the bibliography of their book published in 1961, list 554 titles. Every year, there are many conferences on rotor dynamics, and tens of papers are presented at each.

The author has chosen to provide a list of books dealing strictly with rotordynamics, plus some on bearings and two old books, of historical interest, in which the foundations of rotordynamics were described in relation with the early turbines. The author is sure that many books that should have been included were left out. Proceedings of conferences or collections of papers are also not included, or the list would have been too long. Where translations in different languages exist and an English edition was found, the latter was listed.

G. Belluzzo, *Le turbine a vapore ed a gas*, Hoepli, Milano, 1905.

A. Stodola, *Steam and Gas Turbines*, Mc Graw-Hill, New York, 1927.

A. Palmgren, *Ball and Roller Bearing Engineering*, SKF Industries, Philadelphia, PA, 1959.

F.M. Dimentberg, *Flexural Vibrations of Rotating Shafts*, Butterworth, London, 1961.

O. Pinkus and B. Sternlicht, *Theory of Hydrodynamic Lubrication*, McGraw-Hill, New York, 1961.

A. Tondl, *Some Problems of Rotor Dynamics*, Chapman & Hall, London, 1965.

R.G. Loewi and V.J. Piarulli, *Dynamics of Rotating Shafts*, The Shock and Vibration Information Center, Naval Res. Lab., Washington, D.C., 1969.

L. Buzzi, *Equilibratura*, CEMB, Mandello del Lario, 1971.

G. Schweitzer, *Critical Speeds of Gyroscopes*, Springer, Vienna, Austria, 1972.
H. Schneider, *Balancing Technology*, Schenck, Darmstadt, Germany, 1977.
A.D. Dimarogonas and S.A. Paipetis, *Analytical Methods in Rotor Dynamics*, Applied Science Publishers, London, 1983.
O. Marenholtz, *Dynamics of Rotors*, Springer, Vienna, Austria, 1984.
J. Rao, *Rotor Dynamics*, Wiley Eastern, Delhi, India, 1985.
N.F. Rieger, *Balancing of Rigid and Flexible Rotors*, The Shock and Vibration Information Center, U.S. DoD, Washington, D.C., 1986.
T. Someya (Editor), *Journal Bearing Databook*, Springer-Verlag, Tokyo, Japan, 1988.
J.M. Vance, *Rotordynamics of Turbomachinery*, Wiley, New York, 1988.
M.S. Darlow, *Balancing of High Speed Machinery*, Springer, New York, 1989.
M.J. Goodwin, *Dynamics of Rotor-Bearing Systems*, Unwin Hyman, London, 1989.
M. Lalanne and G. Ferraris, *Rotordynamics Predictions in Engineering*, Wiley, New York, 1990.
R.V. Braembussche (Editor), *Vibration and Rotordynamics*, von Karman Institute for Fluid Dynamics, Lectures Series, 1992.
D. Childs, *Turbomachinery Rotordynamics: Phenomena, Modeling and Analysis*, Wiley, New York, 1993.
E. Krämer, *Dynamics of Rotors and Foundations*, Springer, Berlin, Germany, 1993.
C.W. Lee, *Vibration Analysis of Rotors*, Kluver Academic Publishers, Dordrecht, 1993.
F.C. Moon, *Superconducting Levitation*, Wiley, New York, 1994.
G. Schweitzer, H. Bleuler, and A. Traxler, *Active Magnetic Bearings: Basic, Properties and Applications of Active Magnetic Bearings* V/D/F, Zürich, Switzerland, 1994.

References

[1] G. Genta and E. Brusa, "On the role of non.synchronous rotating damping in rotordynamics," *International Journal of Rotating Machinery*, vol. 6, no. 6, pp. 467–475, 2000.

[2] A. Argondizza, S. Carabelli, and A. Tonoli, "Finite element model reduction for rotating systems," *Proc. 7th Int. Symp. on Magnetic Bearings, Zurich*, pp. 51–56, 2000.

[3] G. Genta and C. Delprete, "Acceleration through critical speeds of an anisotropic, nonlinear, torsionally stiff rotor with many degrees of freedom,," *Journal of Sound and Vibration*, vol. 180, no. 3, pp. 369–386, 1995.

[4] E. Brusa and G. Genta, "Fast-spinning multibody spacecrafts seen as free rotors: Stability considerations," *Space Forum*, vol. 2, no. 57, pp. 319–344, 1999.

[5] G. Genta, C. Delprete, and E. Brusa, "Some considerations on the basic assumptions in rotordynamics," *Journal of Sound and Vibration*, vol. 227, no. 3, pp. 611–645, 1999.

[6] H. Lamb and R. Southwell, "The vibrations of a spinning disk," *Proc. of the Royal Society of London*, vol. 99, pp. 272–280, 1921.

[7] G. Genta and A. Tonoli, "A harmonic finite element for the analysis of flexural, torsional and axial rotordynamic behaviour of discs," *Journal of Sound and Vibration*, vol. 196, no. 1, pp. 19–43, 1996.

[8] G. Genta and A. Tonoli., "A harmonic finite element for the analysis of flexural, torsional and axial rotordynamic behaviour of bladed arrays, journal of sound and vibration," *Journal of Sound and Vibration*, vol. 207, no. 5, pp. 693–720, 1997.

[9] M. Silvagni, G. Genta, and A. Tonoli, "Non-axisymmetrical 3d element for fem rotordynamics," *ISCORMA-2, Gdansk, Poland*, 2003.

[10] G. Genta and S. Carabelli, "Noncolocation effects on the rigid body rotordynamics of rotors on active magnetic bearings," *Proc. 7th Int. Symp. on Magnetic Bearings, Zurich*, 2000.

[11] S. Crandall and J. Mroszczyk, "Conservative and nonconservative coupling in dynamic systems," *I. Mech. E. Conf. Publ. 1988-7, 567-572, I. Mech. E., London*, 2003.

[12] W. Rankine, "Centrifugal whirling of shafts," *The Engineer*, Apr. 9 1869.

[13] A. Föppl, "Das problem der laval'shen turbinewelle," *Civilingenieur*, pp. 332–342, 1895.

[14] G. Belluzzo, *Le turbine a vapore ed a gas.* Hoepli, Milano, Italy, 1905.

[15] A. Stodola, *Steam and Gas Turbines.* Mc Graw-Hill, New York, 1927.

[16] H. Jeffcott, "The lateral vibration of loaded shafts in the neighborhood of a wirling speed-the effect of want of balance," *Phil. Mag.*, vol. 37, no. 6, pp. 304–314, 1919.

[17] G. Genta, *Vibration of Structures and Machines.* Springer-Verlag, New York, 1998.

[18] L. Meirovitch, *Dynamics and Control of Structures.* Wiley, New York, 1990.

[19] D. Inman, *Vibration with Control, Measurement, and Stability.* Prentice-Hall, Englewood Cliffs, NJ, 1989.

[20] W. Campbell, "Protection of steam turbine disk wheels from axial vibration," *Trans. ASME*, no. 46, pp. 31–160, 1924.

[21] A. Muszynska, "Rotor instability," *Senior Mechanical Engineering Seminar, Carson City, Nevada*, 1984.

[22] F. Ehrich and D. Childs, "Self-excited vibration in high performance turbomachinery," *Mechanical Engineering*, May 1984.

[23] S. Crandall, "Rotordynamics," in *Nonlinear Dynamics and Stochastic Mechanics* (W. Kliemann and N. S. Namachchivaya, eds.), CRC Press, Boca Raton, FL, 1995.

[24] J. D. Hartog, *Mechanical Vibration.* McGraw-Hill, New York, 1934.

[25] S. Crandall, "The role of damping in vibration theory," *Journal of Sound and Vibration*, vol. 11, no. 1, pp. 3–18, 1970.

[26] T. Caughey, "Vibration of dynamic systems with linear hysteretic damping (linear theory)," *Proc. Fourth U.S. Nat. Congr. Appl. Mech., ASME, New York*, pp. 87–97, 1962.

[27] M. Dimentberg, *Flexural Vibrations of Rotating Shafts.* Butterworth, London, 1961.

[28] J. Lund, "Stability and damped critical speeds of a flexible rotor in fluid film bearings," *Journal of Engineering for Industry*, vol. 96, pp. 509–517, January 1974.

[29] J. Melanson and J. Zu, "Free vibration and stability analysis of internally damped rotating shafts with general boundary conditions," *Journal of Vibration and Acoustics*, vol. 120, pp. 776–783, January 1978.

[30] G. Genta and F. D. Bona, "Unbalance response of rotors: a modal approach with some extensions to damped natural systems," *Journal of Sound and Vibration*, vol. 140, no. 1, pp. 129–153, 1990.

[31] N. Myklestadt, *Fundamental of Vibration Analysis.* Mc Graw-Hill, New York, 1956.

[32] M. Prohl, "A general method for calculating critical speeds of flexible rotors," *Journal of Applied Mechanics*, January 1945.

[33] G. Cowper, "The shear coefficient in timoshenko's beam theory," *Journal of Applied Mechanics*, pp. 335–340, June 1966.

[34] O. Zienkiewicz, *The Finite Element Method in Engineering Science.* McGraw-Hill, London, 1971.

[35] M. Petyt, *Introduction to the Finite Element Vibration Analysis.* Cambridge University Press, Cambridge, 1990.

[36] K. Bathe, *Finite Element Procedures in Engineering Analysis.* Prentice-Hall, Englewood Cliffs, NJ, 1982.

[37] J. Archer, "Consistent matrix formulations for structural analysis using finite element techniques," *AIAA Journal*, vol. 3, no. 10, pp. 1910–1918, 1965.

[38] G. Genta, "Consistent matrices in rotordynamics," *Meccanica*, vol. 20, pp. 235–248, 1985.

[39] J. Rao, *Rotor Dynamics.* Wiley Eastern, Delhi, India, 1985.

[40] R. Henshell and G. Warburton, "Transmission of vibration in beam systems," *International Journal for Numerical Methods in Engineering*, vol. 1, pp. 47–66, 1969.

[41] G. Genta, "Whirling of unsymmetrical rotors: a finite element approach based on complex co-ordinates," *Journal of Sound and Vibration*, vol. 124, no. 1, pp. 27–53, 1988.

[42] S. Timoshenko, *Vibration Problems in Engineering.* Wiley, New York, 1974.

[43] E. Nestorides, *A Handbook on Torsional Vibration.* Cambridge University Press, Cambridge, 1958.

[44] W. K. Wilson, *Torsional Vibration Problems.* Chapman & Hall, London, 1963.

[45] G. Duffing, *Erzwungene Schwingungen bei ver anderlicher Eigenfrequenz.* F. Vieweg u. Sohn, Braunschweig, 1918.

[46] A. Palmgren, *Ball and Roller Bearing Engineering.* SKF Industries, Philadelphia, PA, 1959.

[47] T. Someya, *Journal Bearing Databook.* Springer-Verlag, Tokyo, Japan, 1988.

[48] O. Reynolds, "On the theory of lubrication and its applications to mr. towers' experiments," *Phil. Trans. Soc., London*, vol. 177, pp. 154–234, 1886.

[49] O. Pinkus and B. Sternlicht, *Theory of Hydrodynamic Lubrication.* McGraw-Hill, New York, 1961.

[50] F. Ocvirk, "Short bearing approximation for full journal bearings," *NACA TN 20808*, 1952.

[51] Barwell *AIAA Journal*, vol. 3, no. 10, pp. 1910–1918, 1956.

[52] A. Tondl, *Some Problems in Rotor Dynamics.* Czechoslovak Academy of Sciences, Prague, 1965.

[53] S. Earnshaw, "On the nature of molecular forces which regulate the constitution of luminiferous ether," *Trans. Cambridge Phil. Soc.*, vol. 7, pp. 97–112, 1842.

[54] W. Braunbeck, "Free suspension of objects by electric and magnetic fields," *Zietschr. Phys.*, vol. 112, pp. 753–763, 1939.

[55] F. Moon, *Superconducting Levitation.* Wiley, New York, 1994.

[56] G. Schweitzer, H. Bleuler, and A. Traxler, *Active Magnetic Bearings: Basic, Properties and Applications of Active Magnetic Bearings.* V/D/F, Zürich, Switzerland, 1994.

[57] G. Adiletta, A. Guido, and C. Rossi, "Chaotic motion of a rigid rotor in short journal bearings," *Nonlinear Dynamics*, vol. 10, pp. 251–269, 1996.

[58] G. Capone, "Descrizione analitica del campo di forze fluidodinamico nei cuscinetti cilindrici lubrificati," *L'energia elettrica*, vol. 3, pp. 105–110, 1991.

[59] P. Hughes, *Spacecraft Attitude Dynamics.* Wiley, New York, 1986.

[60] P. W. Fortescue and J. Stark, *Spacecraft Systems Engineering.* Wiley, New York, 1991.

[61] S. Timoshenko and J. Goodier, *Theory of Elasticity.* McGraw-Hill, Auckland, 1970.

[62] W. Soedel, *Vibrations of Shells and Plates.* Dekker, New York, 1994.

[63] W. Iwan and T. Moeller, "The stability of a spinning elastic disk with a transverse load system," *Journal of Applied Mechanics*, Sept. 1976.

[64] A. Love, *A Treatise on the Mathematical Theory of Elasticity.* Dover, New York, 1944.

[65] M. Endo, K. Hatamura, M. Sakata, and O. Taniguchi, "Flexural vibration of a thin rotating ring," *Journal of Sound and Vibration*, vol. 82, pp. 261–272, 1984.

[66] S. Huang and W. Soedel, "Effects of coriolis acceleration of the free and forced in-plane vibrations of rotating rings on elastic foundation," *Journal of Sound and Vibration*, vol. 82, pp. 261–272, 1987.

[67] Y. Chen, H. Zhao, Z. Shen, I. Grieger, and B. Kröplin, "Vibrations of high speed rotating shells with calculations for cylindrical shells," *Journal of Sound and Vibration*, vol. 160, no. 1, pp. 137–160, 1993.

[68] J. Padovan, "Natural frequencies of rotating prestressed cylinders," *Journal of Sound and Vibration*, vol. 31, no. 4, pp. 469–482, 1973.

[69] D. Thomas, "Dynamics of rotationally periodic structures," *International Journal for Numerical Methods in Engineering*, vol. 14, pp. 81–102, 1979.

[70] M. Geradin and N. Kill, "A new approach to finite element modelling of rotors," vol. 1, pp. 52–64, 1984.

[71] M. Rodino, "Elementi finiti armonici bi e tridimensionali per lŠanalisi dinamica di rotori," *Politecnico di Torino, Thesis*, 2001.

[72] G. Jacquet-Richardet, G. Ferraris, and P. Rieutord, "Frequencies and modes of rotating flexible bladed disc-shaft assemblies: a global cyclic symmetry approach," *Journal of Sound and Vibration*, vol. 191, no. 5, pp. 901–915, 1996.

[73] P. Lancaster, *Lambda-Matrices and Vibrating Systems.* Pergamon Press, New York, 1966.

[74] L. Meirovitch, *Computational Methods in Structural Dynamics.* Sijthoff and Noordhoff Int. Publishers, 1980.

[75] J. Meisel, *Principles of Electromechanical Energy Conversion.* Krieger, 1984.

[76] J. Williams, *Fundamentals of Applied Dynamics.* Wiley, New York, 1995.

[77] H. Schneider, *Balancing Technology.* Schenck, Darmstadt, 1974.

[78] M. Darlow, *Balancing of High Speed Machinery.* Springer, New York, 1989.

Index

Mechanical Engineering Series

J. Angeles, **Fundamentals of Robotic Mechanical Systems: Theory, Methods, and Algorithms, 2nd ed.**

P. Basu, C. Kefa, and L. Jestin, **Boilers and Burners: Design and Theory**

J.M. Berthelot, **Composite Materials: Mechanical Behavior and Structural Analysis**

I.J. Busch-Vishniac, **Electromechanical Sensors and Actuators**

J. Chakrabarty, **Applied Plasticity**

K.K. Choi and N.H. Kim, **Structural Sensitivity Analysis and Optimization 1: Linear Systems**

K.K. Choi and N.H. Kim, **Structural Sensitivity Analysis and Optimization 2: Nonlinear Systems and Applications**

G. Chryssolouris, **Laser Machining: Theory and Practice**

V.N. Constantinescu, **Laminar Viscous Flow**

G.A. Costello, **Theory of Wire Rope, 2nd ed.**

K. Czolczynski, **Rotordynamics of Gas-Lubricated Journal Bearing Systems**

M.S. Darlow, **Balancing of High-Speed Machinery**

J.F. Doyle, **Nonlinear Analysis of Thin-Walled Structures: Statics, Dynamics, and Stability**

J.F. Doyle, **Wave Propagation in Structures: Spectral Analysis Using Fast Discrete Fourier Transforms, 2nd ed.**

P.A. Engel, **Structural Analysis of Printed Circuit Board Systems**

A.C. Fischer-Cripps, **Introduction to Contact Mechanics**

A.C. Fischer-Cripps, **Nanoindentation, 2nd ed.**

J. García de Jalón and E. Bayo, **Kinematic and Dynamic Simulation of Multibody Systems: The Real-Time Challenge**

W.K. Gawronski, **Advanced Structural Dynamics and Active Control of Structures**

W.K. Gawronski, **Dynamics and Control of Structures: A Modal Approach**

G. Genta, **Dynamics of Rotating Systems**

K.C. Gupta, **Mechanics and Control of Robots**

J. Ida and J.P.A. Bastos, **Electromagnetics and Calculations of Fields**

Mechanical Engineering Series

M. Kaviany, **Principles of Convective Heat Transfer, 2nd ed.**

M. Kaviany, **Principles of Heat Transfer in Porous Media, 2nd ed.**

E.N. Kuznetsov, **Underconstrained Structural Systems**

P. Ladevèze, **Nonlinear Computational Structural Mechanics: New Approaches and Non-Incremental Methods of Calculation**

P. Ladevèze and J.-P. Pelle, **Mastering Calculations in Linear and Nonlinear Mechanics**

A. Lawrence, **Modern Inertial Technology: Navigation, Guidance, and Control, 2nd ed.**

R.A. Layton, **Principles of Analytical System Dynamics**

F.F. Ling, W.M. Lai, D.A. Lucca, **Fundamentals of Surface Mechanics: With Applications, 2nd ed.**

C.V. Madhusudana, **Thermal Contact Conductance**

D.P. Miannay, **Fracture Mechanics**

D.P. Miannay, **Time-Dependent Fracture Mechanics**

D.K. Miu, **Mechatronics: Electromechanics and Contromechanics**

D. Post, B. Han, and P. Ifju, **High Sensitivity Moiré:** Experimental Analysis for Mechanics and Materials

F.P. Rimrott, **Introductory Attitude Dynamics**

S.S. Sadhal, P.S. Ayyaswamy, and J.N. Chung, **Transport Phenomena with Drops and Bubbles**

A.A. Shabana, **Theory of Vibration: An Introduction, 2nd ed.**

A.A. Shabana, **Theory of Vibration: Discrete and Continuous Systems, 2nd ed.**